HANDBOOK OF BUSINESS AND CLIMATE CHANGE

ELGAR HANDBOOKS IN ENERGY, THE ENVIRONMENT AND CLIMATE CHANGE

This series provides a definitive overview of recent research in all matters relating to energy, the environment, and climate change in the social sciences, forming a comprehensive guide to the subject. Covering a broad range of research areas including energy policy, the global socio-political impacts of climate change, and environmental economics, this series aims to produce prestigious, high quality works of lasting significance. Each *Handbook* will consist of original contributions by leading authors, selected by an editor recognized as an international leader within the field. Taking an international approach, these *Handbooks* emphasize both the widening of the current debates within the field, and an indication of how research within the field will develop in the future.

Titles in the series include:

Research Handbook on Communicating Climate Change
Edited by David C. Holmes and Lucy M. Richardson

Handbook of Security and the Environment
Edited by Ashok Swain, Joakim Öjendal and Anders Jägerskog

Handbook of Sustainable Politics and Economics of Natural Resources
Edited by Stella Tsani and Indra Overland

Research Handbook on Energy and Society
Edited by Janette Webb, Faye Wade and Margaret Tingey

Handbook on Trade Policy and Climate Change
Edited by Michael Jakob

Handbook of Critical Environmental Politics
Edited by Luigi Pellizzoni, Emanuele Leonardi and Viviana Asara

Handbook on Climate Change and Disasters
Edited by Rajib Shaw

Handbook of Business and Climate Change
Edited by Anant K. Sundaram and Robert G. Hansen

Handbook of Business and Climate Change

Edited by

Anant K. Sundaram

Clinical Professor of Business Administration, Tuck School of Business, Dartmouth College, USA

Robert G. Hansen

Norman W. Martin 1925 Professor of Business Administration, Tuck School of Business, Dartmouth College, USA

ELGAR HANDBOOKS IN ENERGY, THE ENVIRONMENT AND CLIMATE CHANGE

Cheltenham, UK • Northampton, MA, USA

© The Editors and Contributors Severally 2023

All rights reserved. No part of this publication may be reproduced, stored in a retrieval system or transmitted in any form or by any means, electronic, mechanical or photocopying, recording, or otherwise without the prior permission of the publisher.

Published by
Edward Elgar Publishing Limited
The Lypiatts
15 Lansdown Road
Cheltenham
Glos GL50 2JA
UK

Edward Elgar Publishing, Inc.
William Pratt House
9 Dewey Court
Northampton
Massachusetts 01060
USA

A catalogue record for this book
is available from the British Library

Library of Congress Control Number: 2022948481

This book is available electronically in the Elgaronline
Business subject collection
http://dx.doi.org/10.4337/9781839103001

Printed on elemental chlorine free (ECF)
recycled paper containing 30% Post-Consumer Waste

ISBN 978 1 83910 299 8 (cased)
ISBN 978 1 83910 300 1 (eBook)

Printed and bound in the USA

Contents

List of contributors and reviewers   vii
Acknowledgements   x

1   Introduction to the *Handbook of Business and Climate Change*   1
      Anant K. Sundaram and Robert G. Hansen

PART I   THE BUSINESS CASE FOR CLIMATE CONCERNS

2   Business and climate change   8
      Anant K. Sundaram

3   The end of combustion?   38
      David Hone

PART II   KEY INDUSTRIES: IMPACT AND RESPONSE

4   Banks and climate change risk   58
      Edwin Anderson, Ilya Khaykin, Alban Pyanet and Til Schuermann

5   The patchwork quilt: business complexities of decarbonizing the electric sector   89
      Scott G. Fisher, Bruce A. Phillips and Mark W. Scovic

6   Implications of fully decarbonizing the electric industry for business: Icarus or Daedalus?   120
      Bruce A. Phillips, Scott G. Fisher and Mark W. Scovic

7   Climate change and the insurance industry – risks and opportunities for transitioning to a resilient low carbon economy   145
      Maryam Golnaraghi

8   Climate change and aviation   187
      Vincent Etchebehere

9   Leaders and laggards: how have oil and gas companies responded to the energy transition?   208
      Julia Hartmann, Andrew Inkpen and Kannan Ramaswamy

PART III   CORPORATE STRATEGY AND LEADERSHIP IN THE CLIMATE ECONOMY

10   Climate change communication strategies   231
      Paul Argenti, Posie Holmes and Marloes Smittenaar

11	Corporate strategy and climate change: a nonmarket approach to environmental advantage *Thomas C. Lawton and Carl J. Kock*	251
12	Owens Corning: environmental footprint reduction as the foundation for building a net-positive future *Frank O'Brien-Bernini and Amanda Meehan*	271
13	Climate preparedness for business resilience *Janet Peace and Kristiane Huber*	294

PART IV FUNCTIONAL PERSPECTIVES AND CORPORATE PRACTICE

14	The equity value relevance of carbon emissions *Peter M. Clarkson, Jody Grewal and Gordon D. Richardson*	326
15	Getting to 2050: transparency for setting and reaching supply chain climate goals *Suzanne Greene and Alexis Bateman*	340
16	Commodity supply chain management and climate change: a case study of the palm oil industry *Yinjin Lee and Alexis Bateman*	359
17	Carbon pricing *Robert G. Hansen*	379
18	Shifting consumers' decisions towards climate-friendly behavior *Rishad Habib and Katherine White*	405

PART V CLIMATE FINANCE

19	Mainstreaming climate action in public and private investments: mobilizing finance towards sustainable investments through the bond markets *Heike Reichelt, David P. Allen and Scott M. Cantor*	430
20	Green bonds: investor, issuer and climate perspectives *Christa Clapp, Keith Lee and Anouk Brisebois*	458
21	Cost of capital and climate risks *Gianfranco Gianfrate, Dirk Schoenmaker and Saara Wasama*	480
22	ESG investing *Anant K. Sundaram*	503

PART VI THE FUTURE

23	Reflections on the future *Arranged and edited by Anant K. Sundaram and Robert G. Hansen*	526

Index 539

List of contributors and reviewers

CONTRIBUTORS

David P. Allen, External Affairs Officer, Climate Change Group, World Bank, Washington DC, USA.

Edwin Anderson, Partner, Oliver Wyman, New York, NY, USA.

Paul Argenti, Professor of Corporate Communication, Tuck School of Business, Dartmouth College, Hanover, NH, USA.

Alexis Bateman, Research Affiliate and Visiting Lecturer, Massachusetts Institute of Technology Center for Transportation and Logistics, Cambridge, MA, USA.

Anouk Brisebois, Analyst, CICERO Shades of Green, Oslo, Norway.

Scott M. Cantor, Senior Financial Officer, World Bank, Washington DC, USA.

Christa Clapp, Partner and Co-founder of CICERO Shades of Green, Oslo, Norway.

Peter M. Clarkson, Professor of Accounting, UW Business School, University of Queensland, Brisbane, Australia and Beedie School of Business, Simon Fraser University, Vancouver, British Columbia, Canada.

Vincent Etchebehere, Vice President of Sustainable Development and New Mobility, Air France, Paris, France.

Scott G. Fisher, Partner, NorthBridge Group, Concord, MA, USA.

Gianfranco Gianfrate, Professor of Finance, EDHEC Business School, Nice, France.

Maryam Golnaraghi, Director – Climate Change and Environment, The Geneva Association, Zurich, Switzerland.

Suzanne Greene, Research Affiliate, Massachusetts Institute of Technology, Cambridge, MA, USA.

Jody Grewal, Assistant Professor of Accounting, Rotman School of Management, University of Toronto Mississauga, Toronto, Ontario, Canada.

Rishad Habib, Assistant Professor in Marketing, Ted Rogers School of Management, Toronto Metropolitan University, Ontario, Canada.

Robert G. Hansen, Norman W. Martin 1925 Professor of Business Administration, Tuck School of Business, Dartmouth College, Hanover, NH, USA.

Julia Hartmann, Professor of Sustainable Supply Chain Management, EBS Business School, Wiesbaden, Germany.

Posie Holmes, Research Assistant, Tuck School of Business, Dartmouth College, Hanover, NH, USA.

David Hone, Chief Climate Change Advisor, Shell International Ltd., London, UK.

Kristiane Huber, Officer, Flood-Prepared Communities, The Pew Charitable Trusts, Washington DC, USA.

Andrew Inkpen, Seward Chair in Global Strategy, Thunderbird School of Global Management, Arizona State University, Phoenix, AZ, USA and Copenhagen Business School, Denmark.

Ilya Khaykin, Partner, Oliver Wyman, New York, NY, USA.

Carl J. Kock, Professor of Strategy, IE Business School, Madrid, Spain.

Thomas C. Lawton, Professor of Strategy and International Business and Director of the Global Competitiveness Institute, Cork University Business School, University College Cork, Ireland, and Professor of Strategy and International Business, Surrey Business School, University of Surrey, UK.

Keith Lee, Senior Analyst, CICERO Shades of Green, Oslo, Norway.

Yinjin Lee, Research Scientist, Singapore Institute of Manufacturing Technology (SIMTech), Singapore.

Amanda Meehan, Innovation and Sustainability Communications, Owens Corning, Toledo, OH USA.

Frank O'Brien-Bernini, Vice President and Chief Sustainability Officer, Owens Corning, Toledo, OH, USA.

Janet Peace, Head of Advisory Service, Anew Advisory Services LLC, Washington DC, USA.

Bruce A. Phillips, Senior Advisor, NorthBridge Group, Concord, MA, USA.

Alban Pyanet, Partner, Oliver Wyman, New York, NY, USA.

Kannan Ramaswamy, William D. Hacker Chair Professor of Management in the Department of Global Business, Thunderbird School of Global Management, Arizona State University, Phoenix, AZ, USA.

Heike Reichelt, Head of Investor Relations and Sustainable Finance, Capital Markets and Investments, World Bank Treasury, Washington, DC, USA.

Gordon D. Richardson, Professor of Accounting, Rotman School of Management, University of Toronto, Ontario, Canada.

Dirk Schoenmaker, Professor of Banking and Finance, Rotterdam School of Management, Erasmus University, Rotterdam, Netherlands.

Til Schuermann, Partner and Co-Head of the Americas Finance, Risk and Public Policy, Oliver Wyman, San Francisco, CA, USA.

Mark W. Scovic, Analyst, NorthBridge Group, Concord, MA, USA.

Marloes Smittenaar, Communications Manager, Boehringer Ingelheim, The Hague, Netherlands.

Anant K. Sundaram, Clinical Professor of Business Administration, Tuck School of Business, Dartmouth College, Hanover, NH, USA.

Saara Wasama, Research Affiliate, Erasmus Platform for Sustainable Value Creation, Rotterdam School of Management, Erasmus University, Rotterdam, Netherlands.

Katherine White, Senior Associate Dean and Professor, Marketing and Behavioural Science Division, Sauder School of Business, University of British Columbia, Vancouver, British Columbia, Canada.

EXTERNAL REVIEWERS

Paul Argenti, Tuck School of Business at Dartmouth

Christa Clapp, CICERO Shades of Green

Ian Cochran, University of Edinburgh Business School

Olivier Fainsilber, Aviation Sector Expert

Anthea Lavallee, Hubbard Brook Research Foundation

Kevin Leahy, Climate and Energy Policy Expert

Verónica León-Bravo, School of Management at Politecnico di Milano

Ella Mae Matsumara, University of Wisconsin-Madison

Rebecca Reczek, Fisher College of Business at Ohio State University

Irving Schenkler, NYU Stern School of Business

Til Schuermann, Oliver Wyman

Kevin Stiroh, The Federal Reserve Bank of New York

Corey Stock, Tuck Class of '22

Swenja Surminski, London School of Economics

Gregory Unruh, George Mason University

Mark Way, The Nature Conservancy

Acknowledgements

We would like first, and most of all, to thank the 39 authors and co-authors who, along with the two of us, contributed the 23 chapters that comprise this *Handbook*. We began the project around the time that the pandemic was just starting to permeate our collective lives, and were not quite sure how the timeline would evolve as events unfolded. Yet, despite some slowing that resulted in a few months' delay relative to original plans, our contributors helped to deliver what we believe is, collectively, a wonderfully substantive volume that the two of us, as editors, are pleased and proud to be able to put out there. We are deeply appreciative of the fact that our authors helped to keep us more-or-less on track with our originally intended timeline.

The quality of the articles in this compendium also reflects critiques and advice from our reviewers, whose main focus was to nudge the authors with suggestions to improve their work. This *Handbook* would neither have been possible nor have achieved the quality that it did, were it not for their considered, thoughtful feedback. We therefore would like to give special thanks to those who reviewed chapters for us (in many instances, more than one).

Daniel Mather, Commissioning Editor of Edward Elgar Publishing's well-known *Research Handbooks* and *Original Reference* series across the social sciences, approached us with the idea of doing this *Handbook*, and we jumped at the opportunity. We thank Daniel for his guidance, advice, and encouragement throughout. We are grateful to Nina Booth and Pete Waterhouse for their remarkable skills in managing the process of copyedits, and greatly appreciate their careful attention to detail. Vic Froggett (along with Nina Booth) helped us with the process of designing the cover for the *Handbook*, and we are thankful for that.

Additionally, Anant would like to thank the hundreds of Tuck MBA students from whom he has shamelessly gleaned insights on this topic, arising from 13 years of teaching his class, *Business and Climate Change*. When first offered in 2009, it was not obvious that the course – the first of its kind in a US MBA program, as far as we know – would survive for long, let alone prove to be the precursor to a topic of such sweeping breadth and relevance to business. What made it all the more apropos, and came full circle from the standpoint of this collaboration, is the fact that Bob, as (then) Senior Associate Dean of Tuck, was the one who originally proposed the idea of the course to Anant, and provided the resources for its development.

Anant K. Sundaram and Robert G. Hansen
Tuck School of Business at Dartmouth, Hanover, New Hampshire, USA

1. Introduction to the *Handbook of Business and Climate Change*

Anant K. Sundaram and Robert G. Hansen

When we decided to pull together the *Handbook of Business and Climate Change*, the first such book on this huge topic, as far as we know, our aim was to produce a collection of articles that would provide an up-to-date summary of the state of knowledge in the business and climate arena, with a view to taking a deeper dive into a few particularly important questions: How do businesses perceive the issue of climate change? How are they affected by it, and how are they engaging with climate change? Why might they be doing so? What impact are they having, at what cost? And finally, what might the future hold? We also set the goal that the material in the *Handbook* should be accessible and valuable to not only academics and students, but also to practitioners in companies, NGOs, and governments.

This was a formidable task, as the issues raised at the intersection of business and climate change are complex, diverse, and touch all aspects of a company's operations. Businesses are, at once, not only a major source of carbon emissions, but are themselves impacted by climate change. Simultaneously, they are crucially important in developing and deploying solutions to address the problem. Climate change brings myriad risks to the world of business but, on the other hand, opportunities abound, with potentially large rewards for those companies and entrepreneurs who can help make the needed energy transition cheaper, cleaner, faster, and safer. Companies are dealing with climate change as an issue that cuts across all aspects of their business, from finance to marketing and operations, and many companies recognize it as a strategic issue that commands the attention of their top leadership. Stakeholders – shareholders, consumers, employees, citizens – expect companies to be involved, to be a part of the solution, and demonstrably so. While there is some commonality across industries and sectors, heterogeneity of impact and response is a dominant feature.

All of this implies that we had on our hands a broad, deep and complex set of topics and issues that needed to be considered.

We wanted to ensure that there was an appropriate mix of contributions from both academics in business schools and practitioners in the world of business who have been struggling with these issues, in some instances for decades. We felt that this mix would be especially valuable for the *Handbook*, as corporations and markets have been at the forefront of developing many of the innovations and practices that increasingly seem commonplace today. The inclusion of practitioners would also bring in case studies, which we believe can be particularly useful in getting into the details of how companies – the entities, to repeat, that are at the frontlines of developing and deploying solutions to the problem – are framing these issues, what they have actually done, and the challenges that they see remain. Finally, given the global nature of climate change, we wanted to ensure that we had contributors and hence, ideas from across the globe.

With these objectives in mind, and following this introduction, the *Handbook* begins with "Part I. The Business Case for Climate Concerns," where Anant Sundaram of the Tuck School of Business at Dartmouth provides a broad-ranging and detailed definition and exposition of the issues that climate poses for business and why business has to be seriously engaged. He details the carbon emissions[1] of businesses, by industry and geography, and evaluates the magnitude of the risk: Sundaram estimates that the corporate sector accounts for approximately 70 percent of total *direct* carbon emissions in the US, 60 percent in the EU, and nearly 80 percent in China. He discusses the current state of emissions disclosure by companies as well as the changes likely to be coming to this disclosure and reporting arena. Sundaram then lays out a framework for linking corporate value creation to carbon emissions and climate change that would be useful to have in hand as one approaches the rest of the *Handbook*. It should perhaps be required reading for any analyst working in the nexus of business and climate. Using his framework, Sundaram estimates an implicit liability for carbon-related damages – assuming a carbon price of $60 per metric ton – for 1,050 of the largest publicly traded US companies on the order of $2.6 trillion. Following Sundaram is a provocative chapter by David Hone of Shell: "The end of combustion?" – with an emphasis on the question mark. Hone describes how the pace of change in the energy system has increased dramatically, but he highlights the world's remaining dependence on fossil fuels and what would be required to limit warming to 1.5°C or even 2.0°C. Hone pithily describes the daunting scale of the challenge:

> Meeting the most aggressive goal of the Paris Agreement will require very substantial near-term action, probably at a rate that defies all previous change in the energy system and from a starting point that has two hundred years of legacy infrastructure behind it.

In "Part II. Key Industries: Impact and Response," our contributors look at some of the specific industries and companies that are being (and will continue to be) impacted by climate change. Edwin Anderson, Ilya Khaykin, Alban Pyanet and Til Schuermann, all of the consulting firm Oliver Wyman, tackle the case of the banking sector. They break climate risks into two categories following the Taskforce on Climate-related Financial Disclosure (TCFD) guidelines, physical risk and transition risk, focusing on how banks can frame these risks from the standpoint of their core business, and what they can do to prepare. Scott Fisher, Bruce Phillips and Mark Scovic of the energy consulting firm NorthBridge Group take two chapters to cover the implications of climate change for the US electric power industry. Globally, electricity generation accounts for about 40 percent of emissions, so the industry is crucial to achieving significant reductions. In the first of their chapters, Fisher, Phillips and Scovic cover the history of regulation of the industry through to today, with an emphasis on the notion – as well as the downsides – of a "patchwork quilt" of regulations. This patchwork quilt covers a host of regulatory policies, including renewable portfolio standards, feed-in-tariffs, carbon pricing, tax credits and other subsidies, and net metering. An important implication for investment in this sector is the need to understand the local, state and national regulatory structures – and the ability to manage the attendant risks. In their second chapter, Phillips, Fisher and Scovic look to the future, focusing on three broad business challenges facing the industry: deploying already commercialized technologies at scale; commercializing advanced technologies; and managing the existing fleet of generating and distribution assets. To meet the challenges, the electric industry will need strategic foresight, financial resources, and prudent risk management. Phillips et al. use Greek mythology to help understand the challenges, tradeoffs and risks: Will the future of the electric industry resemble what happened to Icarus, who flew too close to the sun and burned up, or to Daedalus, who walked the fine line between flying too high and too low?

The next chapter of Part II covers another industry in which climate change will have a large impact – insurance and re-insurance. Maryam Golnaraghi of The Geneva Institute details the impacts on both the asset and liability sides of insurance companies' balance sheets, and the actions that companies are already taking, both individually and collectively, to manage the risks. Insurers are doubly affected by climate change, as it affects both the asset and liability sides of their balance sheets, but this also means that insurers are in a unique position to affect the future course of how businesses deal with climate issues. With regard to liabilities, Golnaraghi looks at a variety of insurer activities to manage the risk, including risk modeling and pricing, new and alternative risk transfer products including so-called parametric insurance, and incentives for policyholders to build climate resilience into their properties and operations. On the assets side, she discusses the challenges that insurers face in rearranging their investment portfolios to reflect the energy transition.

Aviation, a hard-to-abate industry, has been more recently highlighted for its contribution to climate change, not just from fuel-based carbon emissions but also from those telltale white contrails. Vincent Etchebehere of Air France lays out the facts as we know them on aviation's contributions to emissions from fuel combustion as well as the effects of contrails, and describes how the world's airlines as well as global industry regulators are working to reduce their warming impact, focusing in particular on both the opportunities for and limitations of sustainable aviation fuels, and alternatives such as hydrogen-based fuels and carbon offsets. He also delves into the specific actions being undertaken by the Air France-KLM Group, noting that

> ...the decarbonization challenge remains high for the sector, one whose demand is expected to continue to grow significantly in the future. ...The sector currently sees SAFs [sustainable aviation fuels] as the main lever to achieve this, but overcoming feedstock availability constraints, scaling up production, and enabling decreases in SAF prices will be critical.

Part II concludes by examining the fossil fuel industry itself and how the world's oil and gas companies have responded to the energy transition. Perhaps not surprisingly, strategic responses are quite heterogeneous. Julia Hartmann of the EBS Business School, and co-authors Andrew Inkpen and Kannan Ramaswamy (Thunderbird School of Management of Arizona State University) detail the economic and societal forces driving change, and the heterogeneity that resulted. Specifically, they ask, as these firms prepare for the energy transition that lies ahead, will they proactively engage in the search for viable renewable energy alternatives or will they continue to reactively focus on oil and gas with the expectation that the world will continue to need fossil fuels for many more decades?

Leadership in, and the communication of corporate climate strategy is another area that we felt merited inclusion in the *Handbook*. Many corporations have clearly elevated issues of climate change to one of strategic importance and have attempted to communicate the seriousness of the issue and the company's responses to all of their stakeholders. "Part III. Corporate Strategy and Leadership in the Climate Economy," takes on these topics, beginning with Paul Argenti (Tuck School of Business at Dartmouth), Posie Holmes and Marloes Smittenar writing on climate change communications strategies: the need for such communications, a framework for thinking about how to communicate about climate change most effectively, and the risks of getting it wrong. They argue that sustainability reports alone are not enough to effectively communicate a company's climate strategy and emphasize the importance of communications by top leadership and the use of collectives and collaborations. They discuss the risk from being accused of "greenwashing" and how to avoid that, with the building of trust with constituencies being critical. With regard to climate strategy for companies, Thomas Lawton (Cork University Business School in Ireland) and Carl Kock (IE Business School in Spain), in "Corporate strategy and climate change," examine how companies can develop an overall corporate strategy that includes environmental issues, with a special focus on combining both market and non-market aspects of strategy. Market-based strategies operate through the usual mechanisms of competition such as price, quality, scale and scope, while non-market strategies include social, political and regulatory mechanisms. Diversity and dispersion in corporate responses to environmental issues is a dominant feature noted by Lawton and Kock, with a continuum from being reactive/defensive to proactive. This theme of diversity and dispersion occurs throughout this *Handbook*.

Frank O'Brien-Bernini and Amanda Meehan of Owens Corning, a global buildings and industrial materials manufacturer in another hard-to-abate sector and a company perhaps best known for its fiberglass insulation and shingle roofing, discuss the evolution of the company's sustainability journey since the early 2000s. O'Brien-Bernini and Meehan chronicle that journey, focusing on how goal-setting, continual benchmarking and commitment to transparency helped Owens-Corning meet its goals. Over the years, the company's goals have evolved and expanded in service of an aspiration to become a net-positive company, i.e., one whose increased 'handprint', or the positive impacts of its people and products, is greater than its environmental footprint. Part III closes with a chapter by Janet Peace of Anew (a provider of strategic insight and advice to companies on how to mitigate environmental impact) and Kristiane Huber, who address the topic of building business resilience to climate change. They focus in particular on physical risks that could result from climate change and how such risks can affect a company. They also address whether most companies are prepared, offering suggestions for corporate actions, including identifying business opportunities associated with changes in the climate.

Part IV of this *Handbook*, "Functional Perspectives and Corporate Practice," looks at business and climate change from the perspective of specific practices that companies can take to enhance their understanding and management of the risks and opportunities that climate change poses. Peter Clarkson of the University of Queensland Business School, Australia, Jody Grewal and Gordon Richardson (both of the Rotman School of Business at the University of Toronto, Canada) set the stage by reviewing the evidence on the relevance of carbon emissions for equity valuations. A negative relationship between carbon emissions and firms' valuations is documented, and the authors go on to discuss how disclosure of carbon emissions, especially mandated disclosure, can help investors assess climate risk and in turn incentivize companies to reduce emissions. Following this chapter are two contributions on supply chain management and climate goals of companies: Suzanne Greene and Alexis Bateman of MIT begin with an overview of supply chain and climate. They argue that, due to a lack of action from national governments to address climate change, many companies have assumed the mandate to reduce their overall environmental impact. Incorporating research on the state of supply chain sustainability as well as disclosures, they explore current practices and discuss the state of the industry, to lay out some of the best practices in measurement and standards, transparency, goal-setting, and supplier engagement. The following chapter, by Yinjin Lee of the Singapore Institute of Manufacturing Technology and Alexis Bateman of MIT, uses the global palm oil industry as a case study in supply chain management. Palm oil is a widely-used ingredient in many consumer products, with myriad global companies sourcing the oil. Practices by these companies in pursuing sustainable palm oil are, once again, heterogeneous. Similar to the taxonomy developed by Lawton and Kock, Lee and Bateman categorize palm oil sourcing companies according to their overall profile in how actively they deal with sustainability in their supply chain.

In the chapter that follows, Robert Hansen of the Tuck School of Business at Dartmouth College looks at one increasingly widespread practice to help corporations manage carbon emissions and risks, namely internal carbon pricing. This is the use by companies of an internal price on carbon emissions to better understand potential costs and risks; to affect corporate decisions, particularly with regard to investments; and to incentivize and communicate the company's mitigation efforts. Across firms, there is a great deal of diversity in the level of prices used as well as in the range of decisions potentially affected by the internal price. Internal carbon prices are usually "shadow" prices, so they do not typically directly affect financial flows within the company. While Hansen notes the value to companies of the practice, he casts a critical eye on the system-wide effects, particularly with regard to heterogeneity of carbon prices used and possible redundancy with public policies such as economy-wide carbon pricing and renewable energy standards for electric generation.

Part IV wraps up by looking at a topic not yet broached: consumer behavior with respect to climate change. Rishad Habib of Ryerson University and Katherine White of the University of British Columbia discuss the importance of engaging consumers in climate strategy, presenting a framework to change behavior in the key sectors of transportation, food, energy, waste disposal, and material purchases. Their framework is based on five consumer behavior factors – social influence, habit, individual self, feelings and cognition, and tangibility – and they document the use of the framework in six different settings.

We concluded that the intersection of finance and climate change was rich enough to warrant a separate part of its own. "Part V. Climate Finance," has four contributions, covering both corporate and public finance. Heike Reichelt, David P. Allen and Scott M. Cantor, all of The World Bank, start off by documenting the immense scale of investment needed to achieve the world's climate goals: $1.6 *trillion* to $3.8 *trillion* (in today's dollars) annually from now to 2050. They lay out the elements of the world's "climate finance" system, in particular, the role of traditional capital markets in addition to public finance to meet this challenge. The emergence of "green bonds," first issued in 2008 by the World Bank, and the related institutional structure are discussed in detail along with how the broad architecture of the world's financial system should change in order to meet climate change challenges. Christa Clapp, Keith Lee and Anouk Brisebois of CICERO Shades of Green, a company that gives independent assessments of green and sustainable bond frameworks, dive deeper into the topic of green bonds, detailing the nature of the market for such bonds and the important issues that bond issuers, investors, and regulators must consider. While the green bond market is growing rapidly and is expected to continue to grow, Clapp et al. are careful to point out why green bonds are not a silver bullet, raising questions regarding how they actually contribute to emissions- and climate impact-reductions.

Gianfranco Gianfrate of the University of Nice and Dirk Schoenmaker and Saara Wasama of Erasmus University survey the literature on the relationship between a company's environmental performance and a key financial measure – the company's cost of capital, i.e., its cost of equity and cost of debt. Academic research supports a finding that better environmental performance leads to a lower cost of capital, but the studies suffer from issues of how to measure environmental performance and, of course, at best can only document a relationship, not causation. The empirical relationship appears to be specific to certain countries and industries. This is clearly an area that will benefit from additional research, maybe most important from the theory side, as it is unclear as to why environmental risks should be a risk factor in the cost of capital rather than something that affects cash flows. Finally in Part V, Sundaram takes up a topic of increasing relevance, the rapidly-growing and already sizeable phenomenon of investing in ESG (environment, social and governance) assets, of which climate-investments are a substantial part. His chapter synthesizes and unpacks the considerable amount of academic and practitioner evidence, as well as controversies, which have equally rapidly emerged. Sundaram examines key characteristics and components of ESG, the fast-paced growth in the market, ESG scores and rankings, and the link between corporate investments in attractive ESG attributes and their financial and stock market performance, as well as the link between ESG-themed portfolios and their risk-return performance. The chapter closes with an assessment of implications for corporations as well as buyers and sellers of this form of investing, and whether it is time to bring oversight to an increasingly messy (and sometimes opaque) part of business and portfolio management.

Our final Part is titled "The Future." We simply asked our contributors to tell us what they thought the future might hold for the links between business and climate change. We suggested that they consider what might be one or two developments – these could be technological, market-based, policy-based, something else altogether – that will transform our ability to deal with climate change in the next decade or two. Is there something specific, around the corner, which makes them feel optimistic – or pessimistic? How do they think the world of regulation and corporate practice will play out: Will we reach a global, coordinated set of policy solutions or will we end up with a hodgepodge of disparate, uncoordinated individual corporate actions, carbon taxes, trading schemes, subsidies, and so forth? It should be interesting to look back in a decade or two to compare these thoughts to the reality of what actually came about.

Overall, we hope that readers find this collection to be a valuable resource for getting a summary of the state of knowledge on business and climate change, and for digging deeper into issues such as leadership, strategy and communication; measurement and disclosure of emissions and related valuation impacts; supply chain management; climate and finance; consumer behavior; impacts on key industries; and management practices that can help companies achieve their climate-related goals. If this volume enhances research efforts at the intersection of business and climate change, and, even more saliently, if it contributes even in some small way to the process of resolving one of the world's most pressing problems, we will be very satisfied.

NOTE

1. Sundaram gives some definitions which will be useful for readers to have in mind as they proceed through this *Handbook*. Quoting from his background science section in Chapter 2:
"GHG" is the precise term for the gases that cause atmospheric warming. The three main GHGs are carbon dioxide [CO_2], methane [CH_4] and nitrous oxide [N_2O]). In what follows, I will typically use "CO_2" or sometimes just "carbon" as short-hand references for GHG where it is unlikely to cause confusion. GHG emissions are measured on equivalents based on the "global warming potential" (GWP) of one metric ton (or "tonne," abbreviated hereafter as "t") of a GHG relative to the GWP of one tonne of carbon dioxide (tCO_2) measured over a 100-year period. In other words, emissions are measured on their CO_2-equivalents, or CO_2e, for short.

PART I

THE BUSINESS CASE FOR CLIMATE CONCERNS

2. Business and climate change
Anant K. Sundaram

INTRODUCTION

Climate change is a vital issue of our time. The science and the evidence that greenhouse gas (GHG) emissions cause warming, resulting in discernible impacts on various aspects of climate and the earth's ecosystems, are now well-documented. Businesses[1] are decisively implicated: As producers of goods and services that consumers want and need, they are the most important source of emissions via their control of the world's production processes. Conversely, the effects of climate change, as well as society's responses to solve the problem not only pose risks to businesses, but also simultaneously create opportunities to deploy the talent, the technologies, and the resources to address the problem.

This chapter explores various aspects of this cause-and-effect association. It first summarizes evidence on emissions, looking at where we are today, how we got here, and where the world is expected to head temperature-wise under scenario-based forecasts of the future. It then examines the role of businesses as the source of GHG emissions in the world's three largest-emitting geographic areas, namely the US, the EU and China. This, in turn, leads to a discussion of the role of both GHG sources (inflows) and sinks (outflows), and to a framework that defines key components of a global 'climate economy,' components that are pivotal to address the issues created by climate change.

Following that, the chapter examines the current state of corporate emissions measurement, reporting and management, drawing upon examples of firms that have implemented climate strategies, and looks at impending changes in climate-related reporting. The chapter closes with a framework for how companies should think about linking climate change to their business model and, more broadly, their value-creation process.

A BRIEF BACKGROUND ON THE SCIENCE, THE EVIDENCE, THE FUTURE[2]

The Scientific Evidence

Climate science is not new. The greenhouse-like properties of the atmosphere were studied by French scientist Joseph Fourier in the 1820s, and the heat-trapping role of carbon dioxide (CO_2) was reported by American scientist Eunice Foote in 1856 and then by British physicist John Tyndall in 1859. In 1896, Swedish chemist Svante Arrhenius quantified the effect of changes in CO_2 in the atmosphere on the earth's surface temperature (Rodhe et al., 1997). Concentration of CO_2 in the atmosphere at that time was estimated to be approximately 300 parts per million (ppm). Since then, CO_2 concentration has steadily increased, crossing 410 ppm in 2019.

In its sixth 'Assessment Report' published in August 2021, the Intergovernmental Panel on Climate Change ('IPCC', and this publication is referred to hereafter as IPCC-AR (2021)) summarizes research on attributions of excess temperature (relative to emissions during the pre-industrial period 1850–1900) from human-caused GHG emissions, adjusted for the effects of other human-caused factors as well as exogenous influences. Such attribution analysis for the decade 2010–2019 shows coefficients of +1.5°C from GHGs, –0.4°C from other human drivers such as aerosols, land use and ozone (all of which create a net cooling effect), and roughly 0.0°C from exogenous influences such as solar and volcanic activity, and internal variability. Together, these numbers imply an attributed human influence to global warming of 1.1°C (i.e., 2°F) during the decade 2010–2019 relative to pre-industrial times (IPCC's "Summary for Policymakers," referred to hereafter as IPCC-SPM (2021): see Fig. SPM.2, p. 8).

Net emissions – i.e., total emissions minus emissions absorbed by land and oceans – since 1970 exceed the cumulative total during the preceding 220-year period,[3] currently adding 2.3–2.5 parts per million (ppm) of CO_2 concentration to the earth's atmosphere annually (Lindsey 2020). Similar substantial increases in concentration trends are observed for CH_4 and N_2O. However, their measurements do not go as far back as those for CO_2, since these two gases dissipate faster in the atmosphere than CO_2, which stays in the atmosphere for many hundreds of years (Buis, 2019). In 2018, the latest year for which comprehensive emissions data are available as of the time of writing, the world emitted 49 billion metric tons of CO_2e ($GtCO_2e$),[4] with CO_2 accounting for 74 percent, CH_4 for 17 percent and N_2O for 7 percent, the three together thus accounting for 98 percent of all global GHG emissions (ClimateWatch, 2021); see Figure 2.1.

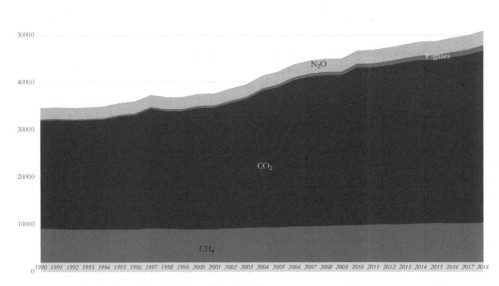

Source: Constructed from dataset in ClimateWatch (2021).

Figure 2.1 *Global GHG emissions (MtCO2e) by major gases: 1990–2018*

Emissions/Temperature Scenarios and Forecasts

AR6 develops five 'illustrative' scenarios for future "shared socio-economic pathways" (SSPs), without taking an explicit position on any one or the other being more likely or desirable.[5] The scenarios are referred to as SSP*x-y*, where 'SSP*x*' refers to the Shared SSP describing the socio-economic trends underlying the scenario, and '*y*' (= 1.9, 2.6, 4.5, 7.0, or 8.5) to the level of radiative forcing resulting from that scenario by the end of the 21st century. Increasing '*y*' implies higher annual emissions and hence, higher forecasted temperatures.

Table 2.1 (IPCC-AR 2021, Table 1, pp. 17–18) summarizes the IPCC's temperature forecasts for various future time periods. The "best estimate" forecasts for end-century (2081–2100) that most closely correspond to the widely referred-to policy goals of "1.5°C" and "2°C" by various global bodies and inter-governmental agreements[6] are SSP1-1.9 (1.4°C) and something between SSP1-2.6 and SSP2-4.5 (1.8–2.7°C). The "best estimates" for 2081–2100 are usually the temperature forecasts commonly associated with – and widely reported with respect to – global warming.

Table 2.1   *IPCC AR6's best estimate and 'very likely' range of global temperature forecasts under different SSPS for near/medium/long terms*

Scenario	Near term, 2021–2040 Best estimate (°C)	Near term, 2021–2040 Very likely range (°C)	Mid-term, 2041–2060 Best estimate (°C)	Mid-term, 2041–2060 Very likely range (°C)	Long term, 2081–2100 Best estimate (°C)	Long term, 2081–2100 Very likely range (°C)
SSP1-1.9	1.5	1.2 to 1.7	1.6	1.2 to 2.0	1.4	1.0 to 1.8
SSP1-2.6	1.5	1.2 to 1.8	1.7	1.3 to 2.2	1.8	1.3 to 2.4
SSP2-4.5	1.5	1.2 to 1.8	2.0	1.6 to 2.5	2.7	2.1 to 3.5
SSP3-7.0	1.5	1.2 to 1.8	2.1	1.7 to 2.6	3.6	2.8 to 4.6
SSP5-8.5	1.6	1.3 to 1.9	2.4	1.9 to 3.0	4.4	3.3 to 5.7

Source: IPCC-AR (2021, Table 1, pp. 17–18).

Researchers, policymakers, NGOs and others work backwards from such end-of-century temperature forecasts to derive corresponding implied *mid*-century GHG emissions forecasts that result from these SSPs. Using data from IPCC-AR (2021; Table 4), implied 2050 emissions forecasts – for CO_2, CH_4, and N_2O combined – which correspond to these three commonly-used SSPs are:[7] 2.3 $GtCO_2e$ for SSP1-1.9, 18.2 $GtCO_2e$ for SSP1-2.6, and 43.9 $GtCO_2e$ for SSP2-4.5. Given current annual emissions of approximately 50 $GtCO_2e$, therefore, achieving a 1.5°C (point estimate) target by end-century will require near-complete abatement of net global annual emissions flows by mid-century,[8] while a 2.0°C target will require a 60–65 percent cut, requiring net zero emissions by (roughly) 2075. Soon thereafter, the world will be required to get to – and continue with – net *negative* emissions, i.e., we would need to get to a world in which global carbon removals will have to equal or exceed carbon inflows.[9]

Collateral Impacts

AR6 also makes assessments of impacts of warming on collateral climate-related phenomena and 'extreme' risks, such as those related to Arctic sea ice area, sea-level rise, and weather/

climate extremes (IPCC-SPM 2021, pp. 9–14). The following is a summary of those assessments (with the IPCC's expressed level of confidence in parentheses):

- During the decade 2011–2020, annual average **Arctic sea ice area** reached its lowest level since at least 1850 (*high confidence*).
- **Sea level** has risen faster since 1900 than over any preceding century in at least the last 3000 years (*high confidence*).
- **Oceans have warmed** faster over the past century than since the end of the last deglacial transition, around 11,000 years ago (*medium confidence*).
- There has been a long-term increase in **ocean acidification** (*high confidence*).[10]
- **Heat extremes** including heatwaves have become more frequent and more intense across most land regions since the 1950s, while **cold extremes** including cold waves have become less frequent and less severe (*virtually certain*).
- The frequency and intensity of **heavy precipitation** events have increased since the 1950s over most land areas where good data are available (*high confidence*); human-induced climate change is the main driver (*medium confidence*). Human-induced climate change has contributed to increases in **agricultural and ecological droughts** (*medium confidence*).
- **Tropical cyclone activity** has increased over the past four decades (*medium confidence*), has shifted northward (*medium confidence*), but there is *low confidence* in longer term trends on tropical cyclones.

In summary, the IPCC says that evidence is strong on the impact of climate change for Arctic sea ice area decrease, sea level rise, ocean acidification, heat/cold extremes and heavier precipitation, but that the evidence is less strong for droughts, tropical cyclone activity, and ocean warming.

What about the future of these collateral impacts? The IPCC predicts (IPCC-SPM 2021, pp. 19–25) that in the case of those collateral impacts where the evidence is strong, conditions will continue to get worse with high levels of certainty. Additionally, the IPCC predicts that permafrost thawing and seasonal snow cover loss will be amplified. However, in the case of those collateral impacts where the evidence is less strong (e.g., droughts, cyclone activity) extreme effects may be regional rather than global.

CORPORATE SHARE OF EMISSIONS: US, EU, AND CHINA

Corporate reporting of emissions is voluntary in most parts of the world. The exception is the European Union (EU), where an emissions trading system (ETS) – typically a 'cap-and-trade' system that imposes a limit ('cap') on the quantity of emissions and allows firms that emit less than the cap to sell their excess emissions to those that exceed the cap – covers a sizeable segment of the corporate sector and therefore compels reporting. As a result, comprehensive global firm-level data are not possible to obtain.[11] However, guidelines under the United Nations Framework Convention on Climate Change (UNFCCC) require member countries to report their "inventory" of direct GHG emissions once every few years, using a (substantially) common methodology.[12] I use country-level GHG inventory reports of direct emissions by various entities to estimate, in broad terms – and admittedly with estimation error – the share of corporate emissions as a percentage of total direct economy-wide emissions.

12 *Handbook of business and climate change*

The current requirement from UNFCCC is for countries to document their emissions inventory through 2019, and for the reports to be submitted before end-2021. Many countries and regions have already reported, including the US and members of the EU. However, China – now the largest emitter in the world – was a holdout, having provided data only through 2014 at the time of writing.[13] Together (and, unfortunately, using only estimates for China), in 2018, these three geographic areas were estimated to account for nearly half of all global GHG emissions, with China accounting for a 26 percent share, followed by the US at 13 percent, and the EU (+UK) 9 percent (Ge, Friedrich and Vigna, 2020).[14]

GHG emissions and sinks (removals) under UNFCCC guidelines are reported by categories of GHG: CO_2, CH_4, N_2O, and HFCs/PFCs/SF_6/Related (see Note 2 for a description of these gases and their global shares). By far the largest category is CO_2, within which, combustion of fossil fuels for energy dominates. Table 2.2 summarizes my analysis of the share of corporate direct emissions in the US, the EU and China, by GHG-type.

Table 2.2   *Corporate (direct) emissions in the US, the EU, China (2019)**

GHG-type and sector	Attribution to 'Corporate'	US emissions ($MtCO_2e$)	EU emissions ($MtCO_2e$)	China emissions** ($MtCO_2e$)
CO_2: Fossil fuel combustion – electric power	100%	1,606.0	824.0	4,679.7
CO_2: Fossil fuel combustion – industrial processes	100%	822.5	494.4	4,010.8
CO_2: Fossil fuel combustion – commercial***	100%	249.7	131.8	535.3
CO_2: Fossil fuel combustion – transportation	50%	908.6	445.0	480.3
CO_2: Non-energy use	0%	0.0	0.0	0.0
CO_2: Other****	33–100%	252.5	303.2	1,547.6
CH_4: Fossil fuel systems	100%	244.1	39.9	NA
CH_4: Industrial (incl. waste treatment)	100%	145.7	131.3	3.4
CH_4: Other (primarily farming- or energy-related)	10%–33%	27.0	13.2	79.5
N_2O	10%	45.7	25.5	4.6
HFCs/PFCs/SF_6/related	100%	181.5	106.0	1.0
Total corporate & industrial emissions (range)	–	4,507.1 to 4,589.6	2,410.6 to 2,514.7	10,807.4 to 11,342.2
Total emissions of the country (or region)	–	6,558.3	4,067.0	14,093.0
Corporate emissions as percent total	–	68.7% to 72.1%	59.3% to 61.8%	76.7% to 80.5%

Sources: US: Analysis by author based on: "Inventory of US GHG Emissions and Sinks 1990-2019", EPA 430-R-21-005, 2021 (EPA, 2021); EU: "Annual EU GHG Inventory 1990–2019 and Inventory Report 2021," EEA/PUBL/2021/066, 2021 (European Environment Agency, 2021); China: "The People's Republic of China Second Biennial Update Report on Climate Change," (Chinese Ministry of Ecology and Environment, 2018).
* '$MtCO_2e$' refers to million tCO_2e; For questions or further details on methods and assumptions used, contact the author.
** Data for China are based on emissions proportions from the latest available inventory extrapolated on the basis of estimated country-level emissions in 2019 of 14,093 $MtCO_2e$ (see Note 13).
*** For China, what is shown here under "Commercial" is actually categorized as "Other Sectors" in their report (75 percent attribution).
**** For China, what is shown here under "CO_2: Other" is actually categorized as "Industrial Processes" in their report.

As Table 2.2 shows, businesses account for the vast majority of direct emissions in all three geographic areas. According to the calculations above, corporate direct emissions account for approximately 70 percent of total emissions in the US, 60 percent in the EU, and nearly 80 percent in China.[15] In turn, a substantial proportion of corporate emissions, over 85 percent, is from CO_2. Compared with the US and the EU, which produce three-quarters of their direct emissions from combusting fossil fuels for energy consumption, China emits a larger, approximately 85 percent, share.

FRAMING CORPORATE RESPONSES TO CLIMATE CHANGE

Key Elements of the 'Climate Economy'

Consider a bathtub into which water flows in at a rate faster than it drains.[16] The predictable outcome is that the tub will overflow, causing water damage. There are only a handful of ways to fix the problem: turn down the faucet so that the rate of inflow becomes less than the rate of outflow, or find the means to bathe with less water (or using means other than water), or increase the size of the drain so that the rate of outflow matches or exceeds the rate of inflow. The analogy to GHG emissions should be obvious. Emissions are currently growing at a rate faster than the ability of land and sea systems to absorb them, with the unabsorbed portion ending up in the earth's atmosphere. CO_2, almost 75 percent of GHG emissions, stays in the atmosphere for a very long time. This cumulative buildup of emissions in the atmosphere – the increasing concentration from the overflow which, as noted before, is currently happening at a rate of 2.3–2.5 ppm annually – causes global warming, in turn resulting in impacts on climate and the earth's ecosystems.

Working off the bathtub analogy, the only three solutions to the warming-causing effects of emissions are equally obvious: one, we can become more carbon-efficient in our combustion of fossil fuels (i.e., produce the same output by combusting lower amounts of fossil fuels thus emitting less carbon); two, we can find alternatives in the form of non-carbon sources of energy (i.e., invest in non-fossil fuel-based sources of energy, such as solar, wind, nuclear, hydro, geothermal, and so forth); or three, we can capture the emitted carbon and store it or find alternative uses for it (i.e., invest in carbon capture, storage and alternative use technologies, sometimes abbreviated as 'CCSU'). A necessary additional aspect of the climate economy, one that is required to sustain solutions to climate change, is for society to put "a price on carbon" to give emitters the incentive to invest in emissions abatement technologies by internalizing the costs of the externalities arising from their actions. Thus, we can think of these responses to addressing GHG emissions – carbon efficiency, non-carbon sources of energy, CCSU, and carbon price – as the four key elements of the emerging 'climate economy.'

Since a vast majority of the world's energy needs – well over 80 percent – is met via carbon-emitting sources[17] (i.e., coal, petroleum and natural gas), investments in carbon efficiency are tantamount to investments in energy efficiency. With respect to non-carbon sources of energy, there is currently a finite set of choices, none perfect and each with its own pros and cons (Barrett 2009): solar (photovoltaic (PV) or concentrated solar power (CSP)), wind (onshore or offshore), nuclear (increasingly 3rd, 3+th and 4th generation technologies), hydro (increasingly 'small' hydro rather than investments in multi-gigawatt hydroelectric projects), biofuels (e.g., woodchips, crop residues, ethanol), geothermal, and, somewhat more distantly in time, tidal/ocean energy. CCSU technologies are still at early stages of implementation to address the problem in any meaningful way. The major choices with CCSU are: capture directly from where the energy is produced and transport to store elsewhere in land or at sea,

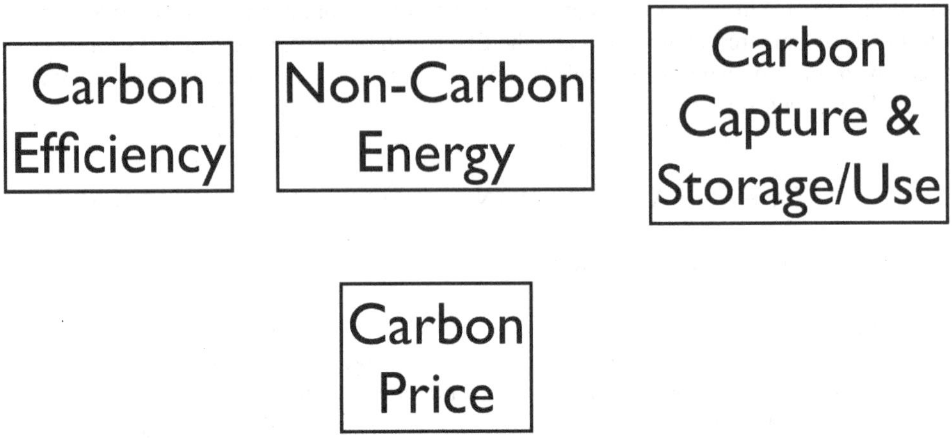

Figure 2.2   Elements of the climate economy

biomass capture and storage (e.g., tree-planting), ocean fertilization and increasing ocean alkalinity, or industrial air capture (Barrett, 2009). Alternative use technologies (e.g., converting the captured carbon into concrete that can be used in construction) are still in early stages of development and are small-scale.[18]

The role of carbon pricing has been discussed extensively in the economics literature – see, for example, Gillingham and Stock (2018), and Hansen (2023, Chapter 17 in this volume). It is addressed here only in passing. Tol (2009, p. 39) notes that the optimal carbon price would be a "…Pigouvian tax[19] that would be placed on carbon…" that is equal to the marginal damage cost to society from carbon emissions, often referred to as the "marginal social cost of carbon."

If market-based solutions (as opposed to mandates) are the approach used, such a price could be directly imposed as a tax by the government on every ton of GHG emitted, or it could emerge in a so-called "cap-and-trade" system whereby policymakers restrict quantities emitted and allow those who have excess carbon (the more carbon-efficient firms) to sell to those who are looking for carbon to buy (the less carbon-efficient firms). Such a tax can also be self-imposed by firms as an "internal carbon price" to change intra-corporate behaviors with respect to emissions. In other words, an external carbon price imposed by governments can coexist with internal carbon prices set by companies.

Based on a survey of approximately 5,600 companies, the Carbon Disclosure Project (CDP) finds that 853 firms globally report some type of internal carbon pricing initiative, and another 1,159 "…anticipate doing so in the next two years" (Carbon Disclosure Project 2021, p. 7). While some firms direct or set-aside capital budgeting dollars to projects that enhance energy efficiency, half the firms use an internal carbon price as a "shadow" price, i.e., one where no actual fee is assessed, but rather, the price is used as a means to do a "what-if" calculation to get managers and divisions to adapt to the possibility of a future carbon-priced world.

Corporate Approaches to Emissions Reductions

Many firms now undertake initiatives to improve the energy/carbon efficiency of operations directly under their control, e.g., by focusing on energy used in raw material extraction and use, and making physical investments (e.g., buildings, machinery, inbound and outbound transportation and logistics systems) more energy-efficient. Others claim "low-carbon design innovations" in product redesign and packaging, or via reorienting research & development (R&D) efforts towards greater energy efficiency in products and processes. Firms also work on the upstream and downstream parts of their value chain to address energy efficiency of activities that, even though only indirectly under their control, are seen as a consequence of the product or service that they produce and sell. Approaches here include improving supplier efficiencies, better supplier sourcing with an eye to energy efficiency, reducing the energy consumed during product use, and making product end-of-life (including recycling) decisions easier and more efficient for consumers. (The chapters in this volume by O'Brien-Bernini and Amanda Meehan 2023, for Owens Corning, and by Etchebehere 2023, for Air France, provide numerous specific examples of such approaches by companies; see also Apple's climate strategy in Box 2.2 below.)

Five sectors account for all GHG emissions in an economy: Industry, Electricity, Agriculture, Residential & Commercial Buildings, and Transportation. The emissions share of each of these sectors in 2019 (latest available data at the time of writing) is:[20] Transportation emits the most in the US with a 29 percent share of emissions, followed by Electricity and Industrial Processes at about a quarter each, Buildings 13 percent and Agriculture 10 percent.

Box 2.1 provides a snapshot of the array of carbon efficiency technologies that are being implemented (or gaining traction) in these five sectors.

BOX 2.1 ENERGY/CARBON EFFICIENCY EXAMPLES IN VARIOUS SECTORS

Sector	Examples of carbon efficiency technologies
Industrial processes	Power control systems, power management systems, industrial automation, predictive maintenance, energy storage technologies, better waste-water treatment, employee incentives and initiatives to enhance energy efficiency (e.g., shop-floor performance, commuting, travel), supplier incentives and initiatives (e.g., sourcing more energy-efficient raw material inputs), enabling customer/consumer behavioral changes.
Electricity	Enhanced power plant efficiency via advanced software and tools, lowered transmission & distribution losses, smart grids, smart meters, microgrids/micro-generation, fuel substitution (e.g., from coal to natural gas), combined heat & power, energy storage technologies.
Agriculture	Reforestation, avoided deforestation, tall grass planting, conservation tillage, soil erosion control, soil restoration, carbon mineralization, developing resilient (e.g., heat-resistant) crops.
Residential/commercial buildings	Efficient lighting (e.g., switching to LEDs and CFLs; light sensors), efficient heating/ventilation/cooling (HVAC) systems, sun-proofing, energy efficient appliances, better insulation, building electronics (software and control systems), better building standards (e.g., LEED), natural light/heat, building 'fabrics' such as aerogel insulation, double shell building systems and active windows, geothermal heat pumps.
Transportation	Electric vehicles, hybrids (including plug-in hybrids), hydrogen vehicles, more efficient batteries and storage technologies, modal switch (e.g., airplanes to trucks to rail), synthetic hydrocarbons, air capture vehicles, synthetic biology, digitization for efficient movement of freight.

Companies also now routinely invest in non-carbon sources of energy to power their operations. Investments in solar and wind projects have become common, with varied types of contracting approaches used: direct investments, co-equity investments via joint ventures, and long-term renewable energy contracts (see the chapters in this volume by Fisher, Phillips and Scovic, 2023, and Phillips, Fisher and Scovic, 2023, for more details on such contracts). Firms also work with their suppliers and employees to incentivize non-CO_2 transportation modes.

Where efficiency or non-CO_2 opportunities are unavailable or insufficient, firms directly invest in or enable carbon removal projects (i.e., offsets), often via "nature-based" solutions including tree- and tall grass-planting, investing in forest management practices and in soil carbon restoration. Offsets for employee travel are not uncommon. Finally, as mentioned above, hundreds of companies globally report using some form of internal carbon prices, but whether those amount to more than shadow prices is still up for debate. Figure 2.3 summarizes.

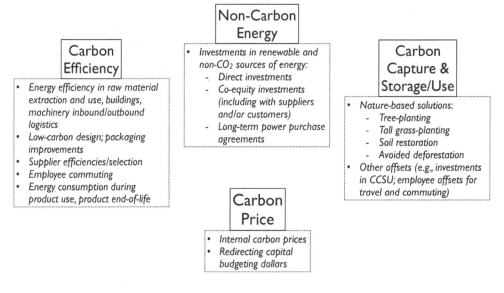

Figure 2.3   Corporate approaches to emissions reductions

Box 2.2 discusses the example of Apple's approaches for carbon reduction (see also the chapters in this volume by O'Brien-Bernini and Meehan, 2023, and Etchebehere, 2023).

BOX 2.2   APPLE'S CARBON STRATEGY[21]

In Fiscal year 2019, Apple – currently the most profitable and valuable company in the world, with a market value approaching $3 trillion at the time of writing, and annual revenues (in the prior four quarters as of the time of writing) of $350 billion – reported that its carbon footprint across the entire value chain totaled 25.1 $MtCO_2e$. The company has little in direct emissions (also called 'Scope 1' emissions, defined and explained in more detail in the next Section "Measurement and Reporting of Corporate Emissions"), with just 50,000 metric tons directly emitted. The company has become carbon-neutral for all of its purchased energy, which consists almost entirely of electricity (also called 'Scope 2' emissions, see next section). Thus, nearly all of its GHG emissions were in upstream and downstream parts of its value chain (also called 'Scope 3' emissions). In Apple's case, its Scope 3 emissions consist of manufacturing (which is almost fully outsourced), product transportation, product use and disposal/recycling, and employee travel and commuting.

Apple's manufacturing accounted for 76 percent of value-chain emissions in 2019, followed by product use (14 percent), transport (5 percent), and employee commuting and travel (3 percent). As of 2020, Apple has set a goal of achieving "total carbon-neutrality", i.e., neutrality across its entire value-chain by the year 2030. Apple provides numerous examples of specific carbon-reduction strategies. The following lists Apple's major strategies, in key categories. (The company gets all of its claimed reductions externally verified.)

Energy Efficiency and Low-Carbon Design

- By increasing efficiency and using recycled and other low-carbon aluminum in products, Apple claims to have decreased its carbon footprint by 4.3 million metric tons in 2019 relative to 2015.
- Apple pursues material use efficiency and use of low-carbon materials for its products. For example, using a forging method in the manufacture of steel for its iPhone 11 Pro, Apple claims it was able to reduce steel use by 30 percent, resulting in savings of 1 kg of CO_2e per iPhone. (Considering that Apple sells over 200 million iPhones per year, this innovation for its iPhone-related steel alone would represent a 200,000 tCO_2e annual emissions reduction). By using a hollow extrusion process for aluminum, the company claims it reduced its need for the metal by 50 percent in its Apple Watch Series 5, thus reducing the CO_2 impact of its aluminum by 25 percent.
- Apple claims to have reduced average energy use in its products by 73 percent in the prior 11 years.
- Internal energy efficiency programs in its data centers, retail outlets and office buildings avoided 264,000 MWh of electricity consumption annually by 2019, avoiding 7,500 tCO_2e per year.
- Energy efficiency initiatives in its supply chain avoided 779,000 tCO_2e in 2019. Apple created a "green" fund to initiate $100 million in efficiency-related investments for its Chinese suppliers.

Non-Carbon Sources of Energy (incl. Supplier Initiatives)

- Between 2013 and 2019, Apple invested $2.5 billion on projects to generate renewable electricity. Through direct ownership (12 percent of projects), equity investment (4 percent) and long-term renewable energy (84 percent) contracts, over four-fifths of the renewable energy that it sources for its direct energy needs were from Apple-created projects by 2019. The company estimates that such investments avoided nearly 900,000 tCO_2e in emissions annually.
- In 2015, Apple created a "Supplier Clean Energy Program" to help suppliers reduce emissions in its supply chain. By FY2019-end, Apple had worked with suppliers to install 2.7GW of renewable energy capacity. Its suppliers consumed 5.7 million MWh of renewable energy annually from these investments, the equivalent of avoiding 4.4

18 *Handbook of business and climate change*

> million tCO$_2$e of emissions annually. The company says it had supplier commitments for another 5.1GW which, together with the previous 2.7 GW, are expected to avoid 14.3 million tCO$_2$e annually, i.e., three-quarters of its total supply chain emissions.
> - Apple works with governments locally and globally to create carbon accounting, verification, regulatory and monitoring systems and infrastructure.
>
> **CCSU/Offsets**
>
> - To achieve its total carbon-neutrality goal, Apple forecasts the need to annually remove 10 million tCO$_2$e in emissions using CCSU and carbon offsets by 2030.
> - Apple works with NGOs such as Conservation International, the Conservation Fund, and World Wide Fund for Nature on mangrove restoration in Colombia, savanna conservation in Kenya and sustainable forestry in China. The company expects its existing initiatives to have the potential to remove 1 to 2 million tCO$_2$e of emissions annually, and plans additional projects in the future.
> - Apple does not report having any internal carbon pricing initiatives.
>
> Altogether, Apple estimates that, relative to 2015, these actions led to the company abating 10.7 million tCO$_2$e annually across its entire value chain by 2019. When compared with its 2019 total value-chain emissions of 25.1 million tCO$_2$e, it implies that Apple would have had business-as-usual (BAU) emissions of 35.8 million tCO$_2$e in the absence of its GHG-reduction strategies. In other words, the company claims it had achieved a 35 percent reduction compared with BAU emissions in 2019 (relative to 2015). Of the 10.7 million tCO$_2$e avoided, energy efficiency accounted for 40 percent; supplier-related efforts (in the suppliers' own facilities as well as their renewable energy investments from Apple's involvement) accounted for 59 percent. All of Apple's other initiatives combined, including CCSU, accounted for just 1 percent.

MEASUREMENT AND REPORTING OF CORPORATE EMISSIONS

Emissions Measurement

As an oft-quoted adage goes, "if you can't measure it, you can't manage it."[22] While it is arguable whether the claim is universally true, it certainly seems apropos in the climate context. One thing is obvious: it would be difficult, if not impossible, to model the price of something – such as carbon emissions – without knowing its quantity, just as it would not be possible to cap emissions in a cap-and-trade system without being able to measure quantity emitted.

A somewhat unheralded development that substantially advanced our ability to manage emissions is the measurement protocol developed in the late 1990s by the Greenhouse Gas Protocol Initiative (often referred to as "GHGProtocol.org" or "GHGProtocol"). GHGProtocol was created as a joint effort between two NGOs, World Resources Institute (WRI) and World Business Council for Sustainable Development (WBCSD), working with a group of companies looking for guidance to measure carbon footprints. The goal of this effort was to create an internationally accepted GHG accounting/reporting standard for businesses, and promote its adoption globally (GHGProtocol, 2004). Methodologies developed by GHGProtocol have now become the most widely-used emissions measurement and GHG accounting protocol in the world.[23]

The protocol categorizes emissions into three types, based on the "scopes" or operational boundaries of the emitter (see GHGProtocol, 2004, pp. 17–33 for more detail). Scope 1 emissions, also called "direct emissions" arise from sources that are owned or controlled by the company. Examples would include boilers, furnaces or vehicles that are owned by the company that combust fossil fuels that the company purchases.

Indirect emissions, "…a consequence of the activities of the company but occur at sources owned or controlled by another company…" (GHGProtocol, 2004, p. 25), are split into two categories, Scopes 2 and 3: Scope 2 emissions are indirect GHG emissions from purchased electricity consumed by a company, where the emissions actually happen outside the boundary of the company. (In other words, a company's Scope 2 emissions are the Scope 1 emissions of the utility that generates the power.) Scope 3 emissions are indirect GHG emissions associated with the rest of the value chain, from raw material extraction, to processing, to manufacture and assembly, to inbound logistics – all of which could be construed as a company's supply chain – to outbound logistics, to the wholesale/retail chain, through to customer use and finally, customer disposal/product recycling and recovery.

Companies have developed their own, now increasingly widely-used parlance that redefines these boundaries. For example, firms refer to "Operational Emissions," a phrase that typically comprises Scope 1 *plus* Scope 2 emissions, *plus* emissions associated with employee travel and commuting.[24] Similarly, firms refer to emissions across the value chain, i.e., Scopes 1+2+3 emissions, as "total emissions," "value-chain emissions" or "life-cycle emissions" (see Box 2.2, "Apple's Carbon Strategy"). Linking back to the prior discussion (see Figure 2.3, for example), note that reductions in any or all of Scopes 1, 2 and 3 can be achieved via carbon efficiency, non-carbon energy, and CCSU.

Corporate Emissions Reporting in the US (and Comparisons to Europe)

CDP gathers data from companies on their Scopes 1, 2 and 3 emissions, as well on whether and how they manage their emissions, e.g., whether the companies get their emissions measurements externally verified, whether they set absolute or relative emissions targets, whether managers are incentivized on the basis of achieving those targets, whether there is Board/senior management oversight of the company's climate strategies, and additional data such as energy use. Figure 2.4 summarizes evidence as of 2017, based on data that CDP reported in 2019, for the S&P500 group of companies.[25]

In 2017, 318 (or 64 percent of) S&P 500 companies reported their emissions publicly or privately to CDP.[26] However, less than 50 percent made public their Scope 1 (47 percent) and Scope 2 (46 percent) emissions data. Public reporting of Scope 3 emissions was less prevalent, with just one-fifth reporting. About 70 percent of the companies reporting Scopes 1 and 2 emissions said they got their disclosure externally verified. Nearly all of the companies reporting Scopes 1 and 2 said they had either a Board-level committee or a senior manager overseeing climate issues. Nine out of ten of those reporting said they incorporated emissions management into managerial incentive compensation systems.

Reporting rates are higher in Europe. The Intercontinental Exchange Inc. (ICE), a well-known provider of data, infrastructure, and technology for global financial markets, reports that in the STOXX 600 Europe Index (similar to the S&P 500 in the US), 74 percent of EU companies

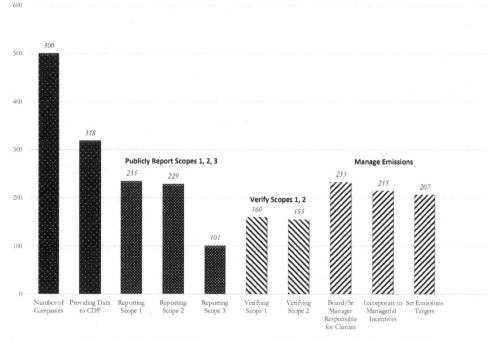

Source: Analysis by author from CDP 2019 corporate reports database (data are for the year 2017).

Figure 2.4   *Voluntary climate reporting and management by S&P 500 companies (2017)*

report both Scopes 1 and 2 emissions (Intercontinental Exchange, 2021). In a separate analysis of CDP data for 2015, I found that 87 percent of the largest 300 EU stocks were reporting Scope 1 emissions, with 86 percent saying they had Board or senior executive oversight, 80 percent saying they set emissions reduction targets and had managerial incentives, and 74 percent that they externally verified Scope 1 emissions.[27]

Data Quality Issues: Scopes 1 v. 2 v. 3

Despite concerns arising from the fact that emissions disclosure is mostly voluntary and from potential biases that therefore arise in selection, coverage, estimation methodologies and external verification, we can make some broad generalizations about the quality of reported data.

Given the wide availability of regularly updated, easily accessible emissions conversion factors for major fossil fuel types, the measurement and reporting of Scope 1 emissions from combusting fossil fuels is relatively straightforward to do. It is easy, even for a small firm, to obtain, maintain, and update data on its direct emissions.[28] All it requires is knowing how much of each type of fossil fuel the firm consumes, then using the available conversion factors. Methane (CH_4) and nitrous oxide (N_2O) are less of an issue for most businesses since, recalling earlier analysis, over four-fifths of GHG emissions in the corporate sector comprise CO_2.

Scope 2 emissions are similarly straightforward to calculate and report. In most countries (again, more so in the advanced economies), there are governmental environmental agencies, such as the US Environmental Protection Agency (EPA), that maintain and update the average emissions content of every kWh of electricity produced and transmitted, broken down by grid regions. Despite some issues that can arise – for example, fuel mix can differ depending on the time of day when electricity is produced, or emissions content of sub-regions may differ from the average for a larger region of which it is a part – estimates are close enough to be useful, on average. As an example, via its "eGRID" initiative – which stands for Emissions & Generation Resource Integrated Database – the US EPA provides data on environmental characteristics of electric power generated in the US, including emissions for 27 sub-regional grids.[29] As with Scope 1, calculating a firm's Scope 2 emissions does not require much more than knowing the number of kilowatt-hours of electricity the firm consumes from each of its grids and then using emissions conversion factors associated with its grids.

Accurate disclosure of Scope 3 emissions – i.e., emissions in the upstream (e.g., supplier) and downstream (e.g., customer) parts of the firm's value chain reflecting the life cycle of the product or service – remains a challenge. First, despite efforts by standard-setters, there is, as yet, no clearly agreed-upon categorizations that firms use to classify components of their value chain for emissions reporting purposes, even within the same industry. As a result, it is not only difficult to undertake comparisons of quantities of Scope 3 emitted among peers, but it is nearly impossible to make comparisons of firm strategies in tackling their upstream and downstream emissions (see Box 2.3 for the case of Scope 3 emissions reporting by leading technology companies).

BOX 2.3   SCOPE 3 EMISSIONS REPORTING BY LEADING TECHNOLOGY COMPANIES

Consider the case of seven well-known technology companies that compete against each other across a wide range of products and services: Apple, Microsoft, Facebook, Amazon, Alphabet, Dell and HP. Using a commonly-used framework of categories of classifying Scope 3 emissions, namely manufacturing,[30] product transportation, product use, end-of-life, and 'other' Scope 3 emissions, three out of the seven – Apple, HP, and Microsoft – provide reasonable detail under each. Dell does not separate out information on transportation or end-of-life, while Amazon and Alphabet do not separate out information on product use or end-of-life. Facebook's reporting is the opaquest, with no information provided on manufacturing, product transportation, product use or end-of-life; rather, the company reports just one number under a category called "other" (Bolduc and Sundaram, 2020, Exhibit 5, p. 17).

Given such inconsistencies, it is pointless to attempt peer comparisons on Scope 3 emissions across these companies. It is a problem that many companies recognize, but given the lack of clear standards and the mostly voluntary nature of the disclosure, most companies make-do. Some just refuse to report. For example, energy company EOG Resources Inc. noted in its 2019 sustainability report that it "… does not believe that it is able to calculate Scope 3 emissions with the accuracy and rigor typically required…" for data it makes public (Eaglesham and Shifflett, 2021), and hence chooses not to.

Second, even when data are made available, firms frequently resort to Scope 3 estimation methodologies and conversion factors that are stale, or whose provenance is ill-explained. Numbers rounded off to the nearest one hundred-thousand are common.[31] Third, even when better data or more robust conversion factors subsequently become available, it is not obvious whether companies (or data providers) go back and change numbers for prior years, and if they do, how they do so. Without such adjustments and explanations, data will not be easily comparable across time. Credible time series analysis becomes difficult, if not impossible.

The cases of Apple and Microsoft – two companies that publicly report a great deal of climate data and that have set aggressive total (i.e., Scopes 1+2+3) carbon-neutrality goals – are illustrative. In its 2020 environmental report (data for 2019 emissions), Apple notes its reported 25 MtCO$_2$e would have been 23 MtCO$_2$e instead, i.e., 7 percent lower, had it used the same methodology in 2019 as it did in 2018 (Apple 2020, p. 13).

Similarly, in its 2018 environmental report (data for 2017 emissions), Microsoft (2018) reported 14 MtCO$_2$e of purchased goods & services and manufacturing Scope 3 emissions. However, its 2021 report (Microsoft, 2021, p. 73) revises that 2017 number downwards to 5.76 MtCO$_2$e, i.e., a reduction of nearly 60 percent. The change in methodology that led to this drastically lower number is not explained. Moreover, in estimating its revised 2017 Scope 3 emissions in 2020, the company notes that its estimates (in 2020) are based on "cradle-to-gate" emissions factors by sector from UK Defra's "UK Defra Table 13 – Indirect Emissions from the Supply Chain, March 2014" (without explaining the terms "cradle-to-gate" or "UK's Defra"): in other words, data reported by the company in 2021, as a part of its 2020 emissions calculations, the data for which happened to get revised for the year 2017, used 2014 conversion factors (Microsoft 2021, p. 82; see also Eaglesham, 2021).

In summary, while Scopes 1 and 2 emissions reporting is generally credible and can be externally validated with some degree of accuracy, the quality, provenance, methodologies and content of Scope 3 reporting leave a lot to be desired among even sophisticated companies, even in advanced economies such as in the US.

That said, the good news is that if every company were to report its Scope 1 emissions – a requirement that would *not* be particularly burdensome even for small businesses, as noted above – by geographic location depending on where its operations are located, that would take us a considerable way towards building a comprehensive picture of aggregate corporate emissions since, after all, total emissions of the corporate sector in a country/region are just the sum of all corporate Scope 1 emissions in that country/region.

Emissions Data and Rankings Providers

Aside from certain regions (e.g., the EU) and some countries/states where there exists some type of an emissions trading system (ETS) – and even then, where only a subset of firms are covered – most corporate emissions reporting is voluntary. Unlike other common types of disclosure (e.g., financial disclosure), emissions reporting is not 'required' disclosure. As a result, questions can be raised regarding various aspects of reported emissions data: selection (i.e., who chooses to report and why), coverage (i.e., what they report), methodologies (which protocols, which conversion factors for converting energy into emissions from which date and source), and whether the data disclosed are verified by an arms-length third-party (Clarkson, Grewal and Richardson, 2023, in this volume, discuss some of these issues; see also Busch, Johnson, Pioch and Kopp, 2018, and Kalesnik, Wilken, and Zynk, 2020).

For about two decades now, some large, publicly-traded corporations globally have been voluntarily disclosing emissions, mostly using GHGProtocol's standards, to two well-known NGOs: CDP and the Global Reporting Initiative (GRI). The number of firms reporting has grown steadily over the past two decades. For example, the number of firms reporting to CDP grew from a few hundred in the early 2000s to many thousands by 2020.[32]

A handful of data vendors use self-reported CDP/GRI data, along with corporate sustainability reports of those who do not report to CDP/GRI to *estimate* emissions for many thousands of publicly-traded firms across multiple sectors and countries. Examples include S&P's Trucost (which claims 15,000 companies in its database; S&P Trucost, 2020), Thomson's Refinitiv (">10,000 companies"; Refinitiv, 2021), MSCI's ESG Carbon Metrics (8,500 companies; MSCI, 2018), and Morningstar's Sustainalytics ("thousands of companies"; Sustainalytics, 2021).[33] These data are increasingly widely used in empirical work making its way into well-known journals (see, for example, Bolton and Kacperczyk, 2021).

Methodologies used to estimate data for firms that do not report their emissions are, at best, partially described, and even when described, few examples are provided. A careful examination of whether and how major data providers disclose emissions estimation methodology for non-reporting firms (S&P Trucost, 2020; Refintiv, 2021; MSCI, 2018; Sustainalytics, 2021) leads to the following conclusions. (1) Most of the estimation methods for non-reporting companies rely on emissions multiples based on dollar sales or number of employees. (2) None provide an example of a calculation for an actual company, nor do they reveal on their websites any information on how many firms actually report versus how many data points are estimated. The latter is a crucial piece of information since a buyer might assume that most are *actually reported* numbers, which they are not.[34] (3) None of the publicly available methodology descriptions provide information on how sectors, industries, or regions/countries are defined.

In summary, it appears that the quality of public emissions data from well-known databases is currently, at best, a work-in-progress, and should be used with considerable caution and caveats.

A Rapidly Evolving Disclosure Landscape

Following a 5–4 ruling by the US Supreme Court in 2007 that classified CO_2 and five other GHGs as "pollutants" thereby giving the US EPA the authority to regulate these gases,[35] the EPA instituted a GHG reporting program ('GHGRP') in October 2009. The program covered all facilities in the US that emitted more than 25,000 tCO_2e. That size threshold meant that most small businesses were exempted from these regulations. The EPA estimated that at the start in 2009, GHGRP covered approximately 8,000 emitting facilities in the US, accounting for 85–90 percent of total US GHG emissions (EPA, 2009).

In the following year, the US Securities and Exchange Commission (SEC) issued a "2010 Climate Change Guidance" noting that publicly traded companies in the US might be required to provide information on climate change-related risks and opportunities in disclosures related to business description, legal proceedings, risk factors and management discussion & analysis of financial condition and operating results (Securities and Exchange Commission, 2010). Pressure for ever greater disclosure of climate-related risks and opportunities is continuing to intensify. In December 2020, a subcommittee at the SEC issued a preliminary recommendation to broaden corporate disclosure to include environmental, social and governance

(ESG) factors. In March 2021, the SEC asked its staff to "…evaluate disclosure rules with an eye toward facilitating the disclosure of consistent, comparable, and reliable information on climate change" (Lee, 2021; HLS Forum on Corporate Governance, 2021). Following a public commenting and review process, in March 2022, the SEC proposed rules to enhance and standardize climate-related disclosures for investors. The main thrust of the rules are as follows (Securities & Exchange Commission, 2022):

> The proposed rule changes would require a registrant to disclose information about (1) the registrant's governance of climate-related risks and relevant risk management processes; (2) how any climate-related risks identified by the registrant have had or are likely to have a material impact on its business and consolidated financial statements, which may manifest over the short-, medium-, or long-term; (3) how any identified climate-related risks have affected or are likely to affect the registrant's strategy, business model, and outlook; and (4) the impact of climate-related events (severe weather events and other natural conditions) and transition activities on the line items of a registrant's consolidated financial statements, as well as on the financial estimates and assumptions used in the financial statements.
>
> For registrants that already conduct scenario analysis, have developed transition plans, or publicly set climate-related targets or goals, the proposed amendments would require certain disclosures to enable investors to understand those aspects of the registrants' climate risk management.
>
> The proposed rules also would require a registrant to disclose information about its direct greenhouse gas (GHG) emissions (Scope 1) and indirect emissions from purchased electricity or other forms of energy (Scope 2). In addition, a registrant would be required to disclose GHG emissions from upstream and downstream activities in its value chain (Scope 3), if material or if the registrant has set a GHG emissions target or goal that includes Scope 3 emissions. These proposals for GHG emissions disclosures would provide investors with decision-useful information to assess a registrant's exposure to, and management of, climate-related risks, and in particular transition risks. The proposed rules would provide a safe harbor for liability from Scope 3 emissions disclosure and an exemption from the Scope 3 emissions disclosure requirement for smaller reporting companies. The proposed disclosures are similar to those that many companies already provide based on broadly accepted disclosure frameworks, such as the Task Force on Climate-Related Financial Disclosures and the Greenhouse Gas Protocol.

The proposed rules are, at the time of writing, already a matter of wide public debate (see, for example, Eaglesham, 2022), and will likely be bogged down in litigation if and when formally adopted.

The EU has a longer history of climate regulation, because of its introduction of a cap-and-trade system in 2005, subsequent to the ratification of the UN's Kyoto Protocol in the 2002. Phase 1 of EU's emissions trading system (EU-ETS) was piloted during 2005–2007, covering 12,000 facilities primarily in the energy sector and among large industrial polluters emitting 2.2 $GtCO_2e$, accounting for roughly 50 percent of the region's GHG emissions (Bayer and Aklin, 2020). Phase 2 lasted from 2008–2012, Phase 3 from 2013–2020, and a Phase 4 (2021–2030) includes a broader set of sectors compared with Phases 1–3. Sectors and firms covered by the EU-ETS have been subject to GHG reporting requirements since 2005.[36]

In 2014, the European Commission issued the "Non-Financial Reporting Directive" (NFRD) requiring large firms to start annually disclosing, by 2018, information on a wide range of non-financial aspects of their operations: environmental (including climate-related) sustainability efforts, treatment of employees, human rights, anti-corruption/bribery initiatives, and Board diversity. However, following uneven success of these efforts – resulting from only a minority of companies providing comprehensive or reliable information (see, for example, Monciardini, Tapio and Tsagas, 2020) – the EU all but abandoned NFRD, revising its approach in 2021 to introduce a new, more far-reaching framework called the "Corporate Sustainability Reporting Directive" (CSRD).

CSRD will require all listed companies and large private companies in the EU, as well as subsidiaries of non-EU companies whose revenues in the EU reach certain size thresholds, to report on environmental, social and governance (ESG) aspects of their operations, their ESG targets and their progress toward achieving targets, as well as auditing and external assurance requirements. While the implementation is a work in progress and still a few years away, two key elements of CSRD bear a close watch for the degree of detail required, as well as for pushing the boundaries of corporate disclosure. The first is the notion of "double-materiality", the idea that firms must provide information not only on how climate change is material to their business models, but also the reverse, i.e., how the firm's operations are material to climate change (Harrison and Bancilhon, 2021). The second is compatibility with the so-called "EU Taxonomy," which lists disclosure requirements for financial products marketed as "green" or as being consistent with environmentally sustainable activities.[37] Also being considered by the EU are new supply chain-related reporting obligations.

These developments bear scrutiny for the likely costs – both direct and indirect – that will be imposed on firms, as well as the collateral set of standard-setting, advising, auditing and assurance businesses that will inevitably emerge so as to be able to make it all happen. Firms could face non-trivial compliance and verification costs. Potential conflicts between the US and the EU, resulting from their fundamentally different approaches to corporate disclosure, are highly likely. The actual implementation of these regulations may take much longer than expected.

LINKING CLIMATE CHANGE TO CORPORATE VALUE: A FRAMEWORK

Linking Climate Initiatives to Corporate Value: Cash Flow and Risk

Before we develop a framework for how businesses can think about making corporate financial decisions related to climate investments, it is useful to recapitulate the core set of ideas that underlie financial decision-making for any investment. A common – often unstated – presumption that is beneath this decision-making approach is that the key goal of a for-profit business is to maximize the (long-run) value of the business for its owners. What is this "value"? In financial terms, it is defined as the value today (i.e., the "present value") of the firm's expected future cash flows discounted back to the present using a discount rate (i.e., a "cost of capital") that reflects the risks of the firm's cash flows. All else being equal, the higher the expected future cash flows or the lower the risk (and therefore, lower the cost of capital) for a given set of cash flows, the higher the value of the firm.

Cash flows, more precisely called "free cash flows," are defined as revenues net of operating expenses (after-tax), from which investment expenditures required to generate those cash flows are subtracted. Cost of capital invokes the idea that when investment decisions are made, investors forego the opportunity to put the capital to use into a *risk-equivalent* asset elsewhere, and therefore that foregone return becomes the benchmark 'cost of capital' to judge the worth of the investment at hand. The cost of capital is thus the *opportunity cost of using capital*. The singular driver of the cost of capital, in turn, is the perceived operating and financial risk of the asset: the higher the risks, the higher the expected return for investors, and therefore, the higher the firm's cost of capital.

Irrespective of the type of decision that a firm makes, a simple and universal fact is that a financial decision becomes a part of its planning and resource allocation process only if it materially adds to the determinants of the firm's value. Thus, in this framework, value is created only by the following four ways:

1. *Increase revenue* for a given level of operating costs, investment efficiency and risk; or
2. *Decrease operating costs* for a given level of revenue, investment efficiency and risk; or
3. *Increase investment efficiency* for a given level of revenue, operating costs, and risk; or
4. *Decrease risks* for a given level of revenue, operating costs and investment efficiency.[38]

There is no other means by which something becomes strategic or part of the business model for a firm, regardless of what it is.[39] Thus, the challenge for a firm contemplating climate-related investments in either mitigating impact on climate (e.g., managing its emissions to respond to transition risk from climate change) or adapting to climate-induced risks (e.g., building resilience to physical risks from climate change) is to ask what investments are required,[40] then link those to the four determinants of firm value, revenue, operating costs, investment efficiency and risk.

There are only two ways to increase revenue: increase the price at which a product is sold or increase the quantity sold. For example, by becoming (or being seen as) a more climate-friendly corporation and thereby changing customer perceptions of a business model, are customers willing to pay more for that product relative to that of competitors that are seen as less climate-friendly? Will becoming a net-zero company result in the opportunity to brand the product or position the company more favorably in the customers' or the public's eyes and thereby charge a higher price? On the quantity front, are customers willing to buy more, or does climate action allow the firm to take market share away from less climate-friendly peers?

As of yet, no convincing evidence has emerged in the aggregate in the energy-consuming sectors that climate action and investments have produced revenue increases for firms. In other words, for companies that are not energy producers, there is little-to-no evidence that consumers have been willing or required to pay more for climate-friendly products, nor that climate action has led to gaining share at the expense of less climate-friendly competitors. In the energy-producing sector, it depends to a large extent on whether a firm is a producer of carbon-based or non-carbon-based energy. Evidence is compelling that there has been a shift in the consumption mix away from fossil fuels, especially coal. Non-carbon sources, especially wind, solar and biofuels, have gained in share while the share of nuclear and hydro has fallen. Yet, the share gain of non-carbon sources – correspondingly, the share loss of fossil fuels – in the total energy mix is not as impressive as we might expect, laying bare the challenges for the task ahead. Between 2000 and 2019, world energy consumption grew by about 50 percent (Ritchie and Roser, 2020). Yet, the share of fossil fuels fell only slightly, from 86 percent in 2000 to 84 percent in 2019 (and the share of nuclear plus hydro also fell, from 13 percent to 11 percent), while the shares of wind, solar, and biofuels increased from less than 1 percent in 2000 to 5 percent in 2019.[41] While the relative rate of growth in renewables is impressive, they still command only a small fraction of the world's energy output in absolute terms.

On the operating cost front, there could be possibilities for reductions in variable costs (e.g., cost of raw material and of other inputs that scale with output) and fixed costs (e.g., overheads, R&D expenses, advertising expenses, inputs that do not scale, or if they do, scale in discrete steps with output). Carbon-reduction strategies can result in lower variable costs if they result in more efficient fossil fuel use, lower packaging costs and lower inbound and outbound shipping/logistics costs. In industries where energy inputs play a significant role, such as utilities, transportation & logistics (including warehousing) and heavy industry, the impact on variable costs can be meaningful. Where carbon taxes or other forms of price on carbon exist, being carbon-efficient will lower a company's tax bill. On the fixed costs side, being seen as a less risky company – see below for more on risk – will lower the cost of insurance.

Climate-friendlier companies may be viewed as more attractive employers, resulting in higher levels of employee motivation and productivity, lowering the share of labor costs embedded in the firm's products. More subtly, there may be opportunities to link divisional/regional success in carbon mitigation to performance incentive systems. A culture of higher productivity resulting from efficiencies in fossil fuel use can create spillover effects, for example, by creating a similar efficiency lens via which employees approach consumption of other inputs from the natural environment, such as minerals, metals, water, air, flora and fauna. There could also be second-order impacts, whereby a firm compels competitors to match its actions, thus raising rivals' costs, creating greater pricing and profitability leeway for itself.

On the investment front, as discussed in the section "Corporate Approaches to Emissions Reductions", companies invest in product redesign (e.g., lower energy consumption during product use, disposal and recycling, or lower emissions associated with inbound/outbound transportation and warehousing for the product), packaging redesign (e.g., lower, eliminated or offset fiber content, mitigating the need for deforestation), re-orienting research & development (R&D) towards creating more carbon-efficient products and processes, or through investments in resilience building (see Peace and Huber, 2023, in this volume). Investments in renewable energy to offset Scopes 1 and 2 emissions, as well as co-investments with suppliers for Scope 3 emissions are becoming common. Many firms use internal carbon prices hoping to influence internal capital flows or change mindsets.

On the risk front, climate action has now become an important component of managing investor perceptions of risk. Firms face many sources of climate-related risks. The most commonly used categorization of climate risk, arising from the work of the Taskforce on Climate-related Financial Disclosure (TCFD), is "physical risk" and "transition risk" (TCFD 2017):

- *Transition risk*: These are the risks associated with transitioning to a low-carbon economy, which TCFD (2017, p. 5) defines as "… extensive policy, legal, technology, and market changes to address mitigation and adaptation requirements related to climate change," noting that, "[d]epending on the nature, speed, and focus of these changes, transition risks may pose varying levels of financial and reputational risk to organizations."
- *Physical risk*: These are risks resulting from climate-driven events and longer-term shifts in climate patterns. These include "…direct damage to assets and indirect impacts from supply chain disruption… [financial performance] affected by changes in water availability sourcing, and quality; food security; and extreme temperature changes in organizations' premises, operations, supply chain, transport needs, and employee safety" (TCFD 2017, p. 6).

From the standpoint of investors, if something is a source of risk,[42] a premium will be demanded to bear that risk. Thus, if GHG emissions are a source of risk, higher-emitting companies will be expected to compensate investors with a carbon premium. In other words, if carbon risk matters, investors will expect higher-emitting firms to produce *higher returns*. In turn, the cost of capital will be higher for such firms, which would, all else equal, lead to lower firm value.

Recent evidence does, indeed, bear out this hypothesis. Bolton and Kacperczyk (2021) find that stocks of firms with higher carbon emissions (and higher changes in carbon emissions) earn higher returns in the cross-section, after controlling for standard asset-pricing return predictors such as firm size, book-to-market value ratio (see also Gianfrate, Schoenmaker and Wasama, 2022 on the impact of carbon risks on cost of capital, and Sundaram, 2023 on the risk-return performance of ESG stocks, in this volume). Bolton and Kacperczyk find, for example, that a one standard deviation increase in the level of, and change in Scope 1 emissions is associated with a 1.8 percent and 3.1 percent increase in annual returns, respectively, an increase that is statistically and economically significant.[43] The authors conclude that the observed higher return most likely results from investors demanding a carbon risk premium.

Figure 2.5 summarizes the discussion above.

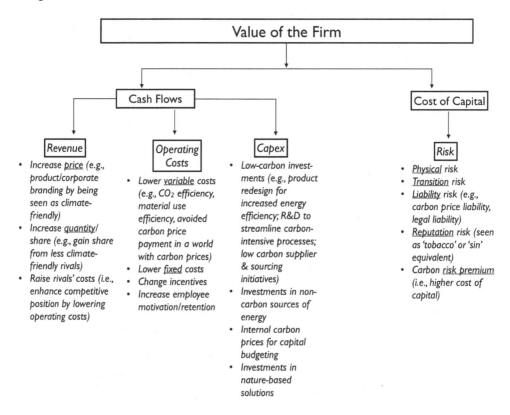

Figure 2.5   Firms' climate actions and impact on firm value

Is it possible to estimate the valuation impact of carbon prices on the corporate sector? If so, how might we assess the market value of the contingent liability that would be associated with firms' emissions in a carbon-priced world? Using evidence from Bolton and Kacperczyk (2021), Box 2.4 attempts to do so, with a rough-and-ready valuation impact analysis for the US equity market.

BOX 2.4 VALUATION OF US EQUITY MARKET LIABILITY IN A CARBON-PRICED WORLD

In 2017, approximately 1,050 publicly-traded US companies had average Scope 1 emissions of 1,922,550 tCO_2e, i.e., approximately 2 GtCO_2e in emissions (Bolton and Kacperczyk, 2021, Table 4, Panel B)). Assume that:

(i) Firms will be long-lived;
(ii) Long-run growth rate in cash flows will be offset by naturally-occurring improvement in emissions efficiencies; Scope 1 (direct) emissions for this group of firms thus continues to be 2 GtCO_2e;[44]
(iii) These firms approximate the Russell 1000 index of the largest 1,000 US firms, which accounts for between 90 percent and 95 percent of the market value of all publicly traded US companies;
(iv) The carbon price per tCO_2e and the discount rate for the firm's cash flows remain constant over time.

Under these assumptions, we can then derive a rough-and-ready estimate of the carbon price-contingent market value liability of firms, using the so-called "constant-growth model" of valuation. The constant-growth model says that, if "C" is the annual expected cash flow and the expected growth rate in C equals zero (which results from the assumption that the long-run growth rate in cash flows will be offset by naturally-occurring improvement in emissions efficiencies), and the discount rate is "r", then the present value of the cash flows is just C/r.

If the carbon price on a firm's direct emissions is, say, $60 per t$CO_2$e, and 2 Gt$CO_2$e is the annual emissions, the expected annual pre-tax cash outflow for this group of firms would be $120 billion. If we assume a (reasonable) 6.5 percent cost of equity capital in 2021 for the Russell 1000 group of firms (2 percent risk-free rate and an equity risk premium of 4.5 percent), the (pre-tax) equity value of the liability would then be $120 billion ÷ 0.065, i.e., or a little over $2.6 trillion. Given the aggregate equity market value of US firms of approximately $45 trillion (at the time of writing), and assuming the Russell 1000 represents 92.5 percent of US equity market value and a corporate tax rate of 25 percent, the value penalty for US firms would be 1.95 trillion ÷ 45.0 trillion, or 4.3 percent.

While this might, at first glance, seem like a somewhat small impact on firm value, it is important to note that this penalty is not uniformly distributed. Rather, it will be concentrated in a handful of hard-to-abate sectors: utilities, oil & gas, materials and industrials. These four sectors together account for only 16 percent of the market value of the US equity market at the time of writing, but are responsible for upwards of 90 percent of direct emissions (Bolton and Kacperczyk, 2021).

The implication is that these four sectors, which constitute the backbone of the manufacturing economy and, more broadly, the consumption of most physical goods (as well as energy) in the economy, would face a *25 percent market value penalty*. The consequences of such a penalty could be severely disruptive for energy consumers and for the global economy.

CONCLUSION

Climate change is a mega-issue of our time. Unlike most other such issues that confront us, climate change is unique in that businesses are implicated front-and-center. Businesses in high-emitting economies are the largest group of emitters. Yet the key solutions to slow climate change – carbon efficiency, non-carbon sources of energy, carbon capture and storage/reuse – also lie squarely within the purview of businesses, since they are the ones that will deploy the talent, the technologies and the resources to make this happen.

Collaterally with a carbon market that results in a carbon price that would be required to enable and sustain solutions to slow climate change, carbon efficiency, non-carbon sources of energy, carbon capture and storage/reuse are the four main pillars of an emerging global climate economy, an economy whose contours are already taking shape in meaningful ways.

The scale of the challenge, to constrain warming of the earth to 1.5–2°C, which requires the global economy to transition from one that emits 50 GtCO_2e annually today to one that is net-zero by the second half of this century, is immense. Equally, the scale of financing, planning and coordination required over the course of the next many years – supported by pragmatic, quickly-convergent global policies – cannot be overstated. It could well be the most consequential socio-economic experiment we run as a society in the next few decades, rivaled perhaps only by the industrial revolution – which, while substantially improving lives and life expectancies globally, is also (ironically) the progenitor of the climate problem.

Whether, and the speed at which, corporations can help address climate change remains to be seen, but rapidly- and steadily-growing evidence from the last two decades speaks to a seriousness-of-purpose amongst thousands of businesses globally, one that is compelling in its power to give us hope.

NOTES

1. I use the term "business" broadly, to refer to any entity that, regardless of ownership type, conducts business by providing a tangible good or rendering a service. It thus includes publicly traded and privately held for-profit firms, as well as non-profit firms and government entities. I also use the term interchangeably (and admittedly, somewhat loosely) with "corporations," "firms," and "companies" throughout. I am grateful to Robert Hansen, Anthea Lavallee and Corey Stock for their thoughtful comments, which improved the chapter.
2. 'GHG' is the precise term for the gases that cause atmospheric warming. The three main GHGs are carbon dioxide [CO_2], methane [CH_4] and nitrous oxide [N_2O]). In what follows, I will typically use "CO_2" or sometimes just "carbon" as shorthand references for GHG where it is unlikely to cause confusion. GHG emissions are measured on equivalents based on the "global warming potential" (GWP) of one metric ton (or "tonne," abbreviated hereafter as "t") of a GHG relative to the GWP of one tonne of carbon dioxide (tCO_2) measured over a 100-year period. In other words, emissions are measured on their CO_2-equivalents, or CO_2e, for short. These three gases account for 98 percent of emissions inflows (ClimateWatch 2021). CO_2 is the most prevalent GHG, accounting for nearly 75 percent of annual emissions. The CO_2es for these three GHGs are: CO_2 = 1×, CH_4 = 25×, and N_2O = 300×. In addition, there are a handful of so-called "F-gases" whose CO_2es are in the many thousands, even though their overall contribution to warming is small.
3. IPCC-SPM (2021, p. 9) notes, with "high confidence" that global temperatures have increased faster since 1970 than any other 50-year period in the prior 2,000 years.
4. Burning fossil fuels – coal, petroleum and natural gas – for energy consumption accounted for most of it.

5. There are three other scenario forecast models that are used by industry practitioners: one is the "IEA Sustainable Development Scenarios" or SDS (IEA 2021); the second is a project between Shell and MIT's Integrated Global System Model, called the "Shell Energy Transformation Scenarios" or SETS (Paltsev, Sokolov, Haigh, Hone and Morris, 2021); and the third is "Network for Greening the Financial System" NGFS by a consortium of central banks (Network for Greening the Financial System, 2020). For more on SDS and SETS, see Hone (2023) in this volume; for more on NGFS, see Anderson, Khaykin, Pyanet and Schuermann (2023) in this volume.
6. The 1.5°C target is commonly attributed to the Paris Climate Agreement in 2015 which specifies a somewhat vague goal of keeping "…warming well below 2°C above pre-industrial levels and pursuing efforts to limit the temperature increase to 1.5°C…" (United Nations, 2015). An IPCC Special Report published in 2018 (IPCC 2018) took the idea further and examined its implications in more detail. The 2°C target is attributed to a 1975 conjecture by economist William Nordhaus at Yale (CarbonBrief, 2014).
7. The data can be found at https://catalogue.ceda.ac.uk/uuid/ae4f1eb6fce24adcb92ddca1a7838a5c.
8. IPCC-SPM (2021, pp. 15–16).
9. In its "Emissions Gap Report", analysis by United Nations Environment Program (2021, Figure 3.1, p. 20) forecasts "illustrative pathways" to net zero emissions for a 1.5°C world will be achieved between 2050 and 2060, while the time frame would shift to between 2070 and 2080 for a 2°C world.
10. Oceans are estimated to absorb approximately one quarter of the CO_2 emissions. When CO_2 combines with water, chemical reactions result in the reduction of the pH level of the water, thus enhancing acidification.
11. A handful of data providers have now emerged that provide firm-level data for thousands of firms worldwide. Most of their data are estimates for a large set of companies based on disclosures by a smaller set – usually just a couple of thousand – of individual firms that report voluntarily to NGOs or via their company sustainability reports. Such extrapolations are typically based on multiples of some financial variable (such as revenues) or using sectoral and country-level input–output tables. See below for a more detailed discussion.
12. Created in 1992 (and operating since 1994), UNFCCC is the UN entity that is tasked with supporting the global response to climate change. Section 4 of the Convention says that member countries must "…develop, periodically update, publish and make available …national inventories of anthropogenic emissions by sources and removals by sinks of all greenhouse gases … using comparable methodologies" (United Nations 1992, p. 5). UNFCCC is the entity that is responsible for hosting the annual UN climate summit called the "Conference of Parties" (COP), and is the entity to which IPCC reports.
13. The data I report for China in Table 2.2 are based on emissions proportions derived from their latest UNFCCC report, which is from 2014 (but only published in 2018), which I extrapolate to 2019 on the basis of its aggregate 14.1 $GtCO_2e$ estimated for 2019 (Larsen et al., 2021).
14. India (7.1 percent global share of emissions), Russia (5.4 percent) and Japan (2.5 percent) are the next three largest emitters. These three countries, plus China, the US, and the EU(+UK) together account for 63 percent of global emissions.
15. Indeed, it seems reasonable to speculate that the corporate role in other major emitting countries – India, Russia, Japan, etc. – is not too different, most likely within the range of the numbers derived above.
16. This metaphor is attributed to Linda Sweeney and John Sterman at MIT.
17. Of the world's total energy consumption of approximately 162,000 Terawatt-hours (TWh) in 2019, coal accounted for a 27.0 percent share, oil 33.1 percent, and natural gas 24.2 percent, totaling 84.3 percent (Ritchie and Roser, 2020).
18. There are also 'geo-engineering' solutions that can lower the intensity of solar rays reaching the earth or alter the reflectivity of the earth's surface, i.e., 'albedo' (Barrett 2009). Examples of the former include mirrors in space to deflect sunlight or spraying large quantities of sulfate particles into the air to mimic the cooling effects from sunlight absorption of a volcanic explosion, while examples of the latter include large-scale planting of "shiny" crops or cloud-seeding to create whiter clouds that reflect sunlight.

19. A Pigouvian tax is a tax on entities that impose negative externalities – i.e., cause adverse side effects – on society through their private actions. The goal of the tax is to disincentivize the entity from imposing such externalities by undertaking efforts to reduce actions that cause the adverse side effects.
20. US EIA, https://www.epa.gov/ghgemissions/sources-greenhouse-gas-emissions.
21. This Box draws upon Apple's 2020 "Environmental Progress Report (Covering Fiscal Year 2019)" (https://www.apple.com/environment/pdf/Apple_Environmental_Progress_Report_2020.pdf) and the Tuck School classroom case "Green Apple: Environmental Sustainability Reporting and Climate Strategy at Apple" (Bolduc and Sundaram, 2020).
22. It is attributed to management consultant Peter Drucker, whose actual quote was, "If you can't measure it, you can't improve it." That version applies as well!
23. For example, GHGProtocol reports that, by 2016, 92 percent of Fortune 500 companies had adopted its standards (https://ghgprotocol.org/about-us).
24. Note that, although companies often assume responsibility for their "operational" emissions, their Scope 2 emissions are the Scope 1 emissions of its electricity generator. Similarly, employee commuting and travel are Scope 1 emissions of the employee or of the transportation companies that the employee uses to commute/travel.
25. The S&P 500 is a widely-used market value-weighted stock market index representing the 500 most valuable companies in the US. Together, this group of companies accounts for over 80 percent of the market value of publicly traded companies in the US stock market. The actual number of stocks in the index is slightly larger than 500 since some companies have multiple categories of shares.
26. Governance & Accountability Institute (2021) reports that as of 2020, this proportion had grown by only one percentage point, to 65 percent.
27. Interestingly, 2015 data for the largest 100 companies in China show only 10 percent reporting and Board/senior executive responsibility, just 5 percent setting targets and 8 percent with managerial incentives, and just 2 percent getting Scope 1 emissions verified. However, there is little doubt that these percentages have increased substantially since 2015 as China, starting in June 2021, has begun to operate an emissions trading system that purports to cover ~40 percent of its emissions (International Carbon Action Partnership, 2021).
28. For the US, see the emissions conversion factors here, for example: https://www.epa.gov/sites/default/files/2020-04/documents/ghg-emission-factors-hub.pdf; for global conversion factors, the IPCC maintains an "emissions factor database": see https://www.ipcc-nggip.iges.or.jp/EFDB/main.php
29. eGRID (US) data can be found here: https://www.epa.gov/egrid. Emissions factors for specific types of fossil fuels can be found here: https://www.epa.gov/climateleadership/ghg-emission-factors-hub.
30. This is also sometimes categorized as "purchased goods and services, and capital goods."
31. Such rounding is somewhat peculiar in a context in which Scopes 1 and 2 emissions are typically reported with single-digit levels of precision. For example, Microsoft reports 2019 Scope 1 emissions of "113,412" tCO_2e and Scope 2 emissions of "275,375" tCO_2e, but reports "6.3 million" tCO_2e in manufacturing Scope 3 emissions, a substantially larger number by many multiples, compared with the sum of its Scopes 1 and 2 emissions (Bolduc and Sundaram 2020, pp. 16–17).
32. CDP (2021), https://www.cdp.net/en/companies/companies-scores.
33. CDP itself, previously mostly a source of primary data, now competes with other data providers in this market and says that they provide data on approximately 5,000 companies. As with the other data providers, their estimates are also built using reported data for a smaller set of companies (Carbon Disclosure Project, 2020).
34. In most cases, the datasets do have a flag for this item at the level of an individual company's data. Thus, it would be relatively easy for them to tally and reveal this information – by year – in their methodology documents, if they chose to do so.
35. *Massachusetts v. Environmental Protection Agency*, 549 U.S. 497 (2007)
36. While prior studies have pointed to mixed success of the EU-ETS resulting from low carbon prices (and considerable volatility in prices), more recent evidence suggests that there have been modest reductions in GHG emissions in the EU relative to modeled baselines, with particularly substantial reductions in electricity generation emissions (Bayer and Aklin, 2020).

37. See the chapters in this volume by Clapp, Lee and Brisebois (2023) and Reichelt, Allen and Cantor (2023) for additional insights related to these developments. See also Gallagher, Bancilhon and Berrutti (2021).
38. The expected long-run growth rate of cash flows also matters since, under the assumption that return on an investment exceeds its cost of capital, a higher rate of long-run cash flow growth will lead to higher value. But this idea of growth in cash flows is subsumed under (1), (2) and (3) above.
39. Damodaran (2020) calls this the "It" Principle: "…for 'it' … to affect value, 'it' has to affect either the cash flows (through revenue growth, operating margins and investment efficiency) or the risk in those cash flows (which plays out in the cost of equity and capital)." See also Henisz et al. (2019).
40. See the fourth section above for a list of the most salient types of actions firms take. The concepts of "physical" and "transition" risks are explained in greater detail below.
41. Ritchie and Roser (2020 – "Energy"). Published online at OurWorldInData.org. Retrieved from: https://ourworldindata.org/energy; see the chart titled "Energy Consumption by Source, World."
42. In the theory and practice of finance, any risk that matters to investors is called a "systematic" risk. It is a measure of relative risk, i.e., risk that affects a firm's cash flows relative to the risk exposure of the average firm. For example, even if transitioning to a carbon-priced world negatively impacts cash flows of a firm, a higher-than-average carbon-intensive firm will be affected correspondingly more negatively, while a lower-than-average carbon-intensive firm will be affected less negatively, perhaps even positively.
43. Considering that the "market risk premium" in US markets, a catch-all measure of systemic market risk, is in the range of 4–5 percent at the time of writing (see for example Damodaran, 2021, who maintains and updates data on US and global equity risk premia), 1.8 percent and 3.1 percent are meaningfully high.
44. This comes from the assumption that the expected long-run increase in cash flows will be offset by the expected annual decrease in emissions.

REFERENCES

Andersen, E., Khaykin, I., Pyanet, A., & Schuermann, T. (2023). Banks and climate change risk. In A. K. Sundaram & R. G. Hansen (Eds.), *Handbook of Business and Climate Change*. Edward Elgar Publishing.

Apple. (2020). *Environmental progress report (covering fiscal year 2019)*. https://www.apple.com/environment/pdf/Apple_Environmental_Progress_Report_2020.pdf

Barrett, S. (2009). The coming global climate-technology revolution. *Journal of Economic Perspectives*, 23(2), 53–75.

Bayer, P., & Aklin, M. (2020). The European Union emissions trading system reduced CO_2 emissions despite low prices. *Proceedings of the National Academy of Sciences*, 177(16), 8804-8812. www.pnas.org/cgi/doi/10.1073/pnas.1918128117

Bolduc, M., & Sundaram, A. K. (2020). *Green Apple: Environmental sustainability reporting & climate strategy at Apple*. Case No. BCC-2020. Tuck School of Business at Dartmouth.

Bolton, P., & Kacperczyk, M. (2021). Do investors care about carbon risk? *Journal of Financial Economics*. https://doi.org/10.1016/j.jfineco.2021.05.008

Buis, A. (2019, October 9). The atmosphere: Getting a handle on carbon dioxide. *NOAA Global Climate Change: Vital Signs of the Planet Blog*. https://climate.nasa.gov/news/2915/the-atmosphere-getting-a-handle-on-carbon-dioxide/

Busch, T., Johnson, M., Pioch, T., and Kopp, M. (2018). *Consistency of corporate carbon emission data*. University of Hamburg/WWF. https://ec.europa.eu/jrc/sites/default/files/paper_timo_busch.pdf

CarbonBrief. (2014, December 8). *Two degrees: The history of climate change's speed limit*. https://www.ngfs.net/sites/default/files/medias/documents/820184_ngfs_scenarios_final_version_v6.pdf

Carbon Disclosure Project. (2020, January). *CDP full GHG emissions dataset – Technical annex III: Statistical framework*. https://6fefcbb86e61af1b2fc4-c70d8ead6ced550b4d987d7c03fcdd1d.ssl.cf3.rackcdn.com/comfy/cms/files/files/000/003/028/original/2020_01_06_GHG_Dataset_Statistical_Framework.pdf

Carbon Disclosure Project. (2021). *Putting a price on carbon: The state of internal carbon pricing by corporates globally*. https://cdn.cdp.net/cdp-production/cms/reports/documents/000/005/651/original/CDP_Global_Carbon_Price_report_2021.pdf?1618938446

Chinese Ministry of Ecology and Environment. (2018, December). *The People's Republic of China second biennial update report on climate change*. https://unfccc.int/sites/default/files/resource/China%202BUR_English.pdf

Clapp, C., Lee, K., & Brisebois, A. (2023). Green bonds: Investor, issuer and climate perspectives. In A. K. Sundaram & R. G. Hansen (Eds.), *Handbook of Business and Climate Change*. Edward Elgar Publishing.

Clarkson, P., Grewal, J., & Richardson, G. (2023). Equity value relevance of carbon emissions. In A. K. Sundaram & R. G. Hansen (Eds.), *Handbook of Business and Climate Change*. Edward Elgar Publishing.

ClimateWatch. (2021). Global Historical Emissions. https://www.climatewatchdata.org/ghg-emissions?breakBy=gas&chartType=area&end_year=2018&start_year=1990

Damodaran, A. (2020, September 21). Sounding good or doing good? A skeptical look at ESG. *Musings on Markets*. https://aswathdamodaran.blogspot.com/2020/09/sounding-good-or-doing-good-skeptical.html

Damodaran, A. (2021). *Damodaran online*. https://pages.stern.nyu.edu/~adamodar/

Eaglesham, J. (2021, September 3). Companies are tallying their carbon emissions, but the data can be tricky. *The Wall Street Journal*. https://www.wsj.com/articles/companies-are-tallying-their-carbon-emissions-but-the-data-can-be-tricky-11630661401

Eaglesham, J. (2022, March 22). New rules put net zero pledges under scrutiny. *The Wall Street Journal*. https://www.wsj.com/articles/new-rules-put-net-zero-pledges-under-scrutiny-11647950909

Eaglesham, J., & Shifflett, S. (2021, August 10). How much carbon comes from a liter of coke? Companies grapple with climate change math. *The Wall Street Journal*. https://www.wsj.com/articles/climate-change-accounting-for-companies-looms-with-all-its-complexities-11628608324

Etchebehere, V. (2023). Climate change and aviation. In A. K. Sundaram & R. G. Hansen (Eds.), *Handbook of Business and Climate Change*. Edward Elgar Publishing.

EPA. (2009, October 30). Greenhouse gas reporting program. United States Environmental Protection Agency. https://www.epa.gov/ghgreporting/10302009-rule

EPA. (2021). *Inventory of US greenhouse gas emissions and sinks 1990-2019*. United States Environmental Protection Agency. EPA-430-R-21-005. https://www.epa.gov/ghgemissions/inventory-us-greenhouse-gas-emissions-and-sinks-1990-2019

European Environment Agency. (2021, May 27). *Annual European Union greenhouse gas inventory 1990-2019 and inventory report 2021.* EEA/PUBL/2021/066. https://www.eea.europa.eu/publications/annual-european-union-greenhouse-gas-inventory-2021

Fisher, S., Phillips, B., & Scovic, M. (2023). The patchwork quilt: Business complexities of decarbonizing the electric sector. In A. K. Sundaram & R. G. Hansen (Eds.), *Handbook of Business and Climate Change*. Edward Elgar Publishing.

Gallagher, E., Bancilhon, C., & Berruti, G. (2021, April 15). *Six things business should know about the EU Taxonomy.* https://www.bsr.org/en/our-insights/blog-view/six-things-business-should-know-about-the-eu-taxonomy

Ge, M., Friedrich, J., & Vigna, L. (2020, February 6). *4 charts explaining greenhouse gas emissions by countries and sectors*. World Resources Institute Insights. https://www.wri.org/insights/4-charts-explain-greenhouse-gas-emissions-countries-and-sectors?hootPostID=2186095b540e9882f3d179982067abefa0f7d846d24e12938bcbcf4ff0f5a6d4

GHGProtocol. (2004). *The Greenhouse Gas Protocol: A corporate accounting and reporting standard.* Revised edition. https://ghgprotocol.org/sites/default/files/standards/ghg-protocol-revised.pdf

Gianfrate, G., Schoenmaker, D., & Wasama, S. (2023). Cost of capital and climate risks. In A. K. Sundaram & R. G. Hansen (Eds.), *Handbook of Business and Climate Change*. Edward Elgar Publishing.

Gillingham, K., & Stock, J. H. (2018). Cost of reducing greenhouse gas emissions. *Journal of Economic Perspectives*, 32(4), 53–72. DOI: 10.1257/jep.32.4.53

Governance & Accountability Institute, Inc. (2021). *Sustainability reporting in focus: 2021 S&P 500 and Russell 1000.* https://www.ga-institute.com/research/ga-research-collection/sustainability-reporting-trends/2021-sustainability-reporting-in-focus.html

Hansen, R. G. (2023). Carbon pricing. In A. K. Sundaram & R. G. Hansen (Eds.), *Handbook of Business and Climate Change*. Edward Elgar Publishing.

Harrison, D., & Bancilhon, C. (2021, February 9). *Double and dynamic: How to enhance the value of your materiality assessment.* https://www.bsr.org/en/our-insights/blog-view/double-and-dynamic-how-to-enhance-the-value-of-your-materiality-assessment

Henisz, W., Koller, T., & Nuttal, R. (2019, November). Five ways that ESG creates value. *McKinsey Quarterly.* https://www.mckinsey.com/~/media/McKinsey/Business%20Functions/Strategy%20and%20Corporate%20Finance/Our%20Insights/Five%20ways%20that%20ESG%20creates%20value/Five-ways-that-ESG-creates-value.ashx

HLS Forum on Corporate Governance. (2021, September 1). *The SEC's upcoming disclosure rules. Harvard Law School.* https://corpgov.law.harvard.edu/2021/09/01/the-secs-upcoming-climate-disclosure-rules/

Hone, D. (2023). The end of combustion? In A. K. Sundaram & R. G. Hansen (Eds.), *Handbook of Business and Climate Change*. Edward Elgar Publishing.

International Carbon Action Partnership. (2021, November 17). *China national ETS.* https://icapcarbonaction.com/en/?option=com_etsmap&task=export&format=pdf&layout=list&systems%5B%5D=55

IEA. (2021). *World Energy Model*. IEA, Paris. https://www.iea.org/reports/world-energy-model

Intercontinental Exchange, Inc. (2021, June 9). *Expansion of ICE ESG reference data shows broad differences in ESG reporting between Europe and the US.* https://ir.theice.com/press/news-details/2021/Expansion-of-ICE-ESG-Reference-Data-Shows-Broad-Differences-in-ESG-Reporting-Between-Europe-and-the-U.S/default.aspx

Kalesnik, V., Wilkens, M., & Zink, J. (2020). *Green data or greenwashing? Do corporate carbon emissions data enable investors to mitigate climate change?* https://papers.ssrn.com/sol3/papers.cfm?abstract_id=3722973

IPCC. (2018). *Global warming of 1.5C.* Intergovernmental Panel on Climate Change. https://www.ipcc.ch/site/assets/uploads/sites/2/2019/06/SR15_Full_Report_Low_Res.pdf

IPCC-AR6. (2021). *Sixth assessment report*. Intergovernmental Panel on Climate Change. https://www.ipcc.ch/assessment-report/ar6/

IPCC-SPM. (2021). *Climate change 2021: The physical science basis – Summary for policymakers*. Intergovernmental Panel on Climate Change. https://www.ipcc.ch/report/ar6/wg1/

Larsen, K., Pitt, H., Grant, M., & Houser, T. (2021, May 6). *China's greenhouse gas emissions exceeded the developed world for the first time in 2019*. The Rhodium Group. https://rhg.com/research/chinas-emissions-surpass-developed-countries/

Lee, A. H. (2021, March 15). *Public input welcomed on climate change disclosures*. Securities and Exchange Commission Statement. https://www.sec.gov/news/public-statement/lee-climate-change-disclosures

Lindsey, R. (2020, August 14). Climate change: Atmospheric carbon dioxide. *NOAA Understanding Climate*. https://www.climate.gov/news-features/understanding-climate/climate-change-atmospheric-carbon-dioxide

Microsoft. (2018). *2017 data factsheet: Environmental indicators*. https://www.google.com/url?sa=t&rct=j&q=&esrc=s&source=web&cd=&ved=2ahUKEwjQ_sWwt6X1AhWxjIkEHYq_AC4QFnoECA0QAQ&url=http%3A%2F%2Fdownload.microsoft.com%2Fdownload%2F0%2F0%2F6%2F00604579-134B-4D0E-97C3-D525DFB7890A%2FMicrosoft_2017_Environmental_Data_Factsheet.pdf&usg=AOvVaw332RJKNQW2sVlxC1UAxl_2

Microsoft. (2021). *2020 environmental sustainability report: A year of action*. https://www.microsoft.com/en-us/corporate-responsibility/sustainability/report

Monciardini, D., Tapio, J., & Tsagas, G. (2020). Rethinking non-financial reporting: A blueprint for structural regulatory changes. *Accounting, Economics, and Law: A Convivium*, 10(2), 1–43. https://doi.org/10.1515/ael-2020-0092

MSCI. (2018, January). *MSCI carbon footprint index ratios methodology*. Morgan Stanley Capital International. https://www.msci.com/documents/1296102/6174917/MSCI+Carbon+Footprint+Index+Ratio+Methodology.pdf/6b10f849-da51-4db6-8892-e8d46721e991

Network for Greening the Financial System. (2020, June). *NGFS climate scenarios for central banks and supervisors*. https://www.ngfs.net/sites/default/files/medias/documents/820184_ngfs_scenarios_final_version_v6.pdf

O'Brien-Bernini, F., & Meehan, A. (2023). Owens Corning: Environmental footprint reduction as the foundation for building a net-positive future. In A. K. Sundaram & R. G. Hansen (Eds.), *Handbook of Business and Climate Change*. Edward Elgar Publishing.

Paltsev, S., Sokolov, A., Haigh, M., Hone, D., & Morris, J. (2021, February). *Changing the global energy system: Temperature implications of the different storylines in the 2021 Shell energy transformation Scenarios. Joint Program Report Series Report 348*. (http://globalchange.mit.edu/publication/17589)

Peace, J., & Huber, K. (2023). Climate preparedness for business resilience. In A. K. Sundaram & R. G. Hansen (Eds.), *Handbook of Business and Climate Change*. Edward Elgar Publishing.

Phillips, B., Fisher, S., & Scovic, M. (2023). Implications of fully decarbonizing the electric industry for business: Icarus or Daedalus? In A. K. Sundaram & R. G. Hansen (Eds.), *Handbook of Business and Climate Change*. Edward Elgar Publishing.

Refinitiv. (2021). *Refinitiv ESG carbon data and estimate models*. https://www.refinitiv.com/content/dam/marketing/en_us/documents/fact-sheets/esg-carbon-data-estimate-models-fact-sheet.pdf

Reichelt, H., Allen, D., & Cantor, R. (2023). Mainstreaming climate action in public and private investments: Mobilizing finance towards sustainable investments through the bond markets. In A. K. Sundaram & R. G. Hansen (Eds.), *Handbook of Business and Climate Change*. Edward Elgar Publishing.

Ritchie, H., & Roser, M. (2020). *Energy*. Our World In Data. https://ourworldindata.org/energy

Rodhe, H., Charlson, R., & Crawford, E. (1997). Svante Arrhenius and the greenhouse effect. *Ambio*, 26(1), 2–5.

S&P Trucost. (2020). *Trucost climate and environmental data: Unlocking essential ESG intelligence*. https://www.spglobal.com/marketintelligence/en/documents/trucost_general_fi_brochure_20200602_final.pdf

Securities and Exchange Commission. (2010, February 2). *Commission guidance regarding disclosure related to climate change*. Release Nos. 33-9106; 34-61469; FR-82. https://www.sec.gov/rules/interp/2010/33-9106.pdf

Securities and Exchange Commission. (2022, March 21). *The enhancement and standardization of climate-related disclosures for investors.* Release Nos. 33-11042; 34-94478; File No. S7-10-22. https://www.sec.gov/rules/proposed/2022/33-11042.pdf

Sundaram, A. K. (2023). ESG investing. In A. K. Sundaram & R. G. Hansen (Eds.), *Handbook of Business and Climate Change.* Edward Elgar Publishing.

Sustainalytics. (2021, April). *The carbon risk ratings methodology: Version 2.0.* https://connect.sustainalytics.com/carbon-risk-rating

TCFD. (2017, June). *Recommendations of the Taskforce on Climate-related Financial Disclosures: Final Report.* https://assets.bbhub.io/company/sites/60/2020/10/FINAL-2017-TCFD-Report-11052018.pdf

Tol, R. (2009). The economic effects of climate change. *Journal of Economic Perspectives*, 23(2), 29–51.

United Nations. (2015). *Paris agreement.* https://unfccc.int/sites/default/files/english_paris_agreement.pdf

United Nations. (1992). *United Nations framework convention on climate change.* https://unfccc.int/resource/docs/convkp/conveng.pdf

United Nations Environment Program. (2021). *The heat is on: A world of climate promises not yet delivered – Emissions gap report 2021.* https://www.unep.org/resources/emissions-gap-report-2021

3. The end of combustion?
David Hone

In July 1912 in Australia, the *Braidwood Dispatch and Mining Journal* published a short article on the global use of coal.

> **COAL CONSUMPTION AFFECTING CLIMATE**
> The furnaces of the world are now burning about 2,000,000,000 tons of coal a year. When this is burned, uniting with oxygen, it adds about 7,000,000,000 tons of carbon dioxide to the atmosphere yearly. This tends to make the air a more effective blanket for the earth and to raise its temperature. The effect may be considerable in a few centuries.

This story had its roots in the late 19th century work of Svante Arrhenius, a Swedish chemist who linked the average surface temperature of the Earth to the level of carbon dioxide in the atmosphere.

At the time, most economies were developing on the back of coal, a fossil fuel. It was the use of coal that had supported the rise of industry in Germany, Great Britain, the US and Australia. One hundred years later, as the 21st century unfolds, the same is true in China and now India. Coal is a relatively easy resource to tap into and make use of. It requires little technology to get going but offers a great deal, such as electricity, railways (in the early days), heating, industry and very importantly, iron ore smelting and cement production. For both Great Britain and the US, coal provided the impetus for the Industrial Revolution. In the case of the latter, very easy-to-access oil soon followed, and mobility flourished, which added enormously to the development of the continent. Society has developed on the back of combustion.

But is society now on a pathway that will bring the combustion era to a close? Many would have us think so. They point to the rapid deployment of wind and solar and the emergence of the electric vehicle and argue that the investment tide has turned.

Nevertheless, the energy system remains dependent on coal, oil and gas with 81 percent[1] of global primary energy requirements coming from these three fuels in 2018. Thirty years earlier, when Dr James Hansen testified on the issue of rising surface temperature to the United States Senate Energy and Natural Resources Committee, and as the first shoots of the energy transition began to appear in the form of small solar arrays and modest clusters of wind turbines, the same three fuels met 82 percent[2] of global energy needs. Perhaps this isn't surprising, given that energy system change tends to be on a multi-generational timescale.

In the early 1990s as the United Nations Framework Convention on Climate Change (UNFCCC) was being created following the Rio Earth Summit, very few people were attuned to the reality of increasing levels of carbon dioxide in the atmosphere and a changing surface temperature. The media didn't pay much attention to Arctic sea ice extent and unusual heat waves were treated as just what they seemed to be, unusual. But in the 21st century the landscape has shifted. There is a growing and deeper awareness in society that the surface temperature of the planet is rising. Whilst many policymakers are now targeting net-zero emissions by 2050, there has been little discussion around the impending consumption of the 1.5°C carbon budget or the distinct possibility of the first 1.5°C year well before the 2020s conclude. This did change to some extent in the Glasgow Climate Pact at COP26 with the statement '... that carbon budgets consistent with achieving the Paris Agreement temperature goal are now small and being rapidly depleted.'

The required rate of change in the energy system to limit warming to 1.5°C or even 2°C exceeds the rate at which aggressive change might be expected to occur and certainly exceeds the historical rate of change, which does not bode well for the goals of the Paris Agreement. Yet broad technological change over the course of this century so far has exceeded all others, which brings hope of a successful outcome. These issues will be discussed further in this chapter.

EXPLORING 1.5°C

The temperature data for 2021 (Figure 3.1)[3] shows that the warming trend continues, with 2020 being the second warmest year recorded. The year 2016 is the warmest in the modern era data series, but it was also a year with very strong El Niño phenomena in the Pacific Ocean, which tends to temporarily raise the global average surface temperature. 2020 was a year where a cooling trend (La Nina) emerged late in the year, which makes the near tie with 2016 temperatures (and an actual tie in the EU dataset) indicative of the rising temperature trend.

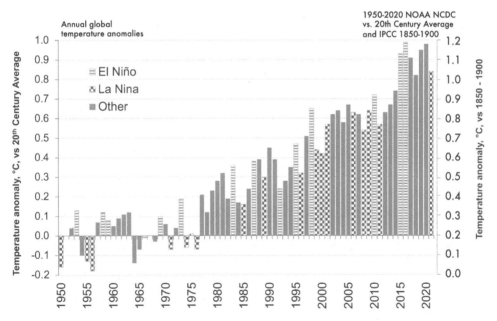

Source: US NOAA National Centers for Environmental Information.

Figure 3.1   *Global average surface temperature rise*

Following the release of the Intergovernmental Panel on Climate Change (IPCC) Special Report on 1.5°C in 2018 (SR15), there is better agreement between institutions on warming so far and the appropriate baseline years to use for reference. A standard baseline of 1850–1900 is emerging as representative of pre-industrial times.[4] On that basis, the 2020 global average surface temperature rise, on a trend basis and not in a particular year, is estimated[5] to be nearly 1.2°C above pre-industrial levels, or just 0.3°C below the 1.5°C threshold of the Paris Agreement. The global average surface temperature was 1.09°C higher in 2011–2020 than 1850–1900.

After the very strong El Niño of 2016 and the elevated global temperature that resulted, a simple analysis of the temperature record shows that the variability we normally associate with the annual temperature data vanishes when only looking at years with a similar El Niño Southern Oscillation status; for example, those years in the last 60 where a very strong El Niño condition prevailed. The Oceanic Niño Index was at or above 2.0 in 1966, 1973, 1983, 1998 and 2016.

A plot (Figure 3.2)[6] of the average surface temperature anomaly for those years reveals a straight line, with a high regression coefficient of 0.9987 and a slope of 0.022. This aligns very closely with the IPCC SR15 finding that warming is progressing at a rate of 0.2°C per decade.

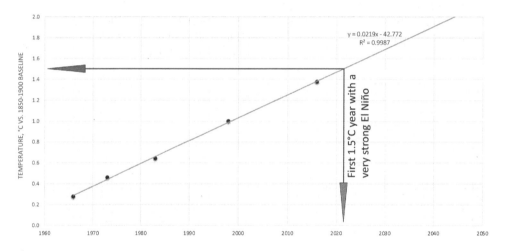

Figure 3.2   *Extrapolating temperature forward assuming a very strong El Niño year*

A further finding is revealed by extrapolation of the trend line. Projecting forward with the same slope sees the line crossing 1.5°C in 2021–2022. This means that should we have a very strong El Niño from now onwards (but recognizing they are infrequent), the year in which it occurs will most likely be the first year in which surface warming equals or exceeds 1.5°C.

This is not the same as reaching 1.5°C as described in the Paris Agreement, given that once the El Niño has ended the temperature will almost certainly dip below this threshold for several years. But a 1.5°C year won't go unnoticed and may well create even greater pressure on the need for change. Perhaps it will be the signal that jolts even wider societal consciousness.

Nevertheless, based on the 2020 temperature assessment of 1.17°C and a 0.2°C per decade rate of change, the rise in average surface temperature will likely pass 1.5°C in 2037/2038.

The goals of the Paris Agreement require the signatory countries to act to hold the increase in the global average temperature to well below 2°C above pre-industrial levels and to pursue efforts to limit the temperature increase to 1.5°C above pre-industrial levels. To achieve such an outcome, countries should aim to reach global peaking of greenhouse gas emissions as soon as possible and achieve a balance between anthropogenic emissions by sources and removals by sinks of greenhouse gases in the second half of this century.

The text of the Paris Agreement is not specific about the precise definition of 2°C or 1.5°C, nor does it instruct the world to achieve a particular emissions outcome by 2050. The only timeline given is for a balance between sinks and sources to occur in the second half of the century, i.e., between 2050 and 2100. Specifically, Article 4.1 of the Agreement states:

> In order to achieve the long-term temperature goal set out in Article 2 [. . . well below 2°C above pre-industrial levels and pursuing efforts to limit the temperature increase to 1.5°C above pre-industrial levels . . .], Parties aim to reach global peaking of greenhouse gas emissions as soon as possible, recognizing that peaking will take longer for developing country Parties, and to undertake rapid reductions thereafter in accordance with best available science, so as to achieve a balance between anthropogenic emissions by sources and removals by sinks of greenhouse gases in the second half of this century, on the basis of equity, and in the context of sustainable development and efforts to eradicate poverty.

This balance is now widely referred to as net-zero emissions which means that anthropogenic carbon dioxide emissions are precisely matched with carbon dioxide removals from the atmosphere, for example via carbon capture and geological storage. Net-zero emissions requires a steep overall reduction in emissions, with the matching step left for any remaining emissions.

The details were left to the IPCC and the request by the Parties to the Agreement for the IPCC to prepare a special advisory report. The IPCC SR15 was delivered in October 2018 and has set the scene for much of the energy transition pressure now visible across society. Since the release of the report, 1.5°C has supplanted 2°C as the goal of the Paris Agreement, due in large part to the sobering assessment by the IPCC of the impact differences between the two.

THE RATE OF TRANSITION

In broad terms, limiting the rise in global average surface temperature means limiting the cumulative amount of carbon dioxide released to the atmosphere. An approximate linear relationship of 185 Gt of carbon dioxide emitted per 0.1°C of warming has been established.[7] That's about 5 years of emissions assuming 2019/2020 levels continue. To limit warming to a given level, there is therefore a limit on the cumulative carbon dioxide release prior to the point of net-zero emissions, when net carbon dioxide emissions to the atmosphere cease. This limit is the carbon budget. How we consume the available carbon budget depends on the emissions trajectory going forwards, but the current pathway has society consuming the available carbon budget very quickly, which is why emissions need to fall rapidly in the near term.

It is possible to exceed the carbon budget and then meet it again in the future by assuming future net removal of carbon dioxide from the atmosphere through very large-scale reforestation or through industrial scale processes involving air capture of carbon dioxide in combination with geological storage. However, such a pathway may mean exceeding 1.5°C for some years or decades before eventually stabilizing at or below 1.5°C of surface temperature warming. This is known as an overshoot[8] pathway. A working assumption amongst many energy and climate modelers is that the temperature goal pertains to 2100, in which case overshoot could take place prior to that. However, the Paris Agreement makes no mention of 2100 so a strict interpretation could point to warming never exceeding 1.5°C.

In the IPCC Sixth Assessment Report Working Group 1 (AR6-WG-I) released in August 2021, one table stood out. Specifically, Table SPM.2 addresses the issue of the remaining carbon budget and its uncertainties. The table notes that for the 50th percentile point of the distribution of the transient climate response to cumulative emissions for 1.5°C, the remaining carbon budget from 1 January 2020 is 500 Gt carbon dioxide. This means that if cumulative carbon dioxide emissions exceed 500 Gt from that date it is about as likely as not that warming relative to 1850–1900 will exceed 1.5°C.

Greta Thunberg has cited the 67th percentile point on numerous occasions, as have other climate commentators. Adhering to this carbon budget (400 Gt from 1 January 2020) means there is a higher level of certainty of not exceeding 1.5°C, specifically – in the parlance of the IPCC – it is unlikely to be exceeded versus about as likely as not for the 50th percentile point. Many in society are looking for this additional certainty. Again, the Paris Agreement is not specific on the issue.

However, as each day passes, the notional carbon budget available for a 1.5°C outcome shrinks. At the start of 2022, two years after the baseline date set by the IPCC AR6-WG-I and accounting for annual anthropogenic emissions of some 40 Gt per year, around 80 Gt of the available budget had been consumed through an ongoing and largely unchanged set of human activities. Even the COVID-19 pandemic and the global shutdowns that ensued have put but a mere dent in the process of producing and consuming fossil fuels, digging up limestone and eliminating forests for agriculture. That leaves a 320 Gt budget at 67 percent, a vanishingly small portion of what was already a very small amount.

However this number is viewed, reductions must come at an extraordinary rate to meet it. At the time of the Paris Agreement, a ~600 Gt budget, as it was then for 1.5°C, meant a reduction in emissions of around 50 percent by 2030. The mathematics behind this are relatively simple. The budget is equated to the area under the emissions curve, which at 40 Gt in 2016 and zero in 2050 and a step down to 20 by 2030, 10 by 2040 and near zero by 2050, means cumulative emissions from 2016 to 2050 of 620 Gt.

The Glasgow Climate Pact is now targeting a reduction of 45 percent in carbon dioxide emissions by 2030 relative to 2010. In 2010, emissions were some 38 Gt, which therefore requires a reduction to 21 Gt by 2030. In 2019 emissions had risen to some 41 Gt. If emissions are to fall by 20 Gt then, in under a decade, the global energy system will need to look very different. To meet such a goal there are changes that can be made and reductions that are unlikely. This isn't just a case of every sector and every activity reducing by 45 percent. Some may even grow just to force reductions in others! Table 3.1 outlines a notional calculation which meets a goal of 21 Gt global carbon dioxide emissions in 2030.

The table isn't a forecast but is indicative of what would have to happen to achieve a near 50 percent reduction in global emissions, taking into account some practical considerations. The 2020s reduction effort falls on three main sectors: electricity generation, passenger vehicles and ending deforestation. That does not mean no changes elsewhere, they just won't be very visible by 2030 although significant efforts will have to be made to prepare a range of industrial and transport technologies for deployment in the 2030s and the ongoing reductions that will be required. Time must be allowed for technologies to permeate and become deployable. For example, hydrogen may play a role decarbonizing heavy freight vehicles through fuel cell technology, but no such truck currently exists as a commercial proposition. By the late 2020s such trucks will likely exist, but the number on the road by 2030 will still be small.

In addition to the actions in Table 3.1, the emissions of other gases will need to fall. Methane emissions should come down globally as the oil and gas industry contracts in this scenario, but agricultural methane will require specific attention. That probably points to diet change – notably a shift away from red meat towards chicken, pork and plants. Such a move would also help with the deforestation task.

Achieving all the steps in Table 3.1 is unlikely, but it could also happen through a partial implementation of the measures outlined and a broad cessation of emission intensive activities and consumerism by much of the global population. However, as of 2021 there is not a single national policy framework in operation that is aimed at delivering significant demand reduction. Irrespective, this illustration is not meant to challenge the proposition, only outline what it means.

Over 2020 and 2021, we have all had to learn quickly about exponential rates of change as COVID-19 spread throughout the world, but as has been seen with the virus itself, exponential patterns typically stop, and some other pattern of change emerges. Some mechanism intervenes in the process, typically starving the exponential process of its means for reproduction.

With the urgency around rising levels of carbon dioxide in the atmosphere, there is a need to see renewable energy grow exponentially and quickly back out fossil fuels. In 2016, American futurist Ray Kurzweil forecast the dominance of solar PV in a little over a decade. Kurzweil, as reported by Solar Power World,[9] said the following:

> In 2012, solar panels were producing 0.5 percent of the world's energy supply. Some people dismissed it, saying, "It's a nice thing to do, but at a half percent, it's a fringe player. That's not going to solve the problem," Kurzweil said. "They were ignoring the exponential growth just as they ignored the exponential growth of the Internet and genome project. Half a percent is only eight doublings away from 100 percent.
>
> Now it is four years later, [and solar] has doubled twice again. Now solar panels produce 2 percent of the world's energy, right on schedule. People dismiss it, '2 percent. Nice, but a fringe player.' That ignores the exponential growth, which means it is only six doublings or [12] years from 100 percent.

There is no doubt that solar PV is growing very rapidly, but by 2019 global installed capacity was already falling short of the two-year doubling requirement if 2011 is the baseline. In 2011, global deployment was 70 GW installed capacity and in 2019 it was approaching 600 GW, which is closer to doubling every three years, or a growth rate of around 30 percent per year. But can this rate of change persist? It would imply 25 TW installed capacity in 2034, from which enough electricity could be generated to meet all global requirements (say 50 percent above 2020 levels), assuming it could be stored and channeled to the necessary services at the right time and place. After another six years (i.e., 2040), all global energy requirements could conceivably be met, however the demand for energy may not be entirely in the form of electricity. In 2020 electricity made up around 20 percent of final demand, with the balance being predominantly hydrocarbons.

In practice, ongoing exponential deployment of solar is unlikely to happen. If the rate of deployment means doubling capacity every two to three years, then in the year prior to full deployment, PV module manufacturing capacity would need to reach 30,000 GW per year, some 150 times 2020 capacity. However, in the 2040s new PV module demand would drop to a fraction of this as global energy demands would have been met, leaving thousands of

Table 3.1 An example calculation for a 45 percent reduction in carbon dioxide emissions by 2030 relative to 2010

Sector	Considerations	Actual 2019 CO₂ emissions	2030 CO₂ emissions scenario	Discussion
Electricity generation	Clear commercial alternatives to gas and coal are available, so an accelerated transition is possible. This is the sector that has to do the early heavy lifting in terms of emission reductions.	12.4	3.0	The world currently has 2000 GW of coal capacity, the vast majority of which would have to shut down in the 2020s, replaced by wind and solar. As grid batteries are not yet a mainstream technology for managing lengthy periods of intermittency, backup would likely come from existing natural gas facilities so as to keep emissions to a minimum. Some 8000 GW of renewable capacity might be needed (not forgetting that electricity demand will also increase because of transport demands), with solar and offshore wind overlapping in many regions to provide 24/7 coverage. In 2019 the world had 600 GW each of solar and wind, generating about 800 TWh and 1500 TWh, or 8–9 percent of global electricity demand. Capacity increased by 100 GW solar and 60 GW wind between 2018 and 2019, so this would need to increase by about 30% per annum year on year throughout the 2020s. It means that by 2030, the global solar capacity build rate would need to be around 1000 GW per year.
Residential	Current emissions are from natural gas, heating oil and LPG in homes for heating and cooking, which may be slow to dislodge.	2.0	1.0	Over the span of a decade, nearly half of all households may have to replace boilers and incentives could force a change to electricity for cooking.
Freight road transport	Unlikely to change in the short term due to the lack of a clear alternative technology. Hydrogen fuel cell trucks will only make an impact in the 2030s and beyond.	2.4	2.5	
Passenger road transport	Change is now possible in this sector, but EV manufacture would have to be ramped up very rapidly. In 2020, global battery electric vehicle (BEV) production exceeded 2 million per year and there were about 7 million BEVs on the road, compared with total vehicle manufacturing capacity of 80 million and over a billion internal combustion passenger vehicles in use.	3.6	1.0	To reduce emissions to 1 billion tonnes per year by 2030, some 600–700 million BEVs would need to be produced, which means a complete switch to BEV manufacture by all the major automakers before 2025. The supply chain required to feed such a shift in manufacturing would also require a rapid increase in mining activities for key battery materials such as cobalt, lithium and nickel. The expansion of existing mines, the opening of new mines and the construction of sufficient processing facilities to refine the ores would all need to be complete by 2024, which means an almost immediate and sharp ramp up in global activity, well beyond that currently underway for the already planned battery factories.

Sector	Considerations	Actual 2019 CO_2 emissions	2030 CO_2 emissions scenario	Discussion
Aviation		0.9	0.5	The only major change possible in the short term is a reduction in use. The pandemic may ensure this happens, but if not, society will have to make difficult choices.
Shipping		0.8	1.0	May increase in the short term due to the quantity of materials needed to move throughout the world to rebuild the power generation sector and manufacture batteries. A new generation of ships won't appear until the 2030s.
Industry, agriculture and services		3.8	3.5	May increase in the short term due to materials needed for the energy transition. However, a shift to further electricity use might counter this trend.
Heavy industry		3.9	4.5	Will likely increase in the short term due to supply of materials for the energy transition. New low emission processing technologies won't be in place within a decade.
Land use change	Government policy applied rigorously throughout the world ought to be able to end deforestation within a decade, given sufficient impetus and attention.	5.0	0.0	This task is both the easiest and most difficult at the same time, in that the decision to chop down a tree is an entirely human impulse based on need. It is entirely possible for everyone to decide to stop deforestation, but hardly plausible that they will. Deforestation is closely linked with human development, the provision of food and growing populations.
Cement		2.0	2.0	Will hardly change despite efforts to reduce emissions at individual facilities, given the quantity of cement required for transition construction.
Other Sources		4.0	2.0	Largely related to the energy use in processing and delivery of fossil fuels to market. As fossil fuel use declines in this story, this category should decline proportionately.
Total		40.8	21.0	

factories stranded and millions of installers out of work. In reality, the market would see this glut coming and investment would taper long before, leading to a longer transition. This then challenges the other goal behind the requirement for very rapid deployment of new energy infrastructure; the need to limit warming to 1.5°C.

The rate at which new energy system technologies can be developed and deployed is critical to the overall timing of the transition. Historically, technologies pass through a period of early exponential growth before reaching meaningful scale, say 1 percent or more of global energy supply. This first transition can take 20 to 30 years.[10] It can then take at least the same time again for the technology to find its place in the energy system, with growth beginning to plateau (see Figure 3.3). [11]

For wind and solar, we are perhaps entering the second phase, but for up-and-coming new energy carriers such as hydrogen, to be deployed and used in a very different setting to previous industrial use, the journey is only just beginning.

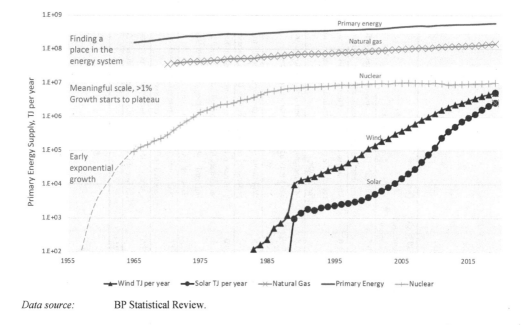

Data source: BP Statistical Review.

Figure 3.3   Primary energy system development over time

The same issue will face the industry that must supply critical materials for the energy transition, the mining industry. The possible demand for lithium, cobalt and nickel in the 2020s and 2030s implies a quadrupling or more of supply in a decade, or double to triple the historical trend.[12]

NET ZERO EMISSIONS

With the challenge posed by limiting emissions to a strict carbon budget in the near term, it is perhaps not surprising that the political focus has shifted to the call for net-zero emissions in 2050. A new reality has also emerged in recent years that supports this course; we are now headed in that direction anyway.

Thinking back even a decade, the concept of net-zero emissions barely registered in the political consciousness. In the Copenhagen Accord of 2009 there is only mention of deep cuts in global emissions being required according to science, and as documented by the IPCC Fourth Assessment Report. In that report, where some of the scenarios show emissions falling to zero late in the second half of the century, net-zero wasn't a key message for policymakers. Rather, the key IPCC message in 2007 in relation to mitigation in the long term, i.e., after 2030, was:

> In order to stabilize the concentration of GHGs in the atmosphere, emissions would need to peak and decline thereafter. The lower the stabilization level, the more quickly this peak and decline would need to occur. Mitigation efforts over the next two to three decades will have a large impact on opportunities to achieve lower stabilization levels.

By the time the IPCC SR15 appeared in 2018, the key mitigation message was very different.

> Reaching and sustaining net zero global anthropogenic CO_2 emissions and declining net non-CO_2 radiative forcing would halt anthropogenic global warming on multi-decadal time scales. The maximum temperature reached is then determined by cumulative net global anthropogenic CO_2 emissions up to the time of net zero CO_2 emissions and the level of non-CO_2 radiative forcing in the decades prior to the time that maximum temperatures are reached.

Moreover, a timeline to 2050 was proposed in SR15 for reaching net-zero emissions.

The history for this change perhaps started with the simple recognition that climate change is a stock problem and that the stock will only stop growing (and therefore stop the problem getting worse) when the flow into the atmosphere is the same as the flow out, i.e. net-zero. The science community has always known this, but the concept has taken some time to register more widely.

The impact of emissions on a given ecosystem, be it chemicals into water or carbon dioxide into the atmosphere, can be described in one of two ways.

1. If the material being emitted remains in the environment for a short while before it breaks down, is deposited somewhere or leaves with the main flow through the system (e.g., river water), then the impact that it has is largely related to the rate at which it flows into that system at any given time. This is a flow problem and the rate at which the material is emitted on a daily, weekly or yearly basis is all-important.
2. If the material is very slow to be removed and doesn't break down, it will then tend to accumulate, and its impact will grow and grow. Even a very small discharge may eventually cause a problem. If the emission finally stops, the problem will at least stop getting worse, but it won't get any better until the material is removed, either through some natural process or by intervention. This is a stock problem and the key determinant here is the total amount of emissions over time. The instantaneous rate of emissions is far less important.

Within society there is endless confusion between stock and flow, so it is not surprising that this stretches into the climate issue. Perhaps the COVID-19 pandemic has helped improve this understanding with the recognition that atmospheric carbon dioxide levels continued to rise even though anthropogenic emissions fell somewhat in 2020. In any case, with society having done little to arrest the flow of carbon dioxide into the atmosphere, the timing for net zero has been brought forward from late in the century to around 2050, based simply on stock considerations. Both companies and policymakers are now focused on the actions required for net zero, given that 2050 is, in many cases, within their long-term planning horizon window for major capital investment. There is growing pressure on all parties to do something, which has led to the declaration of a net-zero emissions goal from many countries, with presumably more to come.

The effort required to achieve an outcome of net-zero emissions by 2050 is possibly beyond the plausible rate of energy system change, which might raise the question of why countries are targeting this goal (apart from the fact that it is perceived to be critical). One answer is perhaps because now, versus just a few years ago, we are heading there anyway; it is just a question of when the goal is reached.

Offering perspective on such long-range questions is a discipline and one that sits well with the Massachusetts Institute of Technology (MIT). Since the early 1990s the MIT Joint Program on the Science and Policy of Global Change (MITJP) integrates natural and social sciences to produce analyses of the interactions among our global environment, economy, and human activities, and the potential impact of policies and social trends that affect these relationships.

A 2020 analysis released by MITJP[13] helps to frame the question of when the world might reach net-zero emissions. Rather than offer a calculation of the steps required to reach net-zero emissions by a particular date, say 2050, the analysis shows how societal, political and technology shifts are taking us to near (net) zero emissions as an inevitable outcome. The report illustrated how a cascade of growing pressures is operating across society, starting with the physical reality of a rising average surface temperature. While political trends, such as populism or leaning to the left or right, tend to come and go over time and social norms shift around as the decades pass, the temperature trend is essentially a monotonic increasing function. As such, the influence it has on our consciousness will only grow over time. The cascade can be simplified as follows:

1. Climate changes:
    a. Global surface temperature continues to rise, and impacts become more apparent.
    b. Sea level keeps rising with visible consequences.
2. Concern rises:
    a. Voter pressure on cities, states and countries to develop 'green' policies.
    b. Shareholders pushing companies to take on net-zero emission goals and targets.
3. Local and national governments pursue (piecemeal) interventions:
    a. Ongoing actions under the UNFCC under the banner of the Paris Agreement and the emergence of net-zero emissions (NZE) as a framing concept.
    b. Incentives and mandates drive down the cost of new energy technologies and lead to further uptake.
    c. Large established NZE policy frameworks continue to operate (e.g. EU, California) and some new NZE policy frameworks emerge (e.g. China by 2060).
4. Technology marches on:
    a. Renewable energy access becomes cheaper.
    b. Developments in physics, chemistry and materials sciences (e.g. PV, storage).
    c. Rapid and broadening digitalization of society.

5. The market rules:
    a. Financial markets distance themselves from fossil fuel investments, but particularly coal, and climate-related financial disclosures bring increasing transparency.
    b. Demands by businesses and consumers for lower carbon footprint products and some preparedness to pay for this.
    c. Development of markets to support low-carbon investment (e.g. nature-based solutions).
    d. Alternatives to coal, oil and gas becoming increasingly competitive.

The ongoing combined effect of these pressures gives rise to a scenario of continuous change and transition. The resultant MIT 'Growing Pressures' scenario is built on a series of simple premises; for example, if by 2050 the push-back by financial markets in combination with the falling cost of renewables means that new coal fired power station development ceases globally, then by about 2100 at the latest coal fired generation of electricity will have ceased (because the power stations built up to 2050 would have been largely decommissioned by then). An overview of the premises is shown in Figure 3.4, set against the Growing Pressures emissions trajectory.

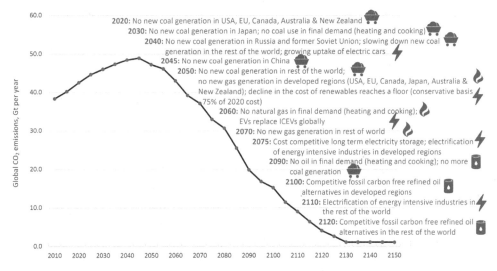

Source: MIT Joint Program. Used with permission.

Figure 3.4   Progression towards net-zero emissions in the Growing Pressures scenario

The premises are not meant to represent the fast pace changes needed to limit warming to 1.5°C, but an assessment of events that are now seemingly locked into our collective energy system future because of the growing pressures. This then establishes a new baseline from which to think about mitigation actions and to assess the progress that is being made towards a better outcome.

With net-zero emissions looking more like an inevitable outcome than an aspiration, the framing of the climate issue may also change. In the IPCC Fifth Assessment Report, readers were presented with a series of impact risk tables that gave the impression of a binary outcome of either 2°C or 4°C, i.e., society could act and limit warming to 2°C or accept the consequences of 4°C of warming.

But the Growing Pressures scenario limits warming to around 2.8°C (central estimate), effectively eliminating the IPCC central outcome of 4°C. In less than a decade the framing of the climate issue has moved from being somewhat unbounded in terms of temperature rise, to one that is bounded between central estimates of 1.5°C and 2.8°C.

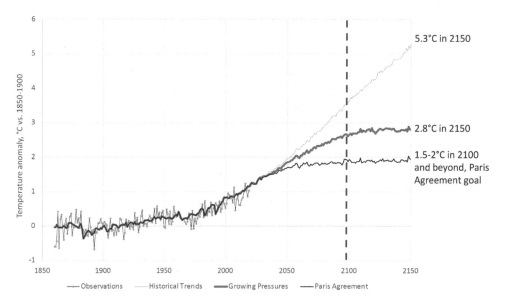

Source: MIT Joint Program. Used with permission.

Figure 3.5 MIT assessed scenario outcomes

This finding is not an argument for just letting events play out; 2.8°C would have serious consequences in terms of adaptation. Rather, the finding illustrates that change is underway and highlights the steps needed to accelerate that change. It also strengthens the hand of the UNFCCC as they encourage adoption of a net-zero emissions goal by as many countries as possible.

Returning to this analysis in a decade hence might see the boundaries contract further. Scenarios that continue historical trends of unfettered fossil fuel use no longer seem relevant when a shift toward a low-carbon society is already under way. The task in front of society is now about the pace of change, not whether change can happen. That pace can be increased, for example through widespread use of policy instruments such as carbon pricing.

THE FUTURE ENERGY SYSTEM AND THE QUESTION OF DEMAND

Four energy system changes are commonly seen in long term energy system scenarios that seek deep decarbonization. Both the International Energy Agency (IEA) NZE 2050 scenario

(June 2021) and the Shell Sky 1.5 scenario (February 2021) are typical examples. The changes are:

1. Zero emission electricity from wind, solar and hydro (with nuclear also playing a role in some scenarios).
2. Broadening use of electricity in society, with a rapid shift from the relatively modest role that electricity plays in final energy now at around 20 percent, to a level exceeding 50 percent or even 60 percent. This means electrification of energy services in transport (electric vehicles), homes (electricity for cooking and heating) and industry (using electricity for process heat, for example electric arc furnaces).
3. Using clean electricity to produce hydrogen, which in turn can be used for direct reduction of iron ore (instead of metallurgical coal), for combustion in processes requiring high heat loads, for aviation in jet engines, and in conjunction with a fuel cell for applications that can run on electricity but are challenged by the energy density of batteries (e.g., long haul trucks).
4. Creating synthetic 'drop-in' hydrocarbon fuels from biomass or via direct synthesis from carbon and hydrogen for use in legacy energy service delivery that will continue unchanged for many decades (e.g., aviation).

Critically, carbon dioxide sinks also play a role in balancing overall emissions to limit warming. A sink is any reservoir, natural or otherwise, that accumulates and stores some carbon for an indefinite period and thereby lowers the concentration of carbon dioxide in the atmosphere. The biosphere is a natural sink in that carbon accumulates in vegetation, such as trees. Fossil carbon can also be returned to the lithosphere where it originated through geological storage of carbon dioxide.

Many scenarios see significant change in land use through afforestation and reforestation, which in turn acts as a carbon sink. The Shell Sky 1.5 scenario requires large scale reforestation to remove carbon dioxide from the atmosphere and help correct the overshoot of 1.5°C during the 21st century.

Both the IEA NZE2050 and Shell Sky 1.5 scenarios also make use of carbon capture and geological storage, either through direct application within industry or as indirect removals to balance ongoing emissions from fossil fuel use. An indirect removal is the industrial capture and removal of carbon dioxide from the atmosphere in conjunction with geological storage, but at a site not directly linked with the emissions. Direct air capture of carbon dioxide (DACCS) and combining geological storage with bioenergy use (BECCS) are both indirect removal examples. Natural sinks through activities such as reforestation can also act as an indirect removal. But the role of balancing emissions and the recognition that net-zero emissions is likely long before zero emissions might, if ever, be achieved is a debated issue. Some believe that society should push fossil fuels out of the energy mix as fast as possible and that this should be the focus of any transition, whereas others see some ongoing role for fossil fuels well into next century to meet particular energy service needs.

In any case, large-scale use of carbon capture and storage also has implications for policy. The use of carbon capture and geological storage described above will require carbon prices of $70–$200 per tonne of carbon dioxide for deployment,[14] depending on the opportunities available and the technologies used.

Within IPCC SR15, four archetype scenarios (P1, P2, P3 and P4) are presented and each one makes use of sinks. The reasons for needing to do this fall into two categories.

1. The point at which net-zero emissions is required (i.e., 2050) comes before all anthropogenic emissions can be stopped, so remaining emissions must be balanced with sinks to remove the same amount of carbon dioxide from the atmosphere as added by remaining sources.
2. By the time net-zero emissions is reached, the carbon dioxide load in the atmosphere has exceeded that associated with 1.5°C of warming, so sinks are required to lower atmospheric levels at a faster rate than natural decline, ensuring a 1.5°C or lower outcome by the end of the century.

All four scenarios are based around the same carbon budget, or total cumulative emissions of 420 Gt from 1.1.2018 consistent with providing a 66 percent chance of limiting warming to 1.5°C.

In attempting to align with a very limited carbon budget, the P1 scenario imagines a world of falling overall energy use and a very rapid shift away from fossil fuels. There is no use of geological storage of carbon dioxide and a modest reliance on natural sinks, although in doing so the world must shift away from net-deforestation by the 2040s. By contrast, the P4 scenario sees increasing demand for energy and, as a consequence, a much more difficult decarbonization journey that involves considerable use of sinks in the second half of the century. Notably, the P4 scenario exceeds the carbon budget by quite some amount with peak emissions to the atmosphere of 900 Gt, before reining in the amount to 200 Gt by 2100 through large-scale use of BECCS. This results in P4 being a so-called overshoot scenario, in that the world exceeds 1.5°C during the century before returning to 1.5°C and below by the end of the century.

A deeper dive into the four stories reveals an interesting trend shown in Figure 3.6 – an almost linear relationship linking cumulative sink capacity to 2100 with the growth in final energy demand (measured as the increase in 2050 demand over 2010 demand).

This trend points to two aspects of the energy transition which are, in turn, related to the reasons for needing sinks outlined above;

1. As energy demand increases and emissions rise from sectors of the economy where no large-scale mitigation solutions currently exist, perhaps a sector such as aviation, greater and greater sink capacity will be required.
2. With energy demand rising quickly and a starting dependency on fossil fuels to meet that demand, the carbon budget that equates to 1.5°C is exceeded well before the point of net-zero emissions, which then means extensive sink capacity is required to remove this carbon from the atmosphere and meet 1.5°C later in the century.

The end of combustion? 53

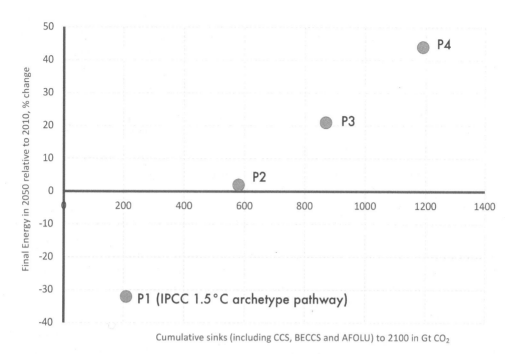

Source: IPCC SR15.

Figure 3.6   The relationship between future energy demand and sinks in SR15

That then brings into focus the question of energy demand. Over the last 30 years (to 2019), world total primary energy grew by 68 percent. This derives from GDP growth of 180 percent, and significant total energy service growth. In a discussion on energy growth, it is important to focus on non-OECD countries, because that is where the relationship of GDP growth to energy service growth is strongest. OECD countries are not far from the point where energy service growth is matched by energy service efficiency improvement and hence energy demand stays flat. In the last 30 years, non-OECD primary energy growth was over 80 percent of the global total and in the next 30 years we may see non-OECD growth making up almost all of the global energy demand increase. But this is the portion of the world where population increase continues in many countries and where access to energy is growing at a rapid rate. This in turn creates new demand for energy services. In short, we are more likely to underestimate energy demand than overestimate it.

There are also other reasons for continued energy demand growth.

1. Efficiency rebound on a macro scale is widespread and often not recognized. For example, in the 1960s air-conditioned cars hardly existed but today, you typically don't buy a car without air conditioning. So, in the 1960s there were perhaps 50 million upscale cars globally with expensive inefficient air conditioners, but now there are over one billion cars with cheap efficient air conditioners. That's a great deal more energy consumption today for air conditioning than in the 1960s.

2. Society continues to find and adopt new and interesting ways to consume energy. When energy is available, it gets used. Think how much more energy intensive our life is compared with that of our parents or grandparents. One example: millions of people around the world are starting to adopt and make use of Bitcoin, but the media reported in early 2021 that Bitcoin 'mining' at various sites around the world collectively consumes more electricity than the Netherlands.
3. Adoption of new technologies can lead to explosive growth in energy services. The internet and all the servers and devices connected to it is one such example. It didn't really exist 30 years ago, at least not as a consumer service.

The IPCC scenarios demonstrate the considerable sink capacity required for a 1.5°C outcome, unless energy demand can be curtailed in the near term and rapid solutions found and deployed for current hard-to-abate sectors.

The shape of the future energy system therefore becomes more complex, requiring the rapid deployment of clean electricity, the development and commercialization of hydrogen as an energy carrier, widespread electrification of energy services, a synthetic fuel pathway and, to meet the Paris Agreement goal, the eventual drawdown of carbon dioxide from the atmosphere at an annual rate exceeding the current emissions of the United States.

The two scenarios discussed above, IEA NZE2050 and Shell Sky 1.5 are at near opposite ends of the P1 to P4 spectrum.

The IEA NZE2050 scenario looks at the period from 2020 to 2050 and presents a proposal for the problem of containing emissions to a 500 Gt carbon budget (the 580 Gt IPCC number less 80 Gt for the years 2018 and 2019). Apart from recognizing that the land-based system is likely to reduce this budget by a further 40 Gt over the period (i.e. reducing it to 460 Gt), the analysis limits itself to the energy system and the changes that would be required to meet a 460 Gt cumulative emissions constraint within a 30 year time frame. As is the case in similar scenarios, including Sky 1.5, the IEA analysis assumes rapid electrification of final energy (e.g. electric cars instead of gasoline) and makes use of renewables and nuclear power generation to produce the electricity. They introduce hydrogen as an energy carrier, make use of bioenergy and include carbon capture and storage (CCS) where fossil fuel remains in use. For the latter, the IEA deploy CCS directly on facilities that use fossil fuels and indirectly through direct air capture for fossil fuel use in applications such as aviation.

All the above steps are well understood, but even given rapid and stretching deployment rates of all technologies, these steps are unlikely to contain emissions to less than 500 Gt in under 30 years because of the expected increase in overall energy use. As such, the IEA have curtailed energy demand to meet the carbon constraint. In the 1.5°C Scenario primary energy demand falls from 612 EJ in 2019 to 543 EJ in 2050, a drop of over 10 percent. Efficiency will certainly help deliver such an outcome and the use of renewable electricity gives the story a big boost as the thermal losses in power stations vanish, although new losses emerge through the use of transmission and storage. But the big story is the widespread assumption of behavioral change across society to reduce energy demand. The chosen measures include the following;

1. Ridesharing in all urban car trips.
2. Reducing motorway/freeway speeds to below 100 kph/60 mph.
3. Increasing temperatures in air-conditioned vehicles and buildings.
4. Lowering temperatures in heated buildings.
5. Replacing short flights with high-speed rail.

6. Limiting long haul air travel to 2019 levels.

With low energy demand and high deployment rates of new energy technologies it then becomes possible to resolve the carbon budget within a 30-year time frame.

At the other end of the spectrum is the Sky 1.5 scenario, which tackles the problem of a limited carbon budget in a very different way. Sky 1.5 looks at the period from 2020 to 2100, an 80-year time frame, and starts with an expectation of rising global energy demand, even assuming significant energy efficiency improvements across society. The growth in population and the demand for energy services, including significant new demand from developing countries for basic services, cannot be contained and energy demand rises. This immediately poses a challenge in that rising demand more quickly consumes the available carbon budget at the front end, when alternative energy technologies and sinks have not been deployed.

Sky 1.5 also recognizes that for many energy technologies, the 2020s still remain a period of development and limited deployment and even for more mature technologies such as solar and wind, a period of early growth where change on a global scale will remain limited.

The solution to this approach is to accept that, at least in the short term, the carbon budget may be consumed and the temperature it is linked to (1.5°C in this case) potentially surpassed for a period of time. The subsequent rapid deployment of sink capacity, both manmade in the form of air capture with geological storage and natural as reforestation, then offers the possibility of a period of net-negative emissions later in the century, to redress the imbalance and reduce cumulative carbon dioxide emissions and therefore temperature. This is the approach that Sky 1.5 takes, as do the IPCC P3 and P4 scenarios. Sky 1.5 takes this approach out of necessity, in that the scenario reaches a limit on energy technology deployment and does not foresee a fall in energy demand.

The two scenarios are attempting to answer the same question but take very different approaches to doing so. There is no right or wrong here, just different ways of solving a tough problem.

NAVIGATING THE FUTURE

Over one hundred years ago the *Braidwood Dispatch and Mining Journal* set out in clear, albeit brief terms, the issue of climate change. Through most of the 20th century, society collectively put the issue to one side with development and growth taking priority, although many of the technologies necessary to address the problem are now in the process of early deployment and others are in the early stages of commercial readiness.

Meeting the most aggressive goal of the Paris Agreement will require very substantial near-term action, probably at a rate that defies all previous change in the energy system and from a starting point that has 200 years of legacy infrastructure behind it. Some countries may well reach net-zero emissions by 2050 and there may be deep wells of change in certain sectors on a global or at least regional basis, such as for passenger road transport. But complete change in every sector across every country by 2050 would be an unexpected outcome.

Nevertheless, a complete transition is in the offing and the timeline, while uncertain, is diminishing. Somewhere between 2050 and 2130 anthropogenic greenhouse gas emissions will come close to net-zero and the ongoing rise in surface temperature will stop. Combustion may not have ceased at that point, but it will have been substantially curtailed. The task will then be adapting to that warmer world.

NOTES

1. IEA World Energy Balances 2020 Edition.
2. IEA World Energy Balances 2020 Edition.
3. US NOAA National Centers for Environmental Information.
4. IPCC SR15 1.2.1.2 – Choice of reference period.
5. https://www.globalwarmingindex.org
6. Berkeley Earth Temperature Data, own analysis.
7. Allen, M. R., Frame, D. J., Huntingford, C., Jones, C. D., Lowe, J. A., Meinshausen, M., and Meinshausen, N. (2009) Warming caused by cumulative carbon emissions towards the trillionth tonne. *Nature*, 458, 1163–1166, doi:10.1038/nature08019.
8. IPCC SR15 terminology.
9. https://www.solarpowerworldonline.com/2016/03/futurist-ray-kurzweil-predicts-solar-industry-dominance-12-years/
10. Kramer, G. and Haigh, M. (2009) No quick switch to low-carbon energy. *Nature*, 462, 568–569. https://doi.org/10.1038/462568a
11. Figure based on analysis by Kramer, G., and Haigh, M. (200) No quick switch to low-carbon energy. *Nature*, 462, 568–569. https://doi.org/10.1038/462568a
12. Battery trends, Shell Climate Change Blog, David Hone, August 2021. https://blogs.shell.com/2021/08/02/battery-trends/
13. Morris, J., Hone, D., Haigh, M., Sokolov, A., and Paltsev, S. (2020) Future energy: In search of a scenario reflecting current and future pressures and trends. *Joint Program Report Series Report* 344, November, 24 p. (http://globalchange.mit.edu/publication/17501)
14. Shell Sky Scenario (2018).

PART II

KEY INDUSTRIES: IMPACT AND RESPONSE

4. Banks and climate change risk

Edwin Anderson, Ilya Khaykin, Alban Pyanet and Til Schuermann

INTRODUCTION

The average human generates about five metric tons of CO_2 each year. If you live in the US, it is more than three times at 16 metric tons (Boden, Marland and Andres, 2017). To put this into perspective, if you were to drive a Ford Mustang from Chicago to Los Angeles, it would contribute one metric ton of CO_2, and a quarter of that will remain in the atmosphere for well over a thousand years. Atmospheric concentration of CO_2 is now higher than at any time in at least the last two million years (IPCC, 2021). CO_2 is the major contributor to climate change, resulting in an increase of global average surface temperature of more than 1°C since the late 19th century, but with an accelerating trend since 1980 (WMO, 2021). This increase has led to a wide range of climate impacts, including sea level rise of 15–25 cm since 1900, although that rate of increase too has accelerated at least three-fold since about 2006 (Hsiang and Kopp, 2018; IPCC, 2021). Tropical storm frequency and intensity is increasing in both the Atlantic and Pacific; drought is spreading in Africa and the American west with impact on food supply and knock-on effects such as increased wildfires (IPCC, 2021).

Banks play an important role in the economy: they intermediate between borrowers and savers, those that demand and supply capital for households, firms and even governmental entities. As a result, banks touch every part of the economy, which makes them more exposed to systematic risk than any other industry or sector (Schuermann and Stiroh, 2006). The physical risks described above have a direct impact on banks. Real estate lending, whether residential or commercial, becomes riskier in coastal areas such as Florida and the Gulf states, and increased exposure to wildfire makes it more difficult for households and firms to obtain flood and fire insurance, directly increasing credit risk through reduced recovery value of the collateral. Rates of warming and rates of ocean rise are all increasing, and tipping points can well occur in the melting of glaciers and polar icecaps, and the thawing of permafrost in the arctic tundra, releasing large quantities of trapped carbon and methane (Steffen et al., 2018).

In addition to these physical risks come the risks associated with transitioning to a lower carbon economy. Without significant changes to our economies, global temperatures are projected to be another 2–3°C warmer by the end of the century (NGFS, 2020a). To avoid such dramatic increases and the resulting damage, our collective carbon footprint will have to decline drastically. This implies correspondingly dramatic changes in behavior, policy and technological innovation. How smoothly those changes occur determines the degree of transition risk, and it is the transition risks that likely will have the more dramatic and near-term impacts on banks' risk profiles. In particular, the scenarios that are especially useful from a risk management perspective explore sudden and abrupt changes, either because of much delayed policy actions (which, when they do happen, will have to be that much more draconian) or because of very disruptive (but beneficial) technology changes. Each would command

Banks and climate change risk 59

a massive shift in relative pricing (and thus relative risk) of carbon intensive and low carbon technologies, and the industries and firms that depend on them.

Every institution is confronted with two challenges: what is the impact of climate change on my institution, and what contribution is my institution making to climate change? Our chapter focuses mostly on the first, and largely through the risk management lens. Of course, activities such as "green" lending facilitates the transition to a low carbon economy and, thus, can also reduce risk to the bank. In this way, the two perspectives, often referred to as the double materiality concept,[1] are interrelated. According to the European Commission's 2019 *Guidelines on reporting climate-related information*, "these two risk perspectives already overlap in some cases and are increasingly likely to do so in the future".

So what are banks doing to prepare? The most immediate action would be to understand and assess exposure to the different sources and channels of climate risk. These exposures would impact the ability of borrowers to repay their loans and the value of the collateral pledged against those obligations, though that is hardly the only channel. It is critical not to paint this risk picture with too broad a brush and classify entire industries and sectors into either "brown" or "green". To take an obvious example from the auto sector, suppose a manufacturer has both internal combustion engines (ICE) and electric vehicles (EV). Is it brown or green? There are a few companies in this sector that produce only EVs, of course. Thus, even apparently "brown" firms could actually become better credits if they are able to adapt more nimbly. Such changes require capital expenditures which need to be financed – for instance, by banks. The devil (and angel) is in the details, and it takes work to uncover the nuanced risks and opportunities.

How far along are banks in this journey? Not far. In 2018, Oliver Wyman and the International Association of Credit Portfolio Managers (IACPM) conducted a survey of 45 leading banks and found that the majority had not explicitly captured climate risk in the credit rating process (Oliver Wyman and IACPM, 2019). Just 16 percent had captured this source either indirectly or directly, the latter at just five percent. But the last two to three years have seen a marked uptick in awareness and activity at banks as they respond to rising pressure from their boards, investors (both equity and debt), and increasingly from regulators. Climate risk is being incorporated into risk management frameworks, reflected in strategy and risk appetite and is increasingly featured in their public disclosures. Indeed, it was the pressure from the Financial Stability Board's (FSB) Task Force on Climate-related Financial Disclosures (TCFD)[2] stemming from a set of recommendations made public in 2017 that contributed to an increasing share of mind to this issue.

Regulatory attention is shifting to climate change risk at an accelerating rate. The Bank of England conducted a climate stress test for insurers in 2019 and for banks in 2021. The Federal Reserve joined the Network for Greening the Financial System (NGFS) in December 2020, a network that includes over 90 other central banks and supervisors covering all major economies in an effort to share best practices on how the financial sector can support the transition towards a sustainable economy. The Basel Committee for Banking Supervision (BCBS) at the BIS established the Task Force on Climate-related Financial Risks (TCFR) in February 2020, placing climate risk squarely on the supervisory agenda globally.

More recently, the United Nations Environment Programme – Finance Initiative (UNEP-FI) has convened the Net-Zero Banking Alliance (NZBA).[3] The Alliance is an industry-led group of banks committed to aligning their lending and investment portfolios with net-zero emissions by 2050. Forty-three banks from 23 countries, comprising $28.5 trillion of assets have signed on as of April 2021. In addition to making a commitment towards net-zero by 2050, banks in the Alliance also commit to setting interim targets for 2030 and to disclosing portfolio emissions and sector-specific emissions intensities. A number of banks who are not members of the NZBA have also made similar commitments towards net-zero portfolio alignment. The Glasgow Financial Alliance for Net Zero, an umbrella organization including the NZBA, brings together similar alliances across the financial sector including insurers, asset owners and asset managers that are committed to net zero alignment.

The rest of the chapter will proceed as follows. The next section sketches out the basic dynamics of climate change; introduces the distinction between physical and transition risk; describes the channels of impact on banks; presents evidence in the literature on how different asset classes are already impacted by climate change risk; provides a discussion of the price of carbon, a common summary measure used for policy decisions; presents a discussion of the many sources of uncertainty arising from the modeling of climate change and its risks; and discusses the use of scenarios as a way to organize around the complexities and account for some of the uncertainties. The third section presents a more detailed analysis, with the help of examples to illustrate the impact of transition and physical risk on the dominant risk type in banks. The fourth section presents a discussion of policy developments relevant for banks, and the final section concludes.

CLIMATE CHANGE DYNAMICS AND CLIMATE RISKS

The basic dynamics of anthropogenic climate change seem deceptively simple but are quite complex. Human economic activity has increased the emission of greenhouse gases (GHGs), mostly carbon, sufficiently to result in an increase of the average global temperature by about 1°C since the beginning of industrialization, about the mid-19th century. The rate of temperature increase has accelerated since 1980. This temperature increase has a significant impact on our environment, including sea level rise, increased frequency and severity of weather events such as storms, increased drought in areas already susceptible to drought, and an accelerated rate of species extinction (IPCC, 2021). The changing environment has a mostly adverse impact on humans and our economies, from agricultural productivity to property damage and reduced labor productivity, especially in equatorial countries.[4] This damage, in turn, has an adverse impact on banks who, by virtue of their intermediation function, are broadly exposed to variation in economic output. For overviews written more for an economics and business audience, see Hsiang and Kopp (2018) and Auffhammer (2018).

Any policy consideration for mitigating the emission of GHGs and aiming for a transition towards a lower carbon economy is complicated by two effects. First, because carbon stays in the atmosphere for hundreds of years, many of the changes underway are irreversible at reasonable time scales and thus reductions in carbon emissions will have to be quite significant to make a difference. All transition scenarios present a carbon budget to be spent, judiciously, along the path to a lower carbon economy (Hone, 2023). Second, the impact of temperature increase will likely be nonlinear and subject to tipping points. For example, Steffen et al. (2018) consider a set of self-reinforcing negative feedback loops that could push the earth to a "hothouse" scenario impossible to prevent even with human emission reductions. The mechanisms include accelerated melting of glaciers[5] (Hugonnet et al., 2021), shrinking of the polar icecap in the Arctic, Greenland and Antarctica; weakening of the Atlantic meridional overturning circulation which is a major mechanism for heat redistribution of our planet (e.g. the Gulf Stream)[6] leading to catastrophic climate change in Northern Europe; and thawing of the permafrost in Siberia releasing methane, a GHG with 25× more warming potential than CO_2. As we will discuss in more depth later, both effects will increase the incentive to act sooner in accelerating our transition to a low carbon economy.

Figure 4.1, adapted from Nordhaus' 2018 Nobel lecture, captures this dynamic from dramatically increased economic growth through industrialization to rising CO_2 emissions inducing climate change, which in turn results in economic damage and which then motivates (either gradually or suddenly) climate change policies to facilitate a transition to a lower carbon economy.

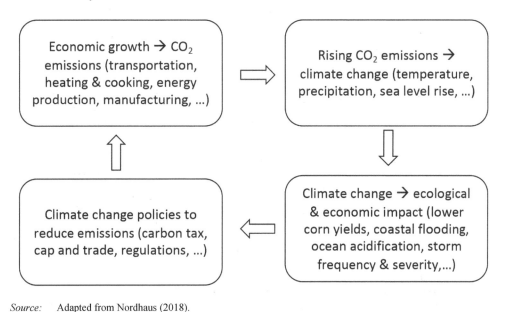

Source: Adapted from Nordhaus (2018).

Figure 4.1   Economic growth and climate change dynamics

Banks' exposure to climate change risk, both physical and transition risks, manifests through the traditional risk types used in bank risk management (BCBS, 2021a) as laid out in Table 4.1.

Table 4.1   The potential effects of climate risk on typical risk types

Risk type	Potential effects of climate risk (physical & transition risks)
Credit	Reduced ability of borrowers to service debt (increase in probability of default, PD) and reduction in value of pledged collateral (increase in loss given default, LGD)
Market	Reduction in value of securities and other financial assets due to large, sudden price adjustments; climate risks could also lead to a change in correlations between assets and changes to market liquidity for some assets
Liquidity & funding	Banks' access to stable sources of funding; change in the price of funding, wholesale or deposit driven
Operational	Increased exposure to damage to physical assets; increased legal and regulatory compliance risks associated with climate sensitive businesses
Reputational	Increased reputational risks due to changing investor and consumer sentiments
Legal	Exposure to litigation triggered by insufficient action to mitigate carbon emissions, breaches of environmental law and regulations, failure to disclose climate related risks, damages from climate related risks and events
Strategic	Missed opportunities in shifting business activities in slow or failed response to climate change

Source: BCBS (2021a) with authors' addition.

In this chapter, we will focus our attention mostly on credit risk which is the dominant risk type in banks, though we also touch on market and strategic risks.

Physical and Transition Risk

When analyzing the dynamics of migrating to a low(er) carbon economy, it is useful to distinguish two sources of risk: physical risk and transition risk.[7] Physical risk captures the wide set of physical impacts that climate change brings about, such as increasing temperatures, sea level rise, increased frequency and severity of storms, drought, and so on. Transition risk captures the impact from a range of behavioral, technological, and policy changes required in the transition to a low(er) carbon economy. In general, the higher the tolerance for physical risk, the lower the transition risk – and vice versa. Physical risks manifest gradually, tipping points notwithstanding, but by the time they do arrive, unambiguously, only radical behavioral change coupled with significant technological advances are likely to have an impact.[8] In that scenario, transition risk is very high since it will require sudden, large scale shifts – which are quite costly. Indeed, the reference scenarios designed by the NGFS (2020a, 2021) and by regulators such as UK's Prudential Regulation Authority (PRA) and France's Autorité de Contrôle Prudentiel et de Résolution (ACPR) include disorderly transition scenarios. How these two risks interact and are traded off are central features in climate scenario analysis.[9]

We illustrate this trade-off with the set of six scenarios in Figure 4.2 that were developed by the NGFS in 2021 to evaluate resilience to different climate transition scenarios. There is a clear trade-off between the two risk types as the left panel in Figure 4.2 shows. The right panel displays the CO_2 emissions trajectories that correspond to each of the six scenarios. In order to avoid ending in a "hot house world" of a 3°C average global temperature increase by 2100, we need more action on climate than just policies currently pledged (NDC scenario: Nationally Determined Contributions). To stay just below 2°C and do so in an orderly fashion ("Below 2°C") requires a gradual but immediate increase in stringency of climate policies. To arrive at the objective set by the Paris agreement, namely 1.5°C (United Nations Framework Convention on Climate Change, UNFCCC, 2015), in an orderly fashion requires very well coordinated policies that enable reaching global net-zero CO_2 emission by 2050 ("Net Zero 2050"). If those policies are not well coordinated, the transition will be disorderly ("Divergent Net Zero"). These policies all require an immediate decrease in emissions of GHGs. If emissions do not decrease until 2030, very strong policies will be required to keep temperature increases to below 2°C ("Delayed Transition").[10]

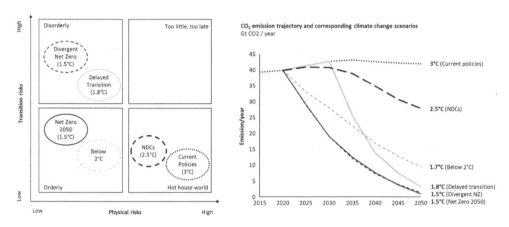

Source: NGFS (2021).

Figure 4.2   *Trading off physical and transition risk with alternative 2100 average global temperatures – illustration using NGFS 2021 scenarios*

The timelines of the NGFS climate changes scenarios go out to 2050, and climate scientists usually consider 2100 as the target horizon for their analysis. Banks have a shorter decision horizon than other financial intermediaries such as insurers. Insurers, as well as pension funds, which often have a 50+ year horizon (e.g. for life insurance, retirement plans), naturally worry about long horizon risks and were early movers in considering the impact of climate change. Banks have decision horizons closer to three to five years driven by the typical tenor and duration of lending. That said, many residential mortgages have a 30-year term (though an average duration closer to seven years), and many commercial real estate loans have a ten-year term with a balloon payment at the end (and interest only payments along the way). Thus, an underwriting decision today would naturally consider the question of refinancing at the end of the ten years, and would have to consider the exit price the next investor will pay, who will also be looking ten years forward – extending the analysis more than a decade forward. Finally, investment banks typically have a shorter decision horizon, particularly those with a significant sales and trading business.

The more immediate challenge for financial institutions is managing transition risk from changing consumer preferences, accelerating policy actions, targeted technological innovations such as improved battery storage and declining prices for solar panels which dramatically change the relative economics of carbon intensive vs. low carbon energy. The implementation of a carbon tax, for instance, could severely impact the profitability of companies based on their direct emission of GHGs, energy consumption, or related upstream or downstream emissions of GHGs. In this way, attention to transition risk helps bring the climate change risk problem closer to the here and now.

That transition to a low carbon economy will be costly, and banks will play a central role in helping with the financing of the transition, either through lending or facilitating the issuance of debt and equity securities. To quote the well-known climate scientist Johan Rockström, director of the Potsdam Institute for Climate Impact Research, "[w]e need bankers as well as activists."[11] The borrowers and counterparties will be households (e.g. financing of solar panels and electric cars), firms (investing in greener technologies), as well as national, state and local governments. McCollum et al. (2013) estimate that such a transition would require investment in low carbon energy and energy efficiency of about $1.1 trillion annually. More recently, the International Energy Agency puts the estimate closer to $4 trillion (IEA, 2021).

Done right, such financing will not only reduce climate risk but also the credit risk faced by the banks themselves. Consider the case of an energy company with a portfolio of mixed power generation technologies, but one of those uses coal. The company is a client of the bank: it has existing loans and loan commitments, in addition to other services such as transaction banking, cash management, and perhaps even capital markets advisory. Without any changes in the company energy production mix, it faces headwinds in a transition scenario and thus becomes a higher credit risk to the bank; we expand on this idea below in the next section.

So what should the bank do? One could consider broadly three options. First, the bank could decide to drop the company as a client altogether, perhaps driven by a view that the credit and reputation risks are too high to warrant continuing the relationship. Second, the bank could help the energy company spin off the businesses that use coal. In a transition scenario, there may be few buyers for such businesses, but much like a good bank/bad bank scenario for banks that are weighed down by sizable nonperforming assets, a clean break allows management to subsequently focus on the "good" or "clean" assets and business activities.[12] Note that while this option may reduce the resulting risk to the bank, if the coal assets simply shift ownership, this does not necessarily contribute to the transition to a low carbon economy.

The third option is for the bank to work with the energy company to develop a more nuanced transition plan to exit coal, short of selling the whole business. This plan could include financing of a mix of technologies to lessen the impact of the coal plant in the short and medium term while new, cleaner, replacement energy generation can be financed and constructed. Many of these greener replacement technologies tend to be newer and are often riskier, which may lend itself more to equity financing. For many larger institutions, this presents green finance opportunities for their associated investment banks.

Approaches that take a blunt view of "green" and "brown" often miss important dimensions of the problem and may be quite counterproductive. Such approaches include blunt asset categorization schemes and many "negative selection" asset management products, including certain ESG funds. When such measures lead to divestment, they may just push those assets into more opaque areas, like private holders, who may have lower concerns about environmental impact than their original holders. Many relatively "brown" technologies are still needed in the economy (cement, for example), so investment strategies that reward "best in class" companies will help support "less brown" companies – in contrast to a negative selection strategy that would impact all firms in a sector, and therefore effectively reward the firm in the strongest fiscal position with the least dependence on third party financing.

What hopefully becomes clear is the scale and scope of modeling that any climate change related risk analysis requires. We need an architecture that considers physical, economic and social dynamics jointly. This is a tall order, and we turn to that issue next.

Integrated Assessment Models

As their name implies, integrated assessment models (IAMs) aim to capture, in one model architecture, the dynamics illustrated in Figure 4.1. They integrate modeling insights from different disciplines from the physical and social sciences. As an illustration, consider the representative model structure depicted in Figure 4.3 based on the REMIND integrated assessment model from PIK.[13] There are two sets of modules reflecting the two risk types, transition and physical risk. There are, in turn, two transition risk modules. The first captures a range of socio-demographic and economic variables such as population growth, economic growth, energy and land use, as well as policy changes. The second covers emissions and climate forcing dynamics around greenhouse gases, land use changes and atmospheric changes. On the physical risk side, there are climate projections about temperature, precipitation, extreme weather and so on, and their impact on water availability, agriculture and food production, sea level rise, and economic damages. IAMs are used to evaluate policy alternatives that would have an impact on the production of GHGs, technological change (e.g. support for renewable energy technologies, battery storage), and consumer behavior (e.g. purchase of electric vehi-

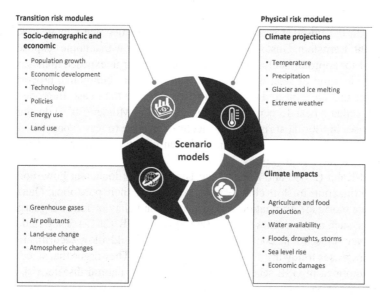

Source: Adapted from Potsdam Institute for Climate Impact Research (PIK); Oliver Wyman.

Figure 4.3   Representative model structure of IAMs

cles or home solar panels). These models form the basis of policy reports such as those produced by the important IPCC (Intergovernmental Panel on Climate Change), a UN body for assessing the science related to climate change. The Fifth Assessment Report was published in 2014, and the Sixth is expected in 2022 with an initial working group report released in August 2021 (IPCC, 2021).

A popular model used, for instance, by the Interagency Working Group (IWG) (2010), is Nordhaus' Dynamic Integrated model of Climate and the Economy (DICE).[14] The IWG also considered two other IAMs, namely Anthoff and Tol's climate Framework for Uncertainty, Negotiation and Distribution (FUND)[15] model, and Hope's Policy Analysis of the Greenhouse Effect (PAGE) model (Hope, 2006). Another model that is commonly used is the Global Change Analysis Model (GCAM; Calvin et al., 2019)[16], a partial equilibrium model of the interaction between five systems: energy, water, agriculture and land use, economy, and climate. The NGFS, in its development of climate scenarios to be considered by banking supervisors and regulators (NGFS 2020a, b, 2021), used GCAM plus two other models, both of which are general equilibrium models: MESSAGEix-GLOBIOM (Krey et al., 2020; Fricko et al., 2017)[17] and REMIND-MAgPIE.[18] These two IAMs were also used to evaluate the impact of transition risk on banks (UNEP-FI, 2018)[19] and were used by the NGFS to develop their scenarios (NGFS 2020b, 2021). In the work of UNEP-FI, Oliver Wyman and a working group of 16 global banks developed a methodology for scenario analysis of bank lending portfolios. For a comparative discussion of several of these models (and others), see Gillingham et al. (2018), NGFS (2020b), Riahi et al. (2017), and van Vuuren et al. (2011).

Impact on Bank Assets: Mortgages, Commercial Debt, Municipal and Sovereign Debt

There is mounting evidence that typical bank assets – loans, securities – are already impacted by climate risks. Real estate credit dominates bank balance sheets making up about half of total bank credit in mid-2021,[20] and there is growing evidence that climate risk is impacting mortgage credit. Bernstein, Gustafson and Lewis (2019) show that homes exposed to sea level rise (SLR) sell for approximately 7 percent less than similar unexposed homes, where similar is same zip code, time of sale, distance to coast, elevation, bedrooms, property and owner type. Properties that are expected to be inundated sooner, in their case after one foot of global average SLR, sell at a near 15 percent discount. Keys and Mulder (2020) document that sales volume of houses in Florida at risk of SLR declined relative to less exposed houses by 16–20 percent from 2013–2018. They show this is driven largely by demand and not supply (e.g. changing lender behavior) effects.

Ouasad and Kahn (2019) and Keenan and Bradt (2020) document how mortgage lenders are using securitization to shift climate risks away from their portfolios. Ouasad and Kahn (2019) examine mortgage origination and securitization behavior following large ($1 billion) disaster events, such as hurricanes Katrina (2005) and Sandy (2012). They find that the share of mortgages very close to the conforming limit (which would allow securitization via Fannie and Freddie) increases following a billion-dollar disaster.[21] They argue that since unobservable flood risk is unpriced in GSE securitization underwriting, natural disaster risk increases the returns to securitization and pushes lenders to originate within the conforming limit to subsequently securitize. Similarly, Keenan and Bradt (2020) show that concentrated local lenders transfer risk from high-risk coastal areas (SE Atlantic, Gulf coast) through increased securitization. Both articles make clear the data challenges that come with appropriately assessing climate risk.

Issler et al. (2020) look at mortgage performance in the wake of severe California wildfires in 2018–2019. They find that delinquency and foreclosures double after a wildfire, but the increase is less if the fires are large – because large fires receive more fire-fighting attention. With increased risk of wildfires, property insurance is becoming more expensive and, in some cases, harder to obtain at all. This reduction in availability has a direct impact on the credit risk of mortgages for properties in areas prone to wildfires (and those areas are expanding).

Bolton and Kacperczyk (2020) find that investors are requiring higher returns of firms with higher total carbon emissions after controlling for a standard set of risk factors, implying that carbon exposure is being priced into equity returns. Ilhan, Sautner and Vilkov (2020) use options prices to show that costs for protecting against large downside risks for carbon intensive firms are higher than other types of firms.

Investing in or lending to "green" firms is a natural but coarse and rather long horizon way to hedge against climate change risk. Motivated by this hedging concern, Engle et al. (2020) create a climate news index based on textual analysis on climate change news in the *Wall Street Journal* to allow for hedging against short term climate change risk. Huynh and Xia (2021) used this index to show that climate news risk is priced into corporate bonds. Bonds with a higher climate news beta earn lower future returns since investors are willing to pay higher prices (earn lower returns) for bonds issued by "green" firms. Jiang, Li and Qian (2019), using syndicated loan data, show evidence of SLR risk priced into lending spreads at longer maturities (more than five years). Importantly, this relationship is absent for shorter maturity loans since SLR is not a short horizon risk. Firms with more tangible, harder to relocate assets are more sensitive to SLR risk. Moreover, this SLR relationship is concentrated among banks that have more experience with SLR risk, and that SLR risk sensitivity is higher at times of climate news spikes as measured by the Engle et al. (2020) climate change news index.

The syndicated loan market provides a rich laboratory for exploring the pricing of climate change risk since it reaches a wider set of borrowing firms that may not all have easy access to the corporate bond market. Zhou, Wilson and Caldecott (2021) analyze a sample of 12,072 loan deals between 2000 and 2020 for 5033 borrowers across 118 countries from just the energy and electric utility sector to better understand the change in borrowing costs as climate awareness grows. They distinguish between energy production (coal mining, oil & gas production and services, biofuels) and power generation (coal power, gas power, solar PV,[22] on- and offshore wind). They find that loan spreads for coal mining companies have increased on average by 65 percent when comparing 2000–2010 to 2010–2020. Loan spreads for oil & gas production and services increased by 30 percent and 32 percent respectively. Meanwhile, spreads for biofuels declined by 11 percent. On the power generation side, average loan spreads rose for all technologies except for offshore wind which fell by 4 percent; coal experienced the largest increase of 54 percent. However, when focusing just on the average over 2007–2010 compared with 2017–2020, loan spreads for all renewables declined, led by offshore wind at −24 percent. Coal power again saw the largest increase by 38 percent, while gas increased by just 7 percent, showing the importance of gas (a relatively lower carbon intensive fossil fuel) for power generation in the transition to net zero.

The municipal debt market presents another interesting environment to test whether climate change risk is priced since municipalities do not have the choice to pick up and move as corporations do. Painter (2020) shows that bonds issued by municipalities in locations more at risk for SLR command higher yields (lower prices), and that this effect is concentrated in longer dated bonds but absent at shorter maturities. The impact is economically meaningful: a 1 percent increase in climate risk for a county increases annualized issuance costs by 23 basis points. Goldsmith-Pinkham et al. (2021), in related work, show that such climate change risk amounts to a 3 to 8 percent reduction in the present value of local government cash flows.

Pricing for climate change risk is evident even in the sovereign debts market. Cevik and Tovar Jalles (2020) find that countries that are more resilient to climate change have lower bond yields and spreads relative to countries with greater vulnerability to risks associated with climate change. This effect was stronger for developing economies. Using data from seven Caribbean countries, Mallucci (2020) shows that extreme weather events such as hurricanes and chronic physical risks restricts governments' ability to access sovereign debt markets.

Note that most of these findings are not specifically attributable to either transition or physical risk. They find that exposure to climate risk matters, and valuation and pricing is impacted by news about emerging physical risk and/or the evolving sources of transition risk (public policy, technological innovations, changing consumer preferences). For instance, exposure to SLR comes with exposure to policies that would respond, either with haste or delay, to the threat of SLR. Both would impact valuations. To make progress on the assessment and analysis of climate change risk, it will be useful to make that separation between transition and physical risk.

Appropriately pricing climate change risk, whether transition or physical, is clearly hard and still very much evolving. Major events, such as hurricane Sandy, result in repricing of climate sensitive assets, and a shift in consumer and investor preferences can have an impact on spending and investor behavior. Indeed, the *Financial Times* has begun reporting on 'green bubbles' as the interest in climate friendly assets is outstripping its supply.[23] Moreover, climate sensitive assets may be especially vulnerable to sudden repricing as climate change risks are re-assessed in light of emerging evidence and the possibility for sudden and dramatic policy actions, especially for countries who are late movers. Carney calls this phenomenon appropriately a "climate Minsky moment" (Carney 2018).[24]

The Social Cost of Carbon

The social cost of carbon (SCC) is the most common "price" used by economists to quantify economic damage from climate change. Such a price is necessary to assess policy alternatives by comparing the benefit of the policy to its cost, meaning that any carbon tax or similar should aim to make the price of carbon equal to the social cost of carbon. Auffhammer (2018, p. 34) defines SCC as

> an estimate of the discounted present value of damages from one additional ton of CO_2 equivalent emitted at a certain point in time. This social cost of carbon is increasing over time, as later emissions result in larger damages due to the elevated stock of greenhouse gases in the atmosphere.

Since a carbon tax (tied to the SCC) is arguably the most versatile policy variable in considering alternative climate transition scenarios used for risk management, we provide a brief discussion here.

An upward sloping term structure of the price of carbon has important implications on investment and hedging strategies. Climate abating investments will reduce the likelihood of adverse climate events and thus provide valuable insurance against such bad states of the world; they are a hedge against future climate disasters. Giglio et al. (2021) develop a model which endogenizes the likelihood of climate disaster. The faster the economy grows, the higher the probability of adverse climate events. But conditional on a climate event, the economy adapts and is now more resilient to continued climate change. Therefore, hedging against climate disaster in the near term is more valuable than hedging long term cash flows which are less affected due to adaptation along the way.

Tightly coupled to the SCC is the choice of discount rate which reflects economic agents' preferences, especially their degree of risk aversion. We can summarize as follows. The greater the risk aversion, the higher the perceived cost of future bad outcomes, the higher the willingness to pay for insurance to mitigate those outcomes (or make them less likely), the lower the discount rate attached to future bad outcomes (since, if the discount rate were high, the present value of future damage is low), and thus the higher the SCC. With the possibility of climate disasters, the value of that insurance increases, raising the SCC and reducing the discount rate (Weitzman, 2014). Commonly used discount rates for policy purposes are 3 to 5 percent (Greenstone, Kopits and Wolverton, 2013).

The range of SCC estimates in the literature is astoundingly wide. Howard and Sterner (2017) conduct a meta-analysis of about 20 studies to arrive at an estimate for current SCC of $74 (in 2020 dollars[25]) with a 95 percent confidence interval of $17 to $144. When taking into account the possibility of catastrophic risks, the estimate increases to $90 with an upper 95 percent bound at $225. Gillingham and Stock (2018) review over 50 studies and report ranges from <$10 to over $1000 per metric ton of carbon. Most of the estimates are above typical values of the SCC used for policy purposes. For example, Auffhammer (2018) uses an approach proposed by Greenstone, Kopits and Wolverton (2013) which was based on the Interagency Working Group on Social Cost of Carbon (2010) under the Obama administration, to arrive at a social cost of carbon for 2020 at $51, for 2030 at $61, and 2050 at $84, all using a discount rate of 3 percent.

Since the horizons for climate risk related analysis are so long, it is challenging to find suitable long-term interest rates off the shelf to use as references for discounting. In a very interesting paper, Giglio et al. (2015) study the price difference between 100-year ("leasehold") and perpetual ("freehold") contracts on otherwise similar real estate assets across a number of countries. The price difference between these contracts captures the present value of owning real estate in 100 years and is thus directly informative about discount rates over those long horizons. The authors estimate the 100-year discount rates for risky real estate assets to be about 2.6 percent. In follow up work, Giglio et al. (2021) show that this 100-year discount rate for real estate should be an upper bound on the discount rate used for hedging investments in climate change abatement, a level that is lower than the rates typically used in policy analysis. Because the term structure of discount rates for this hedging investment is upward sloping, the discount rate used for shorter horizons should be even lower. Giglio et al. (2021) argue that the data is consistent with a model that allows for climate disasters (tail events), feedback between the economy and climate disaster risks (where faster, more polluting economic growth raises the probability of climate disaster) and allows for economic adaptability; faced with growing climate disaster risk, the economy adapts by making more climate risk mitigating investments (see also Giglio, Kelly and Stroebel, 2020).

Uncertainty

Modeling the impact of human economic activity on global climate, the impact of climate change back on the economy, and the impact of possible policy changes, technological innovations and changing consumer and investor preferences on climate and the economy is a daunting task. It is, perhaps, in the language of economists, the ultimate general equilibrium problem. Each major component (physical, socioeconomic, technological change) has complicated dynamics on its own, but their combination and interaction raise the level of complexity yet further. Moreover, the modeling horizon is much longer than is typical in economics and finance: the standard horizon discussed in climate change is to the year 2100 (and often beyond), with intermediate stopping points at 2030 and 2050. Any risk management strategy needs to confront the myriad sources of uncertainty in climate impact modeling.

These complex systems are captured in IAMs which integrate the different components (physical, socioeconomic, etc.) into one model architecture; see the discussion above. Our understanding of the climate and economic system is very far from complete and subject to significant uncertainty. Any risk analysis of climate change impact on banks needs to account for such uncertainty as best as possible. Barnett, Brock and Hansen (2020; BBH) propose a simple but powerful framework where uncertainty has three possible components:

(a) *Risk*: uncertainty *within* a model, for instance the addition of random shocks to allow for the presence of idiosyncratic risk, or the acknowledgment that parameters are estimated with noise. Here the model is presumed to correctly describe the dynamics, probabilities are known (we can characterize the shocks), but the outcomes are uncertain.

(b) *Ambiguity*: uncertainty *across* models, reflecting the fact that there is a host of competing models that have been developed to describe the physical and socioeconomic dynamics of climate change. Here there is uncertainty about which models are better (in some unambiguous sense), and thus which ones should receive more weight for decision making.

(c) *Misspecification*: uncertainty *about* models, acknowledging that any model is an abstraction of reality, and different models make different choices about the degree of abstraction, meaning what is salient to model and what can be abstracted away.[26] These models have flaws, but we don't know with precision how many nor how impactful.

The discussion above on the social cost of carbon and the wide range of estimates already illustrate some of these uncertainties. Indeed, BBH point out that these three uncertainties are multiplicative – they tend to compound, in part driven by possibly significant nonlinearities in climate and economic systems.[27] Gillingham et al. (2018) examine uncertainties in baseline trajectories from six IAMs with such nonlinearities in mind. Specifically, they consider model and parametric uncertainty for population growth, total factor productivity (essentially productivity underlying the economy), and climate sensitivity. Economic variables such as total factor productivity[28] exhibit more uncertainty than climate variables. The uncertainty of productivity growth compounds greatly over the 21st century and induces an extremely large uncertainty about output, emissions, concentrations, temperature change, and damages by the end of the century. They find that parametric (within model, or risk in BBH) dominates structural (across model or ambiguity in BBH) uncertainty. This result suggests increased effort and focus should be placed on reducing uncertainty of the impact of climate and especially economic variables rather than building a wider variety of models.

Cai and Lontzek (2019) explicitly extend the scope of IAMs to include climate and economic uncertainties and risks in a rational expectations model using the DICE modeling framework. They show that, while the SCC increases over time (we know that it has an upward sloping term structure), its standard deviation grows faster than its mean! There are substantial risks driven by small but nontrivial probabilities of very bad outcomes. In their benchmark model simulation, the median SCC in 2100 is $286, but with a 10 percent chance of exceeding $700 and a 1 percent chance of exceeding $1200. Their results feed into the tipping point literature and further motivate an insurance approach to investing in climate risk mitigation strategies. Specifically (Cai and Lontzek, 2019, p. 2689), "[i]nsurance has a negative expected rate of return, but we buy it because of its option value. Similarly, simple discounting rules motivated by a deterministic model are not valid in a dynamic stochastic context because they would ignore the real option aspect of R&D investments." See also Cai, Lenton and Lontzek (2019), Pindyck (2020) and Lemoine (2021).

Climate Scenario Analysis

Faced with such complexity of models and outcomes, the research and policy community has arrived at a core set of scenarios that capture key features of climate change (both physical and transition risks) and span the bulk of outcomes explored in the literature. All of the scenarios go through the year 2100, though some can be extended beyond. There are broadly two complementary types: representative concentration pathways (RCPs) which describe different levels and paths of GHGs and resultant radiative forcing[29] induced climate change (such as temperature increase) (van Vuuren et al., 2011); and shared socioeconomic pathways (SSPs) which describe alternative socioeconomic developments of energy and land use, emissions, population growth and carbon intensity (and size) of the economy (Riahi et al., 2017). While there is no one-to-one mapping between the two, the five SSPs were developed with the four original RCPs in mind.

There are four commonly referred to RCP scenarios, named according to the radiative forcing levels in the year 2100 expressed in units of watts per square meter: RCP 2.6, 4.5, 6, and 8.5.[30] Recently two more RCPs, 1.9 and 3.4, were added. The RCPs were the result of collaboration of IAMs, climate modelers and terrestrial ecosystem modelers. RCPs provide relevant outputs required for climate modeling and atmospheric chemistry modeling such as emissions of GHGs, also reflecting assumptions about technological changes, climate policy and, implicitly, preference and risk aversion choices. Note that temperature is not an output of the RCPs; that is, an output of climate models that use RCP information as inputs. These RCPs were used in the Fifth Assessment Report by the IPCC (2014).

- **RCP 2.6** is the most stringent scenario requiring the most significant mitigations. It requires that CO_2 emissions peak in 2020 and go to zero by 2100, and methane (CH_4) declines by half from 2020 to 2100. This scenario is broadly consistent with temperature increases to under 2°C by the end of the century.
- **RCP 4.5** and **RCP 6.0** are considered intermediate scenarios, with emissions peaking around 2040 and 2080 respectively. CO_2 levels must decline to half their peak levels by 2100. These scenarios are consistent with global temperature increases to between 2°C and 3°C by 2100.

- **RCP 8.5** is a high baseline scenario that is highly energy intensive where emissions continue to rise throughout the 21st century as a result of high population growth and a lower rate of technological development. It is consistent with global temperature increases of 4°C to 5°C.

These four scenarios were chosen to span 90+ percent of outcomes covered in the extant literature.

The Paris Agreement stipulates to "hold the increase in the global average temperature to well below 2°C above pre-industrial levels and pursuing efforts to limit the increase to 1.5°C" (UNFCCC, 2015). That policy objective allows for low to moderate physical risk in a transition to a low carbon and sustainable economy. This transition will require significant changes in behavior achieved only by major policy initiatives that are adhered to. Only RCP 2.6 is consistent with this policy objective. Alternatively, a "business as usual" scenario where no further changes from current policies are implemented, results in very high physical risk but, of course, low transition risk. RCP 6.0 is commonly considered a BAU scenario, with RCP 8.5 as a more pessimistic alternative.

In addition, there are the more recent scenarios of RCP 1.9 and RCP 3.4. The first was added after the Paris Agreement to represent pathways consistent with a 1.5°C warming limit.[31] The second is an intermediate scenario that is considered to be more plausibly achievable than RCP 2.6 (Pielke, Burgess and Ritchie, 2021).

The five SSP narratives are summarized in Table 2 in Riahi et al. (2017). Each of the scenarios considers challenges (low to high) of mitigation and adaptation. They describe five distinct pathways about socioeconomic developments in the absence of explicit *additional* policy measures to accelerate mitigation and enhance adaptive capacity. Riahi et al. (2017) made use of six IAMs to arrive at these scenarios.

- **SSP1 – Sustainability**: considers a coordinated and gradual shift towards a sustainable path.
- **SSP2 – Middle of the Road**: a continuation of the current historical path with uneven development and income growth, and slow progress towards international sustainable development goals.
- **SSP3 – Fragmentation**: assumes a resurgence of nationalism with countries focused on achieving internal energy and food security goals, and slow economic development.
- **SSP4 – Inequality**: considers increased inequality where the gap widens between an internationally-connected society and a fragmented collection of lower-income societies. The former society diversifies its energy mix including low-carbon energy sources.
- **SSP5 – Conventional (Fossil fuel) Development**: global markets are increasingly integrated, and fossil fuel use increases driving rapid economic growth. There is faith in the ability to manage climate change, including by geo-engineering if necessary.

With six RCPs and five SSPs, not all combinations are feasibly arrived at by IAMs. For example, SSP3, which describes a world where national and regional rivalries impeded global coordination, is incompatible with low RCP scenarios. RCPs consistent with the Paris Agreement (RCP 2.6) or better are feasible for all IAMs only with SSP1, "sustainability", while SSP3, "fragmentation", renders that climate goal out of reach.[32]

More recently, in part thanks to the coordination efforts of the central banking and bank regulatory and supervisory community by the NGFS, a core set of three pairs of reference scenarios have become the standard for climate scenario analysis by financial intermediaries, banks in particular, and their regulators (NGFS, 2020a, 2021). As introduced in the second section, they focus on two versions of achieving the objectives of the Paris Agreement (consistent with RCP 2.6) – one orderly, the other a disorderly transition – in addition to two "Hot house world" scenarios, which assume little to no incremental policy actions. In addition, they also include two scenarios that go beyond the below 2°C limit of the Paris Agreement and are focused on achieving net zero emissions by 2050, one in an orderly manner and another disorderly. While very different temperature end-states and transition paths, all NGFS scenarios use the same socioeconomic narrative based on SSP2, a world where the fundamental trends for economic, social and technological trends do not deviate significantly from historical patterns. The NGFS made use of three IAMs to generate the specific transition pathways: GCAM, REMIND-MAgPIE and MESSAGEix-GLOBIOM (NGFS, 2020a, 2021).

From a risk management perspective, the scenarios that describe an orderly transition with low physical risk (bottom left quadrant in the left panel of Figure 4.2) are likely to be the least interesting and relevant. Disorderly transition to a low carbon economy is both more likely and more disruptive, and certainly more costly, and therefore worthy of considerable attention. The pessimistic Hot House scenarios – high physical risk but low transition risk – are also relevant for risk managers. In contrast to more traditional stress testing in financial institutions which features a baseline along with the stress scenario as a basis for comparison, the "baseline" scenario is less obvious in climate change risk. Much of this difference is clearly driven by the much longer horizon for climate change – decades as opposed to two to five years. Practically this means that several (more than two) scenarios need to be explored to gain proper and useful insights into the impact of climate change on a bank's risk profile.[33]

IAMs continue to play a very important role in furthering our understanding of the impact of anthropogenic climate change. However, they come with limitations. The more aggressive RCPs require significant negative CO_2 emissions through carbon absorption (e.g. trees, ocean) or carbon capture and storage (CCS) technologies. While CCS technologies exist today, they are not yet available at any scale needed to make an impact at a global level. Thus many of the scenario paths can only come about through the invention and development of new technologies.

From a business or economics perspective, an important limitation is the lack of attention paid to prices. IAMs focus on volumes: coal, oil, cars, airplanes, solar panels, and so on. Prices play a critical role in clearing any market by allowing supply and demand to meet. Prices reflect changes in incentives and behaviors to clear the market for carbon (both demand and supply) and the associated elements such as energy production (coal, oil, solar), transportation (airplanes, ICE and EV), and so on.[34]

DEEP DIVE INTO CREDIT RISK

Credit risk is the dominant risk in nearly all banks. This risk comes most obviously from traditional bank lending to households and firms, but also from national, state and local governments. Even for global investment banks, credit risk dominates, emanating from counterparty risk in trading (e.g. through derivatives), as well as through extension of credit from securities financing (e.g. facilitating short positions) and margin lending – in addition to commercial bank lending activities. In this section, we focus our discussion on the ways that credit risk exposure is impacted through transition and physical risks that come with climate change. For further discussion of climate-related risk drivers and transmission channels, including the range of methodologies to address them, see UNEP-FI (2018; 2020a, b; 2021) and BCBS (2021a, b).

We present three examples to bring to life the impact of climate change on bank credit portfolios. The first two examine transition risk by considering two sectors: oil and gas upstream companies (so companies that deal with the exploration and initial production stages), and metals and mining. The first sector presents a relatively straightforward assessment of climate change impact while the second tries to highlight some of the subtle complications which make it hard to simply classify any sector, or companies therein, as either low carbon ("green") or carbon intensive ("brown"). Our third example examines the impact of physical risk. Our examples are based on the methodology we developed in numerous exercises including our work with UNEP-FI (2018) as well as our product developed with S&P Global, Climate Credit Analytics (CCA).[35]

The literature overview in the previous section provided evidence of how climate change is impacting the assets that banks typically hold, either directly or in the form of pledged collateral. Following the financial crisis, banks and their supervisors have developed a rich toolkit to evaluate the impact of stress scenarios on bank financial health. The idea was to consider real world scenarios – drop in GDP, rise in unemployment, drop in home prices and stock markets, widening of credit spreads – that were plausible but stressful and then model their impact on bank balance sheets and income statements to assess their resilience.[36] While not exactly the same, the approach in assessing climate change impact on banks also follows the path of scenario analysis, recognizing however that (a) the horizons are much longer than traditional stress testing (which are typically two to five years), and (b) the mechanisms at play are far more complex and varied, as illustrated in our earlier discussions of IAMs and uncertainties and scenario analysis.

In considering how to translate climate scenarios into an impact on bank financials, one quickly encounters a key challenge of granularity or resolution: how detailed (granular, high resolution) should this mapping and modeling be? A more top-down approach can be done at either the fully macro or a somewhat more granular sector level (to allow for cross-sector heterogeneity with respect to climate change risk), while a more granular bottom up approach goes directly to the borrower level impact analysis (UNEP-FI, 2018 and 2021). In this section, we pursue the more detailed bottom up approach to also account for significant within-sector, and thus cross-borrower heterogeneity. Results from borrower level analysis can be used to better inform sector level impact.

The basic idea is illustrated in Figure 4.4. We begin with a particular climate scenario using an IAM such as illustrated in Figure 4.3, or as specified by, say, the NGFS reference scenarios. This scenario generates transition scenario outputs such as energy demand and supply (and thus market clearing prices), land use, emissions, a carbon price and macroeconomic impacts. These, in turn, drive business performance for a given business, say an oil and gas upstream company that focuses on exploration and production, via price and volume, unit cost, capital expenditure, and impact on asset values (some assets could end up stranded). With the scenario-conditional business drivers in hand, we can generate scenario adjusted financial statements. This step is important because it allows us to use traditional credit analysis combining a set of risk factors in a credit rating tool to arrive at impact on the building blocks in credit risk,[37] namely the probability of default (PD) and loss given default (LGD); exposure at default (EAD) is likely less relevant.[38]

Banks and climate change risk 75

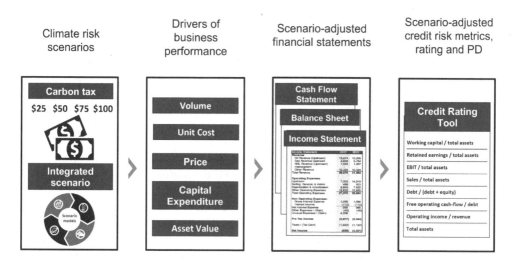

Source: Authors' illustration.

Figure 4.4   Linking climate scenarios to credit risk models

Transition risk example 1: oil & gas upstream companies[39]

Upstream companies deal with the exploration and initial production stages of the oil and gas industry. They may be companies that directly extract these hydrocarbons, or they may assist such companies with a variety of services such as provision of oil drilling equipment, feasibility studies, rig operators, and so on. Consider the impact of a carbon tax that is levied on producers as it manifests through the five drivers of business performance.[40]

- **Volume**. Some of the additional costs borne by the producers can be passed onto the consumer, which will reduce demand as we travel up the demand curve for oil and gas; how much will depend on things like market power. Not all producers will be affected the same way. High-cost producers will likely decrease production sooner and/or more than lower cost producers and are therefore more likely to exit sooner.
- **Unit cost**. The marginal cost of extraction will be impacted by the cost of emissions. Carbon emissions are generated during the actual extraction and production process, and, of course, when the hydrocarbons are used directly (e.g. in cars). This shifts supply curves upward to reflect the additional costs of emissions from the carbon tax. There would be some modest decrease in the cost of oil field services due to a decline in demand, but not nearly enough to make up for the increase coming from emissions.
- **Price**. Clearly the price paid by the consumer will increase (see Volume discussion above), but the profit margin to the producer will become smaller. How much will depend on price and demand elasticities.
- **Capital expenditure**. CapEx are investments needed to maintain current and to enable future production. CapEx is expected to decrease reflecting the lower demand. CapEx is correlated with the price of crude which provides a modeling strategy.
- **Asset value**. Some of the reserves may become uneconomic and will need to be written off as stranded assets, impairing the balance sheet.

Taken together, total demand for oil declines and the oil producer would experience an adverse impact on profitability, with a corresponding deterioration in the balance sheet. In addition, some reserves on the balance sheet may become uneconomic and will have to be written off as stranded assets, further adversely impacting the balance sheet. As a result, the company's credit quality would deteriorate, either as measured through a formal credit scoring model or through an expert driven credit rating process. Importantly, however, not all upstream oil and gas companies will be equally impacted; there will be big losers (tar sands, shale requiring fracking), moderate losers (light crude producers), and in some cases a few winners, at least for a while (natural gas producers in certain scenarios).[41]

Clearly not all elements in the modeling chain described in Figure 4.4 can be pinned down with equal empirical rigor. Much as in the IAMs themselves, the analysis of impact on company financials and then credit quality involves expert judgment. The use of judgment is well established in credit analysis where empirical analysis and models serve to support a final credit judgment, e.g. the assigning of a credit rating by the rating agencies. Disorderly transition scenarios are expected to generate far more adverse credit impacts than orderly transitions, all else equal, because companies will not have sufficient time to pursue a gradual adaptation path which would be less costly and carbon policy will need to be much more aggressive to achieve comparable outcomes (i.e., higher carbon prices).

Transition risk example 2: metals & mining

The companies in the metals and mining sector are engaged in the location and extraction of metal and mineral reserves, including iron ore, bauxite for aluminum, precious metals such as gold, silver and platinum, and minerals such as coal and precious gemstones such as diamonds. This sector also produces metals that are critical for the transition to a lower carbon economy such as lithium for battery production.

Since energy is required for the actual mining of metals and minerals, the introduction of a carbon tax will raise production costs for all. To understand the net impact of a transition scenario on a metal and mining company and to analyze more fully the impact of a carbon tax, it is useful to divide the production into three groups. For each group we comment on the impact of the main business drivers laid out in Figure 4.4 and used in the prior transition risk example.

- **Coal**: There are, broadly, two types of coal: metallurgical coal, used primarily in steel making, and thermal coal, used primarily for power generation. Since coal is a highly carbon-intensive fuel, it is likely to experience sharp declines in demand, though more for thermal than for metallurgical coal as there are few alternatives for the production of steel and other alloys. High-cost producers are expected to decrease production sooner and/or more than lower cost producers. Capital expenditures are expected to decline, though perhaps not as much as the overall production volume as less carbon intensive and more environmentally friendly extraction processes are implemented. Similar to the oil & gas sector, coal mining companies will suffer from the cost of stranded assets.
- **Energy transition metals & minerals**: A group of mostly metals that are critical inputs to technologies expected to play a vital role in the transition to a low carbon economy, especially in the production of batteries used for EVs and large-scale power storage. Examples are lithium, cobalt, nickel, manganese, copper and the rare earth metals. Demand is expected to increase significantly, especially in more aggressive transition scenarios (which would command a higher carbon tax). Unit cost of production will increase

as a result of a carbon tax, shifting the supply curve upward. Since there are few if any substitutes for these metals and minerals, the price elasticity of demand is expected to be low and producers are expected to be able to price through the carbon tax impact. Capital expenditures are expected to increase commensurately, also impacted by some of the environmental motivations described above.

- **Other metals & minerals**: For other metals and minerals, some of them may play an important role in either (or even both) low carbon or carbon intensive technologies. This group includes precious metals and minerals such as gold, silver, platinum and diamonds, non-precious metals such as aluminum, zinc, iron, and lead, and more exotic metals such as titanium, molybdenum and uranium. Uranium is used in nuclear energy production which may play a large role in the pursuit of low-carbon power production. Overall, however, demand for this group is expected to increase with economic growth, though how much will depend a lot on the growth path. Similar to energy transition metals and minerals, there are few if any substitutes, resulting in low price elasticity of demand with associated pricing power.

To illustrate some of the complications, consider coal producers. Pure coal mining companies are expected to perform very poorly and thus become increasingly credit risky in a transition scenario. Diversified miners will be able to offset some of the revenue losses from coal with other metals and minerals expected to benefit from a transition to a low carbon economy. Thermal coal is considered more of a regional market in contrast to metallurgical coal, which is a global commodity. This distinction matters because the phasing out of coal for electricity generation is expected to be quite uneven across the world, happening earlier in developed regions such as Europe and North America, and more slowly in the developing countries.

Note that several of the metals and minerals in the second and third group come with significant political risk since extraction is concentrated in a few countries. For instance, about 60 percent of the world's cobalt supply comes from the Democratic Republic of Congo, and about 80 percent of manganese resources are in South Africa.

Physical risk example: real estate assets and climate adjusted flood risk
The insurance and reinsurance industry, and related catastrophe risk modeling firms, have a long history of assessing exposure to natural disasters such as water and wind damage resulting from storms and floods. These firms have been adjusting their models to account for increasing physical risk due to climate change. Because of this long history, catastrophe models used in insurance provide a useful starting point for evaluating the impact of physical risks on bank portfolios.

Figure 4.5 provides an illustration of the approach to capturing physical risk impacts on bank portfolios. As a starting point, the modeling needs to consider the relevant geographical exposures and the corresponding perils such as floods, subsidence and hurricanes applicable for these exposures and geographies. Second, catastrophe models are peril specific and would need to be conditioned on the requisite climate scenarios, typically expressed through the corresponding RCP. Once the models are applied to the locations and properties (step 3 in Figure 4.5), the outputs of the catastrophe, or other climatic models are integrated to account for peril dependencies – floods often come with landslides – (step 4), then translated into a financial impact and aggregated across perils (step 5). This translation is achieved through vulnerability functions that relate physical outcomes (e.g. flood depth) into a financial impact (e.g. property damage) resulting in a risk-adjusted property value (step 6). Damages are then translated into the corresponding credit impacts, for example relating damages to the corresponding impact on PDs and LGDs of the loan (step 7).

Illustration of select framework components

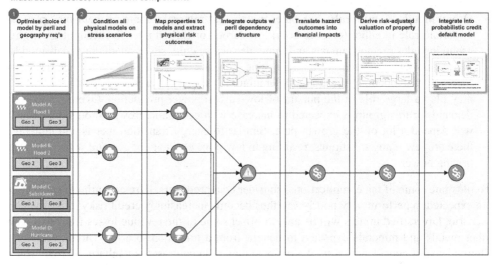

Source: Authors' illustrations, Marsh.

Figure 4.5   *Approach to capturing physical risk impacts on bank portfolios*

This analysis can imply a significant undertaking for banks since bank risk managers are typically less familiar with these types of physical risks and the corresponding catastrophe and climate models. Moreover, there are embedded assumptions about how perils are (or are not) correlated which might well change with climate change.

These three case studies hopefully highlight and bring to life some of the challenges banks face when assessing their exposure to and impact from climate change risk. They require nuanced and detailed understanding of production, demand and market dynamics of their counterparties, climate knowledge and an appreciation for political risk, and an analysis horizon that is longer than typical for risk managers. And all of these aspects place new demands on data, a topic we turn to next.

Data challenges

Operating such risk analysis machinery requires a particular fuel: data. Data is almost universally considered as the biggest challenge when performing climate risk analysis, including stress testing. In addition to traditional portfolio data (such as current ratings, PD, LGD, EAD), five types of data are typically required to perform climate stress testing. First, financial statements are necessary as a starting point to conduct the financial analysis, as illustrated above in Figure 4.4. Second, industry specific data such as cost curves for oil and gas or metals and mining (as demonstrated in the examples above) provides important information to identify which company or which part of the business will be most impacted. Third, emission data allows one to assess the carbon intensity of companies and their potential exposure to transition risk. Fourth, physical risk data, in particular the location of assets, is critical to assess potential damages or failures from various perils, and these are not typically collected as part of credit underwriting for many asset classes today, and other asset classes may be hard to extract despite their being collected at the point of origination. Finally, transition plans provide information on companies' climate-related strategy and are particularly relevant given the time horizon of the climate stress testing exercises (e.g. most companies in carbon intensive sectors will not stay idle while their business model is disappearing).

These data can be sourced from internal databases, external providers, or directly from counterparties. Unsurprisingly, traditional portfolio data is the easiest to source, though some issues of mapping between databases may arise. Financial statements of borrowers are already collected by banks (e.g. to onboard and rate counterparties in the first place), though are not always readily available or collected in a consistent manner, especially for smaller counterparties or across jurisdictions. In practice, some institutions prefer sourcing financial statements from third party providers when possible. Industry specific data is often already available to banks, though by the business teams rather than the risk teams, creating more need for cross-department collaboration.

Physical risk data presents another set of challenges: banks must have good data on the precise location of assets, which can be difficult even for mortgage portfolios and especially for commercial lending where borrowers may have operations in numerous locations. Ideally, data also includes relevant descriptive information on the nature of the asset and features relevant for resiliency, e.g. number of floors, age of building, flood defenses, etc. A further complication for wholesale portfolios involves measuring the impacts of physical risks on supply chains, as data on the suppliers of borrowers is unlikely to be available. Finally, transition plans are not available in structured databases and must be collected (and assessed) by engaging with clients, creating a significant burden on organizations.

POLICY AND REGULATORY DEVELOPMENTS

Climate risk is rising rapidly on the regulatory and central banking agenda, likely stimulated by Mark Carney's September 2015 speech, "Breaking the tragedy of the horizon – climate change and financial stability" (Carney, 2015). If climate change risks destabilizing the economy, then climate risk bears monitoring for both monetary policy and financial stability reasons. Bolton et al. (2020), in a more expansive discussion, make the point that central banks and bank supervisors effectively have little choice but to confront the challenges faced by a country, and thus its banking system, in its transition to a low carbon economy.[42] There will be disruption to the economy, and because there is no guarantee that this transition will be smooth, there is ample potential for systemic instability. Of particular interest may be the possibility of a "climate Minsky moment" (Carney, 2018) where there is a sudden repricing of assets as a result of an equally sudden re-assessment of climate risks and what it would take to address them.

The Bank of England was an early mover, publishing a report on the impact of climate change on the insurance sector in 2015 and the banking sector in 2018 (Bank of England, 2015, 2018). With the formation of the NGFS in 2017, central banks and banking supervisors have a forum where they can exchange ideas and emerging practices on how to assess and manage climate change risk. The use of scenario analysis and climate stress tests is rapidly gaining traction. For example, the Bank of England initiated a stress test for insurers in 2019 and for banks in 2021. As discussed in the second section, the NGFS published a set of scenarios in 2020 and 2021 that span a range of plausible paths, exploring both transition and physical risks that the industry and its regulators could anchor on.[43]

Bolton et al. (2020) use the Basel three-pillar framework for capital adequacy to consider how climate risk could be incorporated into microprudential banking regulation and supervision.

- **Pillar 1** (minimum capital requirements): regulators and supervisors could simply change risk weights assigned to assets by considering climate risk exposure. "Green" assets would thus receive lower risk weights than "brown" assets.
- **Pillar 2** (supervisory review of risk management practices): regulators and supervisors could, at their discretion, impose higher capital requirements for banks with relatively poor practices in their risk management of climate change exposure. This would be accompanied by gradually increasing supervisory expectations of banks' risk management capabilities.
- **Pillar 3** (disclosure requirements): regulators and supervisors could require disclosure of climate-related risk exposure for more effective market discipline.

With the formation of the TCFD in 2017 and its high level guidance on assessing and disclosing consistent information about climate related risks (and opportunities), the groundwork for the "Pillar 3" approach was set, and its implementation for transition and physical risk was laid out, respectively, in UNEP-FI (2018, 2020a). Several countries have already committed to mandatory TCFD disclosures, e.g. New Zealand by 2023, UK by 2025 (with many requirements in place by 2023), and Hong Kong by 2025.[44] Bank regulators and supervisors have not (yet) gone so far as to actively change risk weights (Pillar 1), but they have started to probe banks' abilities to appropriately identify, measure and manage climate related risks (Pillar 2), including through the use of scenario analysis and stress testing. Note that a stress test can be interpreted as a highly bespoke set of risk weights arising from the specifics of the scenario. For instance, a stress scenario will almost certainly be punitive to "brown" assets and sectors as well as, say, real estate lending exposed to climate risks, with an effect not unlike a higher risk weight on such assets.[45]

In addition to these microprudential channels, central banks may also be called upon to flex their macroprudential and monetary policy muscles. Bolton et al. (2020) point to a number of avenues where central banks could be called on to act. For example, climate change could result in supply shocks to commodities and agricultural products impacting aggregate price levels (and thus inflation), with even the possibility of stagflationary effects. Sudden repricing of assets (the climate Minsky moment) could force central banks' hands to engage in purchases of affected assets to stabilize markets, not unlike actions taken by central banks in early 2020 in the COVID-19 pandemic. These actions could extend into green quantitative easing to dampen the disruptive effects of the transition to a lower carbon economy. Importantly, such activities need not be driven by any climate policy per se, meaning a central bank acting on behalf or in support of climate policy of the government, risking their independence. Rather, their hand could be forced within their existing mandates of price and financial stability while maintaining their independence.

Climate stress testing

Stress testing has become an important part of the toolkit for risk managers and bank regulators since the great financial crisis of 2008–2009. The US, EU and UK conduct annual stress tests of their largest banks and include a broader set of banks every other year, largely to assess their resilience to severe but plausible financial and economic shocks. The scenario horizon is typically two to five years to reflect typical maturity profiles of bank balance sheets. But climate scenarios operate at somewhat longer time horizons, and the stress testing machinery that banks and their regulators have built up over the past decade likely will not be directly applicable to exploring bank resilience to climate scenarios.[46]

Many regulators, such as the PRA, ACPR, APRA, ECB, HKMA, MAS,[47] and Bank of Canada are planning or have already executed so-called climate stress tests. While the regulatory requirements are still in flux, some common features of climate stress tests are beginning to emerge. Most regulators choose to use the NGFS scenarios, either directly or versions consistent with the NGFS scenarios. Climate scenarios typically extend for about 30 years to 2050, a common milestone in climate change analysis. While stress tests against financial and economic shocks proceed in either annual or quarterly time steps, such granularity would be over-engineered for the 30-year horizon used in climate scenarios; most regulators use a five-year interval.

The ECB released results of a top-down climate stress test in September 2021 (Alogoskoufis et al., 2021). Their exercise encompassed about four million corporates worldwide and about 1,600 consolidated banking groups in the euro area. While they report that the effects of climate risk would increase moderately, this is true only on average. For corporates and banks most exposed to climate risk, whether transition or physical risk, the impact is potentially very significant, once again highlighting the importance of accounting for substantial heterogeneity in the impact of climate change risk.

The Bank of England's PRA biennial exploratory scenario provides a useful example of what kinds of questions need to be addressed (Bank of England, 2021). We provide a summary in Table 4.2 by highlighting a set of five questions. The questions touch on basics of risk management, such as climate risk exposure identification (who, where, how big) across transition and physical risk, questions on quantification and modeling, and of course questions on data. While the precise direction of travel by regulators remains very much in flux, banks will need to increase their climate risk awareness and demonstrate an ability to answer what are basic and fundamental risk management questions.

Finally, Jung, Engle and Berner (2021) present a market-based approach for assessing the impact of a climate shock on banks – a climate stress test. They extend the SRISK methodology of Acharya, Engle and Richardson (2012) and introduce a measure called CRISK, systemic climate risk, which is the expected capital shortfall of a financial institution conditional on a climate stress scenario. They use a stranded asset portfolio as a proxy for transition risk which allows them to estimate time varying climate betas. The authors explore the impact of the collapse of fossil-fuel prices in 2020 on banks' capital positions. They find that the CRISK measure during 2020 roughly aligns with the size of currently active loans made to the US firms in the oil and gas industry.

Table 4.2   Key questions and requirements to respond for the Bank of England's 2021 climate stress test

Key questions	Specific requirements
Where are the key climate risks within your portfolios?	• Key climate risks identified across a 30-year time horizon, with both physical and transition risks accounted for • Comprehensive coverage of banking book
How do you measure your exposure to climate risk? How are transition risks and physical risks interlinked?	• Granular expectations for analysis of major counterparties, including name level assessment for minimum Top 100 corporate exposures • For the retail portfolios, minimum 4-digit postcode granularity for UK physical risk • Three climate scenarios to be analyzed
How do key climate risks challenge your business model and how would you strategically act to mitigate them?	• Impact of climate risk on business model to be analyzed • List of potential participant management responses to be developed (and sized)
How do key climate risks challenge your client's business models and how well prepared are they to mitigate them?	• Client engagement/research for top counterparties to understand • Perspectives on business model challenges due to climate risks • Adaptation plans to mitigate climate risks and business model challenges
Where are your key data and capability gaps? What do you need to enhance your understanding of the climate risk across your portfolios?	• Limitations to be documented, requiring comprehensive view on climate risk data requirements and specific plans to close the gaps in the data set • Climate risk data requirements should be coherent across bank initiatives (i.e. risk, finance, sustainability etc.)

Source: Authors' compilation based on Bank of England (2021).

CONCLUDING REMARKS

In 1977, William Nordhaus made a presentation at the annual meeting of the American Economic Association on the economic impact of climate change, an important step on a path that would ultimately lead him to receive the Nobel Prize in economics in 2018.[48] This was not the first time that the Nobel committee considered this topic worthy of recognition. In 2007, the IPCC and Al Gore were awarded the Nobel Peace Prize "for their efforts to build up and disseminate greater knowledge about man-made climate change, and to lay the foundations for the measures that are needed to counteract such change".[49] Banks as financial intermediaries will play a critical role in the transition to a low carbon economy and thereby be exposed to all the risks that come with that transition.

It is hard to overstate the scale and scope of the problem and of the challenges that confront us in this transition. Central banks who help manage systemic risks in a modern economy – such as financial crises – have recognized that climate change risk may be the ultimate systemic risk because it impacts the whole planet at a time scale that makes it very hard to just borrow from the future, essentially through the printing of money, to address the crisis of the present. As a result, they are starting to exert pressure on banks to assess and begin to disclose their exposure to climate change risk as captured by the four core elements promulgated by the TCFD:[50]

- Governance: the organization's governance around climate-related risks and opportunities.
- Strategy: the actual and potential impacts of climate-related risks and opportunities on the organization's business, strategy, and financial planning where such information is material.

- Risk management: the organization's approach to identifying, assessing and managing climate-related risks.
- Metrics and targets: the metrics and targets used to assess and manage relevant climate-related risks and opportunities where such information is material.

These four elements are shaping how banks are preparing their organizations on how to manage their role in the transition to a lower carbon economy. With the experience of the global financial crisis of 2008–2009, and the more recent COVID-19 pandemic, banks have built up better processes to improve their resilience to a variety of shocks, including a wide (and widening) range of stress scenarios to test against. Banks play a vital role as shock absorbers,[51] and preparation for perhaps the most impactful shock we are confronting will require a lot of work.

ACKNOWLEDGEMENTS

The authors of this chapter wish to thank Kevin Stiroh, Johannes Stroebel and the editors Bob Hansen and Anant Sundaram for their very helpful comments and suggestions. A brief editorial, "Managing the Risk of Climate Change," based on an earlier version of this chapter, appeared in the *Journal of Risk Management in Financial Institutions* in 2021, Volume 14, Issue 2. All remaining errors are the authors'.

NOTES

1. See European Commission's 2019 *Guidelines on non-financial reporting: Supplement on reporting climate-related information* for a definition of the concept of double materiality. https://ec.europa.eu/finance/docs/policy/190618-climate-related-information-reporting-guidelines_en.pdf
2. https://www.fsb-tcfd.org/
3. https://www.unepfi.org/net-zero-banking/
4. These are clearly broad and sweeping statements and abstract from differences across regions, levels of economic development and rates of technological change. See Cruz and Rossi-Hansberg (2021) for a study that addresses the importance of heterogeneity in modeling climate impact.
5. Hugonnet et al. (2021) document that glacier melting has accelerated in the first two decades of the 21st century, contributing to about 21 percent of the observed sea level rise.
6. Caesar et al. (2021).
7. https://assets.bbhub.io/company/sites/60/2020/10/FINAL-2017-TCFD-Report-11052018.pdf
8. Note that asset prices could well be subject to sudden and abrupt repricing as a result of sudden and very large changes in the market's beliefs about the evolution of physical risks in a kind of Minsky moment. See also the discussion in the fourth section.
9. As noted in Table 4.1, each of these risk categories includes the associated legal risks including potential third-party lawsuits such as those from shareholders.
10. Since scenario analysis plays such an important role in climate change risk assessment, we devote a whole section below to the topic.
11. Interview by Jonathan Watts in *The Guardian*, 29 May 2021. https://www.theguardian.com/environment/2021/may/29/johan-rockstrom-interview-breaking-boundaries-attenborough-biden
12. See, for example, Mackenzie's piece in *Bloomberg*, "Think hard before you invest in a 'climate bad bank'" (18 June 2021).
13. PIK: Potsdam Institut für Klimafolgenforschung (Potsdam Institute for Climate Impact Research).
14. https://sites.google.com/site/williamdnordhaus/dice-rice
15. http://www.fund-model.org/

16. https://jgcri.github.io/gcam-doc/overview.html
17. https://docs.messageix.org/projects/global/en/latest/
18. https://www.pik-potsdam.de/en/institute/departments/transformation-pathways/models/remind
19. UNEP-FI: United Nations Environmental Programme – Finance Initiative, is a partnership between UNEP and the global financial sector to mobilize private sector finance for sustainable development.
20. Federal Reserve H.8 data.
21. Fannie Mae and Freddie Mac are restricted by law to purchasing single-family mortgages with origination balances below a specific amount, known as the "conforming loan limit." Loans above this limit are known as jumbo loans. See https://www.fhfa.gov/DataTools/Downloads/Pages/Conforming-Loan-Limits.aspx.
22. PV: photovoltaic; electricity production through solar panels.
23. Nauman, B., "'Green bubble' warnings grow as money pours into renewable stocks", 19 February 2021, *Financial Times*.
24. A Minsky moment, named after the economist Hyman Minsky, describes a period of sudden and rapid instability in the market following a prolonged period of prosperity. The turning point is driven by a sudden re-evaluation of risks which results in a sudden collapse of prices for risk assets.
25. Auffhammer reports the numbers in 2007 dollars. We have updated them to 2020 dollars using the CPI, specifically the CPIAUCSL series from FRED: https://fred.stlouisfed.org/series/CPIAUCSL.
26. Barnett, Brock and Hanson (2020) make the interesting observation that some climate models are so complicated on their own that researchers have developed approximations called "emulators" to capture the most salient features.
27. Depending on the nature of the nonlinearity and their sign, some errors could well offset instead of compound.
28. Total factor productivity is the total economic productivity of all factors such as capital and labor. Specifically, it is measured by the ratio of total output (such as GDP) to total inputs.
29. Radiative forcing is the difference between incoming and outgoing energy in the Earth's climate. With increased GHGs, this (positive) difference grows.
30. See Table 2 in van Vuuren et al. (2011).
31. https://climatescenarios.org/primer/mitigation/
32. See discussion at https://climatescenarios.org/primer/mitigation/
33. See also BCBS (2021b) and our discussion in the fourth section.
34. See also the discussion on the challenges to meet the growing demands in the business and finance community for reliable and practically useful climate information in Fiedler et al. (2021).
35. www.spglobal.com/marketintelligence/en/solutions/climate-credit-analytics
36. For a detailed description of stress scenario impact on commercial, investment and custody banks, see Cope et al. (2020).
37. The modeling of losses is typically broken down into these three components such that expected loss (EL) is just their product: $EL = PD \times LGD \times EAD$. See also Table 4.1.
38. EAD is more relevant when considering and testing the potential implications of a bank's planned management actions over the long term; for instance, when a bank makes a commitment to divest from a specific industry by a certain date.
39. This section is an abbreviated version of Oliver Wyman (2020).
40. To be sure, in the long run all taxes are borne by the consumer.
41. See also Baldassarri et al. (2020) for additional discussion of sector-wide impacts of transition risks.
42. See, for instance, Scott et al. (2017) for the Bank of England, ECB (2019), Brunetti et al. (2021) and Board of Governors (2020), Litterman et al. (2020).
43. For a recent summary of supervisory climate stress tests, see UNEP-FI (2021).
44. New Zealand: https://www.beehive.govt.nz/release/nz-becomes-first-world-climate-reporting; UK: https://www.gov.uk/government/publications/uk-joint-regulator-and-government-tcfd-taskforce-interim-report-and-roadmap; Hong Kong: https://www.reuters.com/article/us-hong-kong-regulator-climate-change/hong-kong-sets-new-climate-disclosure-rules-aligns-with-global-standard-idUSKBN28R0Y5
45. For a discussion of stress testing and risk weighting, see Schuermann (2020).
46. See also BCBS (2021b) for a discussion of climate scenario analysis vs. stress testing, where the latter is largely a special case of the former.

47. PRA: Prudential Regulation Authority (UK); ACPR: Autorité de Contrôle Prudentiel et de Résolution (France); APRA: Australian Prudential Regulation Authority; ECB: European Central Bank; HKMA: Hong Kong Monetary Authority; MAS: Monetary Authority of Singapore.
48. See Nordhaus (1977) and Nordhaus (2018).
49. https://www.nobelprize.org/prizes/peace/2007/summary/
50. Financial Stability Board (2017)
51. Banks as shock absorbers is a term used by Timothy Geithner in his 2014 book on the global financial crisis, *Stress Test*.

REFERENCES

Acharya, V.V., R. Engle and M. Richardson (2012). Capital shortfall: A new approach to ranking and regulating systemic risks. *American Economic Review: Papers & Proceedings*, 102(3), 59–64.

Alogoskoufis, S., N. Dunz, T. Emambakhsh, T. Hennig, M. Kaijser, C. Kouratzoglou, M.A. Muñoz, L. Parisi, and C. Salleo (2021). ECB economy-wide climate stress test: Methodology and results. ECB Occasional Paper No. 281, September

Auffhammer, M. (2018). Quantifying economic damages from climate change. *Journal of Economic Perspectives*, 32(4), 33–52.

Baldassarri, G., H. von Högersthal, A. Liu, H. Tomičić and L. Vidovic (2020). Carbon pricing paths to a greener future, and potential roadblocks to public companies' creditworthiness. *Journal of Energy Markets*, 13(2), 1–24.

Bank of England (2015). The impact of climate change on the UK insurance sector. September. https://www.bankofengland.co.uk/-/media/boe/files/prudential-regulation/publication/impact-of-climate-change-on-the-uk-insurance-sector.pdf

Bank of England (2018). Transition in thinking: The impact of climate change on the UK banking sector. September. https://www.bankofengland.co.uk/-/media/boe/files/prudential-regulation/report/transition-in-thinking-the-impact-of-climate-change-on-the-uk-banking-sector.pdf

Bank of England (2021). Guidance for participants of the 2021 Biennial Exploratory Scenario: Financial risks from climate change. June. https://www.bankofengland.co.uk/-/media/boe/files/stress-testing/2021/the-2021-biennial-exploratory-scenario-on-the-financial-risks-from-climate-change.pdf

Barnett, M., W. Brock and L. P. Hansen (2020). Pricing uncertainty induced by climate change. *Review of Financial Studies*, 33(3), 1024–1066.

Basel Committee for Banking Supervision (BCBS) (2021a). Climate-related risk drivers and their transmission channels. https://www.bis.org/bcbs/publ/d517.htm

Basel Committee for Banking Supervision (BCBS) (2021b). Climate-related financial risks - measurement methodologies. https://www.bis.org/bcbs/publ/d518.htm

Bernstein, A., M. Gustafson and R. Lewis (2019). Disaster on the horizon: The price effect of sea level rise. *Journal of Financial Economics*, 134(2), 253–272.

Board of Governors of the Federal Reserve System (2020). *Financial Stability Report*. November. https://www.federalreserve.gov/publications/files/financial-stability-report-20201109.pdf

Boden, T., G. Marland, and R. J. Andres (2017). Global, regional, and national fossil-fuel CO_2 emissions (1751–2014) (V. 2017), Oak Ridge, TN: Carbon Dioxide Information Analysis Center (CDIAC). Oak Ridge National Laboratory (ORNL). http://dx.doi.org/10.3334/CDIAC/00001_V2017.

Bolton, P., M. Despres, L.A. Pereira da Silva, F. Samama and R. Svartzman (2020). The green swan: Central banking and financial stability in the age of climate change. BIS. Available at https://www.bis.org/publ/othp31.htm

Bolton, P. and M.T. Kacperczyk (2020). Do investors care about carbon risk? *Journal of Financial Economics*. Available at https://ssrn.com/abstract=3398441

Brunetti, C., B. Dennis, D. Gates, D. Hancock, D. Ignell, E.K. Kiser, G. Kotta, A. Kovner, R.J. Rosen, and N.K. Tabor (2021). Climate change and financial stability. FEDS Notes. https://www.federalreserve.gov/econres/notes/feds-notes/climate-change-and-financial-stability-20210319.htm

Caesar, L., G. D. McCarthy, D. J. R. Thornalley, N. Cahill, and S. Rahmstorf (2021). Current Atlantic Meridional Overturning Circulation weakest in last millennium. *Nature Geoscience*, 14, 118-120.

Cai, Y. and T. S. Lontzek (2019). The social cost of carbon with economic and climate risks, *Journal of Political Economy*, 127(6), 2684–2734.

Cai, Y., T.M. Lenton, and T. S. Lontzek (2019). Risk of multiple interacting tipping points should encourage rapid CO_2 emission reduction. *Nature Climate Change*, 6, 520–525.

Calvin, K. et al. (2019). GCAM Documentation. http://www.globalchange.umd.edu/gcam/. College Park, MD: Joint Global Change Research Institute.

Carney, M. (2015). Breaking the tragedy of the horizon – climate change and financial stability. Speech at Lloyd's of London, London, 29 September. Available at https://www.bis.org/review/r151009a.pdf

Carney, M. (2018). A transition in thinking and action. Remarks at the International Climate Risk Conference for Supervisors, The Netherlands Bank, Amsterdam, 6 April. Available at: https://www.bis.org/review/r180420b.pdf

Cevik, S. and J. Tovar Jalles (2020). This changes everything: climate shocks and sovereign bonds. IMF Working Paper 20–79.

Cope, D., C. Hsu, C.D. Lively, J. Morgan, T. Schuermann and E. Sekeris (2020). Stress testing for commercial, investment and custody banks. In J. Doyne Farmer, A. Kleinnijenhuis, T. Schuermann and T. Wetzer (eds), *Handbook of Financial Stress Testing*. Cambridge University Press. Working paper version is available at http://ssrn.com/abstract=3530185

Cruz, J-L and E. Rossi-Hansberg (2021). The economic geography of global warming. NBER Working Paper 28466.

Engle, R., S. Giglio, H. Lee, B. Kelly and J. Stroebel (2020). Hedging climate change news. *Review of Financial Studies*, 33, 1184–1216

European Commission (2019). Guidelines on reporting climate-related information. https://ec.europa.eu/finance/docs/policy/190618-climate-related-information-reporting-guidelines_en.pdf

Fiedler, T., A.J. Pitman, K. Mackenzie, N. Wood, C. Jakob and S.E. Perkins-Kirkpatrick (2021). Business risk and the emergence of climate analytics, *Nature Climate Change*, 11, 87–94.

Financial Stability Board (2017). Recommendations of the Task Force on Climate-related Financial Disclosures. June. https://www.fsb-tcfd.org/recommendations/

Fricko, O., P. Havlik, J. Rogelj, Z. Klimont, M. Gusti, N. Johnson, P. Kolp, M. Strubegger, H. Valin, M. Amann, T. Ermolieva, N. Forsell, M. Herrero, C. Heyes, G. Kindermann, V. Krey, D. L. McCollum, M. Obersteiner, S. Pachauri, S. Rao, E. Schmid, W. Schoepp, and K. Riahi (2017). The marker quantification of the Shared Socioeconomic Pathway 2: A middle-of-the-road scenario for the 21st century, *Global Environmental Change*, 42, 251–267.

Geithner, T.F. (2014). *Stress Test: Reflection on Financial Crises*. New York, NY: Crown Publishers.

Giglio, S., M. Maggiori, and J. Stroebel (2015). Very long-run discount rates, *Quarterly Journal of Economics*, 130(1), 1–53

Giglio, S., B. T. Kelly, and J. Stroebel (2020). Climate finance, forthcoming. *Annual Review of Financial Economics*. Available at SSRN: https://ssrn.com/abstract=3719139

Giglio, S., M. Maggiori, K. Rao, J. Stroebel, and A. Weber (2021). Climate change and long-run discount rates: Evidence from real estate, *Review of Financial Studies*, 34(8), 3527–3571.

Gillingham, K., W. Nordhaus, D. Anthoff, G. Blanford, V. Bosetti, P. Christensen, H. McJeon, and J. Riley (2018). Modeling uncertainty in integrated assessment of climate change: A multimodel comparison. *Journal of the Association of Environmental and Resource Economists*, 5(4), 791–826.

Gillingham, K. and J.H. Stock (2018). The cost of reducing greenhouse gas emissions. *Journal of Economic Perspectives*, 32(4), 53–72.

Goldsmith-Pinkham, P. S., M. Gustafson, R. Lewis, and M. Schwert (2021). Sea level rise exposure and municipal bond yields. Available at SSRN: https://ssrn.com/abstract=3478364

Greenstone, M., E. Kopits, and A. Wolverton (2013). Developing a social cost of carbon for US regulatory analysis: A methodology and interpretation. *Review of Environmental Economics and Policy*, 7(1), 23–46.

Hone, D. (2023). The end of combustion? In A. Sundaram and R. Hansen (eds), *Handbook of Business and Climate Change*. Edward Elgar Publishing.

Hope, C. W. (2006). The marginal impact of CO_2 from PAGE2002: An Integrated Assessment Model incorporating the IPCC's five reasons for concern. *The Integrated Assessment Journal*, 6(1), 19–56.

Howard, P.H and T. Sterner (2017). Few and not so far between: A meta-analysis of climate damage estimates. *Environmental Resource Economics*, 68(4), 1–29.

Hsiang, S. and R. E. Kopp. (2018). An economist's guide to climate change science. *Journal of Economic Perspectives*, 32(4), 3–32.

Hugonnet, R., R. McNabb, E. Berhier, B. Menounos, C. Nuth, L. Girod, D. Farinotti, M. Huss, I. Dussaillant, F. Brun, and A. Kaab (2021). Accelerated global glacier mass loss in the early twenty-first century. *Nature*, 592, 726–731.

Huynh, T.D. and Y. Xia (2021). Climate change news risk and corporate bond returns. *Journal of Financial and Quantitative Analysis*, 56(6), 1985–2009.

Ilhan, E., Z. Sautner, and G. Vilkov (2020). Carbon tail risk. *Review of Financial Studies*. Available at: https://ssrn.com/abstract=3204420

Interagency Working Group on Social Cost of Carbon (2010). Social cost of carbon for regulatory impact analysis under Executive Order 12866. February. United States Government. http://www.whitehouse.gov/sites/default/files/omb/inforeg/for-agencies/Social-Cost-of-Carbon-for-RIA.pdf

Intergovernmental Panel on Climate Change (IPCC) (2014). AR5 Synthesis Report: Climate change 2014. Available at https://www.ipcc.ch/report/ar5/syr/

Intergovernmental Panel on Climate Change (IPCC) (2021). Climate Change 2021: The physical science basis. Available at https://www.ipcc.ch/report/ar6/wg1/

International Energy Agency (2021). Net zero by 2050: A roadmap for the global energy sector. May. https://www.iea.org/reports/net-zero-by-2050

Issler, P., R. Stanton, C. Vergara-Alert, and N. Wallace (2020). Mortgage markets with climate-change risk: evidence from wildfires in California. Available at SSRN: https://ssrn.com/abstract=3511843

Jiang, F., C.W. Li, and Y. Qian (2019). Do costs of corporate loans rise with sea level? Available at SSRN: https://ssrn.com/abstract=3477450

Jung, H., R. Engle, and R. Berner (2021). Climate stress testing. Federal Reserve Bank of New York Staff Reports, no. 977.

Keenan, J. M. and J. T. Bradt (2020). Underwaterwriting: from theory to empiricism in regional mortgage markets in the US. *Climatic Change*, 162, 2043–2067.

Keys, B. J. and P. Mulder (2020). Neglected no more: Housing markets, mortgage lending, and sea level rise. NBER Working Paper 27930.

Krey, V., P. Havlik, P. N. Kishimoto, O. Fricko, J. Zilliacus, M. Gidden, M. Strubegger, G. Kartasasmita, T. Ermolieva, N. Forsell, M. Gusti, N. Johnson, J. Kikstra, G. Kindermann, P. Kolp, F. Lovat, D. L. McCollum, J. Min, S. Pachauri, Parkinson S. C., S. Rao, J. Rogelj, H. and Ünlü, G. Valin, P. Wagner, B. Zakeri, M. Obersteiner, and K. Riahi (2020). MESSAGEix-GLOBIOM Documentation – 2020 release, Technical Report, International Institute for Applied Systems Analysis (IIASA), Laxenburg, Austria. Available at https://pure.iiasa.ac.at/id/eprint/17115

Lemoine, D. (2021). The climate risk premium: how uncertainty affects the social cost of carbon. *Journal of the Association of Environmental and Resource Economists*, 8(1), 27–57.

Litterman, R. et al. (2020). Managing climate risk in the US financial system. Report of the Climate-Related Market Risk Subcommittee, Market Risk Advisory Committee of the U.S. Commodity Futures Trading Commission. https://www.cftc.gov/sites/default/files/2020-09/9-9-20%20Report%20of%20the%20Subcommittee%20on%20Climate-Related%20Market%20Risk%20-%20Managing%20Climate%20Risk%20in%20the%20U.S.%20Financial%20System%20for%20posting.pdf

Mallucci, E. (2020). Natural disasters, climate change, and sovereign risk. International Finance Discussion Papers 1291. Washington: Board of Governors of the Federal Reserve System.

McCollum, D., Y. Nagai, K. Riahi, G. Marangoni, K. Calvin, R. Pietzcker, J. van Vliet, and B. van der Zwaan (2013). Energy investments under climate policy: a comparison of global models. *Climate Change Economics* 4(4).

Network for Greening the Financial System (2020a). Guide to climate scenario analysis for central banks and supervisors. June. https://www.ngfs.net/sites/default/files/medias/documents/ngfs_guide_scenario_analysis_final.pdf

Network for Greening the Financial System (2020b). NGFS Climate Scenario Database – Technical Documentation. June. https://www.ngfs.net/sites/default/files/ngfs_climate_scenario_technical_documentation_final.pdf

Network for Greening the Financial System (2021). NGFS climate scenarios for central banks and supervisors. June. https://www.ngfs.net/en/communique-de-presse/ngfs-publishes-second-vintage-climate-scenarios-forward-looking-climate-risks-assessment

Nordhaus, W.D. (1977). Economic growth and climate: The carbon dioxide problem. *American Economic Review*, 67(1), 341–346.
Nordhaus, W. D. (2018). Climate change: The ultimate challenge for economics. Nobel Prize lecture. Available at https://www.nobelprize.org/uploads/2018/10/nordhaus-lecture.pdf
Oliver Wyman (2020). Assessing credit risk in a changing climate: transition-related risks in corporate lending portfolios. Ch.3 in *Case Studies of Environmental Risk: Analysis Methodologies*. NGFS, September. https://www.ngfs.net/sites/default/files/medias/documents/case_studies_of_environmental _risk_analysis_methodologies.pdf
Oliver Wyman and IACPM (2019). Climate change: Managing a new financial risk. Available at https:// www.oliverwyman.com/our-expertise/insights/2019/feb/climate-risk-and-the-financial-impact.html
Ouazad, A. and M. E. Kahn (2019). Mortgage finance in the face of rising climate risk. NBER Working Paper Working Paper 26322.
Painter, M. (2020). An inconvenient cost: The effects of climate change on municipal bonds. *Journal of Financial Economics*, 135(2), 468–482.
Pielke, R., M. G. Burgess, and J. Ritchie (2021). Most plausible 2005-2040 emissions scenarios project less than 2.5°C of warming by 2100. *SocArXiv*. March 23.
Pindyck, R. A. (2020). What we know and don't know about climate change, and implications for policy. NBER Working Paper No. 27304.
Riahi, K., D. P. Van Vuuren et al. (2017). The Shared Socioeconomic Pathways and their energy, land use, and greenhouse gas emissions implications: An overview. *Global Environmental Change*, 42, 153–168.
Schuermann, T. (2020). Capital adequacy pre- and postcrisis and the role of stress testing. *Journal of Money, Credit and Banking*, 52(S1), 87–105.
Schuermann, T. and K. J. Stiroh (2006). Visible and hidden risk factors for banks. Federal Reserve Bank of New York Staff Report 252.
Scott, M., J. van Huizen and C. Jung (2017). The Bank of England's response to climate change. Bank of England *Quarterly Bulletin* 2017 Q2. https://www.bankofengland.co.uk/quarterly-bulletin/2017/q2/ the-banks-response-to-climate-change
Steffen, W., J. Rockström, K. Richardson, T. M. Lenton, C. Folke, D. Liverman, C. P. Summerhayes, A. D. Barnosky, S. E. Cornell, M. Crucifix, J. F. Donges, I. Fetzer, S. J. Lade, M. Scheffer, R. Winkelmann, and H. J. Schellnhuber (2018). Trajectories of the Earth System in the Anthropocene. *PNAS*, 115(33), 8252–8259.
United Nations Environmental Program Finance Initiative (2018). Extending our horizons: Assessing credit risk and opportunity in a changing climate. Part I: transition risk. April. https://www.unepfi.org/ wordpress/wp-content/uploads/2018/04/EXTENDING-OUR-HORIZONS.pdf
United Nations Environmental Program Finance Initiative (2020a). Charting a new climate: Assessing credit risk and opportunity in a changing climate. Part II: physical risk. September. https://www .unepfi.org/publications/banking-publications/charting-a-new-climate/
United Nations Environmental Program Finance Initiative (2020b). Beyond the horizon: new tools and frameworks for transition risk assessments from UNEP FI's TCFD Banking Programme. September. https://www.unepfi.org/publications/banking-publications/beyond-the-horizon/
United Nations Environmental Program Finance Initiative (2021). The climate risk landscape: A comprehensive overview of climate risk assessment methodologies. February. https://www.unepfi.org/ wordpress/wp-content/uploads/2021/02/UNEP-FI-The-Climate-Risk-Landscape.pdf
United Nations Framework Convention on Climate Change (2015). *Paris Agreement. United Nations Framework Convention on Climate Change*. Bonn, Germany. Available at: unfccc.int/sites/default/ files/english_paris_agreement.pdf
Van Vuuren, D. P., et al. (2011). The representative concentration pathways: An overview. *Climatic Change*, 109, 5–31.
Weitzman, M.L. (2014). Fat tails and the social cost of carbon. *American Economic Review: Papers & Proceedings*, 104(5), 544–546.
World Meteorological Organization (WMO) (2021). *United in Science 2021 Report*. Available at https:// public.wmo.int/en/resources/library/united-science-2021
Zhou, X., C. Wilson and B. Caldecott (2021). The energy transition and changing financing costs. Oxford Sustainable Finance Programme working paper. https://www.smithschool.ox.ac.uk/research/ sustainable-finance/publications/The-energy-transition-and-changing-financing-costs.pdf

5. The patchwork quilt: business complexities of decarbonizing the electric sector

Scott G. Fisher, Bruce A. Phillips and Mark W. Scovic

INTRODUCTION

The economic growth and well-being of societies across the world are heavily dependent upon the adequacy of their electric systems. Electricity enables light, heat, cooling, the production of goods, communications over long distances, data access and computations, modern healthcare, mass media entertainment, and transportation, all at the flip of a switch or the turn of a key. The development and operation of a robust, reliable, and affordable electric system are as complex as they are important. They require large investments, access to natural resources, technological expertise, and well-crafted public policies and rules.

The electric sector is also central to addressing the world's climate challenge. Power generation accounts for 40 percent of energy-related carbon dioxide (colloquially, "carbon") emissions worldwide, highlighting the importance of making substantial changes to the electric sector to address climate-related issues (International Energy Administration, 2021).[1] The electric sector can also be leveraged to reduce greenhouse gas emissions from other sectors of the economy. For example, the transportation sector accounts for 23 percent of global energy-related carbon emissions (International Energy Administration, 2021), and significant emissions reductions could be achieved by a shift to cleaner electricity generation coupled with electrification of parts of the transportation sector. Further, regulators and electric utilities are increasingly developing and implementing plans to improve the resiliency of electric systems, to protect against the greater frequency of hurricanes, extreme heatwaves and cold spells, and rising water levels.

These factors contribute to a complicated web of challenges and opportunities for electric sector businesses. The complexity of these challenges and opportunities is amplified by greatly varying regional economic conditions, access to capital, natural resources, government structures and policies, pre-existing electric system structures, and societal needs. The landscape for electric sector businesses therefore varies across countries and, on a more granular basis, across localities within a single country. Despite this regional variation, several similar dynamics and considerations apply across much of the world, albeit certain factors may be more prevalent in one locality versus another.

This chapter and the chapter that follows examine these issues in detail. This chapter discusses the complexity and varying conditions in the electric sector, what we refer to as a "patchwork quilt," focusing on the sector's recent history and current structure, often using the United States (US) as a case study. The following chapter takes a longer view, discussing the prospect of a transition to a fully decarbonized economy-wide energy system that will last several decades, and the unprecedented investment opportunities and challenges that this presents for business.

In this chapter, the "Electric Industry Structure and Evolution" section provides a brief overview and history of the electric industry, including structural, technological, and greenhouse gas emission developments. Over the last several decades, the US electric industry has been shaped by several important trends including the restructuring of traditional vertically-integrated utilities in some regions of the country to allow for wholesale and retail competition, the growth of combined-cycle generating technologies and innovations in natural gas production that have driven down the cost of natural gas as a fuel for the generation of electricity, and, most recently, the substantial cost declines and the rapid growth of wind, solar, and battery technologies. At the national level, these developments have led to the significant expansion of natural gas-fired and renewable generation, the decline of coal-fired generation, and reductions in the emissions of carbon dioxide and other air pollutants. Still, as discussed in the next chapter, limiting global warming to a level between 1.5°C and 2.0°C above pre-industrial levels will likely require fundamental changes to the electric industry and massive expansion of sector infrastructure.

The subsequent section of this chapter, "Today's Complex Patchwork Quilt," covers the varied policies and market-related factors that define the opportunities and challenges for potential investors and businesses active in the electric sector. This section provides insights about the nature, dynamics, and interplay of the various types of climate and clean energy policies that comprise the patchwork quilt, and the associated considerations for businesses. There is great variation in the electric sector at the regional level across the US and the world. Major contributors to this variation include differing natural endowments of coal, natural gas, hydro, solar and wind resources, and differing state and local priorities for energy and environmental policy. These have led to the patchwork quilt of conditions in the regional electric markets, especially in countries such as the US which lack stable energy and climate policies at the federal level. The quilt has two overarching ramifications for businesses.[2] First, the quilt can contribute to large investment opportunities, but these are highly dependent on regional market and policy conditions, some of which can be local and complex. Second, in many cases, the patchwork quilt is accompanied by increased business risks due to the interplay between different regulators' policies, the uncertainty about future policy changes, and other factors, including the emergence of new technologies and changing commodity prices. These can lead to financial risks for investors and unnecessary costs to satisfy climate goals.

Using this chapter as a foundation, the next chapter looks at the energy transition that is widely expected to occur over several decades. This transition will entail decarbonizing today's carbon-emitting fossil-fired generation or replacing it with carbon free electricity, as well as massive electricity system expansion with substantial investment opportunities to electrify and decarbonize portions of the transportation, industry, and space conditioning sectors of the economy. The transition is deeply uncertain in its form and timing due to ongoing uncertainties about federal and state policy, technology pathways, and the pace at which it will be possible to deploy new energy infrastructure. For companies in the electric industry, this points to three broad challenges: how to deploy carbon-free technologies that have already been commercialized, how to demonstrate and commercialize new advanced clean energy technologies, and how to manage the existing fleets, particularly fossil-fired and nuclear generating plants.

To succeed in this environment, businesses need strategic foresight, financial resources, and prudent risk management. Companies must be able to understand and manage commercial risks arising from uncertain state regulation, federal policy, technology, supply chain, and project development forces.

ELECTRIC INDUSTRY STRUCTURE AND EVOLUTION

The Electric Grid

The electric grid, or electric system, is an interconnected network that produces, transmits and consumes electricity. It is often thought of as being composed of four primary segments: the *generation* of electricity at power plants, long distance *transmission* of electricity across high voltage power lines, *distribution* on local, low voltage power lines, and *customer* use of electricity to power lighting, appliances, commercial processes, industrial machines, and other activities.

The generation and consumption of electricity must be precisely matched on an instantaneous and locational basis to avoid blackouts and shorter service interruptions. To accomplish this, the organizations operating the US electric grids direct the operation of power plants in real time to meet electric demand at the lowest operating cost in an approach known as "security-constrained economic dispatch," which the US Department of Energy defines as "an area-wide optimization process designed to meet electricity demand at the lowest cost, given the operational and reliability limitations of the area's generation fleet and transmission system" (US Department of Energy, 2007).

In the centralized wholesale power markets of the US, the system operator holds daily competitive "energy market" auctions to select which power plants generate electricity, offer backup reserve capacity if needed to respond to unanticipated events, and provide ancillary services to the grid, with market prices that settle on an hourly or more granular basis.[3] The operation of the system is quite complex, as demand is constantly changing, different generation resources have different operating constraints, transmission constraints limit power flows across the system, and the availability of generation resources and power lines can change due to factors such as weather and equipment failures. These factors, as well as significant differences between the dispatch costs of various types of generation resources, contribute to market volatility and uncertainty for competitive generation companies and other businesses.

Some system operators also administer "capacity markets" to ensure that there are always adequate quantities of supply resources available. Winning bidders in a capacity auction administered by the system operator must guarantee that specified megawatt quantities of supply will be available in the upcoming year associated with the auction. Supply resources in capacity auctions may include new generation resources, existing generation resources, upgrades to existing generation resources, demand response (guarantees from consumers that they will reduce their electricity use when called upon), energy efficiency, and transmission upgrades.

Industry Restructuring

Bolstered by state regulation and New Deal legislation during the first half of the 20th century, the US electric industry expanded from a series of largely urbanized and separate electric systems to one with long-distance transmission interties connecting adjoining utilities, electric cooperatives and public power authorities, and with more extensive service to rural areas. From that era up to the 1980s and 1990s when large portions of the electric industry were restructured, the predominant industry model revolved around vertically integrated generation–transmission–distribution utilities each supplying power to customers in a defined geographic region with retail rates set by a state regulator and interstate transmission overseen by federal regulators.

The industry enjoyed steady growth and remained largely unchanged structurally until the energy crisis and OPEC oil embargo of the 1970s, which resulted in supply shortages of fossil fuels and large swings in their prices. In response to that, the Public Utility Regulatory Policies Act (PURPA) was enacted in 1978. Among other policies intended to promote energy efficiency, solar and wind technologies, and nuclear power, PURPA designated cogeneration facilities[4] and smaller renewable generation projects owned by independent power producers as qualifying facilities (QFs), which allowed them to sell electricity to utilities at pre-established prices based on the utility's "avoided cost," which was basically the incremental cost that the utility would incur if it instead generated the electricity itself or purchased it from another source. While largely unintended, PURPA had the effect of demonstrating that non-utility power suppliers could develop and operate large-scale power plants, thereby encouraging further regulatory reforms in the 1980s and 1990s. In 1992, Congress passed the Energy Policy Act (EPACT) to facilitate competition in wholesale electricity markets by opening transmission lines to non-utility generators and expanding the definition of qualifying facilities. EPACT essentially forced utilities to open their transmission systems, giving non-utility generators more freedom to sell their electricity to buyers located large distances from their plants.

In 1999, the Federal Energy Regulatory Commission (FERC) issued Order 2000, encouraging owners of transmission facilities to pool their transmission assets under the control of a Regional Transmission Operator (RTO). While Order 2000 leaves flexibility as to how the RTOs may be structured, the primary purposes of each RTO are to control and monitor the transmission system. Common practice is for the RTO to administer a wholesale electricity market across its footprint and coordinate system-wide dispatch of electricity generators based on market bids. The intent of establishing these organized electricity markets was to provide stronger incentives to improve the economic efficiency of the electric sector and lower the cost of electricity service for consumers relative to the cost that would be achieved under traditional regulation without these markets. Many regions of the US are now served through RTOs (US Federal Energy Regulatory Commission, 2015) (see Figure 5.1).

All the RTOs in the US are regulated by the FERC, apart from the Electric Reliability Council of Texas (ERCOT), which as a single-state transmission organization avoids certain aspects of federal regulation. The transmission systems in regions without RTOs, mostly in the western and southeast states, are overseen by vertically integrated electric utilities and public power authorities.

Often driven by state regulations, many utilities in RTOs have sold their power plants to third parties who compete in the wholesale markets. These utilities rely on wholesale market purchases to satisfy their customers' electricity supply needs. While the local utility still is responsible for the distribution of electricity supplied to all customers in the utility's service area, in several states, each customer now may choose among many companies to provide the electricity supply at the retail level. These competitive retail providers provide an agreed-upon price structure that the customer pays, assume responsibility for the wholesale market cost to supply the customer, and manage the associated risks accordingly. Despite the similar purpose of the different RTOs, the structure and operations of each regional market differs. This contributes to the patchwork quilt theme discussed later in this chapter.

The patchwork quilt: business complexities of decarbonizing the electric sector 93

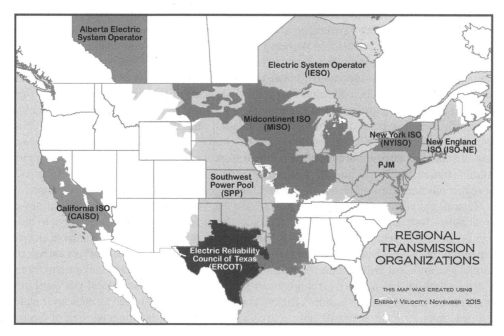

Source: US Federal Energy Regulatory Commission.

Figure 5.1   Regional transmission organizations

The Changing Generation Mix and Declining Emissions

Most electricity generation in the US is fueled by natural gas, coal, nuclear, and renewable energy technologies including hydropower as well as non-hydropower renewables such as wind and solar (US Energy Information Administration, 2021g) (see Figure 5.2).

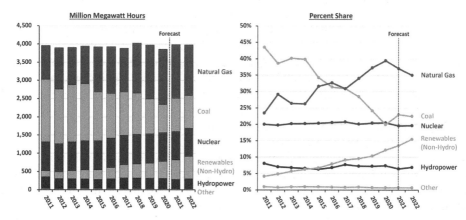

Source: US Energy Information Administration.

Figure 5.2   US electricity generation by fuel type

From 2011 to 2021, generation from non-hydropower renewables, predominantly wind and solar, grew from less than 5 percent to almost 15 percent of US generation; natural gas-fired generation grew from less than 25 percent to roughly 35 percent; and coal-fired generation declined from approximately 45 percent to less than 25 percent. The drivers of these trends are explained later. Not surprisingly, US electric sector carbon emissions have declined, due to the growth in zero-carbon generation such as wind and solar, the growth in natural gas-fired generation, which has lower carbon emission rates than coal-fired generation, and the decline in coal-fired generation with its high carbon emission rates. US electric sector carbon emission levels in 2019 were 33 percent below 2005 levels, dropping the sector to be the second largest source of carbon emissions behind the transportation sector (US Energy Information Administration, 2020b). While this represents meaningful progress, it falls short of the pace of reductions needed to achieve many stated climate goals, as discussed in the next chapter.

Natural gas and coal-fired power plants have the capability to provide reliable power around-the-clock, and many natural gas-fired plants can quickly increase their output to meet rapidly increasing demand in peak hours. This capability makes these plants valuable electric generation resources. But without controls to capture and sequester their carbon emissions, natural gas and coal will remain the dominant sources of electric sector carbon emissions. Since natural gas has a lower carbon content than coal, and because natural gas combined cycle generating technology is more efficient than conventional coal-fired technology, natural gas generation typically emits 40 percent to 50 percent of the carbon on a per unit of electric energy basis relative to conventional coal.

Nuclear, hydropower, wind, and solar generation are the main sources of electricity in the US that do not directly emit carbon.[5] As of 2021, nuclear generation supplies the largest share of electricity among these generation types. However, most of the country's nuclear fleet is aging. Many nuclear plants have reached or are approaching their original plant licensing lives. Some, especially those without long-term contracts or government guarantees of cost recovery, struggle to earn enough revenue to continue operation. Hydropower also provides a significant share of electricity. Many hydroelectric plants can be operated to provide around-the-clock electricity or be dispatched in response to changing demand (a.k.a. "load") and market conditions. But opportunities to develop new hydroelectric generation plants are limited due to the difficulty of permitting new reservoirs. Solar and wind facilities are increasingly used to produce energy whenever they can because of their low variable operating costs, their lack of carbon emissions, and the availability of financial credits for electricity generated from wind. But the variability of their output, and the possibility of low or zero output during extended periods of relatively calm winds and cloud cover, create significant challenges for these technologies alone to reliably serve peak electric loads.

Growth of Natural Gas-fired Electric Generation

Two developments drove natural gas generation to become the US's most widely relied upon fuel source for power generation, allowing natural gas generation to displace coal generation and reduce electric sector carbon emissions. The first was the commercialization of combined cycle technology, and the second was the "fracking revolution."

The natural gas-fired combined cycle generating technology was widely commercialized in the last quarter of the 20th century. With this generating technology, the hot gas exhausted from a gas turbine, itself driving a generator to produce electricity, flows into a heat recovery steam generator, which boils water to produce steam that drives a steam turbine to generate additional electricity, increasing total system efficiency. This allows more power to be produced with less fuel, resulting in lower fuel costs and less carbon emissions than older generation natural gas plant technologies and coal-fired alternatives.

In the first decade of the 21st century, natural gas generation received another boost as the fracking revolution unfolded. Specifically, advances in certain technologies that extract natural gas from underground shale rock formations substantially changed the economics of natural gas production in the US, where shale rock formations are plentiful. The most prominent of these extraction technologies is a production procedure that is known as hydraulic fracturing (a.k.a. "fracking"), which involves drilling underground to the shale rock formation and injecting a fracturing fluid mixture of sand, water, and chemicals into the rock to fracture it, thereby releasing natural gas and other hydrocarbons. While fracking technology had existed for decades, it had not been commercially viable on a widespread basis. However, advances in another technology, horizontal drilling, combined with the development of more effective mixtures for fracturing fluids, suddenly changed the economics of fracking. As its name implies, horizontal drilling is the process of drilling horizontally after reaching the depth at which the shale rock formation lies, and horizontal drilling provides access to larger sections of geologic formations. These advances led to an increase in US natural gas production of more than 50 percent from 2008 to 2018 (US Energy Information Administration, 2021h), and a decline in US natural gas prices of more than 50 percent over that period (US Energy Information Administration, 2021i).

These developments have greatly contributed to a reduced cost of natural gas-fired generation relative to coal-fired generation, allowing natural gas generation to displace coal generation and reduce electric sector carbon emissions. The availability of low-cost natural gas supplies has also enabled the US to become a net exporter of energy with implications for global energy trade and diplomacy.

Commercialization of Wind, Solar and Batteries

While the deployment of natural gas-fired generation substantially increased in recent decades, more recent cost reductions in wind, solar, and batteries indicate a major surge in the deployment of these technologies going forward. In part because of these cost reductions, wind and solar technologies are expected to have a key role in achieving climate goals, as discussed in the next chapter.

Significant declines in the costs of wind and solar generation technologies have helped spur increasingly rapid deployment of these technologies. The US average levelized cost[6] of onshore wind projects has declined by over 60 percent, from $90/MWh in 2010 to $33/MWh in 2020 (Wiser et al., 2021). The cost of utility-scale solar photovoltaic (PV) generation has declined even more. The US average levelized cost of this type of generation declined from about $220/MWh in 2010 to $34/MWh in 2020, a cost reduction of 85 percent (Bolinger, Seel, Warner, & Robson, 2021). These cost declines, coupled with state renewable procurement mandates and voluntary purchases by large technology and consumer products corporations, have driven rapid increases in the construction of wind and solar generation. In aggregate in 2020, wind and solar generation represented 80 percent of the total US generation capacity additions, as compared to about 30 percent in 2010 (Bolinger, Seel, Warner, & Robson, 2021). Along with the development of natural gas-fired power plants and their increasing utilization rates displacing coal-fired generation, solar and wind deployment have been significant contributors to US electric sector carbon emission reductions.

Similarly, the US is undergoing a substantial increase in battery capacity on the electric grid. By the end of 2010, large-scale battery storage systems[7] installed in the US totaled about 50 MW of power capacity.[8] By the end of 2019, this value had increased by a factor of 20, to over 1,000 MW. Furthermore, the US Energy Information Administration projects the amount of installed large-scale storage battery to be over 12,000 MW by the end of 2023, an amount roughly 12 times that installed as of the end of 2019 (US Energy Information Administration, 2021b). These increases are driven by both significant battery cost reductions and the adoption of policies at the state and federal levels to promote battery development for the services that batteries can provide. With the capability to store and release electricity over a four-hour period or somewhat longer, batteries can address short-term supply deficits and serve particularly high peak demands. Batteries also can increase the amount of electricity that is sourced from environmentally friendly generation resources such as wind and solar generation by charging batteries from these generation resources during hours when electricity demand levels are low compared with the availability of these resources and discharging the batteries during hours when demand levels are high. They are also adept at providing "ancillary services" that stabilize grid operations. However, battery technology performance and costs will need to further improve substantially from today's levels for batteries to be suitable to address longer-term seasonal variations in solar and wind output or multi-day periods of calm winds and cloudy skies.

Another related trend is the adoption of distributed energy resources (DERs). DERs are generation resources or controllable loads that are directly connected to a local distribution system or connected to a host facility that is directly connected to the local distribution system. DERs include rooftop solar generation, small-scale wind generation, backup generators, and batteries connected to the distribution system or installed on the customer's side of the utility electric meter. In certain circumstances, DERs can defer or avoid the need for expensive grid investments, and they can improve the reliability of customer service by avoiding or mitigating service interruptions from distribution-level power outages. Depending on the fuel type used, DERs may also displace the need for more emissions-intensive centralized power plants that are powered by fossil fuels.

Climate and Energy Policies

The technical advances, cost reductions, and increased deployment of natural gas, solar, wind, battery, DER, and other technologies are the result of private sector investments and government policy both overseas and in the US at the federal and state levels. The next section, "Today's Complex Patchwork Quilt," identifies and addresses the types of climate and clean energy policies that are often applied in the electric sector. It presents specific examples, and provides insights regarding policy dynamics and the associated ramifications for businesses. The patchwork quilt of energy and climate policies that exists today has helped to reduce harmful emissions. However, as discussed in the next chapter, the pace of overall reductions to date falls short of that needed to achieve many stated climate goals, and varying policy priorities and unnecessary costs due to the lack of coordination and alignment may undermine the prospects of satisfying these goals.

In the US, this lack of coordination and alignment is partially driven by the failures of federal efforts in recent decades to establish uniform national emission or clean energy requirements. These include the US government not adopting the Kyoto Protocol treaty negotiated in 1997, the Senate's rejection of the American Clean Energy and Security Act of 2009 (also known as the Waxman-Markey Bill), the holdup in the court system of President Obama's Clean Power Plan, President Trump's withdrawal of the US from the Paris Agreement in 2017, and Congress's failure to enact the Clean Electricity Performance Program proposed in the House in 2021. While the US rejoined the Paris Agreement under President Biden, strong government programs and coordination will be required over several decades for the US to achieve the goals of the Paris Agreement or other ambitious climate goals.

TODAY'S COMPLEX PATCHWORK QUILT

Today's global and US electric sector landscape is a "patchwork quilt" of regional market conditions and climate and energy policies. The patchwork quilt has supported various levels of clean energy resource deployment and electric sector greenhouse gas emission reduction, and it often complicates business opportunities.

The quilt is partially defined by regional differences in the abundance of natural resources, the topography of the electricity grid, and the structure of the electricity markets, all of which shape the opportunities and risks in each region. In addition, the government policies intended to support clean energy and address the changing climate have also become a major part of the quilt and the business opportunities that it presents. Regional and political differences regarding climate concerns and the preferred approaches to address these concerns lead to policies that vary locationally and across different levels of government.

The patchwork quilt has two overarching ramifications for businesses.[9] First, while the quilt can contribute to large investment opportunities, these are highly dependent on regional market and policy conditions, some of which can be quite local and complex. Second, in many cases, the patchwork quilt is accompanied by increased business risks due to the interplay between different regulators' policies, the uncertainty about future policy changes, and other factors including the emergence of new technologies and changing commodity prices. The nature of many of the policy approaches and the inconsistencies between policies also can lead to unnecessary costs associated with satisfying climate goals.

The US provides an exemplary case study of the patchwork quilt, given the size of its economy and carbon footprint, its geographic variation in natural resources, and its frequently polarized political environment.

Case Study: The Patchwork Quilt in the United States

The mix of electricity generation in operation in each region of the US, and the mix that is being constructed in each region, provide a starting point to understand the patchwork quilt that characterizes the US electricity landscape. Figure 5.3 depicts the aggregate regional sources of grid-connected electricity generated across the US as of 2020[10] (US Energy Information Administration, 2021c, 2021f).

98   *Handbook of business and climate change*

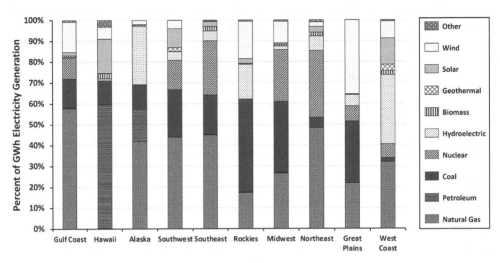

Source: Based on data from the US Energy Information Administration.

Figure 5.3   *Regional electricity generation mix across the US (2020)*

As the figure shows, as of 2020, most of the power generated in almost every region of the US was generated by fossil fuels, predominantly natural gas and coal, with petroleum being the dominant fuel for electricity generation in Hawaii. Despite the national expansion of natural gas-fired generation in recent years as described previously, coal-fired generation still held a significant share as of 2020, especially in the coal-rich Rockies, Midwest and Great Plains regions. Among the clean sources of generation, nuclear power is the largest generation source, especially in the Northeast, Southeast and Midwest regions. Substantial amounts of power are also generated by hydroelectric sources in the Northwest (part of "West Coast") and Alaska where rivers and waterways are plentiful. Wind power has become a significant source of electricity in the Great Plains states where winds are especially strong and project sites are more abundant due to relatively low population densities.

The picture is quite different when trends in new construction are studied instead of looking at a snapshot of existing generation. Figure 5.4 shows the compositions of the aggregate regional capacities of grid-connected electricity generation resources built during 2018–2020[11] (US Energy Information Administration, 2021e, 2021f).

The patchwork quilt: business complexities of decarbonizing the electric sector   99

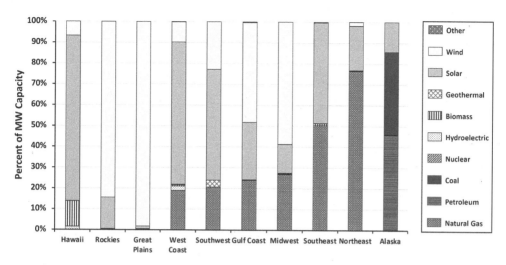

Source: Based on data from the US Energy Information Administration.

Figure 5.4   Regional electricity generation builds across the US (2018–2020)

In many regions, renewable power generation such as wind and solar comprised the majority of generating capacity built during this period. Wind power development has represented the largest source of new power on a megawatt capacity basis in the mid-American regions such as the Great Plains, Rockies, Midwest and Gulf Coast, where wind intensities are quite high. Similarly, solar power development has been substantial in areas that experience high levels of insolation such as Hawaii, the Southwest, the Southeast, and California (part of "West Coast"). State policies in these regions are also supportive of solar power. The confluence of strong solar energy policy support, solar cost reductions, and high levels of insolation can drive solar adoption to levels at which the need for flexible and controllable electricity supply increases quickly and significantly as the sun sets. California and Hawaii have both experienced this dynamic to a relatively significant degree. As discussed in the next chapter, transforming the electricity grid to handle increasing levels of renewable and distributed generation resources will require substantial investments in the coming decades.

Natural gas was the dominant fuel for new capacity resources in the Northeast during 2018–2020, due largely to substantial natural gas resources unlocked in Pennsylvania by the fracking revolution and the relative lack of other resources in much of the Northeast. The New England states do not have especially productive conditions for solar and wind, except offshore wind, which is beginning to emerge in the US as costs decline and states enact supportive policies. Natural gas-fired generation represented roughly half of the new capacity additions in the Southeast, as natural gas resources are abundant in the nearby Gulf Coast. Alaska's megawatt capacity additions were much smaller than the additions in other regions. While Alaska relied heavily on fossil fuels for its additions, solar power was a part of this mix.

The patchwork quilt of market conditions and climate-related policies also contributes to a variation in the electricity rates that electricity users pay across the US, as shown in Figure 5.5 (US Energy Information Administration, 2021d).

100   *Handbook of business and climate change*

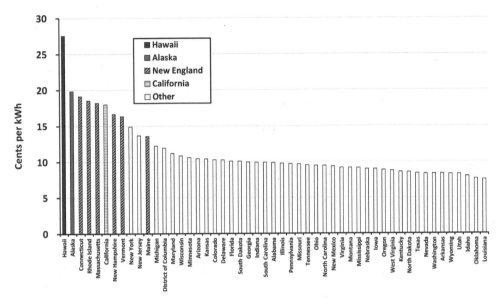

Source: US Energy Information Administration.

Figure 5.5   Average retail price of electricity across the US (2020)

Hawaii and Alaska have the highest average electricity rates in the US. Hawaii's electricity costs are driven by the limitations of the resources on the islands that comprise Hawaii and the costs of importing fuels across the ocean. Due to Alaska's dispersed population, many Alaskan communities are not connected to a larger grid that leverages significant quantities of generation resources (Institute for Energy Research, 2013). Even the Railbelt, the grid that serves many of Alaska's more urban populations, is isolated from the electric grids in Canada (US Energy Information Administration, 2021a).

After Hawaii and Alaska, the average rates in the New England states are among the highest in the US. These relatively high rates are driven by a dependency on the imports of natural gas to fuel many of New England's power plants, coupled with limited pipeline capacity into the region (ISO New England, 2021). Some parties also attribute New England's relatively high electricity rates to the costs of policies adopted in the region to support the growth of renewable generation resources and implement energy efficiency programs (Brown, 2017), as well as policymakers' reluctance to expand natural gas pipeline infrastructure (Consumer Energy Alliance, 2017). Furthermore, while some parties have argued that electricity rates in New England could be lowered by proposed electric transmission projects that would import larger amounts of power from Canada to New England, such as the Northern Pass project and the New England Clean Energy Connect, these proposals have been controversial and rejected for a variety of reasons, including their cutting through wildlife habitats, their impacts on landscapes, their disproportionate benefits across the states they traverse, general attitudes toward their project sponsors, and concerns about associated hydroelectric generation resources being located on unceded First Nations lands (Iaconangelo, 2021).

California's electricity rates are also among the highest in the US. Drivers of California's relatively high rates include above-market generation contracts and utility-owned generation that in part support the state's clean energy goals, costs associated with wildfire risk mitigation and damages, the costs of the state's energy efficiency and other public purpose programs, and authorized utility rates of return on equity that are higher than in other parts of the US (Energy Institute at Haas, UC Berkeley, 2021; California Public Utilities Commission, 2021). While California's electricity rates are among the highest in the US, the upward effect on overall electricity bills is mitigated by the fact that California's electricity use per capita is among the lowest in the US, driven by the state's mild climate and its aggressive energy efficiency programs (California Public Utilities Commission, 2021).

Electricity rates for California utility customers without solar generation systems on their premises have been further increased by California's utility rate structures. California's utility rate structures incorporate electricity bill reductions for customers who generate solar power on their premises, and these bill reductions have been adequately large to support the substantial growth of customer-sited solar generation systems in the state. However, as explained in a 2021 independent study prepared for the California Public Utilities Commission, under California's current rate structures, the electricity bill reductions that California utility customers are provided by generating solar power on their premises are in part funded by increases in other customers' rates (Verdant Associates, LLC, 2021). This has been a contentious issue in California, and similar rate design issues have been contentious in other states where the growth of customer-sited solar generation systems has been relatively high or where such systems are disproportionately adopted by utility customers with higher incomes, leaving disproportionately lower-income utility customers to bear additional cost burdens. This is further discussed in the later section, "Retail Ratemaking and Net Metering." The variation in rate designs between utility service areas, as well as uncertainty about how rate designs may be changed in a particular jurisdiction, adds to the patchwork quilt of business opportunities and risks.

Patchwork Quilt Policies Affecting Business

As the regional study of the US electricity generation mix and retail rates indicates, government policies play a significant role in defining the business opportunities in the electric sector. The influence, heterogeneity, and complexity of policymaking has increased with climate concerns, as policymakers at various levels of government respond to concerns about climate change and clean energy deployment with widely varying priority levels and preferred approaches. Investors in electricity generation projects and related technologies must understand the landscape of existing and potential climate-related policies applicable to a given location before committing to projects. The unique combination of policies that exists in a particular region, or that may be implemented in the region in the future, can be the key to whether an electric sector business venture or a business that uses large quantities of electricity is a money-maker or is destined for losses. Some relevant climate-related policy support is federal but much of the support is state or regional. The most prominent types of policies which electricity businesses and major electricity consumers must monitor and consider are discussed below.

Mandated portfolio standards

Government-mandated portfolio standards typically require that a specified percentage of the overall electricity generation be produced by a certain type of generation. In the US, these portfolio standards are typically established at the state level and support renewable forms of generation, and they are commonly referred to as Renewable Portfolio Standards (RPS).[12] A given state's RPS may specify different percentage requirements for different generation types (e.g., solar, wind, etc.), with requirements as to where qualifying generation resources must be located (e.g., in the respective state, in the respective RTO, etc.). Each retail supplier in the state (e.g., the regulated electric utilities and competitive retail suppliers) must show regulators that it is supplying electricity from the required generation types in amounts at least equal to the respective required percentages of its overall supply. Most RPSs incorporate a Renewable Energy Credit (REC) system in which qualifying generators are granted a certificate for each megawatt-hour that they generate. They can sell these certificates to retail suppliers, thereby allowing the retail suppliers to verify satisfaction of their requirements with regulators. Market prices for RECs can vary significantly from one state to another, based on the cost to build and operate the specified generation types, as well as supply and demand factors such as the percentage requirements and whether generation resources located in a region larger than the state may qualify for the state's RECs. With over half of the US states adopting RPS programs, a 2021 Lawrence Berkeley National Laboratory report cited RPS policies as a "key driver" of renewable energy growth, noting,

> Roughly half of all growth in US renewable electricity (RE) generation and capacity since 2000 is associated with state RPS requirements, though that percentage has declined in recent years, representing 23% of all US RE capacity additions in 2019. However, within particular regions – namely, the Northeast and Mid-Atlantic – RPS policies have remained a dominant driver for RE growth. (Barbose, 2021)

A benefit to mandated portfolio standards is that they help to directly ensure the development of specified quantities of certain types of generation that are deemed favorable for climate or other policy reasons. However, as explained in the later section, "Preferential Technology Deployment Policies," one of the downsides of RPS programs is their cost inefficiencies in meeting climate goals. To mitigate this issue, some states have built upon their RPS programs by adopting a Clean Energy Standard (CES). While there is no universally accepted definition of a CES, the term frequently refers to a standard that is generally technology-agnostic, and that includes clean or low-emission generation resources that would not be qualifying resources in an RPS, such as nuclear, or coal or natural gas coupled with carbon capture infrastructure, as well as generation resources that typically would qualify in an RPS.

Feed-in tariffs

A feed-in tariff is a policy mechanism that pays owners of qualifying generation technologies a pre-established price for each unit of electricity. Where feed-in tariffs are used, they are generally offered for renewable generation resources. Prices under feed-in tariffs are typically structured as fixed prices per megawatt-hour, or they may be structured to pay the difference between a fixed benchmark price and the prevailing wholesale market price while allowing the resource receiving the feed-in tariff price to sell its output into the local electricity market. The prices are often locked in over a contract period of 10 to 25 years. Costs are typically recovered from retail electricity customers through grid charges or other surcharges. Feed-in tariffs are currently quite rare in the US, but they were widely used in the 1980s to procure output from generators qualifying under PURPA. Feed-in tariffs are used more frequently outside of the US. Worldwide, feed-in tariffs were in place in 83 jurisdictions at the national, state or provincial levels as of the end of 2020 (REN21, 2021).

Feed-in tariffs provide benefits to renewable generation investors because they remove energy price risk from the investment equation. However, this risk is effectively shifted to electricity customers, which can result in costly situations if technology costs drop after feed-in tariff contracts are signed. This risk is further compounded by the fact that the feed-in tariff price is determined administratively, rather than directly through competitive market forces. This leads to an increased likelihood that the price paid for the electricity under the feed-in tariff is excessively high, especially because a policy objective of a feed-in tariff is to set price levels high enough to stimulate deployment of the chosen technologies. One of the more controversial examples of electricity feed-in tariffs is that stemming from Germany's Renewable Energy Act, which was first introduced in 2000. Under this Act, a long-term, fixed price was provided for the megawatt-hours generated by each qualifying renewable energy project (Raikar & Adamson, 2020). These prices were typically set for 20 years, they often were set at levels above market prices, and they are recovered from charges imposed on electricity customers (Raikar & Adamson, 2020). By 2016, German citizens were paying over six euro cents per kilowatt-hour, or approximately €23 billion annually, in support payments under the program (Thalman, 2015). Some critics claim that the high power costs in Germany, which are at least partially attributable to these support programs, have impaired the ability of German businesses, and especially its smaller businesses, to compete (Wilkes & Parkin, 2018).

In addition to the risk of paying an unnecessarily high price for the targeted type of generation, feed-in tariffs can increase society's overall electricity costs as investors may choose to ignore opportunities to develop lower cost generation projects, even if they provide similar or greater environmental benefits, and instead develop the specific types of projects that are eligible for the tariffs. This may also disrupt the competitive wholesale electricity markets, especially if the feed-in tariffs support excess supplier entry, disadvantaging competitive generation projects that are not supported by the tariffs. Consequently, policymakers must be careful about the design of feed-in tariffs, as they involve tradeoffs from a societal perspective.

Competitive solicitations for long-term contracts
New investments in generating plants that are made on a "merchant" basis, which is without regulatory or contractual assurance of full cost recovery, face the risk that revenues will be adversely affected by decreases in future electricity market prices without commensurate decreases in the project's fuel costs. As will be explained later, this is an especially significant risk for climate-friendly resources such as solar or wind generation, which have no fuel costs. Lower electricity market prices could be caused by many factors including decreases in the prices of fuels, such as natural gas, which power the generation units that often determine wholesale market prices, by low load growth, or by policies or cost reductions that support or increase the market's supply of new generation resources, including other climate-friendly resources such as solar, wind, hydroelectric, or nuclear generation.

As a result of these factors, most investors in renewable generation projects have especially large appetites to mitigate the risks associated with ongoing market exposure. For example, according to the American Wind Energy Association, 73 percent of the new wind capacity installed in the US in 2019 was either contracted with a long-term contract (a.k.a. "power purchase agreement," or "PPA") or was built by a utility that has regulated cost recovery mechanisms (American Wind Energy Association, 2020). Furthermore, of the remaining 27 percent, only 3 to 4 percent did not have some form of financial hedge in place (American Wind Energy Association, 2020).

Competitive solicitations for long-term PPAs can address this issue and provide the same level of financial certainty to generation investors as feed-in tariffs while harnessing the forces of competition to better ensure that the buyer receives the lowest price for the electricity. As of 2020, approximately 50 countries throughout the world use these types of solicitations to procure renewable energy, and the use of this approach continues to grow (Raikar & Adamson, 2020).

Many US states also have established solicitations for long-term PPAs for renewable energy. For example, to meet its mandate of supplying 70 percent of its electricity consumption from renewable generation, the State of New York relies heavily on annual solicitations administered by the New York State Energy Research and Development Authority (NYSERDA) and the New York Power Authority (NYPA) for contracts with durations of up to 20 years (New York State Energy Research and Development Authority, 2020c). New York's solicitations result in the procurement of contracts for solar, hydroelectric, land-based wind, offshore wind generation, as well as contracts for utility-scale storage (New York State Energy Research and Development Authority, 2020b). By 2020, New York's annual solicitation in aggregate constituted the largest competitive procurement of renewable energy in US history to date (Sylvia, 2020), garnering contracts for over 4,600 MW of renewable capacity (New York State Energy Research and Development Authority, 2020a). Massachusetts has also utilized solicitations for long-term contracts as a significant step toward meeting its climate goals, including a solicitation that resulted in a 20-year contract for almost ten million annual megawatt-hours of hydroelectric generation from Hydro-Québec, which was approved by the Massachusetts Department of Public Utilities in 2019 (Walton, 2019). The development of the US offshore wind generation market is also being driven by competitive solicitations for long-term contracts. For example, Massachusetts, Rhode Island, Connecticut, New Jersey, Maryland, and New York have conducted or plan to conduct such solicitations (Beiter et al., 2020; Kuffner, 2020). States or regulated utilities that administer long-term PPA solicitations for renewable energy may include the associated RECs (from the state's RPS) in the purchase under the contract. The RECs may then be resold into the market, or they may be used by the regulated utility to satisfy its share of the RPS requirement.

Competitive solicitations for long-term contracts entail tradeoffs. Long-term contracts mitigate the energy price and policy risks faced by investors in generation projects, reduce the financing costs of the projects, support development of the targeted projects, and speed commercialization of clean energy technologies. Furthermore, unlike feed-in tariffs that administratively set a long-term price that may be unnecessarily high, solicitations harness competitive forces to set the price received, thereby reducing the chance of locking into needlessly high prices. Still, retail electricity consumers, who ultimately cover the costs of contracts executed by governments or regulated utilities, assume the risk that the prices in the long-term contracts turn out to be above prevailing market levels in the future. This procurement system may also disadvantage competitive generation projects that are not provided the same long-term cost recovery guarantees, making it more costly to finance projects that are not supported by the long-term contracts. Policymakers must consider these tradeoffs when deciding whether and how to use this procurement and contracting method.

Carbon pricing

The idea behind carbon pricing mechanisms is simple: since the world's climate challenge is largely driven by carbon dioxide (a.k.a. "carbon") emissions, parties that emit carbon should be charged an amount equal to the damages that their carbon emissions impose on society. As of 2021, 64 carbon pricing mechanisms were in operation around the world, covering over 20 percent of global greenhouse gas emissions and generating $53 billion in revenue (The World Bank, 2021).

Carbon pricing mechanisms generally come in one of two forms: a carbon tax and a carbon cap-and-trade program (a.k.a. "emissions trading system," or "ETS"). A carbon tax is self-explanatory, as it is a predetermined price that is imposed by a government per ton of carbon dioxide emitted. If this price reflects the true cost to society of the emitted ton of carbon dioxide, the tax is imposed on all sources of carbon emission, and the tax can be administered cost-effectively, then it will reduce emissions in a cost-effective fashion.

In contrast to a carbon tax, which directly mandates the price of carbon, a carbon cap-and-trade program directly mandates the maximum quantity (a.k.a. "cap") of carbon that can be emitted in a given period such as a year, and it allows a market system to determine the price of an emitted ton, given the mandated quantity. On a periodic basis, emission "allowances," which in aggregate match the carbon cap, are sold or allocated by the government entity administering the cap-and-trade program. Parties can then buy and sell the allowances, so that parties can purchase the necessary allowances to legally emit a commensurate amount of carbon as part of their operations. Typically, the carbon cap is programmed to decrease over time so that carbon reduction goals can be achieved. Under some cap-and-trade programs, parties can bank unused allowances for emissions in future periods. By allowing for trading of the allowances, the resources with the lowest costs to reduce emissions are the most likely to operate, resulting in a greater likelihood that the targeted carbon reductions are achieved in a cost-effective manner. Some cap-and-trade programs also incorporate offset credits, which credit carbon emitters for emission reductions achieved through other activities. In the US, a group of Mid-Atlantic and Northeastern states jointly formed a cap-and-trade program known as the Regional Greenhouse Gas Initiative (RGGI) in 2009, and California adopted a cap-and-trade program that connected with Québec's cap-and-trade program in 2014.

Tax credits

Tax credits can provide significant support for the development of clean generation resources. In the US, tax credits are offered at the federal, state, and local levels. The US federal government offers income tax credits in the form of Production Tax Credits (PTCs) and Investment Tax Credits (ITCs). The PTC, which provides tax credits based on the quantity of electricity produced, was first enacted by the 1992 Energy Policy Act. This Act established a ten-year PTC of $0.015 per kWh with an inflation-adjustment mechanism for wind and closed-loop biomass generation resources brought online before July 1, 1999. The PTC has been extended multiple times, its value has changed over time, and the list of qualifying technologies has expanded to include open-loop biomass, small irrigation power, municipal solid waste, qualified hydropower, marine, and hydrokinetic generation. In past years, the PTC has been a key deciding factor in determining how rapidly wind generation is deployed. For example, uncertainty about whether the government would extend the PTC resulted in a 92 percent reduction in US wind power capacity installations in 2013 versus the previous year (American Wind Energy Association, 2014).

The ITC was first established by the Energy Policy Act of 2005 (PL 109-58), providing for a 30 percent tax credit on the eligible property of residential and larger solar energy systems placed in service in 2006 and 2007 (Raikar & Adamson, 2020). As with the PTC, the ITC has been extended several times and its value has changed over time. Solar, fuel cell, small wind, offshore wind, and geothermal projects all qualify for varying levels of ITC.

One of the complicating factors with tax credits is that they generally can be monetized only if the entity claiming the credit is subject to federal income taxes and has sufficient profitability to produce a tax liability to absorb the credits. Consequently, developers often must seek partners to engage in complex tax equity agreements to realize the value of the credits. This adds to the cost of project development and acts as a barrier to enter the clean energy development space. Developers face legal and accounting fees to engage in and administer such agreements, and the tax equity partners require returns for their participation. A 2017 study by Woodlawn Associates found that tax equity investors require between 7.5 percent and 9.5 percent in after-tax returns for unleveraged projects, net of their tax benefits (Lutton & You, 2017), and most tax equity investors will only join in funds intended to finance at least $75 to $100 million of investments within one year, which is more capital than most developers can deploy (Lutton & You, 2017).

Since tax credits are typically offered to only certain types of resources, they may entail cost inefficiencies in meeting climate goals, and they can disadvantage other types of resources that may be able to reduce emissions more cost effectively. This basic dynamic is further explained in the later section, "Preferential Technology Deployment Policies." In the US, the PTC has received criticism from some parties along these lines, due in part to the fact that it is credited based on the megawatt-hours generated. With this structure, the PTC provides an economic incentive for certain types of qualifying generation resources to offer negative prices into the daily wholesale electricity markets because they will receive the tax credit if they generate. For example, a wind generator's marginal cost to generate is very close to zero because it incurs no fuel costs. With the PTC available to the wind generator, the generator has an incentive to offer a negative price into the market, of a magnitude that is up to the pre-tax value of the PTC that it receives for generating.[13] During hours in which wind generation is the marginal type of resource, this can reduce market clearing electricity prices to these negative values. As more wind generation is built, increasing the number of hours in which wind generation sets a negative market price supported by the PTC, the opportunities to profitably deploy other types of clean energy resources could be significantly diminished, and existing clean energy resources may retire prematurely without additional financial support. Furthermore, these negative market clearing prices could encourage excessive consumption of electricity and lead to additional costs incurred to expand the electric system to accommodate this incremental demand. These issues are likely to be more significant as resources that qualify for the PTC, such as wind generation, increase their market share. Consequently, while tax credits such as the PTC can be useful measures, they must be applied carefully, and they should be periodically reassessed for their range of impacts.

Accelerated tax depreciation benefits

Accelerating the tax depreciation schedules of investments is another policy mechanism employed by governments to support the development of certain types of projects over time. Value from accelerated depreciation is reaped by the project owner due to the time value of money. If an asset with an expected 25-year life is assumed to depreciate for tax purposes over only five years instead of 25 years, then the aggregate tax deduction associated with the depreciation of the asset over its life occurs earlier in time, providing a higher present value to develop the project.

In the US, at the federal level, renewable energy projects qualify for Modified Accelerated Cost Recovery System (MACRS) treatment. Qualifying assets with long expected useful asset lives may depreciate for tax purposes over much shorter periods, sometimes 5–6 years. In addition, certain projects may qualify for "bonus depreciation," in which a certain percentage of the capital cost is depreciated for tax purposes in the first year of the project's life.

Other direct subsidies, taxes and low-cost financing arrangements

Governments at federal or local levels may provide other direct financial subsidies or low-cost financing programs for selected energy technologies. Some such programs have creative structures, such as PSE&G's Solar Loan Program, which offers loans to help qualified entities finance a portion of their solar generation investment. PSE&G administers solicitations for parties seeking loans and it grants loans based on the lowest Solar REC (SREC) prices bid. The loans can be repaid with cash or SRECs, in which the SRECs are valued at the higher of the bid price or the market SREC price (PSE&G, 2016). This allows the most competitive bidders to finance their solar investments in a way that provides them a guaranteed minimum SREC price for the life of the loan, while retaining upside SREC pricing potential.

Government levies on certain technologies can also materially change investment opportunities and patterns on a locational basis. Wyoming's tax on wind power provides an excellent example. As of 2009, Wyoming led its peers in terms of wind power development. However, the 2012 imposition of a $1 per megawatt-hour state tax on wind-energy output effectively stymied wind development in the state. From 2009 to 2018, the cumulative installed wind capacity in three nearby states with similar wind development costs, Montana, Colorado, and New Mexico, increased 415 percent, 205 percent, and 204 percent, respectively. In comparison, Wyoming's cumulative installed wind capacity in the same period increased only 28 percent (Cotting & Horwath, 2019).

Wholesale market design

Policies pertaining to the structure of the wholesale electricity markets also influence the opportunities and risks for generation investors and other businesses. Investors in merchant or partially merchant generating projects rely upon wholesale market prices for their project revenues, so they must consider both the opportunities for market compensation under the existing market designs and the possibility of changes in the market designs over time. For example, a developer of an intermittent renewable generation resource such as a wind or solar resource must consider whether a given electricity market of interest includes a "capacity" product and, if so, then to what requirements would the generation resource owner be subjected, and to what extent would the project's compensation for providing capacity be limited due to the intermittent nature of the resource.

Similarly, the economic viability of grid-scale batteries, which can facilitate the integration of intermittent generation resources into the grid, may hinge on the pricing structures for the services that batteries can provide to the grid. Policy changes in the frequency regulation market administered by the PJM Interconnection (PJM)[14] provide a notable case in point. In 2012, to help maintain the stability of its transmission system, PJM introduced a fast-responding frequency regulation product designed to compensate generation resources that can quickly adjust power output but are limited by the time they can sustain that output. Between 2012 and 2016, utility-scale battery capacity in PJM increased from 38 MW to 274 MW, which at the time was almost half of the entire utility-scale battery capacity in the US (Fletcher & Marcy, 2018). More than 90 percent of this PJM battery capacity was providing frequency regulation service (Fletcher & Marcy, 2018). The market collapsed, however, when PJM subsequently changed its policy by placing a cap on fast responding frequency regulation, imperiling the income streams for many of the utility-scale battery projects (Maloney, 2018). This led to subsequent protracted litigation and market design uncertainty.

Another example of how market design influences business opportunities and risks is related to the anticipated significant growth of solar and wind generation. Since solar and wind generators incur no fuel costs, their marginal costs to generate are very close to zero, and the prices they bid into the daily wholesale electricity markets, which are used to develop hourly market prices, are based on these very low marginal costs. Furthermore, growth of solar and wind generation is expected to increase the frequency in which these resources are the marginal resources in the wholesale electricity markets, and therefore increase the hours in which the bids submitted by owners of these resources set the market prices. Consequently, substantial growth of solar and wind generation could create significant downside market price and revenue risks for generation developers and existing generation resources. This raises questions about potential changes in today's market designs to accommodate the growth of these clean energy resources, and how such changes may affect the profitability and opportunities for generation developers and other businesses affected by electricity market prices.

Indeed, the design of the wholesale market in a region, and the uncertainty about design changes over time as the grid transitions to a more climate-friendly system, can make or break the profitability of new and existing projects. Wholesale market designs are continually evolving, with significant changes possible over the coming decade as emerging climate-friendly technologies become increasingly cost competitive and regulators grapple with changes in market design to accommodate and compensate these emerging technologies. For example, in the US, FERC Order 2222, issued in September 2020, is intended to remove barriers preventing distributed energy resources from competing on a level playing field in the organized capacity, energy, and ancillary services markets (Participation of Distributed Energy Resource Aggregations in Markets Operated by Regional Transmission Organizations and Independent System Operators, 2020; US Federal Energy Regulatory Commission, 2020). The market designs that are adopted in response to the need identified in this order are likely to vary across regional markets, to evolve over time, and to impact the opportunities afforded to distributed energy resource projects while affecting the market landscape for all market participants.

Retail ratemaking and net metering
Policies regarding retail ratemaking and net metering can determine the financial viability of distributed generation projects, such as solar generation located on a customer's premises. Before the turn of the 21st century, customer meters that could read both a customer's peak

demand during a monthly period as well as the total kilowatt-hour usage over that period were typically only installed at larger customers' premises, as these types of meters were considered too expensive to install at residential customers' premises given their lower loads. Consequently, larger customers' rates often included a significant charge based on peak demand, because these customers had the meters to accommodate such charges and it was generally accepted that this type of charge is aligned with drivers of several of the electric system's costs, especially because distribution capacity must be built to accommodate peak power flows. In contrast to the rate structures for larger customers, the rate structures for residential customers generally did not include a peak demand charge due to the limitations of these customers' meters. Instead, residential rate structures generally included charges based on the customer's kilowatt-hour usage over a month.

This conventional rate structure was not very controversial until the costs of distributed solar generation systems dropped significantly in the first 20 years of the 21st century, accelerating the adoption of residential solar systems in many regions. These systems are often interconnected "behind the meter," and for ratemaking purposes they are often treated as reductions in customer usage. The amount of the electrical output of these solar systems, over some periods of time, can be similar in magnitude to the overall electric usage of the customer itself. As a result, residential customers with solar generation systems and with large portions of their bill based on their metered kilowatt-hour usage can avoid paying a substantial portion of what they would otherwise be charged for use of the distribution system. Net metering policies in certain jurisdictions also allow these customers to receive credits for any net positive amounts of electricity that they export onto the grid, credits that are often similar in magnitude to the customer's overall kilowatt-hour-based rate.

Proponents of maintaining this form of residential rate structure (and net metering treatment) argue that these bill reductions and credits help to compensate the residential solar generation owner for the climate benefits that its solar generation provides. Critics argue that since residential solar generation owners still use the distribution system to receive and export electricity, this approach to ratemaking allows them to avoid paying their fair share of distribution system costs, causing other customers who do not have or are unable to site solar generation on their premises to pay more for their electricity. Critics also argue that lower-income customers are less likely to have the financial or physical means to install solar generation systems on their property, so the burden of the cost shift is disproportionately shouldered by lower-income customers. The advent of "virtual net metering" and community solar programs, which treat shares of electricity generated from independently developed solar generation facilities located elsewhere as if they were located behind the meter of residential customers who subscribe to a share of the facility's output, have provided customers without viable solar generation sites on their premises the opportunity to receive the benefits of the rate structure and net metering treatment. However, critics argue that this exacerbates the problem of customers allegedly not paying their fair share of distribution system costs, leaving even more costs for others to pay. Controversy surrounding these issues has caused some jurisdictions to adopt alternative approaches.

To the businesses investing in residential solar generation, profitability can be highly dependent upon a given jurisdiction's residential rate structure, how the net metering of the excess electricity exported to the grid is credited, and how these policies may change in the future. Furthermore, these issues are relevant to all power plant owners and developers because the design of these policies can affect the amount of residential solar generation in the market and therefore it can affect the wholesale market prices that generating plant owners receive for their plants' outputs.

Siting, permitting and interconnection policies
Policies regarding the siting, permitting and interconnection of new generation resources of all types also can affect business opportunities and carbon reductions. For example, the use of coastal waters by offshore wind projects can involve significant permitting requirements, onshore wind projects can face permitting challenges due to their relatively large geographic footprints and their visibility, natural gas-fired generation and pipelines increasingly face siting and permitting challenges, and long-distance transmission lines have been difficult to site and permit for many years.

Regulators also must decide how to allocate the costs to interconnect generation resources between the resource owner and other customers making use of the transmission and distribution system. Policymakers in different jurisdictions may allocate these costs differently. With clean energy resources becoming an increasing share of new interconnections, some policymakers with aggressive climate goals may modulate the allocation to lower the direct cost to developers, while other policymakers may not. Policy decisions along these lines can affect the financial viability of a project.

Generation project developers also benefit from clear and reliable estimates of the interconnection costs they will incur, and from timely completion of interconnection. Uncertain estimates of the interconnection fees for a given project, especially if changes to estimates can occur after the interconnection agreements are signed, can pose significant financial risks for developers. Similarly, development risks arise if there is a notable chance that projects will be stalled indefinitely in an interconnection queue.

Thematic Challenges for Businesses and the Achievement of Climate Goals

The patchwork quilt presents several thematic challenges for businesses and the achievement of climate goals. These include preferential technology deployment policies, emissions leakage due to uneven climate policy, competing federal and state policies, magnified investment risks for capital-intensive technologies, and policy and market uncertainty.

Preferential technology deployment policies
Policies advancing a limited set of preferred technologies constrain development opportunities for other technologies that are not provided preferred treatment, as compared with adopting a broader technology-agnostic policy approach to decarbonization. For example, while RPS programs typically mandate new wind and solar, they often omit or include much less aggressive mandates for other clean energy generation technologies such as preserving conventional nuclear generation, expanding hydroelectric plants, or developing fossil carbon capture and sequestration (CCS) facilities. Similarly, such preferential policies do not support the development of earlier-stage, emerging, climate-friendly technologies such as small modular nuclear reactors and low-carbon oxy-combustion processes.

When one type of clean generation is financially supported, it is provided an economic advantage over other, unsupported types of clean generation, some of which otherwise may be more cost effective than the supported type. This penalizes the unsupported clean generation types in two ways. First, the unsupported types of clean generation are less likely to be viable investments due to their ineligibility to receive financial support. Second, the financial support may result in additional market entry from the supported generation type, increasing the overall supply in the market and lowering regional energy market prices for some time. The lower market prices may further reduce the likelihood of deploying unsupported clean generation types, and they may increase the likelihood that existing resources of the unsupported clean generation types retire earlier. In addition to potential overall cost inefficiencies, this leads to other types of generation filling the resultant gaps, types which could be higher emitting resources.

The US nuclear industry provides a prime example of this dynamic. Excluding nuclear generation from policies, such as the RPS programs in many states which provide financial support to other climate-friendly generation resources, can undermine the decarbonization of the electric sector. As of the beginning of 2013, the US had 105 nuclear power reactors (Scott & Comstock, 2019). However, seven reactors retired between the beginning of 2013 and the beginning of 2019, with 12 more expected to retire by 2025 and only two new reactors expected to be built (Scott & Comstock, 2019). Generation from greenhouse gas emitting resources often, if not always, replaces a part of the retired nuclear generation. For example, according to the Edison Electric Institute (EEI), the 2018 retirement of the Oyster Creek Nuclear Generating Station resulted in annual additional carbon emissions of 3.1 million tons, as natural gas and coal generation replaced the lost generation (Fisher, 2021). Similarly, EEI reported that the 2014 retirement of the Vermont Yankee Nuclear Power Plant was accompanied by a 2.9 percent increase in New England's carbon emissions the following year (Fisher, 2021). In response to impending nuclear retirements, some states, such as New York, New Jersey, and Illinois, have responded with compensation programs for nuclear plants that were on the brink of closure (Magill, 2019). But proactive adoption of policies that are broader in scope both geographically and technologically, to include nuclear as well as other types of clean generation technologies, could provide greater certainty to electricity generation investors and facilitate the achievement of climate goals.

Similarly, generation resources that employ carbon capture and sequestration have not been deployed at least in part due to policies that have favored other climate-friendly technologies over this type of system. For example, the failure to complete the Hydrogen Energy of California project, an integrated gasification combined cycle power plant with carbon capture and sequestration, was due at least in part because it could not obtain a long-term PPA approved by the California Public Utilities Commission (Reicher, Brown, & Fedor, 2017).

Emissions leakage owing to uneven climate policy
Climate policies that are adopted in one state or jurisdiction but not adopted in another also present risks for investors and often reduce the efficacy in meeting climate goals.

One such example relates to what is commonly known as carbon "leakage." Suppose that one state adopts a policy, such as a carbon tax or a tradable emissions allowances system. These policies are generally designed to favor clean or low-carbon emitting resources, as the costs faced by the carbon-emitting resources in the state are increased, encouraging carbon-emitting resources to be displaced by carbon free or lower carbon emitting resources. This displacement reduces overall carbon emissions. However, if a neighboring state does not adopt a similar policy, and electricity generators in the neighboring state are part of the same wholesale electricity market without transmission constraints between them, then carbon-emitting generation resources in the neighboring state will also be advantaged by the policy. These resources will not face the higher costs of the carbon policy while their competitors in the state that adopted the policy will, and they will have a financial incentive to increase their generating output and sell it into the state that adopted the carbon policy. Consequently, the policy creates unintended incentives for carbon-emitting resources to be built in the neighboring state, reducing the clean energy support and overall emission reductions intended from the carbon policy. In this way, the efficacy in meeting climate goals can be notably diminished.

The magnitude of these effects can be substantial. For example, a 2020 study issued by the Pennsylvania Department of Environmental Protection estimated the carbon reductions associated with the implementation of an Executive Order by the Governor of Pennsylvania to participate in RGGI (Pennsylvania Department of Environmental Protection; ICF Incorporated, L.L.C., 2020). Despite the fact that four of the six states that border Pennsylvania also participated in RGGI,[15] the study estimated that 54 percent of the reductions in Pennsylvania emissions from 2022 to 2030 would be offset by higher emissions elsewhere in PJM (Pennsylvania Department of Environmental Protection; ICF Incorporated, L.L.C., 2020). Similarly, a 2015 study published in *The Energy Journal* estimated that, if cross-border mechanisms to minimize leakage are not effectively enforced, California's cap-and-trade program could increase out-of-state emissions by 45 percent of the domestic reduction (Caron, Rausch, & Winchester, 2015). Cross-border mechanisms can mitigate these types of leakage levels, but development and enforcement of effective mechanisms can be complicated in practice. Regional cap-and-trade programs are helpful to mitigate the effects of climate change, but the achieved climate benefits are often notably reduced by the absence of a more broadly applied program.

Competing federal and state policies
Unsynchronized or competing federal and state policies, which may arise when parties with differing policy viewpoints are in power at the federal and state levels, can diminish or nullify the efficacy of policies intended to achieve climate goals, add to the costs to satisfy climate goals, impair or add risk to climate-friendly investment opportunities, and threaten the financial viability of existing climate-friendly resources. The magnitude of this problem has been emphasized by business leaders in the electric sector, including Exelon CEO Chris Crane:

> [The United States] for too long has separated environmental policy from an energy policy, and competing at federal levels with state levels has made it very difficult for markets to be formed efficiently and made it very difficult for predictability of investments going forward...[L]eadership at the federal level should either come up with a common policy or get out of the way, have the states be allowed to work with the RTOs and design the markets [that] the state desires. (Christian, 2017)

One high-profile example of the issues that can arise from lack of alignment at the federal and state levels is the incongruity between a December 2019 US FERC order regarding capacity resources serving loads in the PJM region (Order Establishing Just and Reasonable Rate, 2019), and the intent of state policies designed to support clean energy resources. PJM's original Minimum Offer Price Rule (MOPR) placed floors on the prices that certain generation resources could offer into PJM's capacity market, to prevent bidders who are net buyers of capacity in the market, such as load-serving utilities, from exercising buyer-side, or "monopsony," market power. This safeguard was designed to prevent net buyers of capacity from artificially suppressing overall capacity market prices by offering their generation capacity into the capacity market at prices below their costs, and to prevent the net buyers from having their contracted electricity suppliers do the same. FERC's 2019 MOPR Order modified the rules of the original MOPR. Specifically, the 2019 MOPR Order required PJM to expand the application of the offer price floor to all new and existing capacity resources that receive or are eligible to receive state subsidies, subject to certain exemptions.[16]

Per the 2019 MOPR Order, the minimum offer price of any capacity resource subject to the MOPR is based on an estimate of the resource's costs going forward, without reductions to account for the state subsidies that the resource receives. Many states in the PJM region have established policies that effectively provide subsidies designed to compensate low-emitting or clean generation resources for the positive value that they provide for the earth's climate owing to their low greenhouse gas emissions. These include renewable portfolio standards, clean energy standards, mandates to conduct competitive solicitations for clean energy resources, and other policies. By setting a minimum offer price that reflects a resource's costs without a reduction to account for the state subsidies that the resource receives, the 2019 MOPR Order threatened to offset the climate-based compensation from such state policies because it jeopardized the ability of clean energy resources to offer a sufficiently low price for the resource to be a winning bidder in the capacity market and receive the associated capacity market revenues. All else equal, this reduced the profit opportunities for new clean energy investments, and it increased the possibility of early retirements of existing nuclear plants that are major contributors to greenhouse gas reductions. Furthermore, requirements on these clean energy resources to bid higher prices would likely cause capacity market clearing prices to be higher than they otherwise would be, thereby providing additional financial support to resources that emit greater quantities of greenhouse gases and increasing capacity costs for consumers.

The FERC subsequently took measures to mitigate and nullify the relevant provisions in the 2019 MOPR Order, after the composition of the FERC had changed once President Biden took office. After taking action in 2021 to reverse changes enacted by the 2019 MOPR Order, two FERC commissioners stated, "[The 2019 MOPR Order] was really an effort to strip away any the [sic] influence of disfavored state policies on capacity prices…" (Statement of Chairman Glick and Commissioner Clements, 2021). While the FERC policy has changed since the time of the 2019 FERC MOPR Order, this presents a prime example of how lack of coordination or alignment at the federal and state levels can jeopardize climate-friendly business opportunities and the achievement of climate policy goals.

Magnified investment risks for capital-intensive technologies

Without regulatory or financial hedges against revenue uncertainty, clean generation projects with high upfront capital requirements relative to their total lifetime project costs, such as solar and wind generation, can face magnified risks. As shown in Figure 5.6,[17] solar and wind generation generally entail higher capital costs than natural gas generation, as a percent of total lifetime project costs (US Energy Information Administration, 2020a; National Renewable Energy Laboratory, n.d.; Murphy et al., 2021).

The relatively high capital costs of solar and wind generation are counterbalanced by relatively low operating costs over the project's lifetime, largely because these technologies do not incur fuel costs. However, since investors in solar and wind generation must make proportionally large upfront financial commitments to cover the capital costs associated with project construction, they are likely to be more exposed to the uncertainty of future market revenues than some other technologies with higher emissions, absent means to mitigate the revenue risk.

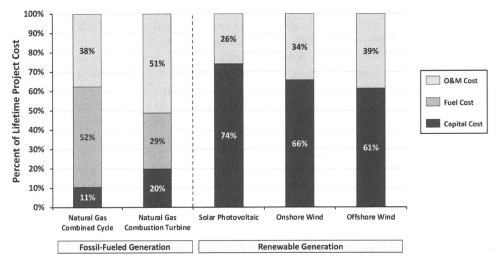

Source: Based on data from the US Energy Information Administration and the National Renewable Energy Laboratory.

Figure 5.6   Illustrative cost breakdown of various generation technologies

Furthermore, without reasonable opportunities for clean energy investors to mitigate their revenue risks, the ability to achieve decarbonization goals may be jeopardized. The 2017 Stanford University study, *Derisking Decarbonization: Making Green Energy Investments Blue Chip*, indicates that the world's institutional funds will need to be tapped to a greater extent to make the estimated multi-trillion-dollar annual investments needed globally to keep global warming below 2°C, but the conservative risk profile of the vast proportion of these funds does not match the higher-risk nature of most unhedged clean energy projects (Reicher, Brown, & Fedor, 2017).

Policy and market uncertainty
Another key business challenge with the patchwork quilt stems from its lack of stability. Individuals and political parties can change quickly at both the local and federal government levels. This turnover is often accompanied by different views about clean energy and climate change. This can lead to new policies or the rollback of old policies, which in turn affects the market landscape and business opportunities. When this policy uncertainty is added to the other market and business risks that an electricity project investor faces, the overall risk to the entity can be substantial. Abigail Ross Hopper, President and CEO of the Solar Energy Industries Association, identifies uncertainty, especially about policy change, as a "key" to the development of solar generation in the electric sector:

> [SEIA member companies] want to know what the rules of the road are, and then they will go out and innovate and adapt and build businesses based on that... Long-term certainty in terms of either tax policy, carbon policy – whatever kind of policy you want to call it – but long-term certainty is really key to that rapid deployment of solar and solar-plus-storage. (Wagman, 2021)

Uncertainty affects both the prospects for new projects and the economics of existing projects. The economics of a project can be affected by a policy change even if the policy does not directly address that specific type of project. For example, policy changes that alleviate emission requirements on fossil-fired generation resources lower the costs incurred by such resources, reducing the opportunities for more climate-friendly generation projects. The lower costs of the fossil-fired generation, resulting from the less stringent emission requirements, could provide a competitive edge to fossil-fired resources to meet a region's electricity demands. In addition, the lower costs of the fossil-fired generation could lower wholesale electricity market prices that may drive the revenues for a clean generation project.

Takeaways for Businesses: Selective Opportunities and the Need for Risk Management

The patchwork quilt of regional market conditions and climate and clean energy policies that characterizes much of the electricity landscape has two overarching ramifications for businesses in the electric sector.

First, the patchwork quilt can contribute to large investment opportunities for businesses, but these are highly dependent on regional market and policy conditions, some of which can be quite local and complex. Opportunities are shaped by regional market structures and natural resource endowments. They are also dependent upon the potential for technology cost reductions, especially in the technologies that are the most conducive to a given region, which may include emerging technologies such as solar and wind power, batteries, electric vehicles, and other technologies on the horizon such as small modular nuclear reactors and hydrogen. Some of these technologies have already been discussed, while others will be discussed in the next chapter.

Importantly, the business opportunities in any location are also highly dependent upon the policies that apply to that location and the policies in neighboring locations that can affect the overall system. Such policies may include mandated portfolio standards, feed-in tariffs, centralized contracting, carbon pricing, tax and subsidization policies, wholesale market design, retail rate design, and siting, permitting and interconnection policies. Business challenges and economic inefficiencies are especially likely when policies favor certain technologies over other potentially more promising technologies, the unevenness of policies across a geographic region creates distorted incentives, or federal and state policies compete with one another. As policymakers at various levels of government respond to concerns about climate change, with widely varying priority levels and preferred approaches, policy considerations can be the deciding factor between a profitable investment and one that is destined to lose money.

The second overarching ramification for businesses in the electric sector relates to the complexity of the risks that businesses face, and the resultant need to assess and manage these risks. Projects in the electric sector face risks associated with obtaining the necessary siting and permitting approvals, acquiring satisfactory financing arrangements, future market conditions, and changes in government policies. Fossil-fired projects face risks associated with increasing public sentiment to implement new policies that are detrimental to generation resources that emit harmful pollutants, while investors in more climate-friendly technologies face the risk that supportive policies will be rolled back as federal or local government priorities change. Projects without regulatory or financial hedges against market uncertainty face magnified risks. This is especially true for clean generation technologies such as solar and wind, given their high upfront capital costs relative to their total lifetime project costs, which require solar and wind developers to make proportionally large upfront financial commitments. Furthermore, the low variable costs of solar and wind generation could result in significant electricity market price declines as the deployment of these technologies expands, adding downside market risk for generation resources without revenue hedges. While regional variations in economic conditions, natural resource endowments, and societal needs mean the patchwork quilt is to some extent inevitable, policy approaches that are better coordinated and more uniform would reduce today's investor risks and help reduce carbon emissions in a more cost-effective manner.

With this current state of the electric sector as a foundation, the next chapter takes a longer view, discussing the prospect of a transition to a fully decarbonized economy-wide energy system that will last several decades, the linchpin role of the electric sector including massive expansion to help decarbonize other sectors of the economy, the deep uncertainties associated with technology pathways, policy development, scaling deployment and consumer acceptance, and three broad challenges for businesses in the electric sector.

NOTES

1. In addition to these direct greenhouse gas emissions from the generation of electricity, some processes that support the electric sector, such as the extraction, processing and transport of the natural resources used in today's electric systems, also emit greenhouse gases.
2. Affected businesses include both energy sector investors as well as businesses that consume electricity and therefore are exposed to electricity costs that are affected by electric sector developments.
3. As will be discussed later, the system operators that administer these auctions are often referred to as "Regional Transmission Operators."
4. Cogeneration facilities are electricity generation facilities that capture the waste heat, usually in the form of steam, and apply it for another purpose such as an industrial process, often through a sale of the steam to another party.
5. The electricity generation processes for these technologies do not directly produce carbon emissions, but carbon may be emitted in the manufacturing and development of these generation plants.
6. The "levelized cost" of a generating plant is a measure of the average price of electricity, in real dollars, which would be required to recover the costs of building and operating the plant during its life, and to achieve a fair return on the investment. The levelized cost reflects the plant's capital costs (often including the cost of any required transmission system upgrades), operations and maintenance costs, fuel costs, financing costs, and an assumed technology-specific capacity factor, but it generally does not include the costs associated with the use of the electricity transmission and distribution system to deliver the electricity to consumers. The quoted levelized cost values are expressed in 2020 dollars.
7. Large-scale battery storage systems are defined as those that are connected to the grid and have a nameplate power capacity greater than 1 MW.
8. Power capacity refers to the maximum amount of power output a battery can provide in any instant.
9. Affected businesses include both energy sector investors as well as businesses that consume electricity and therefore are exposed to electricity costs that are affected by electric sector developments.
10. Northeast: CT, DE, DC, ME, MD, MA, NH, NJ, NY, PA, RI, VT; Southeast: AL, FL, GA, KY, NC, SC, TN, VA, WV; Midwest: AR, IL, IN, IA, MI, MN, MO, OH, WI; Gulf Coast: LA, MS, TX; Great Plains: KS, NE, ND, OK, SD; Rockies: CO, ID, MT, WY; Southwest: AZ, NV, NM, UT; West Coast: CA, OR, WA. Values are calculated by summing EIA's reported net generation values (EIA-906, EIA-920, and EIA-923) and EIA's small scale PV estimates.
11. Northeast: CT, DE, DC, ME, MD, MA, NH, NJ, NY, PA, RI, VT; Southeast: AL, FL, GA, KY, NC, SC, TN, VA, WV; Midwest: AR, IL, IN, IA, MI, MN, MO, OH, WI; Gulf Coast: LA, MS, TX; Great Plains: KS, NE, ND, OK, SD; Rockies: CO, ID, MT, WY; Southwest: AZ, NV, NM, UT; West Coast: CA, OR, WA. Values are calculated by summing EIA's reported operable generator values (2020 Form EIA-860 Data – Schedule 3) and increases in EIA's small scale PV estimates.

12. Alternatively, these types of programs are sometimes called, "Alternative Energy Portfolio Standards."
13. The PTC is quoted in after-tax dollars because it is a tax credit. Therefore, the pre-tax value of the PTC is the quoted PTC value divided by one minus the tax rate.
14. PJM is a regional transmission organization that operates an electric transmission system serving all or parts of Delaware, Illinois, Indiana, Kentucky, Maryland, Michigan, New Jersey, North Carolina, Ohio, Pennsylvania, Tennessee, Virginia, West Virginia, and the District of Columbia.
15. New York, New Jersey, Delaware, and Maryland were a part of RGGI, while West Virginia and Ohio were not.
16. Exemptions applied to most existing renewable generation resources, demand response, energy efficiency, and energy storage resources, but not to new renewable or existing nuclear generation resources.
17. This figure depicts the costs of the electricity generation only, and not the fully delivered costs of electricity. Percentages are calculated based on estimated costs expressed in real dollars.

REFERENCES

American Wind Energy Association. (2014). *AWEA U.S. Wind Industry Fourth Quarter 2013 Market Report.* American Wind Energy Association.

American Wind Energy Association. (2020). *Wind Powers America Amended Annual Report 2019.* American Wind Energy Association.

Barbose, G. (2021). *U.S. Renewables Portfolio Standards – 2021 Status Update: Early Release.* Lawrence Berkeley National Laboratory.

Beiter, P., Heeter, J., Spitsen, P., & Riley, D. (2020). *Comparing Offshore Wind Energy Procurement and Project Revenue Sources across U.S. States.* NREL.

Bolinger, M., Seel, J., Warner, C., & Robson, D. (2021). *Utility-Scale Solar, 2021 Edition.* Lawrence Berkeley National Laboratory.

Brown, M. (2017, August 20). *Sorting Out Our Perverse Electricity Markets.* Retrieved from CommonWealth Magazine: https://commonwealthmagazine.org/energy/sorting-perverse-electricity-markets/

California Public Utilities Commission. (2021). *Utility Costs and Affordability of the Grid of the Future: An Evaluation of Electric Costs, Rates, and Equity Issues Pursuant to P.U. Code Section 913.1.*

Caron, J., Rausch, S., & Winchester, N. (2015). Leakage from Sub-national Climate Policy: The Case of California's Cap-and-Trade Program. *The Energy Journal*, 167–190.

Christian, M. (2017, February 13). *Exelon CEO: Feds Should Form Common Power Policy or 'Get Out of the Way'.* Retrieved from S&P Global: https://www.spglobal.com/marketintelligence/en/news-insights/trending/lbx4kdqv0kxhfy6sulaf9a2

Consumer Energy Alliance. (2017). *Families, Communities and Finances: The Consequences of Denying Critical Pipeline Infrastructure.* Consumer Energy Alliance.

Cotting, A., & Horwath, J. (2019, April 10). *How Wyoming Went from Leader to Laggard in Wind Energy.* Retrieved from S&P Global: https://www.spglobal.com/marketintelligence/en/news-insights/trending/WDrrAH2joStLEQyVTq5BaA2

Energy Institute at Haas, UC Berkeley. (2021). *Designing Electricity Rates for an Equitable Energy Transition.* Next 10.

Fisher, E. S. (2021, February 12). *Edison Electric Institute Letter to the New Jersey Board of Public Utilities.* Retrieved from New Jersey Board of Public Utilities: https://publicaccess.bpu.state.nj.us/DocumentHandler.ashx?document_id=1235396

Fletcher, F., & Marcy, C. (2018, February 28). *The Design and Application of Utility-Scale Battery Storage Varies by Region.* Retrieved from US Energy Information Administration: https://www.eia.gov/todayinenergy/detail.php?id=35132

Iaconangelo, D. (2021, November 12). *$1B Transmission Smack Down May Upend Northeast Renewables.* Retrieved from E&E News - Energy Wire: https://www.eenews.net/articles/1b-transmission-smack-down-may-upend-northeast-renewables/

Institute for Energy Research. (2013, August 26). *Alaska: An Energy and Economic Analysis*. Retrieved from Institute for Energy Research: https://www.instituteforenergyresearch.org/fossil-fuels/gas-and-oil/alaska-an-energy-and-economic-analysis/

International Energy Administration. (2021, March 25). *Global Energy-Related CO_2 Emissions by Sector*. Retrieved from IEA: https://www.iea.org/data-and-statistics/charts/global-energy-related-co2-emissions-by-sector

ISO New England. (2021). *Natural Gas Infrastructure Constraints*. Retrieved from ISO New England: https://www.iso-ne.com/about/what-we-do/in-depth/natural-gas-infrastructure-constraints

Kuffner, A. (2020, October 27). *RI Set to Double Down on Offshore Wind Power*. Retrieved from *The Providence Journal*: https://www.providencejournal.com/story/news/2020/10/27/ri-set-double-down-offshore-wind-power/3745858001/

Lutton, J., & You, S. (2017). *Tax Equity 101: Structures*. Retrieved from Woodlawn Associates: https://woodlawnassociates.com/tax-equity-101/

Magill, J. (2019, May 24). *Nuclear Retirements in Northeast Expected to Create Gas Demand*. Retrieved from S&P Global: https://www.spglobal.com/platts/en/market-insights/latest-news/electric-power/052419-nuclear-retirements-in-northeast-expected-to-create-gas-demand

Maloney, P. (2018, March 21). *Project Finance Getting More Viable for Energy Storage, Moody's Says*. Retrieved from Utility Dive: https://www.utilitydive.com/news/project-finance-getting-more-viable-for-energy-storage-moodys-says/519701/

Murphy, C., Mai, T., Sun, Y., Jadun, P., Muratori, M., Nelson, B., & Jones, R. (2021). *Electrification Futures Study: Scenarios of Power System Evolution and Infrastructure Development for the United States*. Golden, CO: National Renewable Energy Laboratory.

National Renewable Energy Laboratory. (n.d.). *Utility-Scale Energy Technology Capacity Factors*. Retrieved from National Renewable Energy Laboratory: https://www.nrel.gov/analysis/tech-cap-factor.html

New York State Energy Research and Development Authority. (2020a). *2020 Solicitation*. Retrieved from New York State Energy Research and Development Authority: https://www.nyserda.ny.gov/All-Programs/Programs/Clean-Energy-Standard/Renewable-Generators-and-Developers/RES-Tier-One-Eligibility/Solicitations-for-Long-term-Contracts/2020-Solicitation-Resources

New York State Energy Research and Development Authority. (2020b). *Clean Energy Standard*. Retrieved from New York State Energy Research and Development Authority: https://www.nyserda.ny.gov/all-programs/programs/clean-energy-standard/renewable-generators-and-developers/res-tier-one-eligibility/solicitations-for-long-term-contracts

New York State Energy Research and Development Authority. (2020c, September 29). *New York State Energy Research and Development Authority*. Retrieved from NYSERDA Seeks to Acquire Approximately 1.6 Million New York Tier 1 Eligible Renewable Energy Certificates Annually: https://portal.nyserda.ny.gov/servlet/servlet.FileDownload?file=00Pt000000P00roEAB

Order Establishing Just and Reasonable Rate, EL16-49-000 & EL18-178-000 (Consolidated) (US Federal Energy Regulatory Commission December 19, 2019).

Participation of Distributed Energy Resource Aggregations in Markets Operated by Regional Transmission Organizations and Independent System Operators, Docket No. RM18-9-000; Order No. 2222 (U.S. Federal Energy Regulatory Commission September 17, 2020).

Pennsylvania Department of Environmental Protection; ICF Incorporated, L.L.C. (2020). *Pennsylvania RGGI Modeling Report*. Pennsylvania Department of Environmental Protection.

PSE&G. (2016, February). *PSE&G Solar Loan Program*. Retrieved from PSE&G: https://nj.pseg.com/-/media/pseg/global/gathercontentdocuments/7-4-7howtoapply/solarloan_commercial_brochure.ashx

Raikar, S., & Adamson, S. (2020). *Renewable Energy Finance*. London: Elsevier.

Reicher, D., Brown, J., & Fedor, D. (2017). *Derisking Decarbonization: Making Green Energy Investments Blue Chip*. Stanford University.

REN21. (2021). *Renewables 2021 Global Status Report*. REN21.

Scott, M., & Comstock, O. (2019, March 21). *Despite Closures, U.S. Nuclear Electricity Generation in 2018 Surpassed Its Previous Peak*. Retrieved from US Energy Information Administration: https://www.eia.gov/todayinenergy/detail.php?id=38792

Statement of Chairman Glick and Commissioner Clements, Docket No. ER21-2582-000 (U.S. Federal Energy Regulatory Commission October 19, 2021).

Sylvia, T. (2020, July 21). *New York Launches Largest Renewable Solicitation in U.S. History*. Retrieved from *PV Magazine*: https://pv-magazine-usa.com/2020/07/21/new-york-launches-largest-renewable-solicitation-in-us-history/

Thalman, E. (2015, October 15). *German Green Power Levy to Rise in 2016 to New Record*. Retrieved from Clean Energy Wire: https://www.cleanenergywire.org/news/german-green-power-levy-rise-2016-new-record

The World Bank. (2021, May 25). *Carbon Prices Now Apply to Over a Fifth of Global Greenhouse Gases*. Retrieved from The World Bank: https://www.worldbank.org/en/news/press-release/2021/05/25/carbon-prices-now-apply-to-over-a-fifth-of-global-greenhouse-gases

US Department of Energy. (2007). *Economic Dispatch of Electric Generation Capacity*.

US Energy Information Administration. (2020a). *Capital Cost and Performance Characteristic Estimates for Utility Scale Electric Power Generating Technologies*. Washington, DC: US Department of Energy.

US Energy Information Administration. (2020b). *U.S. Energy-Related Carbon Dioxide Emissions, 2019*. Washington, DC: US Department of Energy.

US Energy Information Administration. (2021a, January 21). *Alaska State Energy Profile*. Retrieved from US Energy Information Administration: https://www.eia.gov/state/print.php?sid=AK

US Energy Information Administration. (2021b). *Battery Storage in the United States: An Update on Market Trends*. Washington, DC: US Department of Energy.

US Energy Information Administration. (2021c, September 15). *Electricity: Detailed State Data*. Retrieved from US Energy Information Administration: https://www.eia.gov/electricity/data/state/

US Energy Information Administration. (2021d, October 7). *Electricity: Electric Sales, Revenue, and Average Price*. Retrieved from US Energy Information Administration: https://www.eia.gov/electricity/sales_revenue_price/

US Energy Information Administration. (2021e, September 9). *Electricity: Form EIA-860 Detailed Data with Previous Form Data (EIA-860A/860B)*. Retrieved from US Energy Information Administration: https://www.eia.gov/electricity/data/eia860/

US Energy Information Administration. (2021f, November 29). *Electricity: Form EIA-861M (formerly EIA-826) Detailed Data*. Retrieved from US Energy Information Administration: https://www.eia.gov/electricity/data/eia861m/#solarpv

US Energy Information Administration. (2021g). *Short-Term Energy Outlook: December 2021*. Washington, DC: US Department of Energy.

US Energy Information Administration. (2021h, May 28). *U.S. Dry Natural Gas Production*. Retrieved from US Energy Information Administration: https://www.eia.gov/dnav/ng/hist/n9070us2A.htm

US Energy Information Administration. (2021i, May 28). *U.S. Natural Gas Citygate Price*. Retrieved from US Energy Information Administration: https://www.eia.gov/dnav/ng/hist/n3050us3A.htm

US Federal Energy Regulatory Commission. (2015, November). *File:RTO v1.jpg*. Retrieved from Wikimedia Commons (Public domain from FERC, sourced via Wikimedia Commons): https://commons.wikimedia.org/wiki/File:RTO_v1.jpg

US Federal Energy Regulatory Commission. (2020). *FERC Order No. 2222: A New Day for Distributed Energy Resources*. Washington, DC: US Federal Energy Regulatory Commission.

Verdant Associates, LLC. (2021). *Net-Energy Metering 2.0*.

Wagman, D. (2021, February 9). *'The Strategy Is to Go Big.' A Conversation with SEIA's Abigail Ross Hopper*. Retrieved from *PV Magazine*: https://pv-magazine-usa.com/2021/02/09/the-strategy-is-to-go-big-a-conversation-with-seias-abigail-ross-hopper/

Walton, R. (2019, June 27). *Massachusetts Regulators Approve State's Largest Clean Energy Procurement*. Retrieved from Utility Dive: https://www.utilitydive.com/news/massachusetts-regulators-approve-states-largest-clean-energy-procurement/557752/

Wilkes, W., & Parkin, B. (2018, September 24). *Germany's Economic Backbone Suffers from Soaring Power Prices*. Retrieved from BNN Bloomberg: https://www.bnnbloomberg.ca/germany-s-economic-backbone-suffers-from-soaring-power-prices-1.1141809

Wiser, R., Bolinger, M., Hoen, B., Millstein, D., Rand, J., Barbose, G., . . . Paulos, B. (2021). *Land-Based Wind Market Report: 2021 Edition*. Lawrence Berkeley National Laboratory.

6. Implications of fully decarbonizing the electric industry for business: Icarus or Daedalus?
Bruce A. Phillips, Scott G. Fisher and Mark W. Scovic

INTRODUCTION AND SUMMARY

For more than a century, the US electric industry has been a large, complex, and regionally varied sector of the nation's economy. And now, a large and growing number of businesses, policy and other observers see the electric industry as a foundational element of what is likely to be a decades-long transition to a fully decarbonized economy-wide energy system. This energy transition presents unprecedented investment opportunities for companies in the US electric sector, as well as large management challenges and financial risks.

The previous chapter of this book, "The Patchwork Quilt: Business Complexities of Decarbonizing the Electric Sector," reviewed the varying economic, technology and policy dynamics in the electric sector, focusing on the industry's recent history and current structure. This chapter builds on that by taking a longer-term view, examining the scale of new energy infrastructure needed to achieve a fully decarbonized economy-wide energy system and the unprecedented investment opportunities and challenges this presents for businesses in the electric sector.

Over the next several decades, the energy transition will entail replacing or decarbonizing nearly all of today's carbon-emitting fossil generation which still provides roughly 60 percent of all US electric energy. It is also expected to require a doubling of electric generation and a tripling of electric capacity. Even after realizing the savings from ambitious energy efficiency programs, this expansion will be needed to electrify and decarbonize portions of the transportation, industry, and space conditioning sectors of the economy. The resiliency of electric infrastructure will also need to be improved to withstand what most climate scientists expect to be increasingly frequent and severe weather events.

At the same time, the form and timing of the energy transition is unknown due to persistent and intrinsic uncertainties with federal and state climate and energy policy, the pace and scale at which new clean energy infrastructure will be deployed, and long-term evolution of low-carbon technology pathways. These uncertainties are compounded, as detailed in the previous chapter, by the complex regional "patchwork quilt" of natural resource endowments, economic conditions, regulatory systems, and policy priorities currently across the US.

For companies in the electric industry, the energy transition and its uncertainties point to three broad types of business challenges.

1. *Deploying commercial technologies at scale.* The first challenge is how to rapidly scale the deployment of carbon-free technologies that have already been largely commercialized in some regions of the country such as wind, solar and batteries. This is likely to rest on a mix of new government policies and business initiatives including federal or state policies to "pull" these technologies into the marketplace, transmission policy reforms, new financing

structures and business models, expanded supply chains, and reformed corporate practices and government policies to speed siting and permitting.
2. *Commercializing advanced technologies.* The second business challenge is how to demonstrate and commercialize advanced clean energy technologies such as carbon capture, zero-carbon liquid fuels, advanced nuclear and firm renewables, which are likely to be needed to fully decarbonize the economy in a reliable and affordable way. To be successful, these technologies will need to achieve technical milestones and economic benchmarks established by competing clean energy technologies, recognizing how each will operate in an integrated electric system. Given the large capital costs, timeframes and risks involved, patient financing from both the private and public sectors will likely be needed.
3. *Managing existing generation fleets.* The third challenge is how to manage the existing electric generating fleets: whether to secure financial support for marginally economic nuclear plants or plan their retirement and decommissioning; whether to retrofit coal plants with carbon capture equipment or plan for their retirement; and whether to retrofit gas plants with carbon capture equipment, retire them, or maintain them primarily as low-utilization resources to balance the variability of wind and solar output and provide electric system reliability.

To address these challenges, businesses will need strategic foresight, financial resources, and prudent risk management. The ability to manage commercial and organizational risks arising from uncertain government policies, technology competitiveness, and deployment forces will be a core competency.

The remainder of this chapter is structured around the following sections:

1. Rapidly growing support for the energy transition.
2. Recent technology, policy and emission trends in the US electric industry.
3. Growth opportunities for the electric industry.
4. Uncertainties in the energy transition.
5. Implications for electric businesses and decision making.

RAPIDLY GROWING SUPPORT FOR THE ENERGY TRANSITION

A large and growing number of policymakers, business leaders and industry analysts see the electric industry as being in the initial stages of an economy-wide transition to a cleaner, fully decarbonized energy system that will also be more resilient in the face of what most climate scientists expect to be increasingly frequent flooding, wildfires and extreme storms. These and other impacts of climate change have been examined in a series of widely reported and closely examined reports issued by the Intergovernmental Panel on Climate Change (IPCC), the most recent of which, as of this writing, was released in August of 2021 (IPCC, 2021). These developments have led to the adoption of various policy and corporate goals, such as limiting global warming to a level between 1.5°C and 2.0°C above preindustrial levels and achieving "net-zero" greenhouse gas emissions by mid-century.

The electric sector is typically seen as a foundational element of this energy transition because of its large carbon footprint, the availability of proven and relatively low-cost zero-carbon generating technologies, and the opportunity to rely on the electric sector to decarbonize other sectors of the economy. This will fundamentally transform the electric sector over the coming decades and present both unprecedented growth opportunities and investment risks that will need careful management.

In the years leading up to 2021, dozens of states, electric utilities and other corporations in the US responded to growing concerns over climate change impacts and the economic opportunities stemming from clean energy deployment by adopting ambitious clean energy and carbon emission reduction goals.

Building on numerous state Renewable Portfolio Standard (RPS) policies, these include, as of early 2021, 11 states with 100 percent clean energy or net-zero carbon emission goals and 29 electric utilities with pledges to reduce carbon emissions by 80 percent to 100 percent relative to historic levels. These states and utilities are distributed widely across the country including not only the two coasts, but also many entities in the Southwest, Midwest and Southeast regions (Place, 2020).

Beyond the state level, the Regional Greenhouse Gas Initiative (RGGI), a market-based cap-and-trade program covering the electric sector in 11 states, has continued to expand its geographic coverage as new states join the program (RGGI, 2021). Another example of regional action is the Transportation and Climate Initiative (TCI), which is a regional collaboration of Northeast and Mid-Atlantic states and the District of Columbia, established to reduce carbon emissions from the transportation sector (TCI, 2021).

In the private sector worldwide, over 260 corporations, mostly outside of the electric sector, have committed to procuring their electric requirements from zero-carbon sources (RE100, 2020). And in the US, in 2019, 100 major corporations from the consumer products, high technology, raw material, and heavy industry sectors procured over 20 million megawatt hours (MWh) of renewable energy, which represents more than 60 percent of their electricity consumption (RE100, 2021). This includes roughly 10 GW of new capacity in 2020, which is about one third of the total amount of new generating capacity installed nationally in that year. Further, a growing number of major US and international financial institutions including BlackRock, Morgan Stanley, Barclays, TD Bank, Citigroup, and Bank of America have announced goals to achieve net-zero greenhouse gas emissions in their operations and financing activities.

As discussed in more detail later in this chapter, fully or nearly decarbonizing the US electric sector will require fundamental changes to the electric industry and a dramatic, unprecedented expansion of sector infrastructure. At a minimum, it will require decarbonizing or replacing essentially all of today's coal and natural gas-fired electric generation, which represents more than half of the nation's electric energy and generating capacity. It would further entail substantially expanding the electric sector to supply the zero-carbon energy needed to electrify non-electric sectors of the economy, as will be described later.

These state and private sector efforts are very unlikely, by themselves, to lead directly to full decarbonization of the US economy by about mid-century; supportive federal policies will also be required to achieve that goal. But the growing number and diversity of state and private sector efforts, along with the increasing commercial competitiveness of zero-carbon generating technologies, suggests that the US electric sector will continue to reduce its carbon emissions over time and improve the resiliency of its infrastructure.

RECENT TECHNOLOGY, POLICY AND EMISSION TRENDS IN THE US ELECTRIC INDUSTRY

National trends in the electric industry, including technology deployment, fuel prices, federal and state policies, generation mix, and emissions, along with how these differ across the US, provide important context for the prospects and risk of the energy transition.

Since the mid- to late-2000s, carbon emissions from the US electric sector have been declining, and as of 2019 (the last year before the pandemic reduced electric demands), electric sector carbon emissions were 33 percent below 2005 levels (US Energy Information Administration, 2020). While this reduction is well short of that needed to achieve many ambitious climate goals, such as net-zero, it is nonetheless a substantial decline relative to historic levels, and the downward trend is relevant going forward.

The reduction in electric sector carbon emissions has been primarily driven by a shift in the mix of electric generation (with low- or zero carbon generation displacing higher carbon emitting coal generation) and greater energy efficiency. According to one analysis of the period between 2007 and 2017, the growth of relatively low-carbon natural gas-fired generation, the increase in wind and solar generation, and energy efficiency were responsible for 34 percent, 25 percent and 25 percent of emission reductions respectively (Goff, 2017).[1]

The emission reductions associated with gas generation displacing coal generation were made possible by the commercialization of natural gas "fracking" technologies including (1) advanced drilling technologies and computational power to more efficiently control the direction of underground drilling, and (2) the use of water, sand, and chemicals to hydraulically fracture deep geologic formations to release natural gas. The emergence of these technologies in the 2000s coincided with roughly a 50 percent decline in natural gas prices between 2008 and 2012 and further declines since then (US Energy Information Administration, 2021a). These price reductions materially reduced the cost of natural gas relative to coal, allowing for gas generation to displace coal generation and reduce emissions in the electric sector.[2]

Substantial declines in the costs of wind and solar photovoltaic (PV) generation technologies have helped spur increasingly rapid deployment of these technologies, displacing fossil generation and further reducing sector carbon emissions. The US average levelized cost of onshore wind projects declined by over 60 percent, from $90/MWh in 2010 to $33/MWh in 2020 (Wiser et al., 2021). The cost of utility-scale solar PV projects declined even more. The US average levelized cost of this type of generation declined from $220/MWh in 2010 to $34/MWh in 2020, a cost reduction of 85% (Bolinger, Seel, Warner, & Robson, 2021).

Importantly, the technical advances and cost reductions that have enabled these drivers of emission reductions are the result of a mix of private sector investments and government policy both in the US and overseas. In the US, at the state and regional levels, supportive policies have included a large and growing number of RPS and Clean Energy Standard (CES) programs requiring electric companies to procure a minimum amount of renewable or clean energy each year, typically increasing over time. They also include two major greenhouse gas emission cap-and-trade programs, one in California and the other across the RGGI region. While several widely recognized proposals to establish federal carbon emission regulations have not been approved (including for example the Waxman-Markey Bill of 2009 and more recently the Obama administration's Clean Power Plan), the federal government has successfully advanced the development and commercialization of a wide range of low-carbon technologies through fundamental research at the Department of Energy's national laboratories and the Advanced Research Projects Agency – Energy (ARPA-E) program,[3] and through accelerated tax depreciation and tax incentive policies.

These technological and policy developments have led to major changes in the US generation mix and the emission reductions previously cited. Since 2011, non-hydropower renewables have grown from less than 5 percent of US generation to almost 15 percent of US generation, natural gas-fired generation has grown from less than 25 percent to roughly 35 percent (now the largest single source), and coal-fired generation has declined from around 45 percent to less than 25 percent (US Energy Information Administration, 2021b). In recent years, wind and solar have represented more than half of total US electric generating capacity additions (Bolinger, Seel, Warner, & Robson, 2021).

As detailed in the previous chapter, these national trends look somewhat different when examined on a regional level. Regional variations are driven by differing natural endowments of coal, natural gas, hydro, solar and wind resources, and differing state priorities for energy and environmental policies. Electric generation fueled by natural gas has increased the most in regions such as the Northeast and the Southeast, which have ready access to low-cost supplies from producing regions including from the states of Pennsylvania and Texas. The growth of wind generation has been the greatest in the Great Plains, Rockies, Midwest and Gulf Coast states. The growth of solar has been the greatest in the Southwest, West Coast, Hawaii and Southeast. The largest declines in coal generation have occurred in the Midwest and Northeast where relatively high-cost coal has been competitively disadvantaged relative to lower cost natural gas. These changes reflect the patchwork quilt of regional market and policy conditions in the electric sector across the country discussed in the previous chapter.

Looking at recent federal policy, the Biden administration has set an ambitious climate and clean energy policy agenda during its first year. As of this writing in late 2021, the United States has rejoined the Paris Agreement and established several greenhouse gas emission goals: to reduce economy-wide emissions in 2030 by 50 percent or more relative to 2005 levels, to eliminate electric sector carbon emissions no later than 2035, and to reach net-zero economy-wide greenhouse gas emissions no later than 2050. The administration also launched a "whole-of-government" approach to achieve these goals. This involves a mix of legislation, such as the bipartisan Infrastructure Investment and Jobs Act (IIJA) which passed Congress and was signed into law in 2021. It also involves executive and regulatory actions including, for example, a directive for federal agencies to procure clean electricity and the revival of the Department of Energy's loan program office for innovative energy projects (The White House, 2021a, 2021b). The IIJA legislation, among other provisions, established new policies to support the development of zero-carbon fuels and carbon capture technologies, reduce methane emissions, maintain the country's nuclear fleet and invest in new electric transmission (CATF, 2021). More ambitious climate legislation is pending as of this writing and further executive and regulatory action is expected.

GROWTH OPPORTUNITIES FOR THE ELECTRIC INDUSTRY

As mentioned, fully or nearly decarbonizing the US electric sector will require replacing essentially all of today's coal and natural gas-fired electric generation or at least eliminating the carbon emissions from that generation. This will be a large lift. As of 2020, fossil generation in the electric power sector still represented 58 percent of the nation's electric energy

Implications of fully decarbonizing the electric industry for business 125

and 64 percent of its generating capacity (US Energy Information Administration, 2021c). In addition, decarbonizing other sectors of the nation's economy, including transportation, industry and buildings, is widely expected to be accomplished by electrifying many energy services in those sectors with zero-carbon or low-carbon electricity. This process, sometimes referred to as "beneficial electrification," would substantially expand the demand for electricity. New electric loads would come from charging electric vehicles, powering electric industrial boilers (often for process heat), using electrolysis to produce hydrogen (a zero-carbon fuel used in a variety of industrial applications), installing residential and commercial water heaters, installing heat pumps for space heating, and deploying direct air capture systems (a process for capturing carbon dioxide from ambient air for sequestration underground or for use in manufacturing processes).

To illustrate, one scenario for expanding electric service over time in the home heating and transportation sectors, drawn from a recent nationwide study, would involve the following. Residential space heating would transition from two-thirds fossil fuel today (predominantly oil and natural gas) to almost entirely electric heat pumps (80 percent) and electric resistance heating (20 percent) by 2050. Residential water heating would transition from roughly one-half fossil fuel today to a mix of electric heat pumps (60 percent) and electric resistance heating (40 percent) by 2050. In the transportation sector, cars and light-duty trucks would transition to almost all electric vehicles by mid-century. Medium and heavy-duty trucks would transition to a mix of technologies, with about 80 percent being powered by either electricity or hydrogen fuel cells (Sustainable Development Solutions Network, 2020).

Looking at the aggregate impact of these changes on the electric sector, a review of studies conducted between 2015 and 2021 reveals total US electric generation would need to more than double (a 137 percent increase) relative to today's levels by 2050 (Figure 6.1).[4] At the same time, US electric capacity would need to roughly triple (a 278 percent increase).[5] These increases would be needed even after accounting for ambitious improvements in the efficiency with which energy is used by consumers, and opportunities to charge batteries and electric vehicles during night-time periods of low electric demand.

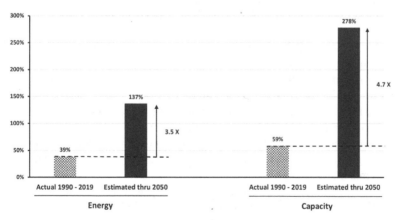

Source: Based on data presented in The NorthBridge Group (2021).

Figure 6.1 Estimated increase in the electric sector growth compared with history

This increase suggests a rate of expansion for the US electric sector far beyond that experienced over the last 30 years: about 3.5 times recent history on an energy basis and over four times that on a capacity basis.[6]

Not surprisingly, this unprecedented expansion would drive large investment opportunities across a wide range of electric generation, energy storage and transmission technologies. The dollar value of the investments required is magnified by the fact that many low- and zero-carbon generating technologies are more capital intensive than the technologies they will replace. One recent study estimated that the US would need to make cumulative capital investments in generation and related infrastructure of close to $2.6 trillion by 2030 and $10 trillion by 2050 (Larsen et al., 2020). These correspond to average annual capital investments over the 30-year period of about $300 billion to $350 billion. To put this in context, US investor-owned electric utility capital expenditures in 2019 were $125 billion (Edison Electric Institute, 2020). Consequently, capital investments would need to increase several-fold over the 30-year period.

Several recent studies of deep decarbonization pathways in the US point to a common set of technologies and investment priorities for the years leading up to 2030 and then diverge over the following decades as technical, economic and market uncertainties have a larger impact on modelling results and possible futures. The common investment priorities for the decade of the 2020s mostly include building out renewable generation (solar PV, onshore wind, and offshore wind), maintaining existing low- and zero-carbon capacity (including nuclear and natural gas plants), expanding transmission infrastructure to facilitate inter-regional transfers of power, and improving the resiliency of the existing electric grid system.

The largest of these investment priorities has been estimated to include the following items (Larson et al., 2020; Williams et al., 2021):

- Building 500 GW to 600 GW of new onshore and offshore wind and solar PV capacity (which is roughly three times the currently installed US wind and solar capacity and 50 percent of the total US installed electric capacity of all fuel types as of the end of 2020).
- Building almost 200,000 GW-km of new transmission (roughly a 60 percent increase over current transmission capacity) to connect new renewable generation supplies with load centers and improve the resiliency of the electric grid.
- Building 5 GW to 15 GW of new battery storage, which is five to 15 times the battery storage currently installed in the US.
- Investing to maintain existing nuclear power plants (that are determined to be safe to operate) to provide dependable, around-the-clock zero-carbon generation and existing natural gas-fired generating capacity to maintain system reliability.
- Capturing and sequestering carbon emissions from five to ten natural gas-fired power plants and carbon-emitting industrial facilities; and constructing an interstate pipeline system (roughly 20,000 km in scale) to transport captured carbon dioxide from point of capture to major storage basins.
- Building end-use energy infrastructure to electrify light-duty vehicles (estimated at 50 percent of new vehicles sold), and heat pumps for space heating (estimated at 25 percent of residences).

UNCERTAINTIES IN THE ENERGY TRANSITION

While holding out the prospect of enormous investment opportunities for businesses, a US energy transition that achieves net-zero economy-wide greenhouse gas emissions by mid-century is a daunting challenge as it rests on a rapid and fundamental restructuring of the country's entire energy sector, a collection of large, diverse and complex industries. It should not be surprising then that the transition is uncertain, both in its timing and its eventual technological direction.

There are at least three main sources of uncertainty. These stem from challenges related to: (1) establishing ambitious and sustainable government policies consistent with mid-century net-zero emission goals; (2) rapidly scaling deployment of electric infrastructure and consumer acceptance of new technologies to electrify other sectors of the economy, such as transportation and home energy services; and (3) anticipating the mix of variable renewable (wind and solar) technologies and firm clean energy technologies needed for the electric sector to be fully decarbonized in a reliable, affordable, socially acceptable and timely manner.

Each of these challenges will influence what the transition means for US businesses.

Ambitious and Sustainable Policies

The progress made to date in improving the performance, reducing the cost and advancing deployment of low-carbon and zero-carbon generating resources is in part due to federal and state policies. Supportive government policies will continue to be needed going forward to rapidly deploy proven technologies and commercialize new advanced technologies. One question is whether these government policies will be sufficiently ambitious and stable over time to achieve net-zero economy wide emissions by mid-century, and this will of course depend on the degree of public support for these policies.

At the federal level, technology innovation policies including fundamental research performed at the Department of Energy's national laboratories and ARPA-E, along with many decades of collaboration among government, universities, research institutions, industry, and entrepreneurs, have made the US the global leader in the development of new electric technologies (Breakthrough Energy, IHS Market, Energy Futures Initiative, 2019; National Academies of Sciences, Engineering, and Medicine, 2021). These programs and other federal policies, including tax incentives for solar, wind and carbon sequestration that help "push" new emerging technologies into the commercial marketplace, have often had bipartisan political support. However, policies to "pull" new technologies into the marketplace, such as carbon taxes, cap-and-trade programs, and greenhouse gas emission limits, have not had sufficient support at the federal level to be established. The existence, stringency, design and predictability of such push and pull policies will heavily influence the pace and character of the transition but are subject to considerable uncertainty.

Policy analysts generally agree the most economically efficient policy approach to reduce carbon emissions would involve a nationally uniform carbon price, which could take the form either of a tax on carbon emissions or market-based emission credit price such as in a cap-and-trade program. Either type of policy could be designed to achieve a desired level of decarbonization over time in a predictable manner and achieve related policy goals (Phillips & Reilly, 2019). But despite several efforts stretching back over 20 years, federal policymakers have yet to develop sufficiently broad agreement to pass these types of policies into law. This period includes proposals put forward by Republican presidential candidates in the 2000s, the American Clean Energy and Security Act of 2009 (also known as the Waxman-Markey bill) and President Obama's Clean Power Plan in 2015.

In recent years, a wide range of policy proposals have been advanced at the federal level that rely heavily on sector-specific policies and elements of industrial policy. Examples include the Green New Deal, which was put forward to address climate change through industrial policies linked to non-environmental policy goals, and sectoral policies such as a clean energy standard (a market-based system requiring a certain percentage of electric sales from zero- and low-carbon generating technologies), expanded clean energy investment and production tax incentives, reform of the centralized regional wholesale electric markets to support clean energy deployment, vehicle emission standards, government procurement guidelines to increase the amount of low-carbon electricity purchased by federal agencies, and use of federal power marketing administrations or other government entities to more directly support development of clean energy and electric transmission.

As mentioned earlier, the Biden administration established an ambitious agenda for clean energy and climate policy. In its first year, the administration pursued that agenda in part by initiating several executive and regulatory actions including directing federal agencies to procure carbon free electricity and reviving the Department of Energy's loan program for innovative clean energy projects. It also secured passage of the bipartisan Infrastructure Investment of Jobs Act which includes support for clean energy technologies as listed earlier in this chapter. Another administration initiative, the Build Back Better Act, has provisions to develop and deploy clean energy including extended and reformed tax incentives for renewables, carbon capture, hydrogen, advanced nuclear, existing nuclear, and high-voltage transmission. That Act was passed by the House as H.R. 5376, but as of this writing in late 2021, appears unlikely to be passed by the Senate. If the Build Back Better Act does not go forward in some form, other executive and regulatory actions will be needed to achieve all the administration's climate goals.

Going forward, regardless of the precise mix of federal clean energy and climate policies, the states are widely expected to continue to play a prominent role in the energy transition. This could involve increasing the stringency or accelerating the timelines of state clean energy standard programs, expanding the scope of carbon pricing programs (such as RGGI and California's cap and trade program), expanding the use of targeted procurement programs for clean energy resources such as offshore wind, and adopting analogous programs for the transportation sector. Programs such as these will evolve and likely expand over time on an incremental basis as they have done in recent years. A key point of uncertainty is whether these types of programs will be largely restricted to states with relatively low greenhouse gas emissions or instead be adopted more broadly by other states including those with higher emissions.

Importantly, both federal and state policies will only be sustainable over the coming decades if they are compatible with the provision of energy services to consumers and businesses that are reliable, affordable to all customers, and aligned with other societal priorities including employment and social equity; for example, supporting the "just transition" of local communities that are today largely dependent on fossil fuel production and consumption. Federal policies will also need to be sufficiently flexible to conform with the differing resource endowments, public priorities, and electric regulatory systems in major regions of the country.

If state and federal policies are not adequately aligned in these ways, public pressure will inevitably mount to delay, revise or reverse some policies. At the state level, the announcements in 2021 that Massachusetts will not participate in the regional Transportation and Climate Initiative and that Virginia will withdraw from RGGI illustrate the uncertain nature of policy support for some climate initiatives.

Similarly, the clean energy and emission goals established by electric utilities, large energy-consuming corporations, and financial institutions will only be sustainable over time and achievable if they can be achieved while also meeting other customer, investor, and societal goals.

Scaling Deployment and Consumer Acceptance

With a doubling or tripling in the size of the electric sector and need for perhaps $10 trillion of capital investment over three decades, another source of uncertainty is whether the industrial and consumer changes needed to achieve this transformation can be practically scaled in a timely way.

There are several elements to this, most importantly: siting and permitting electric infrastructure, expanding industrial supply chains, and gaining widespread consumer acceptance.

The first of these involves the ability to site, permit and gain local community support for new clean energy generating plants and infrastructure at a pace several times higher than experienced in the last three decades. This becomes particularly daunting when accounting for the land use and community impacts of wind, solar and transmission projects which are widely expected to comprise a large majority of new energy infrastructure. According to one recent study, by mid-century onshore wind and solar could span almost 600,000 sq km of land area, roughly the size of Illinois, Indiana, Ohio, Kentucky, Tennessee, Massachusetts, Connecticut, and Rhode Island combined (Larson, et al., 2020). At the same time, the nation's transmission system may need to be doubled or tripled (The NorthBridge Group, 2021). The land use issue is particularly important for pathways that rely heavily on variable renewable technologies because these technologies are less energy dense than some other generating technologies and rely on long distance transmission to connect resource rich regions with load centers. Policy and corporate business practice reforms to better engage local communities and share the economic benefits of clean energy development could help address these issues, but are only now beginning to be explored in the US.

The most likely alternative technology pathways to extensive wind and solar development would involve greater deployment of more energy dense generating technologies, such as fossil plants with carbon capture and advanced nuclear technologies. However, these are technologies that have limited social license in many regions of the country today. These two factors, the extensive land and community impacts of variable renewables and the social opposition to carbon capture and nuclear technologies, may make rapid scaling of clean energy infrastructure a difficult and uncertain enterprise.

Another challenge involves rapidly scaling supply chains for electric energy-related industrial and consumer goods. These include raw material procurement, manufacturing and transportation of electric plant and equipment, workforce training and construction services. As in recent years with the imports of solar panels from China and offshore wind technology from Europe, the scaling of supply chains in future years will depend to some extent on international trade policies, and the extent to which manufacturing is located overseas or in the US. The ability to rapidly scale these systems will influence the opportunity of the US to capitalize on its world-class ability to develop advanced energy technologies and use the energy transition to grow the nation's export markets.

The third challenge revolves around the changes in consumer home energy systems, widespread market penetration of electric vehicles, conversion of industrial fossil boiler systems to electricity and use of zero-carbon fuels in industrial manufacturing processes. The most recent economy-wide analyses of deep decarbonization pathways often assume consumers and industry will be willing to rapidly adopt these new technologies and systems and accept that some of this new electric demand will be curtailable or interruptible when inadequate electricity supply is available to serve those electric demands. In essence, by the time full economy-wide decarbonization is achieved, these analyses often assume the technical operations of the electric grid will be largely reversed relative to how it has been operated historically: instead of managing firm, generating resources (power plants physically capable of producing energy on a reliable and regular basis throughout the year) to satisfy firm "on-demand" electric loads as is currently done, large amounts of flexible consumer and industrial electric load will be managed within the available limits of variable electric supplies sourced predominantly from variable wind and solar moderated with battery storage.[7] For some industrial processes that might otherwise operate on a regular, around-the-clock basis throughout the year, this implies operating in a new manner, with plant utilization at any point in time possibly limited by the availability of electricity. This in turn could influence the daily and seasonal time patterns of workforce requirements at these facilities. For the public, this new approach to operating the electric system, with flexible loads managed within the limits of variable supplies and battery systems, suggests consumers will need to be comfortable with their electric vehicle charging and home appliance use being subject to the availability of electric supplies, at least during some times of the year.

These industrial and consumer scaling challenges may well be overcome in one form or another through time. But the magnitude and speed of the anticipated changes add uncertainty to the pace of the energy transition, and the potential for consumer dissatisfaction creates risk for policy reversals.

These challenges could also influence the mix of clean energy resources eventually developed. For instance, local community resistance to the siting of onshore wind or solar PV, or marine community opposition to offshore wind, could slow or limit the deployment of those technologies and shift the mix of new generation required to meet climate goals to other types of clean energy that might be more readily sited and permitted. And consumer or industrial reluctance to have some of their electric usage curtailed or remotely managed would increase the need for clean energy technologies that are reliably firm rather than variable.

Alternative Technology Pathways

Recent studies, as described before, point to the need for many common types of investments through 2030. These include, in particular, wind, solar and transmission. However, the mix of new electric investments needed after 2030 is less certain. The same studies that point to a common set of investments over the next decade also point to several differing long-term technological pathways for the electric sector starting in the 2030s and beyond. These technology pathways generally fall into two groups, variable renewable-dominant systems and systems with a more diversified mix of technologies.

Variable renewable-dominant systems can be thought of as electric systems where wind and solar supply roughly 85 percent or more of total electric energy and are complemented by much smaller quantities of firm resources such as existing hydro, existing nuclear, and natural gas-fired electric energy (fueled either by natural gas without carbon capture controls or by zero-carbon liquid fuels of various types). In contrast, more diversified electric systems are comprised of variable renewables complemented with larger quantities of existing hydropower and nuclear along with new firm clean energy technologies to help maintain system reliability and provide zero-carbon energy. Examples of new firm clean energy technologies include natural gas-fired electric generation with carbon capture and sequestration, fossil-fired oxy-combustion technologies with carbon sequestration, advanced nuclear electric generation including small modular "shipyard-manufactured" fission reactors or fusion reactors, hydropower, geothermal power, and zero-carbon liquid fuel-fired electric generation such as hydrogen.[8]

While wind and solar generation have become cost-competitive sources of energy in many regions, deep decarbonization studies have shown that a diverse mix of clean energy technologies, including both variable renewables and firm electric generating technologies, will be needed to achieve full decarbonization. Firm electric technologies are important for several reasons. They help maintain reliable electric service for customers requiring on-demand electricity by generating electricity when supplies from wind and solar are insufficient (Lott & Phillips, 2021). They reduce the overall system-wide cost of electric service at high levels of decarbonization when the incremental cost of adding additional variable renewables (along with complementary transmission and batteries) rises to the point where they exceed the incremental cost of adding firm clean energy generation. By expanding the tool kit of clean energy options, they provide flexibility to achieve ambitious decarbonization goals if the siting, permitting, scaling and consumer acceptance challenges discussed earlier slow the pace of rapidly deploying variable renewable resources. And finally, CCS technologies are expected to be particularly valuable in helping decarbonize some industrial processes that currently use natural gas as a feedstock.

The eventual pathway that emerges in the US is uncertain because of the many technological, economic, and social unknowns at play. Continued cost reductions and performance improvements in onshore wind, offshore wind, solar PV, and battery technologies are widely expected by industry and technology experts. Advanced natural gas-fired oxy-combustion technologies with carbon sequestration are also now being demonstrated commercially in the US. Downscaling of nuclear fission technologies coupled with modular shipyard construction could make forms of that technology cost competitive in some markets, perhaps as "drop in" replacements for coal-fired boilers in some of today's coal power plants. This would have the benefit of helping to sustain local communities in coal-heavy regions of the country. Other advanced technologies such as long-duration energy storage or "deep hot rock" geothermal could be demonstrated and become commercially competitive. The degree of success of any of these technical and economic developments is uncertain. The uncertainty is compounded by the social dimensions of siting and permitting new energy infrastructure, and how these differ across alternative generating technologies and regions of the country. Difficulty siting new infrastructure at scale could focus development efforts on repurposing existing coal and gas power plant sites with advanced low carbon generating technologies, and relying on existing transmission, rail, and pipeline corridors to expand electric transmission capacity.

The eventual decarbonization pathway in the US will also likely vary from region to region of the country depending on the relative strength or weakness of the area's wind and solar endowments, ability to build new long-distance transmission, the availability of other natural resources (such as hydro, biomass, and access to geologic sequestration), and existing energy infrastructure (including nuclear, hydropower, and pipelines for fuel transportation).

These many considerations lead to a range of estimates for the mix of electric generation that may eventually be seen in a fully decarbonized electric sector. Figure 6.2 illustrates this point by summarizing the mix of US electric generation estimated in seven national, deep decarbonization studies. The average generation mix across the studies corresponds to 65 percent variable renewables (such as wind and solar) and 35 percent firm generation (such as nuclear, fossil with carbon capture and storage (CCS) and firm renewables such as hydropower). Some of the most recent studies estimate energy shares for wind and solar of 85 percent or more, reflecting the increased cost competitiveness of wind and solar technologies. But, collectively, the studies point to a wide range of outcomes, with the potential for somewhat smaller shares for wind and solar resources (down to around roughly 50 percent) and correspondingly larger shares for firm clean energy technologies. The range of these modelling results illustrates the potential for alternative pathways to develop over time (The NorthBridge Group, 2021).

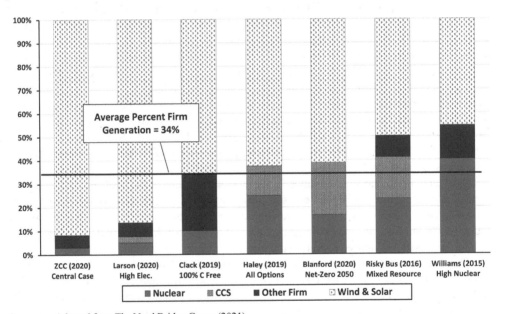

Source: Adapted from The NorthBridge Group (2021).

Figure 6.2   2050 Generation Mix in low-cost scenarios from national studies examining multiple decarbonization pathways

IMPLICATIONS FOR ELECTRIC BUSINESSES AND DECISION MAKING

The fundamental challenge for businesses and investors in the electric sector is how to capture the enormous opportunities for growth while at the same time managing the complex localized commercial risks that arise out of the policy, scaling, and technology uncertainties just discussed. Policy uncertainty, at both the federal and state levels, dampens the financial incentives of the private sector to invest in low-carbon technologies at the massive scale required to meet ambitious emission goals. Policy uncertainty also preserves the opportunities that the owners of today's carbon-emitting fossil power plants have to continue to operate and earn financial returns, at least for the time being. The scaling uncertainties with permitting infrastructure, supply chains, and consumer acceptance raise the cost and financial risks of rapid business expansion and building market share. And the uncertainty around technology pathways makes it more difficult for early investors to demonstrate new technologies and build the first-of-a-kind facilities needed to improve performance, reduce costs and commercialize those technologies. It also raises the risk to investors that, absent long term fixed-price contracts that pass market risk to other parties, capital investments will turn out to have lower than anticipated rates of return. All these uncertainties are made more complicated by the patchwork quilt of market and regulatory conditions currently in place across the country.

To address these challenges, businesses in the electric sector will need to consider three questions: (1) how to rapidly deploy technologies that have already been largely commercialized; (2) how to demonstrate and commercialize new advanced clean energy technologies; (3) how to manage the existing fleets of fossil and nuclear generating plants. Each of these is addressed in the following sections.

Deploying Commercial Technologies at Scale

Over the next ten years, through at least 2030, most decarbonization in the electric sector is anticipated to come from deploying zero-carbon generating technologies that have already been at least partially commercialized in parts of the country – solar PV, onshore wind and offshore wind – along with complementary electric infrastructure such as transmission, batteries and distributed energy resources. As mentioned earlier, to achieve net-zero US economy-wide emissions by 2050, the amount of wind and solar capacity alone that would need to be deployed by 2030 could approximate roughly half of today's total US electric capacity. This level of deployment corresponds to an average of 50 GW to 60 GW annually, which is close to three times the wind and solar installed annually between 2015 and 2020. Further, the nation's transmission system would need to expand by more than 50 percent.

Achieving this ambitious level of deployment during the 2020s will hinge on addressing the policy and scaling issues discussed before, most importantly new federal or state "pull" policies, interstate transmission development and reforms to facilitate the siting and permitting of new facilities.

Wind and solar

If solar and wind capacity is to expand by the magnitude estimated in recent studies, it would represent the single largest growth opportunity associated with the energy transition in the US over the next decade. But for businesses involved in wind and solar development, rapidly scaling deployment raises siting and permitting, contracting, procurement, financing and transmission access questions, especially in regions of the country with competitive wholesale power markets.

Non-utility developers of solar and wind looking for rapid expansion of their portfolios may need to consider the mix of long-term, fixed-priced power purchase agreements (PPAs) and merchant projects (which place greater risk of cost recovery on investors rather than utility ratepayers). For many years, the comparatively large up-front capital costs of wind and solar facilities (relative to total project costs), wholesale market price uncertainty and lack of ambitious carbon pricing or clean energy standards in many regional markets have led most developers to seek long-term, fixed-priced contracts from utilities or corporate buyers. This practice allocates financial risk to ratepayers and buyers, reduces investor risk exposure, and allows developers to finance projects with a lower cost of capital. This is the commercial standard today in the US.

Continued technology advances that allow zero-carbon investments to be commercially competitive with other sources of wholesale electricity without policy supports (such as a CES, carbon taxes or tax incentives) would allow clean energy developers to invest on a merchant basis without long-term, fixed-priced PPAs. This merchant approach to financing investments, which is the commercial norm in most non-electric sectors of free-enterprise economies, would be most practical for relatively large well-capitalized companies in a position to accept greater financial risks. Eliminating the need to secure new PPAs for each project could simplify the contracting and financing process for investors and speed the deployment of clean energy technologies. But in the absence of such technological advances or new ambitious and sustainable energy policies, continued reliance on long-term fixed-priced PPAs is likely to be the default commercial pathway in most parts of the US.

For renewable energy developers in regions of the country with strong wind and solar technical potential looking to sell output in other regions of the country with demands for clean energy, another commercial question is whether and how to bundle wind or solar projects with new interstate transmission. The economics of these PPAs will be affected by the technical structure of wholesale electric markets (e.g., whether they are bilateral, energy-only, or energy and capacity markets), the design of renewable energy and clean energy credit products, and the way these systems evolve in future years.

For vertically integrated utilities with their own generating assets, another question is whether and how much wind and solar capacity might be owned as opposed to contracted from independent developers. These decisions will be influenced by several considerations, including the utility's relative capability in project development, tax law considerations and competitive procurement policies established by their state regulators.

And finally, as wind and solar technologies achieve high levels of market penetration and displace other sources of electric generation with higher variable operating costs, the wholesale price of electric energy during hours of high wind and solar production will decline. During hours when wind and solar displace all other sources of electric generation, wholesale prices will likely drop to zero, eliminating the economic incentive for continued production. This concern has led policy analysts to explore ways that price formation practices in centralized wholesale power markets might be reformed to provide better economic incentives to deploy and operate clean generating technologies in deeply decarbonized power markets.

Transmission

The 50 percent expansion of the nation's transmission system over the next 10 years, and the doubling or tripling estimated in some studies by mid-century, point to very large investment opportunities for transmission developers. But new long-distance transmission facilities have in the past been notoriously difficult and slow to develop, often taking well more than a decade for successful projects (and resulting in failure for others).[9] There are several reasons for this, including: the diffused nature of the reliability and economic benefits which makes it difficult to assign cost responsibility to ratepayers and users of the system; local community, incumbent generator and environmental opposition to project development; a lack of centralized regional and national transmission planning; and the multiple layers of local, state and federal regulatory approval required for inter-state projects.

Some of these hurdles could be overcome by co-locating new capacity with existing transmission lines and other infrastructure rights-of-way (railroad, highways, pipelines, etc.), reforming federal and state regulations to speed and reduce the uncertainty associated with project review, adopting "best practice" project development processes to more effectively engage, compensate and build support from local communities, and expanding the use of independent merchant project business models (as opposed to traditional utility projects with socialized cost recovery from ratepayers).[10]

The outcome of efforts to speed the deployment of interstate transmission will likely have an impact on the mix of variable renewable and firm clean energy that emerges over time. For instance, long-distance transmission lines from the Great Plains states and Southwest to population centers in other regions could facilitate development of wind and solar in those supply regions, while their absence would require more new clean energy generation sited closer to load centers and perhaps shift the generation mix towards other technologies.

Batteries

The 80 percent decline in the cost of lithium-ion battery technology that occurred in the years leading up to 2020, deployment of wind and solar generation, appreciation of the technical services that batteries can provide to the electric grid, and policy support from some state regulators all point to an expectation of robust battery investment growth going forward. The federal government recently estimated that a total of over 12 GW of large-scale battery storage capacity would be installed by the end of 2023, more than a ten-fold increase since 2019 (US Energy Information Administration, 2021d).

At the same time, the relatively short duration of today's batteries' storage capability, generally four to six hours, makes them expensive options to overcome the seasonal variability of solar and wind generation, and may limit their longer-term market penetration. In the long term, success is likely to hinge in part on business model and regulatory reforms, including the adoption of long-term, fixed-priced contracts and regulated cost recovery that allow for capital costs to be financeable and commercially viable in diverse market regimes.

Distributed energy resources

Distributed energy resources include a varied mix of generating, storage and other technologies such as rooftop solar, batteries, backup conventional electric generation (including internal combustion engines), advanced metering, and demand management electronic controls. All these are located on the utility's distribution system (and not on the transmission system) or directly on a customer's premises. In many cases, the distributed energy resource is configured to be "behind the meter," meaning that the resource's output is treated (from the standpoint of utility planning and operations) as a reduction in the customer's metered load and grid-level electric demand. While not all distributed energy resources are low- or zero-carbon technologies, some, such as rooftop solar, do not directly emit greenhouse gases when they produce electricity. Utilities can deploy distributed energy resources in specific locations as a solution to manage localized system congestion. Customers can deploy distributed energy resources to actively manage their electricity usage and improve the reliability of their electric service. When the costs of distributed resources are less than the portion of utility rates that are avoided by them, customers can reduce their electricity bills through adoption of distributed energy resources. The value of demand management services and improved reliability to customers is important in evaluating the economics of distributed energy resources because the cost of many distributed energy resources is higher than the cost of corresponding central station generating technologies, which often benefit from greater economies of scale (Lazard, 2020). For these resources to be widely adopted by utility customers, these benefits will need to outweigh the cost premium. While generally considered to be a smaller part of the overall industry-wide energy transition, distributed energy resources represent a promising business opportunity for many utilities, equipment suppliers and project developers, particularly if regulatory reforms seeking to facilitate their adoption are successful.[11]

Commercializing Advanced Technologies

Looking ahead over the next several decades, the second type of business challenge is how to commercialize advanced technologies that are likely to be valuable and perhaps essential to achieve net-zero emission goals on an economy-wide basis but are not yet demonstrated or widely available on a commercial basis. These include advanced technologies to provide clean firm capacity and energy (such as next-generation small modular reactors, generation IV fission reactors, fusion reactors, advanced low-carbon fossil-fueled oxy-combustion and deep hot rock geothermal technologies), store electricity over extended periods of time (such as thermal energy storage), capture carbon dioxide from fossil fuel-fired electric generation plants and industrial processes, and produce hydrogen through electrolysis. Capturing carbon dioxide and producing hydrogen through electrolysis or steam methane reformation processes on a widespread basis will also require extensive interregional transport, storage and sequestration networks.

While some deep decarbonization analyses suggest these technologies may not need to be widely deployed until the 2030s or later, developing and commercializing them before then is important because of the long periods of time required to fully demonstrate, commercialize and deploy new technologies, the need for supporting infrastructure such as pipeline and storage networks and, if widespread deployment of renewables and transmission is delayed, the need for other advanced technologies to be deployed more quickly.

The commercialization challenge for businesses looking to advance these technologies is to demonstrate them at pilot and commercial scales, improve their performance and drive down costs to commercially competitive levels, develop supply chains, establish workable business models, and gain consumer acceptance. Successful technologies will need to exceed competitive benchmarks based not only on today's technologies but also tomorrow's advanced technologies with lower costs and improved performance. They will also need to be designed to operate and compete in integrated electric systems and competitive wholesale markets;

this is likely to put a premium on the ability to complement the variability of wind and solar resources by dispatching them when wind and solar are not generating output. The federal ecosystem of fundamental research, technology advancement and project demonstration can be of great value in this process.

The commercialization of advanced technologies is an area of great activity in the electric sector. Many important technological advances have occurred in recent decades including the commercialization of natural gas-fired combined cycle technology, onshore and offshore wind turbines, solar PV panels, and lithium-ion batteries. There is much ongoing public and private sector activity including work at the DOE national laboratories, the federal ARPA-E program, early project developments supported by federal tax incentives to sequester carbon dioxide captured from electric generating and industrial emission streams,[12] and early commercialization of advanced oxy-combustion generating technologies. Further, more than a dozen companies are working on advanced nuclear technologies, some of which are in the stages of licensing next-generation technologies (Nuclear Innovation Alliance, 2021).

But the energy sector, with its long technology development cycles and high capital requirements, also presents particularly difficult innovation challenges. Over the years, many high-profile development efforts have not come to fruition as planned. Examples include the federal government's FutureGen project in Illinois, the Southern Company's Kemper project, the Massachusetts Cape Wind project, the Texas Clean Energy Project (TCEP), and several nuclear plants proposed in the early 2000s during a period when many analysts expected a federal carbon price would be established and before directional drilling and fracking drove down the price of natural gas and wholesale electric prices.

While failure is an inevitable element of the innovation process, valuable lessons are often learned that can lead to later success in subsequent efforts and, over time, the rate of success can be improved. For early-stage technology innovators, patient capital to support piloting and project development activities over a 10-year or longer period, in contrast to the three- to five-year investment horizons of many investors today, could be immensely valuable. This would provide organizational and development stability through a series of technology readiness stages. For early-stage investors, providing that patient capital with a strategic view of technology characteristics that will be commercially valuable in deeply decarbonized energy markets can help guide the design and development of recent technologies in ways that will make them viable in competitive marketplaces, not just in an engineering lab. This may involve considering the value of designing generating technologies to have greater operational flexibility (for instance, to be "dispatched" up or down depending on the availability of wind and solar generation), sourcing equipment and components from overseas markets, initially demonstrating technologies in other countries before deploying them at scale in the US, and building the potential for export demand into business plans. The federal government, working in concert with technologists and investors, could offer expanded research and development funding through ARPA-E and the national laboratories, and support early commercial stage demonstration projects (Gates, 2021; National Academies of Sciences, Engineering, and Medicine, 2021; Breakthrough Energy, IHS Market, Energy Futures Initiative, 2019).

Managing Existing Generation Fleets

The third challenge for business involves managing the nation's existing generation assets, in particular the large fleets of coal, natural gas, and nuclear power plants. The lack of sustainable energy policies at the federal level is a major source of risk for the owners of these power plants. Owners of merchant nuclear plants, those whose costs are not recovered through regulated rates, may view the lack of sustainable federal energy polices as wholesale markets not compensating them for the environmental value of zero-carbon generation. Owners of carbon-emitting coal and natural gas-fired power plants may see this as an opportunity to continue to operate their assets for an indeterminate number of years.

Nuclear

As of the end of 2020, the US had a fleet of close to 100 nuclear reactors that generated slightly over half of all the zero-carbon energy produced during that year. The development of a new cohort of nuclear plants, once considered promising before the advent of natural gas fracking and the decline of wholesale electric prices in the 2000s, is – as of 2021 – limited to the two reactors under construction at the Vogtle nuclear station in Georgia. The decline in wholesale electric prices across the country has contributed to the retirement of seven reactors between the beginning of 2013 and the beginning of 2019, with 12 more expected to retire by 2025 (US Energy Information Administration, 2019). It has also imperiled the continued economic operation of other nuclear plants in regions with competitive wholesale markets, a risk that may grow over time with continued market penetration of renewable technologies that depress wholesale market prices. These concerns have pushed nuclear plant owners, policymakers concerned with maintaining employment and local tax bases, and environmental groups focused on preserving zero-carbon generation, to band together in support of state policies that would enable continued plant operations (Clemmer, Richarson, Sattler, & Lochbaum, 2018). As of 2020, the states of New York, Illinois, Connecticut, and New Jersey have all adopted nuclear zero-emission credit (ZEC) systems to extend the operating lives of some plants in those states. These state initiatives along with recently passed federal policy supports are likely to be a major determinant of nuclear plant lives and carbon emissions in competitive market regions of the country. Most recent analyses of pathways to net-zero economy-wide greenhouse gas emissions envision the continued operation of the majority of today's nuclear fleet. Any development of new reactor capacity in the US beyond the two now under construction will likely hinge on major advancements and cost reductions with next-generation technologies. As noted earlier, several companies are working on advanced reactor designs, and several are in the process of licensing next-generation technologies.

Coal

Over 30 percent of the US's coal-fired electric capacity was retired in the years leading up to 2020. However, coal still supplied 20 percent of US electric generation in 2020 and, as of 2021, US DOE forecasts estimate relatively modest declines in coal generation over the next five to ten years absent new policies.[13] At the same time, many deep decarbonization analyses forecast steep declines and the eventual elimination of this source of electricity by 2030 or soon thereafter.

To date, the uncertainty over new emission and energy policies has led owners of many coal-fired electric plants to continue to operate, albeit often with marginal economics and the risk of not recovering ongoing investments in plant and equipment or of transferring their ownership to investment funds able and willing to bear the risk of continued operations. The decisions of plant owners to continue to invest and operate plants or move towards closure hinge on numerous business considerations that vary by region of the country, owner expectations for near term market conditions and competitor behavior, and form of power plant ownership. These include owner views of the likelihood and timing of federal or state clean energy or carbon pricing policies; other government policies that may tighten conventional air pollution, water and waste regulations which would increase power plant costs; commercial opportunities to repurpose brownfield power plant sites for the production or support of low-carbon energy and other uses; the "real option value" of keeping economically marginal coal plants open because of the potential for higher natural gas and wholesale electric energy prices or near term advances in post-combustion CCS technologies; the incentive to continue to operate rate-regulated power plants with unrecovered investment costs still on their books in order to fully recover costs (unless otherwise addressed by securitization policies as in states such as Colorado and New Mexico); the impact of must-take coal supply contracts and plant site remediation costs triggered by closure; and the interest of state policymakers and local communities in supporting local employment, tax bases and a "just transition" for communities currently hosting coal-fired power plants and related fuel facilities. All these contribute to complex and situation-specific plant management decisions.

Natural gas

Now the single largest source of capacity, at roughly 40 percent, and energy, at roughly 35 percent, in the US electric sector, natural gas power plants in some ways face similar issues as the owners of coal plants since the carbon emissions from natural gas generation will eventually need to be largely or fully eliminated to achieve net-zero emission goals. At the same time, the nation's fleet of gas-fired combined cycle and combustion turbine plants has continued to grow each year in response to the need for new electric capacity and energy. Proposals to develop new gas-fired electric generation raise questions for investors about the risk of stranded costs, that is, that these long-lived investments may be prematurely retired at some point in the future before their initial investment costs are fully recovered.

However, to a greater extent than coal, electric generation from natural gas may have an important continuing role in a deeply decarbonized US economy. This could come about as the result of advances in post-combustion carbon capture or commercialization of oxy-combustion technologies that would allow natural gas-fired power plants to generate energy with very low or perhaps even zero-carbon emissions. One example of this could be the Allam-Fetvedt cycle oxy-combustion technology, developed by 8 Rivers Capital and NET Power, which announced plans in 2021 to build their first commercial scale projects in Colorado and Illinois, with the Colorado project targeted for operation by 2025 (Bloomberg, 2021). Alternately, some currently operating combined cycle and combustion turbine plants (with upstream transport and storage facilities) might be retained indefinitely to provide reliability to the electric grid while being operated just 5 percent to 10 percent of the hours during the year. In this decarbonization pathway, these generating facilities may be fueled by natural gas (with the carbon emissions offset by other means such as use of land carbon sinks or direct air capture) or zero-carbon liquid fuels such as hydrogen. The technical or economic feasibility of such a low-utilization system of fuel production, transportation and generation has not yet been widely explored in the public literature.

Regulated Utilities and Diversified Electric Companies

The challenges just discussed apply to "pure play" companies and investors that focus exclusively on a single line of business such as developing central station solar PV plants, offshore wind plants or new transmission lines.

But many of the larger corporations in the US electric sector are diversified enterprises that need to consider the impact of initiatives in one line of business on other parts of their business. Vertically integrated electric utilities have generation, transmission and distribution assets and will consider some of the issues outlined earlier about scaling commercial technologies and managing existing generating fleets in tandem with one another. Many utilities also often provide electric service in neighboring states, making them regulated by more than one state public utility commission with perhaps differing policy goals. For all these types of diversified companies, the challenges outlined before may be particularly complex to navigate.

To illustrate, rate regulated utilities are accountable to their state regulatory commissions and policymakers who are typically concerned with diverse policy objectives, including service reliability, affordable electric rates, employment and local economic development, and ensuring a sustainable environment, among other goals. Clean energy and climate goals are part of this, but not the only important policy goal for many of these regulators. Utilities with ambitious clean energy and greenhouse gas emission reduction goals, including net-zero emission goals, take these other policy goals into account in their corporate planning efforts and weigh their tradeoffs. In the absence of a federal requirement mandating ambitious clean energy deployment or emission reductions, regulated utilities in states with supportive government policies will generally see more clean energy investment and better prospects for full cost recovery of prudent investments than utilities in states with less supportive government policies. The reverse may also often occur. A special case involves demonstrating advanced technologies with new and unfamiliar cost and performance risks. While clearly important to achieving climate goals from a national policy perspective, the technical and performance risks associated with demonstrating advanced technologies coupled with conventional cost-based electric utility regulation typically create a financial disincentive for regulated utilities and their owners to invest in early-stage commercialization projects.[14]

Regulated utilities providing electric service from a single integrated generating system to adjoining states with state regulators that have conflicting policy priorities may face particularly difficult planning decisions and tradeoffs. Generating investments made to comply with the requirements or goals of one service area is likely to influence and perhaps raise the cost of providing electric service to an adjoining service area. For example, building new carbon-free generation to meet an RPS or CES requirement in one state could increase the utility's overall cost of generation, accounting for both the direct cost of the new resource and its impact displacing generation from other power plants. In some situations, total generation costs are allocated between the service areas based on their relative loads, leaving both service areas paying for a share of the costs resulting from one state's policy. This could lead to the need to separate the utility's integrated electric system on a jurisdictional basis, so that separate planning and system operations could be carried out in each state, with the costs resulting from each state's policies appropriately recovered from each state.

For corporations with both rate regulated and competitive lines of business, investments in transmission to move electricity from one region to another may influence the operations, emissions, economic value, and operating lives of generation assets. Large investments in distributed energy resources such as rooftop solar and batteries could similarly influence the value of central-station generation.

CLOSING

Looking ahead over the coming decades, climate change is widely expected to spur a global transition to a cleaner, decarbonized energy system that must also be more resilient in the face of increasingly frequent extreme weather events.

To meet mid-century decarbonization goals, a step-change increase in the pace of clean energy deployment will be needed, leading to a doubling or tripling of the electric sector while fundamentally transforming its structure and operations.

For businesses in the electric sector, these changes will spur diverse and unprecedented investment opportunities, several times greater than in recent years. But at the same time, the energy transition also presents complex financial and organizational risks. These stem in large part from today's patchwork quilt of market and policy conditions across the country, the uncertainty of establishing sustainable national policies consistent with ambitious net-zero emission goals, the difficulty of rapidly scaling deployment and gaining consumer acceptance of new clean energy technologies, and the unknowns associated with the long-term mix of electric generation, storage and consumer technologies.

This points to three challenges for companies in the US electric industry. The first is how to rapidly scale deployment of wind, solar, long-distance transmission and other technologies that have already been largely commercialized in some regions of the country. The second is how to rapidly commercialize advanced clean energy technologies such as carbon capture, zero-carbon liquid fuels, next-generation nuclear and firm renewables, which will be needed to fully decarbonize the economy in a reliable, affordable and socially acceptable way. The third is how to manage the existing generating fleets: to secure financial support for marginally economic nuclear plants or wind down their operation; to plan the retirement of coal plants or retrofit them with carbon capture; and to retrofit gas plants with carbon capture, retire them, or maintain them as low utilization generating resources to ensure electric reliability.

To address these challenges and succeed in this complex environment, successful electric businesses will need strategic foresight, the ability to critically assess regionally varied investment opportunities, and skills to manage large commercial and organizational risks. Core competencies will include constructively contributing to federal and state energy policies, developing new supply chains, reforming project siting and permitting practices, analyzing rapidly evolving electric markets and customer acceptance trends, assessing the commercial promise of emerging technologies, and managing commercial risks through contractual means, regulatory processes and financial scale.

Final note: In Greek mythology, Daedalus was a skilled craftsman who made wings for himself and his son, Icarus, so they could escape from the island of Crete. Daedalus was careful not to fly too high or too low and made the trip safely. But, despite his father's warning, Icarus flew too high. The heat of the sun melted the wax used to make the wings, causing Icarus to fall into the sea and drown.

NOTES

1. Discussions of carbon emissions in this chapter generally refer to the emissions directly resulting from power plant operations, not the larger technology "lifecycle" emissions that also include fuel production, fuel transportation, power plant manufacturing and material disposal.
2. The increase in US natural gas spot prices during 2021 has been forecast by the US EIA to be moderate in the coming years and is not expected to materially reverse this long-term trend.
3. ARPA-E is an agency in the United States Department of Energy tasked with promoting and funding research and development of advanced energy technologies.
4. Electric generation is also referred to as electric energy.
5. Electric capacity is a measure of the maximum amount of energy that may be generated at a point in time. Generating technologies with relatively low utilization rates or capacity factors, such as solar and wind plants, require relatively large amounts of capacity to produce the same amount of cumulative electric generation as other technologies with higher capacity factors. Because of this, the proportional increase in capacity required to achieve deep decarbonization is typically larger than the proportional increase in generation.
6. Based on analysis by the NorthBridge Group of modeling studies examining economy-wide deep decarbonization pathways in the US. The estimated increase in energy is the average of cases from nine national studies reporting total energy requirements. The estimated increase in capacity is the average of cases from six studies reporting total capacity requirements. (The NorthBridge Group, 2021)
7. Some amount of firm clean electric generation would also be available to serve other types of electric demand that is unwilling or unable to be managed in a more flexible way.
8. Hydrogen may be produced from natural gas through steam reformation with carbon capture or electrolysis with electricity supplied from zero-carbon generation.
9. See *Superpower: One Man's Quest to Transform American Energy* by Russell Gold (2019) for a history of the Clean Line Energy proposals to develop and transfer large scale wind energy from the Great Plains to the east.
10. New York's Accelerated Renewable Energy Growth and Community Benefit Act is one example of efforts to expedite the siting and permitting of clean energy infrastructure.
11. For example, FERC Order 2222, issued in September 2020, was intended to remove barriers preventing distributed energy resources from competing on a level playing field in the organized capacity, energy, and ancillary services markets run by regional grid operators.
12. This tax incentive is often referred to by the relevant section of the federal tax code, 45-Q.
13. According to the US EIA, the increase in natural gas prices during 2021 has caused gas-fired generation to decline and coal-fired generation to increase, but this effect is expected to be reversed within the next two years and be followed (absent new policies) by an extended period of modest declines in coal generation.
14. Several electric utilities have incurred large financial disallowances associated with their efforts to develop new nuclear facilities (for example, The Vogtle and Sumner nuclear plants) and low-carbon fossil plants (for example, the Kemper coal CCS and Edwardsport IGCC projects).

REFERENCES

Bloomberg. (2021, April 15). *U.S. Startup Plans to Build First Zero-Emission Gas Power Plants*. Retrieved August 20, 2021, from Bloomberg Green: https://www.bloomberg.com/news/articles/2021-04-15/u-s-startup-plans-to-build-first-zero-emission-gas-power-plants

Bolinger, M., Seel, J., Warner, C., & Robson, D. (2021). *Utility-Scale Solar, 2021 Edition*. Lawrence Berkeley National Laboratory. Retrieved August 20, 2021, from https://emp.lbl.gov/sites/default/files/2020_utility-scale_solar_data_update.pdf

Breakthrough Energy, IHS Market, Energy Futures Initiative. (2019). *Advancing the Landscape of Clean Energy Innovation*. Breakthrough Energy. Retrieved August 20, 2021, from https://www.breakthroughenergy.org/reports/advancing-the-landscape/

CATF. (2021, November 15). *Infrastructure Investment and Jobs Act: A Downpayment for Fullfilling Federal Promises on Climate Action*. Retrieved from Clean Air Task Force: https://www.catf.us/resource/infrastructure-investment-jobs-act-a-down-payment-on-fulfilling-federal-promises-for-climate-action/

Clemmer, S., Richarson, J., Sattler, S., & Lochbaum, D. (2018). *The Nuclear Power Dilemma: Declining Profits, Plant Closures, and the Threat of Rising Carbon Emissions*. Union of Concerned Scientists. Union of Concerned Scientists. Retrieved August 20, 2021, from https://www.ucsusa.org/resources/nuclear-power-dilemma

Edison Electric Institute. (2020). *2019 Financial Review – Annual Report of the U.S. Investor-Owned Electric Utility Industry*. EEI. Retrieved August 20, 2021, from https://www.eei.org/issuesandpolicy/Finance%20and%20Tax/Financial_Review/FinancialReview_2019.pdf

Gates, B. (2021). *How to Avoid a Climate Disaster, The Solutions We Have and the Breakthroughs We Need*. New York: Alfred A Knopf.

Goff, M. (2017, July 13). *How Natural Gas and Wind Decarbonize the Grid*. Retrieved from The Breakthrough Institute: https://thebreakthrough.org/issues/energy/how-natural-gas-and-wind-decarbonize-the-grid

Gold, R. (2019). *Superpower: One Man's Quest to Transform American Energy*. New York: Simon & Schuster.

IPCC, M.-D. V. (2021). *IPCC, 2021: Summary for Policymakers. In Climate Change 2021: The Physical Science Basis. Contributions of Working Group I to the Sixth Assessment Report of the Intergovernmental Panel on Climate Change*. Cambridge University Press.

Larson, E., Greig, C., Jenkins, J., Mayfield, E., Pascale, A., Zhang, C., . . . Swan, A. (2020). *et-Zero America: Potential Pathways, Infrastructure, and Impacts, interim report*. Princeton: Princeton University. Retrieved August 20, 2021, from https://environmenthalfcentury.princeton.edu/sites/g/files/toruqf331/files/2020-12/Princeton_NZA_Interim_Report_15_Dec_2020_FINAL.pdf

Lazard. (2020, October 2020). *Levelized Cost of Energy, Levelized Cost of Storage, and Levelized Cost of Hydrogen*. Retrieved August 20, 2021, from Lazard.com: https://www.lazard.com/perspective/lcoe2020

Lott, M., & Phillips, B. (December 2021). *Advancing Corporate Procurement of Zero-Carbon Electricity in the United States: Moving from RE100 to ZC100*. New York: Columbia University CGEP.

National Academies of Sciences, Engineering, and Medicine. (2021). *The Future of Electric Power in the United States*. Washington, DC: The National Academies Press. Retrieved August 20, 2021, from https://www.nap.edu/catalog/25968/the-future-of-electric-power-in-the-united-states

Nuclear Innovation Alliance. (2021, August 20). *Industry Innovation Leadership Council*. Retrieved from Nuclear Innovation Alliance: https://nuclearinnovationalliance.org/industry-innovation-leadership-council

Phillips, B., & Reilly, J. (2019, February). Designing Successful Greenhouse Gas Emission Reduction Policies: A Primer for Policymakers – The Perfect or the Good? *MIT Joint Program Report Series, 335*. Retrieved from https://globalchange.mit.edu/publication/17200

Place, A. (2020). *State and Utility Decarbonization Commitments*. Retrieved June 17, 2021, from https://www.catf.us/2020/10/state-and-regional-decarbonization-commitments/

RE100. (2020). *RE 100 Annual Progress and Insights Report*. Retrieved August 20, 2021, from file:///C:/Users/BPHILL~1/AppData/Local/Temp/RE100%20Annual%20Report%202020.pdf

RE100. (2021). *RE100 Members' Electricity and Renewable Electricity Consumption by Market*. Retrieved August 20, 2021, from https://www.there100.org/about-us

RGGI. (2021, December 29). Retrieved from The Regional Greenhouse Gas Initiative: https://www.rggi.org/

Sustainable Development Solutions Network. (2020). *Zero Carbon Action Plan*. New York: Sustainable Development Solutions Network (SDSN). Retrieved August 20, 2021, from https://irp-cdn.multiscreensite.com/6f2c9f57/files/uploaded/zero-carbon-action-plan%20%281%29.pdf

TCI. (2021, December 29). Retrieved from Transportation and Climate Initiative: https://www.transportationandclimate.org/

The NorthBridge Group. (2021, April 20). *Review and Assessment of Literature on Deep Decarbonization in the United States: Importance of System Scale and Technological Diversity*. Retrieved from

Clean Air Task Force: https://www.catf.us/wp-content/uploads/2021/06/NorthBridge_Deep_Decarbonization_Literature_Review.pdf
The White House. (2021a, January 27). *Executive Order on Tackling the Climate Crisis at Home and Abroad*. Retrieved August 20, 2021, from The White House: https://www.whitehouse.gov/briefing-room/presidential-actions/2021/01/27/executive-order-on-tackling-the-climate-crisis-at-home-and-abroad/
The White House. (2021b, April 22). *FACT SHEET: President Biden Sets 2030 Greenhouse Gas Pollution Reduction Target Aimed at Creating Good-Paying Union Jobs and Securing U.S. Leadership on Clean Energy Technologies*. Retrieved August 20, 2021, from The White House: https://www.whitehouse.gov/briefing-room/statements-releases/2021/04/22/fact-sheet-president-biden-sets-2030-greenhouse-gas-pollution-reduction-target-aimed-at-creating-good-paying-union-jobs-and-securing-u-s-leadership-on-clean-energy-technologies/
US Energy Information Administration. (2019, March 21). *Despite Closures, U.S. Nuclear Electricity Generation in 2018 Surpassed its Previous Peak*. Retrieved August 20, 2021, from US Energy Information Administration: https://www.eia.gov/todayinenergy/detail.php?id=38792
US Energy Information Administration. (2020). *U.S. Energy-related Carbon Dioxide Emissions, 2019*. Retrieved August 20, 2021, from US Energy Information Administration: https://www.eia.gov/environment/emissions/carbon/pdf/2019_co2analysis.pdf
US Energy Information Administration. (2021a, August 20). *U.S. Natural Gas Citygate Price*. Retrieved from US Energy Information Administration: https://www.eia.gov/dnav/ng/hist/n3050us3A.htm
US Energy Information Administration. (2021b). *Short-term Energy Outlook*. US Department of Energy. Retrieved August 20, 2021, from https://www.eia.gov/outlooks/steo/archives/jun21.pdf
US Energy Information Administration. (2021c, August 20). *Annual Energy Outlook – 2020 Data from Interactive Table*. Retrieved from US Energy Information Administration: https://www.eia.gov/outlooks/aeo/data/browser/
US Energy Information Administration. (2021d). *Battery Storage in the United States: An Update on Market Trends*. Washington, DC: US Department of Energy.
Williams, J. H., Jones, R. A., Haley, B., Kwok, G., Hargreaves, J., Farbes, J., & Torn, M. S. (2021, January 14). Carbon Neutral Pathways for the United States. *AGU Advances, 2*(1). Retrieved August 20, 2021, from https://agupubs.onlinelibrary.wiley.com/doi/10.1029/2020AV000284
Wiser, R., Bolinger, M., Hoen, B., Millstein, D., Rand, J., Barbose, G., . . . Paulos, B. (2021). *Land-based Wind Market Report: 2021 Edition*. Lawrence Berkeley National Laboratory. Retrieved August 20, 2021, from https://emp.lbl.gov/sites/default/files/2020_wind_energy_technology_data_update.pdf

7. Climate change and the insurance industry – risks and opportunities for transitioning to a resilient low carbon economy

Maryam Golnaraghi

INTRODUCTION

As the world grapples with a global pandemic and its aftermath, according to the World Economic Forum, "failure to address climate change" and "extreme weather events" remain among the top most likely and most impactful risks facing humanity (World Economic Forum 2016, 2017, 2018, 2019, 2020a, 2020b, 2020c, 2021). Despite decades of international policy negotiations facilitated by the United Nations' Framework Convention on Climate Change (UNFCCC) (United Nations 2015, 2021a), only in the last few years has the focus of the climate change debate shifted from being mainly a scientific, environmental and corporate social responsibility issue to one of the core drivers of financial instability, socio-economic development, trade and labor. There is now a widespread global recognition of climate change science and the associated socio-economic impacts as set out by the United Nations Intergovernmental Panel on Climate Change (IPCC 2018, 2021). Six years after the Paris Agreement (United Nations 2015), during the 26th Conference of Parties meeting (COP26) in Glasgow, nearly 200 participating governments publicly agreed that holding warming to 1.5°C below the preindustrial average temperature, "rather than" well below 2°C should be the ceiling of global targets (UNFCCC 2021a). However, in reality the development of national plans for an orderly transition to a low carbon future remains stubbornly slow. To reach the climate change goals, and pivot away from carbon-intensive sectors, dramatic changes in business models and everyday life are needed to impact the core and essential sectors of the world economy. Transitioning to a lower-carbon economy would entail extensive public policy, legal, technology, market and consumer behavior changes over time.

As of February 2022, progress has been made in a number of areas by the financial sector, financial and insurance regulators and supervisory bodies, international rating agencies, platforms working with CEOs of corporations in different economic sectors and by a variety of stakeholders using climate litigation to mobilize climate action. The Geneva Association (2021a, 2021b, 2021c) provides a review of such developments and their implications for the insurance industry.[1] For example, the development of sustainable finance frameworks aim to mobilize mainstream finance to provide long-term funding for the decarbonization transition (The Geneva Association 2021a). In parallel, international rating agencies, such as Moody's and Standard & Poor's, are investing in building their in-house capacities in climate risk analytics and increasingly considering climate change risk in their company, municipal and sovereign credit rating practices (Moody's 2018, 2019; Standard & Poor's 2018, 2020). Increasingly, climate litigation has also become a powerful instrument for those seeking to recover perceived climate change-related damages, influence climate policy or change corporate and government behavior (strategic litigation). These developments have profound implications for insurers, which we will further explore in this chapter.

Risk is the central issue to the insurance industry's purpose. For the past 30 years, "property & casualty" (hereafter "P&C") re/insurers have provided leadership in natural catastrophe (hereafter "NatCat") risk modelling and pricing; conducting research and promoting risk reduction and preventive measures (The Geneva Association 2018a, 2018b). Furthermore, by understanding the risks, primary insurers and reinsurance companies (hereafter referred to as "re/insurers") have fostered socio-economic resilience to natural catastrophe risk, amplified by climate change. By offering innovative risk transfer solutions, re/insurers enable the entrepreneurial pathways from start-up to commercialization of the clean and green technologies of the future and incentivize reduction in greenhouse gas (GHG) emissions (e.g., green building insurance). In order to better tackle the challenges of shifting the economy to a low-carbon model, re/insurers also participate in various alliances and partnerships to promote science-based methodologies, share expertise and collaborate to ensure stronger impact on the real economy. As institutional investors, many P&C and life re/insurers are taking steps to integrate climate change in their investment strategies and make investment decisions that support climate mitigation (The Geneva Association 2018a). Re/insurers are also actively involved in initiatives for the development of sustainable finance frameworks that aim to mobilize mainstream finance to invest at scale in transitioning to a resilient low-carbon economy (The Geneva Association 2018a, 2021a).

This chapter is focused on implications (risks and opportunities) associated with climate change for life and P&C (also referred to as non-life) re/insurers, for both sides of the balance sheet (liabilities and assets). Specifically, the next section provides the terminology and approach, the third section highlights major developments related to public policy, financial, regulations, technology, markets, litigation, and climate risk analytics that have implications for re/insurers. The fourth section provides a holistic decision-based analysis of impacts of climate change on re/insurers (life and P&C, for both sides of the balance sheet) and the fifth section offers examples of actions being taken by re/insurers to increase climate resilience and support the transitioning. In the sixth section, we highlight a number of industry-led initiatives and alliances that have been forged to date. The seventh section offers a summary and some suggestions for system-wide changes needed to help scale up and expedite re/insurers' contributions to the transition.

TERMINOLOGY AND APPROACH

This chapter is based on literature reviews, industry surveys and interviews, multi-stakeholder consultations and novel research conducted by the author at The Geneva Association.[2]

Definitions of Climate Risks and Opportunities

Box 7.1 provides a list of key definitions used in this report.

BOX 7.1 DEFINITIONS OF CLIMATE RISK USED IN THIS REPORT

Climate risks are categorized under physical risk, transition risk and litigation risk. Definitions used in this report are based on the Task Force on Climate-related Financial Disclosure (TCFD) and Burger et al. (2020) and Markell and Ruhl (2012).

Physical Risks – These can be divided into two categories:

- *Acute Risk* – Refers to those risks that are event-driven, including increased severity of extreme weather events, such as cyclones, hurricanes, or floods, resulting from climate change.
- *Chronic Risk* – Refers to longer-term shifts in climatic patterns such as sustained higher temperatures, sea level rise or chronic heat waves, resulting from climate change.

Transition Risks – Transitioning to a lower-carbon economy may entail extensive policy, legal, technology, and market changes to address mitigation and adaptation requirements related to climate change. Depending on the nature, speed, and focus of these changes, transition risks may pose varying levels of financial and reputational risk to organizations.

- Policy risks – policy actions around climate change continue to evolve. Their objectives generally fall into two categories: policy actions that attempt to constrain actions that contribute to the adverse effects of climate change or policy actions that seek to promote adaptation to climate change.
- Technology risks are related to technological improvements or innovations that support the transition to a lower-carbon and/or energy efficient economic system. These risks can have a significant impact on organizations' competitiveness, their production and distribution costs, and ultimately the demand for their products and services from end users. To the extent that new technology displaces old systems and disrupts some parts of the existing economic system, winners and losers will emerge. The timing of technology development and deployment, however, is a key uncertainty in assessing technology risk.
- Market risks – while the ways in which markets could be affected by climate change are varied and complex, one of the major ways is through shifts in supply and demand for certain commodities, products, and services as climate-related risks and opportunities are increasingly taken into account.
- Reputation risks – climate change has been identified as a potential source of reputational risk tied to changing customer or community perceptions of an organization's contribution to or detraction from the transition to a lower-carbon economy.
- Litigation risk – climate litigation risk is defined by the TCFD as a "liability risk" under transition risk, yet the implications could go far beyond. Recent years have seen an increase in climate-related litigation claims being brought before the courts by property owners, municipalities, states, insurers, shareholders, and public interest organizations for reasons such as failure of organizations to mitigate impacts of climate change, failure to adapt to climate change, and the insufficiency of disclosure around material financial risks.

148 *Handbook of business and climate change*

> **Litigation risk** – For our purposes, to cover the wider scope of cases, we may refer to the definition from Burger et al. (2020) and Markell and Ruhl (2012):
>
>> Cases brought before administrative, judicial and other investigatory bodies, financial supervisory authorities and ombudsman schemes or in domestic or international courts and organizations, that raise issues of law or facts regarding the science of climate change and climate change mitigation and adaptation efforts. This broader definition was used by The Geneva Association (2021c), which is one of the sources for this chapter.
>
> *Source:* TCFD (2017) and The Geneva Association (2021c).

About the Insurance Industry Value-chain

This section provides an overview of the insurance industry value chain and the insurance business model (Figure 7.1) (The Geneva Association 2018a).

The insurance industry's value chain includes:

- *Policyholders*: Buyers of insurance, e.g. oneself, a car owner, all drivers of a certain vehicle, all people working for a company, a company itself, municipalities and states, etc.
- *(Primary) Insurers*: They enable individuals and collectives to bear risks. For example, most drivers cannot afford the costs associated with casualties of a car accident.
- *Reinsurers*: Reinsurance companies act as insurers for insurance companies. Their important role is highlighted in more detail below.
- *Brokers/agents*: Are intermediaries or "sales agents" between policyholders and insurers to enable back-to-back business, as well as between primary insurers and reinsurers. Maintaining a distribution network and consulting services, brokers offer a critical service as intermediaries in enabling the risk transfer transaction but do not bear the risk.
- *Financial market*: A well-functioning financial market is needed for the production of insurance products. On the one hand, earned premiums need to be reinvested. On the other hand, insurance companies need to raise additional risk capital.

From an underwriting point of view, there are three basic ways of classifying insurance (Skipper, 1998):

- *Life versus non-life (or P&C) insurance*. Life insurance pays benefits on a person's death, living a certain length of time, sustaining disability or injury. Non-life (P&C) insurance generally covers property losses, liability losses and, in some countries, workers' compensation and health insurance payments.
- *Social versus private insurance*. Social insurance is government-administered and emphasizes social equity and income redistribution whereas private insurance is based on individual actuarial equity, with premiums reflecting individual risk characteristics embedded in a portfolio to benefit from diversification.
- *Commercial versus personal insurance*. Commercial insurance is purchased by businesses or other organizations to insure large business-related risks. Personal insurance is purchased by individuals and covers mass risks.

Climate change and the insurance industry 149

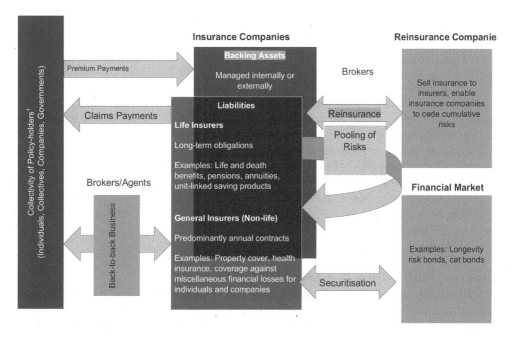

Note: [1] Policyholders: Buyers of insurance, e.g. oneself, a car owner, all drivers of a certain vehicle, all people working for a company, a company itself, municipalities and states, etc.
Source: The Geneva Association (2018a).

Figure 7.1 The insurance industry value chain

Liability side: transferring and carrying of risk
Transferring and carrying risk is at the heart of the insurance business. Insurers assess, price, assume and transfer risk on behalf of their policy-holders. As risk underwriters, insurance companies offer protection to people, businesses, and governments in return for a premium.

The insurance policy is a mutual agreement whereby the insured transfers the risks of an uncertain loss to the insurer by paying upfront a certain fixed amount. Subsequently, in the occurrence of a covered event, the insurance company indemnifies the policyholder (insured). It needs to be noted that the actual insurance product is not the payment in the event of a covered loss. It is rather the guarantee (promise to pay) that losses will be indemnified if the policyholder suffers a loss. The guarantees of the insurance mechanism rely on three methods, including:

i. *Pooling of risks*. By pooling similar risks of different individuals, the uncertain magnitude of the losses becomes controllable. Residual annual fluctuations are offset over time. This is the fundamental role of insurance: organizing the diversification of risks.[3] In this sense, insurance companies achieve the management of their liabilities accordingly. By holding a capital buffer, insurance companies may still endure a period of adverse loss experience.
ii. *Retrocession*. Cumulative and peak risks can be capped by ceding them to a certain portion to reinsurance companies. Correspondingly, reinsurance companies act as insur-

ers for insurance companies. Huge losses are spread across many parties and become bearable.
iii. *Securitization.* The coverage of certain risks can also be financed by placing them in the financial market.

To fulfil promises made to policyholders in short or long-term horizons, insurers need to carefully match their liabilities (or future claims payment) with financial assets (investment portfolio) that closely match the duration of their liabilities, using asset-liability management (ALM) modelling (The Geneva Association 2016).

The claims settlement and general risk management advice are among critical services that insurers offer to their policyholders. Early warning tools, satellite and drone technologies are proving useful to insurance and reinsurance companies to design contingency plans and more efficient claims management processes. On the other hand, insurers offer support to their policyholders by way of raising risk awareness, offering advice on how to reduce or prevent risks (for example retrofits to reduce flood risk of their homes or building back better after a flood) (The Geneva Association 2020a, 2020b, 2020c, 2020d, 2020e, 2020f).

Asset side: liability-driven investment strategy
As institutional investors, managing the assets of insurance companies (or an investment approach) is liability-driven, constrained by regulations and driven by a number of internal and external factors. Insurers invest conservatively. Insurance companies need to ensure that they remain solvent and can make their pay-outs to the policyholders with the highest probability at any time. Insurers have a fiduciary duty to enhance the value of their "policyholder" assets. These fiduciary duties also pose constraints on the industry's investment strategies. On the other hand, insurance regulators impose risk-based capital charges on investments to ensure adequate capital levels to cover insurers' liabilities; the riskier the investment, the higher the capital charge. These vary by country and region. It is important to note that different lines of business are exposed to different risks.

These dictate how financial risks associated with assets and liabilities are managed differently by life and non-life insurers (Asset Liability Management, referred to as ALM). Specifically,

i. *Life insurers are typically "buy-and-hold" investors.* They aim to generate predictable and stable income to match cash flows of long-dated and generally predictable liabilities. Life insurance contract duration can range from ten years to several decades, involving pay out patterns of 20 to 30 years. Life insurers are deeply concerned about the asset-liability mismatch, with interest rate risk being a key issue.
ii. *Non-life insurers are geared towards more liquid investments with shorter time horizons, typically one to three years in duration.* However, in some instances (e.g. asbestos-related) claims are paid out many years later, exposing them to interest rate risk.

MAJOR DEVELOPMENTS RELATED TO CLIMATE CHANGE WITH IMPLICATIONS FOR RE/INSURERS

Since the adoption of the Paris Agreement (UNFCCC 2015), actions within the public and private sectors, for transitioning to a resilient low-carbon economy are slowly gaining momentum. A number of developments with direct and indirect implications for the insurance industry are highlighted in this section.

Public Policy: National Targets and Policy-related Actions on Mitigation and Adaptation

Six years after the Paris Agreement, while governments have submitted their Nationally Determined Contributions (NDCs) to mitigate the impacts of climate change (UNFCCC 2021b), the development of public policies for an orderly transition to a low carbon future has been slow. The IPCC stated with high confidence that, "[a] mix of adaptation and mitigation options to limit global warming to below 1.5°C implemented in a participatory and integrated manner, can enable rapid, systemic transitions" (Intergovernmental Panel on Climate Change, 2019).

On the adaptation side, with more evidence of rising socio-economic costs associated with physical risks of climate, there is a need for a paradigm shift in governments' approaches, from "inaction" or "post-disaster reaction" towards a more coordinated and integrated approach to risk management framework, to build/strengthen socio-economic resilience to extreme weather events (The Geneva Association 2016). Building resilience and addressing physical risks should include investments in risk analysis and risk communication, emergency preparedness empowered by improved early warning systems, ex-ante investments in reducing existing risks and preventing new risks, leveraging risk financing and risk transfer and efforts to build back better after events (The Geneva Association 2017, 2020a, 2020b, 2020c, 2020d, 2020e, 2020f and 2017; The World Bank 2021). Traditional governmental post-disaster financial assistance is proving ineffective and insufficient by dis-incentivizing people, businesses and local governments from taking proactive action to managing their risks (retrofitting homes and infrastructure, protection of private assets) (The Geneva Association 2016, 2020a, 2020b, 2020c, 2020d, 2020e, 2020f).

Increasingly, governments are recognizing the role and benefits of collaborating with the insurance industry and leveraging market-based insurance in carrying and transferring risk. There is increasing evidence that countries with widespread market-based insurance coverage recover faster from the financial impacts of extreme events; it is the uninsured part of losses that drives macroeconomic costs (The Geneva Association 2016; Von Peter et al. 2012). Yet there is a large and, in some places, widening protection gap, indicating that the benefits of risk transfer measures are not being harnessed to their full potential (The Geneva Association 2019a, 2019b). Various collaborations and public–private partnerships are coming into focus such as the Insurance Development Forum,[4] InsuResilience Global Partnership,[5] The Microinsurance Network,[6] The World Bank Group's Disaster Risk Financing and Insurance Program[7] and efforts by the Atlantic Council's The Adrienne Arsht-Rockefeller Foundation Resilience Center to address extreme heat with innovative insurance solutions,[8] to expand market-based insurance to address the widening protection gap, particularly in the most vulnerable nations (The Geneva Association and Insurance Development Forum 2017, and The Geneva Association 2017).

Launch of the Financial Stability Board's Task Force on Climate-related Financial Disclosure (FSB-TCFD)

A pivotal point was the launch of the Financial Stability Board's (FSB) Task Force on Climate-related Financial Disclosure (TCFD) (TCFD 2016), which promotes the need for an orderly and well-planned transitioning to a low-carbon economy to ensure financial stability of the world economy. TCFD raised the need for decision-relevant, clear, consistent and comparable climate information for stakeholder groups. TCFD provided principles-based guidance on climate-related financial disclosures based on defining climate-related risks, opportunities and the need for forward-looking climate risk assessment and scenario analysis (TCFD 2017).[9] Over time, these efforts aim at standardizing decision-relevant disclosure and reporting of climate change risks and opportunities to empower institutional investors and the public sector with a risk- and opportunity-informed approach in funding the transitioning. The recommendations are increasingly being adopted by companies in various sectors (TCFD 2017, 2018, 2019, 2021; The Geneva Association 2021a). Beyond disclosure, implementation of TCFD by the insurance industry has been on the rise, leading to dialogue on climate change risk across the organization to raise risk awareness, and strengthen collaboration to leverage expertise across the company (The Geneva Association 2021a).

Emergence of Sustainable Finance Frameworks to Direct Mainstream Finance to Fund the Transition to a Low-carbon Economy

Sustainable finance initiatives around the world aim to mobilize mainstream finance towards environmentally sustainable investments through both the public and private sectors with the ultimate objective to limit global warming and related risks. These initiatives aim to remove some of the major hurdles investors are facing in investing at-scale in the transition to a low carbon economy. Examples of such initiatives include: EU action plan on Sustainable Finance and the European Commission Renewed Sustainable Finance Strategy,[10] Canadian Expert Panel on Sustainable Finance,[11] and Australia Sustainable Finance Initiative.[12]

Actions by Financial and Insurance Regulatory and Supervisory Bodies

There is growing action by financial services' regulators to understand the potential impact of climate change with implications for the insurance industry (The Geneva Association 2021b; IIF 2021). Much of the regulatory focus to date has been on transparency supported by public disclosure; however, regulatory interest in climate change risk assessments and scenario analysis is growing at the international, regional, national, and sub-national levels. Actions by a wide spectrum of regulators are relevant to re/insurers, given the highly regulated nature of the insurance business, through their investment and underwriting activities. Regulators are undertaking a variety of different approaches to engage with the financial sector and insurance companies, such as setting up industry-led forums and processes, public consultations, conducting surveys and calling for voluntary disclosure of climate risks. The Geneva Association

(2021b) provides examples of actions by various regulatory bodies and highlights a number of challenges associated with the status quo.

These efforts, among others, will have implications on future regulatory requirements for climate risk assessment, risk management and disclosure, and reporting for the industry. Platforms such as the Financial Stability Board (FSB), Network for Greening the Financial System (NGFS), the International Association of Insurance Supervisors (IAIS) and the Sustainable Insurance Forum (SIF) offer mechanisms to enable collaboration among regulatory bodies and other stakeholders, including the insurance industry and other financial institutions. Examples include the Climate Financial Risk Forum (CFRF),[13] established by the Bank of England – Prudential Regulation Authority (BoE-PRA) and Financial Conduct Authority (FCA) and the FSB-TCFD, which has been instrumental in engaging actors from different segments of the financial and insurance sectors as well as supervisory bodies and rating agencies, for development of the TCFD guidelines. The Geneva Association (2021b) provides a detailed analysis and the challenges with the status quo regulatory approaches to conducting climate risk assessment to date.

Actions by International Rating Agencies

In parallel, international rating agencies such as Moody's and Standard & Poor's are investing in building their in-house capacities in climate risk analytics and increasingly considering climate change risk and related mitigation and adaptation measures in their company, municipal and sovereign credit rating practices (Moody's 2018, 2019; Standard & Poor's 2018, 2020). For example, since 2019 Moody's has invested in three data companies, Vigeo Eiris, 427 and Risk Management Solutions; in 2016 Standards & Poor's bought Trucost and in 2022 The Climate Service, Inc.; and in 2020 Morningstar bought Sustainanalytics.[14]

Towards Forward-looking Climate Risk Assessment and Scenario Analysis

Following the launch of TCFD recommendations, the development of climate risk assessment methodologies and tools, such as scenario analysis has gained significant momentum. There is a wide range of initiatives by various stakeholder groups, for example re/insurers, financial institutions, regulatory and standard setting bodies, international organizations, commercial climate and Environmental Social and Governance (ESG) data providers, consulting firms and academia, to develop methodologies for analyzing physical and transition risks with a forward looking approach (UNEP-FI PSI 2020, 2021; UNEP-FI 2019; UNEP-FI and Oliver Wyman 2018; ClimateWise 2019a, 2019b, 2019c, 2019d). Re/insurers are initiating and/or engaging in various intra- and inter-sectoral pilot projects to develop new methodologies, publishing risk reports (CRO Forum 2018, 2020) and developing proposals for appropriate decision-relevant assessments and disclosure of climate change risks, including the TCFD. For P&C re/insurers, the deep knowledge in extreme weather risk modelling has been also instrumental to raising awareness on the asset side; for example, leveraging NatCat risk modelling and expertise has led to a better grasp of the potential impacts on real estate investments (Surminski et al. 2020). There are challenges to integrate climate change related risks into NatCat models as explained in Box 7.2.

> **BOX 7.2 MODELLING THE RISK OF EXTREME WEATHER EVENTS IN PROPERTY INSURANCE – STATUS QUO**
>
> The industry strives to evolve the forefront of NatCat modelling to derive decision-useful quantitative information to inform today's high-level portfolio strategy decisions, including stronger attribution to individual risk trends like climate change. Today's NatCat models are designed to provide decision-useful output for the blend of hazard, exposure, vulnerability, and insurance-cover specifics that are present today. The latter three factors have been the dominant drivers for key natural perils exposure in the past five decades. While highly useful for today's and next year's portfolio management, catastrophe models are limited in providing decision-useful quantitative information over a longer term, impeding progress toward a forward-looking climate change and exposure/vulnerability landscape. Understanding and anticipating the dominant drivers next to climate change will be paramount for conditioning these models for more distant futures.
>
> *Source:* The Geneva Association (2018b, 2021a).

However, overall initiatives remain fragmented and considerable work remains to produce meaningful and decision-useful information. Despite an active and fast growing commercial data provider market (Keenan 2019), this area will evolve and further develop, with an increasing number of solutions becoming available, requiring careful due diligence by users (e.g., insurers, and financial organizations about the viability of these solutions). Furthermore, the quickly evolving nature of climate science as well as other factors that will influence transition efforts need to be carefully considered. Achieving consensus and converging on best practices will take time.

Against this backdrop, in 2020, at the request of its Board of Directors, The Geneva Association established an industry-led 'Task Force on Climate Risk Assessment for the Insurance Industry', involving global P&C and life re/insurance companies.[15][16] The GA Task Force's aim is to advance and accelerate the development of holistic methodologies and tools for conducting meaningful and decision-relevant climate risk assessment and scenario analysis. These efforts aim to shape future innovations and support re/insurers, regulators, and other stakeholders in this space (for more information see the sixth section).

In November 2021, during the United Nations COP26 meetings in Glasgow, The International Financial Reporting Standards Foundation (IFRS) announced that it will develop and provide the global financial markets with high-quality disclosure standards to meet the investors' information needs on climate and other sustainability issues.[17]

New Climate Technologies for Sectoral Decarbonization

To reach the climate change goals and pivot away from carbon-intensive sectors, dramatic changes in business models and everyday life are needed to impact the core and essential sectors of the world economy.

Limiting the global average temperature increase to 1.5°C from pre-industrial levels over the next few decades requires a well-planned whole-of-economy approach – an unprecedented transformation across all sectors of economy and society, affecting the way we produce and consume the goods and services that are essential to our lives. Economies, industries, and businesses are increasingly looking to ways to achieve net-zero emission in their systems, value- and supply-chains.

New technologies, new industries and new infrastructure systems will need to be developed and implemented for ambitious net-zero climate goals to be achieved. The United Nations Framework Convention on Climate Change (UNFCCC) has called for the need to step up, expedite and scale up innovation and technological developments from R&D to commercialization and implementation, to enable the transitioning of economic sectors, such as energy, transportation, heavy industries,[18] buildings/construction, food and agriculture (Golnaraghi 2022).

Apart from scaling up technological developments from R&D to commercialization, new technologies and processes, particularly when implemented at scale, will also have a significant impact, affecting consumers' demand, company competitiveness and related business models. Such developments (e.g. off-shore wind farms, large-scale solar farms, batteries and energy storage systems, hydrogen-based systems, carbon capture and storage technologies) may be accompanied by a variety of technical, operational, safety, environmental, physical climate change risks concerns, among others, which need to be assessed, priced, shared and allocated for large-scale implementation and also to raise private capital. Re/insurers as risk managers and investors play a critical role de-risking and investing in new technologies for decarbonization for commercialization and large-scale implementation (Golnaraghi 2022 and The Geneva Association 2021d).

Latest trends with new market developments indicate a rising number of platforms such as the Mission Possible Partnership and the First Movers' Coalition of The World Economic Forum (WEF). The World Business Council for Sustainable Development (WBCSD) is convening corporations to assess together sectoral net zero transition plans and technological needs for transitioning. Corporations and governments (national to local) are starting to invest in larger scale projects.[19] Governments' new infrastructure spending, some as part of their COVID-19 economic recovery plans, will be injecting significant funds in this space for development of resilient green smart infrastructure systems (Golnaraghi 2022).[20]

Climate Change Litigation

In the absence of clear government and corporate climate change policies for transitioning to a resilient low-carbon economy, globally, courts have increasingly become an instrument for some of those who suffer loss or expect to suffer loss as a result of climate change-related impacts to pursue judicial remedies to recover damages or fund abatement efforts, while others are using litigation as a tool to leverage more ambitious climate policy and actions or to oppose them (The Geneva Association, 2021c).

Between 1986 and 2020, of the 1,727 climate change litigation cases documented, 1,308 were in the US and 419 in other jurisdictions, regional and international courts (Figure 7.2). More than half of cases have been brought since 2015 after The Paris Agreement, and, as of April 2021, the majority of cases have been brought against governments but there is clear evidence that the number of lawsuits against corporate entities – particularly carbon majors – is on the rise (Box 7.3).

BOX 7.3 THREE DISTINCT WAVES OF CLIMATE CHANGE LITIGATION SINCE 1986 TO PRESENT

There are three distinct waves:

- First wave (pre-2007) – cases predominantly in the US and Australia, primarily against national governments to raise environmental standards.
- Second wave (2007–2015) – a surge in cases with expansion to European countries and courts, primarily against governments to accelerate climate policy and tortious cases against corporations for their causal contribution to climate change.
- Third wave (post-2015) – expansion to other jurisdictions, increase in volume and pace, and new types of claims; noticeable increase against major emitters and claims against directors, trustees, and fiduciaries for failure to consider emissions, most prominent cases involve novel causes of action and the application of established legal duties to force more ambitious climate policies.

Source: The Geneva Association (2021c).

A recent report by The Geneva Association (2021c) offers a forward-looking view into the evolving landscape of climate change litigation and offers a number of climate litigation cases to illustrate the nature of this emerging risk. Specifically, four important issues are highlighted here:

1. Climate litigation risk is being amplified by a number of key factors, including: increased physical and transition climate risk; increasing public awareness of the climate crisis; changing requirements for states and corporations triggered by the proliferation of national and international agreements and commitments on climate change; stronger climate commitments from governments, corporates and investors; accessibility to funding for climate litigation; evolving legal duties; and progress with climate change attribution science.
2. Since the adoption of the Paris Agreement, climate litigation against government and corporations has gained pace, increased in volume and expanded in scope and geographical coverage. Importantly, more than half of the total recorded cases dating back to 1986 have been brought since 2015.
3. Climate change litigation has become a global phenomenon, with cross-pollination of ideas, strategies and support across jurisdictions, linked to accessible global data platforms, plaintiffs using cases from different jurisdictions in novel ways, the rising number of precedent cases and a growing network within the legal community.
4. Climate litigation cases take many forms and they are brought by a variety of plaintiffs against corporations and government: litigation cases against companies may be linked to: (i) claims against companies for causal contribution to climate change; (ii) claims against companies, fund managers and/or their fiduciaries for miscommunication or mismanagement of climate risk; (iii) claims against companies, challenging their role in society (duty of care/human rights cases); (iv) cases brought by conservationists against new projects and technologies (wind and solar farm energy producers) alleging, for instance, threats to wildlife and biodiversity. Cases against governments, with implications for corporates may include: (i) claims against governments (and utilities) for

failure to adapt to climate risks; (ii) litigation to challenge approval of projects, policies, and technologies; (iii) litigation to accelerate climate policy which can change the competitive landscape for companies (The Geneva Association 2021c; Setzer and Higham 2021).

This emerging area may have impacts on liabilities, investments and board of directors, officers (Clyde & Co 2018, 2021) and required ongoing monitoring by governments, corporations, their investors and insurers.

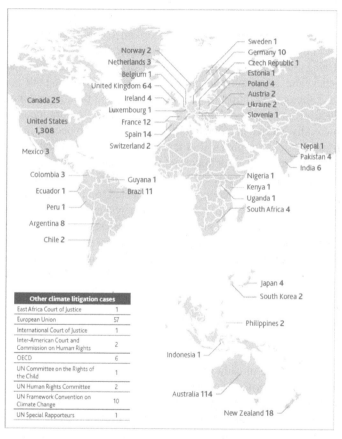

Source: The Geneva Association (2021c).

Figure 7.2   Geographical distribution of 1,727 cases worldwide between 1986 and 2020 (of which 419 cases are outside the US)

ASSESSING CLIMATE CHANGE RISKS ON RE/INSURERS ON BOTH SIDES OF THE BALANCE SHEET

Climate change has been considered among high risk categories by the CRO Forum (2020).[21] Climate change presents varying levels of physical and transition risk to both sides of the balance sheet (liabilities and assets) for non-life and life re/insurers (The Geneva Association, 2021a). The time horizon over which the risk manifests itself is a key factor and varies across the different lines of business and investments, which adds to the complexity of assessing climate risk impacts. The Geneva Association has offered a holistic decision-based overview of how re/insurers are impacted by climate change on both sides of the balance sheet, summarized in Tables 7.1–7.4.[22]

In the first report of a three part series, The Geneva Association (2021a) offers an overview of different types of risks that can be considered under physical risk (i.e. chronic and acute risks) and transition risk (i.e. policy risk, litigation or legal risk, market risk and technological risk) and their relation to re/insurers' risk landscape over two time horizons: (1) the short- to medium-term, or 'business-planning', time horizon (suggested 2020–2030) and (2) the long-term, or 'strategic-planning', time horizon (suggested 2030–2050). This is presented in Table 7.1. Using two time horizons, the report offers the changing characteristics (e.g. severity, frequency and geographic locations) and implications of both acute and chronic physical risks and factors that impact them. Importantly, on the longer timeframes, physical climate risks, biodiversity loss and large-scale degradation of natural ecosystems are deeply intertwined, potentially leading to planetary tipping points which need to be considered. While there has been much focus on physical climate risks and its implications on the insurance sector, as reflected in Table 7.1, it is important to note that over the coming decades, policy, regulatory, technological advancement, market conditions and other aspects of societal transitioning towards low-carbon economies are expected to affect the level of transition and physical climate change risks. When considering the potential effects of climate change on longer-duration lines of business (e.g. mortality protection, retirement savings), beyond understanding the direct impact of climate change on physical risks assumed through underwriting, it is important to establish assumptions for societal progress in combating climate change and how climate change may impact other key drivers (e.g. economic growth, financial market performance).

Liability Side

Non-life (P&C) re/insurers
Non-life (P&C) insurers are already experiencing an evolution in risk exposures as a result of gradual climate change, with examples of such impacts highlighted in Table 7.2 (Geneva Association 2021a). Through extensive investments in NatCat-centric research, the industry has developed a robust natural catastrophe risk management system to understand the present amount of embedded climate change. With the majority of affected property insurance cover offered on an annual basis, P&C re/insurers have the opportunity to monitor gradual changes to the climate risk landscape and consider adjustments to pricing and/or product offerings. While this allows insurers to adjust their prices and offerings on a short term basis, they have to watch for insurability and work on the viability of their business model over longer time horizons (2030–2050).

Importantly, exposure and vulnerability factors, such as population growth, urbanization, an increasing concentration of people and assets in high-risk zones (e.g. along coastlines and flood zones), development choices and supply-chain disruptions (caused by extreme events), distort and could mask any embedded climate change signal (The Geneva Association 2018b, 2020a, 2020b, 2020c, 2020d, 2020f).

Table 7.1 An overview of physical and transition risks and relevant time horizons for re/insurers' decision-making

	Physical risk		Transition risk			
	\multicolumn{2}{l	}{• Can be driven by events or longer-term shifts in climate patterns. • Currently a gradual change, at small annual increments compounding over years, with a low probability of a sudden change.}	\multicolumn{4}{l	}{• May entail policy, legal, technology and market changes to address mitigation and adaptation requirements related to climate change.}
	Chronic risk	Acute risk	Policy risk	Litigation or legal risk*	Market risk	Technological risk
Overview	• Progressive shifts in climate patterns, such as sea-level rise and droughts; cascading effects on food production, water security, migration	• Changes in the nature of extreme weather events, such as wildfires, flooding, storms.	• Includes policy efforts to limit emissions or promote climate-friendly adaptation.	• May arise from insureds' causal contributions to climate change or failure to mitigate the impacts of or adapt to climate change. • May challenge insureds' role in society (duty of care/human rights cases) or force a review of their projects or technologies.	• Includes the impact a changing climate may have on the supply and demand of goods and services.	• Includes the potential for new technology to disrupt or displace existing systems.

160 *Handbook of business and climate change*

	Physical risk		Transition risk	
The risk landscape over the business planning horizon: 2020–2030 (short term)	• The impacts are already present though the rate of annual change is slow, e.g. mean global temperatures are elevated by 1°C over pre-industrial levels, sea-level rise is adversely affecting low lying coasts today, etc.	• Attributing the role of climate change in current extreme events is a difficult, ongoing subject of scientific study. • For major perils, such as hurricanes, a response to anthropogenic climate change can be implied, but evidence of signal is strongly masked by natural climate variability and other man-made changes to the risk landscape. • For secondary perils such as wildfire and local flooding, attribution is already much clearer. • Due to slow gradual change, the climate state of 2030 will not differ significantly from today.	• Public perception supports meaningful near-term policy action; however, action may be inconsistent across the globe and benefits from action may take time to accrue but could have an acute impact on investment portfolios.	• Likely to increase due to increases in: the value of losses and damages from climate change; the scrutiny of action, or inaction, to address climate change; cases that can be used as precedent; substantial policy change in this timeframe.	• Transportation and energy generation are among the markets that will likely experience near-term impacts. • Valuation of assets in investment portfolios may become more volatile and/or experience pressure as carbon-intensive sectors become less productive and/or viable.

	Physical risk		Transition risk
The risk landscape over the strategic planning horizon: 2030–2050 (long term)	• Unmitigated, the negative impacts are expected to increase significantly, including: more and prolonged heat waves and droughts, inundation of coastal real estate, disrupted food production and water scarcity, disruption of ecosystems and loss of biodiversity, the spread of diseases and other health impacts, geopolitical consequences.	• The severity and frequency of perils will change, and in many areas of the world likely increase, e.g. sea-level rise will worsen storm surge risk and cyclone severity: increase in the number and extent of wildfires and local flooding.	• The impact will be highly dependent on the extent of action taken in the short term. • Actions taken in the long term would also take time to accrue but could have an acute impact on asset portfolios.	• The extent of policy action taken (or not) will likely guide how the risk emerges and evolves. • Knock-on effects of climate litigation against governments could shift the dial in certain jurisdictions for corporate clients, directly affecting the corporate strategies, business models and operational decision-making of entire sectors.	• The impact will depend on the timing of transition for various sectors and asset classes, which will be informed by all aspects of transition risk (i.e. policy, legal, market, technological). • In the absence of successful mitigation in the longer term, it is likely that geoengineering solutions will be pursued, with potential unintended consequences.

Note: Using TCFD definition for litigation or legal risks.
*For more information on climate change litigation see The Geneva Association (2021c).
Source: The Geneva Association (2021a).

162   *Handbook of business and climate change*

For P&C re/insurers, the impact of transition risk will be driven by less predictable external forces such as public policy action (policy risk), court rulings (litigation risk), consumer/societal pressures (market risk) and technological advances (technological risk) (Table 7.2). From a policy perspective, actions to reduce carbon emissions could change the scope of business re/insurers are willing or able to underwrite or the price at which they are willing to offer coverage (The Geneva Association 2016, 2018a).[23] While there is a clear trend towards policy action to reduce carbon emissions in a number of major economies, the implications from an insurance perspective will likely be more gradual. However, transition could be quick and therefore the potential risk of a 'cliff effect' cannot be completely discounted. Similarly, the evolving legal environment and the absence of court rulings that establish principles for climate-change related liability create a high degree of uncertainty for how liability insurance (insurer provides protection against liabilities that arise from lawsuits) may be impacted.[24] Similar to policy risk, the impact is expected to be gradual; however, effects are already starting to materialize in certain jurisdictions. The growing expectation that the private sector will take proactive actions to reduce its contributions to carbon emissions or to combat climate change risk (i.e. market risk) must also be accounted for.

While the shift away from carbon-intensive businesses may present some challenges, there are growing opportunities through technological innovations and commercialization in areas such as carbon capture and storage, renewable energy generation and electrification of transportation, among others, requiring many of the same insurance coverages as carbon-intensive sectors and serve as potential underwriting and investment opportunities. The interlinkages of transition and physical risk need to be considered. Broadly speaking and assuming that climate change mitigation actions are insufficient, climate change will introduce longer-term shifts in climate patterns and present chronic risks that impact the geographic locations and subsequently impact pricing and premiums at which P&C re/insurers can offer insurance protection. Unless actions are taken by governments, businesses, and people to reduce and/or prevent the risks, insurance coverage may become prohibitively expensive for the consumer (The Geneva Association 2020a, 2020b, 2020c, 2020d, 2020e, 2020f) and investments may be negatively impacted.

Life re/insurers
For life re/insurers, the climate change challenges vary considerably from that of P&C re/insurers given different time horizons and the nature of the risks underwritten (IFoA 2019a, 2019b). Table 7.3 highlights the implications of physical and transition risk for life re/insurers' decisions over different time horizons for the liability side (The Geneva Association 2021a). Four points need to be highlighted:

1. Given the longer-time horizon of the risks assumed by life re/insurers, the physical risk of climate change is generally not expected to have a material or discernible impact over the short term (2020–2030).
2. The expected implications of transition risk for life re/insurers in the short term is unlikely to be impacted significantly (e.g. the frequency of deaths is unlikely to increase).
3. Over the long term (2030–2050), the impacts of both physical and transition climate change risks are more uncertain for life re/insurers.
4. More research on the impacts of climate change on life exposures is needed – to explore whether climate change impacts are statistically significant in light of the long-term nature of life re/insurers business, noting that assessments should also account for the potential impacts of second-order 'knock-on' effects, such as declines in economic growth, population migration, geopolitical conflict and shifts towards low-carbon business models.

Climate change and the insurance industry 163

Table 7.2 Climate change risk and the decision landscape for P&C re/insurers – liability side

	Physical risk		Transition risk			
	Chronic risk	Acute risk	Policy risk	Litigation risk	Market risk	Technological risk
The risk landscape over the business planning horizon: 2020–2030 (short term)	• Climate change alone is not yet impacting the insurability of NatCat risk, except for certain wildfire zones. • Climate change is an additional modifier or accelerator of known and successfully-managed natural catastrophe risk. Thus, climate change is implicitly embedded in current pricing, risk management and claims experience, taking into consideration annual nature of insurance contracts. • Limited data makes it difficult to isolate what impact climate change will have versus other loss drivers, including natural variability, and increasing and more concentrated asset values in high-risk zones such as cities and coastal areas. • As the risk landscape is constantly evolving on multiple fronts, climate change included, a sustainable product offering needs to address today's risk versus an outdated assessment.		• The impact would depend on the timing and scope of the policy to curb carbon emissions. • Significant action could limit the viability of carbon-intensive industries, impacting associated Insurance lines of business.	• The evolving legal and regulatory environment and absence of court rulings that establish principles for climate change-related liability present a high degree of uncertainty, particularly for Directors and Officers (D&O) coverage. • Professionals' duties of care for some insureds could potentially lead to claims related to physical and transition risks, impacting professional indemnity/errors and omissions (E&O) coverage.	• Market forces could limit the viability of some industries, limiting associated insurance. • Exposure through reputational risk (by association in providing coverage to certain companies and industries).	• Progress in carbon removal and renewable energy presents opportunities; however, prototypical technology is often subject to volatility in results.

164 *Handbook of business and climate change*

	Physical risk		Transition risk			
	Chronic risk	Acute risk	Policy risk	Litigation risk	Market risk	Technological risk
The risk landscape over the strategic planning horizon: 2030–2050 (long term)	• Climate change effects will likely manifest in a worsening claims experience for weather-related perils, requiring the industry to adjust its risk assessment and pricing dynamically and in a forward-looking mode (e.g. what used to be a 1-in-200-years event might become a 1-in-100-years event).* • Insuring some risks/geographies will gradually become less affordable (e.g. home coverage in exposed coastal areas) unless adaptation measures reduce risk levels and prevent new risks. • The insurance business model will not be able to address risk with a certainty of loss, e.g. property on storm surge-exposed coastlines or close to rivers. Insurability will require adaptation measures so that insurance covers the random negative deviation from an expected outcome rather than the expected outcome.		• Long-term impacts will align with those in the short term. • Action will occur across all fronts – policy, legal, market and technological – but the potential impacts are difficult to assess given uncertainties regarding the timing, scope, etc. of change. • Considering the inherent uncertainty and interconnection with physical risks, multiple pathways must be contemplated.			

	Physical risk		Transition risk			
	Chronic risk	Acute risk	Policy risk	Litigation risk	Market risk	Technological risk
landscape over the strategic planning horizon: 2030–2050 (long term)	• Secondary effects will be added; for example, a change or loss in biodiversity affects ecosystems and their natural protection levels, or changed temperature regimes and access to water affect urban settlements, production and supply chains, potentially leading to forced migration. • While not the most likely outcome from a 2020 insights perspective, rapid intensification of climate change and/or exceeding tipping points,** could have a dramatic impact on both chronic and acute risk, where risk adaptation and mitigation become cornerstones to maintain insurability.					

Note: * But not necessarily a worsening of profitability, where premium goes at pace with the cost of risk.
** For example this may result from a subsiding Gulf stream, linked to the changing ocean circulation caused by climate change.
Source: The Geneva Association (2021a).

Table 7.3   Climate change risk and decision landscape for life re/insurers – liability side

	Physical risk		Transition risk			
	Chronic risk	Acute risk	Policy risk	Litigation risk	Market risk	Technological risk
The risk landscape over the business-planning horizon: 2020–2030 (short term)	• Limited data makes it difficult to isolate what impact climate change will have versus other drivers. • While natural catastrophes may increase, the mortality impact has historically been small, owing to early warning and emergency preparedness measures. • Diversification of the type and location of risks assumed also serves to reduce exposure to extreme events.		• Transition risk is generally not expected to have an impact on underwriting but there is potential for assets within the investment portfolio to experience effects (see Table 7.4 for further details on implications for investments). • From a market risk perspective, re/insurers may be directly exposed through reputational risk (e.g. stigmatized by the market for supporting carbon-intensive industries).			
The risk landscape over the strategic-planning horizon: 2030–2050 (long term)	• Prolonged exposure to a warming environment could cause increased cardiovascular and respiratory illnesses, agricultural impacts that adversely affect diet and nutrition, increased spread of infectious disease and more. • While the impact on claims from policies already on the books may become more pronounced, it could take more time before it reaches a level of statistical significance relative to the overall size of an insurer's exposures. • The potential impacts of second-order effects, such as potential declines in economic growth, population migration, geopolitical conflict, etc. should be considered and could increase materiality.		• Long-term implications are like those in the short term. • The interaction between transition and physical risks, which will be highly dependent upon the pace and degree of action to address the various aspects of transition risk, will become more relevant. • Second order effects such as potential declines in economic growth, population migration and geopolitical conflict must also be considered.			

Source: The Geneva Association (2021a).

Asset Side (Both Life and P&C)

Through their investment portfolios, re/insurers are exposed to both physical and transition risks. Some implications are highlighted in Table 7.4 (The Geneva Association, 2021a). The increasing call for transparency alongside increasing regulation on sustainable finance are catalysts for increasing consideration of climate change risk and mitigation actions. When deciding on sectors and geographical allocation and investee selection, re/insurers are increasingly considering the resilience of their portfolio and mitigation actions such as divestment, best-in-class strategies and engagement. On the asset side, efforts have been made to analyze the implications of climate change, particularly related to the transition risks (e.g. risks of stranded assets and devaluation in case of not well planned transition scenarios). Efforts are underway to explore implications of physical climate risk and how this can be captured in the risk assessment approaches.

Table 7.4 Climate change risk and decision landscape for P&C and life re/insurers – asset side

	Physical risk		Transition risk			
	Chronic risk	Acute risk	Policy risk	Litigation risk	Market risk	Technological risk
The risk landscape over the business-planning time horizon: 2020–2030 (short term)	• Insurers are exposed directly through investments in real assets (e.g. buildings, infrastructure) and indirectly through investments in companies (e.g. equity, debt) which are exposed to those risks. • While assets may be increasingly impacted by more frequent or severe natural catastrophes the losses may be offset by insurance protection (e.g. property insurance on real estate).		• The degree to which transition risk impacts re/insurer investment portfolios will be driven by the speed and magnitude of developments across the different facets of this spectrum. • Significant actions or developments could lead to abrupt losses in investment value and drive more expeditious action on the part of re/insurers. • Litigation risk may be most prominent over the next 10 years as the world moves towards transitioning, particularly if the transition is not well planned. Litigation could impact the investment side through: (i) acceleration of the transition and undermining investees' strategies; (ii) financial risk of litigation against investees and related damages claims and costs of defense and (iii) impact on investment portfolios (e.g. devaluation of corporate and sovereign bonds linked to a litigation against corporations and governments).			
The risk landscape over the strategic-planning horizon: 2030–2050 (long term)	• As the effects of climate change accelerate, certain investments and sectors may become less attractive (stranded assets), e.g. real estate in coastal communities, stakes in fossil fuel companies if renewable fuels become viable on a mass scale. • Renewable energy and greening of other key sectors along with investment in resilient and green infrastructure are anticipated to grow significantly. • New technologies (clean, green and carbon capture and storage) will continue to emerge and bring new opportunities for investing but risks of new technologies need to be managed for large scale deployment.					

Source: The Geneva Association (2021a).

ACTIONS BY RE/INSURERS IN SUPPORTING A RESILIENT LOW CARBON TRANSITION

Response of Boards and C-level Executives to Climate Change

In this section we explore the extent to which climate change is on the agenda of the boards and the C-level suite in the insurance industry. This also depends on the size of the company, line of business, and the jurisdiction. The response of the boards and C-suite can generally be characterized in three ways (Geneva Association 2018a).[25] Specifically, companies consider climate change as:

1. *A strategic core business issue being mainstreamed across all aspects of their company*. Driven by Board and C-level considerations, for these companies climate change is a strategic issue with short, medium, long-term implications, impacting corporate policy,

strategy, enterprise risk management and related decisions on both sides of the balance sheet. These companies are undertaking institution-wide measures under the direct supervision of the Board. The company has established inter-departmental mechanisms and/or teams to explore and advance methodologies and tools to produce decision-useful forward-looking climate risk information and is actively assessing climate risks and opportunities and to provide advice to the C-level suite and the Board. These organizations have engaged proactively with the financial and insurance regulatory community to develop exercises for forward looking climate risk assessment, and support shaping future regulations in this area. They are launching and/or engaging in multi-stakeholder coalitions and initiatives to mobilize collective actions and innovations to incentivize and support their clients' transitioning to resilient net-zero. These companies actively reduced their carbon footprint.

2. *A sustainability issue but transitioning into core business.* For these companies, Board and C-level executives are shifting their perspectives about climate change as a sustainability issue linked to corporate social responsibility (CSR) to a more strategic issue. This shift may be motivated by requests from their shareholders, regulators or rating agencies or other external developments impacting the sector. These companies are increasingly exploring how to bring this issue into all aspects of core business decision making. They are establishing inter-departmental mechanisms and/or teams to explore how to conduct forward-looking climate risk information and link these risks and opportunities to their business model and operations and what information they need to provide to the C-level executives and the Board. They are also starting to explore the appropriate governance structure, accountability, and institutional arrangements, and considering external drivers such as public policy, markets, technology.

3. *As a corporate social responsibility and environmental issue.* For these companies, climate change remains a CSR issue, managed internally by the CSR team and/or a sustainability unit that in some cases may report to a dedicated Sustainability Committee at the Board level. In some companies, C-level executives have specific roles and responsibilities, with the CEO driving the topic, primarily focused on reducing the company's carbon footprint and recycling programs. Generally, a lack of climate change-specific core-business considerations may be driven by the governance structure, a lack of resources and expertise, culture, internal silos or a lack of political and regulatory action in their jurisdiction.

It is important to note that this landscape is rapidly changing with more re/insurers taking steps to incorporate climate change into their strategy, policies, enterprise risk management and core practices on both sides of the balance sheet.

Examples of Actions and Challenges for Re/insurers on the Liability Side to Support Transitioning to a Resilient Low-carbon Economy

Range of activities on the liability side

Insurers provide a variety of expertise in catastrophe risk modelling, and risk pricing, along with significant knowledge in risk reduction and preventive measures. They offer innovative and specialized risk transfer solutions in three broad categories, to: (i) build financial resilience to impacts of extreme events; (ii) incentivize reduction of GHG emissions; (iii) enable entre-

preneurial pathways for green and clean technology from start-up to commercialization (The Geneva Association 2018a, 2020a, 2020b, 2020c, 2020d, 2020e, 2020f, 2021a, 2021b, 2021c).
Specifically,

1. The insurance industry is sharing a significant amount of risk knowledge, risk modelling and risk pricing expertise directly or through various partnerships with its clients in the public and private sectors to enable risk awareness and to promote risk-based decision making (The Geneva Association 2018b, 2020a, 2020b, 2020c, 2020d, 2020e, 2020f, 2018b, 2021a, 2021b, 2021c; The Geneva Association and IDF 2017). For the past 30 years, P&C re/insurers have provided leadership in natural catastrophe (NatCat) risk modelling and pricing (Box 7.3). Over the last few years, the insurance sector has actively engaged with the governments, international multi-lateral agencies (e.g., The United Nations) and development banks to extend the development and use of NatCat models for public-sector development planning as well as expansion of social and private insurance markets in emerging and low-income economies (The Geneva Association and IDF 2017; IDF 2020). Larger companies are engaged in a wide range of research initiatives (advancing risk modelling, improving risk information and preventive measures against physical risks). The research may be conducted in-house, be outsourced or conducted at centers of excellence that may be bilaterally or multilaterally funded by the industry.[26]

2. A number of companies have set up "innovation units" or "incubators" or are "funding new centers of excellence in adaptation" to develop new ideas and solutions to help economies tackle underinsurance and improve socio-economic resilience to physical risks of climate. The research is translated into actionable guidelines for their policyholders (e.g., home-owners, commercial business and governments). For example Intact Center on Climate Adaptation (ICCA),[27] AXA Innovation,[28] Swiss Re Institute,[29] Lloyd's Innovation Lab,[30] Munich Re Innovation Hub[31] and innovations units of the largest insurance and reinsurance companies.

3. Insurers offer incentives for risk reduction, for example by means of premium reductions if their policyholders adopt and implement preventive measures (e.g. retrofitting homes against flood or wind damage) (The Geneva Association 2020a, 2020b, 2020c, 2020d, 2020e, 2020f). This issue is being investigated by the academic community, in terms of where this type of incentive works, where it is offered, and why this is not more widespread. In addition, other incentive mechanisms for risk reduction are being explored, for example linking risk reduction retrofits to increased asset valuation.

4. Companies are starting to develop a wide range of products and services to help customers to build climate resilience and reduce GHG emissions. Examples include:
   a. *Traditional and/or alternative risk transfer* products (e.g. parametric insurance) for weather-related extremes, such as tropical cyclones, storms, floods, forest fires, etc. (see Box 7.4 for more details on parametric insurance). These parametric products are also increasingly used for the protection and preservation of nature-based ecosystems (e.g. coral reefs and mangroves), which serve not only as a buffer to enhance resilience to extreme events but also as a sink for carbon. Examples include, innovation in products for protection and restoration of coral reefs and wetlands (The Geneva Association 2022; Beck 2021);

b. *Crop insurance* against climate risks; however, there are a number of challenges to the innovation of broader coverage in this area such as access to risk data;
c. *Micro-insurance products* that are being introduced and other products in low-income countries around the world, currently for small farmers who lack access to traditional insurance;
d. *Services that support issuance of Cat bonds*[32] for customers such as infrastructure-related companies and manufacturers with large production bases;[33]
e. *Issuance of resilience bonds*[34] *and green bonds*[35] to support implementation of major projects for transitioning to resilient low-carbon solutions.
f. *Specialized insurance products for renewable energy* ranging from residential solar systems to micro-hydro turbines and on-shore and off-shore wind farms; and,
g. *"Green buildings" insurance* and products and related incentives.

5. The industry is providing specialized products and services to protect governments' budgets against the impacts of weather-related catastrophes. Examples include:
   a. *Regional pools*. A number of reinsurers have emerged as industry leaders in helping with the design and providing cover to regional sovereign risk-pooling schemes such as the Caribbean Catastrophe Risk Insurance Facility (CCRIF), Pacific Catastrophe Risk Assessment and Financing Initiative (PCRAFI), or the African Risk Capacity (Africa's first parametric natural disaster insurance pool). These primarily are using parametric insurance solutions (The Geneva Association 2017 and 2018a; IDF 2019).[36]
   b. *Disaster expense insurance for local governments* that is provided in some Asian countries to compensate for the expenses needed by the municipality for evacuation of residents in case of a natural catastrophe.
   c. *New initiatives towards protection of publicly owned assets and infrastructure*. General insurers are increasingly engaging with the governments and infrastructure owners to develop approaches for addressing physical climate-related risks for publicly owned assets and ensuring climate resilient infrastructure (The Geneva Association 2019a; Coalition for Climate Resilient Investment (CCRI),[37] IDF 2019).

6. The industry is working to improve its products and services in areas such as business interruption, contingent business interruption and other risks associated with supply chain failures linked to natural catastrophes.

7. A number of companies have developed "specialized energy business units". They are offering products and services to address business risks associated with the complex value chain from start-up to commercialization and roll-out to support the development of renewable energy and other technologies.

8. Primary insurers are constantly striving to increase efficiency and speed up assessments, contract settlement and pay-outs after a disaster, using the latest satellite and near-real-time extreme event forecasting linked to their NATCAT risk modelling technologies.

BOX 7.4 INNOVATION IN PARAMETRIC INSURANCE

Parametric insurance solutions have emerged to address some of the challenges and shortcomings with the traditional "indemnity" insurance. Indemnity insurance has been the traditional mechanism insurers have used for centuries, to indemnify or compensate the inured for losses incurred by damage to an asset, which means that the amount of payout is directly linked to the amount of the incurred loss. Therefore, challenges of indemnity insurance include: need for detailed information about the assets, their vulnerability and exposure to set up the insurance coverage, time and efforts it takes to conduct a detailed assessment of the losses after an event which could delay the payout of the claims and related costs associated with the assessment, particularly if the catastrophe has impacted a large group of insureds over extensive geography.

Parametric insurance on the other hand is constructed based on an independent parameter or sets of parameters (e.g. meteorological, hydrological or seismic) which are closely correlated to the customer's risk. For example, this could be the wind speed of a tropical cyclone. In some cases, to further enhance the design of the contract, insurers use NATCAT models to take into account vulnerability and exposures. Unlike the indemnity insurance, for parametric contracts the claims payouts are triggered automatically once a previously-agreed threshold is reached. The threshold level may be triggered differently, for example, full limit may be made when the index value reaches the pre-defined threshold; on the other hand, the pay-out may be linked to severity levels, for example, a category 3 may trigger 25 percent, category 4 may trigger 50 percent and category 5 cyclone may trigger 100 percent of the pre-agreed payment. These products have become more popular over the last decade, and have been primarily used to enable swift payouts post disasters. They can be custom-made for a client, or be offered as regional facilities for governments.

Parametric insurance solutions' main challenge is "basis risk." This arises when the index does not reflect the actual losses incurred on the ground by the insureds. For example, when the index is triggered when the threshold is reached, (i) the payout may be less than the actual total damages; (ii) in some cases the threshold may not be reached, with no payout, yet the insured may have incurred significant losses; (iii) payouts may be more than incurred losses; or (iv) there may be no losses despite the payout.

Increasingly, some hybrid structures including a combination of indemnity and parametric insurance are being developed and offered, particularly for government when the parametric part addresses the more urgent needs for funding after the disaster, while the indemnity part offers funding for the reconstruction phase.

The Geneva Association and IDF highlight the development of these markets and efforts by the insurance industry in innovating these solutions since 1996.

Source: The Geneva Association (2017) and IDF (2019).

Factors hindering expansion of market-based insurance for building resilience to physical climate risks

There are seven primary factors hindering the expansion of market-based insurance in high-, middle-, and low-income countries, including (The Geneva Association 2018a, 2019a, 2019b):

1. *Limited access to risk information and related risk pricing difficulties.* Pricing of physical climate risks (e.g. weather-related extremes) is difficult due to lack of hazard/vulnerability/exposure data in many regions, the complexities of disasters and volatility of losses. When risk-based premiums are applied, this often stands in direct conflict with affordability of cover (higher risk, higher premium).
2. *Public policy, regulatory and legislative issues.* Public policy and regulations can create the necessary preconditions and the operating environment for the insurance sector. Specifically, government policies, regulation and investments in prevention and risk reduction measures are critical foundations of insurability.
3. *Lack of awareness about insurance* (from ministries of finance to the general public). In general, the insurance business model and how insurance leads to financial resilience at both macro and micro levels are not well understood.
4. *Limited take-up of disaster insurance.* This means that there is a relatively small pool of policyholders even where it is available at an affordable price. The main reasons are lack of risk awareness, limited understanding of insurance mechanisms, lack of a culture of risk management, underestimation of potential impacts, and reliance on other support mechanisms such as post-disaster government hand-outs.
5. *Weakness of domestic insurance market.* For some rural areas and in some middle- and low-income nations this limits access to insurance, distribution channels and pay-out systems.
6. *Regulatory barriers in some countries, which may hinder access to global reinsurance capacity and expertise*, which in turn may be hindered by regulatory barriers in some countries.
7. *Scalability and financial sustainability of insurance programs.* The relationship between the public and private sector is of particular importance in the context of rising losses, where effective Public–Private Partnerships (PPPs) seem to be the only viable option for maintaining insurability.

Opportunities for de-risking new technologies and infrastructure systems to scale up sectoral decarbonization

De-risking New Climate technologies for decarbonization of sectoral transitioning

Assessing, pricing and sharing these new risks associated with untested technologies along the entire life cycle will be fundamental for sectoral adoption, large-scale implementation and raising private capital.

Insurers play a vital role in managing risks of untested technologies. Innovations in insurance products and service are needed for large-scale deployment, where market conditions allow. However, given the massive scale of technological disruptions needed to achieve the climate targets, insurers need to reconsider their status quo approaches. For example, insurers need to build their technical expertise far beyond actuarial expertise in their commercial and specialty lines, by bringing on board technology experts with a deep understanding and hands-on experience with new climate technologies and associated risks. Insurers may also need to move from traditional project-basis coverage to a more integrated approach to addressing client needs (The Geneva Association 2022).

Underwriting and investing in new infrastructure systems
Beyond these, the insurance industry is embarking on supporting strengthening of climate resilient sustainable infrastructure systems. Insuring infrastructure systems against physical climate risks is challenged by a number of factors (The Geneva Association 2018a, 2019a). While critical infrastructure constitutes the backbone of a functional society, almost no consideration has been given in many countries to assessing impacts of natural hazards and climate change related factors. Focus has been mainly on operational risks, such as fire and explosion, but the need for factoring in climate resilience into infrastructure systems and supply chains is coming into focus.[38] In general,

i. There are limited incentives, especially for private operators, to increase climate resilience. However, it is believed that many countries are primarily focusing on decarbonization without paying sufficient attention to ensuring climate resilience.
ii. The extent to which insurers have been underwriting infrastructure risks varies from country to country. Insurers need to have access to high quality data to assess various risks associated with all phases of the infrastructure project throughout the entire life cycle, including their design, construction, operation, and maintenance for assessing and pricing risks, identifying sensitive ecosystems, evaluating the overall resilience of infrastructure projects, anticipating failures, and conducting proactive maintenance and preventive retrofits. Infrastructure data policies are needed to enable access to reliable data for the entire infrastructure lifecycle.
iii. Governments need to establish clear and aligned public policy – supported by effective legislative and regulatory frameworks to mandate climate resilience requirements as pre-requisite for the entire infrastructure lifecycle. Examples include land zoning, updated standards and building codes and investments in other risk-reduction measures such as nature-based systems as buffers (e.g., wetlands, mangroves).
iv. Investments in critical infrastructure also need to consider climate resilience for the entire life-cycle (The Geneva Association, 2019a; Coalition for Climate Resilient Investment).[39]

Actions on the Asset Side

Range of activities on the investment side
Increasingly, climate change is considered as a risk factor and an emerging investment theme for the majority of the re/insurers' CIOs, recognizing the importance of "climate-aware investing" (The Geneva Association 2018a). As previously noted, the insurance industry is increasingly integrating climate change considerations in its investment strategies and processes either as part of its broader sustainability topic or independently as a core business issue. According to The Geneva Association (2018a, 2021a):

i. Re/insurers have had a stronger focus on transition risk due to existing expertise in credit risk, and financial regulatory bodies and methodologies for climate risk assessment of the investment portfolios have primarily focused on transition risk. However, increasingly, they are leveraging internal industry expertise in NatCat modelling and physical climate risk (particularly general insurers) to assess physical risk of their investment portfolios, directly (e.g. implications of extreme weather events on buildings and infrastructure), and indirectly through investment in companies (e.g. debt and equity). On the other hand, considering the interplay of physical and transition risk and incorporating climate change risk considerations are fundamental for making long-term investment decisions as, over time, some of these assets may not be insurable. The industry through the Geneva Association Task Force on Climate Risk Assessment is developing new thinking about approaches to forward looking climate risk analysis.

ii. Overall investment strategies may include:
   a. Not investing in companies with more than 30 percent of their business associated with thermal coal mining or coal power generation.
   b. Making a conscious choice not to divest from fossil-fuel intensive energy companies as these companies are primary investors in green and clean technologies as part of their long-term strategy. Divesting could choke this critical source of funding and potentially compromise the transition to a low-carbon economy (UNEP and Bloomberg New Energy Finance 2017).[40]
   c. Not investing in fossil intensive sectors if they are providing underwriting services to these. Such policies are motivated by their enterprise risk management (ERM) practices.

iii. Investment approaches may include:
   a. Screening fossil fuel intensive sectors,
   b. Best-in-class (inclusionary screening): only include companies that perform best within each sector or industry,[41]
   c. Thematic investments: such as renewables, and constructing a specialized portfolio of related securities, divestment, selling all holdings in a particular sector or industry, such as coal,
   d. Active ownership: use their ownership stake in a company to influence its strategy, operations, governance, and risk management to achieve climate-resilient business strategies.
   e. Engagement with investees is becoming a major strategy for re/insurers – for example to promote and foster speedier decarbonization (transition), relocation of assets that are at risk to extreme events (physical risks), and enhanced risk management approaches.

iv. Many re/insurers require their external asset managers to have integrated climate change and the broader ESG factors in their investment processes and/or principles as part of their due diligence.[42]

Challenges hindering mainstream investing in the low carbon transitioning
There are a number of challenges hindering re/insurers in scaling up their core investments to support transitioning to a low-carbon economy. These can be framed under five areas: (1) financing and market-related factors such as the need for green and transition taxonomies, standards and access to a pipeline of investable grade projects; (2) financial and insurance regulations, particularly in relation to capital charges associated with long-term investing; (3) climate-change related national targets policies and regulatory frameworks for carbon-intensive sectors; (4) risks associated with new technologies for decarbonization; and

(5) transparency and disclosure of investment-relevant climate risk information for corporations for informed investing.

According to a survey by the Geneva Association (The Geneva Association 2018a):

i. The massive green financing gap could be addressed through the development of generally accepted definitions and standards for "green" as an asset class (i.e. taxonomy); expansion of the pipeline of investable-grade opportunities that meet their investment criteria and risk appetite;

ii. There is a need for expansion of the green bond market with appropriate monitoring, as well as other new investment tools and related markets. Cheaper and more widespread green bond funding is needed to drive more investment toward climate-resilient projects. Specifically, there is a need for: (i) more issuance of green bonds coupled with a broader variety of issuers; and (ii) emergence of new instruments (e.g. green loans, green securitizations, transition bonds[43]).

iii. There is a need for methodologies for and expertise in due diligence and monitoring among third-party asset managers.

iv. Risk-based capital charges within financial and insurance regulations could potentially restrain the insurance industry's investing in low-carbon transition with a long-term investment horizon. Insurance companies evaluate investments on a risk-return basis including return on capital. They are increasingly interested in reallocating capital towards long-term investments linked to green investments. However, some respondents noted that international and national financial and insurance regulations on capital charges may potentially restrain their capacity to invest in new technologies for industrial ecarbonization.[44]

v. Fragmentation in climate policies and regulations within and across nations lead to risks, impacting investors' confidence. This is further exacerbated by retroactive policy change, lack of policy or conflicting policy signals (such as fossil fuel subsidies). Greater clarity about national policies, particularly in relation to major carbon emitting sectors, could help to form an informed view about different climate scenarios and build these views into their investment strategies. There is a need for policy incentives to encourage green investment at scale (e.g. tax incentives and subsidies associated with investments in renewable energy or EVs). Lack of appropriate price signals, such as failure to price carbon and natural capital, are also seen as barriers to scaling up green investments.[45]

vi. While the insurance industry has contributed to the growing investments in renewable energy, the green and clean technology markets cannot yet accommodate the scaling up of the risk-adjusted returns that the insurance industry is seeking. As of April 2021, markets for green and clean technologies for disrupting the economic sector's value chains remain volatile, and in general, do not meet their criteria. Most of these new technologies are still in their infancy, and investing in them may be more aligned for those asset managers who may be willing to take higher risks in exchange for higher returns. There is need for more "green" technology-investment opportunities and structures that are close to the insurance industry's risk appetite.

vii. Insurers, as investors, need data to make informed investment decisions. Evaluating the physical, and transition risks associated with transitioning to a low carbon economy starts access to decision-useful climate risk assessment. However, fragmentation and lack of common reporting frameworks are leading to reported fatigue without producing consistent and reliable data and transparency.

EXAMPLES OF MULTI-LATERAL INITIATIVES ENGAGING RE/INSURERS TO TACKLE CLIMATE CHANGE

The insurance industry is actively engaging in many initiatives and multi-stakeholder collaborations to scale up the industry's contributions to supporting the net zero transitioning and expansion of insurance to address the rising protection gap and strengthening societal resilience. A few examples are provided below.

Industry-led Initiatives through The Geneva Association[46] – An International Platform of CEOs of Re/insurance Companies

Through The Geneva Association a number of industry-level initiatives are underway to leverage and scale up the insurance industry's contributions.

The Geneva Association Task Force on Climate Risk Assessment for Insurance Industry
The development of climate risk assessment methodologies and tools, such as scenario analysis, that would produce meaningful and decision-useful information is a work in progress. Despite some actions by stakeholder groups (e.g. re/insurers, financial institutions, regulatory and standard setting bodies, international organizations, commercial data providers, consulting firms and academia), initiatives remain fragmented and considerable work lies ahead because of the quickly evolving nature of climate science as well as other factors that will influence transition efforts. Achieving consensus will take time.

The industry has stepped up its global collaboration through *The Geneva Association Task Force on Climate Risk Assessment for Insurance Industry* (The GA Task Force) to rethink, innovate and develop further forward-looking methodologies for assessing climate risk that could produce decision-useful information for both sides of the balance sheet, not only to support the sector with advancing forward-looking climate change risk assessment tools but also as a contribution to the global developments in producing decision useful information linked to TCFD. Comprising leading experts from the 19 largest P&C and life insurers, The GA Task force is developing an approach that is anchored in the decision-making structure of the companies with a much more cohesive view of decisions for both sides of the balance sheet, taking a short-term and long-term view of the future (The Geneva Association 2021a, 2021b). The results are expected to be published in Q2 2022.

Industry Alliances and Partnerships to Support Net Zero

A number of re/insurers have been actively engaged in and/or led the launch of intra- and inter-industry alliances and platforms to support the net zero transitioning; for example, the United Nations' convened net zero alliances including The Net Zero Asset Owners Alliance,[47] and The Net Zero Insurance Alliance[48].

The Glasgow Financial Alliance for Net Zero (GFANZ) has been launched, chaired by Mark Carney, UN Special Envoy on Climate Action, to enable and facilitate access to mainstream finance at scale to support the transitioning to net-zero.[49]

Public–Private Partnerships to Support Climate Adaptation and Building Societal Resilience to Physical Risks of Climate Change

Nationally and globally re/insures are actively engaging with governments, international development community among others to address the rising protection gap in all countries (University of Cambridge Institute for Sustainability Leadership, 2021; The Geneva Association 2017, 2020a, 2020b, 2020c, 2020d, 2020e, 2020f). Importantly, the Insurance Development Forum (IDF) and InsuResilience Global Partnership,[50] and a number of city-level partnerships (The Geneva Association, 2021e) are among examples of multi-stakeholder partnerships that aim to expand access to insurance to the most vulnerable in emerging and low-income countries.

Among other multi-stakeholder initiatives are the Coalition for Climate Resilient Investment (CCRI),[51] which was established to ensure that physical climate risk is addressed, in areas such as infrastructure investments.

Other United Nations and Industry-led Platforms

Platforms such as the United Nations Environment Program – Financial Initiative's Principles for Sustainable Insurance (UNEP-FI PSI) are launching a number of difference initiatives with the insurance industry.[52] Examples of other industry-led multi-stakeholder platforms are: Munich Climate Insurance Initiative (MCII)[53] and ClimateWise, an industry-led initiative at the Centre for Sustainable Finance at the University of Cambridge.[54]

WAY FORWARD – BREAKING THE SILOS TOWARDS MORE INTEGRATED APPROACHES TO RESILIENT LOW-CARBON AND ENVIRONMENTALLY SUSTAINABLE BUSINESS MODELS

It has become evident that much lies ahead by way of building and strengthening socio-economic resilience to large-scale risks that impact society: people, communities, businesses, governments (at all levels) and the financial sector need to develop the ability to prepare, plan for, prevent, absorb, recover from and more successfully adapt to adverse events in a proactive way. On the other hand, limiting global average temperature increase to meet the global climate change targets over the next few decades requires a well-planned whole-of-economy approach to achieving net zero targets – an unprecedented transformation across all sectors of the economy and society is needed, affecting the way we produce and consume the goods and services that are essential to our lives.

Assessment and proactive management of the climate change risks are fundamental to developing and implementing pathways to meet the ambitious climate change targets: assessing the impacts of acute and chronic physical risks, implications of unclear national to local public policies, strategies and regulations for transitioning of carbon intensive sectors; uncertainties associated with market conditions and response; growing risks of climate change litigation, untested risks associated with new technologies, new processes and new infrastructure systems for decarbonization of carbon-intensive sector and related value chains and supply chains; reputational risks associated with inaction or lack of action, just to name a few.

As economies, industries and businesses increasingly look ahead, new business models need to be developed and implemented with a more integrated view: building resilience, addressing not only climate change but also other large-scale environmental and nature-based risks.

There are growing concerns with implications of societal behaviors (consumption patterns, industrial practices and business models, public sector's development choices and urbanization), leading to large-scale air, soil, water and ocean pollution (e.g. micro plastics), biodiversity loss and destruction of nature-based systems. These in turn are compromising food, water and health systems, with knock-on effects such as mass migration, war and conflict, and reduced economic activity (IPBES 2019; WEF 2021; Dasgupta Review 2021). Deforestation and loss of mangroves and wetlands are directly impacting the resilience of communities to extreme weather events (Beck 2021; Scyphers et al., 2020), increasing transmission of disease from animals to humans and nature's capacity as sources and sinks for greenhouse gases (GHGs). Economic sectors are being impacted by these large-scale man-made environmental risks.[55] In November 2022, The Geneva Association published a report on the implications of nature-based risks and the insurance industry, capturing risks and opportunities for the industry and innovative insurance solutions in this area (The Geneva Association 2022).

As reported by The Geneva Association and the IDF, the world is already experiencing a large (and growing) protection gap in all economies. Beyond the impacts of climate change and large-scale environmental risks, major drivers of insurability over long-term horizons are increasing exposure and vulnerability factors, such as population growth, urbanization, concentration of people and assets in high-risk zones (e.g. along coastlines and flood zones), types of development and supply-chain disruptions (caused by extreme events), which could distort and mask any embedded climate change and environmental signals. Lack of data in capturing these risks and their interconnections may limit our understanding and capturing the underpinning risks, which may lead to incorrect conclusions. On the other hand, societal actions to adapt to and mitigate these impacts of climate change and environmental and nature based risks and reduce societal exposures and vulnerabilities are vital to our ability to manage these risks over time.

As carbon-intensive sectors reduce or suspend their operations in the face of market and public policy pressure, the opportunities to offer these businesses insurance protection will diminish. At the same time, re/insurers themselves are facing market pressure to take proactive action; for example, to no longer underwrite risks for carbon-intensive business (reducing insurance liability exposure) and to no longer invest in them (reduce insurance asset portfolio exposure). While the shift away from carbon-intensive businesses may present some challenges, there are growing opportunities through technological innovations and commercialization in areas such as carbon capture and storage, renewable energy generation and electrification of transportation, among others.

The insurance industry has a unique opportunity by taking a more forward-looking holistic view of the future business models in various sectors, particularly with respect to its de-risking and investing functions. In working with its customers, investees, governments, the broader financial sectors and other stakeholders, it can promote the need to shift from fragmented conversations related to climate change, ESG, sustainability, and environmental and nature-based risks to a more integrated view towards resilient, net zero and environmentally sustainable behaviors and business models. The insurance industry is a critical part of the solution, and despite progress made by the largest insurance, reinsurance companies and cooperatives around the world, much lies ahead by way of innovations in insurance solutions and investment strategies as well as deeper cross-sectoral partnerships and collaboration to drive out-of-the box-thinking and fast track transitioning of the global economy at scale.

NOTES

1. For a detailed literature review please refer to these reports. Note that this is a fast moving area and by the time this chapter is published, other areas may have progressed and not been captured in this report.
2. The Geneva Association is the international strategic think tank of the insurance industry, whose members are CEOs of the largest insurance and reinsurance companies, globally (www.genevaassociation.org).
3. In other words, it is important to underline that the role of insurance is the socialization of risk.
4. For activities of the Insurance Development Forum see: https://www.insdevforum.org/
5. For more information on the InsuResilience Global Partnership see: https://unfccc.int/topics/adaptation-and-resilience/resources/S-N/IRGP#:~:text=The%20InsuResilience%20Global%20Partnership%20aims,capacity%20and%20strengthening%20local%20resilience.
6. For more information on the Microinsurance Network see: https://microinsurancenetwork.org/about
7. For more information see: https://www.worldbank.org/en/programs/disaster-risk-financing-and-insurance-program
8. For more information see: https://www.onebillionresilient.org/post/extreme-heat-resilience-alliance-reducing-extreme-heat-risk-for-vulnerable-people
9. The TCFD has split climate-change related risks into physical and transition risks, offering more clarity to companies to assess the associated financial risk to underpin their decisions.
10. For more information: https://ec.europa.eu/info/consultations/finance-2020-sustainable-finance-strategy_en
11. For more information: https://www.canada.ca/en/department-finance/news/2019/06/expert-panel-on-sustainable-finance-delivers-final-report-finance-minister-joins-international-climate-coalition.html
12. For more information: https://www.sustainablefinance.org.au/
13. https://www.bankofengland.co.uk/climate-change/climate-financial-risk-forum
14. For more information about S&P ratings and latest report cards for companies from different sectors see: https://www.spglobal.com/ratings/en/products-benefits/products/esg-in-credit-ratings#sector-report-cards.
15. The Geneva Association is an international think tank. Its members are CEOs of the largest re/insurance companies (P&C and life), which in total manage US$17.1 trillion in assets; employ 2.4 million people; and protect 1.8 billion people globally.
16. The Board's decision followed two Geneva Association conferences on this topic: (1) How Will Risk Modelling Shape the Future of Risk Transfer?, hosted by SCOR (9 March 2017, Paris) https://www.genevaassociation.org/how-will-risk-modelling-shape-future-risk-transfer; and (2) Advancements in Modelling and Integration of Physical and Transition Climate Risk, hosted by Tokio Marie (11–12 July 2019, London) https://www.genevaassociation.org/climate-change-forum-2019
17. IFRS.org
18. Such as cement & concrete, iron & steel, chemicals & plastics, aluminum, pulp & paper
19. For example: city-level energy efficiency retrofits, to large-scale corporate pilots, new distributed, cheaper, more efficient and smaller sustainable infrastructure systems for energy, water and waste management, to large scale deployment of carbon capture and storage (CCS).
20. For example, the Biden administration's new Infrastructure proposal, which was passed by the US Congress on 6 November 2021 (a US$1.2 trillion package) is the largest US spending in public

works and infrastructure with many elements for decarbonizing the energy and transportation systems in the US.
21. According to CRO Forum 2020, environmental issues are now firmly in the spotlight, dominated by climate change, resource scarcity and pollution of the biosphere. There is growing concern about the consequences of the unchecked emission of greenhouse gases driving climate change, such as the occurrence of more extreme and frequent weather events. The pressure placed on the planet from a growing human population is causing resource scarcity, driven by unsustainable practices in mineral extraction, food and energy production. Anthropogenic activities are also polluting the land and sea with non-biodegradable waste such as plastics, and the air with particulate and gaseous pollutants. All forms of pollution are becoming ubiquitous, with harmful consequences for life on Earth.
22. Tables 7.1 to 7 4 have been reproduced here from The Geneva Association (2021a).
23. Technically, insurance prices are based on risk (along with a number of other factors) and can provide governments, businesses and individuals with reasonably accurate signals as to the impacts and characteristics of the hazards they face. Prices commensurate with underlying risks tend to encourage re/insureds to take mitigation measures which reduce vulnerability across society (The Geneva Association 2016).
24. The Geneva Association is undertaking a study to explore the evolving landscape of climate change litigation and implications for the insurance industry.
25. And additional interviews with C-Suite.
26. Examples include Intact Center for Climate Adaptation (https://www.intactcentreclimateadaptation.ca/), Partners for Actions (https://uwaterloo.ca/partners-for-action/#:~:text=Partners%20for%20Action%20is%20an,flood%20resiliency%20in%20Canadian%20communities), offering innovations in home and community protection and expansion of flood insurance protection. A wide range of innovations with new products and services through innovation labs of large reinsurance companies such as Swiss Re Institute's Quantum Cities https://www.swissre.com/institute/research/topics-and-risk-dialogues/digital-business-model-and-cyber-risk/sustainable-quantum-cities.html.
27. For example, ICCA has developed a number of programs and is working with the government, industry, academia and other key stakeholders to enable a program for home retrofits to reduce risks of floods. It is working on innovative nature-based solutions to build community resilience and supporting the development of new standards for construction and new builds incorporating climate risk considerations (for more information see The Geneva Association 2020f and https://www.intactcentreclimateadaptation.ca/).
28. https://www.axa.com/en/about-us/innovation#tab=axa-next
29. https://www.swissre.com/institute/
30. https://www.lloyds.com/news-and-insights/lloyds-lab
31. https://www.munichre.com/en/company/innovation.html
32. With the rising impacts of extreme weather events, local, provincial and national governments need to manage their budgets to expedite recovery from disasters by covering damages and paying for the cost of reconstruction of public assets and infrastructure. Catastrophe (CAT) bonds are designed to transfer these risks to the capital markets. CAT bonds serve as an insurance policy for the bond issuer, where the principal of the bond is forgiven when a disaster reaches a pre-agreed threshold.
33. To reduce customers' burden of cumbersome administrative procedures associated with Cat bonds issuance as well as to help their customers diversify the hedging methods for catastrophe risks through their advice on setting optimal issuance conditions etc.
34. Resilience bonds are a type of CAT bond that is designed to incentivize cities and other jurisdictions to invest in resilience. These bonds include a resilience rebate that turns avoided "measurable" losses from a risk reduction plan into a revenue stream (Vaijhala and Rhodes, 2018)
35. The proceeds of Green Bonds must be used for climate-related projects that reduce emissions or improve resilience. Green Bonds are increasingly used for integrated low-carbon and resilience projects. However, less than 5 percent of pre-2019 global Green Bond proceeds have funded adaptation and resilience projects (Climate Bonds Initiative 2019, 2020).
36. Caribbean Catastrophe Risk Insurance Facility (CCRIF) (http://www.ccrif.org/); The African Risk Capacity (ARC) (http://www.africanriskcapacity.org/); Pacific Catastrophe Risk Assessment and Financing Initiative (PCRAFI) (http://pcrafi.sopac.org/)

37. https://resilientinvestment.org/
38. An example is the February 2021 winter weather in Texas and the importance of infrastructure resilience investment. https://www.c2es.org/2021/02/winter-weather-in-texas-and-the-importance-of-infrastructure-resilience-investment/
39. https://resilientinvestment.org/
40. For example, the majority of investments in renewable energy and other sources of clean energy are made specifically through these carbon-intensive companies.
41. A more targeted version of this type of investing is to exclude any companies that score below a pre-determined threshold, regardless of their sector.
42. In general, insurers tend to outsource their investment activities to large investment companies that are at the forefront of developments when it comes to climate risks.
43. Transition bonds are designed for carbon-intensive sectors with proceeds used to finance new and/or existing projects for transition towards a reduced environmental impact, such as reduced carbon emissions (Takatsuki and Foll 2019). These bonds are intended for companies and projects that would not qualify as "green".
44. While green bonds are generally issued by large companies or entities that issue liquid debt securities, and as such green bond investors do not have to give up liquidity, there are not any pricing differences with traditional bonds of comparable credit ratings and maturities – green bonds do not trade at a premium compared with their peers. The fact that "green bond" has not been established as an asset class on its own does not allow for different risk capital charges to be applied to them.
45. *Natural capital* is defined as the world's stocks of natural assets which include geology, soil, air, water, and all living things. It is from this natural capital that humans derive a wide range of services, often called ecosystem services, which make human life possible (https://naturalcapitalforum.com)
46. https://www.genevaassociation.org/about-us
47. https://www.unepfi.org/net-zero-alliance/
48. https://www.unepfi.org/net-zero-insurance/
49. GFANZ includes over 160 financial and insurance firms (together responsible for assets in excess of US$70 trillion). GFANZ builds on existing initiatives, including *Net Zero Banking Alliance*, *Net Zero Asset Owner Alliance*, *Net Zero Asset Manager Alliance* and the *Net Zero Insurance Alliance* (later to be established at COP26).
50. https://www.insuresilience.org/
51. https://resileintinvestment.org
52. https://www.unepfi.org/psi/
53. https://climate-insurance.org/
54. https://www.cisl.cam.ac.uk/business-action/sustainable-finance/climatewise
55. *Agricultural sector*: 75 percent of food crop production depend on pollination services. A decline of pollination may generate an annual net loss of US$235-577 billion globally (Global Canopy and Vivid Economics 2020).
    Fishing sector: Ocean acidification and warming may impact biodiversity by disrupting the physiology of marine species, community composition, species distribution, abundance and extinction risks. Nearly 90 percent of the world's marine fish stocks are fished at or beyond maximum sustainable levels (Global Canopy and Vivid Economics 2020).
    Pharmaceutical: 50 percent of prescription drugs are based on natural molecules, 70 percent of cancer drugs are natural products and 75 percent of approved anti-tumor pharmaceuticals in the past 70 years have been non-synthetic, biodiversity loss may lead to the loss of potential existing and new sources of pharmaceutical molecules and drug discoveries.
    Plastic & Packaging sector: Single-use packaging alone has been estimated to be responsible for US$40 billion in externalities (economic costs not borne by those responsible for causing them) (UNEP, 2019). An estimated 1.2–2.4 million tons of plastic flows from rivers into the ocean every year, where it is ingested by hundreds of ocean species (WWF, 2020).
    Tourism sector: Coastal tourism services may be threatened by the degradation and destruction of coral reefs due to ocean warming and acidification and climate change. Fifty percent of the world's coral reef system has already been destroyed (IPBES 2019).

REFERENCES

Beck, M., 2021. Saving coastlines from climate disasters. *LA Times*. April 11, 2021.

Burger, M., R. Horton, and J. Wentz. 2020. The law and science of climate change attribution. *Columbia Journal of Environmental Law* 45(1). https://doi.org/10.7916/cjel.v45i1.4730.

Climate Bonds Initiative. 2019. *Green bonds: The state of the market 2018*. https://www.climatebonds.net/resources/reports/green-bonds-state-market-2018

Climate Bonds Initiative. 2020. *Explaining Green Bonds*. https://www.climatebonds.net/market/explaining-green-bonds

ClimateWise. 2019a. *Physical Risk Framework: Understanding the Impacts of Climate Change on Real Estate Lending and Investment Portfolios – Summary for Decision Makers*. https://www.cisl.cam.ac.uk/resources/publication-pdfs/physical-risk-framework-report-summary.pdf

ClimateWise. 2019b. *Physical Risk Framework: Understanding the Impacts of Climate Change on Real Estate Lending and Investment Portfolios – Full Report*. https://www.cisl.cam.ac.uk/resources/publication-pdfs/cisl-climate-wise-physical-risk-framework-report.pdf.

ClimateWise. 2019c. *Transition Risk Framework: Managing the Impacts of the Low Carbon Transition on Infrastructure Investments Public Report*. https://www.cisl.cam.ac.uk/resources/publication-pdfs/cisl-climate-wise-transition-risk-framework-report.pdf.

ClimateWise. 2019d. *Transition Risk Framework: Managing the Impacts of the Low Carbon Transition on Infrastructure Investments Step-by-Step Guide*. https://www.cisl.cam.ac.uk/resources/publication-pdfs/transition-risk-framework-report-step-by-step.pdf.

Clyde & Co. 2018. *Climate Change a Burning Issue for Businesses and Boardroom*. https://online.flippingbook.com/view/6686/

Clyde & Co. 2021. *Stepping up Good Governance to Seize Opportunities and Reduce Exposure, Climate Change Risk and Liability Report 2021*. https://www.clydeco.com/en/reports/2021/3/climate-change-risk-and-liability-report-2021

CRO Forum. 2018. *The Heat Is On: Insurability and Resilience in a Changing Climate*. https://www.thecroforum.org/wp-content/uploads/2019/01/CROF-ERI-2019-The-heat-is-on-Position-paper-1.pdf

CRO Forum. 2020. *Emerging Risk Initiative Major Trends and Emerging Risk Radar*. https://www.thecroforum.org/wp-content/uploads/2020/06/ERI-Risk-Radar-2020-update.pdf

Dasgupta, P. 2021. *The Economics of Biodiversity: The Dasgupta Review*. https://assets.publishing.service.gov.uk/government/uploads/system/uploads/attachment_data/file/957291/Dasgupta_Review_-_Full_Report.pdf

Global Canopy and Vivid Economics. 2020. *The Case for a Task Force on Nature-related Financial Disclosures*. https://www.globalcanopy.org/sites/default/files/documents/resources/Task-Force-on-Nature-related-Financial-Disclosures-Full-Report_1.pdf

Golnaraghi, M. 2022. Decarbonizing our economy: Factors to extend the scope and accelerate deployment of new climate technologies. *The Actuary Magazine* (January 2022). https://theactuarymagazine.org/decarbonizing-our-economy/.

IFoA. 2019a. *A Practical Guide to Climate Change for Life Actuaries* (Authors: D. Ford, B. Ashton, K. Audley, M. Qazvini, Y. Yuan and Y. McLintoc). https://www.actuaries.org.uk/system/files/field/document/A%20Practical%20Guide%20to%20Climate%20Change%20%20for%20Life%20actuaries%20-%20Oct%20v7a.pdf

IFoA. 2019b. *A User Guide to Climate-related Financial Disclosures*. https://www.actuaries.org.uk/system/files/field/document/J27006_IEMA_TCFD_Guide_V4.pdf

Intergovernmental Panel on Climate Change – IPCC. 2018. *Global Warming of 1.5°C. A Special Report*. Available at: https://www.ipcc.ch/sr15/.

Intergovernmental Panel on Climate Change – IPCC. 2021. *AR6 Climate Change 2021: The Physical Science Basis*. https://www.ipcc.ch/report/ar6/wg1/

Intergovernmental Science-Policy Platform on Biodiversity and Ecosystem Services. 2019. *The Global Assessment Report on Biodiversity and Ecosystem Services of the Intergovernmental Science-Policy Platform on Biodiversity and Ecosystem Services* (IPBES). https://ipbes.net/sites/default/files/2020-02/ipbes_global_assessment_report_summary_for_policymakers_en.pdf

Institute of International Finance (IIF). 2021. *Prudential Pathways: Industry Perspectives on Supervisory and Regulatory Approaches to Climate-related and Environmental Risks.* https://www.iif.com/Portals/0/Files/content/Regulatory/01_21_2021_prudential_pathways.pdf.

Insurance Development Forum (IDF). 2019. *IDF Practical Guide to Insuring Public Assets.* https://www.internationalinsurance.org/sites/default/files/2019-12/2019_09_IDF%20Practical%20Guide%20to%20Insuring%20Public%20Assets%20-LR.pdf

Insurance Development Forum (IDF). 2020. *The Development Impact of Risk Analytics.* http://www.insdevforum.org/wp-content/uploads/2020/10/IDF_Risk_Analytics_11October.pdf

Keenan J., 2019. A climate intelligence arms race in financial markets. *Science* 365(6459), 1240-1243. 10.1126/science.aay8442

Markell, D.L., and J.B. Ruhl. 2012. An empirical assessment of climate change in the courts: A new jurisprudence or business as usual? *Florida Law Review* 64(1), 15–86. https://scholarship.law.ufl.edu/cgi/viewcontent.cgi?article=1013&context=flr

Moody's Investor Service. 2018. *Climate Change Risks Outweigh Opportunities for P&C Re/insurers.* March.

Moody's Investor Service. 2019. *The Impact of Environmental, Social and Governance Risks on Insurance Ratings.* July.

Scyphers S. B., et al. 2020 Designing effective incentives for living shorelines as a habitat conservation strategy along residential coasts. *Conservation Letters.* https://conbio.onlinelibrary.wiley.com/doi/full/10.1111/conl.12744.

Setzer, J. and C. Higham. 2021. *Global Trends in Climate Change Litigation: 2021 snapshot.* (Forthcoming.)

Skipper H.D. 1998. *International Risk and Insurance: An Environmental Managerial Approach.* McGraw-Hill Education.

Standard & Poor's Global Ratings. 2018. *How Environmental, Social, And Governance Factors Help Shape The Ratings On Governments, Insurers, And Financial Institutions.* October. https://www.spglobal.com/en/research-insights/articles/how-environmental-social-and-governance-factors-help-shape-the-ratings-on-governments-insurers-and-financial-institutions

Standard & Poor's Global Ratings. 2020. *ESG Industry Report Card: EMEA Insurance.* February. https://www.spglobal.com/ratings/en/research/articles/200211-environmental-social-and-governance-esg-industry-report-card-emea-insurance-11337605

Surminski, S., M. Westcott, J. Ward, P. Sayers, D. Bresch and C. Bronwyn. 2020. Be prepared – exploring future climate-related risk for residential and commercial real-estate portfolios. *Journal of Alternative Investments* 23(1), 24–34.

Takatsuki, Y., and J. Foll. 2019. Financing brown to green: Guidelines for transition bonds. *AXA Investment Managers.* https://realassets.axa-im.com/content/-/asset_publisher/x7LvZDsY05WX/content/financing-brown-to-green-guidelines-for-transition-bonds/23818

Task Force on Climate-related Financial Disclosures (TCFD). 2016. *Phase I Report of the Task Force on Climate-Related Financial Disclosure.* March. https://www.fsb-tcfd.org/wp-content/uploads/2016/03/Phase_I_Report_v15.pdf

Task Force on Climate-related Financial Disclosures (TCFD). 2017. *Final Report: Recommendations of the Task Force on Climate-related Financial Disclosures and Related Annexes.* June. https://www.fsb-tcfd.org/publications/final-recommendations-report/

Task Force on Climate-related Financial Disclosures (TCFD). 2018. *TCFD: 2018 Status Report.* September. https://www.fsb-tcfd.org/publications/tcfd-2018-status-report/

Task Force on Climate-related Financial Disclosures (TCFD). 2019. *TCFD 2019 Status Report.* June. https://www.fsb-tcfd.org/publications/tcfd-2019-status-report/

Task Force on Climate-related Financial Disclosures (TCFD). 2021. *Implementing the Recommendations of the Task Force on Climate-related Financial Disclosures.* https://assets.bbhub.io/company/sites/60/2021/07/2021-TCFD-Implementing_Guidance.pdf

The Geneva Association, 2016. *An Integrated Approach to Managing Extreme Events and Climate Risks.* M. Golnaraghi, S. Surminski, S. and K. Schanz. https://www.genevaassociation.org/media/952146/20160908_ecoben20_final.pdf

The Geneva Association. 2017. *The Stakeholder Landscape in Extreme Events and Climate Risk*, M. Golnaraghi and P. Khalil. January. https://www.genevaassociation.org/sites/default/files/research-topics-document-type/pdf_public//stakeholder-landscape-in-eecr.pdf

The Geneva Association. 2018a. *Climate Change and the Insurance Industry: Taking Actions as Risk Managers and Investors*. M. Golnaraghi. January. https://www.genevaassociation.org/research-topics/extreme-eventsand-climate-risk/climate-change-and-insurance-industrytaking-action

The Geneva Association. 2018b. *Managing Physical Climate Risk: Leveraging Innovations in Catastrophe Risk Modelling*. M. Golnaraghi, P. Nunn, R. Muir-Wood, J. Guin, D. Whitaker, J. Slingo, G. Asrar, I. Branagan, G. Lemcke, C. Souch, M. Jean, A. Allmann, M. Jahn, D. N. Bresch, P. Khalil and M. Beck. November. https://www.genevaassociation.org/research-topics/extremeevents-and-climate-risk/managing-physical-climaterisk%E2%80%94leveraging

The Geneva Association. 2019a. *Investing in Climate-resilient Decarbonised Infrastructure to Meet Socio-economic and Climate Change Goals*. M. Golnaraghi. December. https://www.genevaassociation.org/research-topics/climate-change-and-emerging-environmental-topics/investing-climate-resilient

The Geneva Association. 2019b. *The Role of Trust in Narrowing Protection Gaps*. https://www.genevaassociation.org/research-topics/socio-economic-resilience/role-trust-narrowing-protection-gaps-research-report

The Geneva Association. 2020a. *Building Flood Resilience in a Changing Climate: Insights from the United States, England and Germany*. M. Golnaraghi, S. Surminski and C. Kousky. June. https://www.enevaassociation.org/research-topics/building-floodresilience

The Geneva Association. 2020b. *Flood Risk Management in the United States*. C. Kousky and M. Golnaraghi. June. https://www.genevaassociation.org/researchtopics/flood-risk-management-united-states

The Geneva Association. 2020c. *Flood Risk Management in England*. S. Surminski, S. Mehryar and M. Golnaraghi. June. https://www.genevaassociation.org/research-topics/flood-risk-management-england

The Geneva Association. 2020d. *Flood Risk Management in Germany*. S. Surminski, V. Roezer and M. Golnaraghi. June. https://www.genevaassociation.org/research-topics/flood-risk-management-germany

The Geneva Association. 2020e. *Flood Risk Management in Australia*. N. Duffy, A. Dyer and M. Golnaraghi. December. https://www.genevaassociation.org/research-topics/climate-change-and-emerging-environmental-topics/flood-risk-management-australia

The Geneva Association. 2020f. *Flood Risk Management in Canada*. M. Golnaraghi, J. Thistlethwaite, D. Henstra, C. Stewart. December. https://www.genevaassociation.org/research-topics/climate-change-and-emerging-environmental-topics/flood-risk-management-canada

The Geneva Association. 2021a. *Climate Change Risk Assessment for the Insurance Industry: A Holistic Decision- making Framework and Key Considerations for Both Sides of the Balance Sheet*. M. Golnaraghi et al. February. https://www.genevaassociation.org/sites/default/files/research-topics-document-type/pdf_public/climate_risk_web_final_250221.pdf

The Geneva Association. 2021b. *Climate Change Risk Assessment for the Insurance Industry: Industry perspectives on regulatory approaches and opportunities for multi-stakeholder collaboration*. M. Golnaraghi et al. June (Forthcoming).

The Geneva Association. 2021c. *Climate Change Litigation – Part 1: The Evolving Global Landscape*. M. Golnaraghi, M. Setzer, J. N. Brook, W. Lawrence and L. Williams. https://www.genevaassociation.org/sites/default/files/research-topics-document-type/pdf_public/climate_litigation_04-07-2021.pdf

The Geneva Association. 2021d. *Future Proofing Technological Innovations for a Resilient Net-Zero Economy*. Conference Summary – A Geneva Association-OECD Conference (12 October 2021) https://www.genevaassociation.org/future-proofing-technological-innovations-resilient-net-zero-economy-summary

The Geneva Association. 2021e. *Future Urban Landscape: An insurance perspective*. K.-U. Schanz. https://www.genevaassociation.org/sites/default/files/research-topics-document-type/pdf_public/future_urban_risk_web.pdf

The Geneva Association. 2022. *Nature and the insurance industry: Taking action towards a nature-positive economy*. (November).

The Geneva Association and Insurance Development Forum. 2017. *Guidelines for Risk Assessment to Support Sovereign Risk Financing and Risk Transfer*. M. Golnaraghi, I. Branagan, S. Fraser, J. Gascoigne, and A. M. Gordon. A request from United Nations Office for Disaster Risk Reduction (UNISDR) as part of "Words into Action Guideline on National Disaster Risk Assessment: Governance System, Methodologies, and Use of Result". Available at: https://www.genevaassociation.org/research-topics/extreme-events-and-climate-risk/guidelines-risk-assessment-support-sovereign-risk

The World Bank. 2021. *Global Partnership on Disaster Risk Financing Analytics: Results and Achievements*. https://www.gfdrr.org/en/publication/global-partnership-disaster-risk-financing-analytics-results-and-achievements

United Nations Environment Program – Finance Initiative. 2019. *Changing Course: A comprehensive investor guide to scenario-based methods for climate risk assessment, in response to the TCFD*. https://www.unepfi.org/wordpress/wp-content/uploads/2019/05/TCFD-Changing-Course-Oct-19.pdf

United Nations Environment Program – Finance Initiative Principles for Sustainable Insurance. 2020. *Using Hindsight and foresight: Enhancing the Insurance Industry's Assessment of Climate Change Futures – A Progress Update*. September. https://www.unepfi.org/publications/psi-tcfd-pilot-progress-update/

United Nations Environment Program – Finance Initiative Principles for Sustainable Insurance. 2021. *Insuring the Climate Transition: Enhancing the insurance industry's assessment of climate change futures*. January. https://www.unepfi.org/psi/wp-content/uploads/2021/01/PSI-TCFD-final-report.pdf

United Nations Environment Program – Finance Initiative and Oliver Wyman. 2018. *Extending our Horizons*. https://www.unepfi.org/wordpress/wp-content/uploads/2018/04/EXTENDING-OUR-HORIZONS.pdf.

United Nations' Framework Convention on Climate Change (UNFCCC). 2015. *Paris Agreement*. https://unfccc.int/sites/default/files/english_paris_agreement.pdf

United Nations' Framework Convention on Climate Change (UNFCCC). 2021a. Draft CMA decision proposed by the President. https://unfccc.int/sites/default/files/resource/Overarching_decision_1-CMA-3_1.pdf

United Nations' Framework Convention on Climate Change (UNFCCC). 2021b. *NDC Synthesis Report*. https://unfccc.int/process-and-meetings/the-paris-agreement/nationally-determined-contributions-ndcs/nationally-determined-contributions-ndcs/ndc-synthesis-report

United Nations Environment Programme (UNEP) and Bloomberg New Energy Initiative (BNEF). 2017. *Global Trends in Renewable Energy Investment* 2017. https://europa.eu/capacity4dev/file/46372/download?token=OAVxO9NT

University of Cambridge Institute for Sustainability Leadership (CISL). 2021. *Risk Sharing in the Climate Emergency: Financial Regulations for a Resilient, Net Zero, Just Transition*. https://www.cisl.cam.ac.uk/files/risk_sharing_in_the_climate_emergency_web1.pdf

Von Peter, G., von Dahlen, S. and Saxena, S. 2012. *Unmitigated Disasters? New Evidence on the Macroeconomic Cost of Natural Catastrophes*, BIS Working Papers No. 394, Basel: Bank for International Settlements (BIS). Available at: http://www.bis.org/publ/work394.pdf.

Vaijhala, S., and Rhodes, J. 2018. *Resilience Bonds: A Business model for Resilient Infrastructure*. Veolia Institute. https://www.institut.veolia.org/sites/g/files/dvc2551/files/document/2018/12/03-02_Resilience_Bonds_a_business-model_for_resilient_infrastructure.pdf

World Economic Forum. 2016. *The Global Risks Report 2016*. http://www3.weforum.org/docs/GRR/WEF_GRR16.pdf

World Economic Forum. 2017. *The Global Risks Report 2017*. http://www3.weforum.org/docs/GRR17_Report_web.pdf

World Economic Forum. 2018. *The Global Risks Report 2018*. http://www3.weforum.org/docs/WEF_GRR18_Report.pdf

World Economic Forum. 2019. *The Global Risks Report 2019*. http://www3.weforum.org/docs/WEF_Global_Risks_Report_2019.pdf

World Economic Forum. 2020a. *The Global Risks Report 2020*. http://www3.weforum.org/docs/WEF_Global_Risk_Report_2020.pdf

World Economic Forum. 2020b. *The Net-Zero Challenge: Fast-Forward to Decisive Climate Action*. http://www3.weforum.org/docs/WEF_The_Net_Zero_Challenge.pdf

World Economic Forum. 2020c. How biodiversity loss is hurting our ability to combat pandemics. https://www.weforum.org/agenda/2020/03/biodiversity-loss-is-hurting-our-ability-to-prepare-for-pandemics/

World Economic Forum. 2021. *The Global Risks Report 2021*. http://www3.weforum.org/docs/WEF_The_Global_Risks_Report_2021.pdf

World Wide Fund for Nature (WWF). 2020. *Living Planet Report 2020: Bending the Curve of Biodiversity Loss*. https://f.hubspotusercontent20.net/hubfs/4783129/LPR/PDFs/ENGLISH-FULL.pdf

8. Climate change and aviation
Vincent Etchebehere

INTRODUCTION

The climate crisis continues to intensify, as highlighted in the Sixth Assessment Report of the Inter-governmental Panel on Climate Change (hereafter, IPCC) from their Working Group 1 (IPCC, 2021). The report links human-caused activities to global warming, highlighting the fact that global temperatures, on average, were 1.1°C higher during the decade 2010–2019 than in the pre-industrial era. The consequences of such warming include rising sea levels, extreme heat waves, floods, severe droughts, and irreversible trends such as glaciers and polar ice caps melting (IPCC, 2021, pp. 20–25). To limit the temperature increase to less than 2°C by the end of this century, we must transform our economies and industries, and must drastically reduce CO_2 emissions.

In 2019, aviation accounted for 2.6 percent of worldwide CO_2 emissions (Delbecq et al., 2022). The sector's carbon footprint has been steadily growing in absolute terms due to the increase in air traffic, while emissions intensity – measured by CO_2 emissions per passenger-kilometer (km) – has decreased. What's more, the sector generates other externalities, such as condensation trails (contrails) that, according to recent scientific studies, could have an impact on climate change of a magnitude at least equivalent to that of its CO_2 emissions, thus doubling its warming effect.

Aviation is one of the hardest-to-abate sectors vis-à-vis carbon emissions. Past reductions levers are insufficient to offset forecast traffic growth. Solutions to further reduce CO_2 emissions have been identified, and some of them have been validated. However, they are yet to be scaled-up to support a decarbonization trajectory compatible with the "well below 2°C target" of the Paris Agreement, or the increasingly widely-accepted 1.5°C transition scenario.

Time is of the essence in the global race to fight the climate crisis. This is particularly true for aviation, which has been increasingly scrutinized for its carbon footprint, as illustrated by the *flygskam* movement that appeared in 2018 in Sweden[1] and then rapidly spread across the world. As stated by the International Air Transport Association (IATA) in December 2021: "Flying sustainably is not an option. And it is more than the industry's license to grow. The freedom to fly at all may depend on eradicating carbon emissions."[2]

This chapter aims to shed light on the impact of the aviation sector on climate change, reviewing existing decarbonization solutions and those under development, as well as depicting the challenges of implementing environmental transition actions, drawing upon Air France's experience.

THE AVIATION SECTOR'S IMPACT ON CLIMATE CHANGE

Aviation can be split into three main categories: *commercial* (both passenger and freight transportation), *military*, and *private* (private jets, leisure, or rescue aircraft).

This chapter focuses on commercial aviation which accounts for 88 percent of the sector's CO_2 emissions, as shown in Figure 8.1.

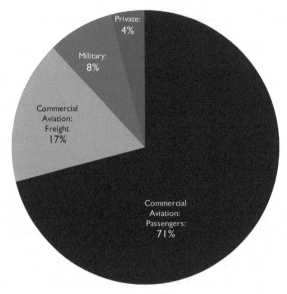

Source: Recreated using data from Gössling and Humpe (2020).

Figure 8.1   *Global distribution of aviation fuel use*

The Aviation Sector's CO_2 Emissions

Carbon dioxide emissions represent approximately 75 percent of global human-caused greenhouse gas (GHG) emissions, the rest being methane (CH_4) at 17.3 percent, nitrous oxide (N_2O) at 6.2 percent and various fluorinated gases at 2.1 percent (Climate Watch, 2021). A vast majority of global CO_2 emissions, about 90 percent, are in turn generated by the combustion of the three fossil fuels, namely oil, coal, and natural gas (Crippa et al., 2018).

Aviation CO_2 emissions are primarily linked to the utilization of jet kerosene, a fossil fuel. In 2019, over 99.9 percent of global aviation fuel needs were dependent on jet kerosene, the other 0.1 percent being linked to the usage of Sustainable Aviation Fuels, or SAFs (World Economic Forum – in Collaboration with McKinsey & Company, 2020). Jet kerosene consumption by aviation will likely be phased out over time in order to reduce CO_2 emissions, but also because oil is a finite resource and its availability will decrease over time. In a sustainable low-carbon world, the final energy – i.e., the energy delivered to consumers for their end-use – will not be produced primarily by the extraction of finite natural resources as is the case today, but rather, it will need to be produced from renewable resources (e.g., hydro, wind, solar, biomass) and nuclear power.

Jet-fuel related CO_2 emissions come from both its combustion in the aircraft, mostly in the air, but also during airplane taxiing and, to a lesser extent, its production, processing, and transport.

In the case of Air France, jet kerosene utilization accounts for over 90 percent of the company's total CO_2 footprint, and it is mostly linked to its combustion (Scope 1 – direct emissions from the company's core activities, accounting for 83 percent of the total) but also its production, processing, and transport (Scope 3 – indirect emissions, mainly from the jet fuel supply chain, 17 percent of the total), as shown in Figure 8.2. Scope 2 emissions, which are emissions embedded in purchased electricity, are quite small.

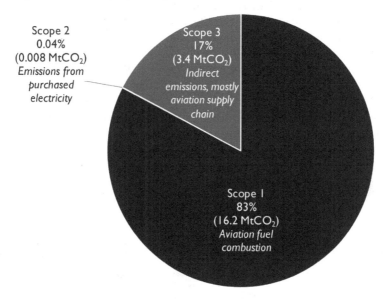

Source: Air France-KLM company documents.

Figure 8.2 Air France Group 2019 CO_2 emissions (estimated total = 20 million tCO_2)

Flight operations, in turn, account for 99.7 percent of Air France's Scope 1 emissions, while ground operations (e.g., engine bench testing, ground vehicles) account for the remaining 0.3 percent.

Scope 3 emissions stem not only from jet fuel-related upstream activities (for example, extraction, production, and distribution of jet fuel), but also from procured goods and services (for example, road deliveries of supplies), passenger travel to and from airports, and employee commuting.

While many airlines have a goal to reach carbon neutrality by 2050, this breakdown shows that aviation decarbonization will essentially come from its ability to limit aircraft fuel consumption, and gradually replace jet kerosene with less-emitting, non-fossil aviation fuel.

The Aviation Sector's Emissions Evolution Over Time

The sector's carbon intensity (i.e., CO_2 emissions per passenger-kilometer) decreased by 85 percent between 1960 and 2018 (Delbecq et al., 2022), one of the highest carbon intensity improvements across industrial sectors. Over the past three decades, it decreased by 1.5 percent per year on average. This improvement has been driven by the use of more modern, thus more fuel-efficient, aircraft. It also reflects an increase in average passenger load factors, a metric of capacity utilization in the passenger aviation sector.

However, annual global aviation CO_2 emissions have increased steadily, from 400 million tonnes in 1990 to 900 million tonnes in 2018 (Meunier & Amant, 2020). This increase in absolute emissions over the last three decades has been driven by worldwide air traffic (both domestic and international) growing at an average annual rate of approximately 5 percent. As reported by the International Energy Agency, the number of air passengers per year globally grew from 1.2 billion in 1990 to 4.6 billion in 2018, which averages out to 4.9 percent per year. In 2018, Asia represented 36 percent of the total number of passengers, Europe 26 percent, and North America 24 percent.

After the unprecedented drop caused by the COVID-19 pandemic, global air traffic is expected, at the time of writing, to return to its pre-pandemic 2019 level sometime between 2025 and 2027. It is then expected to resume steady growth, although at a more modest rate than the pre-COVID growth rate, at just above 3.1 percent per year, according to IATA's December 2021 forecast shown in Figure 8.3.

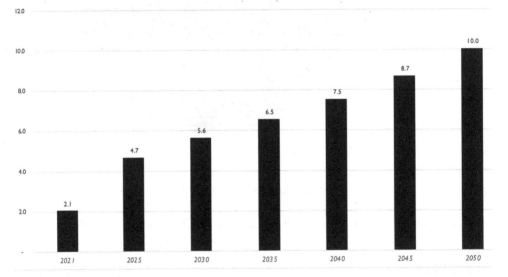

Source: Recreated using data from IATA (2021).

Figure 8.3   Forecasted growth of passenger traffic: 2021–2050 (billions)

The aviation decarbonization challenge for the coming decades will thus consist of deploying carbon intensity improvement levers that will not only offset, but significantly outweigh the expected continuous air traffic growth.

Non-CO_2 Impacts

On top of CO_2, jet kerosene combustion emits other elements such as nitrogen, methane, and soot particles. The latter can lead to the formation of contrails when water vapor condenses as ice. Contrails form when the air is very cold, humid, and supersaturated. They trap and absorb outgoing heat which otherwise would escape into space. Contrails are the most impactful aviation-related non-CO_2 effect as it pertains to global warming.

Aviation's *radiative forcing* (RF), the metric for measuring warming (and hence climate impacts), has been estimated at around 5 percent of global anthropogenic forcing (with a 2 percent – 14 percent uncertainty range). Although non-CO_2 impacts have a larger scientific uncertainty, they appear to account for more than half of aviation's RF impact (Figure 8.4).

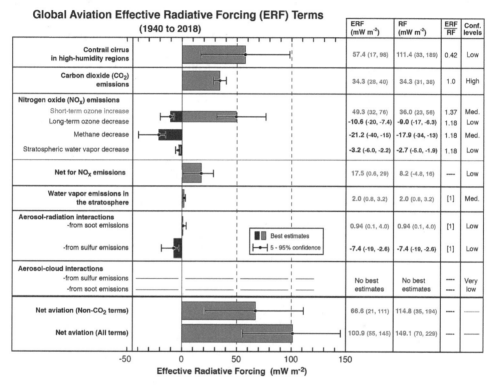

Source: Lee et al. (2021), used with permission.

Figure 8.4   Distribution of effective radiation forcing due to global aviation

Measures to limit the chances of formation of contrails, such as altering aircraft trajectories, are being studied. For example, Air France is collaborating with *Météo France*, the French national meteorological service, to better predict areas of possible contrails formation and consequently alter flight routes. In addition, recent studies show that the usage of sustainable aviation fuels could significantly reduce formation of contrails (Narciso & Melo de Sousa, 2021).

Impact of Climate Change on Aviation

While being a significant contributor to effective radiative forcing, it is worth noting that aviation itself is increasingly impacted by climate change. The main sources of these impacts include:

1. **Flooding of coastal airports:** There are 1,238 airports around the world that are located less than 10 meters above sea level and are facing such threats. The 20 busiest such airports account for 18 percent of worldwide air traffic and 25 percent of air freight. The rise in sea level would place around a hundred of those airports below sea level by 2100. The adaptation costs, such as the construction of sea walls, are estimated at around $57 billion in current dollars (Yesudian & Dawson, 2021).
2. **Wind changes:** Climate change will trigger changes in physical atmospheric conditions, particularly in jet streams. This will lead to increased flight times, but, more importantly, to a rise in en-route air turbulence, leading to increases in physical damages for passengers, crew, and aircraft.
3. **Decrease in aircraft performance:** Rising air temperature reduces aircraft lift as well as engine thrust. Already today, heatwaves at airports can lead to massive flight cancellations and reduction in aircraft maximum take-off weights.
4. **Indirect impacts:** Heatwaves will reduce the attractiveness of some sunnier tourist destinations, thereby reducing travel demand.

Rachel Burbidge, Environmental Policy Officer at Eurocontrol, the European air traffic control organization, has developed a complete list of these impacts, as shown in Table 8.1.

Table 8.1   Potential impacts of climate change on air transport

Climate risk	Impact
Precipitation change	• Disruption to operations, e.g., airfield flooding, ground subsidence • Reduction in airport throughput • Drainage system capacity • Inundation of ground infrastructure (e.g., electrical) • Inundation of ground transport access (passengers and staff) • Loss of local utilities provision
Temperature change	• Changes in aircraft performance • Changes in noise impact due to changes in performance • Heat damage to airport surface (runway, taxiway) • Increased heating and cooling requirements • Increased pressure on local utilities, e.g., water/power (cooling)
Sea-level rise	• Loss of airport capacity • Impacts on en-route capacity due to lack of airport capacity • Loss of airport infrastructure • Loss of ground transport access
Wind changes	• Convective weather: Disruption to operations • Convective weather: Route extensions • Jet-stream: Increase in en-route turbulence • Local wind patterns: Disruption to operations and changes to distribution of noise impact
Extreme events	• Disruptions to operations, route extensions • Disruptions to ground transport access • Disruptions to utilities supply

Source: Burbidge (2018), used with permission.

THE AVIATION SECTOR'S CLIMATE TRANSITION

GHG Emissions Reduction Governance in the Aviation Sector

Delegations from 195 countries signed the Paris Agreement in 2015 at the 21st Conference of the Parties (COP21). The aim was to limit global warming "well below" 2°C compared with pre-industrial levels. Within the Paris Agreement framework, each signatory country provides a Nationally Determined Contribution (NDC) compatible with this goal. Domestic aviation emissions are integrated into those NDCs. International aviation, which accounts for around 60 percent of commercial aviation emissions is, however, not part of those NDCs. This is due to the challenge of accounting for GHG emissions for a flight between two countries, which could be allocated in several different ways: to the departing country, the arrival country, the aircraft registration's country, or even the airline's home country. At the first Conference of Parties (or 'COP,' the annual UN climate summit) in 1995, the United Nations delegated responsibility for controlling civil aviation emissions to the International Civil Aviation Organization (ICAO). In 2010, ICAO set an aspirational goal to stabilize international civil aviation emissions, referring to a "carbon-neutral growth" within ten years. In 2016, it adopted the Carbon Offsetting and Reduction Scheme for International Aviation (CORSIA) to offset, at a country-by-country level, CO_2 emissions in excess of that in 2020, which was considered the reference year (it has since been changed to 2019 due to the COVID-19 crisis).

However, this net emissions stabilization objective appears to be inconsistent with recommendations by the IPCC that call for a drastic 45 percent emissions reduction globally by 2030 versus 1990 levels. A long-term emissions reduction goal for international civil aviation emissions reduction is therefore a high priority item on the ICAO's 41st assembly agenda in September 2022.

Acceleration of Stakeholders' Expectations

In a context of extremely hot summers, natural disasters, and alarming reports by the IPCC, the aviation sector has increasingly been scrutinized with respect to its climate footprint. Airlines have been facing rising expectations from stakeholders:

- **Investors:** Environmental, Social, and Governance (ESG) characteristics have become a new standard for asset managers and owners' extra-financial reporting. Investment strategies thus increasingly rely on ESG characteristics as one of the key valuation metrics.
- **Banks:** The approach to climate risk in banks' portfolios has become more rigorous and systematic. Recommendations of the Task Force on Climate-related Financial Disclosures (TCFD), established by the G20 during COP21 to guide financial and non-financial companies in their public disclosure of the impacts of climate-related risks and opportunities, have been gaining momentum. These recommendations aim to increase the transparency between companies and investors, thereby reducing investment risks, and reconciling short-term financial decisions with the longer-term consequences of climate change. In 2019, the Air France-KLM Group adopted the TCFD metrics for the management of financial risks linked to climate change, and integrated them into its Universal Registration Document, as illustrated in Table 8.2.

Table 8.2   Air France-KLM TCFD principles

	TCFD principles		Implication
1.	Governance	1.1	Describe the Board's oversight of climate-related risks and opportunities
		1.2	Describe the management's role in assessing and managing climate-related risks and opportunities
2.	Strategy	2.1	Describe the climate-related risks and opportunities the organization has identified over the short, medium and long term
		2.2	Describe the climate-related risks and opportunities of the organization's business, strategy and financial planning
		2.3	Describe the resilience of the organization's strategy, taking into consideration different climate-related scenarios, including a 2°C or lower scenario
3.	Risk management	3.1	Describe the organization's processes for identifying and assessing climate-related risks
		3.2	Describe the organization's processes for managing climate-related risks
		3.3	Describe how the processes for identifying, assessing, and managing climate-related risks are integrated into the organization's overall risk management
4.	Metrics and targets	3.1	Describe the metrics used by the organization to assess climate-related risks and opportunities in line with its strategy and risk management process
		3.2	Disclose Scope 1, Scope 2 and if possible, Scope 3 greenhouse gas (GHG) emissions, and the related risks
		3.3	Describe the targets used by the organization to manage climate-related risks and opportunities, and performance against targets

Source: Air France-KLM (2020), used with permission.

- **Public opinion:** The sector is now confronted with an urgent 'societal acceptability' challenge. Air transportation must contend with ever-increasing public pressure at both local and global levels. Challenges range from the *flygskam* movement, to calls for jet fuel taxes in several countries. Furthermore, the COVID-19 crisis resulted in increased questioning of the extent to which societies should focus on economic growth and globalization at the cost of environmental impacts. Aviation is very much in the spotlight on these issues.
- **Corporate customers:** In the global race to fight climate change, companies across all economic sectors are setting climate objectives and emissions reductions trajectories. Airlines' corporate clients seek to reduce emissions linked to their business trips by changing the way they are managing travel. They are shifting to low-carbon means of transport on short routes, but also setting up carbon offsetting programs. Some have even recently considered the financing of Sustainable Aviation Fuels (SAFs). Such initiatives are particularly important for corporate clients whose business travel (Scope 3) emissions are a significant part of their total carbon footprint.
- **Consumers:** With growing environmental consciousness, consumers are now expecting clear information regarding the carbon footprint of the products and services they buy. Many are also now seeking to reduce their individual carbon footprints. A desire for more sustainable mobility is emerging that is translating into fewer trips, longer stays at destination, combining professional and leisure trips, or favoring trains for short distance travels. Amongst the many customer surveys that highlight those changes, one shows, for instance, that one out of seven consumers would opt for a less polluting mode of transport than air travel even if it was less convenient or more expensive (Ipsos survey for the World Economic Forum, 2019).

- **Aviation Employees:** As the sector is being criticized for its large environmental footprint, employees are increasingly self-conscious and wondering why their sector is under attack while it accounts for only 2.6 percent of global CO_2 emissions. This is a major shift away from a time when working for the sector came with a positive halo linked to dreams, glamour, freedom, and discovery. This creates a motivation and pride challenge for current staff but is also limiting for attracting future talent at a time when top graduates seek jobs in sectors at the cutting-edge of the environmental transition. To tackle this talent issue internally, Air France found that staff's initial need was to better understand the climate crisis and environmental challenges, something that had not been part of their prior education. This has, in turn, required training sessions, webinars, and workshops. Air France chose to deploy *Climate Fresk* 3-hour workshops focused on the mechanisms of climate change and based on IPCC findings (Climate Fresk, n.d.).[3]

Fast-growing Commitments and Climate Targets in Aviation

In this rapidly evolving context, the aviation sector has been stepping up its climate commitments at a rapid pace.

In 2009, IATA had set a worldwide target of 50 percent reduction in net CO_2 aviation emissions by 2050, relative to 2005 levels. This went well beyond ICAO's net emissions stabilization goal as from 2020. By 2021, 50 airlines around the globe, including Air France, had made *2050 Net Zero* commitments.[4] Aircraft manufacturers, engine makers, as well as airports, followed suit. In October 2021, IATA set a *Net Zero by 2050* objective, committing the whole sector towards this goal.

Aiming for *net zero emissions* is an ambitious goal in a hard-to-abate sector such as aviation, and we can already see the limitations of such commitments. For example, the aviation sector has yet to agree on a *net zero* standard. Whereas terms such as *net zero emissions*, or *carbon neutrality*, have precise scientific meanings, there has not been a granular scientific definition of what this means for a sector or for individual companies. This requires defining the volume of additional GHG emissions, in other words, a 'carbon budget', to be generated by a given sector to comply with the Paris Agreement's projected trajectory. Most airlines have yet to indicate their precise CO_2 emissions reduction goals by 2050, let alone separate them out from the quantity of offsetting remaining emissions needed.

Another, and more significant, trend in airlines' climate goal setting, has been the commitment to the Science Based Targets initiative (SBTi) (Science Based Targets, n.d.). Such targets provide guidance to sectors, and companies within them, on reducing GHG to a level and at a pace compliant with what the latest climate science deems necessary to meet the goals of the Paris Agreement. With the release of SBTi guidance for aviation in September 2021, many airlines have committed to submitting their short/mid-term GHG emissions reduction objectives to SBTi for approval. In January 2022, 15 airlines or airline groups worldwide, had committed to SBTi, including Air France and Air France-KLM.

In October 2021, SBTi released a so-called *SBT Net Zero* referential. By committing to *SBT Net Zero*, a company will reduce its GHG emissions by at least a certain percentage compared with a reference year, and will use only eligible solutions (e.g., carbon capture) to offset its remaining emissions. As the momentum around *SBTi Net Zero* referential is picking up, one can expect that a growing number of airlines will base their 2050 commitments on this standard. This will help them embark on a credible, rigorous, and acceptable long-term decarbonization strategy.

> ### BOX 8.1 THE AIR FRANCE-KLM GROUP'S SUSTAINABILITY TRAJECTORY
>
> Air France-KLM's ambition is to achieve high standards of performance over the long term by reconciling profitable growth, environmental protection, social progress, and the development of the regions in which it operates.
>
> Air France-KLM has been a signatory of the United Nations Global Compact since 2003 and is committed to applying its ten principles in the areas of human rights, labor, the environment and anti corruption, and to promoting this commitment to its partners. By integrating sustainability into its business and operations, the company strives to contribute significantly, within its scope of influence, to the UN's Sustainable Development Goals.
>
> In 2020, for the 16th year running, the Air France-KLM Group has been included in the Dow Jones Sustainability Indexes (DJSI World and DJSI Europe). The Air France-KLM Group is considered the European leader, and is also ranked the world's number one airline on environmental criteria. The Ecovadis sustainability rating agency awarded the Air France-KLM Group the Platinum Medal, the highest recognition – top 1 percent highest label – in 2020. CDP awarded the Group a score of B- (Management level) for the Climate Change Rating questionnaire. The rating agency ISS-ESG granted Air France-KLM "Prime" status: the Group is the only airline company to figure in the annual ISS ESG ranking of the large global companies deemed to be achieving the highest standards of ESG performance.
>
> The Group's target had been to reduce its CO_2 emissions by 20 percent per passenger-km in 2020 compared with 2011 (including market based measures) but this objective was achieved in 2018 with a reduction of 21.6 percent. In absolute CO_2 emissions terms, the Group reduced its carbon footprint by 6 percent in 2019 relative to 2005.
>
> The Air France-KLM Group's ambition is to reduce its absolute CO_2 emissions by 15 percent by 2030 relative to 2005, and to achieve net zero emissions by 2050. In October 2021, it committed – with the Science-Based Targets initiative (SBTi) – to setting targets in line with the Paris Agreement to limit global warming.
>
> The Air France-KLM group's decarbonization trajectory is mainly based on:
>
> - The Group's airlines fleet renewal with new generation aircraft emitting 20–25 percent less CO_2. Between 2019 and 2021, the Group invested €2.5 billion in fleet renewal.
> - The use of Sustainable Aviation Fuels (SAFs). The Group is working to make these fuels more accessible in terms of quantity and price by creating a sustainable aviation fuel industry in Europe.
> - The search for greater operations efficiency, by favoring more direct trajectories and applying procedures that limit fuel consumption (lighter aircraft, single-engine taxi, continuous descent).
>
> In addition, Air France-KLM is mobilizing the entire sector and is committed to the development of innovative solutions for aircraft design and maintenance, engines, or synthetic fuels, which will ultimately lead to low carbon aviation.

AVIATION'S DECARBONIZATION LEVERS

Defining ambitious GHG emissions reduction goals is a critical step for air sector actors engaging in mitigation of climate change. The biggest challenge, for a sector considered to be one of the most hard-to-abate ones, will then be to fulfill those goals.

As there is no silver bullet, achieving aviation decarbonization will require a combination of GHG reductions levers, with different emissions reduction potential, levels of development, and challenges associated with implementing each of them.

Fleet Renewal

Currently, the most impactful way for airlines to reduce their GHG footprint is to invest in a more fuel-efficient fleet. On average, new-generation aircraft enable a 15 to 25 percent carbon emissions reduction per passenger-km compared with the aircraft they replace. Airline fleet renewal strategies have historically largely been driven by unit-cost considerations (for which fuel efficiency has been the major driver for decades), and by the opportunity to enhance the onboard customer experience. Although it is still too early to demonstrate, decarbonization strategies may well be accelerating the pace of fleet renewals. This effect could be augmented through country-level incentive programs to retire older aircraft, as is presently contemplated in France and at the EU level (Trevidic, 2022).

The COVID-19 public health crisis accelerated the phase-out of older, underperforming aircraft in many airlines worldwide. In the case of Air France, the phasing out of the costly-to-operate Airbus A340, and the exceptionally large Airbus A380, were accelerated. At the same time, the airline started a $1 billion annual investment cycle to add 60 short- and medium-haul Airbus A220s, as well as 38 long-haul Airbus A350s.

It is also worth noting that newer generation aircraft also offer significantly lower noise footprints (34 percent lower in the case of Airbus A220 compared with A319–A320s), a significant environmental externality to which communities living near airports pay attention.

In-flight Fuel Consumption Optimization

With jet fuel accounting for 20–30 percent of total costs, airlines have for decades aimed to optimize in-flight fuel consumption as a means to control costs. Such actions have environmental benefits as well that can lead to a 2 to 5 percent annual reduction in carbon emissions:

- **Weight reductions**: The lighter the aircraft, the less the fuel it consumes. Weight reduction initiatives include, for example, diminishing the weight of seats, galley, and service equipment, paper documentation for cargo and flight decks, and magazines.
- **Flight operations optimization**: Pilots apply, where possible, the most fuel-efficient procedures: Flight Plan precision, speed adjustments and optimized trajectories during the flight, and, on the ground, taxiing with half of the engines shut down. New AI-based tools can help ensure the best application of such fuel-efficiency practices for each flight. For example, Air France cockpit crews use SkyBreathe, developed by Openairlines, a French start-up company.

Sustainable Aviation Fuels

The use of Sustainable Aviation Fuels (SAF) is likely to become the most impactful measure to reduce aviation CO_2 emissions. They enable an 80 percent average reduction in emissions over the entire fuel life cycle, and even up to 100 percent for the most innovative technologies.

Sustainable Aviation Fuels (SAF) are non-fossil aviation fuels produced from:

- Biomass, also called *biofuels* – They are converted from used cooking oils, municipal waste, or forest residues. SAFs being used today are mostly such biofuels.
- Direct carbon capture and green hydrogen, called *power-to-liquid* or *e-fuels*. Those SAFs, allowing the highest emissions reduction potential, are still in research and development phases.

SAFs' chemical and physical characteristics are almost identical to those of conventional jet fuels and can be safely mixed (referred to as *drop-in*) with the latter to varying degrees. This represents a great advantage compared with other aviation decarbonization levers that imply important technological developments which will take decades to emerge. The American Society of Testing and Materials (ASTM) has so far approved seven alternative jet fuels for blending with conventional fossil jet fuel, up to certain limits. The maximum blending rate, for the moment set at 50 percent, will increase in the next decade as aircraft manufacturers and motorists gradually test up to 100 percent SAF incorporation. In 2018, the Boeing ecoDemonstrator flight-test program accomplished the world's first commercial airplane flight using 100 percent sustainable fuels with a 777 Freighter, in collaboration with FedEx. Airbus operated its first A319neo flight with 100 percent sustainable aviation fuel in October 2021.

A key aspect for SAF benefits and acceptability lies in their intrinsic sustainability criteria. These criteria are: a minimal impact on biodiversity, no competition with land used for food production, and no negative implications for limiting access to food resources. The sustainability of first-generation biofuels, used in road transport, has been called into question in the past because of high indirect land use change (ILUC) emissions (Valin (IIASA) et al., 2015). In Europe, the Renewable Energy Directive (RED II) set strict SAF sustainability criteria, particularly to avoid ILUC emissions for biofuels (Renewable Energy – Recast to 2030 (RED II), 2019).

In 2019, close to 200,000 metric tons of SAFs were produced globally, accounting for less than 0.1 percent of the roughly 300 million tons of commercial airlines jet fuel used that year (World Economic Forum – in Collaboration with McKinsey & Company, 2020). SAF utilization by airlines is currently constrained by limited production, and by extremely high costs compared with that of conventional jet fuel. SAF prices currently range from two to eight times the cost of conventional fuel, depending on the pathway. Those costs will inevitably decrease over time with further innovations and scale efficiencies as production rises. Nonetheless a significant price gap will likely remain between conventional jet fuel and SAFs, triggering the question for airlines on how to absorb the cost premium.

Opportunities and Limits of New Aircraft Propulsion Technologies

In recent years, aircraft manufacturers as well as innovative start-ups have announced promising developments in disruptive aircraft propulsion technologies, mainly electric (and hybrid-electric), and hydrogen-based ones. These technologies will likely be first made available to smaller aircraft, such as regional jets and turboprops, for distances not exceeding 1000 km. According to consultants Arthur D. Little, however, aircraft flying routes longer than 1000 km generate 83 percent of worldwide aviation emissions. Introducing such technologies on larger aircraft would imply overcoming significant challenges, such as producing lighter batteries. Currently, for instance, 50 kg of batteries are needed to produce the same amount of energy as 1 kg of fuel. Thus, for example, a 1000 km flight in, say, an Airbus A220, would require around 2.9 tonnes of fuel, or theoretically 145 tonnes of batteries.

The initial step in the emergence of these new propulsion technologies will likely involve hybrid-electric concepts. When combined with a new airframe body, such as a blended wing, they could lead to 40 percent CO_2 emissions reductions. Smaller, fully electric aircraft could be developed in the 2030s, allowing 100 percent CO_2 emissions reductions inflight. Norway, for example, aims to operate all short-haul flights electrically by 2040.

Hydrogen is another promising possibility. It is lighter than jet fuel but needs higher storage volume (around four times for liquid hydrogen). Significantly larger tanks and fundamental changes in the aircraft fuel system are needed. Airbus is contemplating entry into service of hydrogen-based aircraft by the mid-2030s for short and medium distances.

In summary, electric and hydrogen-based technologies are promising developments for future low-carbon air operations, especially on short- and medium-haul flights. However, for long-haul operations that account for most aviation sector emissions, the main decarbonization lever will most likely be the use of SAFs over the next few decades.

Carbon Offsets

Offsetting refers to the action by a company to compensate for its GHG emissions by financing either a reduction in emissions elsewhere or a carbon capture and sequestration program, via natural or artificial carbon sinks. In some regions aviation is integrated into cap-and-trade mechanisms, which allow airlines to purchase CO_2 allowances from other sectors. It is the case for instance in Europe where CO_2 emissions from aviation have been included in the EU emissions trading system (EU ETS) since 2012. Under the EU ETS, all airlines operating in Europe are required to monitor, report and verify their emissions, and to surrender allowances against those emissions. They receive tradeable allowances covering a certain level of emissions from their flights per year.

Carbon offsetting has been seen as an environmentally effective option for aviation, for which abatement costs are very significant, under the concept that an offset, provided it is robust, always represents a tonne of CO_2 which has been avoided or reduced. However, decarbonization standards – such as SBTi – are preventing companies counting offsets in their short-term climate objectives. This is in line with the principle that, in the fight against climate change, the priority for each sector and company is to reduce their emissions "in-house" in a timely manner. As such, carbon offsets can be used as a complement but are frowned upon as a substitute. That said, such offsets will nonetheless be needed to reach aviation's *net zero* objective by 2050.

The effectiveness of offsets is currently a matter of great debate, and offsetting schemes are often considered as "greenwashing," which allows purchasers to avoid efforts that genuinely address climate change. The quality of some projects is regularly called into question.

Effective offsets programs comply with several criteria:

- CO_2 reduction or removal must be 'additional' to 'business-as-usual' activity. Reduction or removal must be permanent and cannot be reversed.
- To quantify the emissions reductions, a baseline must be defined to assess what would have happened if the project had not been implemented. Emissions reductions need to be measured with valid protocols and be externally audited.
- Accounting procedures must track units and avoid double-counting emissions reductions.
- Offsetting projects must comply with local, national, and international laws, and must have safeguards in place to manage environmental and social risks.

Airlines are aligned with market-based initiatives such as CORSIA, which may advance global reforestation. Currently, tree-planting project prices can cost as low as $5 per ton of CO_2 captured but increasing demand will inevitably drive prices up over time. Other offsetting projects include avoiding emissions, for instance by mitigating deforestation or developing renewable energy sources or cooking stoves. Geological sequestration is still expensive and at an early technological phase.

Next to CORSIA frameworks, some airlines have committed to proactively offsetting their CO_2 emissions. Since November 2019, EasyJet has offset 100 percent of the emissions from the fuel on its flights and from their ground operations. Since January 1, 2020, Air France has been proactively offsetting 100 percent of the CO_2 emissions generated by its domestic flights. In cooperation with its partner EcoAct, this offsetting takes the form of participation in certified projects to support reforestation, preserve forests and biodiversity, and develop renewable energies. More and more airlines also offer their individual and corporate customers the opportunity to offset their CO_2 emissions on a voluntary basis, by making CO_2 emission calculators available to customers on their websites. These calculators are directly linked to an emission evaluation system, enabling passengers to offset the carbon emissions associated with their travel.

Air France is partnering with the *A Tree for You* (ATFY), a non-profit organization that connects donors with tree-planting projects around the globe. When purchasing a flight ticket, customers are invited to donate through the *Trip and Tree* program.

New carbon offset schemes are also expected to emerge from new technologies:

- Direct Air Capture (DAC) is a nascent technology that makes it possible to remove CO_2 directly from the atmosphere. It currently relies on large fans that filter air, using a chemical adsorbent to generate a pure CO_2 stream that can be stored or re-used. To have any significant effect on global CO_2 concentrations, DAC would need to be rolled out on a vast scale, along with access to low-carbon energy.
- Carbon Capture Utilization and Storage (CCUS) can capture high proportions of the CO_2 emissions generated by electricity production and industrial processes. It could, for instance, be used to produce SAFs. The CCUS chain consists of capturing the CO_2, recycling it for other industrial purposes, or storing it underground. This technology has been available for many years but has not been widely used due to skepticism as to its ability to be a major part of the world's climate response. One of the main arguments against the use of CCUS technology is that it could drive a prolonged use of fossil energy, rather than pushing investment towards low-carbon and renewable energy. However, the IPCC states that CCUS will be critical to limit global warming to 1.5°C (Masson-Delmotte et al., 2021).

ENABLERS FOR ATTAINING *NET ZERO* EMISSIONS OBJECTIVE IN AVIATION

The air transport sector has rapidly stepped up its climate ambitions to target net zero emissions by 2050. Key decarbonization levers were identified and are the basis for *net zero* trajectories built at airline level as well as at IATA level, as illustrated in Figure 8.5.

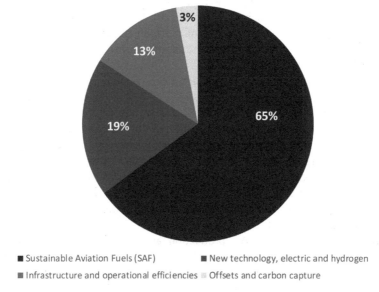

Source: IATA (2021).

Figure 8.5 Decarbonization levers in aviation

However, numerous challenges remain to make *net zero* aviation not only an aspiration but a reality:

- SAFs are considered as the main lever to achieve the *net zero* objective. Nonetheless, this implies sufficient biomass availability (for second generation biofuels) and green hydrogen availability (i.e., hydrogen produced using renewable energy).
- Carbon offsets will be necessary to absorb residual emissions by 2050. Here also, this will depend on aviation access to available natural *sinks* (reforestation projects in particular), as well as DAC and CCUS, which are still nascent.
- New aircraft propulsion technologies will play a role over short and medium distances. This will nonetheless require sufficient access to low-carbon electricity.

Stimulating SAF Adoption

Three conditions will be critical to scale SAF production and to make them economically viable: developing e-fuels, scaling up production and reducing the price gap with conventional jet fuels.

In 2021, The European Commission released an SAF incorporation mandate trajectory within the 'Fit for 55' – a plan to reduce GHG emissions by 55 percent by 2030 in the EU – package for all flights departing from the EU. This trajectory sets an ambitious 63 percent SAF incorporation by 2050, of which 28 percent is for e-fuels. KLM was the first airline in the world to operate a commercial flight with e-fuels in February 2021. However, critical investments in research and development will be needed to see this technology and production pathway mature. Those investments will be even more crucial as aviation cannot rely on biofuels alone. The availability of biomass, necessary for biofuel production, will be much lower than the needs by the different modes of transport (road passengers, road freight, air, ocean shipping) to achieve decarbonization. The use of e-fuels is unavoidable to overcome this shortage in feedstock availability.

The e-fuels production process requires a significant amount of electricity to capture CO_2 and to produce hydrogen (through electrolysis). For e-fuels to achieve the desired GHG emissions abatement over their life cycle, this electricity must be low-carbon. The low-carbon electricity needed to produce e-fuels comes in addition to the one required to produce other alternative energy carriers (electric batteries, ammonia, hydrogen). Based on current growth hypotheses for the different means of transport, nearly 56,000 Terawatt hours (TWh) would be needed by 2050 to produce alternatives to fossil fuels for all means of transportation (in comparison, the current world electricity production is 27,000 TWh). Power generation systems at country and regional levels must be assessed, keeping in mind this energy challenge, and investments planned accordingly.

As mentioned above, current SAF production around the world remains very small. Governments have an important role to play in scaling up SAF production. They can put in place incentives for SAF producers to reduce project risks, to ensure a more competitive business case, and to facilitate the emergence of a sustainable market. Several instruments could be deployed, such as capital subsidies and loan guarantees, feedstock subsidies, tax relief policies, financial market policies (including green bond mechanisms and debt guarantees), accounting policies for accelerated depreciation and amortization schedules, and support for research and development programs.

This has been the approach initiated in the United States through the implementation of various programs:

- The California Low Carbon Fuel Standard (CA-LCFS), updated in 2019, incentivizes SAF production by allowing it to generate credits that can then be sold to other parties.
- The US Sustainable Skies Act, introduced in May 2021, generates credits of $1.50 to $2.00 per gallon for blenders that supply SAF with 50 percent or more lifecycle GHG emission reductions.
- In September 2021, the US federal government announced a new plan to attain SAF production of 3 billion gallons per year by 2030. This plan includes an SAF tax credit aimed at cutting costs for fuel suppliers, and supporting SAF projects with the launch of a new "SAF Grand Challenge" to ramp up domestic production of SAFs (The White House, 2021).[5]

Next to policies aimed at boosting SAF supply, supportive mechanisms could also be considered on the demand side. SAF usage by airlines is unlikely to develop significantly at current prices. The price gap between SAF and conventional jet fuel is expected to diminish as production grows, but cost competitiveness is unlikely to be reached without government policy support.

In particular, an effective policy support would consist of providing direct tax incentives on SAF usage. To ensure this results in the highest climate benefits, the tax incentive might be correlated to GHG emissions reduction, hence favoring SAF technologies that drive the largest reductions.

In the absence of SAF cost competitiveness policies, airlines will be forced to reflect SAF cost premiums into ticket prices. As France introduced an SAF national incorporation mandate of 1 percent as of 2022, the company introduced an SAF contribution to all tickets departing France, covering the premium generated by the policy measure. In view of the high costs represented by airlines decarbonization, ticket prices will likely increase in the future. Without supportive policy this could reach such levels that the most price-sensitive customers would no longer be able to fly. This would result in a severe downturn for airlines and reverse a trend, which started with the appearance of jumbo jets, that made aviation increasingly accessible to consumers.

Lastly, SAF supporting policies ought to be harmonized as much as possible at an international level. An SAF incorporation mandate can be a powerful instrument as it will provide a clear demand signal to fuel producers, thus helping to stimulate production. However, such mandates necessarily increase fuel costs for flights. Since the geographic scope of national mandates is limited to flights taking off from a country, there are potential competitive distortion risks. They could take the form of a traffic shift to surrounding hubs not subjected to the mandate (with lower ticket prices offered), or fuel tankering, whereby an aircraft carries excess fuel in order to reduce or eliminate the need to refuel at its destination. This would undermine policy effectiveness by failing to drive emissions reductions, or could even generate more emissions. The introduction of SAF mandates, if not done homogeneously across all regions can trigger the distortion of competition, reducing the attractiveness of some national/regionals hubs, weakening other airlines or simply encouraging carbon leakage outside national/regional borders.

Accelerating Fuel Efficiency Levers

Incremental fuel efficiency improvements in aviation over time have mostly come from more efficient engines, better aircraft aerodynamics, and reduced weight. The use of composites instead of aluminum to build the latest generation of aircraft has brought weight down, allowing engines to operate more efficiently. Geared turbofan engines, and further advances in design, could drive a further 15–25 percent fuel efficiency improvements over the next two decades. From the mid-2030s new propulsion technologies (electric, hybrid electric, at a later-stage hydrogen) could progressively offer low-carbon transportation for short to medium distances. To accelerate the development of disruptive technologies and reduce aircraft energy consumption, investment support schemes must be maintained, especially for new aircraft shapes, new engines, and new propulsion means. Certification models for disruptive innovations need to be prepared, and mechanisms should be studied to ensure competitive costs for lower-emitting equipment.

Further flight optimization measures could also be deployed, driven by regulators, airports and airlines. Significant CO_2 emissions reductions, for example, can be achieved through traffic and airspace optimization. In Europe the Single European Sky project aims for the better management of air traffic that could lead to 10 percent CO_2 reduction for intra-Europe traffic (Single European Sky to Cut Aviation Emissions, 2011).

Aviation Sector Actors' Transformation

A prerequisite to fully embed this decarbonization challenge relies on the capacity for aviation sector players to adapt their internal processes, governance and decision-making. Until recently, sustainability was often treated by companies as an extra-financial reporting activity, separate from strategic action plans. The situation has changed rapidly and climate and environmental stakes are increasingly integrated into corporate strategy. A key success factor is to ensure that the responsibility for decarbonization is not the "property" of a sustainability department but is shared widely across the organizations. This can only be achieved through a series of actions:

- Acculturation on environmental and climate stakes within the entire organization;
- Employee-training and stronger sustainability knowledge-building in multiple departments (e.g., procurement, finance, operations);
- Introduction of internal carbon pricing so that CO_2 emissions are reflected in business decisions and capital budgeting;
- Robust GHG emissions evaluation methodologies and tools that allow accurate assessments of life-cycle emissions of products and services, as well as modeling of forward-looking CO_2 emissions trajectories and associated costs.

The climate challenge should also prompt airlines to revisit their traditional operating models and value propositions. This is true, for example, on flight connection opportunities. Whereas airlines have traditionally forged alliances with peer operators, collaboration with railway systems can have short-term benefits in terms of overall GHG emissions as well as meeting rising customer expectations. In the case of Air France, the recent past has seen a reinforcement in its cooperation with French railways operator SNCF. In the context of a regulatory obligation to close air routes when a train alternative under 2h 30min exists, Air France has discontinued three domestic routes and implemented code shares on connecting high speed train services. This allows the airline to keep feeding traffic to and from its hub, while meeting a growing customer demand that favors low-carbon alternatives when they exist. Such partnerships between air and rail can drive significant transport-related CO_2 emissions reductions in domestic markets in the short-term. This calls for an internal mindset shift so that rail is no longer perceived as a direct competitor, but rather as a natural partner to deliver a lower-carbon, integrated intermodal transportation system.

CONCLUSION

The air sector has been dialing up its climate goals in recent years, encouraged by increasing stakeholder pressure. Research and development efforts are intensifying towards technological innovations that will ultimately enable a more sustainable, less climate-impacting form of aviation.

Yet the decarbonization challenge remains high for the sector, one whose demand is expected to continue to grow significantly in the future. While aviation emissions have been steadily rising in the past decades, they need to enter a stabilization stage followed by a structural decline stage so as to contribute to the global carbon-neutrality objective by 2050. The sector currently sees SAFs as the main lever to achieve this, but overcoming feedstock availability constraints, scaling up e-fuels production, and enabling decreases in SAF prices will be critical for its large-scale adoption by the world's airlines.

Last, but not least, the environmental transition of the aviation sector must be approached in a systemic manner, not focusing just on climate change. We live in a finite world governed by complex interactions, as evidenced by the Meadows report (Meadows et al., 1972), recently reformulated under the "planetary boundaries" concept, a framework that aims to define the environmental limits of nine major phenomena, including climate change, within which humanity can safely operate (Steffen et al., 2015).

The solution to address climate change cannot apply pressure on another planetary boundary. The development of innovative technologies for aviation, particularly SAFs, must not lead to land-system changes or a worsening of biodiversity integrity.

The transition to more sustainable aviation must fully consider all those dimensions. While technology is likely to facilitate this transition – provided its effects are also consistent with all planetary boundaries – it will not suffice in a context of constrained resources and competition among sectors. As highlighted by the Meadows report 50 years ago, a change in social behaviors, resulting in more sober and reasonable consumption choices, will be critical to ensure a sustainable prosperity for the world.

ACKNOWLEDGEMENT

The author owes a debt of gratitude to Olivier Fainsilber, an Aviation Sector Expert, for the detailed editorial contributions he provided to help to shape this chapter.

NOTES

1. 'Flygskam' is a Swedish word meaning "flight shame". The goal of the movement, which began in 2019, is to encourage people to stop flying so as to lower carbon emissions.
2. https://airlines.iata.org/analysis/2050-net-zero-carbon-emissions
3. Created by the association La Fresque du Climat, the Climate Fresk is a collaborative workshop to collectively understand the implications of climate change and trigger action. It is based on data from the first IPCC working group (Climate Fresk, n.d.).
4. Net zero emissions are achieved when anthropogenic CO_2 emissions are balanced globally by CO_2 removals over a specified period of time (IPCC, 2018).
5. The goal of the Challenge – a joint effort of the US Department of Energy, US Department of Transportation, and US Department of Agriculture – is to reduce costs, enhance sustainability and expand the production and use of SAFs in aviation, while achieving a minimum of 50 percent reduction in life-cycle GHG emissions compared with conventional fuels and meeting 100 percent of aviation fuel demand for SAFs by 2050.

REFERENCES

Air France-KLM. (2020). *Universal Registry Document.* Air France-KLM.
Burbidge, R. (2018). Adapting aviation to a changing climate: Key priorities for action. *Journal of Air Transport Management* (71), pp. 167–174.
Climate Fresk. (n.d.). Association La Fresque du Climat. Paris, France. Retrieved from https://climatefresk.org/

Climate Watch. (2021). *Global Historical Emissions.* https://www.climatewatchdata.org/ghg-emissions?breakBy=gas&chartType=area&end_year=2018&start_year=1990

Crippa, M., Guizzardi, D., Muntean, M., Schaaf, E., Solazzo, E., Monforti-Ferrario, F., . . . Vignati, E. (2018). *Fossil CO_2 Emissions of all World Countries – Technical Report.* Luxembourg: Publications Office of the European Union.

Delbecq, S., Fontane, J., Gourdain, N., Mugnier, H., Planès, T., & Simatos, F. (2022). Aviation and climate: the state of the art. *56th 3AF International Conference on Applied Aerodynamics.* March. https://www.researchgate.net/publication/359541413_Aviation_and_climate_the_state-of-the-art

Gössling, S., & Humpe, A. (2020). The global scale, distribution and growth of aviation: Implications for climate change. *Global Environmental Change,* 65, 102194.

IATA. (2021). *Our Commitment to Fly Net Zero by 2020.* International Air Transport Association. https://www.iata.org/en/programs/environment/flynetzero/#:~:text=IATA%20-%20Fly%20Net%20Zero%20Our%20Commitment%20to,net-zero%20carbon%20emissions%20from%20their%20operations%20by%202050.

IPCC. (2018). *Global Warming of 1.5°C – An IPCC Special Report on the Impacts of Global Warming of 1.5°C above Pre-industrial Levels.*

IPCC. (2021). *Sixth Assessment Report.* Intergovernmental Panel on Climate Change. https://www.ipcc.ch/assessment-report/ar6/

Ipsos survey for the World Economic Forum. (2019). 27-country survey conducted via Ipsos's Global Advisor online survey platform between June 21 and July 5, 2019. Washington, DC, United States of America: World Economic Forum.

Lee, D. S., Fahey, D. W., Skowron, A., Allen, M. R., Burkhardt, U., Chen, Q., . . . Wilcox, L. J. (2021). The contribution of global aviation to anthropogenic climate forcing for 2000 to 2018. *Atmospheric Environment,* January 1, p. 117834.

Masson-Delmotte, V., Zhai, P., Pirani, A., Connors, S., Péan, C., Berger, S., . . . Zhou, B. (2021). *Climate Change 2021: The Physical Science Basis. Contribution of Working Group I to the Sixth Assessment Report of the Intergovernmental Panel on Climate Change.* Cambridge, UK: Cambridge University Press.

Meadows, D. H., Meadows, D. L., Randers, J., Behrens, W., & Club of Rome. (1972). *The Limits to Growth: A Report for the Club of Rome's Project on the Predicament of Mankind.* New York: Universe Books.

Meunier, N., & Amant, S. (2020). Comment construire une aviation durable pour tou·te·s? *Carbone 4: Décryptage Mobilité.*

Narciso, M., & Melo de Sousa, J. M. (2021). Influence of sustainable aviation fuels on the formation of contrails and their properties. *Energies,* p. 5557.

Renewable Energy – Recast to 2030 (RED II). (2019, July 23). Retrieved from EU Science Hub – Reference Regulatory Framework: https://ec.europa.eu/jrc/en/jec/renewable-energy-recast-2030-red-ii#

Science Based Targets. (n.d.). Retrieved from https://sciencebasedtargets.org/

Single European Sky to Cut Aviation Emissions. (2011, July 28). Retrieved from European Commission Eco-Innovation Action Plan: https://ec.europa.eu/environment/ecoap/about-eco-innovation/policies-matters/eu/496_en#

Steffen, W., Richardson, K., Rockström, J., Cornell, S. E., Fetzer, I., Bennett, E. M., . . . Sörlin, S. (2015, January 15). Planetary boundaries: Guiding human development on a changing planet. *Science,* 10.1126(1259855).

The White House. (2021, September 9). Biden administration advances the future of sustainable fuels in American Aviation. *Fact Sheet – Statements and Releases.* Washington, DC, United States of America.

Trevidic, B. (2022, January 18). Une "prime à la casse" pour réduire les émissions de CO2 des avions. *Les Echos.*

Valin (IIASA), H., Peters (Ecofys), D., van den Berg (E4tech), M., Frank, S., Havlik, P., Forsell (IIASA), N., & Hamelinck (Ecofys), C. (2015). *The Land Use Change Impact of Biofuels Consumed in the EU – Quantification of Area and Greenhouse Gas Impacts.* Commissioned and funded by the European Commission. Utrecht: ECOFYS Netherlands B.V.

World Economic Forum – in Collaboration with McKinsey & Company. (2020). *Clean Skies for Tomorrow - Sustainable Aviation Fuels as a Pathway to Net-Zero Aviation.* Geneva: World Economic Forum.

Yesudian, A. N., & Dawson, R. J. (2021). Global analysis of sea level rise risk to airports. *Climate Risk Management*, 31.

9. Leaders and laggards: how have oil and gas companies responded to the energy transition?
Julia Hartmann, Andrew Inkpen and Kannan Ramaswamy

INTRODUCTION

From biomass to coal, coal to oil and gas, and now oil and gas to renewable energy, the world has seen multiple energy transitions. The current transition from oil and gas to renewable energy began several decades ago and will continue for several more. Transitions from one source of energy to a new source proceed through various technological phases that take many years (Rhodes, 2018). For example, the first oil well was drilled in 1859 but it was not until the 1950s that oil made up 25 percent of the world's total primary energy (*The Economist*, 2020a). Transitions also require huge amounts of infrastructure, regulatory changes, and they change the way economies work.

Although renewable energy has been part of the energy mix for decades, it is only recently that forms of renewable energy, such as solar and wind, have become cost effective when compared with fossil fuel energy (US Energy Information Administration, 2021). As the transition occurs, the proportion of the world's energy generated by oil and gas will decrease. Wind and solar energy will likely replace coal and natural gas in the production of electricity. Electric cars and trucks will be powered by renewable energy, reducing the need for motor fuels refined from oil. The demand for oil will peak and then start to decline, raising many questions about the role that oil and gas companies will play in the transition. Oil and gas companies play a central role in the supply of global energy[1] so it is imperative to understand their future role as the transition to renewable energies plays out. Given the large uncertainties associated with the energy transition, how will oil and gas firms position themselves for the future? What strategies will these firms deploy as they prepare for the transition? Will they proactively engage in the search for viable renewable energy alternatives or will they continue to focus on oil and gas with the expectation that the world will continue to need fossil fuels for many more decades?

Given the uncertain landscape for oil and gas firms, this chapter explores how the industry is responding to the energy transition. It discusses broad trends from a study of global oil and gas companies and the approaches they have adopted in addressing the energy transition. We focus on a broad spectrum of adaptation strategies ranging from a complete transition away from fossil fuels, to middle-of-the road approaches that hedge the bets between fossil fuels and renewables, and a business-as-usual approach that seems to wish away the energy transition. In doing so, we paint a rich picture that captures the stresses and strains of the transition, what lies at stake for the supermajors[2] as well as large independents, and the variety of strategic experimentations underway as the industry grapples with its emerging future. We also address some of the shaping forces impacting the firms' strategic choices. These forces include normative and societal pressures originating in environmental health and safety aspirations of the population, and regulatory pressures growing out of governments' desire to spur climate change progress along the transition pathways.

The Oil and Gas Industry's Experience with Renewable Energy

Within the oil and gas industry, there have been attempts by the majors to make inroads into renewable energy. Exxon formed a solar power subsidiary in 1973 called Solar Power Corporation. This was the first company established to specifically manufacture terrestrial photovoltaic (PV) cells in the United States (Jones and Bouamane, 2012). The first PV cells manufactured were used on Exxon's off-shore platforms in the Gulf of Mexico. Exxon sold the business in 1984 to Amoco-owned Solarex. Mobil entered the solar industry in 1974 through a joint venture with Tyco. Both Exxon and Mobil relied on outside entrepreneurs to introduce them to solar technology. Mobil exited the solar industry in 1994 and its Chairman announced, "Our photovoltaics technology is good, but it does not provide us a reasonable business opportunity, either now or in the foreseeable future" (Southerland, 1993).

ARCO (acquired by BP in 2000) was the third US oil company to enter the solar sector (Jones and Bouamane, 2012). ARCO acquired Solar Power in 1977 and became the largest PV manufacturer in the world. ARCO sold the business to Siemens in 1990. Amoco invested in Solarex in 1979 and became 100 percent owner in 1983. Solarex became the world's largest manufacturer of polycrystalline and thin film PV cells and modules, with manufacturing facilities in the United States and Australia (Burr, 1995). Royal Dutch Shell (Shell) entered the solar business in 1973 and in 2001 formed a joint venture with Siemens and E.On that was one of the world's largest PV cell manufacturers. In 2002 Shell acquired 100 percent of the venture (Renewable Energy World, 2002). Shell's solar business was largely closed by 2007.

BP entered the solar industry in 1981 with a joint venture with Lucas Energy, which it bought out a few years later. With BP's acquisition of Amoco in 1999, BP Solar became one of the world's largest PV cell manufacturers with a market share of about 20 percent (Renewable Energy World, 2002). In 2011, BP announced that it was exiting the solar industry, concluding, "solar has evolved into a low-margin commodity market, and in 2011 we began winding down our remaining solar operations as we prepare to exit the business" (BP, 2011:17). Although many of these early forays by the majors into alternative energy investments did not produce lasting impacts, they represented the first wave of cautious diversification into energy sources beyond oil and gas.

Total is an exception among the supermajors because it is the only company to have consistently retained solar energy as part of its portfolio. Total's entry came in 1983 and over the next few years the company was in several joint ventures, including Tenesol and Photovoltech. In 2011, SunPower, a Silicon Valley-based manufacturer of high-efficiency solar cells, solar panels and solar systems acquired PV cell manufacturer Tenesol, by now a wholly-owned Total subsidiary. Total then acquired 60 percent of SunPower's stock (Herndon et al., 2011) with the remaining shares publicly traded. In addition to various other renewable energy investments, in 2019 Total owned 46.74 percent of SunPower. Reflecting its commitment to renewable energies including solar, the company changed its name to TotalEnergies in June, 2021.

The investments in renewable energies by the oil majors in the first wave differ remarkably from the second wave of investments currently underway. The first wave occurred before there was widespread public understanding of climate change and the role played by the oil and gas industry given the GHG emissions that originated from the industry and its products. The first wave was driven primarily by the oil crises of the 1970s and the need to develop alternatives to reduce dependence on imported oil. As those crises receded into the past and the oil industry recovered, renewable energy businesses were shut down. The second wave of diversification into renewable sources of energy can be seen from two perspectives. One, oil and gas firms' investments in renewable energy are a response to an existential threat that has appeared on the horizon. Renewable energy as a competitive threat will reduce the demand for oil and gas, driving oil and gas firms to invest in renewable energy for defensive reasons. Although the demand for oil and gas will likely increase over the next few decades, there is little doubt that the world is transitioning away from fossil fuels. Unlike the oil crises of the 1970s, the energy transition will not be a passing phenomenon. Two, some oil and gas firms see themselves more broadly as energy firms and, therefore, view renewable energy as an opportunity that can be exploited using an existing base of skills.

We believe that unlike investments made during the first wave that were primarily reactive and motivated by growth, the investments of today can be viewed as prudent responses that will open alternative paths to a sustainable future should the core business become untenable. For some oil and gas firms, the investments are options on a different energy future that at this point in time is highly uncertain. For other firms, as we will explain, investments in renewable energy represent the future strategy and a major shift away from fossil fuels. That said, we acknowledge that it is possible that some of the oil and gas firms making renewable energy investments today may reverse course in the future, especially if there is a sustained period of rising oil and gas prices.

THE OIL AND GAS INDUSTRY AND GHG EMISSIONS

Most of the world's human-caused greenhouse gas emissions come from the burning of fossil fuels for energy use – coal, oil, natural gas, and gas liquids such as propane and butane (Harvey, 2021; Kenner and Heede, 2021). In the United States, almost half of the emissions come from oil products, which are primarily used for transportation. Outside the United States the proportion of emissions from transportation is lower but oil and gas are still major contributors to greenhouse gases.

There have been fierce debates about who is responsible for emissions from burning oil and gas. Environmental organizations such as 350.org blame the hydrocarbon companies and want to ban all future oil and gas projects. They argue that the oil and gas companies are responsible for emissions and the world needs to cut off the social license and financing for fossil fuel companies. The oil and gas industry, not surprisingly, takes a different tack. Many of the industry incumbents believe that they are only responding to demand for reliable energy resources from consumers worldwide. Oil and gas firms do not create the demand for energy. The demand comes from the users; if there was no demand there would be no need to produce oil and gas. At some point in the future, the demand will drop or even cease when alternative forms of energy become more widespread and commercially viable.

Regardless of the specific responsibility for greenhouse gas emissions, the products produced by the oil and gas industry are major contributors to man-made climate change. If the world is going to meet the Paris Agreement goals and, eventually, decarbonize, the oil and gas industry will have to play a key role. As a result, the oil and gas industry is under increasing pressure to reduce its emissions.

Some oil and gas firms have taken a very public stance in trying to reduce emissions. BP, for example, has said that its ambition is to become a net zero company by 2050 or sooner and to help the world get to net zero. BP has ten specific aims associated with the net zero ambition. One of those aims is "to increase the proportion of investment we make into our non-oil and gas businesses. Over time, as investment goes up in low and no carbon, we see it going down in oil and gas" (BP, 2019, 7). Others, such as ExxonMobil, have made more modest promises, undertaking to spend more in terms of research on carbon sequestration, or the exploration of the potential use of algae as an input into the energy production process. Still others such as Chevron, have remained focused on operating their fossil fuels business more efficiently without making any major pronouncements about the future.

RENEWABLE ENERGY STRATEGY IN THE OIL AND GAS INDUSTRY

The value chain of the oil and gas industry is typically categorized into three distinct sets of interrelated activities. *Exploration and production* relates to finding and producing oil and gas reserves. *Midstream* encompasses the transportation of crude oil and gas across pipelines, ships, and specialized liquefied natural gas (LNG) carriers. *Downstream* includes refining processes that transform crude oil or gas into component products such as gasoline and other chemical compounds that are sold to consumers or processors who convert these products further. It also covers all the marketing functions involved in the sale of the industry's products. The industry includes both integrated players such as ExxonMobil, Shell and BP that operate across all the three areas of the value chain, as well as specialists who tend to specialize in a narrower set of activities that could fall under one or more of the three core value chain segments.

In developing a comprehensive picture of the renewable energy strategies deployed in the industry, we began by identifying the largest publicly traded companies in all the core segments of the oil industry value chain: integrated companies, as well as exploration and production specialists, midstream transportation and storage firms, and refining and marketing firms based on The Platts Top 250 Global Energy Rankings. This source is widely used by industry professionals as well as investment professionals to assess the overall structure, conduct, and performance of the oil and gas industry. Since the ranking includes many firms that operate in industries beyond oil and gas, such as coal and electricity generation, we only selected 90 oil and gas companies in the list that were active in core segments of the oil and gas industry. Of these, 31 percent were integrated companies, 17 percent were exploration and production specialists, 23 percent were refining and marketing companies, and 23 percent were storage and transport companies. Collectively, these companies represent the single largest block of private capital engaged in the production of oil and gas assets globally and accounted for close to 50 percent of the top rankings in the industry. State-owned companies that were not publicly traded were not included because they often do not disclose operational and strategic details comparable to their publicly traded peers.[3]

Company annual reports were used as source material to measure a company's renewable energy strategy. Consistent with the EIA definition, we used the term renewable energy to include energy that is produced by sunlight, wind, waves, tides, and geothermal heat. We excluded biomass and wood paste from our study, although they do fall within the domain of renewable energy as defined by EIA. None of the firms in our sample engaged in those sub-segments of the renewables business. Areas such as hydrogen production and carbon sequestration had not been commercialized by the oil and gas companies and were excluded. The annual reports form the primary means of communication between an organization and its investors and provide a wealth of information on the current and future strategic priorities of the firm. Investments in renewable energy constitute deliberate strategic choices for an oil and gas firm and represent a significant level of managerial commitment. In turn, management commitment is accompanied by a pattern of action (Ghemawat, 1991). Commitment is defined as willingness to allocate resources and champion activities (Hitt et al., 1990). This definition has the important benefit of capturing the presence or absence of clear alignment between the chosen renewables strategy and related actions to translate the strategy into performance outcomes.

Company annual reports provide evidence of managerial commitment to renewables investments because they delineate both the strategic intent of investments made and management's view of how the investments are expected to play a role in shaping the overall strategy of the company. The annual report data were systematically coded to derive a measure of managerial commitment to renewable energy. The managerial commitment measure included an assessment of both the scope and intensity of focus on renewables as reflected in the mission statement of the company, its strategy as set forth by management in the annual report, as well as specific investments and actions in the renewables sphere outlined in the statutory filings.

Data Extraction and Coding

We analyzed the annual reports for all the firms using 2016 as the base year. The reports formed the source material for content analysis and measure development. We looked at two dimensions of renewable strategy. The *breadth of management commitment to renewables* was defined as variety in terms of renewables technologies, projects, partnerships, and geographies that the firms had implemented or planned to implement. The *depth of managerial commitment to renewables* was defined as the degree to which a renewables strategy had become intertwined with the overall strategy of the firm as reflected in its mission, vision, and strategy statements, as well as its reporting structure.

Several coding schemas were developed and iteratively tested to ensure that relevant content was captured according to established practices for content analysis methods. The final coding scheme was tested on a sub-sample to ensure high inter-rater reliability. The overall group of firms was broken down in sub-groups that were randomly assigned to each of the three authors.[4]

In drawing additional insights into the contextual and organizational factors influencing managerial commitment, we also examined the role of regulatory pressures, and normative pressures arising from a company's external context.

OIL AND GAS COMPANIES AND THE ENERGY TRANSITION LANDSCAPE

Our analysis revealed an array of strategies used by oil and gas companies with respect to their involvement in renewable energy (Figure 9.1).

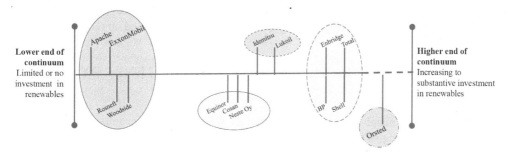

Note: The groupings of firms are only meant to represent similarities in approach to the energy transition and not to illustrate either a good or bad strategy. The position of each firm along the continuum from fossil fuels reliance to a complete renewables portfolio is meant to be illustrative. The companies have been chosen as examples to illustrate a commonly occurring pattern of transition strategy. There are several other firms that would be classifiable in similar groups that are not featured in this chart since it was intended to focus only on example firms.

Figure 9.1   The transition landscape: some exemplar firms and competitive domains

At one end of the spectrum is a set of companies that have either chosen not to invest in the renewable sector or have very limited commercial renewable activities. About one quarter of the companies in our dataset fit into that category. For example, ConocoPhillips, one of the largest independent upstream companies, has limited renewable investments and has said,

> In alignment with our mission to power civilization, and consistent with our positions on sustainable development and climate change, we are evaluating and supporting development of technologies for renewable energy. We are leveraging our expertise, intellectual property and physical assets in pursuit of economically viable, renewable energy business opportunities. (ConocoPhillips, 2018)

Another company toward the left side of the figure is Woodside, one of the world's largest gas producers. In the past few years Woodside has made some small investments in hydrogen, which Woodside has described as the "ultimate clean fuel" (Macdonald-Smith, 2018). The company was looking at two types of hydrogen, blue and green: blue hydrogen is made from natural gas and green hydrogen is produced from water using renewable power.[5]

Moving towards the middle of Figure 9.1 are companies that have made some significant investments in renewable energy. Some of the firms have made a serious commitment to at least one renewable technology. Equinor of Norway, for example, has made a major push into wind energy and has begun constructing and managing wind farms both at home and abroad. Other firms seem to be experimenting and seeking to learn more about the renewable space before they make a major commitment to one or more renewable technologies. Idimetsu Kosan of Japan is actively engaged in wind, solar, biomass, geothermal energy, and hydrogen. The company has a target of 5 GW of renewable energy development overseas by 2030 (Kito, 2020).

Toward the right side of Figure 9.1 are a set of firms that have established renewable energy as a core element in their strategy. Two of these firms, Enbridge and Shell, are profiled in more detail in the next section. Three of the firms are supermajors – BP, Shell, and Total. It is interesting to note that the two US-based supermajors, Chevron and ExxonMobil, are yet to make a major strategic commitment to renewable energy. Finally, on the far right is the Danish company Ørsted, formerly known as Danish Oil and Natural Gas. Ørsted is the rare company that has divested its fossil fuels business and is now completely focused on renewable energy.

OIL AND GAS COMPANIES AND RENEWABLE ENERGY MINI-CASES

To provide deeper insight into company strategies we provide five examples of oil and gas firms and their strategies in the renewable sector:

- ExxonMobil – the largest supermajor with very limited renewable investments.
- Lukoil – a Russian oil company with some investments in renewable power generation.
- Enbridge – a transportation company steadily increasing its renewable power business.
- Royal Dutch Shell – the second largest supermajor, with renewable energy investments since 1973.
- Ørsted – completely divested from fossil fuels.

ExxonMobil

ExxonMobil, the largest supermajor, had a short history with renewable energy several decades ago. The company formed a solar power subsidiary in 1975 and sold most of it by the mid-1980s. ExxonMobil today sees itself "sticking to its knitting" and has publicly stated that it will hew fairly closely to areas where it has expertise, such as GHG emissions reduction and carbon capture and sequestration. The ExxonMobil *Outlook for Energy* acknowledges that an energy transition is underway but provides little discussion of what it means or when peak demand for oil could occur. The *Outlook* says that demand for light duty transportation liquid fuels could peak and be displaced during the outlook period (up to 2040) if electric cars capture 100 percent of the market by 2040. Total liquids demand would still grow because of demand from other sectors such as petrochemicals but the base level would revert back to 2013 levels (ExxonMobil, 2018).

Unlike its peer companies such as BP and Shell, ExxonMobil has not invested in wind and solar energy. When asked about wind and solar in 2018, ExxonMobil's CEO said "We don't have a lot to offer in that space" (Helman, 2018). ExxonMobil's investment in renewable energy is confined to biofuels. The company has spent about $300 million over the past decade on several programs to transform algae and plant waste into biofuels that could be used for transportation. These investments in algae have been largely confined to the company's university partnership with Stanford dating back to 2002, which entails funding in the range of approximately $120 million over 10+ years.[6] ExxonMobil also partnered with Synthetic Genomics, a company established by famous geneticist Craig Venter to explore biofuels and algae, a multi-year investment estimated at $600 million in 2007.[7] These investments pale into insignificance compared with the company's annual average R&D spending, which stood at $1.2 billion.[8] ExxonMobil's CEO has repeatedly mentioned[9] that demand for oil and gas is expected to grow for at least the next two decades as more developing countries improve their living standards. Therefore, he has recast the company's continued emphasis on oil and gas as a moral imperative to support the growth aspirations of the underprivileged countries worldwide, taking the position that the company is better served by focusing on remaining one of the most efficient oil and gas producers in the world.

Lukoil

Lukoil was formed in 1991 when the Government of the Russian Soviet Federal Socialist Republic decided to consolidate three oil production enterprises of Langepas, Uray, and Kogalym as well as several refineries, including those in Perm and Volgograd. Today, Lukoil is the third largest oil producing company in Russia after Rosneft and Surgutneftegas, with facilities spread over 30 countries across the globe.

The year 2010 is the first year in which reference is made to renewable energy in the Lukoil annual report. At that time, Lukoil founded Ecoenergo as a subdivision of the power generation business. The division's main purpose was to specifically focus on renewable energy projects involving wind power, hydroelectric and solar power (Lukoil, 2010). In 2019, Ecoenergo accounted for 5.8 percent of total energy production by Lukoil (Lukoil, 2020). Lukoil operated four hydroelectric power plants in Russia and solar and wind power generation facilities in Russia, Romania and Bulgaria. Lukoil estimated that power generation from renewable sources prevents emission of over 500 thousand tons of CO_2 equivalent per year (Lukoil, 2019). In 2018, Lukoil launched projects in biofuel generation for usage in both marine and road transportation vehicles.

Despite these investments in diverse renewable energy technologies, Lukoil has not formulated a strategy that considers renewable energy a key component. The current strategy covers the period 2018 to 2027 and although it rests on the assumption that the importance of renewable energy and electric vehicles will persist during this period, the company does not specify any strategic goal that would leverage this opportunity. Similarly, in its most recent CSR (Corporate Social Responsibility) report, Lukoil specifies strategic goals related to efficiency, competitiveness, financial performance and social responsibility, but emission or renewable energy related goals are barely mentioned (Lukoil, 2018). One of its goals is: "to leverage climate-related opportunities the Company established a Competence Center on renewables, which evaluates the feasibility of renewable energy projects and generates proposals for further developing this business" (Lukoil, 2019), a fairly low-level of strategic commitment to a renewable energy future.

Enbridge

Enbridge, based in Calgary, was formed in 1949 as the Interprovincial Pipe Line Company and its main business was moving Canadian crude oil. Enbridge's current vision is "to be the leading energy delivery company in North America" (Enbridge Inc., 2020). Enbridge moved nearly two-thirds of Canada's crude oil exports to the US, transported nearly 20 percent of the natural gas consumed in the US, and operated North America's third-largest natural gas utility by consumer count.

Enbridge first invested in renewable energy in the SunBridge wind power project in Saskatchewan in 2002. In the mid-2000s, Enbridge Annual Reports discussed renewable investments as part of CSR. A few years later Enbridge began referring to itself as "one of the world's most sustainable corporations" (Enbridge Inc., 2008). By 2013 renewable energy was discussed in the annual report as a key element in the firm's investment portfolio and the company said "We're investing in renewable and alternative energy technologies that provide attractive returns to our investors while reducing our carbon footprint" (Enbridge Inc., 2013). In 2019 renewable power was reported as one of the company's five business segments. The Annual Report said that management "continue to see significant opportunity in renewable energy, particularly offshore wind. Furthermore, we have tested our existing assets for various energy transition scenarios and concluded that they are highly resilient and can be relied upon for stable cash flow generation well into the future" (Enbridge Inc., 2019). As further evidence of its increasing commitment to renewable energy, renewables was mentioned three times in the 2005 Annual Report and 101 times in 2019.

In 2020 Enbridge had interests in 12 wind farms, four solar energy operations, and a geothermal project, representing more than 1,800 MW of renewable power capacity. The company had a strong presence in Europe with three operating offshore windfarms and four new development projects in offshore France. The company was beginning to invest in self-powering pipeline assets with its own renewable power plants. These investments in renewables suggest that Enbridge is proactively exploring its role in a post-fossil fuels world.

Royal Dutch Shell

Shell first became involved in renewable energy in 1973 with an investment in photovoltaic cells. Shell invested in forestry projects in the early 1980s. In 1997, Shell announced that renewable energy would be one of its core businesses. According to the Group Managing Director, Shell decided to invest in renewables because it is "an energy solution for the future" (Knott, 1997). Despite this statement, a decade later in 2007, renewable energy was not shown as a separate business segment in Shell's financial results. The company said that it was "investing to build a commercially material business in at least one alternative energy technology" (Royal Dutch Shell, 2007). In 2016 Shell formed a New Energies business to pursue three main areas of opportunities: new fuels for transport, such as biofuels and hydrogen; integrated energy solutions, with wind and solar energy which can partner with gas to manage intermittency; and connecting customers with new business models for energy through digitalization and the decentralization of energy systems.

In 2020 Shell was involved in a diverse set of renewable energy investments: wind farms, solar plants, battery storage systems for homes and renewable energy generators, renewable natural gas, and biofuel from sugarcane. It managed over 10,000 MW of power across North America – of which over one-third came from renewable producers. Shell said it hopes to achieve equity returns of between 8 percent and 12 percent from its power business, lower than the 12 percent to 15 percent target for its traditional oil-and-gas business (Petroni, 2019). Although it sees a growing role for cleaner energy forms, its traditional business will still dominate spending plans amid robust global demand for hydrocarbons. Shell said it would spend $2 billion–$3 billion on its New Energies business each year between 2021 and 2025 (Shell, 2020a).

The company said that "Electricity, including from renewable sources, will be a large part of Shell's future as the world moves to lower carbon energy. We want it to become the fourth pillar of our business, alongside oil, gas and chemicals" (Shell, 2020b). In December 2018, Shell announced plans to link executive remuneration to short-term targets to reduce the Net Carbon Footprint of the energy products sold, including customers' emissions from their use of Shell energy products. Collectively, these actions capture the ambition of Shell to play a major role in the evolving world of renewable energy as it foresees an eclipse of its traditional oil and gas prospects over the medium to long term.

Ørsted

In 2017, DONG, Denmark's largest company (majority owned by the Danish state), decided to change course dramatically and announced a sale of all its oil and gas assets. DONG redirected its strategy toward onshore and offshore wind farms and changed its name to Ørsted[10] to signify the move away from the fossil fuels business. In making the strategic change, Ørsted Chairman Thomas Thune Andersen said

> Our vision is to create a world that runs entirely on green energy. We want to spearhead the green transformation. We do so by continuously investing in our competitiveness and core competences to create opportunities for long-term, profitable growth within renewable energy... In the coming years, growth will primarily be driven by our build-out of offshore wind, where we have the largest investment programme in the sector. With the decisions we made in 2017, we completed our strategic transformation from black to green energy. None of the other major energy companies in Europe have come this far in their transformation processes. (Ørsted, 2017)

Ørsted's decision to shed its fossil fuels portfolio was partly precipitated by a crisis in its core fossil fuels operations. Most of its portfolio comprising coal, natural gas, and crude oil was significantly underperforming, forcing a major strategic reorientation. As CEO Henrik Poulsen observed, "The one business where we had some true differentiation was wind" (*The Economist*, 2020b). Ørsted moved quickly to capitalize on its offshore expertise to secure contracts in Britain when large subsidies were on offer. It built on that early experience to establish a sound wind power business that has become very cost efficient, and technology driven, with a distinct advantage in its ability to sense opportunities way ahead of its generalist rivals. Its ability to bring some of its project management expertise garnered in the fossil fuel world to execute projects in renewables in a more cost-efficient fashion also adds to its arsenal of advantages.

In 2020, offshore wind accounted for 83 percent of Ørsted's EBITA and 76 percent of its capital employed. Ørsted also invests in onshore wind and biofuels. Since Ørsted began a radical transformation away from fossil fuels to renewables its stock price has performed well. In the three years to mid-2020 the stock was up more than 150 percent. Ørsted expects to generate a return on its invested capital in the 10 percent range in a few years, which would put it on par with returns that most oil and gas companies typically generate. A strategic reinvention as radical as that of Ørsted is an outlier in the oil and gas business. Although a very bold statement of strategic intent, only time will tell whether the departure from the historic core business was prudent and beneficial to shareholders.

The five case studies presented here exemplify a spectrum of transition strategies that clearly suggests that the drive to invest in renewable energy varies across firms. It is very likely that the differences originate in the unique views of the future that the firms' strategists envision. When confronted with poor performance coupled with the likelihood of an uncertain future, it was perhaps easier for Ørsted CEO Henrik Poulsen to embrace a renewables future. However, in the case of companies such as Chevron and ExxonMobil that have had a strong history of long-term performance, coupled with massive investments in productive reserves, such dramatic moves might be neither prudent nor feasible. Thus, these supermajors might have to adopt a more measured stance towards the energy transition like their European peers Total and Shell. It is also likely that the shaping forces (i.e., normative and regulatory pressures) at play in Europe might be more pronounced than they are in the US, allowing US firms the luxury of a more drawn-out response. We focus more on this issue in the following discussion on the trends in transition strategies over time.

RENEWABLES TRAJECTORIES OVER TIME

To see how oil and gas companies' renewable activities changed over time, we examined renewable activity in 2011 and 2016 for 74 of the 90 companies from our original dataset. We dropped 16 companies because comparable data were not available in the earlier time period. The coding scheme described in the Appendix was used to measure renewable activity in both 2016 and 2011. The *breadth of managerial commitment to renewables* was defined as variety in terms of renewables technologies, projects, partnerships, and geographies that the firms had implemented or planned to implement. The *depth of managerial commitment to renewables* was defined as the degree to which a renewables strategy had become intertwined with the overall strategy of the firm as reflected in its mission, vision, and strategy statements, as well as its reporting structure. Thus, *depth* relates to the intensity of managerial commitment to renewables while *breadth* relates to the scope of such commitment.

The main results are summarized in Figure 9.2.

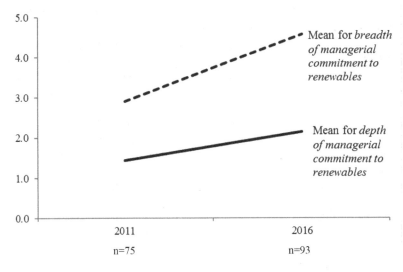

Figure 9.2   *Trends in changes in managerial commitment to renewables over time (2011–2016)*

In 2016, 40 companies had a score of zero in both depth and breadth and this number did not change fundamentally. The "0" scores indicated that the companies in question did not reflect any noticeable commitment to renewables either in terms of intensity of commitment or in terms of the scope of their commitment. The average depth score increased to 2.15 in

2016 from 1.44 in 2011. Along similar lines, the scope of managerial commitment (breadth) increased from 2.91 in 2011 to 4.57 in 2016, The increase in the *depth* score suggests that companies in the sample increased the intensity of their commitment to renewable energies through a combination of moves such as, more clearly defining the renewables business, providing the organizational structure to support such activity, and/or linking renewables to the overall strategy of the firm. Similarly, the increase in the *breadth* score can be interpreted to suggest that the firms in the sample increased their experimentation efforts by exploring a wider range of renewables' technologies such as solar, wind, and geothermal. These changes are discussed in greater detail in the segment that follows.

INTERPRETING THE CHANGES IN RENEWABLES' COMMITMENTS OF OIL AND GAS FIRMS: 2011 VERSUS 2016

Figure 9.2 offers a snapshot of the entire dataset. Although it is notable that both the depth and breadth dimensions of managerial commitment to renewables seem to have increased during the 5-year period, it is important to explore the concomitant granular details of the trend at the level of individual organizations. Tables 9.1 and 9.2 depict the changes in score for both the depth and breadth of managerial commitment to renewables for a sub-sample of firms. These scores were derived by subtracting the 2011 scores from the corresponding 2016 scores. All firms that scored a "0" on both dimensions in 2011 and 2016, and all firms for which there was no change in scores for the breadth and depth dimensions are excluded in Tables 9.1 and 9.2. A positive number represents an increase in depth or breadth of commitment while a negative number reflects a decrease from 2011 to 2016. Our results show that 32 of the sample companies did not change their commitment between 2011 and 2016. However, 30 of the firms increased the depth of managerial commitment to renewables, while 35 increased the breadth of managerial commitment to renewables. Twelve firms decreased the depth of managerial commitment to renewables during the time period while seven of the companies decreased the breadth of their commitment to renewables. Interestingly, 25 of the firms in our sample did not appear to shift their stance with respect to renewables, and received scores of "0" on both dimensions (breadth and depth) for the entire time period covered in the study.

Table 9.1 suggests that for the majority of firms the depth of managerial commitment to renewable energy remained the same or increased during the period.

Many of the large players, especially the European supermajors such as Shell and BP made a firm turn towards renewables during this period while smaller firms in Asia and elsewhere made more incremental changes away from fossil fuels. An interesting picture emerges in looking at the sub-segment of firms that appear to have decreased their commitment measured in terms of depth (i.e., degree to which their renewable activities had become intertwined with their overall strategy). These changes could reflect the period of ferment that Christensen (1997) identified, characterized by rapid swings in technology adoption, experimentation, and consolidation. The decrease in commitment is difficult to untangle ex-ante because the changes might reflect varying forecasts of individual incumbent firms regarding the speed of the ongoing energy transition.

The trend with respect to changes in the breadth of commitment (i.e., variety in technologies, projects, and geographies that firms are pursuing within the renewables domain, see Table 9.2) is quite revealing. It clearly suggests that a large majority of incumbent firms increased the

Table 9.1   Change in depth of managerial commitment to renewable energy, 2011 to 2016

Company	Depth 2011	Depth 2016	Change in depth
Part A: Increases			
Andeavor	0	1	1
BP p.l.c.	3	8	5
Cosmo Energy Holdings Co, Ltd	1	6	5
Empresas Copec SA	4	7	3
Enagás, SA	1	2	1
Enbridge Energy Partners	0	1	1
Enbridge Inc	10	12.5	2.5
Eni S.p.A.	3	7	4
Equinor ASA	1	6	5
ExxonMobil	1	3	2
Formosa Petrochemical Corp	0	0.5	0.5
Galp Energia, SGPS, SA	3	5.5	2.5
Gazprom	0	0.5	0.5
Hèllenic Petroleum SA	1	2.5	1.5
Hindustan Petroleum Corp Ltd	2	6	4
Indian Oil Corp Ltd	3	5	2
Magellan	0	1	1
Neste	5	11	6
NGL Energy Partners	0	3	3
Oil India Ltd	0	3.5	3.5
ONGC	2	3	1
Origin Energy	8	11	3
ORLEN	1	2	1
PTT Plc	0	2.5	2.5
Reliance Industries Ltd	1	2.5	1.5
Shell	4	8	4
Thai Oil Pci	2	7	5
TOTAL S.A.	8	11	3
TransCanada Corp	2	4	2
Vopak	0	0.5	0.5
Average	**2.2**	**4.8**	**2.6**
Part B: Decreases			
Cenovus Energy Inc	1	0	−1
Chevron Corp	2	0	−2
China Petroleum & Chemical Corp	1	0	−1
Husky Energy Inc	1	0.5	−0.5
JXTG Holdings	2	1.5	−0.5
LUKOIL	3	2	−1
Marathon Petroleum Corp	2	0.5	−1.5
Petróleo Brasileiro SA – Petrobras	6	5.5	−0.5
PGNiG	1	0	−1
Suncor Energy Inc	5	4	−1
Valero Energy Corp	6	5.5	−0.5
YPF Sociedad Anonima	2	1	−1
Average	**2.7**	**1.7**	**−1.0**

Table 9.2 Change in breadth of managerial commitment to renewable energy, 2011 to 2016

Company	Breadth 2011	Breadth 2016	Change in breadth
Part A: Increases			
Andeavor	0	2.5	2.5
BP p.l.c.	15	25	10
Cosmo Energy Holdings Co, Ltd	26	27	1
Empresas Copec SA	10	13	3
Enagás, SA	0	3	3
Enbridge Inc	16	29	13
Eni S.p.A.	0	7.5	7.5
Equinor ASA	3	9	6
ExxonMobil	0	0.5	0.5
Galp Energia, SGPS, SA	5	12	7
Hellenic Petroleum SA	11	14.5	3.5
Idemitsu Kosan Co	3	12	9
Indian Oil Corp Ltd	8	9.5	1.5
JXTG Holdings	0	15.5	15.5
LUKOIL	5	10.5	5.5
Magellan	0	2.5	2.5
Marathon Petroleum Corp	0	7	7
MOL Hungarian Oil & Gas Co	2	3	1
NGL Energy Partners	0	4	4
Oil India Ltd	0	10	10
OMV	3	5	2
Origin Energy	5	18	13
Petróleo Brasileiro SA – Petrobras	9	11	2
PGNiG	0	2	2
PTT Plc	0	1	1
Reliance Industries Ltd	0	2.5	2.5
Rosneft Oil Co	0	3	3
Shell	6	14	8
Snam S.p.A.	0	1	1
Thai Oil Pci	4	5	1
TOTAL S.A.	8	14	6
TransCanada Corp	20	27	7
Valero Energy Corp	12	13	1
Vopak	0	3	3
YPF Sociedad Anonima	0	2.5	2.5
Average	**4.9**	**9.7**	**4.8**
Part B: Decreases			
Chevron Corp	4	0	-4
CNOOC Ltd	4	0	-4
Hindustan Petroleum Corp Ltd	6	5	-1
Neste	7	6	-1
ORLEN	3	2	-1
Repsol, SA	5	3	-2
Suncor Energy Inc	10	7.5	-2.5
Average	**5.6**	**3.4**	**-2.2**

variety in the projects that they were involved in. This seems like a rational strategy at the current time given the fact that there are enormous uncertainties associated with the economic viability and scalability of each of the different types of renewables technology that form their project portfolios. Reductions in breadth are not necessarily a negative development because they might reflect the winnowing of options by some companies in favor of a smaller set of technologies as they move into the scalability phase. However, those firms that reduced *both* depth and breadth are of interest because they might be signaling an exit from the search for viable renewable energy strategies. Firms in this group might have either determined that their own long-term prospects in renewable energy were limited or they could be betting heavily on generating the maximum returns possible from doubling down on their investments in fossil fuels, perhaps for the duration that such fuels are in demand. It is also possible that these firms might be playing a wait-and-see strategy, preferring to pursue investments only when the uncertainty abates further.

Two firms (Suncor and Chevron) decreased both their breadth and depth of commitment to renewables. Chevron proposed a bold commitment to renewable energy a few years ago and the company operated several solar and geothermal plants in 2011. However, by 2016, Chevron had sold all of these assets and announced that the returns from renewable energy could not match those of the more traditional oil and gas business (Chevron, 2011, 2016; Elgin, 2014). Suncor called itself "one of Canada's leading investors" in the renewable energy sector in the message to shareholders back in 2011 (Suncor Energy Inc., 2011, 4). This commitment was scaled back in subsequent years. A 2016 statement from the company said that Suncor "continues to advocate for environmental policies and regulations that help us address climate change" (Suncor Energy Inc., 2016, 5). The company sold several wind power plants deemed non-core to better align the asset portfolio with business objectives and to increase the company's liquidity (Morgan, 2016). In 2020, the company had investments in four wind plants and in ethanol production. A previously approved $300 million wind power plant in Alberta was put on hold in 2020.

Nine firms increased their breadth of commitment to renewable energy (or kept it stable) but decreased the depth of commitment over the years 2011 to 2016. Marathon, for example, strategically evaluated algae as a potential fuel source and discussed the potential for collaboration with several Canadian developers in 2011. At the time, the company had not built a production facility but we classified the strategic plan as depth of commitment given the company's clearly stated intent to pursue such a course (Marathon Petroleum Corporation, 2011). In 2016, Marathon did not report any strategic plans to expand the renewables business. However, by then, the company owned a biofuel production facility that produced biodiesel, glycerin and other fuels from by-products (Marathon Petroleum Corporation, 2016). It appears that Marathon implemented its 2011 renewables strategy by 2016 and has no plans to further lift this strategy to the next level.

Lastly, five firms maintained or increased their depth of commitment to renewable energy, but decreased the breadth of commitment. Broadly speaking, these firms have a succinct renewables strategy. However, over time, they found that some types of renewables might not fit their portfolio. Neste from Finland, for example, experimented with solar, geothermal and synthetic energy back in 2011 (Neste, 2011). Today, the company has a global reputation for diesel produced from renewable raw materials. This group represents the emergence of a more focused renewables strategy that encompasses fewer technological areas. This could be the logical outcome of experimentation with a broad range of technologies. From the experimentation comes a more coherent path to renewable energy and the sharpening of focus on a few technologies that could be viable in the medium to long term.

The Changes in Firm-level Commitment to Renewable Energy

The changes in firm-level commitment to renewable energy between 2011 and 2016 help underscore some critical features of the energy transition. A large majority of the incumbent players have deliberately chosen to stay on the sidelines or engage very minimally in the move away from fossil fuels. The actions of these "decliners" seem to be predicated on the belief that the energy transition will unfold over several decades just like the move from coal to natural gas in electricity generation has taken well over a century. Therefore, while some see the onset of monumental change in the energy landscape, others see the case for status quo at least into the medium-term future. Remarkably, some of the firms that seemed to have embarked on a renewables strategy in 2011 have reversed course by 2016.

The frequent entries into different lines of renewable energy and exits in relatively short periods of time might signal continuing experimentation to understand and assimilate the potential future for non-fossil sources. As Christensen (1997) argued, periods of discontinuity are often characterized by an era of ferment and incremental change when incumbents switch back and forth between alternative paths as new technologies and business models are tested and scaled. Since these periods can also be capital intensive, some firms might be willing to trade-off an early mover advantage for more certainty regarding their future direction. The early results of such a process are visible in the set of firms that have jettisoned some technologies or approaches to focus on a smaller set of options and opportunities that they see as being viable. Moves by companies such as Neste of Finland are examples of this process as the company has chosen to focus on bio-diesel derived from non-fossil sources after having tested the waters in solar and geothermal power generation.

The industry snapshot reveals that most of the incumbents have chosen to delay significant strategic changes as the initial period of the energy transition unfolds. While this might appear to be a prudent strategy in the short term, it may hobble the firms from garnering early mover advantages in areas where leaders are already emerging. The argument that "we are an oil and gas firm and have little to contribute to renewable energy" can be countered by firms such as BP, Equinor, Shell, and Total that have proactively entered the renewables sector. Incumbents must also be cognizant of how previous technological transitions have played out. Christensen (1997) found that incumbent firms rarely lead through transitions and most lose their dominance to new entrants despite making substantial strategic investments in the emerging technologies. The new entrant may be more committed to the new technology and its future potential whereas the incumbent is trying to protect cash flow that may be suffering because of the transition away from the older technology. The incumbent may also find itself constrained by its resource allocation process. If a new technology, such as renewable energy, does not meet the needs of existing customers, resource allocation systems may deny capital and human resources to the technology (Christensen, 1997). In this case, the best possible approach for the incumbent is a spinoff that can operate outside the firm's traditional boundaries.

The energy sector has always required a long-term focus and renewable energy is no different. As Shell has said publicly in support of its diverse investments in renewable energy, "We are thinking ahead; where is the future going?" (Reed, 2021). A diverse portfolio of renewable energy investments could be equated with a real options approach. At this stage in the energy transition there remains significant uncertainty about the future size of the market for renewable technologies, customer preferences and acceptance, whether the various technologies will be viable, and whether they can be produced at feasible scale and cost. The various investments made today can be viewed as acquiring the right to expand in the future once the uncertainty is reduced. Properly implemented, a real options approach could give the incumbent firms an initial advantage over capital-constrained new entrants. Eventually, the cost curves and performance outcomes of oil and gas and renewable energy will become more closely aligned, negating the financial advantages of the deep pocketed incumbents.

What will be the fate of those firms (the decliners) that choose to sit out the energy transition? In the near term they will continue to provide a product that retains high demand and provides the majority of the world's energy. Inside these firms there may well be a "business as usual mindset." As industries such as automobiles move into an electrified phase in the 2030s, will they go the way of US coal-fired power plants, video rental stores, and newspapers – crowded out by new technologies, products, and services? Or will they carve out a niche, perhaps as a raw materials supplier for the petrochemical sector? The small number of firms with strategies that reflect proactive strategic positioning are clearly convinced of the eventuality of a non-fossil fuel future. As a result, these firms are going through a process of choosing between alternative renewable paths. As exemplars of this approach, Shell, BP and Total have established investment plans, business organizations, and partnerships for their forays into the world of renewable energy. In the next section we discuss why some firms have embraced a strong commitment to renewable energy while their peers have not.

MOTIVATORS FOR TRANSITIONING TO RENEWABLE ENERGY

In formulating strategy, firms must balance a variety of considerations, including the country-specific environment the firm operates in, the host countries where it operates, the institutional pressures it faces both locally and globally, and its areas of distinctive competence. Institutional theorists (DiMaggio and Powell, 1983; Meyer and Rowan, 1977; North, 1990) argue that firm behavior is shaped by social beliefs, values, norms, constraints, and societal aspirations that are uniquely country specific. These country-specific factors are used by institutional theorists to describe unique patterns or organizational logics that drive how economic transactions are organized. These logics encompass the way in which owners interact with stakeholders, and governments interact with businesses.

Our research found that two country-specific institutional factors influenced management commitment to invest in renewable energy: regulatory pressure and normative social pressure. Regulatory pressures[11] to invest in renewables are strongest in the most developed countries (with some exceptions such as the United States), and especially in Western Europe, and create a "no choice" scenario for firms. Failing to respond to regulatory pressure is rarely a good option as it can lead to legal sanctions and significant financial costs from either penalties or foregone regulatory incentives. This finding underscores the effect that legislation can have on the potential transition from fossil fuels to alternative, renewable sources. However, legislation does not need to be a blunt instrument because it is quite possible that economic incentives may also play a role in eliciting the behavioral changes desired by society. This finding underscores the salience of the role that governments play in either accelerating or delaying the move towards renewable energy sources. While market demand might lie outside the immediate domain of government policy, governments nevertheless do have the ability to induce appropriate behaviors using a combination of incentives and inducements balanced with penalties and punitive actions.

Normative social pressures encompass commonly accepted local social norms and expectations that are an integral part of doing business in the specific country context. These include the value that society places on clean air and water resources, societal expectations for clean energy sources, or the willingness of the local population to embrace renewable energy alternatives. We measured normative environmental pressure using the citizenship rank provided by the US News World Report. We found that managerial commitment to renewable energy increased in countries where the environmental citizenship was more pronounced among the country's population. Since regulatory and normative pressures vary by country, we can expect that firms in some countries will be further along the energy transition path than peers located in settings where such pressures are less pronounced.

THE FUTURE FOR OIL AND GAS FIRMS AND RENEWABLE ENERGY

As energy transition moves forward and climate change concerns take on greater urgency, oil and gas firms will have to decide on their strategic role. Will they continue to focus on oil and gas and stay out of the renewables sector because, as with previous energy transitions, the shift away from fossil fuels will take many decades? Or will they adopt the Shell approach, which is to accept the transition and seek to play a role as a sustainable energy company, competing across a broad portfolio of energy types? According to Shell:

> It's impossible to predict exactly how the transition to lower-carbon energy will play out, but we expect it will mean profound changes for the energy system over the long term ... This will drive change across our portfolio. It is likely to mean more renewable power, biofuels, and electric vehicle charging points; supplying more natural gas; helping advance technology to capture and store carbon safely underground; and helping develop natural carbon sinks like forests and wetlands. We are making investments in these areas today and we expect to do more. We also expect to continue to invest in finding and producing the oil and gas that the world will need for decades to come. (Shell, 2020b)

The oil and gas industry has three major challenges as it gears up to address the energy transition. One is the industry's historical focus on exploitation as the primary means to success. While new technologies have emerged within the industry (e.g., horizontal drilling, hydraulic fracking), they represent incremental rather than radical innovations. The supermajors have built competitive success by improving efficiency across the value chain – from prospecting to refining to petrochemicals. The renewables business represents frame-breaking change. Firms must embrace the exploration of new technologies and business models, something the established players have been reluctant to do after decades of success with incremental change. The pursuit of new knowledge through exploration will become more vital to success than mere exploitation and incremental efficiency improvements. Getting the balance right between an efficiency (exploitation) focus for current projects and simultaneously emphasizing exploration and the search for innovative growth outside the confines of fossil fuels will not be easy.

A second challenge is that the industry has long relied on discovering the next "big elephant" resource base as a preferred way of keeping returns high. The dictum among the supermajors has been "find enough big elephants and we can keep the oil flowing and the profits high." This has forced the industry to weather some deep swings based on the vagaries of commodity prices and a penchant for megaprojects. Success in renewables will likely require a different approach to risk-return tradeoffs. Most of the successful players in wind and solar rely on a broad portfolio of smaller projects, whereas the supermajors have relied heavily on a small set of megaprojects. Although renewables firms also have large-scale projects, none of their capital expenditure outlays match even the modest drilling programs of a supermajor or even a large independent. Thus, in order to enter the renewables segment, oil and gas firms must first get more comfortable with a mindset of short singles and base hits rather than game-changing sixers or home runs.

The third challenge, and perhaps the most difficult one, is that the organizational cultures necessary to succeed in oil and gas and renewable energy are likely to be different. The oil companies of today have well-honed cultures that incentivize conservative decision-making, incremental change, and protecting the status quo so as to minimize commercial and operational risks. While that culture has worked in the past, success in the emerging renewables industry requires a culture and value system that prizes creativity and tolerates failures, both of which challenge the oil industry's well-established touchstones. The willingness to radically transform culture will face the inertial forces of a successful past and a historical sense of invincibility. These inertial forces are bound to favor the well-trodden path and the abundant talent in fossil fuels, which is precisely the wrong recipe for a successful renewables business. The fossil fuel talent in areas such as geology, geophysics, and reservoir management will be less relevant for an emerging renewable energy company, which needs solutions from fields such as biology, artificial intelligence, nuclear design, and chemical engineering. Getting the right people on the bus is a necessity because the wrong people have an incentive to preserve the status quo.

NOTES

1. Oil and gas production constitutes roughly 56 percent of global energy consumption while renewables accounted for 5.7 percent according to the *BP Statistical Review of World Energy* (BP, 2021).
2. The *supermajors* encompass all the large players in the oil and gas business, all of whom are vertically integrated across the global value chain from exploration to retail and chemicals. They include ExxonMobil, Royal Dutch Shell, Chevron, British Petroleum, Eni, and Total.
3. The problem of missing company data was particularly significant in the Middle East and Africa regions where a large number of state-owned companies operate. Therefore, we had to exclude some very large firms such as Saudi Aramco, Kuwait Oil Company, National Iranian Oil Company, Abu Dhabi National Oil Company, and the Nigerian National Petroleum Company, all major producers.
4. Every annual report was independently coded by at least two coders. Independent scores were compared to ensure consistency across coders, highlighting errors of omission which were corrected. The process yielded a mean inter-rater reliability score of $r=0.92$. We tested for stability of the measure over time by evaluating a smaller sub-group of 15 firms based on their 2011 annual reports. The two scores, 2016 and 2011, were correlated at $r=0.81$.
5. More recently, firms have been exploring hydrogen production from natural gas but current technologies do not offer a definitive way to reduce the emissions of GHG in the gas-to-hydrogen production process. Production of hydrogen using the electrolysis route has to contend with a highly

energy intensive process, a conundrum that has not yet been resolved. Given this scenario, despite potential future developments in this arena, our study chose to limit its focus on renewable energy that has already been commercialized and showed competitive levelized costs on a per unit basis of power produced.
6. Exxon Mobil Corporation (2018).
7. Howell (2009).
8. Statista (2021).
9. Exxon Mobil Corporation (2019).
10. The new name gives credit to the Danish scientist Hans Christian Ørsted, who discovered in the 1820s that electric current creates magnetic fields.
11. Regulatory pressure was measured as change in CO_2 equivalent emissions per unit of gross domestic product. This measure assumes that regulatory pressure varies depending on a country's carbon emission levels.

REFERENCES

BP (2011), *Annual Report and Form 20-F 2011*, accessed 15 June 2020 at https://www.bp.com/content/dam/bp/business-sites/en/global/corporate/pdfs/investors/bp-annual-report-and-form-20f-2011.pdf.

BP (2019), *Annual Report and Form 20-F 2019*, accessed 15 June 2020 at https://www.bp.com/en/global/corporate/investors/results-and-reporting/annual-report.html.

BP (2021), *Statistical Review of World Energy 2021*, accessed 15 July 2021 at https://www.bp.com/en/global/corporate/energy-economics/statistical-review-of-world-energy.html.

Burr, M. T. (1995), Entering the pipeline. *Independent Energy*, 25(7), 1–8.

Chevron (2011), *Chevron 2011 Annual Report*, accessed 15 June 2020 at https://www.chevron.com/media/publications.

Chevron (2016), *Chevron 2016 Annual Report*, accessed 15 June 2020 at https://www.chevron.com/media/publications.

Christensen, C. (1997), *The Innovator's Dilemma: When New Technologies Cause Great Firms to Fail*. Boston, MA: Harvard Business School Publishing.

ConocoPhillips (2018), *Renewable energy position*, accessed 10 July 2019 at http://static.conocophillips.com/files/resources/renewable-energy-position_final.pdf

DiMaggio, P. J. and Powell, W. W. (1983), The iron cage revisited: Institutional isomorphism and collective rationality in organizational fields. *American Sociological Review*, 48(2), 147–160.

Elgin, B. (2014), *Chevron Dims the Lights on Green Power*, accessed 15 June 2020 at https://www.bloomberg.com/news/articles/2014-05-29/chevron-dims-the-lights-on-renewable-energy-projects.

Enbridge Inc. (2008), *Enbridge 2008 Annual Report (Sure and Steady)*, accessed 15 June 2020 at http://www.annualreports.com/Company/enbridge-inc.

Enbridge Inc. (2013), *Enbridge 2013 Annual Report*, accessed 15 June 2020 at http://www.annualreports.com/Company/enbridge-inc.

Enbridge Inc. (2019), *Enbridge 2019 Annual Report*, accessed 15 June 2020 at http://www.annualreports.com/Company/enbridge-inc.

Enbridge Inc. (2020), *Our Strategy*, accessed 15 June 2020 at https://www.enbridge.com/about-us/our-strategy.

Exxon Mobil Corporation (2018), *ExxonMobil to join Stanford Strategic Energy Alliance*, accessed 16 July 2021 at https://corporate.exxonmobil.com/News/Newsroom/News-releases/2018/0301_ExxonMobil-to-join-Stanford-Strategic-Energy-Alliance.

Exxon Mobil Corporation (2019), *2019 Barclays Energy Conference transcript of Darren Woods Keynote*, accessed 16 July 2021 at https://corporate.exxonmobil.com/-/media/Global/Files/investor-relations/other-investor-presentations/2019-Barclays-Energy-Conference-transcript-of-Darren-Woods-keynote.pdf?la=en&hash=11134BA9EDFA008D44F2FEB22D238EE780A5D9B5.

ExxonMobil Corporation (2018), *2018 Outlook for Energy: A View to 2040*, accessed 15 June 2020 at https://www.aop.es/wp-content/uploads/2019/05/2018-Outlook-for-Energy-Exxon.pdf.

Ghemawat, P. (1991), *Commitment: The Dynamic of Strategy*. New York, NY Free Press.

Harvey, M. (2021), Historical pathways to climate change. In *Climate Emergency: How Societies Create the Crisis*. Emerald Publishing Limited, pp. 63–92.

Helman, C. (2018), *Over A Barrel: ExxonMobil Preps for the Low-Carbon Future*, accessed 15 June 2020 at https://www.forbes.com/sites/christopherhelman/2018/02/13/over-a-barrel-exxonmobil-preps-for-the-low-carbon-future/#56d0cf6370d8.

Herndon, A., Martin, C. and Goossens, E. (2011), *Total to Buy 60% of SunPower for $1.38 Billion in Solar Bet*, accessed 15 June 2020 at https://www.bloomberg.com/news/articles/2011-04-28/total-to-begin-friendly-tender-for-up-to-60-of-sunpower-shares.

Hitt, M. A., Hoskisson, R. E. and Ireland, R. D. (1990), Mergers and acquisitions and managerial commitment to innovation in M-form firms. *Strategic Management Journal*, 11(4), 29–47.

Howell, K. (2009), *Exxon Sinks $600M into Algae-based Biofuels in Major Strategy Shift*, accessed 16 July 2021 at https://archive.nytimes.com/www.nytimes.com/gwire/2009/07/14/14greenwire-exxon-sinks-600m-into-algae-based-biofuels-in-33562.html.

Jones, G. and Bouamane, L. (2012), "Power from sunshine": A business history of solar energy. *Working Papers – Harvard Business School Division of Research* (12), 1–87.

Kenner, D. and Heede, R. (2021), White knights, or horsemen of the apocalypse? Prospects for Big Oil to align emissions with a 1.5°C pathway. *Energy Research & Social Science*, 102049.

Kito, S. (2020), *Aiming to Become an Energy Co-Creation Company*, accessed 15 June 2020 at https://sustainability.idemitsu.com/en/themes/251.

Knott, D. (1997), Royal Dutch/Shell reinventing itself as total energy company. *Oil & Gas Journal*, 95(47), 29–36.

Lukoil (2010), *Lukoil Annual Report 2010*, accessed 15 June 2020 at https://www.lukoil.com/InvestorAndShareholderCenter/ReportsAndPresentations/AnnualReports/ArchiveAnnualReports1999-2009?wid=wid4Tgn652RYke-bUzKSTJocw.

Lukoil (2018), *Strategy of Achievements and Responsibility. Sustainability Report 2018*, accessed 15 June 2020 at https://www.lukoil.com/InvestorAndShareholderCenter/ReportsAndPresentations/SustainabilityReport.

Lukoil (2019), *Lukoil Annual Report 2019*, accessed 15 June 2020 at https://www.lukoil.com/InvestorAndShareholderCenter/ReportsAndPresentations/AnnualReports/ArchiveAnnualReports1999-2009?wid=wid4Tgn652RYke-bUzKSTJocw.

Lukoil (2020), *Lukoil Power Generation*, accessed 15 June 2020 at https://www.lukoil.com/Business/Downstream/PowerGeneration.

Macdonald-Smith, A. (2018), *Woodside Petroleum wants Pilbara solar to Fuel Hydrogen Market Ambitions*, accessed 15 June 2β2β at https://www.afr.com/policy/energy-and-climate/woodside-petroleum-wants-pilbara-solar-to-fuel-hydrogen-market-ambitions-20180515-h103m2.

Marathon Petroleum Corporation (2011), *Marathon Petroleum Corporation 2011 Annual Report*, accessed 15 June 2020 at http://www.annualreports.com/Company/marathon-petroleum-corporation.

Marathon Petroleum Corporation (2016), *Marathon Petroleum Corporation 2016 Annual Report*, accessed 15 June 2020 at http://www.annualreports.com/Company/marathon-petroleum-corporation.

Meyer, J. and Rowan, B. (1977), Institutionalized organizations: Formal structure as myth and ceremony. *American Journal of Sociology*, 83(2), 340–363.

Morgan, G. (2016), *Ontario Wind Power Assets are Next for Sale after Suncor Sells Off its Lubricants Business*, accessed 15 June 2020 at https://business.financialpost.com/commodities/energy/ontario-wind-power-assets-are-next-for-sale-after-suncor-sells-off-its-lubricants-business.

Neste (2011), *Neste Oil Annual Report 2011*, accessed 15 June 2020 at https://www.neste.com/sites/default/files/attachments/corporate/investors/agm/review_by_the_board_of_directors_2011.pdf.

North, D. C. (1990), *Institutions, Institutional Change and Economic Performance*. Cambridge, UK: Cambridge University Press.

Ørsted (2017), *Ørsted Annual Report 2017*, accessed 15 June 2020 at https://orsted.com/-/media/Aarsrapport2017/Orsted_Annual_Report_2017_Final.ashx?la=en&hash=16E0E6953A3C42EDDD3EAF6BD95CC497.

Petroni, G. (2019), Oil giant Shell's pivot to electricity could bring investors less sizzle. *The Wall Street Journal*.

Reed, S. (2021), *Shell Gets Greener, Even as Climate Advocates Say, 'Go Faster'*, accessed 15 July 2021 at https://www.nytimes.com/2021/06/18/business/shell-renewable-energy.html#:~:text=Royal%20Dutch%20Shell%2C%20though%20still,changes%20have%20to%20come%20quicker.

Renewable Energy World (2002), *Shell Renewables Completes Acquisition of Siemens Solar*, accessed 15 June 2020 at https://www.renewableenergyworld.com/2002/04/29/shell-renewables-completes-acquisition-of-siemens-solar-6428/#gref.

Rhodes, R. (2018), *Energy: A Human History*. New York: Simon & Schuster.

Royal Dutch Shell (2007), *Shell Annual Report 2007*, accessed 15 June 2020 at https://www.shell.com/investors/financial-reporting/annual-publications/annual-reports-download-centre.html.

Shell (2020a), *Royal Dutch Shell plc 2019 Management Day: Shell, Strongly Positioned for the Future of Energy, Provides Strategy Update and Financial Outlook to 2025*, accessed 15 June 2020 at https://www.shell.com/investors/news-and-media-releases/investor-presentations/2019-investor-presentations/management-day-2019.html#:~:text=Ben%20van%20Beurden%2C%20Chief%20Executive,on%20Tuesday%20June%204%2C%202019.

Shell (2020b), *Shell Energy Transition Report*, accessed 15 June 2020 at https://www.shell.com/energy-and-innovation/the-energy-future/shell-energy-transition-report.html.

Southerland, D. (1993), *Mobil Ends Solar Project, Citing Market Weakness*, accessed 10 July 2020 at https://www.washingtonpost.com/archive/business/1993/11/05/mobil-ends-solar-project-citing-market-weakness/25a0a671-00de-44cd-a38c-ffe6881392e7/.

Statista (2021), *Research and Development Expenses of ExxonMobil from 2001 to 2020*, accessed 16 July 2021 at https://www.statista.com/statistics/281239/research-and-development-costs-of-exxon-mobil/.

Suncor Energy Inc. (2011), *Suncor Energy Inc. Annual Report 2011*, accessed 15 June 2020 at https://www.suncor.com/en-ca/investor-centre/financial-reports/archived-annual-reports.

Suncor Energy Inc. (2016), *Suncor Energy Inc. Annual Report 2016*, accessed 15 June 2020 at https://www.suncor.com/en-ca/investor-centre/financial-reports/archived-annual-reports.

The Economist (2020a), *Not-so-Slow Burn. The World's Energy System must be Transformed Completely*, accessed 15 June 2020 at https://www.economist.com/schools-brief/2020/05/23/the-worlds-energy-system-must-be-transformed-completely.

The Economist (2020b), *Ørsted has Helped Boost the Prospects of Offshore Windpower*, accessed 15 June 2020 at https://www.economist.com/business/2019/08/31/orsted-has-helped-boost-the-prospects-of-offshore-windpower.

US Energy Information Administration (2021), *Annual Energy Outlook 2021 with Projections to 2050*, accessed 15 July 2021 at https://www.eia.gov/outlooks/aeo/pdf/AEO_Narrative_2021.pdf.

PART III

CORPORATE STRATEGY AND LEADERSHIP IN THE CLIMATE ECONOMY

10. Climate change communication strategies[1]
Paul Argenti, Posie Holmes and Marloes Smittenaar

WHY COMMUNICATE ABOUT CLIMATE CHANGE?

Communicating about climate change, or what we will refer to throughout this chapter as climate change communication, grows out of activities related to both Corporate Social Responsibility (CSR) and Corporate Communication (CorpComm). CSR acknowledges the intersection of business and society. Climate change, which has been described by economists as an example of market failure (Benjamin, 2007), sits squarely at the intersection of business and society as corporations (as well as individuals) consume and emit carbon, and society (as well as business) faces the repercussions of climate change. Environmental degradation and sustainability are complex topics and, within them, climate change has taken on CSR prominence because the negative effects are increasingly observable and quantifiable, allowing for clear demarcation of progress (or lack thereof). In addition, the implications of climate change are disproportionally averse to minority and underprivileged populations, further cementing climate change within CSR conversations.

CorpComm gives corporations the opportunity, tools, and mechanisms to speak about climate change.

Many companies have grasped that opportunity, and over time communications about climate change have progressed through three distinct phases.

> Schlichting (2013) identified three phases of industry actors' framing of climate change. The first phase in the early to mid-1990s, led by US organizations, used doubt and uncertainty about the existence of climate change. The second phase framed climate change as a socioeconomic consequence, arguing that particularly after the Kyoto treaty, climate sanctions would harm economic progress. In contrast, European industry pioneered the industrial leadership frame – the third phase – acknowledging corporate responsibility and business opportunities while acting on climate change. (Thaker, 2019)

This chapter will focus on the third phase, acknowledging both responsibility and opportunity, and provides guidance on how to craft a successful climate change communication strategy. Following this introduction, the chapter is organized into four overarching sections. The next section will lay the foundation for a discussion of climate communications by making a case for the urgency, immediacy, and growing visibility of corporate climate communications. The third section covers the objectives of climate change communication and key definitions. The fourth section covers a framework for climate change communications and highlights considerations specific to communicating about climate. While climate change communications are fundamentally linked to a corporation's climate change strategy and overall business strategy, we will not delve into those strategies themselves as those topics are covered elsewhere in this handbook. The topics we will cover are grouped into the buckets of corporations, corporation's messages, constituencies, and constituents' messages. The fifth section is a discussion of risks related to climate change communications and will cover greenwashing. In our conclusion, we

232   *Handbook of business and climate change*

will offer recommendations for climate change communications that stem from our analysis of insights from academic research, company actions, and industry leaders.

THE URGENCY BEHIND CLIMATE CHANGE COMMUNICATIONS

As climate change exerts increasing pressure on ecosystems, water, food systems, weather patterns, and other important resources, the social, political, and economic pressure to discuss climate change also increases. Corporate communications about climate change are growing more urgent as they increase in both frequency and prominence; it is critical for corporations to develop the ability to communicate successfully on this topic.

For example, governments, which are an influential constituency for corporations, are signaling the prioritization of climate change in a way that will continue to push corporations to discuss their own carbon footprints, emission plans, risk assessments, and other key topics related to climate change:

- China's President Xi Jinping announced to the UN in September 2020 that China "aim[s] to have CO_2 emissions peak before 2030 and achieve carbon neutrality before 2060" (McGrath, 2020).
- The Biden Administration in the US announced a target for the country to achieve 50–52 percent reduction from 2005 levels in economy-wide net greenhouse gas pollution in 2030 (FACT SHEET, 2021).
- On March 4, 2020 the European Commission proposed the first European Climate Law to "enshrine the 2050 climate neutrality target into law" and "send a strong political signal to our partners and businesses" (European Commission, 2021).

We can also see many examples, especially recently, of corporations that are leaders in their industries issuing wide-reaching communications about climate change:

- Google and Alphabet CEO Sundar Pichai tweeted on December 7, 2020, "Congratulations to @EUCouncil for agreeing to ambitious emission reduction targets. Google is a strong supporter of the #EUGreenDeal and #EUClimatePact, and we stand ready to help Europe in its goal to become a carbon neutral continent" (Pichai, 2021)
- General Motors announced on January 28, 2021 it will be carbon neutral by 2040 in its global products and operations (General Motors, 2021).
- Scandinavian energy provider Ørsted announced in 2017 that it would change its name from DONG Energy, move away from fossil fuels and solely focus on renewable energy (Orsted, 2021).
- Vattenfal, a Swedish producer and retailer of electricity and heat, announced in 2017 it wants to be "fossil free within one generation" (Vattenfal, 2017).
- Apple announced it is committed to being 100 percent carbon neutral for its supply chain and products (Scope 1, 2, and 3 emissions) by 2030 (Apple, 2020).

In addition to these visible instances of corporations communicating about climate change, published research also shows the accelerating nature of climate change communications.

Discussion of climate change topics in earnings calls has increased sharply from 2002 to 2018 as quantified by Sautner et al. (2020). Their findings show the spikes and dips in discussions of climate change topics in earnings calls following large-scale events such as the Paris Agreement or Doha Climate Summit, and it also shows an undeniable upward trend of analyst and investor engagement on talking about climate with corporate leadership, reaching "its peak at the end of [the] sample period" in 2019 (Sautner et al., 2020).

Another study found that

> in financial markets, stocks of carbon-intensive firms underperform firms with low carbon emissions in abnormally warm weather. Retail investors (not institutional investors) sell carbon-intensive firms in such weather, and return patterns are unlikely to be driven by changes in fundamentals. (Choi et al., 2020)

Market analysis is demonstrating clearly the net positive effects of more sustainable business models, leading many firms to eschew previous decades of apprehension and make these long-term investments in their operations. Communication about those investments and operations is imperative.

These trends suggest corporations should expect more conversations about climate change going forward. Despite the upward trend, there is little codified guidance on climate change CorpComm strategies in the academic literature and business publications. As Thaker (2019) pointed out in his recent study, "while an important issue, few studies have focused on corporate communication about science in general, fewer still on corporate communication about climate change in particular" (Thaker, 2019). Given its importance, what is the objective for a climate change communication strategy?

OBJECTIVE OF CLIMATE CHANGE COMMUNICATIONS

The objective of climate change communications is to support the successful execution of a corporation's climate strategy, which should itself be a supporting element of the overall business strategy. When we talk about climate change communications we are talking about any communication activity pertaining to climate change, carbon footprint, emissions, low-emissions transition, or other related topics. We define success in corporate climate change communications as directly or indirectly adding value to the business by supporting strategy execution, long-term financial benefits, reputation preservation or improvement, advantageous constituency engagement, crisis prevention or mitigation, or other value-additive outcomes. This definition is purposefully broad in order to encapsulate the myriad strategies corporations can adopt regarding climate change. It is possible for corporations to have a winning climate change strategy but communicate it poorly and thus miss out on capturing the full value of their strategy. It is also possible for corporations to have inadequate, unwise, or ill-formed climate change strategies but such strong climate change communications that they essentially disguise their poor strategy and defer the need to improve upon it. Many corporations in that camp will eventually face criticism of greenwashing, which will be covered later in the chapter. That being said, these authors strongly advise that both climate change strategies and communication activities should be filtered through the perspective advanced above, namely that climate change is both real and requires action on the part of the corporations to mitigate.

Many corporations view climate change as a component of CSR, and there is evidence that addressing climate within that CSR lens adds value to corporations. Crane and Glozer (2016) argue that communicating about CSR is "a form of stakeholder management for organizations". In addition, it is used for specific goals such as image enhancement and greater legitimacy and accountability for the company or industry. Further, Gupta and Sharma (2009) argue that companies that commit to some sort of righteous initiative also earn more.

As an example of a successful climate communication, Tim Cook famously responded to investors pressing him about the cost of sustainability programs by declaring, "if you want me to do things only for ROI reasons, you should get out of this stock" (Dormehl, 2014). While not a typical climate change communication, this message from Cook was a highly effective way to signal to investors, and other constituencies, a deep commitment for Apple's climate strategy, all the way up to the most senior leaders. Cook made that remark in 2014 and Apple reports their overall carbon footprint then peaked the next year, in 2015, and by 2020 had been reduced nearly 35 percent since that high point (Environmental Progress Report, 2020). Over the same time period, Apple stock rose from an annual average value of about $23 to $95 per share (AAPL Historical Data, 2021).

In contrast, unsuccessful corporate climate change communications are those communication activities (or lack of communication activities) pertaining to climate change that destroy firm value by revealing an insufficient or nonexistent underlying climate strategy, failing to convey strong underlying climate strategies, erode long-term financial value, damage reputation, mismanage constituency relationships, or fail to navigate crisis situations, or other value-destructive outcomes.

We can see an example of an unsuccessful climate change communication in Bank of America's initial communication decisions following the Trump Administration's decision to open the Arctic National Refuge in Alaska for oil and gas production through a drilling leases auction. In this instance, Bank of America's lack of any communication is what was unsuccessful, especially relative to its peer group of large US banks. Following the Trump Administration auction announcement in late 2019, Goldman Sachs revised its environmental policies in December 2019 and became the first large US bank to vow not to finance drilling in the Arctic (Ziady, 2019). Competitors JPMorgan Chase, Wells Fargo, Citigroup, and Morgan Stanley all quickly followed suit and between February and April of 2020 each made a similar announcement (Woody, 2020).

In the following months, Bank of America, the only firm in this set of major US banks that did not announce it wouldn't finance drilling projects in the Arctic, faced pressure from environmental groups and activists. When, in November 2020, Bank of America "said that it won't provide project financing for oil and gas exploration in the Arctic after facing opposition from environmentalists", their head of public policy and strategy felt it was necessary to clarify that, "there's been misunderstanding around our position, but we have not historically participated in project finance for oil and gas exploration in the Arctic. But given that misinterpretation, we've determined that it's time to codify our existing practice into policy" (Nguyen, 2020). Even though Bank of America had a climate-beneficial track record relative to drilling in the Arctic, it faced negative scrutiny from the press and environmental groups because it did not employ a successful communication strategy to support those activities or its underlying climate-change approach.

Preserving the reputation of a firm through a strong climate change communication strategy is beneficial because companies with better reputations enjoy price advantages, competitive advantages, and stability. For example, companies with better reputations can command premium prices, pay lower prices, entice top recruits, have more stable revenues, face fewer risks of crisis, experience greater loyalty (internally and externally), are given greater latitude by constituents (license to operate), have higher market valuation and stock prices, have greater loyalty of investors, and less stock price volatility (van Riel and Fombrun, 2007).

Contribution/Risk Framework:

The role climate change communications play in adding value is demonstrated in Figure 10.1.

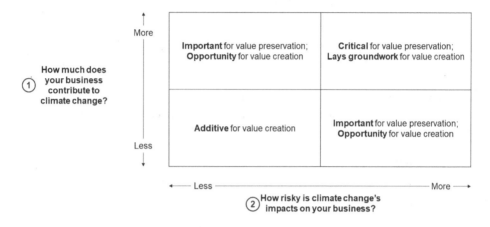

Figure 10.1   Contribution/risk framework

To assess how critical or additive climate change communications will be, corporations should first ask themselves these two essential questions: how much do the business or specific practices contribute to climate change and how much risk does climate change pose for the business or specific practices? The more a business, practice, or decision contributes to climate change, the more constituencies such as government, NGOs, and customers will be urgently looking for answers. The more a business and/or the markets in which it operates are put at risk by climate change (supply chain disruption, increased riskiness of assets, new regulations, etc.) the more constituencies such as employees and investors will be looking for answers.

Communications strategies provide those answers and preserve brand value. In a study of conversations focused on climate change in earnings calls, the sectors with the most climate exposure were Electric, Gas, & Sanitary Services, followed by Construction and Coal Mining – all sectors that both contribute significantly to climate change through emissions and are at risk from disruptions by regulation and physical damage (Sautner et al., 2020). For employee action, in September 2019 over 1,800 Amazon employees walked off the job to protest Amazon's inaction on climate change. The Amazon walkout coincided with The Global Climate Strike (Garcia, 2019).

Determining how much risk climate change presents to a company or activity involves aggregating the effects of various kinds of risks such as physical risks, financial risks, regulatory risks, and reputational risks. The question of risk should also consider business opportunities presented by climate change, such as tailwinds expected in renewable energy industries. For example, climate change might be a financial tailwind for companies such as Tesla as electric vehicles gain momentum, but Tesla is still exposed to risks from climate change that can disrupt supply chains and market activities.

The sliding scale of "more" and "less" are purposefully relative. If the majority of companies in each industry or peer group take actions to reduce their carbon footprint but one company does not, that company (though its footprint has remained constant), now contributes comparably more to climate change relative to its peers.

Climate change communications will help any company reach out to constituents, provide answers, preserve value, and navigate the climate crisis. So, what is the right framework to guide companies considering these communications?

FRAMEWORK TO GUIDE CLIMATE CHANGE COMMUNICATIONS

We propose that corporations designing climate communication strategies should follow the guidance of general CorpComm frameworks and make special note of specific lessons that are important for climate communications. In this section, we will build on Argenti's Corporate Communication Strategy Framework (Figure 10.2) to layer on insights relevant for climate change communications.

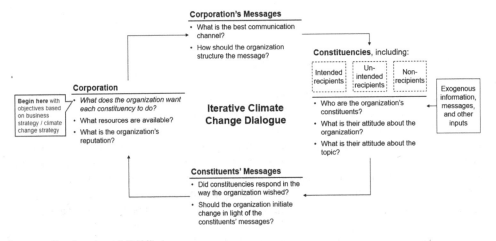

Source: Based on Argenti (2023).

Figure 10.2   Corporate communication model

This framework for CorpComm strategy follows an interconnected structure with four buckets (corporation, corporation's messages, constituencies, and constituents' messages) to create an iterative dialogue about climate change. Corporations have full ownership over decisions and activities in the Corporation and Corporation's Messages buckets, but that's it. Beyond that, messages can be filtered through, amplified by, or even changed entirely by intermediaries. Exogenous information and messages enter the ecosystem to support, muddle, or contradict a corporation's message. And the messages a corporation hears back from constituents might be coming from those who did or did not receive the intended message in the first place, or from constituents speaking directly to other constituents about the corporation. The best climate change communication strategies will understand these dynamics and be prepared to navigate this ongoing, many-headed dialogue.

Corporations

Effective communication strategies need to be firmly rooted in the corporation, its needs, and its abilities. The key questions for corporations to ask themselves about any CorpComm strategy are:

- What does the corporation want each constituency to do?
- What resources are available?
- What is the corporation's reputation?

Specifically for climate change communications, there are several points under the umbrella of Corporation that organizations should incorporate into their climate change communications.

Anchor in strategy

It is essential that the climate change communication strategy is closely aligned with the current state of the corporation's climate strategy and overall business strategy (which should be aligned with each other). As one communication expert at Occidental counseled, "communications can only talk about things that are actually happening" (Schoeb, 2021). Another executive interviewed said: "our communications around climate will evolve following our strategy during the energy revolution" (Industry expert at Total, 2021).

Communications that don't match the realities of a company's actions around climate change can leave the company vulnerable to accusations of greenwashing. For example, PR firm Edelman communicated that it would not work with climate deniers and trumpeted its own environmental work. However, tax filings for the company revealed payments from "the American Fuel and Petrochemical Manufacturers, a major US oil trade organization that even Shell and BP had recently dropped for its aggressive opposition to popular climate solutions" and Edelman soon faced backlash and accusations of greenwashing (Hirja and Taggart, 2021). Edelman broadcast a communication about climate actions that did not match its strategy, which included work with a variety of clients across the sustainability spectrum. We will discuss greenwashing more fully, but the critical lesson here is that when corporations design climate communications they should be firmly rooted in the corporation's underlying strategy and climate actions. The wrong motive, such as mitigating risks or improving reputation, can increase skepticism, which can lead to a negative attitude towards a company, as demonstrated by Elving (2013). A communications executive at oil and gas leader Occidental urged, "communications need to be true, real, and be accurate for where we are as an organization" (Schoeb, 2021), and an executive at a different energy company noted, "you want a communications strategy that is proportionate to where you are. As a business that reflects how you act both internally as well as in public and in industry groups, it is important to make all those things make sense" (Industry Expert One, 2021).

Get the house in order

Another critical step for crafting a climate communication strategy is to first understand the state of affairs as it pertains to climate within its own corporation. Regardless of whether a corporation chooses to share that internal information, it is important for communicators to have that context before shaping the communication strategy. Patagonia is well-known for its climate mission and climate communications, and Patagonia's communications leader stressed that, "it is really fundamental to have your own house in order when you are building a climate communication strategy" (Kenna, 2021).

Sometimes corporations may use government affairs relationships or lobbying activities to support their strategy, and this can be true regarding climate change strategies as well (for developments in favor or against climate change action). This kind of communication activity, in speaking more directly to regulatory or governing bodies, can be effective but corporations should take special care to make sure it is always consistent with climate strategy and public climate change stance or communications. The easiest way to avoid a scandal is to resist scandalous actions in the first place, and lobbying activities revealed to be inconsistent with other climate communications or actions are an open invitation for accusations of greenwashing.

Be proactive

Climate Change is a "predictable surprise" for corporations; there has been ample communication from the scientific community explaining how human activity has driven and continues to drive the proliferation of greenhouse gases in the atmosphere, and detailing the significant challenges we can expect from a climate changed world. Most companies with significant carbon footprints should reasonably assume they will face tough questions from their constituents about their strategy as it pertains to climate change. One of the strongest approaches to avoiding reputational damage is by adopting a proactive approach to examining risks and developing communication strategies that get out ahead of the drivers and impacts of the climate crisis.

Corporation's Messages

After beginning by examining the corporation, the next step in a communications strategy is to delve into the message. Key questions to ask are:

- What is the best communication channel?
- How should the corporation structure the message?

Beyond this general advice, climate change communications have several implications for messages.

Channel decision should support goals

Corporations should make channel decisions depending on audience and topic: who is your audience, what channel should you use to reach them for this topic? What dialogue dynamics are presented by this channel? When it comes to climate change communications, sustainability reports, top leadership, and collectives are three influential channels to consider.

Sustainability reports have grown tremendously in popularity and within those reports corporations are becoming more specific in discussing climate and climate change. A survey conducted by KPMG found that 80 percent of the N100 cohort (representing 5,200 firms globally) report on sustainability as of 2020, up from just 12 percent of the N100 in 1993 (KPMG, 2020). In a 10-year longitudinal analysis of corporate sustainability reporting at 223 companies from Western Europe and North America, Pollach (2018) found the keyword "climate" grew by 238 percent between 2001 and 2010, the largest growth rate of any of the

11 environmental issues evaluated. Incorporating a sustainability report into a communication strategy can leverage already-prepared disclosure information. As a communications expert with experience in both public administration and private corporations advised,

> if you are going to do sustainability work, it makes sense to report it to the public. There is no wrong reason to be more sustainable and it is roundly perceived as good business, both inside and outside business circles; if you are going to do things roundly received as good then why not report it? (Garbow, 2021)

Sustainability reports can be met with criticism, for example if the report is perceived to be inadequate or overly self-serving, but given the trends around sustainability reporting there is also likely to be criticism for corporations that have no reporting available at all.

While sustainability reports appear to be the norm in several industries, sustainability reports alone are not enough for a robust communication strategy because they:

- may not reach all constituency groups,
- have limited ability to use differentiated or local messages for specific constituency groups,
- are generally a passive, inactive, static form of communication.

If the sustainability report conflicts with financial reporting, that is an invitation for scrutiny. In January 2021, *The Guardian* ran an article with the headline "BlackRock holds $85bn in coal despite pledge to sell fossil fuel shares" that publicized apparent disparities between BlackRock's stated sustainability promises and their financials (Jolly, 2021). Released sustainability information should be treated thoughtfully and should align with other sustainability and financial or strategic information being released by the company.

While adopting an additional type of public-facing reporting requires some commitment of time and resources, standardization can reduce the confusion and burden for corporations, and make the released information more valuable for constituents. Although standardization measures have improved as reporting gains momentum, there is not one universal standardized way to report on corporate carbon use. Earlier portions of this book focus more on standardization/disclosure frameworks and organizations, but a few common ones that can support the measurement, disclosure, and management of climate impacts include the Greenhouse Gas (GHG) Protocol, GRI (Global Reporting Initiative), TCFD (Task Force on Climate-related Financial Disclosures), the Carbon Disclosure Project, and the UN SDGs (United Nations Sustainable Development Goals). The adoption of standardizations seen in the N100 does not necessarily imply robust scientific knowledge or sophisticated carbon-reduction strategies; in a study of non-financial reports from 2019, Jagadish Thaker found that in the top 30 Global 500 companies (by revenue), while 43 percent report on the importance/gravity/challenge of climate change and 37 percent identify business responsibility to respond to climate change, only 17 percent include science-based target initiative(s) in their reporting (Thaker, 2019).

Sustainability reporting is becoming commonplace and can support a climate communication strategy, and standardization frameworks or organizations can support a corporation preparing sustainability reports. However, sustainability reports are just one potential channel for climate communications.

Top leadership is another important channel for climate communications because this channel gives legitimacy and authenticity to the message. One expert we interviewed said, "I think that the main issues or topics for climate change should be communicated by top management. In particular, personally by the CEO" (Thaker 2019). Continuing with the example of BlackRock, in his highly influential letter to CEOs, Larry Fink, Founder, Chairman, and Chief Executive Officer of BlackRock, wrote, "companies with a well-articulated long-term strategy, and a clear plan to address the transition to net zero, will distinguish themselves with their stakeholders – with customers, policymakers, employees and shareholders – by inspiring confidence that they can navigate this global transformation" (Fink, 2021). This messaging coming directly from Larry Fink increases its power and credibility; the message becomes a stronger statement about BlackRock's stance and is more influential in its impact on others in business.

Another channel that can be particularly useful for climate communications are collectives or organizations that industries or groups of companies specifically formed to address the topic of climate change. Collectives bring companies together for collective action and engagement. Collectives and collaborations feel like a more trusted source than any one corporation, increasing message believability. Examples of such collectives include:

- The Climate Leadership Council founded in 2017 by AECOM, Allianz, AT&T, Ford, GM, Goldman Sachs, IBM, Johnson & Johnson;
- The Sustainable Apparel Coalition started by Walmart and Patagonia that now has more than 250 brands, retailers, and manufacturer members;
- OGCI: The Oil and Gas Climate Initiative that includes the CEOs from Aramco, BP, Chevron, CNPC, Eni, Equinor, ExxonMobil, Occidental, Petrobras, Repsol, Shell, and Total.

Corporations can use collectives to emphasize their messages, increase influence, and reduce perceived risks from taking on a climate stance alone. In addition, collectives share expertise and amplify climate communications. We heard from an experienced climate leader,

> there is power in the collective and that is really important. Not all companies have the resources or desire to have the capabilities in house to have sophisticated climate policy positions so belonging to those organizations and voicing your concerns through them makes a lot of sense for a lot of companies. (Garbow, 2021)

Companies should, however, be wary of joining collectives that may actually invite greater skepticism, as opposed to invoking greater credibility, as a consequence of the particular actions taken by the collective or the makeup of their membership. For example, outputs from PhRMA (Fabbri, 2018; LaMattina, 2021), the pharmaceutical industry's advocacy association, are frequently met with distrust and skepticism as a consequence of its tight industry ties and perceived lack of objectivity. Corporations collectively taking stands on climate change would do well, then, to learn from other industries' mistakes by ensuring that the combined member makeup is viewed as credible and authentic, as opposed to merely self-serving.

Structure decision should also support goals
Messages can be structured in an infinite number of ways. For climate messages, it is especially important to structure the communication so it is science-based, relatable, and framed in the correct context.

Studies from Yale have shown that anchoring communications about climate change with science-based messaging, especially where there is widespread agreement among scientists (for example, using a phrase such as "97% of climate scientists have concluded that..."), has a powerful effect on how those messages are received:

> communicating the scientific consensus helped neutralize partisan motivated reasoning and bridge the conservative-liberal divide. These findings proved robust across ideology and education levels and build on prior work illustrating that perceived scientific consensus acts as a "gateway" to other key beliefs about climate change (Ding et al., 2011; van der Linden et al., 2015). (van der Linden et al., 2017)

For businesses, electing a science-based approach is a way to reduce politicization, increase buy-in, and open conversations with a wider constituency base beyond those already committed to resolving a human-driven climate crisis. As a communications leader with Patagonia said, "it is important we stay rooted in the science and the facts" (Kenna, 2021).

Even if a corporation does not have all the facts yet, be clear about what you do know and, when applicable, transparent about what you do not know. The IPCC handbook on climate change communications argues for communicators to "lead with what you know" (IPCC, 2018). As one expert urged,

> it is important that you let the general public or other stakeholders know through different channels that you are engaged in the process and working on it. For example, we announced last year that we had the ambition to be net-zero by 2050. Let them know that you are making progress. (Industry Expert at Total, 2021)

Corporations should also structure their messages to be relatable. In their handbook offering guidance on climate change communications, a working group from the Intergovernmental Panel on Climate Change (IPCC) instructs communicators to talk about the real world, not abstract ideas" and "tell a human story" (IPCC, 2018). As one executive said:

> You need to make your climate communications a tangible thing. You can look at dates... and say you are going to change, but for people that is usually too far ahead. You need to make it clear, and small steps so that people can really follow what you are going to do and what the impact is for them. (Industry Expert at Netherlands Wind Energy Association, 2021)

Scientific forecasts of climate change are forward-looking, but to avoid them corporations need to take actions now; the best communication strategies will be able to manage that temporal duality so constituents understand the corporation's strategy, goals, and actions. Another executive noted, "this topic is so complex, and we believe in 'keep it simple' and 'less is more'. Keep it simple is one of our key elements in our strategy" (Industry Expert at Netherlands Wind Energy Association, 2021). To be relatable, be human-centered, specific, and simple. Make sure to avoid jargon that alienates constituencies.

Corporations also need to make decisions about how to frame climate change. Frames are "interpretive storylines that set a specific train of thought in motion, communicating why an issue might be a problem, who or what might be responsible for it, and what should be done about it" (Nisbet, 2009). Nisbet, in his work on frames, argues that audiences rely on frames to make sense of and discuss an issue, so how a corporation frames its climate change communications influences how the audience receives that information.

Nisbet (2009) points to Nordhause and Schellenberger's work that suggests a climate change communication strategy that "involves turning the economic development frame in favor of action, recasting climate change as an opportunity to grow the economy" can speak to a broad coalition of constituencies in a way that lets climate change make sense to them in alignment with what they already care about – this is an important lesson for businesses who also have broad coalitions of constituencies with diverse priorities.

Constituencies

In any communication strategy, there are important questions to ask about constituencies:

- Who are the corporation's constituencies?
- What is their attitude about the corporation?
- What is their attitude about the topic?

As the IPCC (2018) handbook on climate change communication advised, "connect with what matters to your audience." For climate communications, we must highlight a few learnings about various important audiences.

Employees are essential
Having clear and compelling communications with employees and internal constituencies is essential because adapting to climate change will require business changes and risk mitigation that originate within your corporation with your employees. In addition, employees are shown to care about the climate practices of their employers, and sustainability has become a key factor for retaining top talent (Davis-Peccoud, 2013). In our interviews with a communications expert at Patagonia we heard, "really make sure your employees understand why you are talking and what you are talking about – because they will be the first to call you out. Internal communication is essential, so employees are on board and understand why" (Kenna, 2021). Another executive said: "we want our employees to be convinced and to at least know what our position is. Our employees are our ambassadors, and they are one of the most important stakeholder groups when we communicate about climate change" (Industry expert at Total, 2021).

Verifications/certifications can help
Third-party verifications and certifications/statuses can help change constituency attitudes. For example, we heard from an Occidental leader in interviews,

> people wonder if carbon capture is actually going to work, but we have been doing it for 40 years. We got the first MRV (monitoring reporting and verification) plan of any company where the EPA certifies our sequestration efforts. This is a key piece of making sure our communications, what we say, is accurate. (Schoeb, 2021)

Make it local
Climate change communications need to be tailored specifically for different constituency groups and often need to be highly local in their messaging, content, and objective. Besides making content local and easy to understand, leading by example and incorporating culture and social values that are important for constituencies are success factors according to Tan et al. (2008). One interviewee noted, "Content [for climate communication strategy] gets localized for local needs and interests and expectations. No matter what you want to start with,

what is top of mind and what is a priority of your audience?" (Industry Expert One, 2021). For example, Walmart has taken a locally-tailored approach to engaging with utilities corporations, a key constituency for Walmart's intertwined objectives of achieving zero emissions in their operations by 2040 (Climate Change, 2021) and delivering Every Day Low Prices (Our Business, 2021). Walmart works with local utilities corporations and local administrations to create projects and policies in that local market that support Walmart's climate (and overall) strategy, and Walmart's Director of Energy Services is quoted emphasizing how these partnerships are in line with utilities establishing a "customer-centric culture within their customer service and regulatory operations" (Goldman, 2020), presumably a priority for that audience.

Constituents' Messages

The important questions to consider are:

- Did each constituency respond in the way the corporation wished?
- Should the corporation revise the message considering the constituency responses?

For climate, those questions remain critical and are joined by an earnest discussion of how to measure the outcome of climate communications.

Match measurement to goals

Several interviewees argued that measurement methods are entirely subjective to the communication purpose:

- "KPIs [key performance indicators]/goals/objectives that a company is measuring and evaluating the success of depend entirely on the company's strategy. Different companies have different strategies. Whatever you do, your actions and activities are guided by your strategy"(Industry Expert Two, 2021).
- "Our mission in our communication strategy is to convey this message towards our audiences. Our different constituencies. We have KPIs related to how we are carrying our communications. We are committed to climate change, and you can see this, for example, by the amount of content on our website. We do not have KPI's how we want to be perceived. At this point, we are focused on bringing our strategy to our audience" (Industry Expert Three, 2021).
- "The KPIs, goals, and objectives are different for different pieces of communication and campaigns. For example, the Don't Buy This Jacket ad was not intended to sell more jackets; we were seeking to bring attention to consumption and how we all need to lighten our environmental footprint by buying less. With our Facing Extinction campaign, one metric we were looking at was to see how many media interviews we could get for youth climate activists. Sometimes we are looking for the attention of a certain lawmaker – and that is how we measure success. Ultimately, we measure our success based on the outcomes we hope to achieve" (Kenna, 2021).

We argue corporations should design metrics and measurement mechanisms that give them feedback on their specific climate strategy goals. For example, one oil and gas executive talked about the success of making new inroads with constituency groups. He noted, we have been able to have some meaningful conversations with Greenpeace, which would have been unheard of in the past. They are not 100 percent in agreement, but I think it surprises many the things we do agree on… You cannot replace communications and the hard leg work of human interactions. (Industry Expert One).

Often, climate strategy goals are focused on brand trust and reputation. One oil and gas expert said,

the primary objective [in relation to our climate change communication strategy] is around trust. Trust can help us to drive better relations and business outcomes. If we can build trust with the right stakeholders, they will choose our company more often and they will support our goals. The KPIs are really around the measurement of trust.

Trust, highly intertwined with brand reputation, can be approximated, measured over time, and viewed in comparison with competitors through sophisticated measurement tools (Argenti and Calsbeek, 2020).

By following the Argenti communication framework through evaluation of Corporation, Corporation's Messages, Constituencies, and Constituents' Messages, corporations can add specific insights for climate change to a strong CorpComm approach. However, there are risks to keep in mind.

RISKS TO COMMUNICATING ABOUT THE CLIMATE

The most prominent and prevalent risk from climate change communications is "greenwashing". Although some publications (Seele and Geeti et al., 2015) have proposed more precise terminology, greenwashing is the general term used to capture when corporations create or are interpreted to create a false sense of their sustainability. Although climate change communications can be beneficial for an organization, when their constituencies perceive it as greenwashing it can damage a company's reputation in the long run; companies spend years painstakingly building their reputation and greenwashing can damage that valuable asset very quickly.

Accusations of greenwashing, once surfaced, for individual corporations or industries can be persistent over time. For example, an article in *Scientific American* in 2018 argued,

Shell recognized in the 1980s that it played a role in global warming and that the threat from rising temperatures was growing. The research determined that the company generated 4 percent of the world's carbon emissions in 1984, from its production of oil, gas and coal. (Waldman, 2018)

The article went on to note that Shell was "part of the American Legislative Exchange Council, a group that raised doubt about mainstream climate science, until 2015". While we tend to think of greenwashing today as companies promoting false sustainability, the early form of greenwashing seen in this example shows that Shell misrepresented its true environmental impact by concealing or undermining climate change itself. This greenwashing accusation follows the brand today and diminishes the effectiveness of its climate change communications. We can clearly see the ramifications of greenwashing on a Twitter post in which Shell tweeted "What are you willing to change to help reduce emissions?" and included a poll with options "Offset emissions; Stop flying; Buy electric vehicle; Renewable electricity" (Shell, Twitter feed, 2021).

The post received strong backlash from prominent accounts such as US Congresswoman Alexandria Ocasio-Cortez and activist Greta Thunberg; both referencing a legacy of greenwashing. US Congresswoman Alexandria Ocasio-Cortez tweeted in response, "I'm willing to hold you accountable for lying about climate change for 30 years when you secretly knew the entire time that fossil fuels emissions would destroy our planet" (Ocasio Cortez, 2021). Greta Thunberg also responded to the original Shell tweet with a scathing response of, "I don't know about you, but I sure am willing to call-out-the-fossil-fuel-companies-for-knowingly-destroying-future-living-conditions-for-countless-generations-for-profit-and-then-trying-to-distract-people-and-prevent-real-systemic-change-through-endless-greenwash-campaigns" (Thunberg, 2021).

It's highly unlikely that Shell wanted to engage in a public conversation about climate change culpability with Greta Thunberg and Alexandria Ocasio-Cortez over Twitter; in addition to highlighting greenwashing criticism, this example shows the complexity of climate change dialogue when messages leave the corporation's sphere of control. The response was a redirection of the corporation's message, and not at all a match to the corporate objective. In retrospect, Shell might have considered if they wanted to reach specific constituents with this poll, or if they really wanted it going out to the whole Twitterverse.

The best way to avoid greenwashing is to be honest and transparent. Specifically:

- Avoid getting ahead of yourself:
  - Companies communicating about climate change should back up their messages with actions. Otherwise, the company opens itself up to risk through the possibility of backlash, eroded reputation, and destroyed value. "The first risk would be communicating too much, too early. When you are not ready as a company. Communications cannot replace the truth of the business. It must reflect reality. It's not like magic. It needs to be honest and realistic. The more honest you are, the more your communications work in the long run" (Industry Expert at Total, 2021). "There are critics that will say bold communications, like those from Ørsted, are greenwashing, but it's also about intentions of their course. It worked because they followed up with actions" (Industry Expert at Netherlands Wind Energy Association, 2021).
- Focus on building trust with your various constituencies:
  - Trust is a valuable bond that is built over time and allows your constituencies to believe what you say about your climate change strategy and activities are genuine and true. "We believe that creating a trustworthy reputation, can be built on facts. It is important to gain credibility, that is a long-term activity. The backbone is true commitment to the matter, climate change and global warming. Being able to connect this concern we have with the concern of society and our different constituencies" (Industry Expert Three, 2021)

In addition, corporations can guard against greenwashing by engaging the credibility of trusted entities such as:

- Following guidance from regulatory bodies such as the US Federal Trade Commission, which releases Guides for the Use of Environmental Marketing Claims ("Green Guides") (US Federal Trade Commission, 2021).

- Seeking accreditation or certification from reputable groups such as B Corporations, carbon measurement organizations, etc.
- Partnering with respected entities in the space such as activists, NGOs, and nonprofits.

Despite these risks, we feel strongly that the positive benefits of successful climate change communication strategies far outweigh the risks if the companies are genuine in their efforts to help solve the problem of climate change. Brands that use climate change communications particularly well, such as Patagonia, have benefited from "firm messaging over a long period" in addition to "an unbending belief that what it's doing is right", which are two essential ingredients to differentiating genuine climate strategy "at a time when it's difficult to discern which brands are merely saying what they think will help them keep or attract customers" (Chang, 2021). In the next section, we offer a set of summary recommendations for companies to follow based on our research into this topic.

KEY RECOMMENDATIONS AND CONCLUSION

Corporate climate change communication strategies are an essential tool for corporations to realize their climate change strategies and navigate the myriad changes shaking up industries as society reckons with the low-carbon transition needed to resolve the climate crisis. Constituencies are heightening their engagement on the topic of climate change; as scrutiny increases, climate change communications are important to preserve or add value. Corporations should use the CorpComm strategy framework annotated in Figure 10.3 with additional information related to climate change to guide communications.

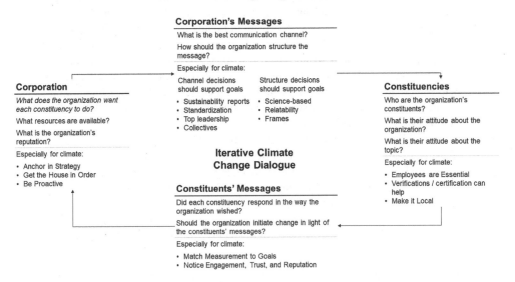

Figure 10.3   Corporate communication model with additional insights for climate change

In addition to those specific recommendations, these authors recommend that corporations think about climate change as a topic that is here to stay. The business community has been gradually ramping up considerations about climate for many years now, but the more recent pace of communication and action has accelerated sharply as climate effects become more prominent, the previously distant forecasted periods become more imminent, and constituency groups express a greater urgency to engage on the topic. As such, we strongly recommend corporations design their climate change communication strategies with the perspective that managing greenhouse gas inputs and emissions will be the new way of doing business. This means having a clear strategy and realistic objectives without getting ahead of where you are as a company (market yourself on the basis of your actions, not your aspirations). Every corporation has a carbon legacy from using resources and therefore every corporation has a story to tell about who it is relative to climate change and who it will be in the face of the climate crisis. Acknowledging that legacy is the only way to move forward as society, governments, and industries attempt to co-create a sustainable future. Be honest and be real. And don't wait for a crisis to start communicating or collaborating with constituencies.

Ask yourself, can you do the things you need to do? Do you have a strategy and is it achievable? What are the capabilities, internal systems or knowledge in your company to support climate communications? How do you connect communicators with climate expertise, internally or externally? If you do not have the right tools, seek the missing pieces. The global transition in the face of climate change is laborious but ultimately necessary, so make sure you have the necessary people, skills, and capabilities to carry you into the future.

Climate communication is an evolving area. Guidelines and recommendations for climate communications will continue to grow and develop as businesses rise to meet this global challenge. Corporations that develop strong climate communication strategies create an opportunity to enhance their reputation, their financial value, and their likelihood of thriving through a global climate revolution and into the long term.

As one interviewee aptly summarized, "The benefit of climate change communications is that you're building a future-proof company. It is necessary for companies to communicate for the survival of the company. You have to have a long-term perspective. Companies certainly will survive the next decade, but you have to look beyond 2030. They need to have a long horizon" (Industry Expert at Netherlands Wind Energy Association, 2021).

NOTE

1. The views expressed by the authors do not necessarily represent those of any of their employers.

REFERENCES

AAPL Historical Data (2021), *NASDAQ*, accessed 27 May 2021 at https://www.nasdaq.com/market-activity/stocks/aapl/historical.
Apple (2020), Apple commits to be 100 percent carbon neutral for its supply chain and products by 2030, *Apple Press Release*, accessed May 27, 2021 at https://www.apple.com/newsroom/2020/07/apple-commits-to-be-100-percent-carbon-neutral-for-its-supply-chain-and-products-by-2030/.
Argenti, P. A. (2023), *Corporate Communication*, 8th ed. New York: McGraw-Hill.
Argenti, P. A. and Calsbeek, R., 2020. What evolutionary biology can teach us about corporate reputation. *Management Business Review*, 1(1), 1–4.
Benjamin, A. (2007), "Stern: Climate change a 'market failure'", accessed 26 July 2021 at https://www.theguardian.com/environment/2007/nov/29/climatechange.carbonemissions.

Chang, A. (2021), "Patagonia shows corporate activism is simpler than it looks", accessed 27 May 2021 at https://www.latimes.com/business/story/2021-05-09/patagonia-shows-corporate-activism-is-simpler-than-it-looks.
Choi, D., Gao, Z., and Jiang, W. (2020), Attention to global warming. *The Review of Financial Studies*, accessed 27 May 2021 at https://academic.oup.com/rfs/article-abstract/33/3/1112/5735304.
Climate Change (2021), Accessed 18 August 2021 at https://corporate.walmart.com/esgreport/esg-issues/climate-change.
Crane, A. and Glozer, S. (2016), Researching corporate social responsibility communication: themes, opportunities and challenges. *Journal of Management Studies*, 53(7), 1223–1251.
Davis-Peccoud, J. (2013), "Sustainability matters in the battle for talent", accessed 27 May 2021 at https://hbr.org/2013/05/sustainability-matters-in-the.
Ding, D., Maibach, E.W., Zhao, X., Roser-Renouf, C., and Leiserowitz, A. (2011), Support for climate policy and societal action are linked to perceptions about scientific agreement. *Nature Climate Change*, 1, 462–465.
Dormehl, L. (2014), "Tim Cook tells profit-obsessed investors to sell their stock", accessed 27 May 2021 at https://www.cultofmac.com/268413/tim-cook-tells-profit-obsessed-investors-sell-stock/.
Elving, W. J. (2013), Skepticism and corporate social responsibility communications: The influence of fit and reputation. *Journal of Marketing Communications*, 19(4), 277–292.
Environmental Progress Report (2020), *Apple*, accessed 27 May 2021 at https://www.apple.com/environment/pdf/Apple_Environmental_Progress_Report_2020.pdf.
European Commission (2021), *The European Climate Law*, accessed 27 May 2021 at https://ec.europa.eu/clima/sites/default/files/eu-climate-action/docs/factsheet_ctp_en.pdf.
Fabbri, A., Lai, A., Grundy, Q., and Bero, L.A. (2018), The influence of industry sponsorship on the research agenda: A scoping review. *The American Journal of Public Health*, accessed 27 July 2021 at https://www.ncbi.nlm.nih.gov/pmc/articles/PMC6187765/.
FACT SHEET (2021), "President Biden sets 2030 greenhouse gas pollution reduction target aimed at creating good-paying union jobs and securing U.S. leadership on clean energy technologies (2021)", White House, accessed 27 May 2021 at www.whitehouse.gov/briefing-room/statements-releases/2021/04/22/fact-sheet-president-biden-sets-2030-greenhouse-gas-pollution-reduction-target-aimed-at-creating-good-paying-union-jobs-and-securing-u-s-leadership-on-clean-energy-technologies/.
Fink, L. (2021), Larry Fink's 2021 letter to CEOs (2021), *BlackRock*, accessed 27 May 2021 at https://www.blackrock.com/corporate/investor-relations/larry-fink-ceo-letter.
Garbow, A. (2021), Avi Garbow, Senior Counselor to the EPA Administrator and Patagonia's Environmental Advocate (on leave), interview by author, February, 2021.
Garcia, A. (2019), "Amazon workers walk out to protest climate change inaction", accessed 27 May 2021 at https://www.cnn.com/2019/09/20/tech/amazon-climate-strike-global-tech.
General Motors (2021), General Motors, the Largest U.S. Automaker, Plans to be Carbon Neutral by 2040, *General Motors Press Release*, accessed 27 May 2021 at https://investor.gm.com/news-releases/news-release-details/general-motors-largest-us-automaker-plans-be-carbon-neutral-2040.
Goldman, M. (2020), "Transforming utility customer service: How Walmart is driving change", accessed 18 August 2021 at https://www.power-grid.com/executive-insight/transforming-utility-customer-service-how-walmart-is-driving-change/#gref.
Gupta, S. and Sharma, N. (2009), CSR: A business opportunity. *Indian Journal of Industrial Relations*, 44(3), 396–401.
Hirja, Z. and Taggart, K. (2021), "A large PR firm pledged to fight climate change. Then it took millions from a notorious fossil fuel trade group", accessed 27 May 2021 at https://www.buzzfeednews.com/article/zahrahirji/edelman-fossil-fuel-pr-climate.
Industry Expert One at International Energy Company, interviewed by author February 2021 (corporation and individual name withheld at interviewee's request).
Industry Expert Two at International Energy Company, interviewed by author February 2021 (corporation and individual name withheld at interviewee's request).
Industry Expert Three at International Energy Company, interviewed by author February 2021 (corporation and individual name withheld at interviewee's request).
Industry Expert at Netherlands Wind Energy Association, interviewed by author, February 2021 (individual name withheld at interviewee's request).

Industry expert at Total, interview by author, February 2021(individual name withheld at interviewee's request).

IPCC (2018) *Principles for Effective Communication and Public Engagement on Climate Change: A Handbook for IPCC Authors*, accessed 27 May 2021 at https://climatecommunication.yale.edu/publications/scientific-agreement-can-neutralize-politicization-facts/.

Jolly, J. (2021), "BlackRock holds $85bn in coal despite pledge to sell fossil fuel shares", accessed 27 May 2021 at https://www.theguardian.com/business/2021/jan/13/blackrock-holds-85bn-in-coal-despite-pledge-to-sell-fossil-fuel-shares.

LaMattina, J. (2021), PhRMA is not effective in improving the image of the pharma industry. *Forbes*, accessed 27 July 2021 at https://www.forbes.com/sites/johnlamattina/2013/02/06/phrma-is-not-effective-in-improving-the-image-of-the-pharma-industry/?sh=1f80ba16e25c.

Kenna, C. (2021), "Global communications and public relations at Patagonia", interview by author, February, 2021.

KPMG (2020), *The Time Has Come: The KPMG Survey of Sustainability Reporting*, accessed 27 May 2021 at https://assets.kpmg/content/dam/kpmg/xx/pdf/2020/11/the-time-has-come.pdf.

McGrath, M. (2020), "Climate change: China aims for 'carbon neutrality by 2060'", accessed 27 June 2021 at www.bbc.com/news/science-environment-54256826.

Nguyen, L. (2020), "Bank of America says it won't finance oil and gas exploration in the Arctic", accessed 27 May 2021 at https://www.bloomberg.com/news/articles/2020-11-30/bofa-says-it-won-t-finance-oil-and-gas-exploration-in-the-arctic.

Nisbet, M. C. (2009), Communicating climate change: Why frames matter for public engagement. *Environment: Science and Policy for Sustainable Development*, 51(2), 12–23, accessed 27 May 2021 at https://research.fit.edu/media/site-specific/researchfitedu/coast-climate-adaptation-library/climate-communications/messaging-climate-change/Nisbet.-2009.-Communicating-CC---Why-Frames-Matter-for-Public-Engagement.pdf.

Ocasio Cortez, Alexandria, Twitter feed, accessed 17 May 2021 at https://twitter.com/AOC/status/1323304992372129792?ref_src=twsrc%5Etfw%7Ctwcamp%5Etweetembed%7Ctwterm%5E1323304992372129792%7Ctwgr%5E%7Ctwcon%5Es1_&ref_url=https%3A%2F%2Fwww.greenmatters.com%2Fp%2Fshell-twitter-poll.

Orsted (2021), We are changing our company name to Orsted. *Orsted News*, accessed May 27 2021 at https://orsted.com/en/media/newsroom/news/2017/10/we-are-changing-our-company-name-to-orsted.

Our Business (2021), accessed 18 August 2021 at https://corporate.walmart.com/our-story/our-business.

Pichai, Sundar, Twitter feed, accessed May 27, 2021 at https://twitter.com/sundarpichai.

Pollach, I. (2018), Issue cycles in corporate sustainability reporting: a longitudinal study. *Environmental Communication*; accessed 27 May 2021 at https://pure.au.dk/portal/files/124885348/Post_print_Irene_Pollach_2018_Issue_cycles_in_corporate_sustainability_reporting.pdf.

Sautner, Z., van Lent, L., Vilkov, G. and Zhang, R. (2020), Firm-level climate change exposure. *ECGI*, accessed May 27 2021 at https://ecgi.global/sites/default/files/working_papers/documents/sautnervanlentvilkovzhangfinal_0.pdf.

Schlichting, I. (2013) Strategic framing of climate change by industry actors: A meta-analysis. *Environmental Communication*, 7(4), 493–511. DOI: 10.1080/17524032.2013.812974

Schoeb, M. (2021), Melissa Schoeb, Vice President Corporate Affairs at Occidental, interview by author, February 2021

Seele, P. and Getti, L. (2015), "Greenwashing revisited: In search of a typology and accusation-based definition incorporating legitimacy strategies", accessed 27 May 2021 at https://onlinelibrary.wiley.com/doi/full/10.1002/bse.1912?casa_token=BCFG8G5aNokAAAAA%3AkbKoMhPbhj7glx4iBgeIbaR2jrZItXFMnE5s76hkyeh7cLYCYvh7mcp0u9w04nSAcDTh3_O5qP1GsdI.

Shell, Twitter feed, accessed 27 May 2021 at https://twitter.com/Shell/status/1323184318735360001?ref_src=twsrc%5Etfw%7Ctwcamp%5Etweetembed%7Ct.wterm%5E1323184318735360001%7Ctwgr%5E%7Ctwcon%5Es1_&ref_url=https%3A%2F%2Fwww.greenmatters.com%2Fp%2Fshell-twitter-poll.

Tan, C. K., Ogawa, A. and Matsumura, T. (2008), "Innovative climate change communication: Team minus 6%, Tokyo", Global Environment Information Centre (GEIC), United Nations University (UNU).

Thaker, J. (2019), Corporate Communication about Climate Science: A comparative analysis of top corporations in New Zealand, Australia, and Global Fortune 500. *Journal of Communication Management*, accessed 27 May 2021 at www.researchgate.net/publication/337893206_Corporate_communication_about_climate_science_A_comparative_analysis_of_top_corporations_in_New_Zealand_Australia_and_Global_Fortune_500.

Thunberg, Greta, Twitter feed, accessed 27 May 2021 at https://twitter.com/GretaThunberg/status/1323355325798506504?ref_src=twsrc%5Etfw%7Ctwcamp%5Etweetembed%7Ctwterm%5E1323355325798506504%7Ctwgr%5E%7Ctwcon%5Es1_&ref_url=https%3A%2F%2Fwww.greenmatters.com%2Fp%2Fshell-twitter-poll.

US Federal Trade Commission (2021), *Guides for the Use of Environmental Marketing Claims* ("Green Guides"), Federal Trade Commission, accessed 27May 2021 at https://www.ftc.gov/policy/federal-register-notices/guides-use-environmental-marketing-claims-green-guides

Vattenfall (2017), Fossil free within one generation. *Annual and Sustainability Report 2017*, accessed May 27, 2021 at https://group.vattenfall.com/siteassets/corporate/investors/annual-reports/2017/vattenfall_annual_and_sustainability_report_2017_eng.pdf.

van der Linden, S., Leiserowitz, A., Feinberg, G., and Maibach, E. (2015). The scientific consensus on climate change as a gateway belief: Experimental evidence. *PloS One*, 10. e0118489. 10.1371/journal.pone.0118489.

van der Linden, S., Leiserowitz, A., and Maibach, E. (2017), Scientific agreement can neutralize the politicization of facts. *Yale Program on Climate Change Communications*, accessed 27 May 2021 at https://climatecommunication.yale.edu/publications/scientific-agreement-can-neutralize-politicization-facts/.

van Riel, C. B. and Fombrun, C. J. (2007), *Essentials of Corporate Communication*. New York: Routledge.

Waldman, S. (2018), "Shell grappled with climate change 20 years ago, documents show", accessed 27 May 2021 at https://www.scientificamerican.com/article/shell-grappled-with-climate-change-20-years-ago-documents-show/.

Woody, T. (2020), "Five major banks refuse to fund Arctic Refuge oil drilling", accessed 27 May 2021 at https://www.wilderness.org/articles/blog/five-major-banks-refuse-fund-arctic-refuge-oil-drilling#.

Ziady, H. (2019), "Goldman Sachs is first big US bank to rule out loans for Arctic drilling", accessed 27 May 2021 at https://www.cnn.com/2019/12/16/business/goldman-sachs-arctic.

11. Corporate strategy and climate change: a nonmarket approach to environmental advantage

Thomas C. Lawton and Carl J. Kock

INTRODUCTION

In previous chapters, authors have acknowledged the business case for integrating environmental sustainability with organizational principles and practices. In this chapter, we argue that such engagement must also be elevated from an operational to a strategic level, as we consider how companies can embed environmental concerns and sustainability initiatives in the design and delivery of corporate strategy. This can result in the development of environmental capabilities that underpin corporate competitiveness. We are acutely aware of the challenges that climate change presents for firms. These challenges include stakeholder pressure to acquire a social license to operate (Demuijnck and Fasterling, 2016), and the development of a climate lens for enterprise management and leadership. But we also see business opportunities to innovate existing products and services and grow entirely new climate-oriented businesses and markets. Moreover, as firms develop such capabilities, this also opens the possibility of further strategic actions that may allow such firms to assert and consolidate an *environmental competitive advantage*.

In focusing on strategic engagement with, and management of, climate change and environmental sustainability, we are moving beyond the market focus of corporate strategy to engage what David Baron (1995a; 1995b) first labeled "nonmarket strategy". Bach and Allen (2010) emphasized that nonmarket strategy recognizes that businesses are social and political entities and not just economic agents. Therefore, nonmarket strategy "considers how managers anticipate, preempt, and respond to actors, influences, and actions emanating from the cultural, social, political, and regulatory arenas" (Lawton, Doh, and Rajwani, 2014, p. 5). Extant work on nonmarket strategy has considered but not always explicitly captured environmental and sustainability thematics. But in recognizing the social agency of firms, this inevitably legitimizes corporate engagement with climate change and related agendas. This is implicit in the fact that a foundational principle of nonmarket strategy is that it deals with externally derived, non-commercial issues, actors and events that impact on the strategic direction and corporate objectives of the business.

This chapter seeks to conceptualize corporate strategy approaches to climate change and consider the development of related environmental capabilities. We therefore look holistically, considering the business opportunities and market initiatives, as well as the strategic regulatory engagement of firms. In the latter, we consider how firms strategically engage and leverage climate change, ultimately, to gain an advantage over the competition. This includes exploring how firms can strategically adapt to stringent regulatory environments and, further, influence the regulatory environment to their advantage through nonmarket strategy, and specifically via lobbying as an instrument of corporate political activity (Hillman, Keim, and Schuler, 2004; Lawton and Rajwani, 2011; Lawton, McGuire, and Rajwani, 2013). An example of proactive regulatory engagement is when a firm that is an environmental leader invests in lobbying to have states set regulation at or near that firm's current environmental capability level. This in turn can severely impact environmental laggards in the same industry, putting them at a competitive disadvantage. In some cases, multiple firms cooperate to advance regulation, and in other (more common) cases, numerous firms may collaborate to delay or prevent regulation through collective action. This may be coordinated via a common trade association (Rajwani, Lawton, and Phillips, 2015; Lawton, Rajwani, and Minto, 2018); or emerge through an advocacy coalition of often diverse organizations with a shared objective (Sabatier, 1988; Jenkins-Smith and Sabatier, 1994).

CORPORATE ENVIRONMENTAL STRATEGY

At a fundamental level, corporate strategy aims to deliver on strategic objectives by establishing and maintaining a competitive advantage that allows a focal firm to earn above average returns or economic rents.[1] In the most basic sense, as Porter suggested in 1980, a firm can become a cost leader by using fewer or less costly inputs, relative to rivals, to create customer value; or it can differentially combine resources in such a way that at least some customers see a distinctly higher value in the resulting products and services, compared with those of other firms. In either case, the firm attempts to widen the difference between the cost of inputs and the customer-perceived value of outputs, thereby adding unique economic value. It actually matters little if cost savings or features that are more valuable to certain customers are established in a market context, where product or process innovation may create differentiated products or more efficient production methods; or a nonmarket context where, for instance, lobbying against environmental regulation may lead to lower costs for a focal firm or higher costs for competitors (Fremeth and Richter, 2011; Capron and Chatain, 2008), or where customers increasingly sensitive to corporate social responsibility (CSR) issues develop preferences for environmentally friendly products or the products of firms with a positive CSR record (Hillman and Keim, 2001). Rather, firms must ensure and assure that the unique value they create indeed is and stays *unique*. If, for instance, a firm lobbies as part of an industry group and is afforded regulatory relief, the same benefits accrue to other firms, and therefore there is no causal effect on competitive advantage. Barney (1991) captured these considerations from a resource-based perspective in the well-known VRIN framework: resources need to be *Valuable* in the sense that they create the above-mentioned added economic value; *Rare*, in the sense that competitors do not have the ability to match the value creation of the focal firm; as well as *Inimitable* and *Non-substitutable* to ensure that the rarity is maintained for some time.

These considerations make it clear that nonmarket – and for this chapter, most crucially, environmental – issues should not just be seen as a constraint or cost imposed upon firms, but also as a potential source of advantage (e.g., Porter and Kramer, 2006). In fact, while early work in the field of environmental strategic management advocated a moral imperative for firms to engage in environmentally friendly practices, a considerable quantity of literature subsequently focused on the question of *does it pay to be green?* And *which contingencies make you greener?* (e.g., King and Lenox, 2002; Cai, Jo, and Pan, 2012; Kock, Santalo and Diestre, 2012; Kurapatskie and Darnall, 2013).

By contrast, only a comparatively smaller number of researchers directly address partial aspects of how firms intentionally use environmental efforts in shaping their competitive strategies. Walsh and Dodds (2017), for instance, investigated how environmental sustainability approaches in the North American hotel industry could help firms assert cost leadership or differentiation advantages. Other authors, conversely, focus on the issue of lobbying. Matsueda (2020), for example, develops a menu-auction model of firm lobbying and explores whether collective or individual lobbying approaches lead to lower costs. Yet other researchers explore how firms can gain a unique advantage over competitors through lobbying, perhaps by even pushing for higher, not lower, regulation (Fremeth and Richter, 2011; Grey, 2018; see also Capron and Chatain, 2008).

EMBEDDING ENVIRONMENTAL ACTIVISM IN NONMARKET STRATEGY

A firm's strategic choices are, to a large extent, prescribed by the competitive environment within which it operates. That competitive environment is typically multidimensional and, for research purposes, is typically dichotomized into market and nonmarket. Baron's seminal definition of the nonmarket environment is that it "consists of the social, political and legal arrangements that structure interactions among companies and their public" (Baron, 1995a, p. 73). For example, the law of contract is an important part of the nonmarket environment that enables companies and their public to contract for the exchange of goods, services, labor, and capital. Variations in contract law between different countries and industries impact the strategic choices of firms. These various social, political, and legal arrangements are collectively referred to as 'regulation'. In industrialized nations, regulation pervades the competitive environment within which firms select and execute their strategies (Shaffer, 1995). Trade policy, competition policy, employment policy, environmental policy, fiscal policy, monetary policy – government policies in general and the particular regulations that they give birth to – have the ability to alter the size of markets through government purchases and regulations affecting substitute and complementary products; to affect the structure of markets through entry and exit barriers and anti-trust legislation; to alter the cost structure of firms through various types of legislation pertaining to multiple factors, such as employment factors and pollution standards (Gale and Buchholz, 1987); to affect the demand for product and services by charging excise taxes and imposing regulations that affect consumer patterns (Wilson, 1985); to effect access to scarce resources (Boddewyn, 1998); and to impact firms' profitability by increasing costs and restricting markets (Schuler, 1996). Consequently, there is substantial interdependence between regulation and the competitive environment within which firms operate (Baron, 1995b; Porter, 1990). Firms therefore take an interest in regulation: an interest in minimizing the cost of existing and proposed regulation upon strategy and business models; an interest in lobbying for regulations which are consistent with and supportive of preferred strategy and business models; and an interest in regulation as a source of competitive advantage. Cumulatively, this constitutes nonmarket strategy: the activities and configurations through which firms strategically manage their political and social environments (Lawton, Dorobantu, Rajwani, and Sun, 2020). For our purposes, and building on the discussion around regulatory engagement, we explicitly add to nonmarket definitions around the strategic management of the environment, and related corporate engagement on climate change (Graf and Kock, 2015; Tashman, Winn, and Rivera, 2015). In doing so, we can distinguish between those firms that purposefully engage, reactively and proactively, and at various points on the spectrum in between. Graf and Kock note that, "firms following a reactive strategy do only what appears to be necessary to build or maintain environmental legitimacy in the eyes of stakeholders" (2015, p. 206). The overall goal is to avoid reputational penalties and regulatory sanctions at the lowest possible cost (Klassen and Whybark, 1999). In contrast, environmentally proactive firms go beyond regulatory rules and aim to outperform in their environmental efforts, exceeding their legal or social expectations (Russo and Fouts, 1997; Sharma and Vredenburg, 1998). Instead of implementing waste control technologies at the end of a production process, for instance, they seek to prevent the generation of new waste from the early production stages on and throughout the whole production process (Klassen and Whybark, 1999).

The goal of this chapter is to align these piecemeal considerations into a more comprehensive overview of different *environmental strategy archetypes*. We do so by highlighting several dimensions that seem likely to systematically shape competitive environmental strategies. Specifically, how aggressive firms are in pursuing their competitive advantage; if they are already proactive in environmental terms; and whether they pursue their environmental competitive strategies alone or in concert with other firms. These key dimensions will be explored in the remainder of the chapter.

ADOPTING A STRATEGIC STANCE: PROACTIVE OR REACTIVE?

So far, we have discussed the basis of competitive advantage, and emphasized the necessity to consider both the market and nonmarket context within which firms compete. But we have been silent on how firms *actually* go about creating strategies to compete. Of relevance for this chapter is how proactive or reactive companies are in attempting to establish their respective positions of advantage. Miles, Snow, Meyer, and Coleman's (1978) classic typology, which remains a relevant and often cited tool, offers pertinent insights here. Specifically, they proposed that firms fall into four strategic types.

1. **Prospectors**: firms that are at the forefront of changes, innovating new products and actively taking advantage of new market opportunities, thereby developing a broad market approach. In the terminology of March's (1991) exploitation-exploration framework, prospectors would seem to focus relatively more resources on the exploration of new areas and less on exploitation of current market opportunities. As such, these firms are aggressively pursuing the establishment of new competitive advantages. The French energy and automation multinational, Schneider Electric, has existed since 1836 but reconstituted its competitive strategy in recent decades to position the corporation at the vanguard of green energy provision. With an explicit and overarching emphasis on efficiency and sustainability, Schneider Electric has been a leader in new market opportunities such as corporate energy efficiency training and consulting, whilst also exploring the sustainability benefits of smart homes.
2. **Defenders**: these firms aim to find secure and stable product niches. Instead of pushing for new product innovations, they are focused on process innovation and cost efficiency.

Accordingly, they conform closely to the notion of exploiting and protecting a current advantage (March, 1991), and will thus be much less aggressive than the previous type when it comes to asserting new sources of advantage. Outdoor clothing and equipment retailer, Patagonia, may illustrate this strategic type, with a long-time focus on their corporate environmental and social footprint, rejection of fast fashion, and early emphasis on recycled fabrics. As the first mover in responsibility and sustainability within their industry, but with competitors now eagerly imitating it, Patagonia strives to defend its position and to continually exploit extant markets and loyal customers through process innovation and the intent to be carbon neutral by 2025.

3. **Analyzers**: these firms share traits of both previous types. Often associated with a 'second-but-better strategy', they take defensive positions in some industries, while being more aggressive in product innovation and seizing opportunities in other markets. These firms may, in fact, oscillate between sometimes being defenders and at other times focusing their resources on prospecting and thus essentially aim to balance exploitation of existing and exploration of new sources of advantage. More aggressive than defenders in establishing new sources of advantages, they also more carefully protect existing markets than prospectors, thereby coming close to what Tushman and O'Reilly (1996) call ambidextrous organizations. In the context of environmental strategy and climate change, many large and diversified conglomerates would fall into this category. For example, energy multinationals such as Chevron or Shell have invested significant resources and made skillful acquisitions in exploring alternative energy operations, particularly in areas including geothermal, solar, and biofuels. But legacy investments in capital assets and managerial capabilities mean that such firms continue to invest in exploiting traditional upstream and downstream oil and gas businesses.

4. **Reactors**: in contrast with the other three types, reactors do not seem to have a consistent strategy and simply (struggle to) respond to changing environments. These companies can seem to lack a clear strategic direction and to merely respond to competitor actions or stakeholder pressures. Here we might find a disconnect between behavior in the market and nonmarket spheres, particularly for firms competing as cost/price leaders. For instance, leading budget airlines, such as Southwest Airlines in the US or Ryanair in Europe, are often prospectors in market strategy, anticipating and aggressively pursuing new market opportunities. But in nonmarket strategy, they are more typically reactors. For instance, on sustainability initiatives and climate change regulatory requirements, they – like more airlines – can lag other companies and even seek to obstruct or delay new environmental policy initiatives.

Overall, the Miles et al. (1978) typology suggests that firms systematically vary in how aggressively they pursue the establishment of competitive advantage, particularly, for our purposes, advantage based on nonmarket capabilities: prospectors lead the charge, followed by analyzers, while defenders, as the name already implies, prefer to hunker down, protect, and incrementally develop their existing area of competence. Reactors, lacking a clear strategy, are unlikely to aggressively charge in any specific direction, as that would require clear strategic intent and therefore, as their name implies, these firms will tend to react to but not initiate change.

Having established that firms differ in their active quest for building new sources of (nonmarket) competitive advantage in a general sense, we turn now to the specifics of the environmental stance of a firm where we can discern a similar, but not identical, dispersion of firms along a continuum from reactive/defensive to proactive when it comes to minimizing the environmental impact of the firms' productive processes.

ENVIRONMENTAL PROACTIVENESS

A comparatively recent phenomenon in environmental management is that some firms, rather than considering environmental efforts as pure costs to be avoided, proactively embrace environmentally friendly practices, and even go beyond compliance in their sustainability strategies. It appears that it is precisely this proactive type of firm that can optimally align corporate environmental and financial performance (King and Lenox, 2002). Hart (1995) and Russo and Fouts (1997) have laid out much of the underlying (natural-)resource-based logic of how environmental firm actions alter capability structures to potentially enhance financial performance as well.

In any case, we take as a key point of departure for our chapter precisely this distinction between firms that are proactive – going beyond what is legally required and often developing new capabilities as a result – and firms that are more reactive, frequently approaching environmental issues through the lens of compliance and cost. Proactive engagement typically suggests a longer-term strategic perspective, with the intent to develop new products, innovate processes, and explore new market opportunities – all of which take time. A reactive mode can indicate a short-term and tactical approach to engaging with climate change. But it can also suggest a viable strategic choice, particularly for firms that are embedded and invested in businesses that run counter to the philosophy that renewables are the only answer in a complex world.

It is also worth noting that, in some instances, one does not exclude the other, i.e., firms – particularly diversified transnational corporations with divergent interests in different places – may be short-term oriented and tactical regarding green strategies in some businesses, and proactively engage in a sustainability strategy elsewhere. Therefore, proactively embracing sustainability is not the only strategy for a firm to pursue. But we do argue that, particularly in advanced economies with rapidly evolving climate action agendas, it makes sense for firms to approach green issues proactively to avoid public scrutiny and penalties, enhance business efficiencies, and capitalize on related customer expectations.

A variety of typologies have been proposed to classify firms' approaches to environmental management issues. Early work by Carroll (1979) and Wartick and Cochrane (1985) on corporate social responsibility, and adaptations to an environmental focus by other authors (e.g., Roome, 1992; Hunt and Auster, 1990) suggested a continuum of firm approaches, with an increasing importance given to social issues that feel very similar in spirit to the Miles et al. (1978) typology discussed above: *reactive, defensive, accommodative,* and *proactive*. Integrating the actual environmental management approaches used, Hart (1995) proposes a similar continuum, starting with an *end-of-pipe* approach, to *pollution prevention or total quality management,* to *product stewardship,* and finally *sustainable development.* The final step is clearly the most desirable from a societal perspective as "…sustainable development aims to minimize the environmental burden of firm growth through the development of clean technologies [and] requires a long-term vision shared among all relevant stakeholders and strong moral leadership" (Buysse and Verbeke, 2003, p. 455).

However, these classifications are not just cross-sectional. Rather, they are reflective of the history of the interaction between firms on the one hand, and environmental stakeholders and regulators on the other. A short review of the evolution that has taken place in the way that companies and shareholders have approached environmental management issues will illustrate this point.

Following the approach of the US Environmental Protection Agency (EPA), we can discern three more-or-less historically successive phases of how firms have responded to environmental issues. In the US context, while concerns about pollution were sparked in the early 1960s by incidents such as oil spills in California and a polluted river in Ohio literally bursting into flames, the establishment of the EPA by President Nixon on December 2, 1970, marks a watershed moment (see 'the origins of EPA' on epa.gov) and the beginning of the first phase. Firms, until then accustomed to treating the environment as both a free source of inputs and a dumpsite for waste materials, were suddenly confronted with an emerging environmental legislation complete with regulations that sought to curb their environmental freeriding. Interpreting such regulations as a mere nuisance that was thrust upon them, firms typically took a reactive or defensive stance, attempting to minimize or avoid the costs associated with compliance. In turn, reactive solutions that usually took the form of end-of-pipe activities such as adding filters, indeed led to higher costs while creating little or no value added for firms, causing a *de facto* negative correlation between environmental investments and financial performance (e.g., Jaggi and Freedman, 1992, for evidence on the late 1970s).

A somewhat more proactive approach emerged in a second phase, as firms strove to find less costly ways to comply. Firms started to conduct internal audits and implement environmental management systems (EMS) that required the redesign of the entire production process to simultaneously comply with regulations and improve operating margins by reducing required inputs and productively using by-products that were formerly wasted. They also embraced the concept of eco-efficiency (DeSimone and Popoff, 1997) to achieve sustainable development through the reduction of production process material and energy intensity and improved durability and recyclability of products. Delmas (2001) and other authors (Melnyk, Sroufe and Calantone, 2003) delivered empirical evidence of the benefits of EMS introductions on increases in self-reported levels of competitive advantage and other measures of corporate performance, but also caution that such positive outcomes are contingent on firms undertaking substantive efforts to go beyond simple compliance and to use the EMS implementation to form strong relationships with internal and external stakeholders (Delmas, 2001), or maximize organizational involvement by actually certifying their EMS within the ISO 14000 family of standards (Melnyk et al., 2003).

Implementing EMS, firms thus became increasingly involved with environmental issues and some started to go *beyond compliance* in what is the currently emerging third phase. Going beyond what regulations require of firms implies that the respective companies have evolved a different approach from earlier phases. Environmental activities are no longer seen as a pure cost but as potential levers for value creation. For instance, through environmentally friendly products for increasingly eco-conscious customers, or the creation of unique organizational capabilities that could lead to cost savings and an enhanced green image. Clearly, only a subset of firms has progressed to a proactive sustainable development approach but that would seem to be precisely why the firms that do are able to create unique added value from their environmental activities and to also benefit financially. Russo and Fouts (1997) explicitly highlight that actions easily imitable by competitors, such as buying off-the-shelf hardware (e.g., filters, more efficient machinery, and so on) to improve a firm's environmental performance do not lead to a sustainable competitive advantage. On the other hand, they suggest that integrating such off-the-shelf items with complex, firm idiosyncratic routines through an environmentally friendly redesign of the entire firm may in fact create rather difficult to match capabilities that can provide long-lasting cost advantages vis-à-vis less environmentally proactive competitors. Empirical results indeed show that higher environmental performance is linked to higher financial performance (Russo and Fouts, 1997; King and Lenox, 2002). Aragón-Correa and Sharma (2003), Christmann (2000), or Hart (1995) have provided additional theoretic and empirical support for this link and identified important moderators such as external contingencies or pre-existing complementary capabilities.

Altogether, the evolution in the approach of firms to environmental issues had its starting point in a reactive end-of-pipe approach that was essentially shared among practically all firms. Over time, some of these firms became more accommodative of environmental concerns and discovered higher efficiencies through preventing pollution in the first place and embracing EMS that are essentially variants of a total quality management approach. Finally, a fraction of these firms is now starting to significantly go beyond compliance to proactively seek advantages from their unique approach to sustainable development. Hence, the extant classification schemes for environmental firm behavior (Carroll, 1979; Wartick and Cochrane, 1985; Hart 1995), on the one hand, seem to reflect firms at different phases of this evolution, but, on the other hand, at the current later stage of this process, we likely also face a distribution of firms across all the proposed categories.

Several scholars have already looked at the reasons why firms would sort themselves into more-or-less proactive environmental performers. Aragón-Correa, Matías-Reche and Senise-Barrio (2004), for instance, report higher environmental commitment in firms that designate a particular manager for environmental issues.

Moreover, an emerging literature on environmental governance suggests that as various stakeholders step up their environmental demands on firms, and even sizable segments of shareholders increasingly believe in a positive link between financial and environmental performance, corporate governance mechanisms are employed to enforce environmental performance by levying pressure on hitherto more hesitant managers (e.g., Kock et al., 2012; Berrone and Gomez-Mejia, 2009; Russo and Harrison, 2005). The stakeholder-agency model pioneered by Hill and Jones (1992) provides one formal framework to appreciate these developments; extending the well-known problem of a misalignment between principals and agents (Fama, 1980; Jensen and Meckling, 1976), and suggesting that various stakeholder groups, other than the owner of firms, also have claims on firms akin to those of principals.

In the next section, our focus is less on explaining why firms are environmentally proactive or not; rather, we will take a firm's current environmental stance – whether reactive or proactive or some blend thereof – and the associated environmental capability structure as given and aim to explore what competitive strategic options are available based on that position.

COMPETITIVE STANCE × DEGREE OF ENVIRONMENTAL PROACTIVITY

Combining strategic and environmental stances allows us to present the environmental strategic archetypes in Figure 11.1.

Corporate strategy and climate change 259

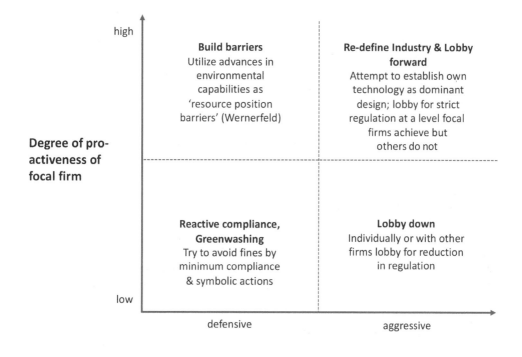

Figure 11.1   Environmental strategic archetypes

On the left-hand side of the horizontal axis we find firms with an overall defensive or reactive strategic posture – the aptly named *defenders* and *reactors* from the Miles et al. (1978) framework – which tend to focus on exploitation of their current market positions and are not overly aggressive in attempting to find new sources of advantage. On the right-hand side of this axis, *prospectors* and *analyzers* are actively pushing to establish new advantages and will therefore aggressively pursue new ideas for creating valuable and unique positions, including significant shifts in the logic of how industries operate, as we will argue below.

The vertical axis maps the degree of environmental proactivity of a firm. At the lower end, firms are in a reactive mode, still primarily considering environmental issues and regulation as a costly imposition on their business, or as an irrelevance to their strategic priorities. These firms think of environmental management in simple end-of-pipe terms and have not built up any specific capabilities to enhance environmental performance, let alone to leverage environmental issues for market positioning (e.g., Hart, 1995; Buysse and Verbecke, 2003). On the other end of that axis, however, firms are proactive and likely have evolved environmental capabilities in a path-dependent fashion as they progressed from a formerly reactive stance through the development of an EMS and finally embracing a proactive, or what Hart (1995) calls 'sustainable development', stance. In fact, Buysse and Verbeke (2003, p. 455), drawing on Hart (1995), emphasize the importance of recognizing the interconnectedness of the different stages, wherein a proactive environmental stance is essentially "…a result of path dependencies and embeddedness. Path dependencies primarily reflect the required sequence of resource accumulation in various individual resource domains to move from one stage to the next. Embeddedness implies the coevolution of various resources and competencies instrumental to a shift in environmental strategy formation." Thus, proactive firms had to orchestrate and build several key capabilities around their environmental management activities. Buysse and Verbeke (2003, p. 455) highlight five such areas where firms need to invest and build:

> 1. conventional green competencies related to green product and manufacturing technologies … 2. employee skills, as measured by resource allocation to environmental training … 3. organizational competencies, as measured by the involvement of functional areas … 4. investments in formal (routine-based) management systems and procedures, at the input, process, and output sides. … 5. efforts to reconfigure the strategic planning process, by explicitly considering environmental issues.

Having gone through this process of investment and internal capability building, proactive firms are likely in a position where they possess valuable and unique skills that could be leveraged to build competitive advantages. Yet, it should be noted that the path-dependent, cumulative build-up of such capabilities also carries the risk of environmental inertia, in that even a high level of environmental skills may become invalidated if the environmental demands, rules and requirements within the firm's organizational environment change. We will discuss some of the resulting implications below.

Reactive Compliance and Greenwashing

The lower left corner of Figure 11.1 depicts a situation where a firm that is defensive and reactive in its overall strategy is also often defensive and reactive in its environmental stance (with notable exceptions, as previously mentioned). Without access to specialized environmental capabilities and lacking the drive to build the same or use environmental issues strategically, the default behavior of such a firm is likely akin to that of firms generally in the first historical phase discussed above, i.e., attempt to minimize the cost of compliance with external regulation while making end-of-pipe investments that are unlikely to create an advantage – or perhaps even simply just pretending to do so. In fact, featuring Kermit the Frog and his famous song lyric, "it's not easy bein' green", a *Financial Times* article (Martin, 2020, among others in the business press) highlighted the practice of pretending to be green while not quite living up to these promises, commonly known as 'greenwashing' (see also Kim and Lyon, 2015; Delmas and Burbano, 2011). In organizational terms, the issue is that of symbolic management or decoupling (Westphal and Zajac, 1998; Zajac and Westphal, 1995), where firms signal to their stakeholders that they are engaged in practices valued by the same, while, in fact, not engaging in any substantive activities towards those ends. Prior studies have delivered evidence for such symbolic management in areas as diverse as corporate governance (Westphal and Zajac, 1998; Zajac and Westphal, 1995), or sense giving for strategic change (Fiss and Zajac, 2006; for a broader review see Schnackenberg, Bundy, Coen and Westphal, 2019), as well as environmental practices. For instance, Rodrigue, Magnan and Cho (2013) report that in many cases environmental governance mechanisms adopted by firms appear to bring few substantive changes in environmental outcomes and thus appear mostly symbolic. Aravind and Christmann (2011) report that firms receiving the environmental management system standard ISO 14001 did not always live up to the requirements of that standard, and only firms that do conform to the spirit of the standard achieve better environmental outcomes.

Mature consumer goods company P&G appears to exemplify this environmental-strategy quadrant, as recently reported (Evans, 2020). The conglomerate is joining several 'nature-based projects as a way to help combat climate change', while at the same time it maintains highly detrimental environmental practices in other areas, such as protecting and restoring tropical trees in the Philippines while continuing to use Canadian climate-critical boreal forest fiber for its throwaway tissue. This inconsistency between environmentally lagging behavior and attempts to portray a green image led some environmental campaign groups to label P&G's strategy as the 'ultimate greenwash' (Evans, 2020). A very established and mature firm in its markets, P&G appears to be searching for, perhaps piecemeal, ways to comply with the environmental requirements imposed by its stakeholders without attempting to go on the offensive aggressively or comprehensively on these issues.

In a similar vein, a lawsuit by Friends of the Earth Netherlands led to a Dutch court order forcing Shell to cut carbon emissions (McFarlane, 2021). While, in the same month (May 2021), Exxon and Chevron management lost shareholder votes resulting in the installation of activist directors and mandates for more aggressive emissions management respectively (Mufson, 2021). In terms of our framework, these are illustrations of environmentally reactive firms whose management literally had to be forced by either activist shareholders or other stakeholders to ramp up their environmental activities.

The reactive–reactive match is a relatively intuitive outcome that illustrates the endogeneity between strategic and environmental stances. It seems a reasonable assumption that if a firm is a defender or strategy-less reactor in the first place, its environmental stance would not develop far beyond reacting to immediate legal requirements. By contrast, for such a firm to muster the energy to develop environmental capabilities or even step outside the behavior of industry peers to become proactive appears rather unlikely.

Build Barriers

While a defensive strategy stance likely begets a defensive environmental approach, there may be occasions where a strategically defensive firm finds itself matched with a proactive environmental approach. This could happen, for example, if a division of a larger firm is more strategically active in the environmental space than the overall company, or when a defender firm acquires an environmental leader that brings to the combination an endowment of environmental capabilities. If the defender is the dominant firm in the combination, a reasonable result would be that the default instinct to defend and consolidate will tend to channel the environmental capabilities to establish a cost leadership or differentiation position, as, for instance, Walsh and Dodds (2017) discuss. The advanced environmental capabilities would in this case be used like any other unique assets to create a defensive resource position barrier (Wernerfelt, 1984) against other firms.

Key to this strategy type is the application of environmental capabilities to strengthen a firm's market position but not to challenge the prevailing logic of that market space. The earlier mentioned French energy and automation multinational, Schneider Electric, illustrates well such a use of environmental capabilities to become best in class in its existing market context. Established in the early 19th century, Schneider has more recently reconfigured its corporate strategy to embody and embed sustainability and strive for environmental advantage.

As this quadrant conceptualizes an uneasy match – between a strategically rather reactive but environmentally proactive focus – one might suspect that this strategy might easily become unstable. For instance, proactive environmental activities may be hampered by an overall reactive strategy stance, pushing the firm towards a more reactive compliance state. In other cases, when a proactive unit is exceptionally strong within a firm, a more aggressive use of the advanced environmental capabilities might supplant the defensive impulses and the original defender might even mutate into an analyzer who uses proactive, and thus perhaps even best-in-class, environmental capabilities, not just to enhance current products or processes, but to shape the competition with more environmentally laggard firms in a more pronounced way, as we discuss next.

Redefine Industry and Lobby Forward

Environmentally proactive firms are still in the minority, which suggests that the complex sets of capabilities they have had to develop to be able to go beyond compliance are comparatively rare. Coupled with an aggressive strategic stance, these firms may be tempted to go beyond a simple improvement of products and processes to actively reshape the competitive landscape for themselves – and their competitors. Both this and the previous strategy-type boast proactive (and arguably industry leading) environmental capabilities. But the key difference is that firms in this quadrant are strategically aggressive and are thus likely to not just rest on being best-in-class but go further and try to change the rules of the game. As mentioned previously, environmental issues are still emerging in many industries. In economic terms, environmental approaches – and related products, services, and processes – will thus tend to be in the early stages of a lifecycle. Utterback and Abernathy (1975) suggested that those early or pre-paradigmatic stages are characterized by experimentation and ongoing changes in the product or basic characteristics to find out what will work or sell to customers. Such experimentation typically ends with a dominant design (Utterback and Abernathy, 1975; Teece, 1986) that incorporates the key drivers of value (for customers) and engineers out most of the problematic issues, thereby creating the product or service that is widely accepted and starts driving significant growth. Looking back, the iPhone, for instance, certainly qualifies as such a dominant design in the phone market. But many environmental markets (green products, services, processes) currently appear to be in an earlier, pre-paradigmatic state where what customers want (or what stakeholders mighty decree as desirable) is still in flux. Moreover, in some cases, new environmental technologies or product markets, such as in the automotive industry, are subject to strong network externalities and self-reinforcing effects with complementary technologies (e.g., cars and refueling stations – neither can exist without the other) that create the possibility of multiple equilibria, as already discussed by Arthur in 1989 (see also Rochet and Tirole, 2006, Cennamo and Santalo, 2013, and others). In other words, it is not clear which of potentially several alternative technologies in the automotive industry – hybrid, electric, natural gas, or hydrogen – will carry the day and eventually come to dominate. And even small initial effects may lead to a lock-in of one technology over the others (Arthur, 1989).

In turn, such a potentially fluctuating state of the industry offers a firm with strong capabilities an opportunity to tip the industry in the direction the firm is already travelling with its path-dependent capability build-up. In other words, the focal firm could try to force a dominant design around its current capability base, thereby placing itself at the forefront of the industry and at the center of the potential fast growth that tends to follow a dominant design. Other firms, by contrast, may be handicapped if their capabilities do not fully match or are simply not applicable to the new requirements (see also Tushman and Anderson's, 1986, concept of competence destroying change). This strongly resonates with Capron and Chatain's (2008) suggestion that firms can gain advantages by improving their own input costs or resource access *or* by increasing the cost basis or worsening resource access for competitors.

Given the importance of the regulator in environmental issues, one form this could take is that of lobbying for a change in the regulation – but a change towards a tougher regulatory environment rather than regulatory relief (Grey, 2018). To the extent that the focal firm already complies with the progressive regulation it promotes, any competitor who does not yet have the same capabilities will be put at a significant disadvantage. Fremeth and Richter provide an excellent example of such a strategy in the form of Hewlett-Packard's (HP) electronic waste (e-waste) recycling initiative:

> The advocating for pragmatic, progressive policy strategy allowed managers at HP to reduce delays in the adoption of specific types of progressive e-waste policies. Enactment of these policies raised rivals' costs given that the rest of the industry was initially not as environmentally responsive as HP. (Fremeth and Richter, 2011, p.150)

Another example of such a strategic redefinition of an industry is provided by Swiss energy company IWB. While IWB was a founding member of Verband der Schweizerischen Gasindustrie (VSG), the association of Swiss gas producers, as IWB's environmental capabilities began outpacing that of the industry, the interests of this firm and the association it co-founded diverged. As IWB puts it on the company website (IWB, 2021), the firm and the industry association were travelling at different speeds. In essence, the association took a more reactive stance limited to the natural gas industry and, by trying to maintain and protect the structure of their existing market, engaged in a form of 'lobbying down'. This caused IWB to leave the association it co-founded, to establish a leading proactive role in the Swiss energy landscape by utilizing a broader range of environmentally friendly technologies, including combined heat and power systems and hydrogen.[2]

In the IWB case, the firm did not so much attempt to lobby for a different set of regulations but rather attempted to change the logic of how the industry operates. The automobile industry offers another concurrent example of how certain actors attempt to change the rules of the game and capitalize on the aforementioned network externalities to, in the extreme, impose their own technology as a dominant design for the entire industry. In particular, the actions of a clearly strategically and environmentally aggressive firm, Tesla, stand out in this regard. At the beginning of this century, it was not apparent to most industry observers which, of a number of technologies, would come to replace internal combustion engines in the global car industry. While most car makers seem to have waited for this uncertainty to resolve, Tesla charged ahead with ever increasing investments in electric and battery technologies. This arguably strongly contributed to shaping the dominant design of the environmental future of the car, focusing much of the industry as well as regulators world-wide on electric engines and batteries to the detriment of other approaches such as natural gas, hydrogen, or hybrids.

Moreover, Tesla's give-away of their (environmental) patent portfolio can be considered a further example of such an aggressive/proactive play, with Tesla arguably hoping that others will embrace their technology – rather than for other car makers to develop their own green tech or converge to a non-Tesla standard – and thereby implicitly accepting Tesla's technology lead. This could be an excellent point of departure if Tesla were to aim for a battery/drivetrain monopoly that other car makers must accept in the mold of 'Intel inside'.

As with the reactive–reactive pairing, the aggressive–proactive combination appears intuitive. The capability building required to become proactive would clearly benefit from the aggressive exploration focus of prospectors or analyzers, making it likely that firms that are environmentally proactive today started out as competitively aggressive firms. Yet we argue that the link is not as strongly endogenous as for the defensive–defensive case, since aggressive firms could channel investments into many different areas and therefore might easily overlook environmental issues, especially when that area is not yet 'hot'. Hence, while it seems likely that most environmentally proactive firms started as aggressive firms, the opposite does not necessarily hold, i.e., many competitively aggressive firms might actually be environmental laggards, which describes the last box that we now turn to.

Lobby-down

If a strategically aggressive firm is an environmental laggard without discernible advantages in the environmental field, an obvious channel for that aggressiveness would again be to attempt to change regulations. However, in this case, the push would be the more common one for regulatory relief. For instance, Grey (2018) describes how Du Pont lobbied for decades against regulation aiming to protect the ozone layer. Similarly, many firms that compete as cost/price leaders, e.g., in air transport or food retail, tend to lobby down in efforts to delay, moderate, or prevent environmental legislation that is likely to add cost and complexity to their business models. Furthermore, showing the coexistence of lobby-forward-style proactive as well as lobby-down firms in the same industry, Walley and Whitehead (1994) provide an example from the US paper industry. In the 1990s, the industry was facing high uncertainty around proposed government regulations mandating chlorine-free paper, specifically, while "…Louisiana-Pacific has started to prepare its organization for chlorine-free paper production, many other industry participants are fighting tooth and nail to undermine proposed legislation" (Walley and Whitehead, 1994, p. 50).

As a final example, we would also locate Toyota in this box, even though that firm has attempted to aggressively build environmental competences around hybrid/hydrogen technologies in the car industry. While the company was clearly proactive with respect to that envisioned future, as the prevailing logic of the industry, and with that, the regulatory interest, turns instead to electric vehicles, Toyota is no longer in the vanguard of the car-industry's green firms (Pander, 2021). This illustrates that proactiveness must be seen in the context of what is required by the actual current environment, which, in turn, is subject to change over time, as indicated with the example of Tesla in this industry. Toyota initially bet on a hybrid/hydrogen future and was proactive with respect to that envisioned future. But as so often happens in disruptive environments that are also characterized by network externalities, the environment appears to have shifted to an e-future and Toyota now finds itself on the reactive end with respect to e-tech; that is, it is lacking in capabilities and technologies for the now dominant technology. More generally, this suggests that if a proactive firm fails to establish a dominant design around its own capabilities, it may see these capabilities literally be destroyed (Tushman and Anderson, 1986) if another firm manages to create a dominant design around a different solution.

In our model, this simply means that Toyota has moved downward on its proactive environmental scale, but it is still a strategically aggressive firm that is now turning its energy to lobbying against e-tech, i.e., to lobby-down to prevent stronger environmental regulation and thereby turning from climate pioneer to a brake on environmental progress (Pander, 2021).

DELIVERING THROUGH INDIVIDUAL OR COLLECTIVE ACTION

In the previous section we discussed how strategically aggressive firms may undertake more pronounced actions, either individually or in concert with like-minded firms. The latter are presumably competitors at a similar level in an industry's value chain, who might share comparable concerns about environmental issues. Often these actions, given the nature of the field, might crystallize into some form of lobbying to change extant regulation for better or worse.

The key question from a competitive perspective is whether such lobbying is indeed undertaken by firms that share common interests and that therefore ought to act together, as common sense might suggest, or by individual firms attempting to decisively deviate from industry norms (Olson, 1965). This characterization partly follows Mancur Olsen's seminal work and his 1965 proposition that individuals in any group seeking collective action will have incentives to free-ride on the efforts of others if the group is working to provide public goods (those that are generally available without the ability to restrict to some subset); but will be disincentivized to do so if the group provides benefits only to contributing participants.[3] The freeriding that Olsen describes is not joining the lobby group but partaking in the same outcomes, without the cost. The lobby forward described earlier, however, would see firms staying outside lobby groups to get stricter rules that only they can currently fulfill, thereby placing all the other firms at a disadvantage.

We strive to understand the various contingencies that explain when and why firms in non-market contexts – specifically the strategic management of climate change – choose to go it alone or pitch in with others. Reasons vary, including organizational scale and scope, industry type, country of origin (and associated regulations and requirements), and brand and market positioning. For instance, going it alone may be preferable if a company is trying to differentiate, or reposition, through its sustainability initiatives. Under the leadership of Paul Polman, the Anglo-Dutch consumer goods multinational, Unilever, sought to place sustainability at the center of corporate and business strategy. The owner of global brands such as Lipton tea and Ben & Jerry's ice cream began the sustainability and responsibility journey more than 20 years ago, under Polman's predecessor as CEO, Niall FitzGerald. Unilever's stated goal of decoupling its environmental impact from its growth has seen Polman adopt a more holistic and long-term approach to performance, warning investors that climate change costs Unilever millions of dollars annually, striving to eliminate deforestation from supply chains, and acquiring companies with eco-friendly brands.

Collective action via trade associations can be preferred both by smaller, resource-constrained companies that cannot afford their own initiatives, and by larger corporates that are either pursuing both individual and collective channels simultaneously, or are keen to shelter behind the collective umbrella of a trade association and therefore avoid specific public scrutiny for their relative lack of engagement on issues such as climate action (Rajwani, Lawton, and Phillips, 2015; Lawton, Rajwani, and Minto, 2018).

In some cases, diversified collective action coalitions can prove more appealing – and more effective – for business. These are similar in form to what Sabatier (1988) and Jenkins-Smith and Sabatier (1994) labeled advocacy coalitions in the public policy sphere, i.e., a varied alliance of actors sharing a common policy goal (such as climate change action). For instance, *Project Gigaton*, an initiative between Walmart, environmental nongovernmental organizations (NGOs) such as the Environmental Defense Fund, and more than 2,300 suppliers, was created to eliminate one billion metric tons (a gigaton) of greenhouse gas emissions from corporate supply chains by 2030.

CONCLUSIONS

This chapter has advanced a systematic framework for the competitive environmental strategies that are both feasible and already in use. Specifically, we have drawn on strategic management concepts and logic to establish the recognizable, and less familiar, strategic positions that firms occupy. This framework can serve as a guide for firms interested in moving beyond being good environmental citizens – i.e., the question of where they fall on the reactive–proactive environmental scale – to leverage proactive environmental strategies in ways that shape their competitive rivalry with other firms.

In that context, it is important to recognize some key features of our framework. As we discussed earlier regarding Figure 11.1, there are a variety of internal and external mechanisms that push firms into a specific quadrant, or cause movement from one to another. For instance, if a firm does not yet have environmental capabilities, the top quadrants cannot be pursued as strategic choices. But being aware of potential strategic options available once environmental capabilities are developed allows firms discretion to plot their future strategic course and potentially target a strategy that creates strong barriers or even redefines the industry landscape. Furthermore, if firms aspire to industry changing environmental strategies, they need to be aware that competitive interdependencies mean that one firm establishing a dominant design and approach might undermine the efforts of other proactive firms. These competitive interdependencies are particularly likely in the environmental domain, which is still relatively nascent and where contestation remains around structure, strategy, and intended outcomes. Conversely, some firms may even choose to be so defensive as to hide their environmental achievements. Kim and Lyon (2015) found that, under some conditions, firms even show undue modesty in not fully reporting their environmental achievements.

In concluding, a note of caution is in order. The essence of the argument underpinning the logic for a proactive corporate strategy approach to climate change and related environmental and sustainability challenges is that firms can earn economic returns that are positive and preemptive. In other words, embrace environmental action before others to develop *valuable* and *rare* (to name the key requirements for earning economic profits according to the resource-based view of Barney, 1991, Peteraf, 1993, and others) environmental capabilities that underpin competitive advantage through facilitating more efficient organizations and/or producing environmentally beneficial goods and services that meet an existing or emerging consumer need. However, that also implies that as other firms jump on the same bandwagon, the rarity of these environmental capabilities – and thus their expected profit potential – decreases. Yet this is what normally happens in the cycle of innovation and imitation. What counts is that currently, across many industries, proactive environmental strategies promise to increase financial returns. Encouraging the discovery and exploitation of these opportunities will engender the market forces of private utility maximization to help in the development and delivery of environmental pursuits.

NOTES

1. Specifically, above average returns or economic rents are essentially the same as economic value added, i.e., firms use resources they take (away) from the rest of the economy and turn these into products or services. If customers value these at more than what the resources cost (where these costs are determined in factor markets, based on what other firms could do with them, i.e., the 'supplier opportunity cost'), then, not only firm profits but also economic value is created, as the firm was able to create a productive combination that – on average – other firms could not create with the focal resources. However, it is important to include *all* resources, including externalities such as pollution, use of clean water, and so on (think via CO_2 trading schemes or other ways to internalize externalities), to determine if firms are truly adding firm and economic value. In fact, if that is not done, then investment decisions are distorted as polluting firms that do not include such externalities in their profit calculation would look better than green firms that use fewer resources or cause less pollution. This would shift economic resources to firms that are less able to create value from these resources.
   In a sense, economic rent – based on accounting for all resources used – requires the most efficient use of resources and is thus a necessary precursor for sustainability.
2. These insights are gleaned from the IWB website and an interview conducted by one of the authors with an IWB executive.
3. Pure public goods are goods that are non-excludable (that is, one person cannot reasonably prevent another from consuming the good) and have no inherent rivalry (one person's consumption of the good does not affect another's, or vice versa). Hence, without selective incentives to motivate participation, collective action is unlikely to occur even when large groups of people with common interests exist.

REFERENCES

Aragón-Correa, J. A., & Sharma, S. (2003). A contingent resource-based view of proactive corporate environmental strategy. *Academy of Management Review, 28*(1), 71–88.

Aragón-Correa, J. A., Matías-Reche, F., & Senise-Barrio, M. E. (2004). Managerial discretion and corporate commitment to the natural environment. *Journal of Business Research, 57*, 964–975.

Aravind, D., & Christmann, P. (2011). Decoupling of standard implementation from certification: Does quality of ISO 14001 implementation affect facilities, environmental performance? *Business Ethics Quarterly, 21*(1), 73–102.

Arthur, W. B. (1989). Competing technologies, increasing returns, and lock-in by historical events. *The Economic Journal, 99*, 116–131.

Bach, D. & Allen, D. B. (2010). What every CEO needs to know about nonmarket strategy. *MIT Sloan Management Review, 51*(3), 41–48.

Barney, J. (1991). Firm resources and sustained competitive advantage. *Journal of Management, 17*(1), 99–120.

Baron, D. P. (1995a). The nonmarket strategy system. *Sloan Management Review, 37*(1), 73–85.

Baron, D. P. (1995b). Integrated strategy: market and nonmarket components. *California Management Review, 37*(2), 47–65.

Berrone, P., & Gomez-Mejia, L. R. (2009). Environmental performance and executive compensation: An integrated agency-institutional perspective. *Academy of Management Journal, 52*(1), 103–126.

Boddewyn. J. J. (1998). Political aspects of MNE theory. *Journal of International Business Studies, 19*(3), 341–363.

Buysse, K., & Verbeke, A. (2003). Proactive environmental strategies: A stakeholder management perspective. *Strategic Management Journal, 24*, 453–470.

Cai, Y., Jo, H., & Pan, C. (2012). Doing well while doing bad? CSR in controversial industry sectors. *Journal of Business Ethics, 108*(4), 467–480.

Capron, L., & Chatain, O. (2008). Competitors' resource-oriented strategies: Acting on competitors' resources through interventions in factor markets and political markets. *Academy of Management Review, 33*(1), 97–121.

Carroll A. B. (1979). A three-dimensional conceptual model of corporate social performance. *Academy of Management Review, 4*, 497–505.

Cennamo, C., & Santalo, J. (2013). Platform competition: Strategic trade-offs in platform markets. *Strategic Management Journal, 34*, 1331–1350.

Christmann, P. (2000). Effects of "best practices" of environmental management on cost advantage: The role of complementary assets. *Academy of Management Journal, 43*(4), 663–680.

Delmas, M. (2001). Stakeholders and competitive advantage: the case of ISO 14001. *Production and Operations Management, 10*(3), 343–358.

Delmas M.A., & Burbano, V.C. (2011). The drivers of greenwashing. *California Management Review, 54*(1), 64–87.

Demuijnck, G., & Fasterling, B. (2016). The social license to operate. *Journal of Business Ethics, 136*(4), 675–685.

DeSimone, L. D., & Popoff, F. (1997). Eco-efficiency. *The Business Link to Sustainable Development, 280*.

Evans, J. (2020). P&G urged to match best in class to avoid 'greenwash' label. *Financial Times*, July 29, 2020. Accessed on August 22, 2021.

Fama, E. F. (1980). Agency problems and the theory of the firm. *Journal of Political Economy, 88*(2), 288–307.

Fiss, P. C., & Zajac, E. J. (2006). The symbolic management of strategic change: Sensegiving via framing and decoupling. *Academy of Management Journal, 49*(6), 1173–1193.

Fremeth, A. R., & Richter, B. K. (2011). Profiting from environmental regulatory uncertainty: Integrated strategies for competitive advantage. *California Management Review, 54*(1), 145–165.

Gale, J., & Buchholz, R. (1987). The political pursuit of competitive advantage: What business can gain from government? In Marcus, A., Kaufman. A., & Beam, D. (Eds.). *Business Strategy and Public Policy*. New York: Quorum, 231–252.

Graf, T., & Kock, C. J. (2015). Environmental performance and non-market strategy. In Lawton, T. C., & Rajwani, T. S. (Eds.) *The Routledge Companion to Non-Market Strategy*. Routledge: London.

Grey, F. (2018). Corporate lobbying for environmental protection. *Journal of Environmental Economics and Management, 90*, 23–40.

Hart S. L. (1995). Natural-resource-based view of the firm. *Academy of Management Review*, 20(4), 986–1014.

Hill, C. W., & Jones, T. M. (1992). Stakeholder-agency theory. *Journal of Management Studies, 29*(2), 131–154.

Hillman A. J., & Keim G. D. (2001). Shareholder value, stakeholder management, and social issues: what's the bottom line? *Strategic Management Journal, 22*, 125–139.

Hillman, A. J., Keim, G. D., & Schuler, D. (2004). Corporate political activity: A review and research agenda. *Journal of Management, 30*(6), 837–857.

Hunt C. B., & Auster, E. R. (1990). Proactive environmental management: Avoiding the toxic trap. *Sloan Management Review, 31*, 7–18.

IWB. (2021). https://iwb.ch/Ueber-uns/Newsroom/Medienmitteilungen/IWB-verl-sst-den-VSG.html. Accessed on August 4, 2021.

Jaggi, B., & Freedman, M. (1992). An examination of the impact of pollution performance on economic and market performance: Pulp and paper firms. *Journal of Business Finance & Accounting, 19*(5), 697–713.

Jenkins-Smith, H. C., & Sabatier, P. A. (1994). Evaluating the advocacy coalition framework. *Journal of Public Policy, 14*(2), 175–203.

Jensen, M. C., & Meckling, W. H. (1976). Theory of the firm: Managerial behavior, agency costs and ownership structure. *Journal of Financial Economics, 3*(4), 305–360.

Kim, E-H., & Lyon, T. P. (2015). Greenwash vs. Brownwash: Exaggeration and undue modesty in corporate sustainability disclosure. *Organization Science, 26*(3), 705–723.

King, A., & Lenox, M. (2002). Exploring the locus of profitable pollution reduction. *Management Science, 48*(2), 289–299.

Klassen, R. D., & Whybark, D. C. (1999). The impact of environmental technologies on manufacturing performance. *Academy of Management Journal*, *42*(6), 599–615.

Kock, C. J., Santalo, J., & Diestre, L. (2012). Corporate governance and the environment: what type of governance creates greener companies? *Journal of Management Studies*, *49*(3), 492–514.

Kurapatskie, B., & Darnall, N. (2013). Which corporate sustainability activities are associated with greater financial payoffs? *Business Strategy and the Environment*, *22*(1), 49–61.

Lawton, T., & Rajwani, T. (2011). Designing lobbying capabilities: Managerial choices in unpredictable environments. *European Business Review*, *23*(2), 167–189.

Lawton, T., McGuire, S., & Rajwani, T. (2013). Corporate political activity: A literature review and research agenda. *International Journal of Management Reviews*, *15*(1), 86–105.

Lawton, T. C., Doh, J. P., & Rajwani, T. (2014). *Aligning for Advantage: Competitive Strategies for the Political and Social Arenas*. Oxford: Oxford University Press.

Lawton, T. C., Rajwani, T., & Minto, A. (2018). Why trade associations matter: Exploring function, meaning, and influence. *Journal of Management Inquiry*, *27*(1), 5–9.

Lawton, T. C., Dorobantu, S., Rajwani, T. S., & Sun, P. (2020). The Implications of COVID-19 for nonmarket strategy research. *Journal of Management Studies*, *57*(8), 1732–1736.

March, J. G. (1991). Exploration and exploitation in organizational learning. *Organization Science*, *2*(1), 71–87.

Martin, M. (2020). ESG: A trend we can't afford to ignore. *Financial Times*, November 26. Accessed online on December 21, 2021.

Matsueda, N. (2020). Collective vs. individual lobbying. *European Journal of Political Economy*, *63*, 101859.

McFarlane, S. (2021). Shell ordered by Dutch court to cut carbon emissions. *Wall Street Journal*, May 26. Accessed online on June 5, 2021.

Melnyk, S. A., Sroufe, R. P., & Calantone, R. (2003). Assessing the impact of environmental management systems on corporate and environmental performance. *Journal of Operations Management*, *21*(3), 329–351.

Miles, R. E., Snow, C. C., Meyer, A. D., & Coleman Jr, H. J. (1978). Organizational strategy, structure, and process. *Academy of Management Review*, *3*(3), 546–562.

Mufson, S. (2021). A bad day for big oil. *The Washington Post*, May 26, 2021. Accessed online on June 5, 2021.

Olson, M. (1965). *The Logic of Collective Action: Public Goods and the Theory of Groups*. Cambridge, MA: Harvard University Press.

Pander, J. (2021). Wie Toyota vom Klimaschutz-Pionier zum Bremser wurde. *Spiegel*, July 29, 2021. Accessed online on August 5, 2021.

Peteraf, M. A. (1993). The cornerstones of competitive advantage: A resource-based view. *Strategic Management Journal*, *14*(3), 179–191.

Porter, M. E. (1980). *Competitive Strategy*. New York: Free Press.

Porter, M. E. (1990). The competitive advantage of nations. *Harvard Business Review*, March-April, 73–93.

Porter, M. E., & Kramer, M. R. (2006). The link between competitive advantage and corporate social responsibility. *Harvard Business Review*, *84*(12), 78–92.

Rajwani, T., Lawton, T., & Phillips, N. (2015). The "voice of industry": Why management researchers should pay more attention to trade associations. *Strategic Organization*, *13*(3), 224–232.

Rochet, J.-C., & Tirole, J. (2006). Two-sided markets: A progress report. *Rand Journal of Economics*, *37*, 645–667.

Rodrigue, M., Magnan, M., & Cho, C. (2013). Is environmental governance substantive or symbolic? An empirical investigation. *Journal of Business Ethics*, *114*(1), 107–129.

Roome, N. (1992). Developing environmental management systems. *Business Strategy and the Environment*, *1*, 11–24.

Russo, M. V., & Fouts, P. A. (1997). A resource-based perspective on corporate environmental performance and profitability. *Academy of Management Journal*, *40*(3), 534–559.

Russo, M. V., & Harrison, N. S. (2005). Organizational design and environmental performance: Clues from the electronics industry. *Academy of Management Journal*, *48*(4), 582–593.

Sabatier, P. A. (1988). Policy change and policy-oriented learning: Exploring an advocacy coalition framework. *Policy Sciences*, *21*(2/3), 129–168.

Schnackenberg, A. K., Bundy, J., Coen, C. A., & Westphal, J. D. (2019). Capitalizing on categories of social construction: A review and integration of organizational research on symbolic management strategies. *Academy of Management Annals*, 13(2), 375–413.

Schuler, D. A. (1996). Corporate political strategy and foreign competition: The case of the steel industry. *Academy of Management Journal*, 39(3), 720–737.

Shaffer, B. (1995). Firm-level responses to government regulation: Theoretical and research approaches. *Journal of Management*, *21*(3), 495–514.

Sharma, S., & Vredenburg, H. (1998). Proactive corporate environmental strategy and the development of competitively valuable organizational capabilities. *Strategic Management Journal*, *19*(8), 729–753.

Tashman, P., Winn, M., & Rivera, J. E. (2015). Corporate climate change adaptation. In Lawton, T. C., & Rajwani, T. S. (Eds.) *The Routledge Companion to Non-Market Strategy*. London: Routledge.

Teece, D. (1986). Profiting from technological innovation: Implications for integration, collaboration, licensing and public policy. *Research Policy*, *15*(6), 285–305.

Tushman, M. L., & Anderson. P. (1986). Technological discontinuities and organizational environments. *Administrative Science Quarterly*, *31*(3), 439–465.

Tushman, M. L., & O'Reilly, C. (1996). Ambidextrous organizations: Managing evolutionary and revolutionary change. *California Management Review*, *38*(4), 8–30.

Utterback, J. M., & Abernathy, W. J. (1975). A dynamic model of product and process innovation. *Omega*, *3*(6), 639–656.

Walley, N., & Whitehead, B. (1994). It's not easy being green. *Harvard Business Review*, *72*(3), 46–52.

Walsh, P. R., & Dodds, R. (2017). Measuring the choice of environmental sustainability strategies in creating a competitive advantage. *Business Strategy and the Environment*, *26*(5), 672–687.

Wartick S. L., & Cochrane, P. L. (1985). The evolution of the corporate social performance model. *Academy of Management Review*, *10*(4), 758–769.

Wernerfelt, B. (1984). A resource-based view of the firm. *Strategic Management Journal*, *5*, 171–180.

Westphal, J.D., & Zajac, E.J. (1998). The symbolic management of stockholders: Corporate governance reforms and shareholder reactions. *Administrative Science Quarterly*, *43*(1), 127–153.

Wilson, J. D. (1985). Optimal property taxation in the presence of interregional capital mobility. *Journal of Urban Economics*, *17*, 73–89.

Zajac, E. J., & Westphal, J. D. (1995). Accounting for the explanations of CEO compensation: Substance and symbolism. *Administrative Science Quarterly*, *40*(2), 283–308.

12. Owens Corning: environmental footprint reduction as the foundation for building a net-positive future

Frank O'Brien-Bernini and Amanda Meehan

INTRODUCTION

Nearly two decades ago, Owens Corning began its sustainability journey with a focus on reducing the environmental – including climate – impact of its operations. Over the years, the company's goals evolved and expanded in service of its aspiration to become a net-positive company; i.e., a company whose increased 'handprint', or the positive impacts of its people and products, is greater than its environmental footprint. Owens Corning has committed to doubling its corporate and product handprint and halving its environmental footprint by 2030, while also working to eliminate injuries and lifestyle-induced diseases, to advance inclusion and diversity, and to make a positive difference in the communities where its employees work and live.

These guiding aspirations are at the core of Owens Corning's current sustainability efforts. In 2019, the company announced its third set of ten-year sustainability goals, to be achieved over the course of the subsequent decade. The 2030 sustainability goals are the most comprehensive and ambitious to date, designed to expand business impact through sustainability, continuing the company's commitment to incorporate environmental, social, and economic initiatives into its global practices and operations.

In recent years, Owens Corning has earned a series of recognitions for its sustainability work. It may seem surprising that an 83-year-old manufacturing company headquartered in the US Midwest is recognized as a leading corporate citizen across environmental, social, and governance (ESG) dimensions, but Frank O'Brien-Bernini, senior vice president and chief sustainability officer, notes that it doesn't happen without unwavering commitment and sustained effort over the years. The company's history of progress in this space has been driven by a determination to realize the power of connecting business strategy to what's good for the world.

The company's history of environmental footprint reduction efforts provides an example of how goal setting, continual benchmarking and a commitment to transparency enable strong collaboration internally and with external partners to deliver meaningful progress. This chapter discusses how Owens Corning approached these challenges and opportunities during the past two decades, with a particular focus on climate change and energy.

About Owens Corning

Owens Corning is a global building and industrial materials company that manufactures and delivers a broad range of high-quality insulation, roofing, and fiberglass composite materials.

Based in Toledo, Ohio, USA, Owens Corning posted 2020 net sales of $7.1 billion. It has been a Fortune 500 company for 67 consecutive years, dating back to the list's inception.

Owens Corning's insulation products conserve energy and improve acoustics, fire resistance, and air quality in the spaces where people live, work, and play. Its durable roofing products and systems protect homes and commercial buildings while enhancing curb appeal. Its fiberglass composites make thousands of products lighter, stronger, and more durable – often in applications where no other material system has been successful. Box 12.1 provides a brief summary of the company's history.

BOX 12.1 A BRIEF HISTORY OF OWENS CORNING

Owens Corning's story begins with innovations in glass manufacturing. In 1931, Games Slayter, a consulting engineer who had invented a method for blowing mineral wool insulation into houses, was hired by William Levis, president of Owens-Illinois Glass Company. In 1932, Dale Kleist, a young researcher who had worked for Slayter, formed a vacuum-tight seal when attempting to weld together glass blocks. A jet of compressed air accidentally struck a stream of molten glass, resulting in fine glass fibers. Kleist refined this serendipitous process by using steam (instead of compressed air) to form glass fibers fine enough to be used as commercial insulation.

In 1935, Owens Corning was formed as a partnership between Corning Glass Works and Owens-Illinois, two of America's leading glassworks companies at the time, and became a separate company in 1938. Although the company still retains the names of its original founders, neither of the founding companies is still involved. Since then, Owens Corning has been at the forefront of most of the major advances in glass fiber technology. The company was listed on the New York Stock Exchange in 1952.

Today, the business is global in scope, with operations in 33 countries, and human in scale, with approximately 19,000 employees and long-standing, local relationships with its customers and communities.

ORIGINS AND MANAGEMENT OF OWENS CORNING'S CLIMATE AND SUSTAINABILITY STRATEGY

Environmental Footprint and Sustainable Products

In the early 2000s, O'Brien-Bernini, who was then leading the company's R&D division, began a low-key campaign to integrate sustainability challenges and opportunities into the business strategy. Glass-melting is done at high temperatures and requires large amounts of energy. With energy costs escalating and the climate impact of human-generated greenhouse gas (GHG) emissions becoming clearer, a critical question was "how could Owens Corning reduce its energy consumption, and the associated GHG emissions in a way that was good for the world and good for business"?

O'Brien-Bernini recalls,

> When I was advocating to create the new role of chief sustainability officer within Owens Corning, my then boss (recently appointed CEO, Dave Brown) replied, "I don't really understand what this is all about, but if you can do your Chief R&D Officer role and the Chief Sustainability Officer role ... then let's do that." And so we did.

In the early years, O'Brien-Bernini says, there was a definite need to focus on educating others. It began with the need to define many new terms. It became quite clear that, above all, the company must define the concept of sustainability in simple language, so that no one could be confused about what it meant. Harking back to the widely adopted definition detailed in the UN's Brundtland Commission Report in 1983,[1] Owens Corning defined sustainability simply as "meeting the needs of the present without compromising the world we leave to the future."

When the company started its sustainability journey in the early 2000s, its corporate leadership saw right away the need to implement a governance structure aligned with existing operational and financial goals, to allocate and influence the deployment of cross-enterprise resources, to communicate up and down the company, to support each other, and to operationalize a "guiding coalition" to drive the changes to which the company was committing itself. This led to the company's first sustainability governance structure, a 'Sustainability Council' chaired by O'Brien-Bernini. Each of the CEO's direct reports designated a member of their leadership team to participate in the Council.

This structure proved to be an effective way to start, with influential business- and functional-leaders taking on accountability for cross-enterprise progress, as well as progress and support inside their own business or function. This model was in place until 2007, when new CEO Mike Thaman made the decision to establish a sole-focused Chief Sustainability Officer and a dedicated Sustainability Team (hereafter, 'Team') to accelerate progress, broaden the agenda and elevate the company's ambitions.

Early Alignment

At first, the engagement of the board of directors in sustainability was primarily educational and operational in nature. Given the industrial nature of the business, sustainability issues would naturally get integrated into product and process R&D, and manufacturing initiatives. The company's first annual sustainability report (issued in 2007, with data for year 2006 (Owens Corning 2007), was designed to give key stakeholders an idea of how Owens Corning was approaching environmental sustainability vis-à-vis its operations.

Rather than rely on guidance from consultants, Owens Corning's sustainability team, using a regularly updated and transparently communicated materiality[2] process, built goals around sustainability objectives that were seen as most important, while also being within the company's capabilities to implement. These objectives have evolved over the years, but they remain consistent in many ways. The company's environmental goals go beyond what is easily attainable, focusing instead on science-based insights on what is needed to mitigate negative impacts. As the goals evolved over time from simply reducing the company's environmental footprint to increasing its positive impacts, Owens Corning has aspired to remain true to its guiding philosophy – articulated as its 'purpose' – of "people and products making the world a better place."

Accountability for realizing this purpose extends today to middle managers, who are responsible for running the daily aspects of the business: designing products, managing facilities, making purchasing decisions, and marketing Owens Corning's products to direct customers, influencers, and consumers. Delivering products at competitive costs is a fundamental imperative for middle managers in the company, and initial efforts to get managers' attention on reducing the company's environmental footprint – while simultaneously meeting operational and financial goals – made buy-in a challenge in the early days of Owens Corning's sustainability efforts.

Mike Thaman increased buy-in by analogizing sustainability initiatives to the safety initiatives that began to sweep corporate America in the 1970s. Once managers saw that safety not only mattered to themselves and to their employees, but also to the company as a whole by improving quality and reducing risks, they began to embrace the idea. In its internal communications, Owens Corning likened sustainability to improving the safety of the natural environment, and the same type of rigorous and repeatable actions that were used to improve operational safety should be employed in the case of sustainability, ultimately making them both personal and habitual throughout the organization.

Managing Capital Expenditure and Costs

Once the foundation for sustainability had been laid, the expectation was that sustainability goals would be treated similarly to any other business goal. Capital expenditure, SG&A, and R&D resources would flow toward those projects that had the ability to meet or exceed "all-in" business goals, i.e., goals that included not only positive financial, but also sustainability impact.

Initially, the CSO, or another member of the senior management team or the Sustainability Council, would sit in on project and budgeting meetings with each division, to ensure that sustainability goals received the same attention as other goals. Over time, divisional managers would be expected to develop the capability to independently manage sustainability goals, without needing as much guidance as they once did (Owens Corning 2014).

Today, Owens Corning applies the principles of Total Productive Maintenance (TPM) in its operations. TPM is a management system designed to help companies improve manufacturing productivity by empowering employees to take an active role in improving production and processes throughout the workplace. The company has found TPM to be an effective way to prevent defects and losses. It was, therefore, a logical progression to extend the TPM framework to help reduce the amount of waste going to landfills, to cut energy use (and hence emissions), and to improve raw material yields. Concurrently, TPM turned out to be collaterally useful in areas such as employee wellness and safety. For example, at one of the company's facilities, applying the TPM methodology has reinforced the principle that all employees own safety and are empowered to directly address concerns. During the rollout of a new safety standard, the plant team used TPM in evaluating hazards and prioritizing improvements, leading to a faster implementation of safe processes.

Building Sustainability into Managerial Performance Evaluation

Early in the implementation process, Owens Corning decided to integrate its sustainability goals into its performance management system. As with any other goal, managers were given specific objectives such as creating a differentiated product position through a product's 'green' attributes; lowering emissions from a certain process so as to obtain a permit for capacity expansion; or meeting a sustainability goal that had been agreed to internally or externally.

Goals related to sustainability were framed similarly to any other goal, be they financial, operational, or otherwise. After two decades of embedding this approach in performance measurement, managers now recognize that progressive leadership responsibilities require that they demonstrate their ability to perform against all-in objectives. Owens Corning has come to call this the "and" expectation: you are expected to meet your financial goals "and" meet the company's sustainability goals.

Ensuring Board Oversight

Owens Corning's board of directors is expected to provide oversight, guidance, and direction on sustainability issues and opportunities that could potentially impact its reputation and long-term economic viability. The entire board monitors Owens Corning's progress regarding sustainability issues and sets expectations for senior management. According to the company's Directors' Code of Conduct, the principles of sustainability include the concepts of personal safety, environmental compliance, product stewardship, and the environmental and social impacts of global operations and the products the company makes and sells.

The board of directors' audit committee is responsible for environmental, health, and safety exposures. According to the Audit Committee Charter: "The committee is responsible to review the impact of significant regulatory changes, proposed regulatory changes and accounting or reporting developments, including significant reporting developments related to the principles of sustainability."

Since the audit committee is responsible for overseeing risk for Owens Corning, they are also responsible for oversight of climate-related risks (and opportunities).

When evaluating a potential director's experience and qualifications, Owens Corning's governance and nominating committee examines a wide range of skills, including sustainability. The committee assesses experience in or management responsibility for furthering sustainable business practices that address environmental, social, or ethical issues. Nine of the company's ten current board members have been judged to demonstrate the skill to further sustainable business practices: see Figure 12.1 for the "skill matrix" for the company's current set of directors.

BOARD OF DIRECTORS SKILL MATRIX

Provided below in a Board of Directors Skill Matrix is a summary of each Director nominee's skills and experience. The categories included in the Matrix are tied to the Company's strategy, and the goal is that the directors collectively possess qualities that facilitate their effective oversight of the Company's strategic plans. While the matrix is useful for determining the collective skills of the Board as a whole, it is not a comparative measure of the value of directors; a director with more focused experience could nonetheless contribute broadly and effectively.

The chart below identifies the principal skills that the Governance and Nominating Committee considered for each director when evaluating the director's experience and qualifications to serve as a director. Each mark ● indicates an experiential strength that was self-selected by each director. In addition, self-selected diversity information is also provided below.

	CHAMBERS	CORDEIRO	ELSNER	FESTA	LONERGAN	MANNEN	MARTIN	MORRIS	NIMOCKS	WILLIAMS
Public Company Management — Experience as an executive officer of a public company or a significant subsidiary, division or business unit.	●	●	●	●	●	●				●
Financial — Would meet definition of audit committee financial expert if serving on Audit Committee.	●	●	●	●	●	●		●		●
Manufacturing — Experience in or management responsibility for a company that is primarily engaged in the manufacture of goods.	●	●	●	●	●	●	●			●
Global Business — Experience working in a globally distributed business and knowledge of different cultural, political and regulatory requirements.	●	●	●	●	●	●	●		●	●
Marketing — Experience in or management responsibility for significant marketing and/or sales operations.	●	●	●	●	●					●
Strategy / Corporate Development — Experience in or management responsibility for developing business strategies or pursuing mergers, acquisitions, divestitures or joint ventures.	●	●	●	●	●	●	●	●	●	●
Technology / Innovation — Experience in or management responsibility for devising, introducing or implementing new technologies, products, services, processes or business models.	●	●	●	●	●		●	●	●	●
Public Policy / Regulatory — Experience in or management responsibility for defining, influencing, or complying with public policy, legislation or regulation.	●		●	●	●			●		●
Sustainability — Experience in or management responsibility for furthering sustainable business practices that address environmental, social or ethical issues.	●	●	●	●	●	●	●		●	●
Gender Diversity			●			●			●	
Racial / Ethnic Diversity		●						●	●	

Source: Board of Directors Skill Matrix from Owens Corning Notice of Annual Meeting of Shareholders and Proxy Statement. https://s21.q4cdn.com/855213745/files/doc_financials/2020/ar/Proxy-2021-03-05.pdf.

Figure 12.1   Board of Directors Skill Matrix

EVOLUTION OF OWENS CORNING'S SUSTAINABILITY GOALS

Starting with Footprint Reduction Goals

In the early 2000s, under O'Brien-Bernini's leadership, a team was formed to holistically address the company's environmental footprint from operations, going well beyond energy-use reduction. The company's approach was – and still is – guided by three principles:

- Greening its *operations*: achieving specific and meaningful environmental footprint reductions.
- Greening its *products*: continuously improving product life-cycle impact – i.e., the total value-chain environmental impact of its products.
- Accelerating *energy efficiency* improvements in the built environment – e.g., through the use of its products in the world.

While the language has evolved, these three core principles continue to serve as the foundation for the aspirations that shape the company's sustainability goals, including the current ones for 2030 (discussed in more detail below).

In its first set of ten-year footprint reduction goals, spanning 2002 to 2012, the Team focused on seven specific aspects, all related to the natural environment and judged to be the most material to the company. The first two focused on conserving physical resources – energy and water – in the production of Owens Corning's products. The other five addressed the abatement of undesirable byproducts from the use of raw materials in Owens Corning's processes: greenhouse gas (GHG) emissions, nitrogen oxide emissions (NOx), volatile organic compounds (VOCs), fine particulate matter (PM2.5), and waste-to-landfill. As described below, these goals were further improved upon in a second stage from 2010–2020. Currently, the company is in its third phase, with a target completion year of 2030.

In the early stages of sustainability goal-setting, Owens Corning used intensity targets, i.e., targets relative to an economic or physical output such as revenue or production – for environmental footprint reduction. This was a common practice among companies at that time, as it was seen as a way of calculating continuous improvement without limiting growth. Moreover, it paralleled the standard way of measuring progress in productivity, e.g., resources or emissions per unit of product, and therefore paralleled how companies measured material or labor costs per unit of product.

Recognizing, however, that climate change doesn't care about emissions intensity, but rather, the flows (and stock) of *absolute* emissions, Owens Corning shifted its climate-related goals from intensity-abatement metrics to absolute abatement metrics. In other words, these would be targets that were independent of other variables. Currently, the company's 2030 GHG goals incorporate Science Based Target Initiative[3] targets – consistent with 1.5°C warming over pre-industrial levels – which requires the reduction of Owens Corning's absolute Scope 1 and Scope 2 emissions by half by 2030 (Owens Corning 2019a).

The Importance of Understanding Stakeholders

Owens Corning is committed to understanding the company's impacts, as well as contextualizing what sustainability means not only for itself but also its various stakeholders. Our view was – and continues to be – that it is only then that we can have confidence that the company's goals are aligned with global needs and with stakeholder expectations.

Every few years, the company completes a materiality assessment to identify sustainability priorities at global and regional levels, allowing the company to set goals aligned with impact, to prioritize investments in future sustainability efforts, to connect the company's organizational strategy with stakeholders' voices, and to facilitate informed decision-making for the future. Ultimately, the company's sustainability materiality assessments are used to answer just one question: what does the world need Owens Corning to be? Addressing this question provides the company with a solid foundation for a continuous, nuanced, informed and ambitious approach to sustainability.

Materiality Assessments

In late 2016, Owens Corning conducted extensive internal and external stakeholder interviews. In particular, the company focused its attention on hearing more directly from employees beyond those in senior leadership roles. We formally interviewed employees throughout the company, mostly outside of headquarters, to understand the issues that were most important to them. The Team simultaneously followed up with external stakeholders who had participated in previous materiality interviews, with the goal of confirming or clarifying previously obtained feedback.

One outcome of this process was a refinement in the company's materiality assessment for sustainability, allowing for greater attention to issues that are of greater salience to the company's employees. For example, the company's surveys highlighted key areas of resonance, such as employee development, safety, wellness, community impact, and waste management. These conversations led to the company reclassifying GHG and toxic air emissions as priority areas. Previously, due to differences in terminology that was in common use inside and outside the company, 'emissions' had been identified as a priority, but the phrase 'climate change' had not been seen as prominent. With conversations and outreach to sharpen the definition of materiality, Owens Corning found that stakeholders were now seeing climate change, greenhouse gas emissions, and toxic air emissions as essentially the same material issue (Owens Corning 2016).

Owens Corning updated its materiality topics in 2019 (Owens Corning 2019b). The following is the list of Owens Corning's 2019 – and currently operational – "Material Topics for Sustainability".

Product handprint

- Circular Economy
- Product Innovation & Stewardship
- Supply Chain Sustainability
- Sustainable Growth

Environmental footprint

- Air Quality Management
- Biodiversity
- Combating Climate Change
- Energy Efficiency & Sourcing Renewable Energy
- Responsible Water Sourcing & Consumption
- Waste Management

Social handprint

- Community Engagement
- Employee Experience
- Health & Wellness
- Human Rights & Ethics
- Inclusion & Diversity
- Living Safely

Global Materiality

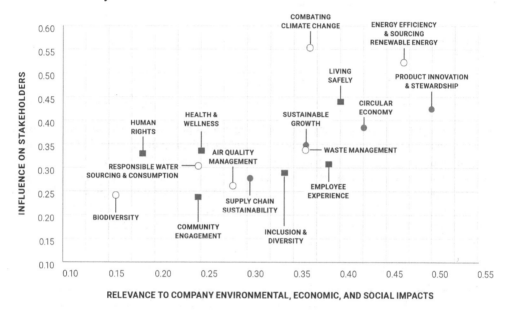

Source: Owens Corning (2019b).

Figure 12.2   Materiality assessment matrix

Owens Corning also decided that, in addition to an overall globally relevant materiality assessment for the company as a whole, the 2019 update to the materiality process would create *regional* materiality assessments for the company's businesses around the globe, with their respective matrix grids (similar to that shown in Figure 12.2).

Further, through the years, the company has partnered with external organizations such as NGOs, trade groups, and research institutions to develop a robust understanding of science and policy goals related to environmental footprint reduction. The Team maintains a matrix of external organizations' objectives and its own, allowing for a nuanced view of where collaboration will be most effective.

The Second and Third Set of Environmental Footprint Reduction Goals

As previously mentioned, the company's first 10-year environmental footprint reduction goals, all intensity-based, were set for 2012 using 2002 as a baseline, covering seven aspects: energy, GHG emissions, NOx, VOCs, PM2.5, waste-to-landfill, and water. The 2012 goals (set in 2002) for each of these aspects were as shown in Table 12.1.

Table 12.1   2012 goals

	2012 reduction target (*Base: 2002*)
Energy	25%
Greenhouse gas emissions	30%
Nitrogen oxide	25%
Volatile organic compounds	25%
Particulate matter	20%
Waste-to-landfill	35%
Water usage	15%

Coming into 2012, Owens Corning had made substantial progress, including meeting six of its seven goals and being within a couple of percentage points on the seventh (GHG). In order to build on this progress, the company announced new goals for 2020, with 2010 as the new baseline. The original seven were consolidated to six, combining NOx and VOCs into one category redefined as "toxic air emissions". In two areas, the extent of the progress achieved by 2015 made it possible for the company to further increase its environmental footprint reduction ambitions mid-cycle. The 2020 goals were set as shown in Table 12.2 (with the two 2015 mid-cycle increases shown in parentheses).

Table 12.2   2020 goals

	2020 reduction target (*Base: 2010*)
Energy	20%
Greenhouse gas emissions	20% *(increased to 50%)*
Toxic air emissions	50% *(increased to 75%)*
Particulate matter	15%
Waste-to-landfill	70%
Water usage	35%

By 2020, Owens Corning had exceeded its targets for energy (actual reduction of 31 percent v. 20 percent targeted), GHG (53 percent actual v. 50 percent target), PM2.5 (39 percent actual v. 15 percent target) and water (43 percent actual v. 35 percent target) goals. The original 50 percent goal for toxic air emissions was achieved (53 percent) but not its 2015 revision to 70 percent; waste-to-landfill remained a challenge, with only a 27 percent reduction relative to its 70 percent goal (Owens Corning 2020). Approaching the end of the second period of goals, Owens Corning worked to establish a third set, with a baseline year of 2018 and a target year of 2030.

In reflection, the progress made against the second set of goals was driven by a mix of process and product innovations – specifically, changes in input materials that reduced emissions, and process improvements that led to greater efficiency and less waste that, in the company's experience, were the largest source of beneficial impacts. We provide specific examples of changes

Owens Corning: environmental footprint reduction 281

in input materials and process improvements below, but before we do so, we turn to a brief discussion that focuses on GHG emissions and energy use reductions goals for 2030.

GHG Emissions and Energy[4]

As mentioned above, the original goal (set in 2002) for GHG emissions was a 30 percent intensity reduction by 2012, and a further (set in 2010) reduction of 20 percent by 2020. The new goal, for 2030, moves to an *absolute* emissions reduction target, aiming for a 50 percent reduction in absolute Scopes 1 and 2 (i.e., direct emissions and emissions embedded in purchased energy, respectively) and a 30 percent reduction in absolute Scope 3 (i.e., emissions in the upstream and downstream parts of the value chain). As an energy-intensive manufacturer in one of the harder-to-abate sectors of the economy, Owens Corning has been focused on reducing GHG emissions from the beginning of its sustainability journey. Initially, Scopes 1 and 2 emissions were prioritized, as the company worked to understand and improve the impact of its own operations. Although the focus was on emissions intensity, the results were substantial on absolute reductions as well – the company has achieved a 60 percent reduction in absolute Scopes 1 and 2 emissions since its peak emissions year 2007, and a 35 percent reduction since the 5.14 million tCO_2e emitted in 2010 (see Figure 12.3).

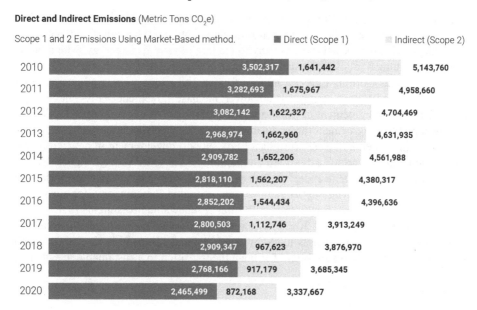

Source: Owens Corning (2020); the 'market-based method' shows emissions the company is responsible for through its electricity purchasing decisions.

Figure 12.3   Owens Corning Scope 1 and Scope 2 emissions: 2010–2020

Owens Corning's 2030 Scopes 1 and 2 goals for GHG are science-based targets externally assured to be in line with the reduction effort needed to limit global warming to 1.5°C by 2050 – this requires a 50 percent reduction in Owens Corning's absolute Scopes 1 and 2 emissions from its 2018 baseline. Additionally, the company has a Scope 3 goal for GHG of a 30 percent reduction, a science-based target in line with what's needed to limit global warming to well below 2°C.

Major contributors to its current GHG emissions are blowing agents used for foam production, combusting fuels in melting processes, and other indirect energy use. The company achieved its initial 30 percent intensity reduction through a change to a new blowing agent in 2009, a large increase in the use of recycled glass to replace virgin raw materials, conversion to more energy and emissions efficient melting processes and glass formulations, and many low-cost/no-cost energy efficiency projects. The company later followed this success by setting a goal to further reduce energy use by 20 percent, from the 2010 baseline, by 2020. (We discuss this more in the next section.)

Owens Corning has also made significant strides in investing in (or partnering for) direct sourcing of renewable energy. An example of such an arrangement, whereby the company entered into large power purchase agreements for wind energy, is described in Box 12.2 below.

BOX 12.2 CASE STUDY – MAKING STRIDES IN WIND ENERGY PRODUCTION

Given the size of Owens Corning's energy demand, the company does not have enough real estate to achieve the emissions reduction it needs via on-site renewables and energy efficiency alone. This reality led the company to explore off-site wind energy, which became economically attractive given the scale needed.

In 2015, Owens Corning took advantage of competitive long-term wind energy pricing by signing a 125 MW wind energy virtual power purchase agreement (VPPA) in Texas, and an additional VPPA agreement in Oklahoma for another 125 MW – totaling a 250 MW virtual power purchase agreement for renewable energy. The deal was structured as a "contract for differences," wherein Owens Corning's PPAs support development of new wind power capacity added to the grid (i.e., assured additionality). These PPAs were signed only two years after its first renewable energy project (a 2.7 MW solar array at a manufacturing plant in Delmar, New York). The key to gaining support within the company, for such a large renewable energy contract, was to start with smaller renewable investments first.

At the time (as of the end of 2015), Owens Corning's wind power purchase agreements added up to the largest purchases made by any industrial company in the world, helping to meet its decarbonization goals and facilitate the transition to a sustainable, clean energy economy.

Project Examples

Fiberglass insulation, first commercialized by Owens Corning in the 1930s, is the most widely used type of insulation in North America today. Residential building insulation saves 12 times as much energy in its *first year* in place as the energy used to produce it. That implies the energy consumed during manufacturing residential building insulation products is saved during just the first 4–5 weeks of product use. The insulation continues to save that amount of energy every month throughout the life of the home in which it is installed.

The company continues to innovate to improve the environmental profile of its insulation products, in keeping with its commitment to expand its product handprint. For example, in 2011, Owens Corning converted all of its residential insulation products to its EcoTouch Insulation with PureFiber Technology (Owens Corning 2012), a product that uses almost entirely natural materials.[5] This product also uses 65 percent recycled content (72 percent of which is post-consumer) and is formaldehyde-free.[6]

In 2020, the company launched a product called FOAMULAR NGX insulation, which is extruded polystyrene (XPS) foam used for insulation. The blowing agent used to make the new product delivers 80 percent less embodied carbon compared with the legacy version of the product, delivering the emissions benefit without a decrease in performance.

The company's new PAROC Natura, a line of stone wool insulation that is fire-safe and durable, uses low-carbon melting technology, renewable electricity, recycled waste materials, and new technologies to reduce the amount of virgin raw material used, thereby offering a product with low CO_2 emissions. Emissions that remain are offset using purchased Verified Emissions Reductions (VERs).[7] The emissions content of this product line is subject to third-party verification, which certifies it as carbon-neutral.

As another example, consider the use of melted glass used in manufacturing insulation. Using crushed post-consumer glass – also called cullet – as raw material not only decreases community landfill waste, but compared with starting with virgin raw materials such as sand, it also lowers energy use. The Glass Packaging Institute, an NGO that advances recycling policies, reports that energy costs drop by about 2–3 percent for each additional 10 percent of cullet used in manufacturing. Owens Corning's use of cullet in glass melted to manufacture insulation products was in the single-digit percentages in the 1980s, but by 2019, the proportion of use of cullet had grown to over 70 percent, resulting in significant savings in energy use.

Similarly, product and process innovation has allowed the company to increase the amount of recycled content in products. As a result, the company's fiberglass insulation now contains between 53 percent and 73 percent certified recycled content, depending on the product. Owens Corning consumed more than 1.3 billion pounds of recycled glass globally in 2020, making the company one of the largest users of recycled glass in the world.

The company is working to further reduce fossil-fuel based primary energy by improving energy efficiency and expanding renewable energy purchases in the US and globally. In 2020, renewable electricity comprised 51 percent of the company's purchased electricity globally. In the US, 59 percent of the company's electricity use (from the grid as well as the PPAs) came from renewables: wind (56 percent), hydro (2 percent), and solar (1 percent). By 2030, the company targets sourcing 100 percent of its electricity from renewables (Owens Corning 2020). Figure 12.4 below shows the progress that Owens Corning has made in sourcing renewable energy during the period 2010–2020 (considering all energy sources, including direct fossil fuel use as well as purchased electricity).

As a final example, consider the company's efforts to abate toxic air emissions from its operations. The company's first toxic air emissions goal was a 25 percent reduction in NOx emissions by 2012. This goal was surpassed by 2007, primarily through the elimination of niter in the insulation batch. Niter was used as a batch oxidizer and to control melt crust. Usage varied depending on raw materials, furnace design, and throughput, but in the early 2000s the company rapidly converted melters based on insights gleaned from glass chemistry and new furnace technology. Through such innovations, niter usage was dramatically reduced despite increased glass melting with the result that the company eventually eliminated *all* niter use.[8]

Overall Energy Usage

	2010	2011	2012	2013	2014	2015	2016	2017	2018	2019	2020
Non-Renewable	9,285,103	9,911,432	9,570,505	9,924,410	9,668,801	9,385,795	9,286,677	8,913,299	9,038,571	8,647,126	7,947,683
Renewable	556,318	589,828	574,326	552,383	592,087	620,467	746,897	1,534,195	1,683,939	1,631,884	1,602,711

TOTAL ENERGY USAGE

PERCENT ENERGY FROM RENEWABLE SOURCES	5.7%	5.6%	5.7%	5.3%	5.8%	6.2%	7.4%	14.7%	15.7%	15.9%	16.8%

Source: Owens Corning (2020).

Figure 12.4   Owens Corning's renewable and non-renewable energy usage – 2010–2020 (MWh)

REPORTING AND TRANSPARENCY

Frameworks and Rankings

Since 2007, the company has viewed annual sustainability reporting as an important tool, both for communicating to stakeholders and as a mechanism for benchmarking and tracking performance over time. The first report, using 2006 data, was considered "a stake in the ground" at the beginning of the company's journey (Owens Corning 2007). Since then, Owens Corning has added disclosures in accordance with other frameworks, to meet a variety of stakeholder needs. There is overlap among the frameworks, but each has a slightly different emphasis and different mechanisms for collecting information. Not all the frameworks require public disclosure, but the company's annual sustainability report is a primary vehicle for those that do.

- **GRI:** The company's 2007 report included a Global Reporting Initiative index for the first time. GRI is a voluntary, international reporting framework that provides guidance to organizations on non-financial reporting. Owens Corning publicly stated with that first index that it was only a baseline, and that the intent was to report more information each year. In 2016, the company began reporting in accordance with the "GRI Standards: Comprehensive Option." This is the more extensive option for GRI reporting requiring additional disclosures on the company's strategy, ethics and integrity, and governance. Owens Corning was among the first US companies to use that standard.
- **CDP:** Formerly known as Carbon Disclosure Project, CDP is a UK-based NGO that works with investors, companies and policymakers to reduce GHG emissions and safeguard water resources and forests. Owens Corning began responding to CDP questionnaires in 2008 for climate change and in 2010 for water. In 2020, the company was named in CDP's "Climate A List" for the fifth year in a row and the "Water Security A-List" for the second straight year. Inclusion on these lists recognizes Owens Corning for its corporate sustainability leadership, through scoring that "measures comprehensiveness of disclosure, awareness and management of environmental risks and best practices associated with environmental leadership, such as setting ambitious and meaningful targets."[9]
- **ISS:** Institutional Shareholder Services assigns participating companies an ESG score, which the organization calls "a scientifically-based rating concept that places a clear,

sector-specific focus on the materiality of non-financial information."[10] Owens Corning uses the ISS framework to help identify gaps in reporting. In 2021, the company earned the following ISS scores: Environmental, 1; Social, 1; Governance, 2. Since the ISS ESG score is based on transparent disclosure, it is used as a key input by other organizations that develop corporate rankings. For instance, 3BL Media uses ISS in calculating the annual ranking of 100 Best Corporate Citizens (a list originally created by *Corporate Responsibility Magazine*). In 2021, Owens Corning became the first company to be ranked number one on that list for three consecutive years – a distinction that the company is particularly proud of because it is the result not only of the efforts and results, but of its commitment to transparency.[11]

- **DJSI:** In 2020, for the 11th year in a row, Owens Corning earned placement in the Dow Jones Sustainability World Index in recognition of its sustainability initiatives. The DJSI World Index is a selective listing of the world's largest companies based on long-term economic, environmental, and social criteria. The company was named 'Industry Leader' for the DJSI World Building Products group for the eighth straight year.

Other frameworks that guide the company's disclosures include the United Nations Global Compact (and the UN Sustainable Development Goals), Ecovadis, TCFD, and SASB. In addition, the company's sustainability team keeps track of questions from investors, and uses the list of questions thus generated as input when deciding on future disclosures.

Learning from Sustainability Rankings

Over the years, the company has earned numerous distinctions and awards related to ESG issues. The company views these as important indications of its satisfying external expectations. As Owens Corning's rankings change over time, the sustainability team works hard to understand the continuously evolving factors that are considered to be indicators of progress. While the company understandably takes pride in its accolades, it believes that the real value comes from the internal benchmarking conversations that lead to better results. For instance, the company didn't earn a spot on the DJSI until 2010, the second year that it submitted to be included in the list. O'Brien-Bernini notes that the learning from the first, failed submission in 2009 was crucial in informing and thus helping to improve not only the company's sustainability initiatives, but also how it communicated with its stakeholders, which contributed to its 2010 successful inclusion in the DJSI.

An important benefit of such rankings is that they spark important internal conversations about where further work might be needed. For instance, knowing that investors or other stakeholders are evaluating the company on a specific item of disclosure can lead to constructive debate about putting resources in place to support that disclosure item. At Owens Corning, it is valued as a way of understanding stakeholder expectations, as an opportunity to do better benchmarking, and as a means to help guide sustainability priorities for the future.

IMPACT OF SUSTAINABILITY ACROSS THE COMPANY

Life Cycle Assessments (LCAs) and Sustainability's Value to the Customer

As part of its product sustainability goals, Owens Corning is committed to evaluating the impacts of its core products throughout their life cycles, and to being fully transparent about its findings. This capability has become central to the company's decarbonization work given the increasingly important metric, "embodied carbon" – that is, the GHG emissions associated with making a specific product – that it uses firm-wide. The company has adopted a two-part methodology to calculate this cradle-to-grave environmental impact.

1. Conduct an LCA according to the ISO 14040, 14044, and 14025, as well as ISO 21930 and EN 15804, followed by a third-party review and verification of appropriate product category rules. LCAs are comprehensive measurements of the environmental footprint of a product at all stages of its life cycle, from the extraction of raw materials, through processing, manufacturing, and product use, and all the way to its eventual end of life through disposal or recycling.
2. Develop an environmental product declaration (EPD) from the LCA and implement continuous and measurable improvements related to those impacts.

By performing LCAs, the company has identified many opportunities for improvement in processes and products. The LCAs have also helped identify high-impact raw materials, enabling Owens Corning to work with suppliers to reduce their footprint, which in turn helps Owens Corning reduce its own. Prior to being introduced in the marketplace, all product packaging and advertising is thoroughly reviewed by the technical services and legal departments, along with each business unit, to ensure compliance with all regulations and codes (Owens Corning 2020).

Owens Corning LCA practitioners have conducted full LCAs on 81 percent of the company's products, including shingles, fiberglass, mineral wool, FOAMGLAS® cellular glass, and extruded polystyrene (XPS) foam insulation, as well as composite glass product offerings such as reinforcements, nonwoven mats, and technical fabrics. In addition, simplified LCAs have been done for an additional 5 percent of the remaining products.

Building Sustainability Momentum

Because Owens Corning approaches sustainability as a company-wide responsibility, there has been a focus for many years on helping business functions see the value of this work. Early on, one team within the sustainability organization focused on understanding the "as is" green and sustainable attributes of the company's building materials, along with benchmarking them versus best-in-class products, regardless of type. They used these insights to focus on how Owens Corning products measure up against those of its competitors, as well as areas where products could improve their green/sustainable attributes.

To accomplish this, the Sustainability Team put together a matrix of green building rating systems, along with programs focused on certification of attributes, as well as transparency in material health.[12] They included programs that may not have been widely accepted at the time, but which they felt could be influential in the future. The Team looked at how green attributes were evaluated in terms of importance, along with negative aspects, such as various red lists of chemicals (i.e., chemicals judged to be problematic by well-known scientific groups and NGOs). Products were scored against green building criteria for each type of building material. Scores could be positive or negative, and had gradations based on whether they met, exceeded, or failed to meet, the criteria.

These sustainability comparisons were christened the "Green Beauty Contest" (GBC). GBC involved an internal competition between Owens Corning's and competitors' products and services, using the green product expectations of their customers (e.g., various green building program point structures). The competition focused on three key objectives, in increasing order of desirability:

1. *Never get "de-spec'd"*: Make sure that no green building program will disallow one of Owens Corning's products on account of its specifications. For example, the LEED certification process may not offer points for insulation containing formaldehyde, so that would have to go.
2. *Achieve "as-equal" spec status*: Assure all products have specifications on par with other products in the marketplace. Essentially, don't get passed over and let a competitor win a bid because they have green attributes that Owens Corning products do not.
3. *Achieve "hard-spec" status*: Because of their performance, durability, safety, brand name, and green attributes, customers will "hard spec" – actively choose, and set as standard – Owens Corning products by name over other products.

Once each product area had been evaluated on the green/sustainability criteria, along with those of the company's competitors' products, results were presented to Owens Corning's leadership. This approach provided an externally informed perspective on the company's products, along with areas for improvement.

The GBC led to several positive outcomes. Importantly, Owens Corning's Sustainability Team became directly engaged with the business units in product strategy and planning processes. The Team implemented an eco-design scorecard for product developers, and integrated discussion of the scorecard in "Product Stewardship" reviews. In addition, adopting an external perspective underscored the importance to business units of many aspects of the sustainability process, such as those emphasizing third-party certification of attributes like low product emissions, high recycled content, as well as giving more importance to LCA of Owens Corning's major product lines.[13]

Integrating Sustainability into the Innovation Process

One way in which Owens Corning has centered sustainability in its operations is by integrating sustainability principles into how the company approaches its innovation process, especially as internal innovation teams develop new products or improve existing ones.

Although the company has had a product stewardship process and a sustainability mapping tool for many years, in 2020 it took a significant step forward by introducing a new framework for its product developers: the "Owens Corning EcoDesign Strategy Wheel." Based on the Okala EcoDesign Strategy Wheel (a brainstorming tool used to explore areas of product development or improvement), this tool integrates stage-specific Design for Environment (DfE) and product sustainability strategies into the innovation process. By providing researchers with a clear view of links to sustainability components of product design, the EcoDesign Strategy

Wheel (Figure 12.5) enables R&D teams to align the product development process with the company's 2030 sustainability goals (Owens Corning 2020).

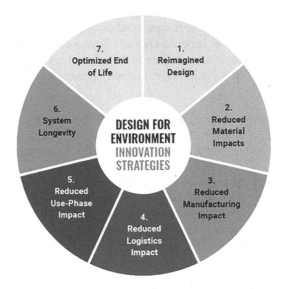

Source: Owens Corning (2020).

Figure 12.5 Owens Corning's "EcoDesign Strategy Wheel"

Carbon Emissions in Budgeting

Early in the company's sustainability journey, and continuing today during its goal setting processes, a shadow price on carbon is used to frame the magnitude of risks and opportunities in the context of a world where carbon emissions are financially valued. However, once the sustainability goals are set, the company leverages these goals to drive action, not the shadow carbon price (while the shadow carbon price tactic has been quite successful in other companies, for Owens Corning it was felt that it would add an unnecessary financial complexity in its project planning and execution). The company does, of course, directly include the forecasted cost of carbon in the financial analysis of projects in regions where there is an Emissions Trading System (ETS) or a carbon tax.

ENTERPRISE COMMITMENT TO 2030 SUSTAINABILITY GOALS

In 2019, as Owens Corning was nearing the end of its second goal period of 2010–2020, the Team took the lead in an intensive process to develop a new set of long-term goals. The Team began with an understanding of stakeholders' priorities, informed by the sustainability materiality assessment discussed earlier, as well as the various benchmarks and frameworks. While some companies may establish goals of this sort as top-down directives, Owens Corning approaches the goal-setting process as a 'change management' initiative. This often takes longer to do up front, but in our experience, collaborating with various levels in the organization throughout the goal-setting process results in a stronger sense of shared ownership in the outcome which, in turn, is beneficial to driving change.

Each year, Owens Corning uses its annual internal 'Sustainability Summit' to advance strategy, execution, communication, and engagement – typically focusing on a few critical issues or opportunities. The company used the 2019 Summit as a forum to gather input on the goals under consideration. During the first two days of the Summit, employees heard about progress on the existing goals and were provided with an overview of relevant megatrends impacting the company, including forecasts of various environmental trends. With that background, working groups used human-centered design techniques to explore ways that Owens Corning could address areas where work was needed. Through these conversations, the sustainability team collected perspectives from employees who work in operational as well as corporate functions, which were then used to create (and then validate) key elements of each goal.

The company's 2030 sustainability goals are broad, and go well beyond climate.[14] Given the focus of this article, however, the most direct climate-related goals include:

- Reducing absolute Scopes 1 and 2 GHG gas emissions by 50 percent, in line with guidance to hold global warming to 1.5°C and reducing absolute Scope 3 GHG gas emissions by 30 percent, in line with guidance to hold global warming to well below 2.0°C. This goal was approved by SBTi.
- Sourcing 100 percent renewable electricity to decarbonize and to reduce product-embodied carbon.
- Maximizing product sustainability, including circular economy strategies for reduced consumption of virgin materials and increased end-of-life reuse and recycling – a powerful (and largely underappreciated) climate action tactic for the company's type of business.

CEO Commitment

In 2019, Brian Chambers assumed the role of CEO of the company. During his transition into the role, numerous discussions were held to address the ways in which Owens Corning could lead in sustainability initiatives, including forming expectations of what might be achieved during Chambers' leadership. The new CEO established two key tenets for the company's 2030 goals. First, the goals must be consonant with what the world expects of Owens Corning. Second, where possible, the company must leverage metrics that have gained scientific consensus with respect to measurement and reporting.

As described above, this had been a major focus of Owens Corning's 2019 Sustainability Summit a few months prior, and Chambers embraced and advocated for this idea. For example, Chambers and O'Brien-Bernini discussed the options for setting a science-based GHG target (via SBTi). While leading climate scientists were advocating for the 1.5°C pathway (vs. the vaguer 'well below 2°C' pathway) the company's analysis indicated that the 1.5°C pathway would require a 50 percent versus a 30 percent reduction in absolute Scopes 1 and 2 emissions.

Chambers' response was indicative of his ambitions in, and commitments to, this area: "Why would we do anything other than challenge ourselves to do what the scientists say must be done?" The conversations that followed led the company's new goal to be set at an ambitious 50 percent reduction in the combined Scopes 1 and 2 emissions. The plan that Owens Corning had set forth to achieve this goal received SBTi verification, with an expectation that any tactical gaps that remained would be resolved over time.

Building Buy-in

Chambers also made clear that these goals must be leadership goals for the whole company, not just those of the Sustainability Team. This, in turn, led to a series of rather direct engagement meetings with all levels of leadership. The process led to an understanding, at the executive committee's level, that they (and their teams) would be accountable for making sure all employees were clear about the specific areas in which their businesses needed to lead for Owens Corning to be able to attain its goals.

As these goals were launched, the CEO stated,

> These goals advance our business priorities by pushing us to discover increasingly more sustainable solutions, [an issue] which is becoming more important to our customers and other key stakeholders. Our employees are energized by the challenge to use their creativity and dedication to achieve these ambitious goals.

Expanding Focus: The Circular Economy

The sustainability team has expanded as the company continues to identify new levers for reducing GHG emissions. With growing scientific evidence recognizing the significant role that the circular economy – i.e., a model of production and resource consumption that uses existing raw materials and natural resources for as long as possible – can play in reducing global CO_2 emissions, Owens Corning declared its aspiration to work towards ensuring that every type of raw material and natural resource extracted to manufacture its products and run its processes remains *indefinitely* in the economy. The company expects this to be a key tenet of its eventual total decarbonization.

In Fall 2020, the company established a dedicated circular economy team to define goals and prioritize projects that align with the company's circular economy ambitions. The team, led by the Chief Sustainability Officer, consolidates and builds on the work already done over the past few decades, and serves as a hub for thought leadership, expertise, and knowledge-sharing. Additionally, the team would collaborate with subject-matter experts and other such teams across the company, as well as industry, governments, and academia. This structure creates shared accountability for meeting goals, ensuring that the circular economy – like other sustainability goals – is not simply a sustainability team initiative but a challenge and an opportunity that is shared by all business units and functions, in all regions.

Owens Corning's circular economy efforts are focused on two areas: manufacturing and end-of-life solutions.

- *Manufacturing*: The work here focuses on meeting the company's 2030 zero-waste goal – by reducing the waste generated by its processes by 50 percent, and then finding ways to reuse or recycle the rest – as well as efforts to create demand for recycled materials throughout the supply chain by expanding the use of recycled materials in the company's manufacturing operations and products, across all businesses.
- *End-of-life solutions*: Owens Corning will seek innovative technologies and business models for its products and materials to be reused and repurposed indefinitely.

The work required here includes partnering internally with R&D, commercial, and corporate development counterparts to shape the vision as well as in its implementation. The company will also engage with external partners to develop end-of-life solutions for its products, as well as in the products where Owens Corning's materials are used (e.g., shingles used in roofing, insulation used in buildings). In our experience, there is a great deal of synergy between these two areas, including opportunities to reclaim customer waste, to deconstruct products at the end of their lives, and to discover uses for those materials as raw material inputs for either the company's operations or other beneficial uses.[15]

Other groups within the company are increasingly connected to the sustainability efforts. For example, with increasing focus on embodied carbon, the company has taken on a science-based target for reducing Scope 3 emissions. Internally, this drives the sourcing and supply chain teams to drive closer collaboration across the company's supply chain. The collaboration is vital to helping Owens Corning increase its product handprint and, through its product stewardship process, ensure its products are safe and environmentally responsible throughout their life cycle. The company expects its suppliers to make efforts to reduce their GHG emissions, as well as to align with other environmental, social, and governance priorities.

CONCLUSION

Although Owens Corning's sustainability journey began with environmental footprint reduction efforts, the company today considers the work to expand its handprint – in both product and social benefits – equally important. This stance has resonated positively with current and prospective employees. The company's recruiters describe increased interest from potential candidates, and many new hires reference Owens Corning's sustainability profile as a factor in their decision to join the company.

We believe, however, that the easy wins are mostly in the past. The sustainability work that companies need to do is in response to challenges that are global in scope, and therefore will require solutions that go beyond any one company. Moreover, transparency and accountability are part of the company's stated approach to attract talent and to collaborate with key partners such as suppliers. For these reasons, we believe it is more important than ever for Owens Corning's employees to be able to "see themselves" in the company's sustainability goals. In the process, the company's sustainability priorities must demonstrate a commitment on their part to connect the 'personal' (i.e., the employee) and the 'collective' (i.e., the organization as a whole).

We believe that the company's long history and considerable accomplishments are merely a baseline for the future. Looking beyond today's goals and results, Owens Corning is committing resources and energy to shape the future and, consonant with Brundtland's original vision of what 'sustainability' means, helping to make the world a better place for the generations that follow.

BOX 12.3 A FEW PERSONAL REFLECTIONS FROM O'BRIEN BERNINI

One of biggest challenges for a chief sustainability officer in any large company can be loneliness, i.e., being the lone voice advocating for work that needs to get done, often on a timescale well beyond normal planning cycles. The CSO is often the only one in the room with the unique, granular perspective on all the implications that a major decision can have on the company's sustainability ambitions. Sharing these perspectives can often create added complications and stress in ensuring initiatives get successfully implemented, especially in an environment in which teams may already be struggling with tight timelines, financial challenges or demanding customer expectations. Even though sustainability is everyone's job, sustainability leaders are the ones who are ultimately accountable to ensure that goals are resourced, on track and achieved.

The most critical competence required of sustainability talent, across the company, is the ability to lead change. To succeed, leaders need the skills and capabilities to set ambitious aspirations that matter and have the confidence to state those aspirations publicly, often before having a defined path to achieve them. In addition, they must be able to commit resources with the resolve to make the necessary progress over the long term, and then have the nerve and integrity to transparently report progress, year after year, whether the results are good or disappointing – recognizing that, most often, it's a mix of both.

After two decades of sustainability leadership, Owens Corning today is experiencing massive cross-enterprise sustainability/business integration. Sustainability has become core to who we are and how we operate, impacting virtually every business discussion and decision. This is deeply energizing for the Sustainability Team and equally, inspiring for all our employees. It's good for the business, and great for the world.

Perhaps the biggest success for sustainability leaders everywhere is that this transformation is not limited to just a few companies. The broader corporate world is embracing the kind of sustainability commitment that has been the core of Owens Corning's purpose and progress for many years. The momentum and energy we share is the best hope for fighting climate change and building a more sustainable future.

ACKNOWLEDGEMENT

We are grateful to Eric Klinger, who helped to pull together the needed corporate data for this chapter.

NOTES

1. In December 1983, Gro Harlem Brundtland, former Prime Minister of Norway, was tasked by the United Nations to chair the World Commission on Environment and Development. Its mission was to re-examine critical environmental and development problems around the world, and to formulate proposals to address them.
2. The concept of materiality – a foundational concept in US securities laws and in corporate disclosure – is one that addresses whether and how particular financial, macroeconomic, social, environmental, etc., issues are important for a company or to the industry to which it belongs. Typically, an issue would be considered material if it is seen as having a significant impact on the firms' earnings and/or risks.
3. Science Based Targets Initiative, or "SBTi" is a partnership between the Carbon Disclosure Project (CDP), the UN Global Compact, World Resources Institute (WRI), and the World Wildlife Fund for Nature (WWF) (https://sciencebasedtargets.org/about-us#who-we-are). SBTi shows companies how much and how quickly they need to reduce their GHG emissions at the firm-level, in order to be consistent with the emissions budget resulting from a 1.5°C global warming target.
4. Owens Corning (2020).

5. It is made with a minimum of 99 percent (by weight) natural materials consisting of minerals and plant-based compounds.
6. Ecotouch PINK Fiberglas Insulation Product Data Fact Sheet: https://dcpd6wotaa0mb.cloudfront.net/mdms/dms/Residential%20Insulation/10013811/10013811-EcoTouch-PINK-FIBERGLAS-Insulation-Product-Data-Sheet.pdf?v=1592540232000
7. VERs certify a reduction in GHGs – typically CO_2 – from independently audited projects using widely-accepted third-party certification standards. Each VER represents the reduction of one metric ton of CO_2-equivalent emissions (tCO_2e).
8. The progress Owens Corning made in the first set of long-term goals (2022–2012) enabled the company not to focus on NOx in its most recent set of global sustainability goals. That is not to suggest it's unimportant or that the company has not continued to make significant progress: as can be seen from its 2020 Sustainability Report (Owens Corning 2020), the company achieved a further 70 percent intensity reduction from 2010–2020.
9. https://www.cdp.net/en/companies/companies-scores
10. https://www.issgovernance.com/esg/ratings/
11. Owens Corning press release: https://newsroom.owenscorning.com/all-news-releases/news-details/2021/Owens-Corning-Earns-No.-1-Ranking-on-100-Best-Corporate-Citizens-List-for-an-Unprecedented-Third-Year-in-a-Row/default.aspx
12. "Material health" refers to understanding how materials used in building construction affect human health and the environment. There are now industry standards and certifications for material health that many suppliers to building construction use.
13. From internal company documents.
14. More about the complete set of our 2030 goals can be found on the company's website: https://www.owenscorning.com/en-us/corporate/sustainability/2030-goals
15. The content in this paragraph is adapted from an unpublished draft of Owens Corning's *2021 Sustainability Report*.

REFERENCES

Owens Corning. (2007). *Sustainability at Owens Corning: 2007 progress report*. https://www.owenscorning.com/owenscorning.com/assets/sustainability/governance/reporting/OwensCorning_Sustainability_07.pdf

Owens Corning. (2012). *Sustainability at Owens Corning: 2012 highlights summary*. https://dcpd6wotaa0mb.cloudfront.net/owenscorning.com/assets/sustainability/governance/reporting/10017732A_Sustainability_Trifold_Brochure_072213-d781fab521e337a8d2020a19967e4842977e9857ac2fc0a6ec2a0fc24a288691.pdf

Owens Corning. (2014). *2014 sustainability highlights*. https://www.owenscorning.com/owenscorning.com/assets/sustainability/governance/reporting/2014_Sustainability_Highlights.pdf

Owens Corning. (2016). *2016 sustainability report: It all adds up*. https://dcpd6wotaa0mb.cloudfront.net/owenscorning.com/assets/sustainability/governance/reporting/2016_OC_Sustainability_Report_FINAL6-14-2017-c606aa00a96047509dd1a8163f5db062c0c508f802b482befadb0ad718fadb30.pdf

Owens Corning. (2019a). *One company, one purpose, one world: 2019 Sustainability Report*. https://www.owenscorning.com/en-us/corporate/sustainability/docs/2020/2019-Owens-Corning-Sustainability-Report.pdf#page=107

Owens Corning. (2019b). *Sustainability materiality*. https://dcpd6wotaa0mb.cloudfront.net/owenscorning.com/assets/sustainability/materiality/Materiality-Report-bf4c1217f8c376af9599261f740b241cbe89dd70b4a593de15dec05a26f03e54.pdf

Owens Corning. (2020). *2020 sustainability report: Beyond today, shaping tomorrow*. https://www.owenscorning.com/en-us/corporate/sustainability/docs/2021/2020-Owens-Corning-Sustainability-Report.pdf

13. Climate preparedness for business resilience
Janet Peace and Kristiane Huber[1]

INTRODUCTION

While the world was focused on a global pandemic in the year 2020, the warmest decade on record, 2010–2019, closed out with the second warmest year on record (National Aeronautics and Space Administration (NASA), 2020). Climate change poses significant risks across the global economy and numerous reports over the last decade (Marsh and McClennan Global Risk Center, 2017; Coppola et al., 2019; Semieniuk, 2020; McKinsey & Company, 2020; S&P Global, 2020; C2ES, 2015) have considered the risks posed to business.

This chapter explores this issue from several directions – how climate change can affect a company; whether most companies are prepared; and, finally, it suggests actions that can be helpful, including identifying business opportunities associated with changes in the climate. Throughout, the focus is on the physical risk associated with climate change and examples from companies are presented to make issues more tangible, illustrate key ideas and suggest best practices.

As has been discussed in previous chapters, the science and the impacts associated with climate change are becoming more prominent and more challenging to ignore. A warmer earth means droughts and heatwaves are more likely; warmer oceans make tropical storms more intense;[2] and more moisture in the atmosphere means any storm is dropping more precipitation (Environmental Protection Agency (EPA), 2021; National Academies of Sciences, 2021). The year 2020 was no exception. It was the sixth year in a row for the US to experience ten or more extreme weather events. The frequency of extreme weather events, such as wildfires, hurricanes, droughts, and flooding events is trending upward, as shown in Figure 13.1.[3]

In 2020 alone, the National Oceanic and Atmospheric Administration (NOAA) reported that extreme weather events across the US imposed a cost of $95 billion on the US, about one-half of a percent of the US GDP.[4] By December of 2021, US damage estimates from NOAA topped $100 billion, with one storm, Hurricane Ida, causing over $64 billion alone (National Centers for Environmental Information (NCEI), 2021). These costs are directly and indirectly borne by American businesses in many forms, such as power outages, crop destruction, damaged buildings and facilities, loss of customers and supply chain disruptions.

And US businesses are not alone in experiencing an uptick in damaging climate related events since the early 2000s. The insurance firm Swiss Re's research institute (2020) reported 202 natural disasters in 2019 with damages of $205 billion ($210, in 2020 USD) worldwide. Similarly, another insurance firm Aon, reported that 2020 had the fifth highest cost on record for natural events, and they too observed that the cost trend is increasing (Aon, 2020). While directly comparing costs between various sources such as NOAA, Swiss Re, and Aon is difficult because of the methodological differences each uses in their calculations, comparing trends within each individual source clearly illustrates an upward cost trend. When multiple models illustrate the same insight, economists like to say that the results are "robust", meaning

Climate preparedness for business resilience 295

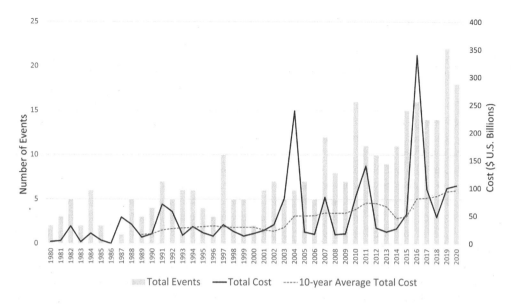

Source: National Centers for Environmental Information (NCEI) (2021).

Figure 13.1 United States billion dollar disaster events 1980–2021

the results are strongly supported (Peace & Weyant, 2008). The increasing nature of these events creates a risk for businesses on a global scale.

As noted previously, many reports have considered the risks posed by climate change to business and the majority have concluded companies are unprepared for the multiple types of risk that a changing climate produces. Being unprepared means companies are less likely to succeed in the face of the impacts and financial implications from climate change.

The alternative is for companies to be climate resilient. Climate resilience has been defined in multiple ways. The Center for Climate and Energy Solutions (C2ES) for example, defined it as the ability to prepare for, recover from and adapt to climate impacts (2019a). The Intergovernmental Panel on Climate Change (IPCC) uses a similar but more expansive definition adding that a response should allow an entity to maintain its essential function, identity and structure while learning, transforming, and potentially reorganizing to cope with the hazards and trends associated with climate change (Lavell et al., 2012). Throughout this chapter, we will consider how companies can be "climate resilient," in other words, have the capacity to adapt to gradual changes and bounce back quickly from physical disruptions related to climate change.

In addition to understanding and managing climate risks, identifying opportunities can round out a company's strategy and help leverage investments in resilience to give them a competitive edge. This chapter reviews the physical risks that climate change poses to businesses, key risk areas that large companies should examine, and considers if companies are generally prepared. In addition, it also offers insights and recommendations on how companies can address barriers, identify opportunities, and increase their resilience to climate change.

BUSINESS RISK

Climate risks faced by business are often divided into two categories – transition risk and physical risk (Campiglio, 2019; Financial Stability Board, 2020; Task Force on Climate-Related Financial Disclosures (TCFD), 2017; World Business Council for Sustainable Development (WBCSD), 2019). Transition risks are those that companies face as society moves to a lower carbon economy. Included in this category are regulatory risk, changing consumer preferences, liability risk, increased competition from businesses facing less carbon-related regulation and from companies providing lower carbon solutions at a lower price (C2ES, 2013a; Center for International Climate Research – Oslo, 2021). To illustrate transition risk, consider that renewable energy sources have been mandated by policymakers, supported by environmental advocates, and have become more cost competitive with traditional fossil energy. As a result, fossil energy sources have lost market share in the electricity sector. The Annual Energy Outlook (AEO) in 2011 reported that renewable generation made up 10 percent of US generation (Energy Information Administration (EIA), 2011) and, in 2021, AEO reported it accounted for 21 percent. This number is expected to expand to 42 percent by 2050 (EIA, 2021). The loss of market share by traditional fossil fuels to lower carbon fuel sources is an example of transition risk associated with a climate-related policy. This policy-related risk is increased by the many local, state, and national governments' greenhouse gas emissions targets and pledges to reach net-zero emission levels by mid-century. These commitments will drive investment away from higher emitting technologies towards lower carbon options and industries.[5]

Much has been written about the transition risk facing companies as policymakers continue to push for incentives and low carbon technology mandates (Center for International Climate Research – Oslo, 2021; Deloitte, 2020; FSB, 2017; Jaffe et al., 2000; Gillingham et al., 2008). As policies continue to expand, this risk will only become more pronounced. Transition risks pose an undeniable threat to companies with large carbon footprints, such as steel or fossil fuel producers. Since most companies, however, routinely review their competition, consumer preferences, policy implications and make business decisions accordingly, transition risk is often front and center in corporate decision-making, especially in recent years.

The same is not true for the other side of the same coin, physical risk. Physical risks are associated with damages that arise from changes to the natural environment. These damages can result from acute or extreme weather events such as hurricanes and wildfires, or chronic stressors such as sea level rise or increasing average temperatures and precipitation.[6] These impacts can severely damage facilities, disrupt employees' lives and commutes, shut off power or otherwise cause significant business interruption and, more broadly, negatively affect regions and supply chains. While most large companies plan for weather events, these decisions are often based on past experience, which is no longer a good predictor of the future. Relying on experience will fail to incorporate how more gradual changes in temperature, sea level, and other climate factors are impacting the environment. More companies need to integrate how future extreme weather can impact their planning, operations, and strategy because costs are real and rising (Box 13.1).

Physical Climate Risks

Climate risks to business are well documented. The World Economic Forum's (WEF) 2021 Global Risks Report, for example, has ranked extreme weather and/or climate inaction among its top two global risks by impact and likelihood, since 2014.

Climate risk from an extreme weather event is related to two core concepts: threat and vulnerability. Threats are the hazards, such as a drought or a hurricane, which can negatively impact an asset. Vulnerability is the potential degree of damage an asset could experience. The probability that the threat could occur, and the severity of the impact are part of what determines the vulnerability of the asset (C2ES, 2020). Sensitivity, exposure, and adaptive capacity are also important to determining how vulnerable any asset is. The IPCC defines sensitivity as the degree to which a system or resource might be affected by climate change. Exposure refers to elements (people, assets, infrastructure, etc.) in an area which could be impacted by a weather-related hazard (Cordona et al., 2012). The adaptive capacity is the ability of an asset, region, or system to adjust and cope with the change. Taking steps to improve adaptive capacity is a cornerstone of improving an asset's resilience. For example, increased precipitation is a threat to assets located in floodplains. The vulnerability of those assets is related to where and how they are built. If they are elevated, the threat still exists, but the adaptive capacity of the asset has increased and its relative vulnerability has decreased. The threat still exists, but the adaptive capacity of the asset and its vulnerability to the threat is reduced. Not building in a floodplain, or at least elevating it, adding plants, and even using porous cement around the building can all improve the resilience of that building to the risk of extreme precipitation (but does not remove the threat).

Broadly speaking, while it is relatively easy to identify vulnerabilities, impacts and costs from past weather-related events (Box 13.1), research suggests assessing how these risks will arise in the future for any specific location or for any specific business can be challenging for a company (C2ES, 2013a, 2015, 2017).

BOX 13.1 EXTREME WEATHER CAN BE COSTLY FOR COMPANIES AND THEIR CLIENTS

While extreme weather directly impacts many business sectors, the three below – Energy, Oil and Gas and Insurance, have received particular attention in recent extreme events and provide critical services the larger economy relies on.

Energy

- Recent studies have suggested a connection between a warming Arctic and the 2021 extreme cold snap in Texas, which triggered massive grid failures across the state (Cohen et al. 2021). That reduced electricity supply coupled with an increased demand for heat caused and wholesale price spikes that rose to $9 per kilowatt hour from an average of 12 cents per kilowatt hour (Hersher, 2021) and may be the most costly disaster in state history (Ferman, 2021).
- Wholesale electricity prices in CA increased and later rolling blackouts resulted from an extreme heat wave in California in 2020. In one hour from 5–6 pm on the 17 August, demand was high and supply limited, resulting in prices more than doubling to $387/

MWh and then further escalating to an average of $800/MWh for the next two hours (*Sacramento Bee*, 2020). According to the California grid operator, CAISO, there just wasn't enough energy to meet the high demand that resulted from the heatwave (S&P Global, 2020b).
- Decreased 2020 snowpack in CA also affected hydroelectricity generation and cost utilities $1.5 billion because of the need to replace hydro-capacity with natural gas (Kathan, 2020).

Oil and Gas

- Hurricane Harvey submerged Houston in 2017 with 60 inches of rain in some areas and caused significant flooding at oil refineries and petrochemical facilities. Almost 4 million barrels of refining capacity per day was knocked off-line (about 25 percent of the US total) and the total storm cost was estimated at $132 billion (2020 dollars) (NCEI, 2021). ConocoPhillips was shut out of its headquarters for over two weeks. To reduce the impacts of flooding, Chevron spent $120 million to raise a dike at its Pascagoula Mississippi refinery and $16.2 million to construct a seawall at its Port Arthur Texas plant (Bhambhani, 2018).

Insurance

- Munich Re has warned that climate change could make insurance coverage unaffordable for many residents in areas of particular risk and noted that damage from CA wildfires in 2019 reached $24 billion. Risks for wildfires, storms and flooding necessitated the upward adjustment of insurance rates in 2019 (Neslen, 2019).
- In 2019, insurers refused to renew over 200,000 home insurance policies, a sharp increase from the prior year. Regulations in several states limit the ability of insurance companies to use past losses to inform rates, triggering the non-renewal of so many policies (Chiglinsky & Chen, 2020).
- The CA Department of Insurance banned insurers from dropping policies in certain areas hit by, and prone to, wildfire. In 2020, the ban was extended for a year. Insurers have said that if the State wants them to continue to offer policies in those areas, they will need to charge higher premiums that reflect the risk (Flavelle, 2020).
- Florida property insurers increased rates by double-digit percentages in 2021, in part, they explained, because of lingering damage from past hurricanes. Many of the larger residential and property insurance companies left the state after Hurricanes Katrina and Wilma in 2005 and the state now relies on small and mid-sized insurers (Barlyn, 2021).
- Canada's largest real estate insurance broker, HUB international, told its Vancouver clients that rates would increase at least 25 percent in 2019 because of increasing climate risk (Littlemore, 2019).

Chief among the obstacles that stand in the way of businesses integrating the future risk associated with climate change is lack of knowledge. Uncertainty about the nature, timing, location, and severity of changes, due to limitations of climate modeling, the complexity of multiple, interrelated risks, was cited as critical during interviews with large companies conducted for C2ES's Weathering the Next Storm report (C2ES, 2015).[7] Companies also said they lacked internal expertise for interpreting climate data and disentangling it from other stresses, such as those caused by urbanization or outdated infrastructure (C2ES, 2013a). Finally, there are few examples of networks for sharing information between companies, communities, and local governments in the US and abroad (although this is changing).

Over the last decade, governments have tried to address the knowledge issue by providing reports, data sets and even tools that can be used to identify and screen significant potential climate hazards. A variety of consulting companies have also stepped-in to provide services to fill the knowledge gap. Many of these downscale climate model data to help identify potential vulnerabilities and localized impacts (e.g., Willis Towers Watson's Acclimatise; Moody's Climate Solutions; The Climate Service). Nevertheless, understanding how these physical risks can impact a company is still an issue and a challenge for decisions about investing in resilience. Preparing for these changes requires better understanding of the risks and either modifications of existing strategies or entirely new processes (C2ES, 2020; Meyer, 2018).

Costs and Benefits can be Challenging to Assess

Decisions on climate risk management are also made more difficult because they need to be based on assessments of vulnerability and probable damage, which are weighed against more certain near-term costs, long-held traditional practices, preference for specific assets, suppliers, or locations. In addition, investments in resilience can also have soft or diffuse benefits that are realized over a prolonged period and accrue to a variety of entities. Such benefits are difficult to define, measure, and report. As Mancur Olson (1965) explained when outcomes have diffused benefits, but the costs are concentrated on a fewer individuals or entities, actions can be more challenging to incentivize and, in this case, justify investments. For example, putting electric wires underground can be very expensive for power companies (and their investors) but doing so can help avoid wildfires that can cause widespread costs to local businesses, electricity outages, loss of life, and impacts to air and water quality. Comparing the probable costs from wildfires and the somewhat, diffuse benefits from avoiding them can make such investments difficult to justify for companies. As Moody's Investor Service reported in 2020, the nature of the risks facing utilities including exposure to floods, fires and heat stress is fairly large and expected to grow over the next 20 years such that "investments in resilience can be worth the cost" to investors but utilities and their regulators need to agree on how to quantify the value of resilience (Utility Dive, 2020) and justify additional charges to ratepayers.

More work is needed to understand and communicate the financial benefit of investing in climate resilience. Understanding the payback for this type of investment has been the focus of several studies (The National Institute of Building Standards (NIBS), 2019; C2ES, 2019b; The Global Commission on Adaptation (GCA), 2019) and all find a significant benefit for each dollar spent on mitigating climate risk. NIBS for example, examined the financial benefit of a variety of broad mitigation strategies that could be used on buildings and related infrastructure (e.g., elevating buildings, flood-proofing basements, changing building codes to account for wind, etc.) (NIBS, 2019). The study found that the payback ranged as high as 12:1 (adopting building codes for higher wind speeds) and as low as 3:1 (federal grants for wildfire prevention in Urban interface) – meaning that for every dollar spent the benefits in terms of safety, preventing property loss and avoiding disruption were significant. The study quantified many of the direct benefits, including productivity losses of damaged facilities and stress on survivors, but many of the more diffuse benefits, such as social impacts and ecosystem impacts, were not included and, as such, benefits per dollar spent are likely much higher. Strategies that improve the resilience of buildings provide an indication of the large value of investing in climate resilience but even so, investment by the private sector has lagged other sources of investment (C2ES, 2019b).[8] Part of the reason for that is that similar studies across other industries are not available. Further, the benefits may not be realized for many years, but the costs are more immediate. Embedding the benefits and the risks within insurance rates could help and is increasingly a focus of the insurance industry (Deloitte, 2019).

Another reason corporate investment in resilience has lagged is competition with other primary business objectives and resources, many of which are often more immediate, tangible and yield more certain returns. Short-term costs and cash flows are often prioritized over investments from which the benefits are over the long-term and based on probabilities. Capital, especially for smaller companies, is limited, and investments in long-lived assets such as facilities or equipment, or even in product development, involves high upfront costs and financial hurdles. Given the uncertainty of the nature and timing of extreme weather events and the lack of detailed, location-specific data on climate change impacts, companies often find it difficult to justify their decisions to invest in resilience. Again, more research expanding on the analyses that NIBS has completed could help to inform and justify private and public resilience investments.

Frameworks for managing climate risk are also emerging. As a recent article published by the Global Association of Risk Professionals (GARP) noted, management of emerging risks "like Climate Change" tends to go through several stages – first, acknowledgment as a substantial threat; second, methodologies/models/frameworks are developed to quantify the risk and its impact; and third, best practices are put into regulations as minimum standards (Folpmers, 2021). The article suggests that climate risk is effectively in the second stage where models and frameworks are being created and, as an example, points to a new framework for managing climate risk in the banking sector by the Basel Committee on Banking Supervision (BCBS, 2021).[9]

CLIMATE CHALLENGES TO BUSINESS

Businesses face all kinds of risk, some of which can cause serious loss of profits or even bankruptcy. Anticipating these risks and preparing for them can hopefully avoid serious losses. Avoiding loss, business interruption or worse, is why many companies have extensive business continuity, disaster response and/or risk planning processes. Anticipating potential disruptions that could impact operations, raw material supplies, and distribution networks, requires that companies stay up to date on where their suppliers are located, raw material costs, relevant policy, and potential legal challenges, among other things. Climate change has implications on all these issues and the following section begins by examining why many current business continuity plans and planning processes inadequately deal with the physical changes of climate change. The section then examines the risk that arises from impacts on supply chains and finally concludes with a brief examination of how fragmented planning outside of a company, and specifically disjointed resilience policy, can impact a company's resilience planning efforts.

Business Continuity Plans

In interviews with corporate representatives conducted in 2015, C2ES found that many companies relied on existing business continuity or risk management processes to deal with climate risks. Many large companies have extensive processes around business continuity, including dedicated "risk management" departments who review operational, legal, compliance and supply chain risks. Business continuity planning helps companies deal with anything that disrupts normal business operations. These plans differ from disaster recovery plans in their holistic (focused across critical business functions) and forward-looking approach to ensure continuation of operations, should a disruption occur. Disaster recovery plans, in contrast, tend to be reactive in approach and only put in place steps for how to recover or restore operations after a disruption, although they can be an important part of a business continuity planning (IBM Services and Consulting, 2020).

Physical changes to climate can disrupt and make business continuity processes and preparations less effective, especially those that rely exclusively on past experience to predict the future. Conventional business continuity planning commonly includes only the hazards the company has faced in the past, look only a few years into the future (three to five), focus on internal operations (rather than the broader environment where they operate)[10] and while they may have business interruption insurance and even include extreme weather events, they usually don't include a comprehensive set of potential indirect impacts that could occur (C2ES, 2015).[11] According to the C2ES interviews, companies often only evaluate whether their infrastructure withstood past extreme weather events and then explore what could be done in the near-term to better protect assets and rebound quickly after another similar event. Part of the reason for this could be where responsibility for risk management typically sits within an organization – sometimes in a finance department, sometimes in legal and sometimes in human resources. Rarely is the sustainability department (where knowledge of climate change might reside), responsible for corporate risk or the development of business continuity plans (C2ES, 2017).

Planning for supply chain disruptions, however, may be changing given the prominence of disruptions caused by the pandemic. Unexpected and indirect impacts became more pronounced in global supply chains in 2020 and 2021. Component shortages, employee shortages, difficulties in transporting goods across regions, led to a variety of unexpected global shortages (Helper and Soltas, 2021). Partners at McKinsey, Kevin Sneader and Susan Lund, suggested that prior to the pandemic, companies often had little visibility into the "deeper tiers of their networks" and CEOs rarely concerned themselves with their supply chains (McKinsey, 2020). They believe that the pandemic may have delivered a "wake-up" call about risks facing supply chains that will extend to climate risks.

As the pandemic has illustrated, not considering potential risks because they have not occurred in the past is problematic. Climate change is similarly not static, and future threats will be unprecedented for communities and the businesses that face them. Houston, for example, historically receives about 50 inches of rain in a year but, in 2017, Hurricane Harvey dropped 60 inches over a few days. While the region has historically been prone to hurricanes, the warmer atmosphere that results from climate change holds more moisture which can result in weather events with increased precipitation. Recent studies are also suggesting climate change may be decreasing the speed at which large storms move, which may help explain why Harvey lingered over Houston for several days (Zang et al., 2020; Princeton, 2020).

Just how much worse are these storms because of human induced climate change? As the Fourth National Climate Assessment (NCA4) reported (2018), scientists have been examining this question for many years and have been able to statistically prove and measure the fingerprints of climate change on many events.[12] Their studies are commonly called "attribution studies" because they seek to identify how much of a storm or event can be "attributed" to human-caused climate change. For instance, studies of Hurricane Harvey found that climate caused precipitation to be up to 37 percent greater (Risser, 2017; Emanuel, 2017; Wang et al., 2018). To be effective, business continuity plans in the region should take these changing conditions into account.

More recently, utilities in Texas were hit by another severe weather event, an extreme cold snap in February 2021, with cold not seen since the 1800s (Box 13.1). The extreme cold increased demand for electricity, impacted supply and caused widespread power outages. Should utilities wanting to factor in climate risk also include this type of extreme in their business continuity plans? The answer is not clear cut. While this type of weather event seems like a rare occurrence, severe cold has hit the region on several occasions in the past (1951, 1983, and 1989) so the cold snap was not without precedent. Additionally, a recent study found that the power outages were caused by a combination of increased electricity demand and extreme cold (Doss-Gollin et al., 2021). Most decarbonization pathways assume increased electrification over time, and as Doss-Gollin et al. suggest, a growing population and future cold snaps could cause similar peak load problems. The dilemma arises because there is a large consensus that the frequency of cold extremes should decrease in most places because of climate change (IPCC, 2021).

So how should a business factor this type of event into their thinking? Because the consequences can be so severe and grow with higher population and electrification for the goal of decarbonization, it seems prudent that a company should (at least) thoroughly explore and understand the risk and continue to monitor the science. As researchers Doss-Gollin (2021) and Cohen (2021) suggest, there could also be a connection between climate change and this type of extreme weather; but like many issues related to climate change, this issue is still being studied.[13]

Nevertheless, there are events so extreme they are difficult to predict and prepare for. Taking climate change into consideration does not mean assessing and investing in resilience solutions for every potential event, including the most unlikely, which could be cost prohibitive. Instead, focusing on more probable events – while understanding that the past is not a good predictor of the future; climate change equates to more extreme events more often; and that the science is evolving – is a suitable place to start.

To examine plausible weather futures, a scenarios approach is often useful. Government resources, such as the Fourth National Climate Assessment (NCA4) put out by the US Global Change Research Program, (USGCRP, 2018) and the US Climate Risk Took Kit (USGCRP, 2016), both illustration future conditions by region and provide a framework for climate scenario planning for those just starting their process for understanding climate risks. More sophisticated users can also access a variety of public and private climate models to help identify risk probabilities in the regions where they operate, such as the potential for flooding, drought or sea-level rise. Local experts (often at universities) can also help refine the models and provide insight into key assumptions. Scenarios can be a very useful tool to help reveal which parts of a business are vulnerable to which types of climate-related impacts and were a key element of the FSB (2017).

Climate scenarios should also consider situations that can arise from indirect impacts, such as when a system-wide failure hampers another interconnected system. Chapter 17 of the NCA4 describes these interconnected impacts as cascading failures.[14] Hurricane Sandy,[15] an Atlantic

Ocean "superstorm" in 2012, for example, had several cascading failures that built upon each other. Jeff Sterba, then CEO of American Water, for example, pointed to the storm's impact on the New York/New Jersey grid and the fact that power outage hindered the ability to move water and use mobile generators. Because the power was out, fuel pumps at gas stations were not working, and gasoline for generators was unavailable. Making matters worse, even if they could have trucked in fuel, once on site there was nowhere to safely store it (Sterba, 2013).

More recently, Hurricane Harvey in 2017 caused power failures, disrupted refineries and water and wastewater treatment plants. Disrupted communications systems also interfered with coordination and evaluation efforts (USGCRP, 2018). Another example of cascading failures is the western wildfires of 2021 that denuded hillsides, contributed to flashfloods, erosion, mudslides, and interrupted primary transportation arteries (Fisher, 2020; Holdman, 2021). These interconnected failures make recovering from these events that much more difficult and provide yet another illustration that past experience with weather is not a good predictor of the future. The likelihood of cascading failures, whether infrastructure or natural systems, is important when assessing potential climate risks and vulnerabilities. Failure to evaluate these risks properly not only impacts business continuity but it can also cause companies to underinvest in resilience. When investments are long-lived and capital intensive, this can be particularly problematic.

Supply Chain and Climate Risk

Experience with Coronavirus has highlighted the fragility of global supply chains. The pandemic has been blamed for transportation holdups in shipping even as the 2021 flooding in central China impacted manufacturers such as Nissan and Apple (Patton, 2021). One news agency, Reuters, headlined a news report with "Global supply chains buckle as virus variant and disasters strike" (Saul et al., 2021). Even before the pandemic, however, companies were experiencing periodic supply-chain disruptions that impacted the flow of raw materials and final products. With about 80 percent of global trade embedded in supply chains, large corporations are becoming aware of risks that could affect their ability to get either raw materials or distribute their products (United Nations Global Compact, 2018). CDP, an organization that tracks corporate climate information, reports that 76 percent of companies responding to their annual survey on supply chains identified ways in which climate change could increase the risk of disruptions to their business (CDP, 2018). And while a sizable percentage of respondents see these risks, CDP (2018) also reports that less than 23 percent have engaged with their suppliers on the topic.

BOX 13.2 SMALL BUSINESSES FACE BIG RISKS

Small- and medium-sized businesses (which often comprise a large portion of the supply chain of larger entities) are often significantly impacted by weather events because they lack the ability to fully evaluate, prepare for, and respond to weather-related stresses.[16] According to the Federal Emergency Management Agency (FEMA), almost 40 percent of small businesses never reopen their doors following a disaster event (FEMA, 2018). Many smaller businesses are not aware of the risks they face from changing climate conditions and may not have plans or adequate insurance coverage in place to respond and recover from known weather events, let alone ones that could be more significant in the future.

Information and awareness barriers, particularly for smaller businesses can make addressing supply chain risks challenging (Box 13.2). Making matters even more difficult is that supply chains are dispersed – connecting various countries and even regions of the same country. While individual entities are identifying risks and adapting, coordination with neighboring jurisdictions can be challenging because of politics, budgets, and regional priorities. Unfortunately, the resilience of an individual region is difficult to isolate from the larger area. For instance, a single event – extensive flooding in Thailand in 2011 – badly damaged global suppliers of parts for the automotive and electronic industries, hurting the bottom lines of Ford, Honda, Toyota, Dell, Cisco, and many other companies. Direct losses from the floods were estimated at $15 to 20 billion (C2ES, 2013a). Disruption to critical systems such as transportation, telecommunications, and electricity transmission in one region can often impact neighboring areas. The saying "a chain is only as strong as its weakest link" is particularly relevant in the climate resilience topic.

Fragmented Policy and Regulation

Exacerbating the physical risks themselves is the variability of policy and preparation between regions. Almost all companies interviewed by C2ES in 2015 said they were concerned about "beyond the fence" risks (including impacts to infrastructure, water, electrical grids, and communication networks). Policy can have significant impacts on these networks and on other issues such as zoning requirements. Across much of the world, adaptation and resilience has been largely pioneered by state and local governments. As risks have become clearer, some private and public entities have been responding, through regulatory approaches, incentives, and education but often these are not coordinated in or between regions. For instance, proposed regulatory reforms in New Jersey would limit public and private development, in any area that could be permanently inundated by sea level rise by the end of the 21st century (New Jersey Department of Environmental Protection, 2020). These regulations are not in place in New Jersey's neighboring states (though some local governments, such as New York City, are also restricting development in areas with significant flood risk (New York City Department of City Planning, 2021)). These rules could impact developers, real estate companies, electric and water utilities who must service these areas and who have interconnected systems that cross these state lines.

It is important to note, however, that government adaptation efforts are not only regulations or restrictions for businesses. More resilient communities and states can be better places to do business (Bailey & Brush, 2020). For instance, local governments that have worked with utilities to ensure continued performance of critical infrastructure and services in disasters, can result in less disruption. Proactive resilience planning can also improve transportation systems and help ensure employees can still get to work. Working with neighboring jurisdictions, however, can ensure that connected infrastructure, such as transportation and communication systems, is protected throughout a region.

An added inconsistency between communities and states preparing for climate change, is the very data and timeframe underlying those decisions. Many climate adaptation and resilience plans rely on climate data downscaled by state universities and government partners that use different methods and produce different results. Similarly, planning timeframes differ, with some regions preparing for mid-century impacts and others preparing for the climate expected in 2100. These inconsistencies can impact prioritization and investment decisions.

A coordinated approach, which includes federal decision makers, all levels of government and the private sector, would be better. While there have been some federal and state climate adaptation efforts, engagement with communities and the private sector has typically focused on raising awareness, sharing information, and responding to disasters but not coordinating policy (Carter et al., 2015). United States policy related to natural disasters, dating back to the early 1800s, and reinforced by Disaster Relief Acts passed in 1950 and 1974, and the 1988 Robert T. Stafford Disaster Relief and Emergency Assistance Act (Stafford Act), focuses on response and recovery, rather than resilience and hazard mitigation (HUD Office of Policy Development and Research, 2015). The Stafford Act covers 75 percent of public losses in disasters and has historically offered insufficient hazard-mitigation funding, discouraging proactive state and local resilience and risk management (Cutter et al., 2013). In places where federal resources and assistance have supported local resilience, this has followed disasters and been connected only to recovery funding; targeting only areas that experienced damaging and disruptive events (The Pew Charitable Trusts, 2018).

In response to the increase in natural disasters, more emphasis has been made to incorporate resilience into recovery spending and investment in pre-disaster mitigation has increased. Since the Disaster recovery Reform Act of 2018, for example, FEMA has prioritized proactive spending on climate resilience in their Buildings Resilient Infrastructure and Communities (BRIC) program (FEMA, 2021). BRIC provides grants to states, territories and tribal governments that have had a major disaster declaration in the past seven years. The program evaluates applications, in part based on how they anticipate and manage future conditions such as climate change, sea level rise and population (Congressional Research Service (CRS), 2021). Additionally, the Biden Administration announced a whole-of-government approach to increase resilience through the activities of several agencies (White House, 2021). A coordinated approach can address the current uneven patchwork of federal, state and community plans and policy that has emerged in the last decade.

ARE BUSINESSES PREPARED? IF NOT, WHAT SHOULD THEY DO?

According to the insurer, Marsh and McLennan (2017) and CDP (2021a), most companies acknowledge climate risks, but many are not building that thinking in their overarching corporate strategies. This finding is consistent with earlier research by C2ES, which found a growing trend over time of large companies assuming their business continuity planning was sufficient but who did not factor future climate change into their process. C2ES also found that while many companies talked publicly about climate risks, few had taken steps to understand these risks enterprise-wide or connect them to strategy (C2ES, 2105).[17] A recent survey of 5000 corporate CEOs by PwC (2021) again confirmed this finding when 60 percent of respondents reported they have not factored climate risk into their strategic risk management activities.

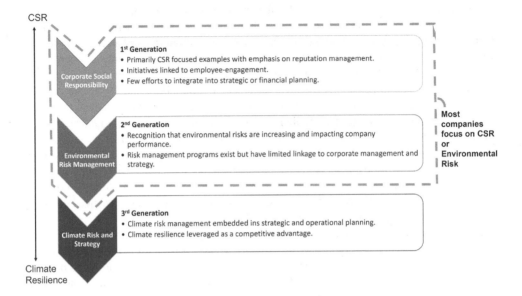

Figure 13.2   Company action on climate resilience

Bringing climate risk into the corporate strategy is exactly what the Financial Stability Board's Task Force on Climate-related Financial Disclosures (TCFD) recommended (FSB, 2017). The TCFD was formed in late 2015 to develop recommendations for what companies should disclose to financial markets about how they were preparing for climate-related financial risks and opportunities. In 2017, the TCFD issued its recommendations with a focus on corporate governance, strategy, risk management, and metrics and targets. Support for these recommendations has been growing. In September of 2018, about 500 companies and other organizations had expressed their support for the TCFD (FSB, 2018), but by 2021 that number had more than doubled to more than 2600 organizations (FSB, 2021). Factoring climate risk into a company's business continuity plan and strategy, however, requires an awareness and understanding of how climate change could affect company operations. This is often done through a vulnerability assessment.

Vulnerability Assessments are Often a First Step

Climate change vulnerability assessments are most companies' first step to screen for the greatest risks (Box 13.3). Ideally, companies should assess potential impacts and their business vulnerabilities for a variety of climate scenarios over shorter and longer time frames and should consider links to other sectors. The DOE Partnership for Energy Sector Resilience, for example, required all utilities who joined to conduct a vulnerability assessment that included the major climate stressors each faced, including flooding, fire, sea level rise, water availability and extreme temperatures (US DOE, 2016). Several lessons emerged from this process which echo themes in this chapter – historic data may underestimate future exposure risks, regulatory processes can impact implementation of resilient solutions and vulnerabilities in one company can cause vulnerability for sectors.

Of course, the energy sector is tightly linked to other sectors up and down their value chain from the inputs used to the sectors using electricity. Identifying these linkages can be particularly useful in a vulnerability assessment as both Pacific Gas and Electric (PG&E) and Sempra Energy point out in their assessments (PG&E, 2016; Sempra, 2016). This type of analysis may be very challenging for large companies, especially the first time. Instead, some companies choose to assess vulnerabilities in a stepwise approach. For example, starting with a narrowly scoped assessment that examines only one region or one impact. Starting with just one asset, one region or one impact can also be used to raise internal awareness and help make the case for a broader more inclusive (and potentially more expensive) evaluation later (C2ES, 2013a).

Even in a more narrowly focused assessment, companies should try to examine their most important assets, critical operations, supply chains and surrounding preparedness, as well as the broader range of potential cascading failures. The DOE's utility partnership found that companies often focus more on assets than vulnerability of operations. While related, assets can be inoperable without being directly impacted by an extreme weather event and, as an illustration, the DOE pointed to lower water availability impacting hydro electricity production or water-cooled generating units (US DOE, 2016). Exelon, for example, identified water scarcity, based on upstream demand and potentially lower water flows, as a potential vulnerability for a new plant and, as a result, could justify investment in an air-cooled rather than water-cooled power plant (C2ES, 2013a). As noted earlier, some impacts, such as upstream water demand, will be outside of a company's region and/or control but nevertheless, in combination with climate stressors could have financial implications for the bottom-line, impacting resource availability, regional preparedness, and employee access.

All vulnerability assessments should use scientifically based, forward-looking assumptions related to the risk from extreme weather events and climate change and incorporate other related factors into the analysis, such as land use, population growth, or competition for scarce resources and raw materials that could also magnify or change those risks over time. The NOAA's Climate Program Office hosts the US Climate Resilience Toolkit (USGCRP, 2016) which has many tools to help with vulnerability assessments and visualization tools for a variety of climate stressors across the US. They also provide links to state climatologists, state climate summaries and flooding projections which can all help identify expected changes to regional climate patterns.

BOX 13.3 VULNERABILITY SCREENING TOOLS

Many screening tools exist to help identify the level of inland flooding, drought, coastal flooding, extreme weather, and wildfire risk. The World Bank hosts a long list of resources including climate data sources, water tools, and reports on sector risks globally on its Climate Change Knowledge Portal. Tools to screen for water risks are particularly useful and WWF's Water Risk Filter and WRI's Aqueduct are referenced by the WBCSD as important tools for informing corporate water stewardship approaches (WBCSD, 2021).

Climate Change should be Incorporated in Standard Risk Management Activities

Once risks are identified, companies, typically, know how to manage them. Generally, one of three options is used – mitigating them, transferring them to someone else or simply accepting them. Mitigating usually requires investments into system hardening, adding redundancy, or moving assets or people out of harm's way to a safer location (Box 13.4). Risk transfer is mostly done through insurance, which essentially pools risk across a variety of clients such that risk is shared and damage in individual events is less costly.

BOX 13.4 WORKING WITH THE NIKE SUPPLY CHAINS

Nike, the global sportswear manufacturer, has had its supply chains affected by climate change on multiple occasions. The 2014 flooding in Thailand, for example, shut down four factories. Conversely, droughts in cotton supplying countries caused cotton price spikes (Bansal, 2016). Nike has, historically, been committed to lean, long-term supply contracts with suppliers and as part of this, it often limits the number of suppliers for any product to create high-volume demand and long-term relationships. Structuring relationships this way has afforded the company greater influence over its suppliers in terms of sustainability practices and worker health and safety. It has also, however, limited their ability to spread their climate risk across different regions (Baskin, 2020). Nevertheless, the company is committed to managing supply chain risk and reported in its 2020 CDP response that they had created a "Supplier Sustainability Council" to address climate change risk management practices amongst key factory partners (CDP, 2021b). This includes factory partners conducting on-site and regional climate change risk assessment to better manage these risks over time. Risks from flooding and droughts, for example, were assessed and Nike required 13 at-risk facilities to develop water-risk mitigation plans (Nike Inc, 2020).

Insurance has a significant role to play with respect to climate resilience. Pricing risk can send a strong signal to change behavior, live in less risk prone areas, and implement standards to protect from extreme weather. For example, offering discounts for activities that lower risk associated with wildfires is one option used by insurer USAA in seven states (Kahn, 2017). A pilot program in Mexico, also recently used an insurance approach to support maintenance of a local reef that protected regional beaches from storm surges (Box 13.5).

BOX 13.5 INSURANCE AND BEACH PROTECTION

The state government of Quintana Roo in Mexico, in partnership with The Nature Conservancy (TNC), and others piloted a conservation strategy for coral reefs that uses an insurance approach to help increase the resilience of local beaches. Essentially, a Coastal Zone Management Trust was developed to accept funds from several sources (including local municipalities) to support the maintenance of Quintana Roo's reef and beaches and purchase an insurance policy. The insurance is an annually renewed parametric policy, a type of insurance where payout is triggered by a set of conditions rather than losses. When wind speeds in the area covered exceeded 100 knots as they did with Hurricane Delta, a payout from the insurance policy facilitated damage assessment, debris removal and initial repairs (and will facilitate additional restoration over time) (The Nature Conservancy, 2020).

Risk acceptance, also known as self-insurance, is yet another approach used to manage risk. Effectively this is doing nothing and accepting the probability or likelihood that a climate-related event is unlikely or that internal resources can be absorbed into the cost of the damage. Across a company's entire operation, a comprehensive risk management approach or business continuity plan will contain elements of all three these options – mitigation, insurance, and acceptance – even if risks are beyond the traditional five-year window of most business continuity plans.

Managing climate risk within a supply chain, however, is more complicated. As shown by the Covid-19 pandemic, disruptions to supply chains can have far reaching implications. Reviewing critical, specialized, and large suppliers for their risk exposure with respect to location, facility, and the degree of specialization is a good first step.[18] Large suppliers, however, may be more prepared to manage climate risk than smaller suppliers (Boxes 13.2 and 13.3). A systematic assessment that prioritizes regions with high climate vulnerability, climate sensitivity of resources, uniqueness of production sites and the number of small suppliers should be considered. Reducing exposure by creating added redundancy or raising inventories where risk is significant can be helpful. Large companies can also directly work with their suppliers to drive resource efficiencies and innovation (Box 13.4). When Diageo's review of water scarcity found issues with the supply of barley used for beer production in Kenya, for example, they worked with local suppliers to develop fields at higher elevations and an alternative crop – sorghum (C2ES, 2015).

Once supply chain risks have been identified, addressing them can require tough decisions about the best options for working with suppliers, investing in system hardening, increasing inventory capacity, creating redundancy in supply chains, and even relocating infrastructure. Resilience planning, however, does not occur in a vacuum. Organizations have multiple goals, including satisfying stakeholders, influencing sustainability practices, addressing traditional pollution issues, and advancing social equity – and they operate under budget constraints. Prioritizing risk reduction efforts to critical assets and where multiple goals can be met, is a strategy often used (C2ES, 2015, 2019b). Take for example, Nike, with its long-term supply relationships. Rather than move away from these regions or suppliers, the company has opted for other strategies that include identification of alternative materials and creating an initiative to help its suppliers reduce their vulnerability.

Risk Disclosure

Public reporting on climate risk, whether in financial reports or separate climate reports is increasing and can be useful for moving the climate conversation from traditional CSR and business continuity planning into corporate strategy (per Figure 13.3). Ilhan et al. (2021) explored the importance of climate risk disclosure to institutional investors and found it as important as financial reporting but also found that reporting needed to improve and be more standardized. Workshops held by C2ES in 2019 and 2020 (C2ES, 2020) explored this issue with members of its business council[19] and key takeaways included:

- Coordination across multiple corporate functions including legal, finance, risk management, and systems planning units was considered necessary for successful engagement on climate risks.

- Executive buy-in can be helped by broadening the climate risk analysis so that business opportunities are also explored. Workshop participants noted that identification of opportunities helped to engage senior leadership teams, which in turn helped build support for a comprehensive corporation-wide climate risk initiative.
- Translating risks and opportunities into financial terms helped management teams justify new investments or strategies compatible with various climate futures.

C2ES also reported that challenges discussed at the workshops included "materiality" of an impact. Materiality is a key threshold issue for financial reporting but not clearly defined for climate risks. Uncertainty about the precise nature of extreme weather events and the long timeframes made assessing its materiality very difficult and ultimately very subjective. In fact, an often-mentioned barrier to acting on climate risk is the perceived mismatch between short-term business decisions and longer-term climate risks. As discussed previously, climate risks are long term, but business continuity plans typically extend five years or less and the success of a business, as reported to shareholders, is done on a quarterly or annual basis.

Nevertheless, many companies have long-lived capital and will invest in more assets that could be increasingly affected by climate change. Take for example, a chemical facility with an average life in North America of at least 30 years, or consider electricity generating stations, many of which are already older than 30 years. Over the next few decades, an annual average temperature increase of about 2.5°F is expected. By late century (from 2071–2100) the average annual temperature could increase by 3°F to 12°F (depending on how climate change is addressed in the coming years) (USGCRP, 2018). The recent (2021) report from the IPCC echoed this prediction and warned of significant impacts and even more extreme weather events over the century. Requiring companies to meaningfully account for and publicly report on their strategies for managing these potential climate risks to their facilities over the longer term is part of the rationale for most mandatory climate disclosure initiatives (Box 13.6).

BOX 13.6 MANDATORY RISK DISCLOSURE

Requiring companies to evaluate and disclose climate risk to investors, as the US Securities and Exchange Commission (SEC) is considering, may also increase investments. In March 2021 the Acting SEC Chair announced it was updating its 2010 guidance on this topic with an "enhancing focus" on climate-related issues (US Securities and Exchange Commission, 2021). Ensuring that companies are transparent about how extreme weather events and environmental regulations could impact a company's bottom line, is consistent with a trend seen in several other countries. In 2015, for example, France adopted the first national law on corporate risk disclosure; in 2019, the EU Commission published guidelines; and in 2020, both New Zealand and the UK announced plans to require mandatory climate risk reporting (Gundlach, 2016; EU Commission, 2019; Jones, 2021; Buchanan, 2020).

Mandatory reporting elements have tended to follow those suggested by TCFD as they sought to provide investors, shareholders, insurers, and others with more in-depth and comparable information on climate related risks, opportunities, corporate governance, and strategies. Over the last decade, corporate support for including more climate related risk information in annual financial documents has significantly increased. Within the S&P Global 100 companies in 2011, for example, just over 30 percent mentioned weather and climate risks in their SEC disclosures. By 2016 that number had increased to more than 50 percent

Climate preparedness for business resilience 311

(C2ES, 2019a). Today, many of these same large companies have signed on as supporters of the TCFD recommendations and many of these have written stand-alone climate risk reports.[20]

Resilience Partnerships

Another strategy that can help with climate resilience is working with others. Companies often point to the difficulty of addressing climate risks that lie outside a company's fence gate, beyond their control – in communities where they operate. Collaboration between a company and a local community or between companies in a specific sector has the potential to reduce costs and make all parties more resilient. PSEG, a large utility in New Jersey for example, was part of DOE's climate resilience partnership for the electricity sector and is now a part of a regional initiative called the New Jersey Climate Change Alliance. Facilitated by Rutgers University, it has created a forum for discussion about the latest science, hazard mapping, best practices, and case studies (New Jersey Climate Change Alliance, 2021). Sempra Energy, similarly, was part of DOE's partnership and works with a variety of partners including a regional collaborative called the San Diego Regional Climate Collaborative. Like the NJ Alliance, it includes academics, non-profit organizations, and business and community leaders who work together to share insights, leverage resources, and identify best practices (San Diego Regional Climate Collaborative, 2021).

Collaboration on a single project, such as a vulnerability assessment, is a good start, but partnerships, like those with PSEG and Sempra, need to reach beyond any one event, meeting, or project, and should be ongoing. Sharing of data is particularly important because while many companies may have evaluated the near- and/or long-term impacts of climate change on their operation, without collaboration, others in the region could be using different data, assuming different timelines, and evaluating disparate impact scenarios, which can pose challenges for coordinated planning and investment. Sempra, for example, has staff meteorologists who monitor drought and fire risk in its region which it shares with others, including county officials. Coordination on data and climate scenarios can help overcome challenges and ensure that when additional information is developed, it is more broadly disseminated (C2ES, 2017).

In addition, large businesses can play a significant role in aiding community resilience – providing services and resources (e.g., communication, technical support and expertise, disaster response), as well as data and tools to smaller businesses, cities, and states. Utility companies, for example, have developed communication systems to coordinate with customers during storms and provide updates on outages, response times, and damage to structures. During Hurricane Katrina, for example, GM allowed rescue workers to use its On-Star system for communications when phone systems went down.[21] Cities and states can develop agreements with companies in the region to provide services and support during extreme events. These types of agreements have been used in Japan, where companies provide emergency shelters, communication and data centers, health services, and other resources during disasters.

The following three climate resilience collaboratives highlight the partners and objectives of each.

- **Gulf Coast Carbon Collaborative**. The US Business Council for Sustainable Development (US BCSD), in cooperation with the New Orleans based Utility, Entergy Corporation, launched the Gulf Coast Carbon Collaborative in December 2019 to create a cross-sector

platform aimed at understanding and reducing the region's climate risk while also reducing regional greenhouse gas emissions (Gulf Coast Carbon Collaborative, 2021).
- **Bay Area Council**. The Bay Area Council came together in 1944 with the aim of coordinating regional economic development and environmental engagement. With a committee dedicated to climate resilience, members have helped raise funds for adaptation projects, coordinate and advocate for water and policy and even participate in a state-wide challenge aimed at building replicable and scalable adaptation projects (Bay Area Council, 2021).
- **Coalition for Climate Resilient Investment (CCRI)**. The CCRI was launched in 2019 by the Global Commission on Adaptation and the World Economic Forum with a goal of forming a global public–private coalition to drive investments in climate resilient infrastructure. CCRI has identified eight areas of focus: Finance and Investment, Food Security and Agriculture, Nature-based Solutions, Water, Cities, locally led Action, Infrastructure, and Preventing Disasters (Coalition for Climate Resilient Investments, 2021).

The federal government can also play a key role in bringing companies together. In 2015, DOE created the Partnership on Energy Sector Resilience. Starting with 17 utilities, the goal was to help climate resilience planning in the electricity sector (Zamuda, 2017). The program started with having each assess their vulnerabilities and progressed to developing tools and helping them with plans that addressed risk. The value of companies working together should not be underappreciated. During DOE workshops companies routinely share what worked, what did not, key strategies (Table 13.1) and even how to talk about these risks to local utility commissions. So useful was this program that companies in other sectors even asked about how such a program could be created for their sector.[22] Unfortunately, for political reasons, the program was shelved, even though companies remained supportive.

Table 13.1   Climate resilience strategies for electricity utilities

Flood protection	Building/strengthening berms, levees, and floodwalls
	Elevating substations/control rooms/pump stations
Wind protection	Upgrading damaged poles and structures
	Burying power lines
General preparedness	Conduct hurricane planning and training
	Manage vegetation/tree trimming
	Coordinating with other groups
	Purchasing or leasing mobile transformers and substations
	Distributed energy, microgrids, islanding capabilities, energy storage
	GIS analysis to find vulnerabilities and plan for new builds
Modernization that improves resilience	Deploying sensors and control technologies
	Installing asset databases/tools, including SCADA system redundancies
Storm-specific readiness	Coordinating priority restoration and waivers
	Securing emergency fuel contracts

Source: Zamuda (2017).

Increasing corporate resilience and moving from CSR and 5-year business continuity plans can be challenging but four key activities can be helpful. First, build awareness of the issue and potential impacts and risks inside the company; next, conduct a vulnerability assessment either across the entire organization or at one location or for one important risk, such as water availability, to clarify specifically how climate risks can impact the company, and how related changes in the region could exacerbate each other. Work with others to ensure consistency of data and understanding. Third, once impacts are identified, take steps to manage them, including supply chain risks which can be more challenging. While identifying these risks, opportunities may also present themselves. Often opportunities are easier to incorporate into corporate strategy than risk minimization activities (C2ES, 2013a). Finally, assess what is working and what needs to change, periodically reassess assumptions about risks and vulnerabilities and adapt strategies as needed. As part of this, publicly report corporate climate risks to share learnings with others. Managing climate risk is a shared responsibility. Of course, these activities should be repeated and refined over time as better information is available and the company changes. Figure 13.3 puts these key activities into a four-step approach.

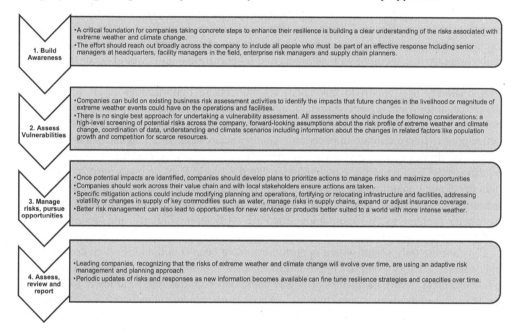

Source: Adapted from C2ES (2013a, 2020).

Figure 13.3   Stepwise approach for managing climate risk

Identifying Opportunities can Help Put a Focus on Climate Risk

Several studies suggest that efforts to build resilience can potentially yield a benefit to the bottom line as well. The Global Commission on Adaptation (GCA) for example, estimates that investing $1.8 trillion globally in five areas could unlock net-benefits worth $7.1 trillion from now until 2030. Areas identified that have a significant benefit-cost ratio greater than 5:1 include early warning systems, making new infrastructure resilient, improving dryland crop production, protecting mangroves and making water resource management more resilient. Strengthening early warning systems was recognized as having the largest benefit–cost ratio at well over 10:1. While it is not clear if this analysis was conducted on a discounted cash-flow basis (to account for the time value of money), it nevertheless helps illustrate the idea that a variety of activities have tangible benefits that exceed the cost to implement. Some researchers also suggest that benefits in this type of analysis undervalue the true societal benefits because diffuse benefits (discussed earlier) are often not included (Asian Development Bank (ADB), 2015).

According to ADB (2015), traditional economic analysis that evaluates the costs and benefit of adaptation/resilience efforts is useful, even though considerable uncertainty remains about the exact nature of climate impacts. It also asserts that economic analyses have always been conducted in the context of uncertainty and further notes that scenarios are one way to deal with this uncertainty.

Examining opportunities as part of climate risk and scenario assessments may also further engage senior company leadership and help encourage the inclusion of climate change into corporate strategy. Preston Chiaro, past President of Rio Tinto Energy, for example, once explained that by reframing a climate risk in terms of the many benefits an action could deliver, greater engagement with suppliers and other business units would occur than if only risk was discussed (C2ES, 2013b).

A relatively new sector to talk about climate related opportunities is IT. Specifically, drone technology and advanced computing technology are being recognized as useful technologies for improving climate resilience. Super computers, for example, are being used to better predict future climate impacts on a smaller scale, such as a neighborhood (Peckham, 2021) and drones can drop small fire-starting "bombs" to create backburns to help combat wildfires.[23] So useful are these technologies that the World Bank is investing $200 billion to help developing countries act on resilience by targeting better systems to gather and analyze information and communicate timely warnings. In a 2019 report, the World Bank and the Consumer Technology Association identified 5G, artificial intelligence, and drones as keys to both disaster prevention and recovery (Consumer Technology Association, 2019).

IT is not the only sector where opportunities have been found and, in fact, companies in every sector identify potential opportunities in their reports to CDP. Table 13.2 highlights a few examples. Many companies primarily name opportunities related to lower carbon technologies, but a few also see opportunities that arise from reducing climate risk. LafargeHolcim, the world's largest cement manufacturer in the world, with operations in 90 countries, for example, reported that it expected more opportunities for concrete because it could be used to withstand the physical changes from climate change (Leonard, 2021). Economic recovery from the Covid-19 pandemic has also connected the dots from federal spending to investments in climate resilience in the Biden administration with planned investments in more resilient infrastructure including more government investment in water and electrical systems (The White House, 2021). Similarly, financing and insurance products are private sector opportunities. Goldman Sachs (2019) described in its report, *Taking the Heat, Making Cities Resilient to Climate Change*, opportunities for catastrophe[24] and resilience bonds,[25] which will likely be needed as companies, investors and cities look for financing options to build resilience against climate change.

Table 13.2   Business opportunities exist in many sectors

Sector	Business opportunities
Agriculture	**Cargill** reported that bio-based alternatives to fossil-based products is a large opportunity and created a dedicated business unit to explore this opportunity.
Infrastructure	**LafargeHolcim** states that climate risk will drive investments in infrastructure and concrete. It also sees an opportunity from increased demand to sell lower carbon concrete as more companies try to reduce their carbon footprint. It expects this demand to increase 5–10% each year eventually reaching over $7 billion.
Electricity	**National Grid** named opportunities over $15 million in benefits associated with investments in low-emissions technologies including the reduction of the potent greenhouse gas SF6. **Schneider Electric** named growth in automation, microgrids, and consulting in energy and sustainability as key opportunities and estimated that an opportunity greater than $22 billion is possible.
Chemicals	**BASF** named new low carbon products and resource efficiency as significant opportunities. Reducing energy is a large opportunity and BASF found over $15 million in energy savings in 2019.
Manufacturing	**Hitachi** reported it can leverage strengths in operational technology, information technology, as well as expertise in research and development to create more than $700 billion in value.
Financial	**ING Group** named the potential for $30 trillion in new investment in energy efficiency and in industrial process changes needed to address climate change. **Goldman Sachs** highlighted insurance products and clean energy finance as opportunities. In 2019, the company reported to CDP that it had deployed $88 billion in clean energy finance since 2012.
Oil and Gas	**Occidental Petroleum** reported that their expertise in carbon capture use and storage was a competitive opportunity to produce low carbon fuels and the creation of low carbon technologies. It created a specific business unit called Oxy Low Carbon Ventures to explore opportunities and estimated a potential positive impact of $380 million opportunity for carbon capture use and storage.

Source: CDP 2020 corporate responses (CDP, 2021).

CONCLUSIONS

The many ways a changing climate affects business and the broader economy combine into a truly unprecedented challenge, and the risks and costs for being unprepared are growing. A business continuity plan that looks out for only five years and is based on past weather events is insufficient for a future with climate change. It is critical for businesses to learn to manage climate risks and that the barriers to do so are addressed by governments, the research and academic community, and by businesses themselves.

While most large companies acknowledge that the climate is changing and mention it in their sustainability reports, most talk about their efforts in terms of how they are transitioning to a lower carbon future. Transition risk is real, especially for companies with large carbon footprints – either in the products they make or the fuels they use. Focusing exclusively on this part of climate risk, however, is not sufficient. Changes to extreme weather and chronic stressors like sea level rise or precipitation patterns are risks today and will only grow in the future. Including physical risks as part of a comprehensive climate strategy is essential. Understanding future risks can also help identify more near-term risks and even opportunities that may not be as apparent.

Several key areas stand out on which businesses need to focus their attention. First, businesses should assess their vulnerabilities and update their business continuity plans by using scenarios to help plan for future climate risks. Vulnerability assessments should consider impacts beyond the company's own operations and include supply chains, with a focus on small suppliers and specialized supply chains with only a few suppliers. Finally, businesses have much to gain from engaging with partners including other businesses, local communities, and universities. Many risks to a company's operation lie outside its control. Working with others can enable more consistency around data and a better understanding about the nature of the risks. Managing climate risk and embracing a more resilient future is a shared responsibility and collective action can result in better solutions at a lower cost to the private and public sectors.

While climate resilience has gained prominence, challenges still exist. Data and analysis that support arguments for investing in climate resilience measures need to expand. While a few sectors have attempted to quantify the value of investing in resilience, more is needed. Cost-benefit analyses are particularly useful to present to senior leadership, regulatory commissions who may oversee system level investments for electric and water utilities and even shareholders. Uncertainty should not prevent economic assessments that consider the benefits and costs of investing in resilience, but it is important to recognize that some benefits are diffuse and difficult to measure quantitatively. Many companies would benefit from improved data on the value of system hardening investments.

To better prepare we recommend the following steps: (1) understand, evaluate, and increase awareness about climate risks; (2) assess climate vulnerabilities of operations, facilities and supply chains using a scenarios approach; (3) embed longer term climate risks into business continuity management plans; (4) partner with others, for instance experts, other businesses, government agencies; (5) invest to mitigate risk; and (6) repeat and refine over time. Climate science and available tools continue to improve and the options for managing risk are expanding quickly. Understanding these risks and opportunities can enable climate change to be part of a company's overarching strategy and investment planning process.

As Larry Fink, CEO of BlackRock, the world's largest asset management company said in his 2020 letter to CEOs, "the evidence on climate risk is compelling investors to reassess core assumptions about modern finance." Further, he predicted that climate change would cause a significant reallocation of capital and, to that end, stated that BlackRock would be "exiting investments that present a high sustainability-related risk" (Fink, 2020). Over the long term, the companies most able to adapt their operations and offerings to these new conditions will likely be the most competitive.

NOTES

1. The views expressed in the chapter do not necessarily reflect those of the authors' employers.
2. According to EPA (2021), storm intensity is a measure of strength, duration and frequency, and is closely related to variation in sea surface temperatures.
3. According to NOAA (Herring, 2020), an extreme weather event is defined as an event lying in the outermost 10-percent of a place's history.
4. In 2021, according to the Bureau of Economic Analysis (BEA, 2021), US GDP was $18.42 trillion in chain adjusted 2012 real US dollars, which equates to $21 trillion in 2020 dollars. Throughout this chapter, where possible, dollars will be reported in real 2020 US dollars.
5. The United Nations Race to Zero Campaign listed on September 1, 2021, 733 cities, 31 regions, 120 countries, 3,067 businesses and 633 Higher Education Institutions with net zero by 2050 goals. https://unfccc.int/climate-action/race-to-zero-campaign#eq-4

6. Physical risk also exists without changes to the natural environment too. Population movements towards high-risk locations like coast areas also increase physical risks.
7. C2ES identified companies in the Standard & Poor's 100 list of the largest 100 publicly traded companies in 2011 as their sample set for companies to interview and this same list of companies was used for each year the data were subsequently updated in reports released in 2015 and 2020.
8. Government investments in resilience come from FEMA, HUD, States and cities and are often associated with disaster recovery initiatives. C2ES (2019a) provides an overview of the primary investments in climate resilience.
9. BCBS is an international body made up of Central Banks and other regulatory authorities from 28 jurisdictions created in 2019 to manage the risks associated with globalization of the financial sector.
10. According to Souter (2019) traditional risk management strategies used by companies don't fully address the broad and long-term threats of a changing climate and further don't consider the environment where they operate.
11. As illustration and for similarity purposes, consider a recent conversation with a large banking company who reported that while their business continuity plan mentioned pandemics, no preparation was made for them, no discussion of cascading impacts was included and no response was ever contemplated (personal communication with the author, 2021).
12. NCA4 is the fourth comprehensive and authoritative report on climate change authored by hundreds of experts and released by the US Global Change Research program every four years.
13. As NCA4 explains, many uncertainties remain related to climate change including but not limited to stability of the Antarctic ice sheet, the connection of wind to a changing climate, where tipping points could exist in ecosystems etc. There is no uncertainty among the scientists writing the NCA4, however, that man-made climate change exists and is getting more pronounced.
14. NCA4 describes cascading failures as unanticipated failures in one system that led to increased risk or failures in another system.
15. Superstorm Sandy affected the Caribbean and the US East Coast, killing nearly 150 people and resulting in estimated damages greater than $70 billion (2020 USD), according to NOAA (NCEI, 2021).
16. Small businesses – defined as those with fewer than five hundred employees – contribute heavily to the US economy. They create two-thirds of net new jobs and account for about 44 percent of US GDP (Kobe and Schwinn, 2018).
17. C2ES tracked public statements from companies identified in the 2011 S&P Global 100, over time (C2ES (2019a).
18. According to the McKinsey Global Institute (2020) supply chains can be broadly grouped as specialized, intermediate and commodity and the more specialized, the more likely it is available from only a few sources and the greater the impact from climate disruptions. In addition, as companies have streamlined their production processes, they have also created unique production sites where that site specializes in the creation of certain products. According to a recent survey, 88 percent of suppliers report they have unique sites for each of their products (McKinsey Global Institute, 2020).
19. The C2ES Business Environmental Leadership Council is made up of 35 mostly Fortune 500 companies across a range of sectors with combined revenues of nearly $3 trillion and 3.7 million employees. Many different sectors are represented, from high technology to diversified manufacturing; from oil and gas to transportation; from utilities to chemicals.
20. As of March 2021, TCFD reported it has more than 1800 supporters in 78 countries.
21. Personal communication with report authors, January 2007.
22. Personal communication, Peace (2016).
23. Referred to as "dragon eggs" these small bombs have been used against wildfires in CA and CO in the last few years (www.thomasnet/insight.com). .
24. Catastrophe (Cat) Bonds are financial instruments that can help manage the financial risk from a natural disaster such as a hurricane. Their goal is not to reduce the physical damage but rather reduce the economic damage associated with the physical damage. They are triggered if a weather event reaches a specific level of damage (either in physical terms or dollar values), the bond sponsor (typically an insurer) keeps a portion of the bond value to pay off losses and investors lose some, or

all, of their investment principal. According to Business Insurance, the Cat Bond market was worth $11 billion in 2020 (Lerner, 2021).

25. Resilience Bonds are a new type of "green or sustainability bond" designed to help entities (governments or companies) pay for infrastructure hardening and climate resilience against disasters. The bonds link insurance coverage that government entities can already purchase (including insurance policies and catastrophe bonds) with capital investments in resilience projects (such as flood walls) that reduce expected losses from disasters. Resilience Bonds fund risk reduction activities via rebates from avoided losses. According to Frederic Asseline of the Asian Development Bank, in 2019 pre-Covid, the "green" bonds market was near $300 billion, representing just 4% of the global bond market.

REFERENCES

Aon. (2020). *Water, Climate and Catastrophe Insight: 2020 Annual Report.* https://www.aon.com/global-weather-catastrophe-natural-disasters-costs-climate-change-2020-annual-report/index.html?utm_source=prnewswire&utm_medium=mediarelease&utm_campaign=natcat21

Asian Development Bank. (2015). *Economic Analysis of Climate-Proofing Investment Projects.* Asian Development Bank. https://www.adb.org/sites/default/files/publication/173454/economic-analysis-climate-proofing-projects.pdf

Bailey, A. & Brush, L. (2020). *The Resilience Factor: The Competitive Edge for Climate-Ready Cities.* Center for Climate and Energy Solutions. https://www.c2es.org/wp-content/uploads/2020/10/the-resilience-factor-competitive-edge-for-climate-ready-cities.pdf

Bansal, A. (2016, November). *Nike's race against climate change.* Harvard Business School, Digital Initiative. https://digital.hbs.edu/platform-rctom/submission/nikes-race-against-climate-change

Basel Committee on Banking Supervision. (2021). Consultative document: Principles for the effective management and supervision of climate related financial risks. Bank for International Settlements. https://www.bis.org/bcbs/publ/d530.pdf

Baskin, K. (2020, February, 11). *Supply Chain Resilience in the Era of Climate Change.* MIT Management, Sloan School. https://mitsloan.mit.edu/ideas-made-to-matter/supply-chain-resilience-era-climate-change

Bay Area Council. (2021). *Water and Climate Resilience.* https://www.bayareacouncil.org/policy/water-resilience

Bhambhani, D. (2018, October 25). *Energy Companies Could Feel the Effects of Climate Change on Their Bottom Line.* Forbes. https://www.forbes.com/sites/dipkabhambhani/2018/10/25/energy-companies-feel-the-effects-of-climate-change-where-it-hurts-the-bottom-line/?sh=2316e5a2199e

Buchanan, K. (2020, October). *New Zealand: Mandatory Climate-related Financial Disclosures Proposed for Financial Sector.* Library of Congress. https://www.loc.gov/item/global-legal-monitor/2020-10-12/new-zealand-mandatory-climate-related-financial-disclosures-proposed-for-financial-sector/

Bureau of Economic Analysis (BEA). (2021, August). *Gross Domestic Product, 2nd Quarter 2021 (Second Estimate); Corporate Profits, 2nd Quarter 2021 (Preliminary Estimate).* https://www.bea.gov/news/2021/gross-domestic-product-2nd-quarter-2021-second-estimate-corporate-profits-2nd-quarter

Campiglio, E. M. (2019). *Climate Risks in Financial Assets.* Council on Economic Policies. https://www.researchgate.net/profile/Emanuele-Campiglio/publication/337102690_Climate_Risks_in_Financial_Assets/links/5dc5232d92851c818036f1d6/Climate-Risks-in-Financial-Assets.pdf

Carter, J. G., Cavan, G., Connelly, A., Guy, S., Handley, J., & Kazmierczak, A. (2015). Climate change and the city: Building capacity for urban adaptation. *Progress in Planning*, 1-66. https://doi.org/10.1016/j.progress.2013.08.001

C2ES, Center for Climate and Energy Solutions. (2013a). *Weathering the Storm: Building Business Resilience to Climate Change.* https://www.c2es.org/site/assets/uploads/2013/07/weathering-the-storm-full-report.pdf

C2ES, Center for Climate and Energy Solutions. (2013b, September). *C2ES Business Resilience Discussion at Climate Week NYC* [Video]. YouTube. https://www.youtube.com/watch?v=vZqxFn5cQaI&t=13s

C2ES, Center for Climate and Energy Solutions. (2015). *Weathering the Next Storm: A Closer Look at Business Resilience.* Center for Climate and Energy Solutions. https://www.c2es.org/document/weathering-the-next-storm-a-closer-look-at-business-resilience

C2ES, Center for Climate and Energy Solutions. (2017). *Guide to Public Private Collaboration of City Climate Resilience Planning.* https://www.c2es.org/wp-content/uploads/2017/05/guide-public-private-collaboration-city-climate-resilience-planning.pdf

C2ES, Center for Climate and Energy Solutions. (2019a). *Business Risks, Opportunities, and Leadership.* https://www.c2es.org/document/business-risks-opportunities-and-leadership/

C2ES, Center for Climate and Energy Solutions. (2019b). *Investing in Resilience.* Center for Climate and Energy Solutions. https://www.c2es.org/site/assets/uploads/2019/11/investing-in-resilience_Brief.pdf

C2ES, Center for Climate and Energy Solutions. (2020). *Implementing TCFD: Strategies for Enhancing Disclosure.* https://www.c2es.org/site/assets/uploads/2020/04/implementing-TCFD-strategies-for-enhancing-disclosure.pdf

Center for International Climate Research. (2021, August). *Shares of Climate Risk* Center for International Climate Research Climate Finance. https://cicero.oslo.no/en/CF-transitional-risk

CDP. (2018). *Closing the Gap: Scaling up Sustainable Supply Chains.* November 2020, from https://www.cdp.net/fr/research/global-reports/global-supply-chain-report-2018

CDP. (2021a). *CDP Disclosure Insight and Action.* https://www.cdp.net/en

CDP. (2021b). *CDP Climate Responses. Nike Inc. Climate Change 2020.* https://www.cdp.net/en/formatted_responses/responses?campaign_id=70692136&discloser_id=852968&locale=en&organization_name=NIKE+Inc.&organization_number=13279&program=Investor&project_year=2020&redirect=https%3A%2F%2Fcdp.credit360.com%2Fsurveys%2F2020%2F6sc15v4h%2F90149&survey_id=68887525

Chiglinsky, K. & Chen C. (2020, December 4). Many Californians being left without homeowners insurance due to wildfire risk. *Insurance Journal.* Retrieved October 2020, from https://www.insurancejournal.com/news/west/2020/12/04/592788.htm

Coalition for Climate Resilient Investments. (2021). *Who We Are.* https://resilientinvestment.org/

Cohen J., Barlow, M., Garfinkel, C.I., & White, I.. (2021). Linking Arctic variability and change with extreme winter weather in the United States. *Science,* 373(6559), 1116–1121. https://www.science.org/doi/10.1126/science.abi9167

Congressional Research Service. (2021). *FEMA Pre-Disaster Mitigation: The Building of Resilient Infrastructure and Communities (BRIC) Program.* https://crsreports.congress.gov/product/pdf/IN/IN11515

Consumer Trade Association. (2019). *Disruptive Tech for Climate Change Resilience.* CES. https://cdn.ces.tech/ces/media/pdfs/white-papers/disruptive-tech-for-climate-change-resilience.pdf

Coppola, M. Krick, T., & Blohmke, J. (2019). Feeling the heat? Companies are under pressure on climate change and need to do more. *Deloitte Insights.* https://www2.deloitte.com/us/en/insights/topics/strategy/impact-and-opportunities-of-climate-change-on-business.html

Cordona, O.D., van Aalst, M.K., Birkmann, J., Fordham, M., McGregor, G., Perez, R., Pulwarty, R.S., Schipper E.L.F., & Sinh, B.T. (2012). Determinants of risk: exposure and vulnerability. In: *Managing the Risks of Extreme Events and Disasters to Advance Climate Change Adaptation.* Intergovernmental Panel on Climate Change. https://www.ipcc.ch/site/assets/uploads/2018/03/SREX-Chap2_FINAL-1.pdf

Cutter, S. L., Ahearn, J. A., Amadei, B., Crawford, P., Eide, E. A., Galloway, G. E., ... & Zoback, M. L. (2013). Disaster resilience: A national imperative. *Environment: Science and Policy for Sustainable Development,* 55(2), 25–29.

Deloitte. (2019). *Climate risk: Regulators sharpen their focus, helping insurers navigate the climate risk landscape.* Deloitte Center for Financial Services. https://www2.deloitte.com/content/dam/Deloitte/us/Documents/financial-services/us-fsi-climate-risk-regulators-sharpen-their-focus.pdf

Deloitte. (2020, January 23). *Financial Risks Stemming from Climate Change: "Challenging the Degree of Resilience into a Constantly Changing Environment".* Deloitte: Risk Advisory: https://www2.deloitte.com/gr/en/blog/risk-advisory/2020/financial-risks-stemming-from-climate-change.html

Doss-Gollin, J., Farnham, D.J., Lall, U., & Modi, V. (2021). How unprecedented was the February 2021 Texas cold snap. *Environmental Research Letters*, 16(6). https://iopscience.iop.org/article/10.1088/1748-9326/ac0278#erlac0278s5

Energy Information Administration. (2011, June 28). *Annual Energy Outlook*. https://www.eia.gov/renewable/annual/preliminary/

Energy Information Administration. (2021, February 3). *Annual Energy Outlook 2021*. https://www.eia.gov/outlooks/aeo/

Emanuel, K. (2017). Assessing the present and future probability of hurricane Harvey's rainfall. *Proceedings of the National Academy of Sciences*, 114(48), 12,681–12,684. doi:https://doi.org/10.1073/pnas.1716222114

Environmental Protection Agency. (2021, April). *Climate Change Indicators: Weather and Climate*. https://www.epa.gov/climate-indicators/weather-climate

EU Commission. (2019). *Guidelines on Reporting Climate Related Information*. European Commission, Directorate General for Financial Stability. Retrieved March 2021, from https://ec.europa.eu/info/publications/non-financial-reporting-guidelines_en#climate

Federal Emergency Management Agency. (2018, October 30). *Stay in Business after a Disaster by Planning Ahead*. https://www.fema.gov/press-release/20210318/stay-business-after-disaster-planning-ahead.

Federal Emergency Management Agency. (2021, July). *Building Resilient Infrastructure and Communities*. https://www.fema.gov/grants/mitigation/building-resilient-infrastructure-communities

Ferman, M. (2021, February 25). Winter storm could cost Texas more money than any disaster in state history. *Texas Tribune*. https://www.texastribune.org/2021/02/25/texas-winter-storm-cost-budget/

Fink, L. (2020). *BlackRock CEO Letter*. Black Rock. https://www.blackrock.com/us/individual/larry-fink-ceo-letter

Fisher, T. (2020, August 21). Wildfires in several states shutting down highways. *Land Line*. https://landline.media/wildfires-in-several-states-shutting-down-highways/

Flavelle, C. (2020, November 5). California bars insurers from dropping policies in wildfire areas. *The New York Times*. https://www.nytimes.com/2020/11/05/climate/california-wildfire-insurance.html

Folpmers, M. (2021, May 14). *The New Basel Framework for Climate Risk Management: Pros and Cons*. Global Association of Risk Professionals. https://www.garp.org/risk-intelligence/culture-governance//a1Z1W000005lC4BUAU

FSB (2017). *Financial Stability Board, Final Report: Recommendations of the Task Force on Climate-related Financial Disclosures*. https://assets.bbhub.io/company/sites/60/2021/10/FINAL-2017-TCFD-Report.pdf

FSB (2018). *Financial Stability Board, Task Force on Climate Related Financial Disclosures: 2018 Status Report*. https://assets.bbhub.io/company/sites/60/2020/10/FINAL-2018-TCFD-Status-Report-092518.pdf

FSB. (2020). *Implications of Climate Change for Financial Stability*. https://www.fsb.org/wp-content/uploads/P231120.pdf

FSB (2021). *Financial Stability Board, Task Force on Climate Related Financial Disclosures: 2021 Status Report*. https://assets.bbhub.io/company/sites/60/2022/03/GPP_TCFD_Status_Report_2021_Book_v17.pdf

Gilliingham, K., Newell, R., & Pizer, W. (2008). Modeling endogenous technological change for climate policy analysis. *Energy Economics*, 30, 2734-2753. https://dukespace.lib.duke.edu/dspace/bitstream/handle/10161/6628/GillinghamNewellPizerEnergyEcon.pdf;sequence=1

Global Commission on Adaptation. (2019). *Adapt Now: A Global Call for Leadership on Climate Resilience*. https://gca.org/wp-content/uploads/2019/09/GlobalCommission_Report_FINAL.pdf

Goldman Sachs Research. (2019). *Taking the Heat: Making Cities Resilient to Climate Change*. Goldman Sachs. https://www.goldmansachs.com/insights/pages/taking-the-heat.html

Gulf Coast Carbon Collaborative. (2021). Gulf Coast Carbon Collaborative. https://carbon-collaborative.org/

Gundlach, J. (2016, March 7). France: New disclosure rules require companies to report on environmental and social factors, including climate risks. *Business and Human Rights Resource Center*. https://www.business-humanrights.org/en/latest-news/france-new-disclosure-rules-require-companies-to-report-on-environmental-social-factors-including-climate-risks/

Helper, S. and Soltas E., (2021). Why the Pandemic Has disrupted Supply Chains. The White House, CEA Blog. June 17. https://www.whitehouse.gov/cea/written-materials/2021/06/17/why-the-pandemic-has-disrupted-supply-chains/

Herring, D. (2020, October 29). *What is an "Extreme Event"? Is there Evidence that Global Warming has Caused or Contributed to any Particular Extreme Event?* National Oceanic and Atmospheric Administration. climate.gov/news-features/climate-qa/what-extreme-event-there-evidence-global-warming-has-caused-or-contributed

Hersher, R. (2021, February 21). *Live Updates: Winter Storms 2021.* National Public Radio. https://www.npr.org/sections/live-updates-winter-storms-2021/2021/02/21/969912613/after-days-of-mass-outages-some-texas-residents-now-face-huge-electric-bills

Holdman, R. (2021, August 13). *Glenwood Canyon Mudslides Caused a "Pothole from Hell" on I70.* CBS Denver. https://denver.cbslocal.com/2021/08/13/interstate-70-mudslides-glenwood-canyon-rain-forecast/

HUD Office of Policy Development and Research. (2015). *History of Federal Disaster Policy.* Evidence Matters. https://www.huduser.gov/portal/periodicals/em/winter15/highlight1_sidebar.html

IBM Services and Consulting. (2020, November 25). *Adapt and Respond to Risks with a Business Continuity Plan (BCP).* IBM. https://www.ibm.com/services/business-continuity/plan

Illam, E., Krueger, P., Sautner, Z. & Starks, L.T. (2021). *Climate Risk Disclosure and Institutional Investors.* European Corporate Governance Institute. Swiss Finance Institute Research Paper No. 19-66, European Corporate Governance Institute. http://dx.doi.org/10.2139/ssrn.3437178 https://www.alliancepolicy.org/site/usermedia/application/6/Kigali_Economic_Report.pdf

Intergovernmental Panel on Climate Change. (2014). *Climate Change 2014: Synthesis Report; Technical Report of the International Panel on Climate Change, Working Groups I, II, and III to the Fifth Assessment Report.* https://www.ipcc.ch/report/ar5/syr/

Intergovernmental Panel on Climate Change. (2021). *AR6 Climate Change 2021: The Physical Science Basis. Contribution of Working Group 1 to the Sixth Assessment Report of the Intergovernmental Panel on Climate Change.* https://www.ipcc.ch/report/ar6/wg1/downloads/report/IPCC_AR6_WGI_Citation.pdf

Jaffe. A., Newell, R.G. & Stavins., R.N. (2000). Technological Change and the Environment. *NBER Working Paper Series.* https://www.nber.org/system/files/working_papers/w7970/w7970.pdf

Jones, H. (2021, March 24). *UK Proposes Requiring Businesses to Disclose Climate Risks by 2022.* Reuters, https://www.reuters.com/article/us-climate-change-britain-finance/uk-proposes-requiring-businesses-to-disclose-climate-risks-by-2022-idUSKBN2BG1HQ

Kathan, J. (2020, August 9). Decline in hydropower hampered by drought will impact utility costs. *The Mercury News.* https://www.mercurynews.com/2020/08/09/decline-in-hydropower-hampered-by-drought-will-impact-utility-costs

Kobe, K., & Schwinn, R. (2018). *Small Business GDP.* US Small Business Administration Office of Advocacy. https://advocacy.sba.gov/2018/12/19/advocacy-releases-small-business-gdp-1998-2014

Lavell, A., Oppenheimer, M., Diop, C., Hess, J., Lempert, R., Li, J., ... Weber, E. (2012). Climate change: New dimensions in disaster risk, exposure, vulnerability, and resilience. In C. Field, V. Barros, T. Stocker, & Q. Dahe (Eds), *Managing the Risks of Extreme Events and Disasters to Advance Climate Change Adaptation: Special Report of the Intergovernmental Panel on Climate Change* (pp. 25–64). Cambridge: Cambridge University Press. doi:10.1017/CBO9781139177245.004

Leonard, L. (2021, August 18). Top 10 cement producers in the world. *Construction Review.* https://constructionreviewonline.com/top-companies/top-10-cement-producers-in-the-world/

Lerner, M. (2021, February 9). Surge in cat bond market set to continue in 2021. *Business Insurance.* https://www.businessinsurance.com/article/20210209/NEWS06/912339653/Surge-in-cat-bond-market-set-to-continue-in-2021

Littlemore, R. (2019, August 1). Facing an insurance hike that could be north of 25 percent, Vancouver condo owners are learning that they can't escape the threat of catastrophic weather in a warming world. *BC Business.* https://www.pressreader.com/canada/bc-business-magazine/20190801/281621011871563

Marsh and McClennan Global Risk Center. (2017). *How Climate Resilient is Your Company.* Marsh and McClennan. https://www.mmc.com/content/dam/mmc-web/Global-Risk-Center/Files/how-climate-resilient-is-your-company.pdf

McKinsey & Company. (2020, January). *Climate Risk and Response: Physical Hazards and Socioeconomic Impacts.* https://www.mckinsey.com/business-functions/sustainability/our-insights/climate-risk-and-response-physical-hazards-and-socioeconomic-impacts

McKinsey Global Institute. (2020). *Could Climate become the Weak Link in Your Supply Chain? Case study.* McKinsey Global Institute. https://www.mckinsey.com/~/media/mckinsey/business%20functions/sustainability/our%20insights/could%20climate%20become%20the%20weak%20link%20in%20your%20supply%20chain/could-climate-become-the-weak-link-in-your-supply-chain-v3.pdf

Meyer, N. (2018). *Best Practices and Challenges: Using Scenarios to Assess and Report Climate Related Financial Risk.* Center for Climate and Energy Solutions. https://www.c2es.org/wp-content/uploads/2018/08/using-scenarios-assess-climate-risk-08-18.pdf

National Academies of Sciences. (2021, August 21). *Global Warming is Making Some Extreme Weather Events Worse.* https://www.nationalacademies.org/based-on-science/climate-change-global-warming-is-contributing-to-extreme-weather-events

National Aeronautics and Space Administration. (2020, January 15). *NASA, NOAA Analyses Reveal 2019 Second Warmest Year on Record.* https://www.nasa.gov/press-release/nasa-noaa-analyses-reveal-2019-second-warmest-year-on-record

National Institute of Building Sciences, Multi-Hazard Mitigation Council. (2019). *Natural Hazard Mitigation Saves: 2019 Report.* National Institute of Building Sciences. https://www.nibs.org/files/pdfs/NIBS_MMC_MitigationSaves_2019.pdf

National Centers for Environmental Information (NCEI). (2021). *Events.* National Oceanographic and Atmospheric Administration. https://www.ncdc.noaa.gov/climate-information/extreme-events

Neslen, A. (2019, March 21). Climate change could make insurance too expensive for most people. *The Guardian.* https://www.theguardian.com/environment/2019/mar/21/climate-change-could-make-insurance-too-expensive-for-ordinary-people-report.

Nike Inc. (2020). *Breaking Barriers, FY20 NIKE, Inc., Impact Report.* https://purpose-cms-preprod01.s3.amazonaws.com/wp-content/uploads/2021/03/30191542/FY20-NIKE-Inc.-Impact-Report1.pdf

New Jersey Climate Change Alliance. (2021). *About Us.* Rutgers University. https://njadapt.rutgers.edu/about-us

New Jersey Department of Environmental Protection (2020). *NJ PACT: Protecting Against Climate Threats.* https://www.nj.gov/dep/njpact/

New York City Department of City Planning. (2021, May 12). *Zoning for Coastal Flood Resiliency.* https://www1.nyc.gov/site/planning/plans/flood-resilience-zoning-text-update/flood-resilience-zoning-text-update.page

Olson, M. (1965). *The Logic of Collective Action: Public Goods and the Theory of Groups.* Second Printing, Harvard University Press, 1971. JSTOR, https://doi.org/10.2307/j.ctvjsf3ts

Patton, D. (2021, July 22). *From Coal to Cars, Chinese Floods Tangle Supply Chains.* Reuters. https://www.reuters.com/business/sustainable-business/coal-cars-chinese-floods-tangle-supply-chains-2021-07-22/

Peace, J. & Weyant, J. (2008). *Insights Not Numbers: The Appropriate Use of Economic Models.* Washington, DC: Pew Center on Global Climate Change. https://www.researchgate.net/publication/241210034_Insights_Not_Numbers_The_Appropriate_Use_of_Economic_Models

Peckham, O. (2021, August 9). *New York Power Authority Employs Supercomputing for Climate Resilience.* HPC Wire. https://www.hpcwire.com/2021/08/09/new-york-power-authority-employs-supercomputing-for-climate-resilience/

PG&E. (2016). *Climate Change Vulnerability Assessment.* San Francisco: Pacific Gas and Electric Company. http://www.pgecurrents.com/wp-content/uploads/2016/02/PGE_climate_resilience.pdf

Princeton. (2020, April 22). Human-caused warming will cause more slow-moving hurricanes warns climatologists. *Science Daily.* www.sciencedaily.com/releases/2020/04/200422151312.htm

PwC. (2021). *A Leadership Agenda to Take on Tomorrow.* https://www.pwc.com/gx/en/ceo-agenda/ceosurvey/2021.html?WT.mc_id=CT3-PL300-DM1-TR2-LS4-ND30-TTA9-CN_CEO-Survey2021-GlobalCEOSurvey-

Risser, M. a. (2017). Attributable human-induced changes in the likelihood and magnitude of the observed extreme precipitation during Hurricane Harvey. *Geophysical Research Letters*, (44), 12,457–12,464. https://doi.org/10.1002/2017GL075888

S&P Global. (2020a). *Climate Change: Energy Transition Risks and Opportunities for European Public Companies' Creditworthiness.* September 2020, from https://www.spglobal.com/marketintelligence/en/news-insights/blog/climate-change-energy-transition-risks-and-opportunities-for-european-public-companies-creditworthiness

S&P Global. (2020b, August 17). California power prices trade near $600/MWH in heat wave. *Market Intelligence.* https://www.spglobal.com/marketintelligence/en/news-insights/latest-news-headlines/calif-power-prices-trade-near-600-mwh-in-heat-wave-59967239

Sacramento Bee. (2020, August 19). California power prices have skyrocketed. Is this normal or more Enron style manipulation? https://www.sacbee.com/news/california/article245048140.html

San Diego Regional Climate Collaborative. (2021). *San Diego Regional Climate Collaborative*: University of San Diego. https://www.sandiego.edu/soles/hub-nonprofit/initiatives/climate-collaborative

Saul, J., Xu, M., & Sun, Y. (2021, July 26). *Global Supply Chains Buckle as Virus Variant and Disasters Strike.* https://www.reuters.com/business/global-supply-chains-buckle-as-virus-variant-disasters-strike-2021-07-23

Semieniuk, G., Campiglio, E., Mercure, J., Volz, U., & Edwards, E.R. (2020). Low-carbon transition risks for finance. *Wires Climate Change.* https://doi.org/10.1002/wcc.678

Sempra. (2016). *Risk Assessment Mitigation Phase, Risk Mitigation Plan, Climate Change Adaptation.* Sempra Energy. https://www.sdge.com/sites/default/files/SDGE-_14_RAMP_Climate_Change_Adaptation_FINAL.pdf

Souter, G. (2019, November 19). *Climate Change Risks Need New Risk Management Approach: Experts.* Business Insurance. https://www.businessinsurance.com/article/20191119/NEWS06/912331736/Climate-change-risks-need-new-risk-management-approach-Experts#:~:text=Traditional%20risk%20management%20strategies%20don,from%20rising%20temperatures%2C%20they%20said.

Sterba, J. (2013, August 1). *Business Resilience Report Launch: Introduction and Discussion* [Video]. Center for Climate and Energy Solutions. https://www.youtube.com/watch?v=uZMgvkvaQf4&t=6s

Swiss Re Institute. (2020, April). *Sigma 2/2020: Natural Catastrophes in Times of Economic Accumulation and Climate Change.* https://www.swissre.com/institute/research/sigma-research/sigma-2020-02.html

Task Force on Climate-related Financial Disclosures. (2017). *Final Report: Recommendations of the Task Force on Climate-related Financial Disclosures.* https://assets.bbhub.io/company/sites/60/2020/10/FINAL-2017-TCFD-Report-11052018.pdf

Task Force on Climate-related Financial Disclosures. (2021). *Support the TCFD Recommendations.* https://www.fsb-tcfd.org/support-tcfd/

The Nature Conservancy. (2020, December 8). *Insuring Nature to Ensure a Resilient Future.* The Nature Conservancy: Perspectives: https://www.nature.org/en-us/what-we-do/our-insights/perspectives/insuring-nature-to-ensure-a-resilient-future/

The Pew Charitable Trusts. (2018). *Natural Disaster Mitigation Spending Not Comprehensively Tracked.* https://www.pewtrusts.org/en/research-and-analysis/issue-briefs/2018/09/natural-disaster-mitigation-spending--not-comprehensively-tracked

United Nations Global Compact. (2018). *Decent Work in Global Supply Chains.* https://www.unglobalcompact.org/library/5635

US Department of Energy. (2016). *A Review of Climate Change Vulnerability Assessments: Current Practices and Lessons Learned from the DOE's Partnership for Energy Sector Climate Resilience.* https://toolkit.climate.gov/sites/default/files/A%20Review%20of%20Climate%20Change%20Vulnerability%20Assessments%20Current%20Practices%20and%20Lessons%20Learned%20from%20DOEs%20Partnership%20for%20Energy%20Sector%20Climate%20Resilience.pdf

USGCRP, (2018). *Impacts, Risks, and Adaptation in the United States: Fourth National Climate Assessment, Volume II.* Reidmiller, D.R., C.W. Avery, D.R. Easterling, K.E. Kunkel, K.L.M. Lewis, T.K. Maycock, and B.C. Stewart (eds). US Global Change Research Program, Washington, DC, USA, 1515 pp. doi: 10.7930/NCA4.2018.

US Global Change Research Program. (2016). *U.S. Resilience Toolkit.* https://toolkit.climate.gov

US Securities and Exchange Commission. (2021, March). *Enhancing Focus on the SEC's Enhanced Climate Change Efforts.* https://www.sec.gov/news/public-statement/roisman-peirce-sec-focus-climate-change

Utility Dive. (2020, April). *As Extreme Weather Spurs Billions in Utility Resilience Spending, Regulators Struggle to Value Investments*. Utility Dive. https://www.utilitydive.com/news/as-extreme-weather-spurs-billions-in-utility-resilience-spending-regulator/576404/

Wang, S. S., Zhao, L., Yoon, J. H., Klotzbach, P., & Gillies, R. R. (2018). Quantitative attribution of climate effects on Hurricane Harvey's extreme rainfall in Texas. *Environmental Research Letters*, 13(5), 054014. https://iopscience.iop.org/article/10.1088/1748-9326/aabb85/meta

World Business Council for Sustainable Development. (2019). *Business Climate Resilience*. https://docs.wbcsd.org/2019/09/WBCSD_Business-Climate-Resilience.pdf

World Business Council for Sustainable Development. (2021). *Global Water Tool*. https://www.wbcsd.org/Programs/Food-and-Nature/Water/Resources/Global-Water-Tool

World Economic Forum. (2021). *The Global Risks Report 2021*. http://www3.weforum.org/docs/WEF_The_Global_Risks_Report_2021.pdf

White House. (2021, August). *FACT SHEET: Biden Administration Announces Nearly $5 Billion in Resilience Funding to Help Communities Prepare for Extreme Weather and Climate-Related Disasters*. https://www.whitehouse.gov/briefing-room/statements-releases/2021/08/09/fact-sheet-biden-administration-announces-nearly-5-billion-in-resilience-funding-to-help-communities-prepare-for-extreme-weather-and-climate-related-disasters/

Zamuda, C. (2017, April). *U.S. Department of Energy's Partnership for Energy Sector Climate Resilience*. Electric Power Research Institute. https://eea.epri.com/pdf/NYSERDA/1.2%20Zamuda%20DOE%20Partnership.pdf

Zhang, G., Murakami, H., Knutson, T. R., Mizuta, R., & Yoshida, K. (2020). Tropical cyclone motion in a changing climate. *Science Advances*, 6(17). doi:10.1126/sciadv.aaz7610

PART IV

FUNCTIONAL PERSPECTIVES AND CORPORATE PRACTICE

14. The equity value relevance of carbon emissions
Peter M. Clarkson, Jody Grewal and Gordon D. Richardson

INTRODUCTION

The focus of this chapter is whether capital markets view the volume of a firm's carbon emissions as value relevant and, if so, the importance of mandated carbon disclosures in facilitating investors' assessment of such emissions. As will be seen, the empirical literature consistently documents an inverse relation between the volume of carbon emissions and firm value, supporting the interpretation that capital markets assess a latent carbon liability commensurate with the quantity of firm-level carbon emissions.[1] As a part of this literature, Clarkson et al. (2015) further establish that the valuation penalty per ton of carbon emissions increases with the firm's relative carbon intensity rank in its sector, suggesting negative valuation consequences associated with carbon intensity rank. They point out that mandated carbon disclosures provide investors with the benchmarking information they need to assess a firm's carbon intensity rank and overall latent carbon liability. Emerging literature (e.g., Grewal, 2021; Tomar, 2021) confirms an additional benefit following from mandated carbon disclosures – realized carbon emissions reductions ('real effects') – which they attribute to enhanced benchmarking information.

We begin by discussing studies that seek to document the firm valuation consequences of carbon emissions and refer to such studies as exploring the valuation 'penalty' attached to a ton of carbon emissions. Early studies in that literature established that investors do attach a valuation penalty to emissions, although, as discussed in the next section, the various studies measure carbon emissions differently. The early studies that we discuss faced the limitation that they assumed the relevant regulatory regime would eventually impose a cost of carbon on firms; as such, the empirical estimates of the valuation penalty in these studies could be confounded by investor expectations of the extent of free carbon allowances should a cap and trade system be imposed. Further, these studies ignored differences across firms that could drive the valuation penalty such as differences in the firm's carbon intensity rank relative to industry or sector peers. Finally, earlier studies relied on self-reported carbon emissions data and thus faced a selection threat when estimating the valuation penalty attached to a ton of carbon emissions.[2]

We discuss extensively how a study by Clarkson et al. (2015) addresses these various design limitations by turning to a setting in which European Union (EU) firms actually face a cost of carbon imposed by the EU cap and trade system and thus the extent of free allowances are known. The EU firms studied by Clarkson et al. faced mandatory Scope 1 emission disclosure requirements at the facility level, alleviating self-selection concerns. The authors develop a measure of carbon intensity rank relative to sector peers in order to control for cross-sectional differences in factors that drive the estimated valuation penalty. Finally, Clarkson et al. (2015) address ambiguity in policy outcomes involving EU versus non-EU emissions for a given EU firm. As we are about to discuss, the Clarkson et al. study demonstrated that investors do attach a valuation penalty to a ton of Scope 1 emissions for the EU firms in their sample; that the valuation penalty is only for Scope 1 emissions in excess of free allowances; that the valuation penalty is declining in their measure of an EU firm's superior carbon intensity performance in its sector; and, finally, that the valuation penalty is lower for the firm's non-EU emissions, relative to its EU emissions, since the penalty estimate for the former reflects investor probabilities assigned to the relevant non-EU regulatory regime eventually imposing a cost of carbon on the firm. Clarkson et al. conclude by proposing the types of mandatory carbon disclosures that would facilitate investors' ability to estimate the valuation penalty for a ton of Scope 1 emissions at the firm level. They assert the following:

> ... at a minimum, investors need the following information in order to estimate latent carbon liabilities: (1) current carbon emissions at the corporate entity level, segregated by regions under different regulatory regimes; (2) the firm's carbon efficiency relative to its sector peers for each sector the company operates in; and (3) other information indicating the firm's ability to pass on increased carbon costs to consumers. (Clarkson et al., 2015, p. 554)

Having documented the valuation relevance of carbon emissions in general and the importance of the mandated carbon disclosures identified in Clarkson et al. (2015) for enhancing the ability of investors, stakeholders and firms themselves to benchmark relative carbon performance, we conclude by discussing a stream of emerging literature which explores the 'real effects', rather than the valuation effects, of mandatory carbon emission disclosures (e.g., Grewal, 2021; Tomar, 2021). The discussed real effects are due to firms learning from their peers about the existence of possible carbon efficiency improvements (an internal 'discovery' effect) as well as increased pressure from investors and other stakeholders on firms to reduce carbon emissions once improved benchmarking information reveals the poor carbon intensity performers within sectors (an external 'pressure' effect). As will be seen, these recent studies confirm this additional benefit to mandated disclosure, namely further reductions in carbon emissions.

The evidence from the studies discussed in this chapter points to a significant market valuation penalty associated with firm carbon emissions, ranging from one-half percent of market capitalization (Griffin et al., 2017); 3.15 percent of market capitalization (Clarkson et al., 2015); 6.57 percent of market capitalization (Chapple et al., 2013); and 7.33 percent of market capitalization (Matsumura et al., 2014). Thus, the documented latent carbon liability is likely to be material to investors. We also discuss the types of mandatory carbon disclosures that would enhance the ability of investors to assess a firm's latent carbon liability, as well as studies confirming the 'real effects' resultant from mandated carbon reporting.

FOUNDATIONAL LITERATURE ON MANDATED CARBON EMISSIONS DISCLOSURE AND VALUATION

The foundational empirical literature on the valuation relevance of a firm's carbon risk exposure uniformly documents a valuation penalty related to the volume of a firm's carbon emissions, consistent with the existence of latent carbon liabilities. This foundational literature is typified by the studies of Chapple et al. (2013), Matsumura et al. (2014), and Griffin et al. (2017).

In conducting their investigations, these studies adopt the volume of the firm's historical carbon emissions (in tons) to proxy for the underlying construct, carbon risk exposure. In this fashion, the maintained assumption of this choice of proxy is that the firm's current volume of carbon emissions reflects the broader set of risk exposures, notably (Griffin et al., 2021):

- future costly and uncertain environmental regulations;
- future uncertain physical climate risks;
- future compliance costs and ligation; and
- future costs of measuring and monitoring emissions.

The carbon emissions data used in these studies has typically been sourced from voluntary disclosures made directly by the firm, primarily in their responses to the Carbon Disclosure Project (hereafter, CDP) survey questionnaire. However, as noted by Matsumura et al. (2014), the CDP surveys only a subset from the global universe of firms and participation in the survey is voluntary. Further, even if a firm chooses to respond, the firm may only provide partial information and/or restrict access to the information provided. In this sense, the foundational studies (as well as many subsequent studies) face a potential self-selection bias.

Another important consideration in carbon valuation studies concerns the 'scope' of the carbon emissions data used to measure the firm's volume of emissions. Scope 1 emissions are those that arise directly from company-owned and controlled resources whereas Scope 2 and Scope 3 emissions are indirect emissions linked to the company's operations.[3] Griffin et al. (2017) conjecture that *ex ante* it is unclear whether indirect (Scopes 2 and 3) emissions will attract the same valuation penalty as direct (Scope 1) emissions, especially if the underlying regulatory regime is targeted towards direct emissions such as in the EU. In this regard, however, the studies cited above consider different compositions, in part driven by their data sources for carbon emissions.

The final design decision that warrants discussion is whether to use an absolute or a relative carbon emissions measure. Absolute carbon emissions are expressed as the volume of emitted greenhouse gases (GHG) (tons of CO_2 equivalent per year), whereas relative carbon emissions (also termed carbon intensity) are expressed as absolute carbon emissions normalized by an activity measure such as revenues, cost of goods sold, number of employees, and so on. Absolute carbon emissions are useful for directly inferring the potential carbon-related latent liability per ton. Alternatively, relative carbon emissions (carbon intensity) can be interpreted as being indicative of the extent to which a firm's business model depends on carbon emissions, and facilitates a comparison of the relative carbon performance and exposure of firms of different sizes and from different sectors (Salo and van Ast, 2009; UNEP-FI, 2013). Both measures are present within the foundational literature, with, for example, Chapple et al. (2013) using a carbon intensity measure and both Matsumura et al. (2014) and Griffin et al. (2017) using the absolute measure.

In early work, Chapple et al. (2013) employ both valuation model and event study methodologies to examine the valuation relevance of carbon emissions in Australia. Their analysis uses a sample of 58 Australian Securities Exchange (ASX) listed firms for the year 2007, a year immediately preceding the proposed introduction of a national carbon emissions trading scheme (ETS). Thus, critical to their research design is the requirement that their carbon emissions data relate only to domestic operations since emissions generated outside of Australia are not subject to domestic regulation. They source their carbon emissions data from two private providers, Citigroup and VicSuper. While the databases of these two providers are developed largely from CDP responses adjusted to reflect only domestic emissions figures, when CDP data are not available, Citigroup estimates carbon emissions in-house while the VicSuper accesses data from Trucost Plc., a research organization that uses a proprietary profiling system to estimate emissions. Thus, although Chapple et al. do not explicitly confront the potential self-selection bias issue, their sample includes firms beyond just those that respond to the CDP survey questionnaire. Finally, while they acknowledge that under the proposed ETS firms may face different valuation impacts relating to direct and indirect carbon emissions, they make no distinction between Scope 1 and Scope 2 emissions following suggestions that the difference in their financial impact is likely to be small (Citigroup, 2008).[4]

Focusing on five distinct information events that they argue will materially alter the probability of an Australian ETS being implemented, Chapple et al. document a significant share price reaction in the predicted direction for two of the events, with the reaction being stronger for high carbon intensity firms. Then, using valuation model methodology, they document the predicted negative association between their carbon emissions measure (carbon intensity) and firm value, with the estimated coefficient on their carbon emissions measure implying a decrease in market value of approximately $26 AUD per ton. Based on mean market capitalization and carbon emissions figures, they determine that this figure implies an assessed market value penalty of 6.57 percent of market capitalization for the high carbon-intensity firms relative to the low carbon-intensity firms in their sample.

Matsumura et al. (2014) base their analysis on a sample of S&P 500 firms from the period 2006–2008. They source their carbon emissions data from firm responses to the CDP survey questionnaire, finding data availability for only 584 out of a potential 1,443 firm-year observations. Matsumura et al. explicitly confront the selection bias by jointly estimating the decision to disclose carbon emissions data and the effect of such emissions on firm value.[5] Their measure of the volume of carbon emissions appears to be based on the total emissions figure reported through the CDP, and potentially represents a mix of various scopes. Specifically, in their footnote 21, Matsumura et al. (2014, p. 704) note that "A number of firms in our sample do not provide carbon emissions broken down into scopes."

Based on these emissions data, Matsumura et al. first documents an economically meaningful negative association between the volume of carbon emissions and firm value, finding that "for every additional thousand metric tons of carbon emissions, firm value decreases, on average, by $212,000."[6] While the study does not discuss the market valuation implications of this $212 USD per ton estimate, based on the descriptive statistics reported in their Table 2, this estimated penalty equates to 7.33 percent of market capitalization based on mean market capitalization and carbon emissions figures, and 1.41 percent based on median figures. They also examine why a firm would choose to disclose its emissions data, given the apparent penalty for doing so. To this end, they document that "the median firm value is about $2.3 billion higher for firms that disclose their carbon emissions compared to firms that choose to not disclose them". Their interpretation is that investors penalize firms for withholding information on carbon emissions.

Most recently, Griffin et al. (2017) revisit the relation between carbon emissions and firm value using a sample of S&P 500 firms from the period 2006–2012. One significant innovation in their study is a methodology to directly confront non-disclosure to the CDP: they develop a carbon emissions estimation model. They then estimate the model parameters based on the emissions data of the firms that publicly disclosed such data through the CDP, and apply these model parameters to estimate the carbon emissions of non-disclosing firms. They base their primary analysis on the aggregate of Scope 1 and Scope 2 emissions, but in further analysis where data by scope are available (Scope 1 and Scope 2), they consider their relative valuation implications.

In the study's pooled sample comprising both disclosing and non-disclosing firms, the authors document a negative association between carbon emissions and firm value, finding for the median S&P 500 firm in their sample period, a "market-implied equity discount of $79 per GHGE [GHG emissions] ton", a figure they note as being substantially lower than Matsumura et al.'s (2014) estimate of $212 per ton. In conjunction, they also confirm that based on median market capitalization and emissions figures, their estimate implies a market value penalty of less than one-half percent of market capitalization. While they further confirm that the penalty relating to Scope 1 emissions is greater than that relating to Scope 2 emissions, they do not provide specific figures on the difference.

When examining the disclosing and non-disclosing firm subsamples separately, Griffin et al. find that, while the coefficient is slightly more negative for the non-disclosing firms in most years, the differences in the estimated penalty per ton of disclosed versus estimated GHG emissions is statistically insignificant in all years but 2012, a finding they interpret as suggesting that market value reflects emissions data developed through channels other than just the self-selected, voluntary CDP disclosure. As such, they conjecture that Matsumura et al.'s finding of an equity valuation discount for not disclosing to the CDP may be due to factors other than the act of disclosing GHG emissions.[7] In conjunction, Griffin et al. also investigate the market reaction to carbon emissions disclosures in 8-K filings, finding evidence that investors respond to the filings and hence confirm 8-K filings as a potential alternative source of information in addition to the CDP. Finally, in supplementary analyses, Griffin et al. address the potential selection bias surrounding the decision to disclose to the CDP by applying the two-stage Heckman approach (as did Matsumura et al.). They continue to find evidence of an equity valuation discount, with the size of the discount being qualitatively similar to that documented in their primary findings of $79 per ton, which they suggest "buttresses the results … that show qualitatively similar [carbon emissions] coefficients for CDP disclosers and CDP non-disclosers."

Lastly, the findings of an assessed market value penalty relating to the volume of a firm's carbon emissions also appear to apply to a firm's cost of borrowing.[8] Here, for example, Jung et al. (2018) find a positive and significant association between the cost of debt and carbon intensity (carbon emissions scaled by sales revenue), but only for firms that do not disclose to the CDP and also that do not make carbon-related disclosures in their CSR reports. The authors interpret this latter finding as suggesting that creditors attach a lower cost of debt penalty to carbon emissions for firms that appear to be proactively managing their carbon exposures. The idea is that, in order for it to be managed, carbon performance must first be measured, and disclosures via either the CDP survey or CSR reports indicate the presence of such measures and carbon awareness more generally. Jung et al. base their study on a sample of 255 firm-year observations for 78 Australian firms from the period 2009–2013. They obtain carbon emissions data under Australia's National Greenhouse and Energy Reporting Act 2007 (NGER Act) which mandates the disclosure of carbon emissions (Scope 1 and Scope 2), and net energy consumption, by entities that meet reporting certain thresholds to the country's Clean Energy Regulator. The Clean Energy Regulator then publishes the data – which are also audited – for each registered corporation. Based on these data, Jung et al. find that a one standard deviation increase in their carbon intensity measure maps onto an increase in the cost of debt ranging between 38 and 62 basis points for the firms in their sample.

In sum, each of the studies described above documents a negative (positive) relation between the volume of carbon emissions and firm value (cost of debt), consistent with the notion that the capital markets assess a latent carbon liability commensurate with the firm's carbon emissions. There is consensus on this finding across the surveyed studies, whether or not the firm participates in the CDP. Griffin et al. conclude that the capital markets can in fact develop carbon emissions estimates through a variety of channels other than the CDP and, as a result, the implication is that market values broadly reflect carbon emissions data, whether disclosed through the CDP or elsewhere.

EXTENSIONS

In the studies discussed above, one of the key research design decisions that the authors have made is to use the firm's total current volume of carbon emissions to proxy for the firm's underlying carbon risk exposure. Equally, each of the studies has been conducted within a regulatory setting without enacted emissions regulation (such as a 'cap and trade' regime), and indeed without mandated carbon emissions disclosure requirements. As such, these studies are largely reliant on emissions data voluntarily disclosed or estimated, rather than verified carbon emissions data broadly across the universe of firms.

In responding to this literature, Clarkson et al. (2015) first discuss the implications of the above described design limitations, and then identify and investigate a setting within which the limitations can be overcome. By considering jurisdictions without enacted regulation, they note that latent carbon emissions liability estimates impound investors' assessment of both the likelihood of future regulation, as well as the form and details of such regulation, inclusive of the possibility of free allowances. They further note that the use of total carbon emissions to proxy for carbon risk exposure implies a uniform valuation impact per ton across the firm's entire volume of carbon emissions, as well as a uniform ability across firms to pass on their carbon-related costs to consumers. In contrast, industry research (e.g., IRRC Institute and Trucost, 2009) argues that the valuation impact of carbon emissions should depend not only on a firm's total emissions but additionally on two important factors: (1) policy outcomes, notably the allocation of free allowances that the firm receives; and (2) the firm's ability to pass on its carbon-related costs.

To directly respond to these challenges, Clarkson et al. (2015) turn to the European Union (EU) where a 'cap and trade' regime in the form of an emissions trading scheme, the EU ETS, has been in place since 2005. They conduct their investigation into the valuation relevance of carbon emissions within the context of the EU ETS based on the set of firms for which carbon emissions data and free allowance allocation data could be obtained over the period 2006–2009.

As they describe, annual verified emissions and allocated allowance data for participating installations in the EU ETS (those meeting reporting thresholds) are recorded by the European Commission in the Community Independent Transaction Log (CITL), with the data being made publicly available in April of the following year. These data are, however, only at the installation level, a constraint that presents one of the critical challenges that Clarkson et al. (2015) face in the conduct of their study. To confirm whether each of the approximately 10,000 installations in the CITL database is a listed entity, or whether it is owned by an immediate or ultimate parent entity that is listed, Clarkson et al. (2015) use the BVD Amadeus

Database, a comprehensive database that records the immediate and ultimate controlling entity of all entities operating in Europe and whether the entity is listed. From this database and using a combination of programming and follow-up hand-matching, they matched each installation to its respective controlling entity. Finally, for each listed entity for each year, they summed the emissions and allowances across all the individual installations under the entity's control to arrive at their aggregated emissions and allowances. Based on this process, their final sample comprises 843 firm-year observations over the period 2006–2009 relating to 221 unique listed firms.

Having developed their sample, Clarkson et al. (2015) then systematically explore each of the factors identified above (the existence of free allowances, cost pass-on ability, and policy outcome ambiguity involving EU versus non-EU emissions) as potential cross-sectional determinants of the latent carbon emissions liability as assessed by the capital market. They start their investigation by reaffirming the existence of a latent liability within their data. Using the sum of Scope 1 carbon emissions for the installations covered under the EU ETS for each of their sample firms, they document an average firm value reduction of €39 per ton (roughly US$51 per ton), an estimate, while highly significant, that is somewhat smaller than the reduction in firm value documented by Matsumura et al. (2014) and Griffin et al. (2017) of US$212 and US$79 per ton, respectively. Following this, as the first core step in their investigation, they disaggregate a sample firm's total emissions into the portion covered by allowance allocations (free permits) and the consequent allocation shortfall.[9] They argue that the portion of emissions covered by free allowances should have valuation implications that are quite different relative to the portion that is not covered (i.e., the shortfall). Based on this disaggregation, they find that the market assesses a significant valuation penalty of €75 per ton (roughly US$97 per ton) of uncovered emissions, a figure that implies a valuation penalty of 3.15 percent of market capitalization based on mean market capitalization and uncovered (shortfall) emissions figures, and a penalty of 1.81 percent of market capitalization based on median figures. In contrast, the market assigns a zero latent liability to covered emissions. Thus, in this regard, they initially confirm the extent of free allowances as an important determinant of the magnitude of a firm's assessed latent carbon liability.

As the second core step in their investigation, Clarkson et al. (2015) extend their valuation model to include measures designed to capture the firm's superior carbon intensity performance in its sector, with carbon intensity measured as Scope 1 GHG emissions divided by sales. They argue that firms with greater carbon efficiency (lower carbon intensity) relative to their sector peers will be able to pass along a greater portion of their carbon-related costs. Regarding pass-on ability, Clarkson et al. (2015) quote IRRC Institute and Trucost (2009, p. 9), which states the following:

> companies that are more carbon efficient than sector peers stand to gain competitive advantage. Carbon pricing could create opportunities for low-emission companies in carbon-intensive sectors. High emitters which find it difficult to fully pass these liabilities on could see profits fall.[10]

It must be noted, however, that a measure of carbon intensity rank relative to sector peers likely captures other negative firm valuation consequences not discussed by Clarkson et al. (2015). For example, Christensen et al. (2019, p. 63) state the following:

> higher transparency regarding CSR and the ability of stakeholders other than investors to benchmark firms against each other at lower costs may increase the societal pressure on poor CSR firms. In the

same way as CSR can build loyalty, poor CSR performance can damage a firm's reputation and create negative publicity. Stakeholders such as social activists, policymakers or consumers can exert pressure through reputational effects like public shaming, boycotts of the firm, or imposing sustainability restrictions along the supply chain.

As further discussed in the next section, Grewal (2021) adapts such arguments to a firm's relative carbon intensity rank relative to industry peers and refers to the above potential threats as arising from "competitive CSR benchmarking". Thus, an EU ETS firm's limited carbon pass-on ability captures only one consequence of a firm's inferior carbon intensity rank in its sectors. More broadly, negative firm valuation consequences for poor carbon performers in their sectors arise from negative reputational effects and the resulting impacts of enhanced external pressure, including pressure from consumers.

The results presented by Clarkson et al. (2015) provide support for their conjecture regarding a firm's carbon cost pass-on ability, as well as the broader negative valuation consequences of an inferior carbon intensity rank discussed above. They find that while the least carbon efficient firms within a sector (based on their industry-year carbon intensity rank) are assessed with a carbon liability, the market does not assign a valuation penalty to firms ranked in the top third of the firms in their sector in terms of carbon efficiency on a sector-year percentile rank basis. Thus, Clarkson et al. (2015) confirm carbon intensity rank as an important determinant of the magnitude of a firm's assessed latent carbon liability.

Finally, Clarkson et al. (2015) consider whether the regulatory regime relating to carbon emissions under which the firm operates has implications for its assessed latent carbon emissions liability. For 189 firm-year observations, a subset of their total sample, they are able to obtain non-EU carbon emissions data for EU firms through the CDP. They investigate the valuation relevance of EU-based carbon emissions relative to those emitted outside of the EU and hence not subject to the EU ETS. Their expectation is that the valuation implication of emissions subject to the EU ETS will differ from that of the same firm's non-EU emissions that are not subject to a 'cap and trade' scheme. Consistent with this expectation, the authors document a valuation penalty of €75 per ton for EU-based emissions not covered by free allowances, but a penalty of only €37 per ton for non-EU emissions, a difference that fits with the probabilistic nature of non-EU emission penalties. In addition, their results also indicate that both the EU and non-EU carbon emission-related liabilities are mitigated for the most carbon efficient firms as captured by their industry-year carbon intensity rank. Thus, Clarkson et al. (2015) confirm that the regulatory regime that a firm faces is an important determinant of its assessed latent carbon emissions liability.

In sum, Clarkson et al. (2015) find that an EU firm's latent carbon liability: (1) relates to the portion of emissions that exceed free allowances within the EU ETS; (2) relates to relative carbon efficiency measured by industry-year carbon intensity rank; and (3) differs for carbon emissions within the EU ETS versus emissions in non-EU ETS jurisdictions. Importantly, their results show that the valuation impact of carbon emissions is not homogenous across firms and industries, but rather investors appear to assess a firm's latent carbon liability within the context of its allocated allowance, its relative carbon efficiency, and the relevant carbon enforcement in different legal jurisdictions. Ultimately, then, as stated in the Introduction, their results lead them to conclude that,

> ... at a minimum, investors need the following information in order to estimate latent carbon liabilities: (1) current carbon emissions at the corporate entity level, segregated by regions under different

regulatory regimes; (2) the firm's carbon efficiency relative to its sector peers for each sector the company operates in; and (3) other information indicating the firm's ability to pass on increased carbon costs to consumers. (Clarkson et al., 2015, p. 554)

As discussed above, a limited carbon pass-on ability represents only one consequence of an EU ETS firm's inferior carbon intensity rank in its sector. More broadly, negative firm valuation consequences for poor carbon performers in their sectors arise from negative reputational effects and the resulting impacts of enhanced external pressure, including pressure from consumers. Nevertheless, if mandatory disclosure requirements include a firm's relative carbon intensity rank in its sectors (item (2) above), investors can attach the appropriate valuation penalty per ton of GHG emissions based on their estimates of the total valuation consequences of a firm's relative carbon intensity rank.

The results of Clarkson et al. (2015) may assist securities regulators and accounting standard setters globally as they deliberate mandatory carbon disclosure requirements for firms, in MD&A and other disclosure channels. We now turn to discuss mandatory carbon disclosure requirements around the globe, and research pertaining to the consequences (benefits) of improved carbon efficiency benchmarking when such mandatory disclosures are made available to investors, other stakeholders, and the firms themselves.

MANDATED CARBON EMISSIONS DISCLOSURE AND CARBON EMISSIONS REDUCTIONS

The preceding sections of this chapter focus on the valuation implications of carbon emissions and how mandated carbon disclosures allow investors to benchmark a firm's carbon intensity ranking and latent carbon liability. In this section, we discuss an emerging literature that examines how mandated carbon disclosures affect firm behaviors and decisions in the real economy ('real effects' as opposed to valuation effects). Importantly, this line of research documents an important additional benefit conferred from mandated carbon disclosures: realized carbon emissions reductions.

A growing number of countries around the world require certain companies to disclose greenhouse gas emissions, including large economies such as the United States and the United Kingdom.[11] The use of so-called 'light-touch' disclosure regulation, in lieu of more explicit regulation that stipulates, prohibits, or taxes firm outputs (e.g., carbon taxation), is gaining popularity as politicians and regulators expect transparency to help curb firms' socially and environmentally undesirable behaviors without having to impose costlier laws.

Theoretical research suggests that mandated reporting elicits changes in firm behavior because improved transparency facilitates monitoring of the reporting firm's behavior and feeds back to the real actions of the firm (Kanodia and Sapra, 2016).[12] As contracting stakeholders of the firm (e.g., investors, government agencies, NGOs, customers, employees, etc.) gain access to improved information, this enhances their ability to exert pressure on the disclosing firm to change its behavior. Firms in turn face increased accountability over the mandated information, as stakeholders exerting pressure expect to see performance improvements over time.

A number of empirical studies confirm these conceptual underpinnings, using settings that examine financial outcomes such as investment efficiency (e.g., Biddle et al., 2009; Graham et al., 2011) as well as non-financial outcomes such as mine-safety violations (Christensen et al., 2017), payments to governments for mineral extraction rights (Rauter, 2020), pollutants (Chen et al., 2018), and carbon emissions (Downar et al., 2021; Tomar 2021; Grewal 2021).

Focusing on carbon emissions, Downar et al. (2021) study firms disclosing under the EU ETS and find that when a new law is introduced in the United Kingdom (UK) – requiring UK firms to disclose carbon emissions in annual financial reports – UK firms reduce their emissions levels and intensity, relative to other EU ETS firms. The authors attribute these reductions to an improvement in transparency, whereby investors and other stakeholders become more aware of and attentive to a firm's carbon emissions once the data are (1) aggregated at the firm-level (making it easier for stakeholders to exert pressure and demand accountability), and (2) disclosed in a highly-disseminated channel (making the information more salient to external users), relative to when emissions were reported at the facility-level in a low-visibility channel under the EU ETS (a government website).

In addition to the real effects (realized carbon emissions reductions) that may occur when firms are mandated to disclose new, improved, or more disseminated carbon data, two recent studies document that mandated carbon reporting also improves benchmarking – both on the part of a firm looking to learn from its peers, and on the part of investors and other stakeholders looking to identify (and influence) poor carbon performers – which may also yield carbon emissions reductions.

In the first study on this topic, Tomar (2021) examines the effects of a disclosure rule enacted by the US Environmental Protection Agency's Greenhouse Gas Reporting Program, requiring US facilities with over 25,000 tons of carbon dioxide equivalent per annum to report facility-level yearly greenhouse gas emissions. Focusing on facilities that did not disclose emissions elsewhere, Tomar documents a 7.9 percent emissions reduction following mandated reporting. Tomar also documents that carbon intensity (measured as carbon emissions divided by cost of goods sold) decreases after mandated disclosure, suggesting that facilities did not simply reduce economic activity in order to cut emissions.

Turning to the mechanism, Tomar conjectures that because the required disclosures are highly granular, and carbon emissions are closely aligned with the industrial production of goods, the mandated information is informative about operations and allows facilities to benchmark their own, relative carbon performance once they can observe their peer's disclosures. Consistent with this channel, Tomar ranks facilities against their peers based on carbon intensities revealed by mandated reporting, and finds that worse performers reduce emissions by a greater extent. The takeaway of the study is that mandated reporting allows firms to utilize their peers' emissions disclosures to identify energy efficiency-improvement opportunities and set benchmarks, which in turn facilitates emissions reductions.

In the other study on this topic, Grewal (2021) examines a regulation in the United Kingdom requiring firms to disclose greenhouse gas emissions. Focusing on firms that already voluntarily disclosed carbon emissions prior the regulation, Grewal finds that these voluntary disclosers reduce carbon emissions levels (by 9 percent) and intensity (by 9.9 percent) following mandated reporting. Grewal's theory is that when the emissions data of previously non-disclosing firms become available, some of the voluntarily-disclosing firms that had good emissions performance (lower levels and intensity) relative to their industry peers before the regulation learn that they are actually relatively poor performers, and try harder to reduce emissions to protect their corporate social responsibility (CSR) reputations. Consistent with her theory of "competitive CSR benchmarking", she finds that the change in a voluntary discloser's emission ranking within its industry after the regulation, predicts its post-regulation emission reductions. She also finds that the relation between worsening rankings and larger subsequent reductions is accentuated for firms that perceive higher climate-change related reputational risks (measured using Carbon Disclosure Project survey responses), further strengthening the theorized channel. Overall, Grewal's study shows that carbon reductions following mandated reporting are attributable to increased stakeholder pressure and reputational concerns, once improved benchmarking information reveals the poor carbon intensity performers in each industry.

To summarize, Tomar (2021) and Grewal (2021) are complementary and provide insight into the different motivations to benchmark that result in emissions reductions after carbon reporting is mandated. In particular, when data are reported in a visible outlet (e.g., financial reports) and presented at the firm-level (facilitating comparisons across firms), competitive benchmarking may inspire firms to cut emissions as it did in Grewal's study, whereas when data are in a less-visible channel (e.g., a government website), and disaggregated (impeding cross-company comparisons), benchmarking to learn from peers may explain emissions reductions as in Tomar's study. Thus, in addition to the importance of mandated carbon reporting identified by Clarkson et al. (2015) to facilitate benchmarking carbon intensity across sectoral peers and the implications for firm valuation thereof, Tomar (2021) and Grewal (2021) show that mandated carbon reporting provides enhanced information for benchmarking, which results in realized carbon emissions reductions.

CONCLUSION

As we transition globally to addressing climate change, more and more countries will soon impose a price on carbon. As this happens, investors will need information to estimate a firm's latent carbon liability in order to enhance the allocative efficiency of stock markets around the globe. Indeed, the resulting valuation 'penalty' attached to a ton of carbon emissions will put pressure on the firm to incur new carbon reduction capital expenditures up to the point where the marginal capital expenditure per ton of carbon emissions reduction equals the valuation penalty per ton of emissions reduction. Thus, providing investors with the mandatory carbon disclosures they need to attach the appropriate valuation penalty to a ton of firm carbon emissions will greatly assist our global transition to a carbon free environment.

In parallel, a critical further benefit to such mandatory disclosures is the fact that the internal discovery and external pressure effects resulting from enhanced benchmarking abilities will lead to further carbon emission reductions. We therefore assert that mandatory carbon emission disclosures is mission critical to our global attempts to address climate change.

NOTES

1. We focus on the valuation penalty per ton of carbon emissions, using a firm valuation approach. An alternative approach would be to explore the impact of investor perceptions of a firm's carbon risk on the firm's stock returns (see, for example, Bolton and Kacperczyk, 2021). The studies that we discuss establish a firm valuation penalty per ton of greenhouse gas emissions. In terms of a firm valuation model based on discounted future cash flows, such a penalty can be attributed to either future cash flow effects ("numerator effects") or equity cost of capital effects ("denominator

effects"). While the equity valuation studies discussed in this chapter do not seek to distinguish between future cash flow effects and equity cost of capital effects, the regulatory policy implications for mandatory greenhouse gas emission disclosures that follow from the firm valuation studies we discuss hold regardless of the source of the valuation penalty.

2. As explained by Matsumura et al. (2014) and Griffin et al. (2017), econometric "selection bias" threats arise when a firm is not randomly assigned to a treatment group. Firm characteristics driving disclosure selection may bias empirical estimation results unless controlled for.

3. Scope 1 emissions are direct emissions from company-owned and controlled resources. Alternatively, Scope 2 emissions are indirect emissions arising from the generation of purchased electricity, steam, heating and cooling consumed by the reporting company, and Scope 3 emissions are all indirect emissions not included in Scope 2 that occur in the company's value chain, including both upstream and downstream emissions (i.e., all remaining indirect emissions linked to the company's operations).

4. If an ETS or carbon tax is implemented in a given country, the distinction between Scope 1 and Scope 2 may not be economically significant in terms of firm valuation, since a price on carbon will be incurred directly by the reporting company (Scope 1 emissions) or indirectly (Scope 2 emissions once the cost of carbon is passed on by the energy provider).

5. Formally, Matsumura et al. (2014) apply what is known as the two-stage Heckman model. Within the context of their study, in the first stage, based on theory, they formulate a model for the decision to disclose (the source of the selection bias). Estimation of the model based on the full sample of firms (both those that disclose to the CDP and those that don't) then yields results that can be used to predict the probability that each firm will disclose. In the second stage, selection bias is then corrected by adding a transformation of this predicted probability as an additional explanatory variable in the valuation model. This allows Matsumura et al. (2014) to "make inferences about the average effect of carbon emissions on firm value for all the firms in the sample, not just for the firms that disclose their emissions."

6. Matsumura et al. (2014) correctly point out that US$212 per ton is not a "price" per ton of carbon but instead represents the impact on firm value, which is the sum of future cash flows discounted over the life of the firm.

7. It seems reasonable, as conjectured by Matsumura et al. (2014), that information asymmetry costs and reputation costs would be reduced through CDP participation. Quantifying the benefits of voluntary GHG emission disclosures is an interesting topic for future research.

8. We discuss the debt cost of capital implications of GHG emissions in order to emphasize that GHG emissions matter to creditors as well as investors.

9. Carbon allowances are issued by a government under an emissions cap-and-trade regulatory program. Each allowance (or emissions permit) allows its owner to emit one tonne of a pollutant such as CO_2e. Under a cap-and-trade system such as the EU ETS, the supply of GHG allowances is limited by the mandated 'cap'. Allowances can be allocated freely by the governing program, be purchased when auctions are held, or be purchased from other entities that have excess. Over time, allowances under the EU ETS are increasingly being provided through auction. A year-over-year reduction in the 'cap' then ensures a parallel drop in total emissions.

10. While the terms "pass-on" and "pass-through" ability are both used in the carbon literature, we use the former term. Smale et al. (2006) discuss factors affecting carbon pass-on ability for firms subject to the EU ETS. They observe (Smale et al., 2006, p. 33) that many of the sectors in the EU ETS are oligopolistic rather than perfectly competitive. Clarkson et al. (2015, footnote 16) offer the following example to support the IRRC Institute and Trucost (2009) argument regarding pass-on ability: "Consider two rivals in an oligopolistic industry sector where the low emitter has 1 tonne of direct emissions per €1000 of sales not covered by free allowances while the high emitter has 10 tonnes per €1000 of sales not covered by free allowances. Assume further that, given the elasticity of demand, both firms can pass the costs of purchased allowances for just 1 tonne of emissions on to customers. In this setting, the high emitter has no pass-on ability for 9 of its 10 tonnes, since price is set by the low-cost producer (the low emitter)."

11. See: https://www.oecd.org/daf/inv/mne/Report-on-Climate-change-disclosure-in-G20-countries.pdf

12. For example, Cheng et al. (2013) show that firms mandated to report internal control weaknesses for the first time under the Sarbanes-Oxley Act improved investment efficiency and Cho (2015) finds that improvements in mandated segment reporting following SFAS 131 increased investment efficiency.

REFERENCES

Biddle, G.C., Hilary, G., & Verdi, R.S. (2009). How does financial reporting quality relate to investment efficiency? *Journal of Accounting and Economics*, 48(2-3), 112–131.
Bolton, P., & Kacperczyk, M. (2021). Do investors care about carbon risk? Forthcoming. *Journal of Financial Economics*, 142(2), 517–549.
Chapple, L., Clarkson, P., & Gold, D. (2013). The cost of carbon: Capital market effects of the proposed emissions trading scheme (ETS). *ABACUS*, 49(1), 1–33
Chen, Y., Hung, M., & Wang, Y. (2018). The effect of mandatory CSR disclosure on firm profitability and social externalities: Evidence from China. *Journal of Accounting and Economics*, 65(1), 169–190.
Cheng, M., Dhaliwal, D., & Zhang, Y. (2013). Does investment efficiency improve after the disclosure of material weaknesses in internal control over financial reporting? *Journal of Accounting and Economics*, 56(1), 1–18.
Cho, Y. (2015). Segment disclosure transparency and internal capital market efficiency: Evidence from SFAS No. 131. *Journal of Accounting Research*, 53(4), 669–723.
Christensen, H., Floyd, E., Liu L., & Maffett, M. (2017). The real effects of mandated information on social responsibility in financial reports: evidence from mine-safety records. *Journal of Accounting and Economics*, 64(2-3), 284–304.
Christensen, H., Hail, L., & Leuz, C. (2019). Adoption of CSR and sustainability reporting standards: Economic analysis and review. *ECGI Working Paper Series in Finance*.
Citigroup, 2008. *Carbon Pollution Reduction Scheme: Impacts Reviewed for ASX100 Companies and More*. Citigroup Global Markets, Australia/NZ Thematic Investing, Equity Research.
Clarkson, P., Li, Y., Pinnuck, M., & Richardson, G. (2015). The valuation relevance of greenhouse gas emissions under the European Union Carbon Emissions Trading Scheme. *European Accounting Review*, 24(3), 551–580.
Downar, B., Ernstberger, J., Reichelstein, S., Schwenen, S. & Zaklan, A. (2021). The impact of carbon disclosure mandates on emissions and financial operating performance. *Review of Financial Studies*, 26(3), 1137–1175.
Graham, J.R., Hanlon, M., & Shevlin, T. (2011). Real effects of accounting rules: Evidence from multinational firms' investment location and profit repatriation decisions. *Journal of Accounting Research*, 49 (1), 137–185.
Grewal, J. (2021). Real effects of disclosure regulation on voluntary disclosers. *Working paper*.
Griffin, P., Lont, D., & Sun, E. (2017). The relevance to investors of greenhouse gas emission disclosures. *Contemporary Accounting Research*, 34(2), 1265–1297.
Griffin, P., Lont, D., & Pomare, C. (2021). The curious case of Canadian corporate emissions valuation. *British Accounting Review*, 53(1).
IRRC Institute and Trucost. (2009). Carbon risks and opportunities in the S&P 500. Retrieved from http://www.trucost.com/publications
Jung, J., Herbohn, K., & Clarkson, P. (2018). Carbon risk, carbon risk awareness and the cost of debt financing. *Journal of Business Ethics*, 150, 1151–1171.
Kanodia, C., & Sapra, H. (2016). A real effects perspective to accounting measurement and disclosure: Implications and insights for future research. *Journal of Accounting Research* 54(2): 623–676.
Matsumura, E., Prakash, R., & Vera-Munoz, S. (2014). Firm-value effects of carbon emissions and carbon disclosures. *The Accounting Review*, 89(2), 695–724.
Rauter, T. (2020). The effect of mandatory extraction payment disclosures on corporate payment and investment policies abroad. *Journal of Accounting Research*, 58(5), 1075–1116.
Salo, J., & van Ast, L. (2009). *Carbon Risks and Opportunities in the S&P 500*. Boston, MZ: Trucost.

Smale, R., Hartley, M., Hepburn, C., Ward, J., & Grubb, M. (2006). The impact of CO_2 emissions trading on firms profits and market prices. *Climate Policy, 6*, 29–46.

Tomar, S. (2021). CSR disclosure and benchmarking-learning: Emissions responses to mandatory greenhouse gas disclosure. *Working paper.*

United Nations Environment Program – Finance Initiative (UNEP-FI), 2013. *Portfolio carbon: Measuring, disclosing and managing the carbon intensity of investments and investment portfolios.* Retrieved from
http://www.unepfi.org/fileadmin/climatechange/UNEP_FI_Investor_Briefing_Portfolio_Carbon.pdg

15. Getting to 2050: transparency for setting and reaching supply chain climate goals

Suzanne Greene and Alexis Bateman

BACKGROUND

Global supply chains are integral to life in the 21st century. By some estimates, as much as 90 percent of global demand cannot be met by local production and roughly 90 percent of goods are produced and delivered through global supply chains (Hult & Tomas 2014). Containerization, increased competition, and deregulation of the shipping industry in the latter part of the 20th century led to the rapid expansion of international trade flows and the transformation of supply chains. With faster, more reliable, and lower-cost transportation, manufacturers and retailers began moving operations to the most economical location for each part of their supply chain (Blume Global 2021).

By the turn of the 21st century, most containers flowing through US ports contained components and partially produced goods rather than raw materials or finished goods, reflecting the fundamental restructuring of global supply chains (Blume Global 2021). However, this expansion of global trade comes with environmental and social costs not captured by the market value of goods and services. One of the costs is the environmental impact of supply chains – particularly the greenhouse gas (GHG) emissions generated in the production, manufacturing, and transportation of materials around the world. While climate change mitigation has continued to emerge as the primary concern of many consumers, investors, as well as governments and corporations, are facing increasing scrutiny and pressure to track and reduce their supply chains' impacts (Bateman et al. 2021). However, companies face key challenges in determining how to account for the full scope of the supply chains to create appropriate goals and take meaningful actions. There is an increasing societal expectation that large and/or multinational companies with significant financial resources and heavy footprints are obligated to account for and reduce impact in their supply chains.

Now is the time to accelerate decarbonization of energy, transportation, and manufacturing systems, getting the world on track to meet 2050 global climate goals to meet the Paris Agreement. It is also the decade to rebuild what has been disrupted by the coronavirus pandemic. In the US, a record flow of capital was channeled into US Environmental, Social, and Governance (ESG) funds in 2020 (Morningstar 2020). At the international level, the 2021 United Nations Climate Change Conference (COP26) marks the first conference since the Paris conference (COP21) where the parties committed to the achievement of more ambitious goals (United Nations 2021).

While governments and non-governmental organizations (NGOs) have traditionally been the driving forces behind environmental initiatives, corporations and other private entities now also recognize the central role of their business operations and global supply chains in reducing climate-warming GHGs (Bateman et al. 2021). With 2050 being the long game, many companies are setting more near-term goals, such as net zero by 2030, or setting interim targets to guide progress until 2050. In the absence of global policy to regulate emissions, there is pressure on companies to act from multiple sources, including governments, mass media, consumers, investors, and others (Meixell & Luoma 2015). In the 2021 State of Supply Chain Sustainability, on average, firms felt more pressure to act on supply chain sustainability between 2019 and 2020. Significant increases in pressure came from investors (+8%), government and international regulatory bodies (+8%), and media (+4%) (Bateman et al. 2021). This increase in pressure across the board signals that there is an increasing awareness of the impact of supply chains on the environment, and an obligation for companies to address and mitigate that impact.

A further, and more recent, addition to the imperative for companies to advance towards emissions reduction, is the increasing role of Environmental, Social and Governance (ESG) ratings for investors. Once thought of as a small, sustainability-minded group of investors, ESG-informed investing and impact investing has grown considerably in the last three years (Carlson 2020). Despite some variability around the quantification of ESG metrics (Berg et al. 2020), more and more investors are using those ratings in making their investment choices; for example, there have been significant shifts from investor groups such as Blackrock (Blackrock 2021). In addition, groups such as the Task Force for Climate-Related Financial Disclosures have been formed to align and advise companies and investors around climate-related problems. This shift is providing an impetus for businesses that are investing for the long term, and highlighting those that will be more resilient to future disruptions related to climate change and, therefore, perform better financially.

Due to a lack of action from national governments to address climate change, many companies have assumed the mandate to reduce their overall environmental impact. Companies may be motivated to take on reduction initiatives for a variety of reasons, including external pressure, existing policies and regulations designed to incentivize reducing emissions, and the competitive advantage inherent in terms of brand benefit of being an environmental forerunner. So far, almost 20 percent of the world's largest companies – representing 61 percent of global GHG emissions, 68 percent of global GDP, and 56 percent of the world's population – have pledged to reach net-zero emissions at some point before 2050 (Black 2021).

This chapter will incorporate research on the state of supply chain sustainability as well as disclosures, exploration of current and best practices, and an overall discussion of the state of the industry. The following sections will explore: the pressure from stakeholders to account for supply chain emissions, different approaches to achieving transparency, leadership practices, and recommendations for future development.

REDUCTIONS OF GREENHOUSE GAS EMISSIONS

Significant progress in curbing GHG emissions was made over the last decade in electricity generation, residential homes, personal cars, and other sectors of the economy. However, almost half of global GHG emissions come from industrial supply chains, and emissions continue to grow (United States Environmental Protection Agency 2020). Industrial supply chains also contain some of the hardest segments of the economy from which to reduce GHG emissions (aka "decarbonizing"), including ocean shipping, air cargo, long-haul trucking, and cement and steel production. Cement and steel are common materials used in modern infrastructure and industrial products; they contribute over 10 percent of global GHG emissions and have no non-carbon alternatives that can satisfy global demand at scale (World Resources Institute 2016). Logistics activities – including freight transportation, warehousing, and ports – contribute over 10 percent of global GHG emissions as well. The demand for freight transportation is projected to triple by 2050 and become the most carbon-intensive sector by 2040 without a widespread and concerted effort to reduce the associated impacts (International Transport Forum & Organisation for Economic Co-operation and Development 2019).

However, many companies have started to make carbon reduction commitments for their own organizations and in general can accurately account for Scope 1 emissions or even Scope 2 emissions as defined by the GHG Protocol, the leading method for corporate carbon accounting. But when companies look toward addressing upstream and downstream Scope 3 emissions, a challenge emerges (see Box 15.1). This challenge lies in both the ambiguity of existing guidelines to comprehensively quantify and set benchmarks for indirect emissions, and in the availability of needed data from suppliers.

BOX 15.1 WHY SCOPE 3?

There are significant challenges that come with the effort to account for Scope 3 emissions including the challenge of gathering accurate data, the risk of double counting, and the push back from industry. While the challenge is immense, there are three counter arguments that drive current efforts and adoption:

1. The risk of climate inaction in the absence of regulation or widespread industry adoption
2. Additional value that comes with supply chain transparency including cooperation and risk management
3. The ability to target hot spots in supply chains for more impactful reduction efforts.

Just as supply chain emissions reduction lags behind other sectors of the economy, disclosure of supply chain emissions significantly lags behind other segments of corporate carbon accounting. Many of the world's largest companies disclose emissions to the CDP (formerly the Carbon Disclosure Project), but less than 20 percent disclosed freight transportation emissions in 2019 (Greene 2020b). Global supply chains are complex networks of business relationships, making it difficult for individual companies to accurately identify and measure sources of GHG emissions. To do so, companies have to commit resources to carbon accounting, life cycle assessment, and other sustainability practices, and they must also work with upstream and downstream supply chain partners. In a competitive business environment, these partners can be hesitant to assess and disclose emissions. Additionally, while supplier codes of conduct and auditing are common practices, adopting new carbon accounting and other sustainability practices requires developing new skills and institutional knowledge, which can be costly. Without immediate returns, companies are hesitant to invest resources in these ways. These investments, however, are essential to shed light on the sources of supply chain emissions and reduce risk for companies that have pledged to reach net-zero emissions. Greater transparency of supply chain emissions is also essential for making investment, procurement, and policy decisions if we are to tackle some of the most complex challenges we face in avoiding a climate crisis.

Complete auditing of supply chain emissions for a particular company allows the company to identify and be accountable for its full impact, make realistic goals, and identify key hot spots in the supply chain that should be addressed in the near term. However, even major advocates in this area say that measuring carbon and other GHGs across product value chains today is almost impossible (World Business Council for Sustainable Development 2020). With the decade of the 2020s marking some of the most significant gains in corporate commitments to address climate impacts, there is significant opportunity to revolutionize supply chain transparency. Some key opportunities include:

- Improving understanding of the value of accurate accounting and reporting up and down the value chain using primary data;
- Building and utilizing consistent methodology to allocate, share, and verify primary emissions data;
- Driving buy-in and adoption of reporting platforms across the value chain to enable ease of information sharing;
- Utilizing technology to monitor and share GHG emissions data;
- Designing systems that utilize open technology standards to share information across technology solutions throughout the value chain;
- Pairing efforts for digitalization and visualization for other purposes, such as risk management and human rights protection, with the drive towards carbon transparency.

STATE OF SUPPLY CHAIN GHG EMISSIONS REPORTING

The disclosure of corporate climate impacts has become an imperative in today's sustainability-conscious society. However, the reporting can vary in quality and is not always standardized. Further, it can be perceived by companies as a costly exercise that does not translate into clear benefits (Sodhi & Tang 2019). In the State of Supply Chain Sustainability 2021 industry survey results, supply chain sustainability is most heavily reported on websites with 87 percent of respondents reporting that they disclose in this format in a regular way. Seventy-eight percent say they disclose in their own CSR or Sustainability Reports. For reporting organizations that have a higher level of rigor and requirements, thus requiring more effort and expertise, only 48 percent of respondents said their organization discloses to organizations such as CDP (formerly known as the Carbon Disclosure Project) (Bateman et al. 2021).

Emissions are typically reported on absolute and intensity bases. Absolute emissions refer to the total emissions generated by one company, facility, or activity during a specific time period, often over the course of one year. Calculating emissions on an intensity basis provides context for absolute emissions values. Here, absolute emissions are allocated to a meaningful metric such as kilograms of a material, product, or revenue, or to activity metrics such as ton-kilometers for transportation. This section will discuss two primary forms of disclosure: annual corporate carbon emissions (an absolute metric) and product carbon footprints (an intensity metric).

Corporate Carbon Emissions

The act of assessing and reporting corporate carbon emissions involves accounting for the direct and indirect emissions related to a company's activities. Following the Greenhouse Gas Protocol's tripartition of emissions scopes, corporate emissions are divided into three scopes (World Resources Institute & World Business Council for Sustainable Development (WRI & WBCSD) 2004, 2011). Scope 1 emissions are considered direct emissions resulting from a company's owned and controlled assets. These could include the emissions from the combustion of fuels used in a company's fleet of trucks, or emissions from a factory's onsite power plant. Scope 2 emissions include the emissions from electricity purchased to power Scope 1 activities, such as electricity used within office buildings or electric cars. Finally, there are Scope 3, or supply chain, emissions – the most diverse and challenging-to-track category. Scope 3 covers a broad range of activities within the upstream and downstream supply chain, such as the production of purchased goods, subcontracted transportation, and the use and disposal of sold products.

Companies typically tally and report these emissions on an annual basis, disclosing them on their websites, corporate sustainability reports, and disclosure platforms (discussed in more detail below). Corporate carbon emissions are often used to judge the performance of a company over time by comparing one year with another to see how the generation of emissions has evolved. The information can be put into context by relating to a performance metric such as revenue or number of employees. This way, a growing company whose emissions may also be growing can show that the carbon intensity of their activities is remaining the same, or decreasing, while the absolute emissions might be growing.

Product Carbon Footprints

While corporate emissions provide an understanding of a company's performance, product carbon footprints make it easier for consumers to understand the climate impacts embedded in a product, process, or service. Carbon footprints overlap with corporate carbon emissions, but they are calculated and allocated in different ways to fulfill the information needs of the end user. As such, there is a separate but related guidance under the Greenhouse Gas Protocol for product carbon footprints, the Product Life Cycle Accounting and Reporting Standard (WRI & WBCSD 2013). As interest grows for product carbon footprint data, and to keep up with digital technology development, the World Business Council for Sustainable Development has released as updated guidance in 2021 on product carbon footprinting, *Pathfinder Framework Guidance for the Accounting and Exchange of Product Life Cycle Emissions* (WBCSD 2021).

Many products have complex life cycles and undergo numerous transformations, as shown in Figure 15.1. The life cycle can be considered in various phases. For example, the journey from raw materials to a product is often referred to as cradle-to-gate, which is the focus of the Pathfinder Framework. However, many products also produce emissions in the use phase, and again as they are transported and processed for disposal. Considering the entire life cycle is referred to as cradle-to-grave, or for recycling or reuse, or cradle-to-cradle, an important consideration for enabling the circular economy, a life cycle assessment (LCA) is a process used to estimate the environmental or social impacts that can occur in the lifetime of a product or service, often covering issues such as carbon emissions, water use, toxicity, and so on

(Klöpffer 2006). LCAs are often leveraged for product carbon footprints and Scope 3 assessments, particularly for purchased goods and services or transportation.

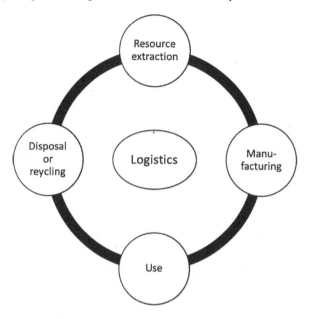

Source: Authors.

Figure 15.1 The life cycle of products can be linear, from extraction to use and disposal, or circular, using recycled materials to create new products. Transportation links the life cycle together

The variety of companies, activities, and processes within a product or company's supply chain can be complicated and challenging to track. This complexity has been a roadblock for comparing one product to another and has limited some green procurement protocols (Cannon et al. 2020). This is why, for example, we do not yet see widespread application of carbon footprint labels – various methods and data sources make it impossible to reliably compare two products. Similarly, companies cannot easily be compared with one another (Lee et al. 2020). However, numerous initiatives have emerged to try to address this issue, offering methodologies and data to guide companies towards more accurate and transparent carbon emissions values, as the next section details.

Methodologies and Standards

To maintain consistency in carbon emissions data, methodologies are followed that define the goals, scope, boundary, data requirements, allocation procedure, and documentation process. Numerous standards, guidelines, and methods exist to provide clarity and structure to corporate and product carbon footprints (Pré 2013). A set of key methods and standards are described here, but many more exist for various regions, products, sectors, or platforms. As time goes on, new methods will surely continue to emerge to fill the gaps in existing guidance, adjust for advances in real-time monitoring, and support new ways of sharing data between or among businesses and with consumers.

While regional accounting guidance exists for corporate carbon accounting, such as the Guidance on How to Measure and Report Your Greenhouse Gas Emissions (Department for Environment, Food and Rural Affairs 2009) for the UK and Bilan Carbone (Agence d l'Environnement et de la Maîtrise de l'Energie 2017) for France, the Greenhouse Gas Protocol dominates the methodological landscape. The Greenhouse Gas Protocol's Corporate Accounting and Reporting Standard sets the baseline, and the Corporate Value Chain (Scope 3) Accounting and Reporting Standard defines accounting for the supply chain. The Greenhouse Gas Protocol provides a broad platform for carbon accounting and has been "built on" by numerous sector-specific methodologies (Greenhouse Gas Protocol 2021) such as the Global Logistics Emission Council Framework, which is described in more detail in Box 15.2. These methods aim to provide more specific details for a specific activity or process, and they have been developed for sectors including logistics, waste removal, concrete, pharmaceuticals, aerospace, and others.

The Greenhouse Gas Protocol was formalized by the International Standards Organization (ISO) as ISO 14064, first established in 2006 and updated in 2018 and 2019. ISO 14064 has three parts. Part 1 defines the requirements for developing and reporting GHG inventories for organizations (International Standards Organization 2018a), Part 2 covers the requirements for quantifying and reporting emissions from projects meant to reduce or remove GHG emissions (International Standards Organization 2019a), and Part 3 defines how GHG inventories and product carbon footprints can be verified and validated (International Standards Organization 2019b). The broad applicability and concrete formalization of the Protocol has led to its widespread adoption by key climate-related initiatives such as CDP, TCFD, Science-Based Targets initiative, and more.

BOX 15.2 CASE STUDY: THE GLOBAL LOGISTICS EMISSIONS COUNCIL FRAMEWORK

Freight transportation is a key element of every supply chain, spanning all geographies and a wide range of technologies, such as ships, planes, trains, and trucks, as well as transfer and storage points such as ports, terminals, and warehouses – leading to a significant and growing climate impact. Adding to the complexity is the web of shippers, logistics service providers, and carriers that coordinate and move diverse goods across a shared set of resources. This complexity transcends to carbon accounting, creating a system of variable calculation methods across transport modes and regions – a puzzle that the Global Logistics Emissions Council (GLEC) was formed in 2014 to address.

Through a partnership of companies, governments and stakeholders, the GLEC collaboratively built a streamlined system for calculating and sharing carbon emissions data across the supply chain. The resulting methodology, the GLEC Framework, was first published in 2016, then updated in 2021, and is built on the principles of the GHG Protocol.

The GLEC Framework has addressed the issue of consistent, reliable carbon accounting for freight transportation and has seen broad adoption from the sector and its customers. Even further, the process of developing the GLEC Framework created an alignment across the supply chain on the need to collaborate to track and reduce carbon emissions.

Getting to 2050: transparency for setting and reaching supply chain climate goals 347

Additional methods and standards exist to define product carbon footprint and LCA principles specifically. The Greenhouse Gas Protocol Product Life Cycle Accounting and Reporting Standard provides climate accounting guidance for products and serves as a partner to the popular corporate carbon accounting guides (World Resources Institute). Within the LCA, ISO 14040, 14044, and 14046 are key standards that define and consolidate the various procedures used to assess environmental and social impacts (Finkbeiner et al. 2006; International Standards Organization 2018a, 2018b, 2019a and 2019b). Product carbon footprints serve as a baseline for carbon footprint labels, which continue to advance as a basis for a green product marketplace. The European Commission continues to lead in this area, and has been advancing new guidance documents as a basis for a carbon labeling scheme (European Commission 2017).

While this section has only scratched the surface of the rapidly-evolving carbon accounting landscape, understanding the key methodologies is a good starting point from which to launch deeper investigations of certain sectors or geographies. Next, we will explore how data, tools and technologies can advance our understanding of carbon emissions.

Types of Data for Supply Chain Climate Accounting

Achieving supply chain transparency will require at least some level of communication with partners along the supply chain; however, lack of data is often highlighted as the biggest challenge in supply-chain transparency (Bateman & Bonanni 2019). Working towards increased accuracy in Scope 3 accounting is an ongoing journey for many companies. These companies move along a spectrum of data sources and increase their commitments where they are most needed and effective as time goes on. For example, companies seeking to set and track progress towards climate goals may need to shift towards the use of more detailed information to understand the performance of their suppliers over time. In general, four tiers of data are used to estimate supply chain emissions (Greene & Lewis 2019); these data tiers range from least accurate to most accurate, as shown in Figure 15.2. With all types of data, it is important to clearly note where the information is used and the source from which it is taken (Blanco & Craig 2009).

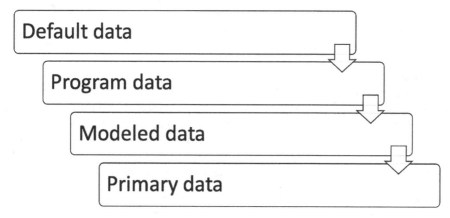

Source: Authors.

Figure 15.2   Types of emissions data: from least accurate, default data, to most accurate, primary data

Default data

Many companies rely on industry average, or default, data to calculate their Scope 3 carbon emissions, representing the average GHGs emitted to manufacture a product or conduct an activity. Often coming from sources such as LCA databases, government bodies, academic publications, or sector-specific initiatives, these data may be tailored for a certain geography or represent an average value worldwide. A commonly-used database is ecoinvent, an inventory of life cycle information for thousands of materials and activities across all sectors and geographies (ecoinvent n.d.). Default data are useful for providing a general overview of emissions and illuminating areas of particularly high emissions, but may be higher or lower than the actual emissions from the production of a product. In this way, default data are regarded as the least accurate form of data – they do not allow a company to accurately track its progress towards its climate goals or procure a verified lower carbon product or service.

Program data

Various government bodies, non-governmental organizations and industry associations offer detailed emissions data for certain activities or companies. These programs offer a neutral platform for data sharing, protect companies hesitant to share sensitive data, and align emissions accounting methods in order to create more reliable and transparent values. These data represent an increased level of resolution and often provide information for the most recent calendar year. An example of such a program is Clean Cargo, an industry group representing container shipping (Business for Social Responsibility n.d.). This program offers annual estimates for container ship emissions for global shipping trade lanes and, for a fee, for specific carriers.

Modeled data

Companies and tool providers can create models that blend supplier, program, and industry average data to provide custom emissions calculations for a specific set of conditions for a company. Through these processes, companies can create custom emissions factors that fit their business model, allowing for simplified consideration of emissions in business decision-making. Models can be created by a company in-house, or through a tool such as EcoTransIT, which leverages default data and actual conditions to improve the accuracy of carbon accounting for freight transportation (Ecotransit World n.d). The model's accuracy will be driven by the data inputs; uncertainty is increased if the main data source is default data.

Primary data

The most accurate data for Scope 3 accounting is primary data collected from suppliers. Primary data can range from precise information communicated about a certain activity or product, to annual average figures calculated and communicated from supplier to buyer. The latter is more common, though with increased transparency emerging as a growing concern within supply chain sustainability, myriad programs and tools are also emerging to advance supplier data collection and real-time monitoring of carbon emissions.

The ubiquity of the sensors and satellites offers the potential for monitoring directly at the source, such as a smokestack or tailpipe. These sensors can serve to cross-check the climate disclosures from companies or countries, taking measurements with or without the knowledge of the emitters (Pan et al. 2021). In one example of permitted monitoring, the Rocky Mountain

Institute's Climate Action Engine promises to bring together measured data, company reports, and modeled data to increase the transparency of the oil and gas industry (Ide & Kirk 2020).

While the wealth of data would be welcome for the increased accuracy provided, new systems will need to be built to manage and share that information in a credible and meaningful way. Data management is part of the digital transformation and is already underway in many global companies. Data sharing is also evolving, with programs such as the World Economic Forum's Global Battery Passport that provides such a framework for sharing data on the battery supply chain, a key element in the clean energy revolution. The Passport will be based on a blockchain-powered platform that collects sustainability data for the materials and processes that make up the battery supply chain, from the extraction phase to the recycling-after-use phase (World Economic Forum 2020). It will be used to share harmonized emissions data across the value chain and to confirm the origin of materials with known social or environmental impacts, such as cobalt, verifying they are being sourced from high-integrity suppliers.

Disclosure Programs and Platforms

Disclosure programs gather and consolidate climate data disclosed by companies, which are in turn used by supply chain partners, investors, and non-governmental organizations to assess the environmental or social impacts of the reporting company. Disclosure programs provide an outlet for climate data, typically with rules that govern the types of data disclosed, the methods used to find that data, and the verification required to use the platform. As such, these systems provide consistency across corporate reports and allow observers to compare a company's activities from year to year. The following section will focus on three common platforms, with the understanding that new platforms and programs continue to emerge and evolve.

The Global Reporting Initiative (GRI) is a non-governmental organization founded in 1997. It published the first global framework for sustainability reporting in 2000, and is the most widely used sustainability disclosure standard (Global Reporting Initiative 2021). The GRI Sustainability Reporting Standards are the most widely used principles for sustainability reporting and are frequently used as a basis for qualitative and quantitative reporting within many corporate sustainability reports. Covering a broad suite of topics within ESG, GRI offers both generic standards and topic-specific standards for water, health & safety, waste, biodiversity, and emissions. The GRI's climate reporting standards draw on Greenhouse Gas Protocol methodologies, adopting unified practices for Scope 3 reporting in its indirect emissions disclosure (Global Sustainability Standards Board 2016).

Founded in 2000, CDP is a platform for environmental disclosures from companies and governments that collects voluntarily reported data on climate, water, and forests from over 9,600 companies and 800 government bodies (CDP Worldwide 2021). Its climate disclosure questionnaire is seen as standard practice for multinational companies. Unlike GRI, which is embedded in corporate sustainability reports, CDP disclosure is done through an external platform, although many companies can choose to re-publish their disclosures on their own websites. CDP's disclosure questionnaire contains questions on Scope 3 emissions following the Greenhouse Gas Protocol. To collaborate further with suppliers, companies can join the CDP Supply Chain program, which allows for direct, confidential customer–supplier communication on annual carbon emissions for the customer, product carbon footprints, and collaboration opportunities.

To avoid duplication of efforts, GRI and CDP published a guidance document in 2017 that shows how CDP's climate change questions are linked with the GRI standards (GRI & CDP 2017). Questionnaire results are scored based on the comprehensiveness of the company's disclosure, its environmental awareness, and the actions it has taken to avert climate risks (CDP Worldwide 2020).

The final platform considered here is the Task Force on Climate-Related Financial Disclosures (TCFD) founded in 2017 by the Financial Stability Board. With disclosure content aimed specifically at an audience of investors, lenders and insurance underwriters, TCFD gathers information on governance, strategy, risk management, and metrics and targets as the information relates to climate change (Task Force on Climate-Related Financial Disclosures 2017). The TCFD acknowledges and seeks to address inconsistency within ESG ratings that make it challenging to act on reported data. TCFD ties into the Sustainability Accounting Standards Board (SASB), a body creating ESG standards specifically for the investor community under the auspices of the Value Reporting Foundation. TCFD and SASB have grown in importance as investors continue to increase pressure for more and better ESG data.

CLIMATE IMPACT REDUCTION STRATEGIES

GHG emissions data informs perhaps one of the most critical corporate sustainability trends of our generation – climate impact reduction. Companies around the world are motivated by internal and external factors to develop strategies that reduce the impacts of their own companies and their supply chains. For many companies, the strategies culminate in setting a climate goal. The Science-Based Targets initiative, a partnership between CDP, UN Global Compact, and World Resources Institute, has emerged as the preeminent goal-setting and verification body. By 2022, over 2,400 companies had committed to an emissions reduction target with the Science-Based Targets initiative (Science Based Targets Initiative 2022). The subsequent section will first define the parameters by which climate goals are typically set, then describe common strategies by which companies meet climate targets for their supply chains.

Climate Goals

Climate goals, or targets, can vary by scope, ambition, timelines, use of offsets, and rigor. Companies may set a target for their Scope 1, 2 and 3 emissions, as well as targets based on emissions intensity. Scope 1 emissions reporting represents the minimum goal that can be set, and Scopes 1–3 represent more comprehensive goals. Ambition refers to the degree to which emissions will be reduced, based on a range of percentages. For example, one company may set a goal of a 30 percent reduction in Scope 1 emissions, while another may set a goal of net zero in which emissions are effectively reduced to carbon neutral. Microsoft has taken its goal even further by pledging to effectively eliminate all present and historic emissions through future actions, neutralizing all emissions since the company's founding (Microsoft 2021).

Timelines for climate targets are typically aligned with the Paris Agreement, meaning goals should be met by 2050 at the latest. However, many companies are setting more near-term goals, such as net zero by 2030, or setting interim targets to guide progress until 2050. For example, The Coca-Cola Company committed to reduce emissions across the entire value chain 25 percent by 2030, from a 2015 base-year (Coca-Cola Company 2022). Target time-

Getting to 2050: transparency for setting and reaching supply chain climate goals 351

lines are typically pinned to a baseline year, in relation to which future reductions are measured, such as the 2015 baseline for Coca-Cola. Additionally, goals can vary depending on the use of carbon offsets versus actual emissions reductions to reach the specified goals. This topic will be described further below. Finally, climate goals can be differentiated by their rigor, such as how well-defined the goal and reduction strategies are, how the goals are publicly communicated and tracked over time, and if they are in line with pathways endorsed by the Science-Based Targets initiative.

For many companies, addressing climate impacts in the supply chain is the thorniest problem. The next section outlines three key strategies companies can adopt within their supply chain, as shown in Figure 15.3.

Source: Authors.

Figure 15.3 *Three strategies for supply chain decarbonization*

Sustainable Sourcing and Procurement

Companies are constantly engaging with other entities in their supply chain through sourcing and procurement practices, and sustainable sourcing is becoming "business as usual" for many global companies (Zimmer et al. 2015). Sustainable sourcing is defined by the International Standards Organization as "the process of making purchasing decisions that meet an organization's needs for goods and services in a way that benefits not only the organization but society as a whole, while minimizing its impact on the environment" (International Standards Organization 2017). Embedding carbon emissions into sourcing and procurement provides a meaningful lens by which companies can include carbon emissions data into decision making, in addition to more traditional concerns such as cost, availability, and quality, as well as other ESG metrics. As low carbon energy sources become increasingly available around the world, there will be a clear distinction among products and services according to their embodied carbon emissions. This will allow for even bigger "gains" by companies, as strategic sourcing can allow companies to reduce their Scope 3 emissions quickly and proactively. For example, as low-carbon steel production technologies scale up, buyers may seek to procure as much supply as possible from the set of producers using low-carbon production techniques. At the same time, sellers of lower-carbon products may be able to charge a price premium.

Companies can leverage standard procurement policies to embed carbon emissions practices (Sancha et al. 2015; Gualandris & Kalchschmidt 2016). First, companies can demand certain low-carbon attributes or disclosures of ESG metrics in a Request for Proposal, soliciting materials or services that fall within a defined carbon intensity level. Companies can also embed sustainability into the contract process, requiring suppliers to disclose carbon emissions or to enact certain reductions during the tenure of the contract. Using concrete carbon accounting and verification methodologies will be key to obtaining the information required by the company at the agreed-upon level of accuracy.

Supplier Engagement and Collaboration

Beyond the procurement and sourcing procedures, companies can leverage engagement with suppliers directly to move towards more lower carbon processes. A company may choose to do this with suppliers of key high-carbon materials or activities vital for their operations. Collaboration could take a number of different forms. A primary collaboration might include training suppliers on how to account for their carbon impact and instructing them on key reduction strategies. Based on initial awareness training and preliminary successes, suppliers are encouraged to provide value for greater impact reductions via cross-supply chain initiatives. In addition, suppliers and buyers may collaborate to deliver products more efficiently. For example, consolidating shipments to send fewer and more carbon-efficient truckloads would result in lower emissions. In another example, buyers and suppliers might collaborate via an industry group or NGO that offers support for low-carbon advancements. In a final example, a buyer and a supplier might collaborate to influence suppliers further upstream in the supply chain, applying pressure to advance the low-carbon transition deeper in the value chain.

Carbon Offsets and Insets

Carbon offsets are certificates that represent the avoidance, reduction, or neutralization of emissions made outside the buyer's value chain. The use of offsets is a common reduction strategy, with as many as 30 percent of companies engaged in supply chain sustainability practices using it as a tool to reduce their impact (Bateman et al. 2021). Carbon offsets are commonly applied to tree plantings, renewable energy, and clean cookstoves (Ecosystems Marketplace 2019). While there are rigorous standards on how offsets must be authenticated, such as avoiding double counting and verification, questions continue to arise on the effectiveness and longevity of some carbon offset projects (United Nations Environmental Programme 2019).

If a company relies on carbon offsets to meet its climate goals, it could be that it effectively does nothing to reduce its own operational or supply chain emissions. In response, groups such as the Science-Based Targets initiative have declared that carbon offsets are not allowed to meet a verified Science-Based Target, but are attempting to define an acceptable level of offsetting for a Net Zero target, making clear the responsibility companies have to make actual emissions reductions within their operations or supply chain. This tension illuminates a potential future challenge around verified carbon neutrality for products or companies, as there are no clear standards for how the use of offsets are declared along the supply chain.

Carbon insets have arisen as a meaningful companion to carbon offsets, representing the purchase of emissions reductions within a company's own value chain, defined alongside carbon offsets in Figure 15.4. While still a relatively new concept, carbon insets represent an opportunity for companies to co-finance supply chain decarbonization and reduce their Scope 3 emissions in a tangible, transparent way. An early example of carbon in-setting was demonstrated by L'Oréal in 2016, wherein the company invested in clean cookstoves for a set of suppliers in Burkina Faso producing shea butter (International Platform for Insetting 2016).

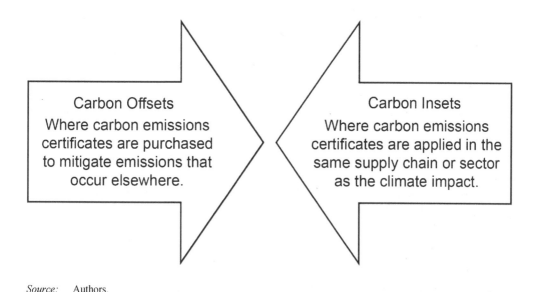

Source: Authors.

Figure 15.4   Defining carbon insets and offsets

It purchased carbon offset certificates related to the new cookstoves in the typical carbon offset marketplace, but since the recipients were in L'Oréal's own supply chain, the offsets could be considered carbon insets.

Many additional opportunities exist to apply carbon insets in a strategic and actionable way, aligning a company's Scope 3 impacts with the sector or value chain from which the impact originated. For example, freight transportation has emerged as a sector where carbon insets could be leveraged for projects that would accelerate the decarbonization of logistics activities, such as scaling up the infrastructure for low-carbon fuels, buying back out-of-date equipment, or retrofitting old engines (Greene & Façanha 2019). A company that has 10 percent of its Scope 3 impact related to freight may choose to spend 10 percent of its funds on carbon insets related to freight. To further expand the use of carbon insets, new methodologies, marketplaces, and certificates may be needed to ensure that projects are transparent and credible (Greene 2020). A starting point for this concept has been advanced by the MIT Center for Transportation & Logistics and Smart Freight Centre, which developed a framework for carbon insets from sustainable aviation fuel (SFC & MIT CTL 2021), an idea that could be expanded to other transportation fuels, feedstocks for plastics, and so on.

FUTURE DIRECTIONS

As more firms are setting large goals around climate targets and net zero carbon goals (Black et al. 2021), increasing supply chain transparency for achieving climate goals is more critical than ever before. With significant targets in place, companies are under substantial pressure to begin gathering the right information about their supply chains to understand their total impact and determine which actions will have the most effect.

According to past research, efforts to effectively measure emissions and manage decarbonization strategies can take a decade or more to become sophisticated (Lubin et al. 2019). We have mapped the journey that some companies have taken to increase supply chain transparency (see Figure 15.2). Many "new entrants" begin by establishing a general policy aimed at reducing climate impacts. Early efforts include disclosing Scope 1 and 2 emissions by using average emissions estimates for each product, service, or activity. These primary actions help tackle more easily achievable goals (low-hanging fruit) and foster early wins by using operational practices that are in the direct control of the company. The achievement of early goals increases motivation to achieve more difficult goals.

Once a firm is ready to advance to the next stage of the decarbonization journey, it is committed for the long haul. This means moving away from a general policy aimed at reducing emissions to the creation of formal targets for short-term reductions. At this stage, comprehensive disclosures that include Scope 3 emissions, in addition to Scopes 1 and 2 emissions, must be provided. These companies also use more sophisticated program data gleaned from NGOs, governments, and industry associates to estimate their emissions. In these ways, companies move beyond "easy wins" into a stepwise plan that includes suppliers and supply chain partners in the reduction journey.

Once a committed company is ready to be numbered among the "Advanced," it needs to be prepared to report on emissions for its complete GHG footprint and, to be most trustworthy, accurate, and verified by a third party. When companies are ready to advance their climate strategy, they are often also ready to partner with others in their industry to make real gains. This could occur through industry associations or other collaborative efforts that scale efforts broadly, such as the Responsible Business Alliance (formerly known as the Electronics Industry Citizenship Coalition) or Climate Collaborative, a consortium of natural products companies cooperating for climate mitigation. Advanced companies engage share partners as well as customers and combine efforts to make reductions across the supply chain. For instance, being in the advanced echelon often includes publication of a detailed plan of execution.

Companies that demonstrate leadership in GHG accounting to set and achieve climate goals are those that boldly commit to setting long-term goals using science-aligned targets and reporting year-to-year goals very transparently. To be confident about committing to science-aligned targets, companies should consider – at a minimum – using modeled data. In some cases, primary data, the most reliable source of data, is being used to estimate supply chain emissions. There are also cross-industry efforts that seek to scale efforts broadly, such as Amazon's Climate Pledge and Walmart's Sustainability Consortium, two corporate developed consortiums channeling corporate efforts to reduce emissions and investment in environmental innovations. These efforts seek to unite players across the spectrum on a shared journey toward emission reductions. Finally, one of the most aggressive efforts a leader in transparency might employ would be tying carbon reduction achievements to business results and employee goals (see Figure 15.5).

In summary, nearly half of global GHG emissions derive from supply chains, and companies within supply chains face increasing pressure from governments, consumers, investors, and the media to reduce their carbon impacts. However, the complexity inherent in supply chains makes achieving transparency – essential for assessing and addressing climate impacts – difficult. Despite the challenge, there is a greater risk in inaction in lieu of broad regulation, widespread standardization, or supply chain resistance. Accounting for and mitigating Scope

3 emissions is one of the most powerful opportunities to rapidly increase the footprints of companies. By addressing Scope 3 emissions, companies are signaling not only that they care

The Journey to Supply Chain Transparency for Achieving Climate Goals

New Entrant	Committed	Advanced	Leader
Establishing a policy around reducing climate impact	Setting targets for incremental short-term reductions	Reporting on emissions for the complete GHG footprint	Setting science-aligned targets
Disclosing initial emissions for business internal operations and purchased electricity (scopes 1 and 2)	Increasingly comprehensive climate disclosure - extending from only scope 1 and 2 to scope 3.	Reporting to third-party verification	Reporting year over year progress
Using average emissions emitted by product, service or activity to estimate supply chain emissions	Using program data provided by NGOs, governments, and industry associations to estimate supply chain emissions	Using modeled data that blend supplier, program, and industry averages to estimate supply chain emissions	Using primary data to estimate supply chain emissions
Identifying early wins and low-hanging fruit	Creating a stepwise plan	Engaging with own industry for shared success	Engaging across industries for shared success
		Publishing a detailed plan for execution	Linking carbon reduction achievements to business results and employee goals

Source: Authors.

Figure 15.5   The journey to supply chain transparency

about their own impact, but that they recognize their supply chains have far-reaching impacts and they have a responsibility to address those impacts in cooperation with supply chain partners.

This chapter has examined current progress in supply chain sustainability, methods used in achieving supply chain transparency, explored several key issue areas that make achieving transparency challenging, and explored various solutions that can enable greater transparency for more accurate monitoring and realization of long-term climate goals.

REFERENCES

Agence d l'Environnement et de la Maîtrise de l'Energie (2017), 'Methodological Guidelines, v8'.

Bateman, A., Betts, K., Cottrill, K., Pang, J., & Suhas Deshpande, A.; Cottrill, K., Grau, A., & McCool, D. (2021), 'State of supply chain sustainability 2021'. MIT Center for Transportation & Logistics and Council of Supply Chain Management Professionals.

Bateman, A. & Bonanni, L. (2019). 'What supply chain transparency really means'. *Harvard Business Review*, https://hbr.org/2019/08/what-supply-chain-transparency-really-means

Berg, F., Kölbel, J. & Rigobon, R. (2020), Aggregate confusion: The divergence of ESG ratings'. Social Science Research Network.

Black, R., Cullen, K., Fay, B., Hale, T., Lang, J., Mahmood, S. & Smith, S.M. (2021), 'Taking stock: A global assessment of net zero targets'. Energy & Climate Intelligence Unit and Oxford Net Zero.

Blackrock (2021). Investing with climate in mind. Retrieved from: https://www.blackrock.com/americas-offshore/en/insights/investing-with-climate-in-mind

Blanco, E. & Craig, A. (2009), 'The value of detailed logistics information in carbon footprints [White paper]'. MIT Center for Transportation & Logistics.

Blume Global (2021). *The History and Evolution of Global Supply Chains*. Retrieved from: https://www.blumeglobal.com/learning/history-of-supply-chain/

Business for Social Responsibility (n.d.), *Clean Cargo*. Accessed 19 April 2021 at https://www.clean-cargo.org/.

Cannon, C., Greene, G., Blank, T.C., Lee, J., & Natali, P. (2020), 'The next frontier of carbon accounting: A unified approach for unlocking systemic change'. Rocky Mountain Institute.

Carlson, D. (2020), 'ESG investing now accounts for one-third of total U.S. assets under management'. Marketwatch.

CDP Worldwide (2020), *The A List 2020*. Accessed 19 February 2021 at https://www.cdp.net/en/companies/companies-scores.

CDP Worldwide (2021), *What We Do*. Accessed 19 February 2021 at https://www.cdp.net/en/info/about-us/what-we-do.

Coca-Cola Company (2022), 'Science-based targets'. Accessed 27 January 2022 at https://www.coca-colacompany.com/sustainable-business/climate/science-based-targets.

Department for Environment, Food and Rural Affairs (2009), 'Guidance on how to measure and report your greenhouse gas emissions (PB13309)'. United Kingdom Government.

Ecoinvent (n.d.), ecoinvent database. https://ecoinvent.org/the-ecoinvent-database/. Accessed 20 April 2021.

Ecosystem Marketplace (2019), 'Financing emissions reductions for the future: State of the voluntary carbon markets 2019'. Ecosystem Marketplace.

Ecotransit World (n.d.), 'Your solution to calculate emissions of global freight transports'. Accessed 1 September 2021 at www.ecotransit.org

European Commission (2017), 'Guidance for the development of product environmental footprint category rules, version 6.3'. European Commission.

Finkbeiner, M., Inaba, A., Tan, R., Christiansen, K., & Klüppel, H. (2006), 'The new international standards for life cycle assessment: ISO 14040 and ISO 14044'. *The International Journal Life Cycle Assessment*, 11, 80–85.

Global Reporting Initiative (2021), *Our Mission and History*. Accessed 19 February 2021 at https://www.globalreporting.org/about-gri/mission-history/.

Global Reporting Initiative and CDP (2017), 'Linking GRI and CDP'.Global Reporting Initiative and CDP.

Global Sustainability Standards Board (2016), 'Global Reporting Initiative 305: Emissions'. Global Sustainability Standards Board.

Greene, S. (2020a), 'Carbon insets for freight transportation'. Smart Freight Centre and DHL.

Greene, S. (2020b), 'Closing the logistics emissions gap: An analysis of emissions disclosure to CDP by corporations worldwide', Smart Freight Centre.

Greene, S. & Façanha C. (2019), 'Carbon offsets for freight transport decarbonization'. *Nature Sustainability*, 2(11), 994–996.

Greene, S. and Lewis, A. (2019), 'Global logistics emissions council framework for logistics emissions methodologies'. Smart Freight Centre.

Greenhouse Gas Protocol (2021), *Greenhouse Gas Protocol Guidance Built on the GHG Protocol* Accessed 10 October 2021 at https://ghgprotocol.org/Guidance-Built-on-GHG-Protocol.

Gualandris, J., & Kalchschmidt, M. (2016), 'Developing environmental and social performance: the role of suppliers' sustainability and buyer–supplier trust'. *International Journal of Production Research*, 54(8), 2470–2486.

Hult, G. & Tomas, M. (2004), Global supply chain management: An integration of scholarly thoughts. *Industrial Marketing Management*, 33, 3–5. 10.1016/j.indmarman.2003.08.003.

Ide, T. & Kirk, T. (2020), 'Making emissions visible: A clear path to insights and action for the oil and gas industry'. Rocky Mountain Institute.

International Platform for Insetting (2016), *Burkina Faso Solidarity-Sourced Shea Butter*. Accessed 19 April 2021 at https://www.insettingplatform.com/case-study/loreal/.

International Standards Organization (2017), 'ISO 20400 Sustainable Procurement'. ISO.

International Standards Organization (2018a), 'ISO 14064-1:2018, Greenhouse Gases – Part 1: Specification with guidance at the organization level for quantification and reporting of greenhouse gas emissions and removals'. ISO.

International Standards Organization (2018b), 'ISO 14067:2018, Greenhouse Gases – Carbon footprint of products – Requirements and guidelines for quantification'. ISO.

International Standards Organization (2019a), 'ISO 14064-2:2019, Greenhouse Gases – Part 2: Specification with guidance at the project level for quantification, monitoring and reporting of greenhouse gas emission reductions or removal enhancements'. ISO.

International Standards Organization (2019b), 'ISO 14064-3:2019, Greenhouse Gases – Part 3: Specification with guidance for the verification and validation of greenhouse gas statements'. ISO.

International Transport Forum & Organisation for Economic Co-operation and Development (2019), 'ITF Transport Outlook 2019'. ITF & OECD.

Klöpffer, W. (2006), 'The hitch hiker's guide to LCA – An orientation in LCA methodology and application. *International Journal of Life Cycle Assessment*, 11, 142.

Lee, J., Bazilian, M., Sovacool, B., & Greene, S. (2020), 'Responsible or reckless? A critical review of the environmental and climate assessments of mineral supply chains'. *Environmental Research Letters*, 15, 103009.

Lubin, D., Nixon, T., & Mangieri, C. (2019), 'Transparency: The pathway to leadership for carbon intensive business'. Reuters News Service.

Meixell, M. J., & Luoma, P. (2015). Stakeholder pressure in sustainable supply chain management: A systematic review. *International Journal of Physical Distribution & Logistics Management*, 45(1/2), 69–89.

Microsoft (2021), *2020 Environmental Sustainability Report*. Microsoft.

Morningstar (2020), 'Sustainable funds U.S. landscape report'. Morningstar.

Pan, G., Xu, Y., Ma, J (2021), 'The potential of CO_2 satellite monitoring for climate governance: A review'. *Journal of Environmental Management*, 277, 111423.

Pré (2013), 'Life cycle-based sustainability – standards & guidelines [White Paper]'.

Sancha, C., Longoni, A., & Giménez, C. (2015), 'Sustainable supplier development practices: Drivers and enablers in a global context'. *Journal of Purchasing and Supply Management*, 21(2), 95–102.

Science-Based Targets initiative (28 January 2022), 'Companies taking action'. Retrieved from https://sciencebasedtargets.org/companies-taking-action).

Smart Freight Centre and MIT Center for Transportation & Logistics (2021), *Sustainable Aviation Fuel Greenhouse Gas Emission Accounting and Insetting Guidelines*. Smart Freight Centre and MIT Center for Transportation & Logistics.

Sodhi, M & Tang, C (2019), 'Research opportunities in supply chain transparency'. *Production and Operations Management*, 28(12), 2946–2959.

Task Force on Climate-Related Financial Disclosures (2017), 'Recommendations of the Task Force on Climate-related Financial Disclosures'. TCFD.

United Nations Environmental Programme (2019), 'Carbon offsets are not our get-out-of-jail free card'. United Nations.

United Nations (2021), 'Secretary-General's remarks briefing to member states by incoming COP26 President (as delivered)'. United Nations.

United States Environmental Protection Agency (2020), 'Inventory of U.S. greenhouse gas emissions and sinks: 1990-2018, Report # 430-R-20-002'. US EPA.

World Business Council for Sustainable Development (2020), 'Value chain carbon transparency pathfinder: Enabling decarbonization through Scope 3 emissions transparency.' WBCSD.

World Business Council for Sustainable Development (2021), *Pathfinder Framework Guidance for the Accounting and Exchange of Product Life Cycle Emissions*. WBCSD.

World Economic Forum (2020), *Global Battery Alliance*. Accessed 18 February 2021 at https://www.weforum.org/global-battery-alliance/action.

World Resources Institute (2016), *World Greenhouse Gas Emissions: 2016*. https://www.wri.org/resources/data-visualizations/world-greenhouse-gas-emissions-2016.

World Resources Institute & World Business Council for Sustainable Development (2004), *Greenhouse Gas Protocol: A Corporate Accounting and Reporting Standard*. WRI & WBCSD.

World Resources Institute & World Business Council for Sustainable Development (2011), *Greenhouse Gas Protocol: Corporate Value Chain (Scope 3) Accounting and Reporting Standard*. WRI & WBCSD.

World Resources Institute & World Business Council for Sustainable Development (2013), *GHG Protocol Product Life Cycle Accounting and Reporting Standard*. WRI & WBCSD.

Zimmer, K., Fröhling, M., & Schultmann, F. (2015), 'Sustainable supplier management – a review of models supporting sustainable supplier selection, monitoring and development'. *International Journal of Production Research*, 54, 1–31.

16. Commodity supply chain management and climate change: a case study of the palm oil industry

Yinjin Lee and Alexis Bateman

INTRODUCTION

As supply chains become increasingly multinational and complex, some of the most significant climate impacts occur deep in the supply chain, far upstream from consumers of the products. The full scope of a product's supply chain, including raw material extraction, processing, manufacturing, transporting, and warehousing, can account for most of the product's embodied carbon. More than 20 percent of global greenhouse gas (GHG) emissions are generated through the production of goods (O'Rourke, 2014).

Businesses often face pressures from governments and non-governmental organizations to mitigate the climate impacts in their commodity supply chains (Lyons-White & Knight, 2018; Thorlakson, 2018). News headlines have taken supply chains and their impacts out of obscurity and into public awareness. The 2021 State of Supply Chain Sustainability Report (Bateman et al., 2021) found increasing pressure across stakeholder groups to improve supply chain sustainability. The report also identified significant media coverage about the clear-cutting of rainforests for commodity production. In response, companies have begun to invest heavily in high-risk commodities to avoid media scrutiny and ensure available production supply.

This study explores the ways companies mitigate the climate impact of their palm oil supply chains. Palm oil is an ingredient that is widely used in personal care and food products, as well as biofuels. In the last few decades, its production has grown at the cost of rapid deforestation, leading to high levels of GHG emissions and biodiversity losses (Carlson et al., 2013, 2017). To research and analyze companies' actions to source palm oil sustainably, we examined existing literature, company disclosures, and NGO reports that were publicly available in 2018. We found that companies' key management practices included third-party standards, compliance instruments, supply chain tracing, commitment, and collaboration. We also found that companies in the palm oil industry were focused on protecting existing forests, ecosystems, workers, and communities and less focused on greenhouse gas accounting. This could be because some of these actions can mitigate climate change more directly than greenhouse gas accounting.

Our research classified companies into four types: Highly Active, Active, Passive, and Inactive in terms of their engagement in ensuring sustainability in their palm oil supply chains. On average, palm oil processors and traders, public, high brand value companies, and companies with more employees were more likely to be Highly Active or Active in managing their palm oil sustainability issues than their counterparts. In general, companies' choice of action seems to be guided by regulatory and authoritative organizations as well as the actions of other companies.

PALM OIL SUPPLY CHAIN IMPACT ON CLIMATE CHANGE

This section briefly discusses the impact of palm oil production on global GHG emissions and describes the palm oil supply chain. Palm oil is the most widely consumed vegetable oil globally; it accounted for 38 percent of global vegetable oil consumption in 2015 (World Wide Fund for Nature, 2016). Palm oil or palm kernel oil is found in approximately half of all products on supermarket shelves, such as hand soap, cereals, and ice cream.

The increasing demand for palm oil has led and will lead to further deforestation and more GHG emissions. Over 75 million tons of palm oil were consumed in 2020, more than three times the amount consumed in 2000 (Statista, 2021). A joint study by Yale and Stanford universities estimated that palm oil plantation expansion would contribute more than 558 million tons of carbon dioxide to the atmosphere in 2020, which is almost as much as Canada's entire 2019 carbon dioxide emissions (Carlson et al., 2013; Ritchie & Roser, 2020).

In general, agriculture is the leading cause of global deforestation, with livestock production responsible for roughly 24 percent of deforested land and agricultural production responsible for roughly 29 percent (Baron et al., 2017). In terms of agricultural deforestation, the largest contributors include soybeans (19 percent), maize (11 percent), palm oil (8 percent), rice (6 percent), and sugarcane (5 percent). Clearing land to produce palm oil causes extensive climate impact; moreover, newly planted palm oil plantations produce double the amount of greenhouse gas as do mature plantations (Cooper et al., 2020).

By far, the greatest forest loss from palm oil production is in Malaysia and Indonesia, which jointly produced approximately 84 percent of the global output in 2016 (United States Department of Agriculture, 2018). Between 1990 and 2015, Indonesia lost 31 million forested acres, roughly equivalent to the size of Germany (Greenpeace, 2016). Malaysian and Indonesian rainforests contain rare or endangered species and are considered some of the most biodiverse forests on earth. Large-scale forest conversion from tropical rainforest to palm oil plantations has led to significant environmental impacts on biodiversity and indigenous species. As a result of the clearcutting of rainforest – there has been a critical loss of habitat for multiple species, including orangutans, which are squeezed into smaller and smaller portions of land from which they need to subsist (World Wildlife Fund, 2021).

As land becomes scarcer, growers are turning to the conversion of peatlands[1] despite the difficulties and expenses involved in making them suitable for palm tree plantations (Cooper et al., 2020). Unfortunately, the conversion of peatlands can release a large amount of GHG (Cooper et al., 2020) and exacerbate forest fires (Azhar et al., 2017; Carlson et al., 2017; Lyons-White & Knight, 2018).

Palm Oil Supply Chain

There is a pressing need to meet the growing demand for palm oil with sustainably produced supplies. However, the complexity of the palm oil supply chain can be a barrier to making and sourcing sustainable palm oil. The palm oil supply chain can be described as having an hourglass shape, as shown in Figure 16.1 (Lyons-White & Knight, 2018). Countless growers and farmers supply palm fruits that are funneled to a smaller number of mills. Products from the mills, such as crude palm oil and palm kernel, are distributed to an even smaller number of processors and traders (PTs).[2] The PTs trade and refine palm products to produce many kinds of oils and derivatives (El-Fegoun, 2015; Hashim et al., 2012). Subsequently, manufacturers use palm oil and palm derivatives purchased from PTs as inputs to their goods, which numerous retailers sell to billions of end-consumers.

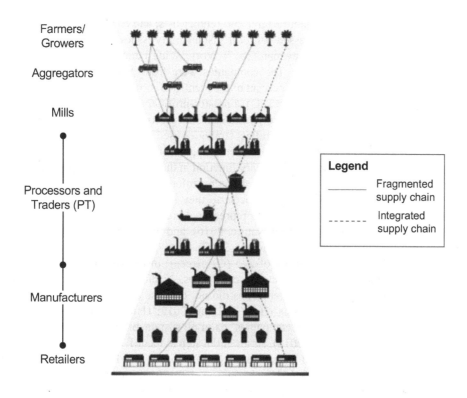

Note: The figure is reproduced with permission from Lyons-White & Knight (2018) and Elsevier.

Figure 16.1   Stages in palm oil supply chains

The top part of the supply chain involves numerous and diverse palm fruit sources, many of which are hard to trace. Some PTs partially integrate their upstream supply chain with their mills and palm oil plantations, although many of them also source from independent farmers and aggregators (Dodson et al., 2020; Lyons-White & Knight, 2018). PTs have limited visibility and control over these independent suppliers (Leegwater & van Duijn, 2012; Nesadurai, 2017). The palm product in a single end-consumer item could be a mixture of palm oil from a large shipment, produced by a few dozen palm oil mills, supplied by hundreds of farmers and plantations (Leegwater & van Duijn, 2012). Addressing sustainability issues at farms and plantations will require accountability from players across the supply chains.

Data

Our research sample included PTs, manufacturers, and retailers (grocery and foodservice companies) (Lee, 2021). Given that no list comprehensively identifies companies with sustainable sourcing practices for palm oil, we focused on companies most likely to have implemented sustainability practices. In total, 275 companies from the consumer goods and foodservice sectors were identified from reports published by non-governmental organizations (NGOs), Deloitte and Forbes top global manufacturer and retailer lists, and ethical consumer websites listing small businesses that addressed palm oil sustainability issues (Deloitte, 2017; Ethical Consumer, 2015; Forbes, 2017; Newman, 2017; World Wide Fund for Nature, 2016, 2017; Zoological Society London, 2017). The companies were from North and South America, Europe, Asia, and Africa. Close to 70 percent of the identified companies were members of the Roundtable of Sustainable Palm Oil (RSPO), a not-for-profit and voluntary group supporting the growth and use of sustainable palm oil products. Our selective sampling approach implies that our results on companies' practices reflect strategies used by the leaders in sustainability; our approach does not purport to represent the entire palm oil industry.

We categorized companies by their primary function in the supply chain stage. We grouped palm oil processors and traders following the conventions used in NGO reports (CERES, 2017; Zoological Society London, 2017) because they served overlapping functions. The retailers and manufacturers in our sample represented roughly 10 percent of global palm oil consumption (World Wide Fund for Nature, 2016), and the PTs controlled more than 87 percent of the global palm oil trade (Nesadurai, 2017).

Of the sampled companies, 179 out of 275 (65 percent) disclosed information related to palm oil sustainability. Of the remaining 96 companies, 27 did not have information available in English for analysis and 69 companies did not disclose any information on palm oil sustainability. Using content analysis (Berg & Lune, 2012; Hsieh & Shannon, 2005), we tabulated data related to the chosen companies' palm oil sustainability from their most recent online information (as of January 2018). As an initial step to identify a list of disclosure items, we referred to the literature, RSPO guidelines, NGO reports, and five leading companies' disclosures (CERES, 2017; RSPO, 2017; Thorlakson et al., 2018; Zoological Society London, 2017). The list of disclosure items (see Appendix) was revised after reading 60 companies' disclosures. Companies were assigned binary scores (1/0) for each disclosure item. For example, if a company stated that they have a supplier code of conduct, it earned a score of 1. A majority (96.6 percent) of the content and materials we used for analysis were published between the years of 2015 and 2018. At least two coders read each document. Independent coding of the documents initially resulted in 85 percent agreement. The coders then discussed their interpretations of the documents to reach 100 percent agreement.

SUPPLY CHAIN MANAGEMENT COMPONENTS

We classified the practices companies used to manage the sustainability of their palm oil supply chains into five components through commonalities in the data, industry and academic expert input, existing literature, and broadly accepted palm oil disclosure guidelines (CERES, 2017; Melot & Delabre, 2017, Newton et al. 2013). The identified components include commitment, external standards, compliance instruments, traceability, and collaboration. These components are interconnected and are not mutually exclusive. In this section, we focus our discussion on the 179 companies that disclosed information about their efforts to source palm oil sustainably.

Commitment

Companies stated their commitment to sustainability in a variety of ways. The sustainability commitments for palm oil are listed in Figure 16.2. This figure shows the percentage of companies committed to addressing environmental and social issues in gray and black, respectively. Although most (9 out of 14) of the commitments focused on environmental issues, social issues, such as workers' rights, were commonly mentioned by the companies. Environmental and social issues can be interdependent. For example, improving farmers' livelihoods usually also included helping them improve yield, which in turn can reduce greenhouse gas emissions for the same volume of palm oil produced.

Ending physically destructive activities that have drastic implications for climate change has typically been companies' priority over accounting for GHG emissions. Making explicit declarations related to reducing deforestation (declared by 64 percent of the companies) was more popular than commitment to reducing GHG emissions (declared by 44 percent of the companies). Other commitments identified the kinds of land areas to be protected, including high conservation value areas, peatlands, and high carbon stock areas that trap large amounts of GHGs. Specific commitments can enhance communications, align expectations, and discourage opportunistic behaviors based on loopholes.

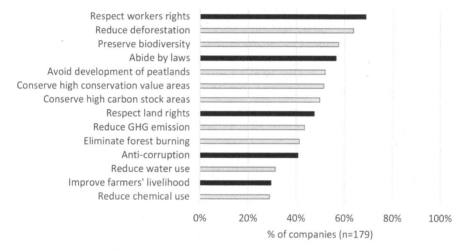

Note: Commitments with gray bars focus on environmental sustainability, while those with black bars focus on social sustainability.

Figure 16.2 *Percentage of firms making each commitment*

External Standards

External standards are awarded to companies by a third-party organization when the companies achieve the sustainability criteria that the third-party organization has established. Sustainability claims verified by independent third parties can be more credible than the claims that companies make by themselves about their products (Sheffi & Blanco, 2018).

A wide range of external standards exists. In the case of palm oil, there are multi-stakeholder, governmental, and NGO-led standards. Multi-stakeholder standards are developed and endorsed by stakeholders from multiple types of organizations, such as NGOs, companies, and governments. The most prevalent external standard is the RSPO certifications and it is a multi-stakeholder standard (Figure 16.3). There are several kinds of RSPO certifications, and they vary by the stringency of the certification process.

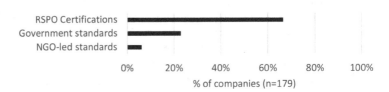

Figure 16.3   *Types of external standards and the percentage of companies using the external standards*

Governmental standards are awarded by the governments of the producing countries. Malaysia and Indonesia, the major global palm oil producers, have sustainability palm oil standard programs. These standards were used by 24 percent of the companies in our sample. We observed that only producers and traders disclosed their use of governmental standards; none of the retailers and manufacturers did so. Most retailers and manufacturers were not located in Malaysia and Indonesia, and they may not be compelled to adopt these national standards. Retailers and manufacturers may also not be interested in the governmental standards because NGOs find that governmental standards' sustainability criteria fall short of RSPO standards (McInnes, 2017). Nonetheless, many NGOs and academics recommend that even RSPO standards be improved to safeguard local communities, workers, and the environment (Carlson et al., 2017; Poynton, 2015).

Lastly, 6 percent of the companies used palm oil with NGO-led standards defined and verified by NGOs. These companies used the Rainforest Alliance certified palm oil. Although NGO-led standards and certifications are popular for commodities that are the key ingredients in the final products, such as coffee (Lee and Bateman, 2021) and chocolate, NGO-led standards were less popular for palm oil. Notably, the Rainforest Alliance phased out its palm oil certification program in 2020, recognizing that their certification process required significant investments from participating companies (Rainforest Alliance, 2021).

In addition to external standards, leaders in sustainable supply chain management were incorporating compliance instruments, which were developed specifically for their supply chains (Bateman et al., 2020; Ernst & Young & United Nations Global Compact, 2016; Gimenez & Sierra, 2013; Pagell et al., 2010). This trend was fueled partly by the ineffectiveness of external standards in ensuring sustainability performance and protecting companies' reputations and partly by a need to customize sustainability requirements suited to their business objectives, such as securing resources (Bager & Lambin, 2020; Starbucks, 2015). NGOs and researchers described this trend as firms "going beyond certifications" (Poynton, 2015; Thorlakson, 2018).

Compliance Instrument

Companies use compliance instruments to ensure that their suppliers meet their internal sustainability standards. This allows the companies and their external stakeholders to hold the suppliers responsible for their impacts on the environment and people (Zimmer et al. 2016; Bai & Sarkis, 2014). Companies begin by determining their internal sustainability standards, sometimes with the help of external advice from NGOs and industry alliances (Consumer Goods Forum, 2015; Poynton, 2015).

Figure 16.4 shows the types of compliance instruments and the percentage of companies using each of them. Fifty-six percent of the companies in our sample worked with approved suppliers that met their internal standards. The approval process either took the form of audits done at the suppliers' facilities or suppliers sharing operational information with buyers.

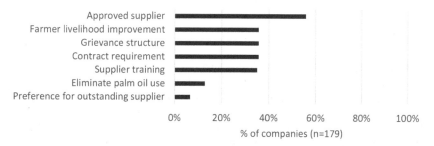

Figure 16.4   Percentage of companies using each compliance instrument

A subset (36 percent) of these companies maintained that they would terminate their contracts with suppliers that do not meet their standards. For instance, in 2018, numerous buyers stopped sourcing from a major palm oil processor, FGV, when alerts of forced labor were identified. In another example, leading manufacturers such as Ferrero and Nestlé terminated contracts with their supplying PTs that violated RSPO standards (Furlong, 2016; Jiang, 2009; Maitar & Skar, 2010; Nash, 2017; Reuters Staff, 2009). Companies who took this firm stance strongly signaled to their suppliers their commitment to source palm oil sustainably.

Thirty-five percent of the companies trained their direct suppliers to understand their internal standards better. Companies provided training documents for the suppliers or took an additional step to provide them with workshops and webinars. Training and engaging suppliers are supported by research to reduce social and environmental violations (Porteous et al., 2015; Villena, 2019). Lastly, only 7 percent of the companies gave preference to suppliers who surpassed the internal standards.

A subset of companies implemented instruments that extended their internal standards to the plantations and farms. Thirty-six percent of companies provided grievance structures that allowed anyone, including NGO representatives or the companies' employees, to report on activities or practices in their supply chains. Companies must publicly address the reports. For example, Unilever published a website listing all reported grievances and their progress in addressing them (Unilever, 2021).

Farmers who lack the information and resources needed to produce palm oil efficiently may clear land by cutting and burning primary forests (Sheffi & Blanco, 2018). In response to this issue, 36 percent of the companies in our sample invested in improving farmers' livelihoods and farming methods. Companies that took this path worked with experts from a combination of NGOs, local governments, and researchers. For example, Proctor & Gamble collaborated with the Malaysia Institute of Supply Chain Innovation and two NGOs, Wild Asia and Proforest, to trace thousands of smallholder farmers in their supply network and share knowledge and resources with these farmers to improve their crop management (P&G, 2018). Transferring knowledge and resources benefits farmers and reduces the climate impacts of palm oil production (Walker et al., 2018).

A final strategy used by 13 percent of our sample is eliminating palm oil altogether. This usually means that they reformulated their products. This alternative is used by a small number of companies, such as soap company Lush Retail Limited and chocolate maker Divine Chocolates. Owing to the premium nature of their products, they could use more expensive and less plentiful alternatives to palm oil, such as coconut oil.

Traceability

A company's supply chain traceability refers to information provided to external stakeholders to identify their direct and indirect suppliers (BSR & United Nations Global Compact, 2014). One of the objectives of tracing and disclosing information about a company's suppliers to the public is that it allows organizations such as journalists and activist NGOs to identify suppliers' locations and verify the suitability of the environmental and social conditions (Doorey, 2011; Lambin et al., 2020). Researchers found that companies in the fashion industry (e.g., Nike and Nudie Jeans Co) were driven to ensure that their identified suppliers met their standards before publishing information about their suppliers (Doorey, 2011; Egels-Zandén et al., 2015). Companies in other industries would likely behave similarly to protect their reputations. However, while traceability is increasingly recognized, it is not widely adopted (Thorlakson et al., 2018). Supply chain tracing is challenging because it requires the industry to standardize information gathering (Gunawan et al., 2020).

Our study observed the information companies in the palm oil industry disclosed about their palm oil supply chains (Figure 16.5). Traceability scope refers to the supply chain stages the companies can trace. Forty percent of the companies discussed their palm oil traceability scope. Companies were more likely to disclose mill, grower, and farmer locations than processor and trader (PT) locations. Only 25 percent of the companies in our study disclosed the percentage of palm oil they could trace beyond their direct suppliers. Companies also discussed the challenges in tracing upstream of the supply chain, citing that aggregators and dealers who acted as intermediaries between farmers and mills were reluctant to share information.

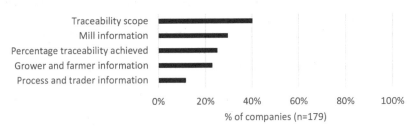

Figure 16.5   *Types of traceability information disclosed*

Collaboration

Practitioners and scholars found that buying firms that proactively work with their suppliers, in addition to using external standards and compliance instruments, can better improve the buying firms' reputations and their suppliers' sustainability performance (Croom et al., 2018; De Marchi et al., 2013; Gold et al., 2010; Pilbeam et al., 2012; Porteous et al., 2015; Poynton, 2015; Zhu & Sarkis, 2007). We identified three kinds of collaborations in the palm oil industry: vertical collaborations among companies downstream and upstream in the supply chain, horizontal collaborations among companies in the same supply chain stage, and collaborations with governments and non-governmental organizations external to the supply chain. These three types of collaboration serve different purposes.

Companies usually shared their internal standards in a supplier code of conduct. In the palm oil industry, supplier codes of conduct facilitated vertical collaboration between buyers and suppliers. Fifty-nine percent of the companies disclosed using a supplier code of conduct with terms related to palm oil sourcing (Figure 16.6). Codes of conduct were always given to the direct suppliers. Sometimes the companies asked their direct suppliers to communicate the same standards to the upstream sources. For example, McDonald's Supplier Code of Conduct explicitly stated that they expected their suppliers to hold the upstream supply chain to the same standards in the code (McDonald's, 2012).

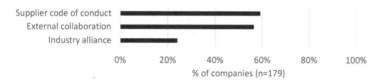

Figure 16.6 Percentage of companies using each type of collaboration approach

Horizontal collaboration with other companies in the same, or similar, stage in the supply chain can help companies increase their influence and power in the industry (Santos & Eisenhardt, 2005; Touboulic et al., 2014) and reduce the costs incurred for each firm to implement changes (Villena, 2019). In the palm oil industry, the Consumer Goods Forum (CGF) serves as a key industry alliance for hundreds of retailers and manufacturers, allowing them to push upstream suppliers for more accountability. In 2010, over 400 CGF companies committed to sourcing palm oil that does not contribute to deforestation by 2020 (Taylor, 2019). Although that commitment was not realized, CGF still serves as a platform for companies to work jointly with NGOs such as Proforest to train suppliers and promote transparency in the upstream supply chain (Consumer Goods Forum, 2021).

Lastly, external collaborations with academics, government, and NGOs help provide expertise and social connections beyond the functions of the businesses (Lambin et al., 2020; Thorlakson, 2018). Fifty-six percent of the companies employed external collaborations. Some of these companies used the help of NGOs to implement compliance instruments, such as supplier codes of conduct. In other cases, they leveraged these organizations' connections with businesses to facilitate their vertical and horizontal collaborations. A prime example of an outcome of a collaboration between companies and NGOs is RSPO, which was firmly supported by Unilever and World Wildlife Fund from the outset (Nesadurai, 2017). We observed that companies also relied on external collaborations to identify buyers' indirect upstream suppliers and train the suppliers to implement more sustainable practices. As in an earlier example, P&G relied on an education institute (Malaysia Institute for Supply Chain Innovation) and two NGOs to trace their upstream supply chain, assess smallholder farmers' sustainability performance, and improve the farmers' livelihoods (P&G, 2018).

COMPANY MANAGEMENT PROFILES

Our study identified four supply chain management approaches through an unsupervised machine learning classification. We first summed all companies' disclosed information according to the five components and applied principal component analysis.[3] We used the identified principal components as input to *k*-means clustering.[4] (See Lee and Bateman, 2019, for more detail regarding the applied classification method.)

The clustering results identified four company management profiles, which we labeled Highly Active, Active, Passive, and Inactive. We classified the 69 companies with no disclosures about palm oil sustainability as Inactive. Figure 16.7 shows the breakdown of commitments and practices in each management profile (omitting the Inactive companies). The percentage of companies with commitments and actions decreased from the Highly Active to the Passive management profiles.

Highly Active companies generally traced part of their supply chain and applied a wide range of external standards and compliance instruments. Most of them had grievance mechanisms (78 percent), provided training to their suppliers (78 percent), and invested in smallholder farmers (78 percent). Eighty-nine percent of the companies reported how far they could trace their supply chain (traceability scope) and 66 percent reported the percentage of traceable supplies. Highly Active companies adopted more compliance instruments and reported more information regarding their supply chain traceability than companies in other profiles. This result suggests that compliance instruments and supply chain tracing can complement each other.

Companies with Active management profiles did more than simply use certified palm oil. Most of them participated in RSPO as members and used formal compliance instruments such as approved suppliers (70 percent) and contract requirements (45 percent) to substantiate elaborated commitments to palm oil sustainability. They collaborated with external groups such as NGOs and governments (65 percent) to increase their supply chain visibility. However, their ability to apply other types of compliance instruments and to trace the supply chain was limited.

There are two types of companies in the Passive management profile. The first type of companies with this profile limited their commitments to reducing deforestation, conserving biodiversity, and protecting workers' rights. Most of them relied on RSPO membership and used RSPO-certified palm oil (49 and 51 percent, respectively). The second type, which accounts for 22 percent of the companies in this profile, chose to eliminate palm oil. Interestingly, they could be considered highly active because avoiding the use of palm oil can be costly. Not using palm oil was a point of product differentiation for many companies that employed this measure. The clustering algorithm classified these companies in the Passive management profile because they disclosed little information about the five components.

Commodity supply chain management and climate change 369

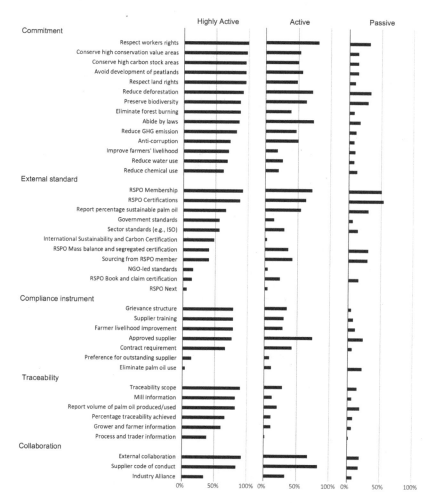

Note: n = 179.

Figure 16.7   *The percentage of companies that had stated specific commitments, implemented external standards, compliance instruments, and collaborations, and disclosed traceability information in each management profile*

In summary, highly active companies made comprehensive commitments and implemented almost all practices. Active companies complied with pressure to use more sustainable palm oil by buying certified products and publishing supplier codes of conduct, but they were not engaging their suppliers as readily as the highly active companies. Passive companies were minimally involved in sourcing sustainable palm oil, including a minority of companies avoiding palm oil in full.

370 *Handbook of business and climate change*

Factors Influencing Management Profiles

To understand how company characteristics and the context of headquarter (HQ) countries influenced their management profiles, we fitted multinomial logistic regression models[5] with different specifications of the independent variables shown in Table 16.1 to the dataset (Lee & Bateman, 2019). Of all the models we explored, Table 16.2 shows the estimation result of the model that fitted the data best. The Inactive company group is the reference alternative. All coefficients are significant at the 89 percent confidence level, and the McFadden adjusted pseudo R^2 is 0.34. The Global Reporting Initiative (GRI)[6] reporting standard serves as a control since companies that used the GRI disclosure guidelines were likely to disclose more information (Thorlakson et al., 2018).

Table 16.1 Description of the independent variables used for multinomial logit regression

Processor and Trader	Equals 1 if the company is a PT, 0 otherwise	Count = 50
Manufacturer	Equals 1 if the company is a manufacturer, 0 otherwise	Count = 112
Retailer	Equals 1 if the company is a retailer, 0 otherwise	Count = 86
High brand value	Equals 1 if company or its product is listed in the Reputation institute and Interbrand list of companies with high brand value, 0 otherwise	Count = 21
Public	Equals 1 if the company is listed as public by Bloomberg.com, 0 otherwise	Count = 132
Number of employees	The larger number of employees as disclosed in their documents or in D&B Hoovers	Average = $6.7 \times 10^4 \pm 1.7 \times 10^5$
Sustainability branding	Equals 1 if the company's disclosures include statements on doing good, brand image and leadership, and meeting demand for sustainable products, 0 otherwise	Count = 34
Stakeholder expectations	Equals 1 if the company's disclosures include statements about meeting stakeholders' expectations, 0 otherwise	Count = 122
Media attention to palm oil sustainability	Percentage of publications about palm oil sustainability, including the keywords "palm oil" and "orangutan", "illegal", "logging", "sustainable", "sustainability", "deforestation", "rainforest", "RSPO" or "biodiversity" out of publications that have the keyword "palm oil" published in the HQ country as recorded by Dow Jones Factiva from 2013 to 2017.	Average = 0.25 ± 0.10
Sustainability competitiveness of HQ country	Sustainable Global Competitiveness Index by World Economic Forum for 2014–2015. It measures countries' preservation of natural resources, quality of social capital, resource efficiency and intensity, wealth of intellectual capital and innovation. The score is divided by 100 so that it ranges from 0 to 1	Average = 0.54 ± 0.06
GRI reporting standard	Equals 1 if company discloses using GRI reporting standard, 0 otherwise	Count = 80

The estimated coefficients in Table 16.2 show the impact of the independent variable on the probability that a company would act with Highly Active, Active, and Passive management profiles. The first three coefficients in Table 16.2 are the constants estimated for each profile with respect to the inactive companies. They are all negative (−17.6, −8.94, and −4.97), implying that, all else being equal, companies would prefer to be inactive and not to disclose any information.

The fourth and fifth coefficients in Table 16.2 are those of the PT dummy variables with respect to Highly Active and Active management profiles. These coefficients being positive means that, on average, PTs were more likely to have Highly Active and Active management profiles than Passive and Inactive management profiles. The sixth coefficient compares the correlation between high brand value companies with High Active, Active, and Passive management profiles against the Inactive management profile. Given that the coefficient for the high brand value company is not specific to the first three management profiles in this best fit model, it means that high brand value companies were equally likely to use any of them. The sixth positive coefficient also indicates that high brand value companies were more likely to be active than non-high brand value companies.

Interpreting the remaining coefficients in the same way, public companies were more likely to be highly active and active than passive and inactive. Compared with companies with few employees, companies with many employees were more likely to be highly active than active, and more likely to be highly active and active than passive and inactive in managing their palm oil sustainability issues. Companies with sustainability branding were likely to use some sustainability practices and were not likely to have an Inactive management profile.

Table 16.2 The estimated coefficients of the multinomial logit regression models

Independent variables	Estimated coefficient	Standard error	p-value
Highly Active Profile (HAP) dummy variable	−17.6	3.28	0.00
Active Profile (AP) dummy variable	−8.94	2.11	0.00
Passive Profile (PP) dummy variable	−4.97	1.55	0.00
Processor and trader (specific to HAP)	6.26	1.45	0.00
Processor and trader (specific to AP)	2.20	1.04	0.03
High brand value	3.94	0.82	0.00
Public (specific to HAP and AP)	0.88	0.46	0.06
Log-number of employees (specific to HAP)	0.48	0.19	0.01
Log-number of employees (specific to AP)	0.21	0.12	0.08
Sustainability branding	1.55	0.69	0.05
Stakeholder expectations (specific to HAP)	1.97	0.84	0.02
Stakeholder expectations (specific to AP)	1.70	0.42	0.00
Media attention (specific to HAP)	12.9	4.19	0.00
Media attention (specific to AP)	7.68	2.47	0.00
Media attention (specific to PP)	5.49	2.22	0.01
Sustainability competitiveness of HQ country	0.65	0.31	0.04
GRI reporting standard (specific to HAP)	2.75	0.70	0.00
GRI reporting standard (specific to AP)	0.70	0.44	0.11

Interestingly, the media attention coefficients are increasingly higher with increasingly active management profiles. This trend suggests that significant media attention on palm oil issues motivated companies to engage actively. To clarify this point, we used scenario analysis to compare the effect of media attention focused on palm oil to the effect of regulation, which was estimated to have the same coefficient for all management profiles. The following scenarios compare the influence of focused media attention and HQ country sustainability competitiveness on private or public companies that used each management profile. Table 16.3 shows that a 25 percent increase in media attention impacts the probability that companies apply Highly Active or Active management profiles more than a 25 percent increase in sustainability competitiveness. While this result is correlative and not causal, it suggests that media attention is a big push for companies to implement a full range of practices. This insight enables future research to explore the impact of media attention versus public awareness on companies' efforts to address supply chain sustainability.

Table 16.3   Results of scenario analyses where media attention and regulation strictness are increased by 25 percent

	Media attention increases by 25%				Sustainability competitiveness increases by 25%			
	Average probability of using each management profile							
	Highly Active	Active	Passive	Inactive	Highly Active	Active	Passive	Inactive
Public								
Before	0.29	0.36	0.14	0.21	0.29	0.36	0.14	0.21
After	0.37	0.43	0.15	0.05	0.29	0.40	0.17	0.13
Change	+0.08	+0.07	+0.01	−0.16	0	+0.04	+0.04	−0.08
Private								
Before	0.11	0.18	0.35	0.36	0.11	0.18	0.35	0.36
After	0.16	0.26	0.44	0.14	0.12	0.21	0.45	0.22
Change	+0.05	+0.08	+0.09	−0.22	+0.01	+0.03	+0.10	−0.14

CONCLUSION

This study explored the ways companies mitigate the climate impact of their palm oil supply chains. First, we identified and discussed five components in companies' management of supply chain sustainability, including commitment, external standards, compliance instruments, traceability, and collaboration. These components are interconnected and not mutually exclusive. We observed that commitments were substantiated by implementing external standards, compliance instruments, and supply chain tracing. Collaborations helped facilitate the first three components.

Based on companies' commitments and practices, we found that companies were more focused on protecting existing forests, ecosystems, workers, and communities and less focused on greenhouse gas accounting. Companies in the palm oil industry sought to address social and environmental issues. Respecting workers' rights and reducing deforestation were the two most common goals. Addressing social issues can be synergistic to reducing climate impact. For example, improving farmers' productivity reduces GHG emissions per volume of palm oil produced (Walker et al., 2018). We inferred that companies mainly committed to actions that would help mitigate climate change more directly (e.g., reducing deforestation) than GHG emission accounting.

RSPO certification, supplier codes of conduct, external partnerships, and supplier approval were the most popular practices. The high popularity of these practices is echoed in other industries (Bateman et al., 2020; Ernst & Young & United Nations Global Compact, 2016). The popularity of RSPO certification may be because its sustainability criteria are focused on palm oil and because its criteria are more robust than other standards, as evaluated by NGOs (McInnes, 2017). Companies' choice of action seems to be guided by authoritative organizations and the actions of other companies.

Second, we identified four supply chain management profiles: Highly Active, Active, Passive, and Inactive. Highly active companies were likely to implement all five supply chain management components. Compared with highly active companies, active companies mostly fell short in practices that required a deep understanding of the supply chain. These practices include, for example, supply chain tracing, providing a grievance structure throughout their supply chain, and improving farmers' livelihoods. Passive companies mainly relied on external standards, including RSPO membership and certification.

We found that, on average, PT, public, high brand value companies, and companies with more employees were more likely to be highly active or active in managing their palm oil sustainability issues. Most of these results were expected, since public, high brand value, and large companies tend to receive more external pressures and have more resources to act than their smaller business counterparts (Gibbon et al., 2008). One surprising finding is that PTs, typically less visible to the public than manufacturers or retailers, were more likely to act. Lee (2021) suggested that the PTs' unexpected behaviors were due to stakeholder pressures and because they had the competitive impetus and resources to do more.

The pressure to disclose supply chain sustainability efforts is an impetus for companies to act (Bateman et al., 2021). The authors recommend that active management of supply chain sustainability can help companies in three ways.

1. *Precautionary action*: Regulation for disclosure and due diligence around raw material supply chains is increasing over time. Companies with existing due diligence and public disclosure around raw materials will be better placed to respond as regulation grows in scope.
2. *Response to stakeholder pressure*: Companies that know and manage their highly scrutinized commodities' supply chains are better able to respond to pressures from their stakeholders. The stakeholders include end-consumer, business customers, investors, NGOs, governmental entities, and regulatory actors. Consistent disclosures of progress achieved in managing supply chains also help build stakeholder trust.
3. *Competitive advantage*: Companies that are early movers, implementing commitments and actions towards supply chain sustainability and disclosing their efforts can enhance their brand value and increase their competitive advantage in their industry.

Companies' approaches to sustainable supply chain management will evolve over the next few years as companies gain experience. This study shows that companies respond to public scrutiny and societal pressure to mitigate the environmental and social impacts of palm oil production and consumption. The type of management approaches that companies use – how those approaches might take form and what they may mean for future supply chain sustainability – could also apply to other commodities that substantially impact climate change.

A few factors limited our study. Similar to other large-scale studies on sustainable sourcing practices and disclosure (Akhavan & Beckmann, 2016; Bager & Lambin, 2020; Thorlakson et al., 2018), we were limited to companies' public disclosures to derive the information that forms the basis for our analysis. Our results are not representative of the entire palm oil industry because our sample focused on companies likely to source palm oil sustainably. Lastly, our analysis is based on correlations, and our findings do not indicate causal relationships.

Our research identified areas for further study. More detail could be collected regarding the geographical and temporal scopes to which companies implement their commitments and practices. For example, some of the commitments were limited to selected geographical areas. Another potential avenue for future research would be to collect data over multiple periods and analyze changes in companies' practices over time. Temporal analysis can examine causal relationships by identifying factors that influence sustainability practices. Lastly, future research can explore the impact of media attention versus public awareness on companies' efforts to address supply chain sustainability.

NOTES

1. Peatland is a type of wetland with organic matter that is not completely decomposed.
2. According to Dodson et al. (2020), there were 6.8 million hectares of oil palm plantations in Indonesia in 2018, 742 mills in 2015, and 78 refineries (operated by producers) in 2017.
3. Principle component analysis is a computation method to reduce the dimensionality of a collection of data points. It transforms the data through eigen decomposition to a new coordinate system with a set of orthogonal coordinates (i.e., components) to capture maximum variance in the data (Hastie et al., 2009).
4. k-means clustering is an algorithm that partitions n observations in k clusters such that each observation falls into its nearest cluster center (Hastie et al., 2009).
5. Multinomial logistics regression is a statistical method to estimate the relationship between more than two categorical dependent variables and a set of independent variables (Ben-Akiva & Lerman, 1985).
6. The Global Reporting Initiative (GRI) is an international, independent standards organization that provides a reporting framework for companies to report on their environmental and social impacts.

REFERENCES

Akhavan, R. M., & Beckmann, M. (2016). A configuration of sustainable sourcing and supply management strategies. *Journal of Purchasing and Supply Management, 25*, 137–151.

Azhar, B., Saadun, N., Prideaux, M., & Lindenmayer, D. B. (2017). The global palm oil sector must change to save biodiversity and improve food security in the tropics. *Journal of Environmental Management, 203*, 457–466. https://doi.org/10.1016/j.jenvman.2017.08.021

Bager, S. L., & Lambin, E. F. (2020). Sustainability strategies by companies in the global coffee sector. *Business Strategy and the Environment*, July, 1–16. https://doi.org/10.1002/bse.2596

Bai, C., & Sarkis, J. (2014). Determining and applying sustainable supplier key performance indicators. *Supply Chain Management: An International Journal, 19*, 275–291.

Baron, V., Rival, A., & Marichal, R. (2017, June 8). *No, Palm Oil is Not Responsible for 40% of Global Deforestation*. Phys Org. https://phys.org/news/2017-06-palm-oil-responsible-global-deforestation.html

Bateman, A., Sheffi, Y., Caplice, C., D'Ambrosio, L., Allegue, L., Barrington, A., Collard, C., Lee, Y. J., & Greene, S. (2020). *State of Supply Chain Sustainability 2020*. MIT Center for Transportation & Logistics and Council of Supply Chain Management Professionals.

Bateman, A., Betts, K., Cottrill, K., Pang, J., & Deshpande, A. S. (2021). *State of Supply Chain Sustainability 2021*. MIT Center for Transportation & Logistics and Council of Supply Chain Management Professionals.

Ben-Akiva, M., & Lerman, S. (1985). *Discrete Choice Analysis*. MIT Press.

Berg, B., & Lune, H. (2012). An introduction to content analysis. In J. J. C. Campanella & D. Musslewhite (Eds.), *Qualitative Research Methods for the Social Sciences* (8th ed., pp. 349–385). Pearson Education Inc.

BSR, & United Nations Global Compact. (2014). *A Guide to Traceability: A Practical Approach to Advance Sustainability in Global Supply Chains*. United Nations Global Compact. https://www.unglobalcompact.org/library/791

Carlson, K. M., Curran, L. M., Asner, G. P., Pittman, A. M., Trigg, S. N., & Adeney, J. M. (2013). Carbon emissions from forest conversion by Kalimantan oil palm plantations. *Nature Climate Change, 3*, 283–288. https://doi.org/10.1038/nclimate1702

Carlson, K. M., Heilmayr, R., Gibbs, H. K., Noojipady, P., Burns, D. N., Morton, D. C., Walker, N. F., Paoli, G. D., & Kremen, C. (2017). Effect of oil palm sustainability certification on deforestation and fire in Indonesia. *Proceedings of the National Academy of Sciences, 115*(1), 201704728. https://doi.org/10.1073/pnas.1704728114

CERES. (2017). *Reporting Guidance for Responsible Palm.* January, 16. CERES.

Consumer Goods Forum. (2015). *Sustainable Palm Oil Sourcing Guidelines.* Consumer Goods Forum.

Consumer Goods Forum. (2021). *CGF Forest Positive Coalition of Action* (Issue March). Consumer Goods Forum.

Cooper, H. v., Evers, S., Aplin, P., Crout, N., Dahalan, M. P. bin, & Sjogersten, S. (2020). Greenhouse gas emissions resulting from conversion of peat swamp forest to oil palm plantation. *Nature Communications, 11*(1). https://doi.org/10.1038/s41467-020-14298-w

Croom, S., Vidal, N., Spetic, W., Marshall, D., & McCarthy, L. (2018). Impact of social sustainability orientation and supply chain practices on operational performance. *International Journal of Operations and Production Management, 38*(12), 2344–2366. https://doi.org/10.1108/IJOPM-03-2017-0180

de Marchi, V., di Maria, E., & Ponte, S. (2013). The greening of global value chains: Insights from the furniture industry. *Competition and Change, 17*(4), 299–318. https://doi.org/10.1179/1024529413Z.00000000040

Deloitte. (2017). Global powers of retailing 2017: The art and science of customers. In *Deloitte.* https://www2.deloitte.com/content/dam/Deloitte/global/Documents/consumer-industrial-products/gx-cip-2017-global-powers-of-retailing.pdf

Dodson, A., Salisbury, C., & Spencer, E. (2020). *Palm Oil Crushers and Refiners: Managing Deforestation Risk through a Supply Chain Bottleneck.* https://www.spott.org/wp-content/uploads/sites/3/dlm_uploads/2020/08/Palm-oil-crushers-and-managing-deforestation-risk-through-supply-chain-bottleneck-SPOTT-2020.pdf

Doorey, D. J. (2011). *The Transparent Supply Chain: from Resistance to Implementation at Nike and Levi-Strauss.* https://doi.org/10.1007/s10551-011-0882-1

Egels-Zandén, N., Hulthén, K., & Wulff, G. (2015). Trade-offs in supply chain transparency: The case of Nudie Jeans Co. *Journal of Cleaner Production, 107*, 95–104. https://doi.org/10.1016/j.jclepro.2014.04.074

El-Fegoun, M. A. B. C. (2015). *Upstream Supply Chain Analysis for Oil Palm.* http://hdl.handle.net/1721.1/102233

Ernst & Young, & United Nations Global Compact. (2016). *The State of Sustainable Supply Chains.* http://www.unglobalcompact.org.au/new/wp-content/uploads/2016/09/UN-GC-EY-building-responsible-and-resilient-supply-chains.pdf

Ethical Consumer. (2015). *Special Report on Palm Oil.* Ethical Consumer.

Forbes. (2017). *Global 2000: The World's Largest Public Companies.* https://www.forbes.com/global2000/

Furlong, H. (2016). *Unilever, Kellogg, Mars Drop Major Palm Oil Supplier after RSPO Revokes its Certification.* Sustainable Brands. https://sustainablebrands.com/read/supply-chain/unilever-kellogg-mars-drop-major-palm-oil-supplier-after-rspo-revokes-its-certification

Gibbon, P., Bair, J., & Ponte, S. (2008). Governing global value chains: An introduction. *Economy and Society, 37*(3), 315–338. https://doi.org/10.1080/03085140802172656

Gimenez, C., & Sierra, V. (2013). Sustainable supply chains: Governance mechanisms to greening suppliers. *Journal of Business Ethics, 116*(1), 189–203. https://doi.org/10.1007/s10551-012-1458-4

Gold, S., Seuring, S., & Beske, P. (2010). Sustainable supply chain management and inter-organizational resources: A literature review. *Corporate Social Responsibility and Environmental Management, 17*(4), 230–245. https://doi.org/10.1002/csr.207

Greenpeace. (2016). *Cutting Deforestation Out of the Palm Oil Supply Chain.* Greenpeace.

Gunawan, I., Vanany, I., & Widodo, E. (2020). Typical traceability barriers in the Indonesian vegetable oil industry. *British Food Journal, 123*, 1223–1248.

Hashim, K., Tahiruddin, S., & Ahmad Jaril Asis. (2012). Palm and palm kernel oil production and processing in Malaysia and Indonesia. In O.-M. Lai, C.-P. Tan, & C. C. Akoh (Eds.), *Palm Oil: Production, Processing, Characterization, and Uses* (pp. 235–250). AOCS Press.

Hastie, T., Tibshirani, R., & Friedman, J. (2009). *The Elements of Statistical Learning* (2nd Edition). Springer.

Hsieh, H.-F., & Shannon, S. E. (2005). Three approaches to qualitative content analysis. *Qualitative Health Research*, *15*(9), 1277–1288. https://doi.org/10.1177/1049732305276687

Jiang, B. (2009). The effects of interorganizational governance on supplier's compliance with SCC: An empirical examination of compliant and non-compliant suppliers. *Journal of Operations Management*, *27*(4), 267–280. https://doi.org/10.1016/j.jom.2008.09.005

Lambin, E. F., Kim, H., Leape, J., & Lee, K. (2020). Scaling up solutions for a sustainability transition. *One Earth*, *3*(1), 89–96. https://doi.org/10.1016/j.oneear.2020.06.010

Lee, Y. (2021). *Sustainable Agri-Food Supply Chains: Consumer Demand and Company Sourcing Practices*. Massachusetts Institute of Technology.

Lee, Y., & Bateman, A. (2019). *Corporate Supply Chain Disclosures and Factors Determining the Disclosure Approaches: A Palm Oil Case Study* (No. 2019-mitscale-ctl-02; SCALE Working Paper Series).

Lee, Y., & Bateman, A. (2021). The competitiveness of fair trade and organic versus conventional coffee based on consumer panel data. *Ecological Economics*, *184*. https://doi.org/10.1016/j.ecolecon.2021.106986

Leegwater, M., & van Duijn, G. (2012). Traceability of RSPO-certified sustainable palm oil. In *Palm Oil: Production, Processing, Characterization, and Uses* (pp. 713–736). AOCS Press. https://doi.org/10.1016/B978-0-9818936-9-3.50027-7

Lyons-White, J., & Knight, A. T. (2018). Palm oil supply chain complexity impedes implementation of corporate no-deforestation commitments. *Global Environmental Change*, *50*, 303–313. https://doi.org/10.1016/j.gloenvcha.2018.04.012

Maitar, B., & Skar, R. (2010). Burger King cancels palm oil contract with rainforest destroyer Sinar Mas. *Greenpeace*. https://www.greenpeace.org/usa/news/burger-king-cancels-palm-oil-contract-with-rainforest-destroyer-sinar-mas/

McDonald's. (2012). *McDonald's Supplier Code of Conduct*. McDonald's.

McInnes, A. (2017). A comparison of leading palm oil certification standards. In *Forest Peoples Programme*, https://www.forestpeoples.org/sites/default/files/documents/Palm%20Oil%20Certification%20Standards_lowres_spreads.pdf

Melot, C., & Delabre, I. (2017). *From Disclosure to Engagement - A Guide to the SPOTT Indicator*. https://www.spott.org/wp-content/uploads/sites/3/2017/05/From-disclosure-to-engagement_A-guide-to-the-SPOTT-indicator-framework.pdf

Nash, J. (2017). Business risks from deforestation. In *Ceres Case Study Series*. https://doi.org/10.1021/cen-v081n041.p002

Nesadurai, H. E. S. (2017). New constellations of social power: States and transnational private governance of palm oil sustainability in Southeast Asia. *Journal of Contemporary Asia*, *48*, 204–229. https://doi.org/10.1080/00472336.2017.1390145

Newman, K. (2017). *A List of Palm Oil Free Products – How You Can Reduce Your Impact*. Travel for Difference.

Newton, P., Agrawal, A., & Wollenberg, L. (2013). Enhancing the sustainability of commodity supply chains in tropical forest and agricultural landscapes. *Global Environmental Change*, *23*(6), 1761–1772.

O'Rourke, D. (2014). The science of sustainable supply chains. *Science*, *344*(6188), 1124–1128.

Pagell, M., Wu, Z., & Wasserman, M. E. (2010). Thinking differently about purchasing portfolios: An assessment of sustainable sourcing. *Journal of Supply Chain Management*, *46*(1), 57–73. https://doi.org/10.1111/j.1745-493X.2009.03186.x

P&G. (2018). *2018 Citizenship Report*. Proctor & Gamble.

Pilbeam, C., Alvarez, G., & Wilson, H. (2012). The governance of supply networks: A systematic literature review. *Supply Chain Management: An International Journal*, *17*(4), 358–376. https://doi.org/10.1108/13598541211246512

Porteous, A. H., Rammohan, S. v, & Lee, H. L. (2015). Carrots or sticks? Improving social and environmental compliance at suppliers through incentives and penalties. *Production and Operations Management*, *24*(9), 1402–1413. https://doi.org/10.1111/poms.12376

Poynton, S. (2015). *Beyond certification*. https://library.oapen.org/bitstream/handle/20.500.12657/25056/9781910174531_text.pdf?sequence=1

Rainforest Alliance. (2021). *New Approach to Palm Sustainability: Phasing Out of Palm Certification in the 2020 Program.* https://www.rainforest-alliance.org/business/certification/new-approach-to-palm-sustainability-phasing-out-of-palm-certification-in-the-2020-program/

Reuters Staff. (2009). *RPT–Unilever Cuts Palm Oil Supplier Ties After Report.* Reuters. https://www.reuters.com/article/unilever/rpt-unilever-cuts-palm-oil-supplier-ties-after-report-idUSGEE5BA0Z320091211?edition-redirect=in

Ritchie, H., & Roser, M. (2020). *CO₂ and Greenhouse Gas Emissions.* Published Online at OurWorldInData.Org.

RSPO. (2017). *Code of Conduct for Members of the Roundtable on Sustainable Palm Oil* (November Issue). https://rspo.org/resources/membership

Santos, F. M., & Eisenhardt, K. M. (2005). Organizational boundaries and theories of organization. *Organization Science, 16*(5), 491–508. https://doi.org/10.1287/orsc.1050.0152

Sheffi, Y., & Blanco, E. (2018). *Balancing Green: When to Embrace Sustainability in a Business (and When Not to).* The MIT Press.

Starbucks. (2015). *Making Coffee the World's First Sustainably Sourced Agricultural Product.* https://stories.starbucks.com/stories/2015/making-coffee-the-worlds-first-sustainably-sourced-agriculture-product/

Statista. (2021). *Palm Oil Consumption Worldwide from 2015/2016 to 2020/2021.* Statista.com

Taylor, M. (2019). *Factbox: Can Global Corporations Meet 2020 No-Deforestation Pledge?* Reuters. https://www.reuters.com/article/us-global-palmoil-factbox-trfn/factbox-can-global-corporations-meet-2020-no-deforestation-pledge-idUSKBN1XP013

Thorlakson, T. (2018). A move beyond sustainability certification: The evolution of the chocolate industry's sustainable sourcing practices. *Business Strategy and the Environment, 27*(8), 1653–1665. https://doi.org/10.1002/bse.2230

Thorlakson, T., de Zegher, J. F., & Lambin, E. F. (2018). Companies' contribution to sustainability through global supply chains. *Proceedings of the National Academy of Sciences,* 201716695. https://doi.org/10.1073/pnas.1716695115

Touboulic, A., Chicksand, D., & Walker, H. (2014). Managing imbalanced supply chain relationships for sustainability: A power perspective. *Decision Sciences, 45*(4), 577–619. https://doi.org/10.1111/deci.12087

Unilever. (2021). *Sustainable Palm Oil.* https://www.unilever.com/planet-and-society/protect-and-regenerate-nature/sustainable-palm-oil/

United States Department of Agriculture. (2018). *Production, Supply, and Distribution.* https://www.fas.usda.gov/

Villena, V. H. (2019). The missing link? The strategic role of procurement in building sustainable supply networks. *Production and Operations Management, 28*(5), 1149–1172. https://doi.org/10.1111/poms.12980

Walker, S., McMurray, A., Rinaldy, F., Brown, K., & Karsiwulan, D. (2018). *Compilation of Best Management Practices to Reduce Total Emissions from Palm Oil Production* (November), Roundtable on Sustainable Palm Oil: https://winrock.org/wp-content/uploads/2019/01/RSPO-Compilation-of-Best-Management-Practices-to-Reduce-Total-Emission-from-Palm-Oil-Production-English-published-on-RSPO-website.pdf

World Wide Fund for Nature. (2016). *Palm Oil Buyers Scorecard: Measuring the Progress of Palm Oil Buyers.* WWF.

World Wide Fund for Nature. (2017). *Palm Oil Buyers Scorecard – Malaysia and Singapore.* WWF.

World Wildlife Fund. (2021). *Palm Oil.* https://www.worldwildlife.org/industries/palm-oil

Zimmer, K., Fröhling, M., & Schultmann, F. (2016). Sustainable supplier management–A review of models supporting sustainable supplier selection, monitoring and development. *International Journal of Production Research, 54*(5), 1412-1442.

Zhu, Q., & Sarkis, J. (2007). The moderating effects of institutional pressures on emergent green supply chain practices and performance. *International Journal of Production Research, 45*(18–19), 4333–4355. https://doi.org/10.1080/00207540701440345

Zoological Society London. (2017). *Hidden Land, Hidden Risks?* (March Issue). spott.org/palm-oil/landbank

APPENDIX

Table 16.A1 Information that was collected from companies' reports and websites classified by the five supply chain management components

Commitment	Compliance instrument
• Respect workers' rights	• Having approved suppliers
• Conserve high conservation value areas	• Having preferred suppliers
• Conserve high carbon stock areas	• Specifying criteria to suspend or exclude suppliers
• Avoid development of peatlands	• Having grievance structure
• Respect land rights	• Providing (non-smallholder farmers) supplier trainings
• Reduce deforestation	• Investing in smallholders
• Preserve biodiversity	• Eliminating palm oil use in products
• Eliminate forest burning	
• Abide by laws	
• Reduce GHG emission	
• Anti-corruption	
• Improve farmers' livelihood	
• Reduce water use	
• Reduce chemical use	
External standard	**Traceability**
• RSPO membership	• Total volume of palm oil product produced/traded/used is reported
• Producing, selling, or using RSPO certified palm oil	• Furthest extent of the supply chain traceable is reported
• Producing, selling, or using RSPO-Next palm oil	• Percentage of supply traceable to the furthest extent of the supply chain is reported
• Using palm oil sourced from RSPO member	• Name of processor and/or trader company
• Percentage of palm oil products handled/traded/processed that are RSPO-certified	• Information of mills that is enough to identify their locations, e.g., name or maps or coordinates
• Whether book-and-claim is one of the dominant supply chain models	• Information of plantations that is enough to identify their locations, e.g., name or maps or coordinates
• Whether mass balance or segregated is one of the dominant supply chain models	
• Producing, selling, or using ISPO and/or MSPO certified palm oil	
• Producing, selling, or using ISCC certified palm oil	
• Producing, selling, or using NGO-Led certified palm oil	
• Being or sourcing from ISO certified sources	
	Collaboration
	• Industry alliance
	• External collaboration: Government, NGO, and academic engagement
	• Supplier code of conduct

17. Carbon pricing
Robert G. Hansen

INTRODUCTION

Carbon pricing is a classic market-based response to the negative externality associated with carbon emissions. If the marginal social cost of carbon emissions is, say, $60 per tonne CO_2 equivalent (tCO_2e),[1] that cost should be part of producers' and consumers' decision-making, if the goal is to promote economic efficiency.[2] To make that happen – to internalize the externality – the public policy remedy is to simply set a carbon price at $60/t$CO_2$e, impose it as an actual cost on emitters of carbon (or set an appropriate cap-and-trade mechanism), and let prices throughout the economy adjust. Activities that generate carbon emissions will become more costly and producers and consumers will adjust demands accordingly. Firms that emit high levels of carbon will find it profitable to engage in (costly) emissions reductions; in carbon-intensive industries, the new equilibrium will yield a smaller size of the industry, implying an exit of resources from such industries. Consumers will be induced to shift their demands to less carbon-intensive goods and services. Research and development of new production and emissions control technologies will be stimulated. With the exception of a possible effect on carbon capture activities, a carbon price per se does nothing to reduce the impact of emissions that were made *before* the pricing policy began, but incentives from that point forward are efficient.

This chapter delves deeper into carbon pricing, with a focus on what is known as *internal carbon pricing*: the practice by companies of using a carbon price within the company, even when public policy has not created an economy-wide carbon price or even required firms to reduce emissions. The next section of the chapter begins with a brief review of the current state of carbon pricing, covering both *external* carbon pricing, that is, mandatory carbon prices that are imposed for an entire economy or region via some form of public policy, as well as the incidence of *internal* carbon pricing within a company on a voluntary basis.[3] The third section turns to the theory behind external carbon pricing and the implications for industries and economies. A system of external carbon pricing is important to understand before moving on to internal carbon pricing, as it sets the standard for efficiency and for the adjustments that will occur in firms and industries, in both the short run and long run. The fourth section introduces internal carbon pricing, reviewing the rationale for voluntarily implementing an internal carbon pricing policy and discussing ways in which carbon prices are used within firms – in particular, which corporate decisions and actions might be affected. This section examines whether the necessary conditions for economic efficiency, as developed for external carbon pricing, are likely to occur under internal carbon pricing. It also considers how internal carbon prices interact with an existing, possibly imperfect, regulatory regime for carbon emissions. While the parallelism between national or regional carbon pricing and internal carbon pricing is intuitively attractive, there are important differences, with implications for efficiency. The fifth section discusses additional evidence on which firms use internal carbon pricing, how those prices are used, and on usage across time, countries, and industries. The final section wraps up with a quick summary and some parting thoughts on the role of internal carbon pricing in what is increasingly a "patchwork quilt" of carbon regulations.[4]

THE CURRENT STATE OF CARBON PRICING INITIATIVES, BOTH EXTERNAL AND INTERNAL

Prevalence of External Carbon Pricing across the Globe

How prevalent across the globe is external carbon pricing, implemented through either a direct carbon tax or a cap-and-trade system? The World Bank's yearly report, *The State and Trends of Carbon Pricing*, and the Bank's dashboard that tracks carbon pricing developments on almost a real-time basis, provide coverage data.[5] For 2021, the Bank reports 64 "carbon pricing instruments" in operation across the globe; instruments are either explicit carbon pricing policies or emissions trading regimes that result in a carbon price.[6] These 64 instruments cover 21.5 percent of all global greenhouse gas emissions. This is an increase from 58 instruments covering 15.1 percent in 2020, with most of the increase arising from China's implementation in 2021 of its emissions trading scheme covering over 2,000 entities in power generation. Even more substantially, in 2011 there were only 21 regimes covering a little over 5 of the global emissions at that time. Figure 17.1 shows the distribution of carbon prices in some of these 64 regimes for 2021.[7] The prices range from a low of $2.61/tCO$_2$e in Japan and $3.18/tCO$_2$e in Mexico to $101.47/tCO$_2$e in Switzerland and $137/tCO$_2$e in Sweden. A majority of prices in the 64 regimes, however, remain below the "carbon price corridor," that range of carbon prices recommended by The World Bank to meet the 2°C goal of the Paris Agreement (the range is $40 to $80/tCO$_2$e). Thus, while there is significant operation of external carbon pricing regimes, and rapid growth in that coverage, carbon prices are low relative to target levels and the variance in prices across countries and regions remains high. There is no de facto global carbon price.

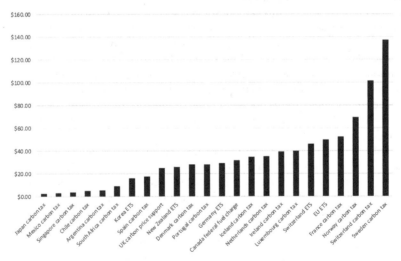

Note: Taxes across countries are not directly comparable due to differences in coverages, exemption, etc. This is an adaptation of an original work by The World Bank. Responsibility for the views and opinions expressed in the adaptation rests solely with the author of the adaptation and are not endorsed by The World Bank.
Source: The World Bank (2021). "State and trends of carbon pricing 2021" (May), World Bank, Washington, DC. Doi: 10.1596/978-1-4648-1728-1. License: Creative Commons Attribution CC BY 3.0 IGO.

Figure 17.1 National and regional carbon pricing, 2021

Prevalence of Internal Carbon Pricing at Companies

Internal carbon pricing is the use by corporations of an assumed price on carbon emissions to assess the economic value of those carbon emissions and to aid in the company's control or management of emissions and related risks and opportunities. Voluntary measurement and disclosure of a company's emissions – whether Scope 1, 2 or 3 – is one thing, but assessing the economic value of those emissions by assigning a price, and using that economic value calculation to affect business decisions, takes voluntary corporate action on environmental issues to another, and potentially more meaningful, level. The extent of the practice might be surprising: according to the most recent survey data from the Carbon Disclosure Project (CDP),[8] 853 companies currently use internal carbon pricing and another 1,159 plan to implement an internal pricing regime within two years; these numbers are out of a total of 5,900 companies responding to CDP's carbon pricing questions. Most companies use a shadow carbon price, meaning that no actual fee is assessed within the corporation. The most-cited objective in using internal carbon pricing is to "drive low-carbon investment." The median internal price as reported by CDP is $25/tCO$_2$e. Figure 17.2 shows the percentage of companies using or planning to use internal carbon pricing (within two years), by industry, for 2018–2020. Not surprisingly, the two industries with the greatest current/planned use are power and fossil fuels. Financial services companies are in third place, but with a high growth rate.

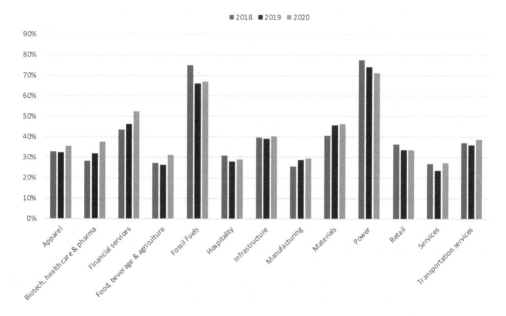

Source: Adapted from Carbon Disclosure Project (2021).

Figure 17.2 Internal carbon pricing by industry, 2018–2020. Percentage of firms reporting that they do or plan to have internal carbon pricing in place within two years

For a timely example of a company adopting internal carbon pricing, at the end of COP26 in 2021, Volvo announced that it would henceforth use an internal carbon price of 1,000 SEK (around 116 USD) on all of its worldwide operations, and that every future car project will undergo a "sustainability sense-check" using that price. The intent is to "…ensure that each car model would be profitable even under a strict carbon pricing scheme, steering project-sourcing and manufacturing-related decisions towards the most sustainable option on the table" (Volvo Cars, 2021).

The fifth section explores additional evidence on which companies use internal carbon pricing as well as why and how they do it. That evidence is better interpreted with an understanding of the theory behind carbon pricing, the subject of the next section.

ECONOMIC THEORY BEHIND CARBON PRICING

Since at least the time of Pigou (1920), economists have supported the idea of pricing externalities to create the conditions under which a market-based economy will yield efficient outcomes. An externality is simply a cost (or benefit) that is borne by society broadly but not borne directly by firms. As such, firms will fail to incorporate the full societal cost of their actions in their decisions on pricing, output, investment, etc., and those decisions will therefore generally be inefficient. With an external cost, the general outcome is that prices are too low and output too high and the reverse for an external benefit. Since Pigou, economists have further developed not only the general theory of pricing externalities but a massive literature on the use of pricing to address the specific problems of carbon emissions and climate change.[9]

In the context of carbon and climate, firms that through their actions cause the release of carbon into the atmosphere are contributing to the societal costs of climate change. They emit too much carbon, and do too little mitigation, as they don't bear the costs. Before we put all the blame onto corporations, we should of course note that consumers in the end bear not only the costs of the externality but they also enjoy the benefits: lower prices and higher quantities of the pollution-producing goods and services. Consumers consume too much of the carbon-intensive goods and services, so that the net value of the last units is negative. Furthermore, it is possible that some consumer groups enjoy the benefits of larger production and lower prices, while other groups bear mostly the costs from climate change.

The intuition of how pricing carbon emissions solves the externality problem is straightforward, but that intuition covers up much of the complexity of the changes in industry equilibrium that need to take place once a pricing regime is instituted. This section reviews the microeconomics of pricing carbon, with a focus on industry-level equilibrium – what happens to prices, outputs, capital stocks, firm profitability, the number of firms – after the levy of an efficient carbon price. This analysis will in turn help to evaluate the likely outcomes and efficiency if firms voluntarily implement their own internal carbon pricing policies, a topic covered in the fourth section.

For simplicity and analytical clarity, I consider a carbon pricing regime created via a direct carbon price, as opposed to a cap-and-trade or emissions trading system. International border adjustment issues are sidestepped by considering only a closed economy, and the timing aspects of the carbon/climate problem are also ignored (generally an optimal dynamic policy will have increasing-over-time carbon prices). Last, subsidy or credit policy for carbon-capture will not be explicitly considered, although that idea arises naturally.

The Efficient Carbon Price and Related Level of Mitigation

Figure 17.3 illustrates the economics behind an efficient carbon price, which equates the marginal cost and marginal benefits of carbon emissions. Carbon emissions for the entire economy are measured on the horizontal axis of Figure 17.3, with the current unregulated level of emissions being noted as Q_C. The marginal cost of emissions, labeled MC, shows the cost (damages) to society from an *additional* ton of carbon released to the atmosphere. The marginal cost of carbon is a relatively straightforward concept: the discounted present value of all the incremental costs imposed from an additional ton of carbon. Its measurement, however, is perhaps not surprisingly difficult, fraught with uncertainty, and heavily debated. Auffhammer (2018) gives a survey of methods for assessing the economic damages from carbon; one thing to note is that most attempts at estimating damages assume a constant marginal cost rather than a rising function, as in Figure 17.3. Auffhammer (2018) reports several recent estimates of the marginal cost, with a wide range from close to zero to almost $100/ton CO_2 (approximately $110/t$CO_2$e). A key parameter in estimating marginal cost is the discount rate, as costs accrue over time, and this accounts for much of the variation in cost estimates. The Biden administration in the US now uses $51/t$CO_2$e. In a new paper, Carleton and Greenstone (2021) state that updating the United States' official estimate of the social cost of carbon using a 2 percent discount rate (and no other updates) yields an estimate of $125/ton or $137.50/t$CO_2$e.

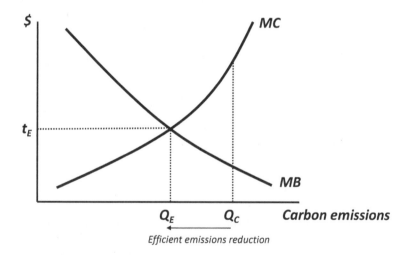

Figure 17.3   An efficient carbon price

The same difficulties in estimation hold true for the second curve of Figure 17.3, MB, which is the marginal benefit of emissions. The marginal benefit of emissions is logically the same as the marginal cost of reducing emissions, for if an entity is allowed to emit, that entity avoids incurring the marginal costs of not emitting. The marginal cost of reducing emissions can include the cost of altering production processes, including the type of fuel used, but it also includes the cost of lost output if firms produce less in order to reduce emissions.[10] Perhaps the best-known estimate of the marginal cost of emissions reduction is from a McKinsey & Co. study authored by Enkvist, Naucler and Rosander (2007). A more recent update is provided by

Gillingham and Stock (2018). These authors note the difference between static and dynamic costs and note some of the cost reductions that have been seen in low-carbon technologies such as solar panels and electric vehicles. Overall, however, both studies support a rising marginal cost of emissions reduction curve as in Figure 17.3, with some reductions coming in only at quite high prices (for example, weatherization assistance programs for housing is estimated to reduce CO_2 at a cost of $350/ton, 2017 dollars).

At Q_C, the marginal cost of emissions exceeds the marginal cost of reducing emissions, representing the inefficiency inherent in an unregulated world. The efficient price is t_E, determined by the intersection of the marginal cost and marginal benefit of emissions. Equivalently, we could say that the efficient level of emissions is Q_E and use that quantity to determine the number of carbon permits to issue in a cap and trade system; the equilibrium market price of such licenses would have to be t_E, the efficient price.

Incidence and Effects of Carbon Price on Downstream Industries

The next conceptual step in implementing a carbon price is to decide at what point in the chain of production it should be levied, that is, the "statutory incidence" question. To minimize administrative costs, it is generally thought that a carbon price should be levied far upstream on energy producers, for instance at the level of an oil refinery, natural gas processor, electric utility, or coal mine.[11] Again, there are numerous issues around this statutory incidence question, including what happens to oil products that are not combusted and the myriad other carbon sources such as pipelines, wells, cement production and agriculture, but these issues are not critical to the general analysis. In the end, the economic incidence of the price should be the same regardless of where it is levied, as is thought to be the case with most taxes; the question of where to levy is mostly if not entirely a matter of minimizing administrative costs.

Assuming that the carbon price is levied upstream on energy producers, we turn to the impact on downstream industries that use energy as an input. I choose to focus on a representative downstream industry – think perhaps of plastics, electricity, fertilizer or chemicals production that uses natural gas as a feedstock – rather than on the primary energy industry that initially bears the statutory carbon price. Importantly, this downstream industry will not have its carbon emissions resulting from usage of the energy input directly priced (to avoid redundancy); all the beneficial economic effects follow from the increase in the price of energy due to the upstream effects on energy producers.

In way of a summary, the main qualitative effects of a carbon price on a downstream industry are given below. There are both short-run and long-run effects; in the short run firms make operational decisions such as substituting lower-carbon fuels and changes in levels of production; in the long-run firms change the scale and nature of their fixed investments. In the long run it is expected that the carbon price will cause exit of resources in the downstream industry, including even the exit of firms. These adjustments, at the industry level, cause the cost of emissions to be reflected in prices of all products that use fossil fuel energy – what was an externality gets internalized. The price changes induce a host of efficient responses, from lower consumer demand to switching of energy sources and production technologies:

- Increases in energy prices increase the marginal cost of production; firms therefore reduce their production, causing industry supply to fall and prices to increase. These results hold

in the short-run and long-run, with greater effects in the long-runs. Higher prices imply less consumption of carbon-intensive products.
- Firms suffer losses in the short-run, with the greatest losses at firms whose costs increase the most. Some firms – the relatively cleaner ones – may even increase profitability, if their costs increase by less than the increase in the output price.
- Some firms and/or assets that suffer losses exit the industry, causing prices to further increase and for consumption of carbon-intensive products to fall further. This is the long-run effect.
- Firms have an incentive to offset energy price increases through improvements in existing technologies, adopting new technologies, switching energy sources, and other cost-saving technologies and innovations. The availability of cleaner technology reduces the implied increase in both short- and long-run prices.
- Firms' incentives to save costs result in the marginal cost of reducing carbon emissions to be equal across all firms, a requirement for efficient economy-wide emissions reductions.
- Any industries downstream of this one will experience similar economic outcomes, as they will have an increase in their input prices.

Figure 17.4 depicts such an industry pre-carbon pricing, in both a short-run and long-run competitive equilibrium.[12,13] Panel A of Figure 17.4 shows the situation of a typical firm while Panel B shows the industry overall. The price of output equals the marginal cost of production for all firms, and there is no incentive for exit or entry: if all firms are identical, then price equals average total cost for all firms; if firms are heterogeneous in capabilities, then the marginal firm has zero profits but inframarginal firms enjoy profits due to their unique resource or capabilities.[14] Since the demand curve represents marginal value to consumers, in equilibrium consumers' marginal value equals price, which in turn equals marginal cost: we have the economically efficient level of production.[15]

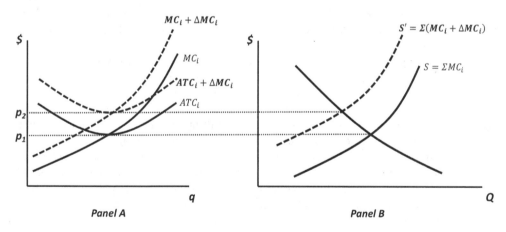

Figure 17.4   Impact of carbon tax on competitive firms and industry

Now imagine a carbon price of C/tCO_2e has been assessed on primary energy producers upstream from our representative industry, so that the energy input to the downstream industry increases in price. For now, assume complete pass-through so that the price per unit of energy increases by $C*(tCO_2e/unit of energy).[16,17]

For the downstream industry, higher energy prices due to the carbon price cause higher costs of production, and in particular a higher marginal cost of production. Firms will adapt and attempt to minimize the impact of the higher energy prices. The increase in firms' marginal costs of production depends upon a host of conditions, perhaps the most important being: (1) the energy intensity of production; (2) the ease of increasing energy efficiency in production; and (3) the ease of switching from high-carbon to low-carbon energy. These conditions will differ by firm and by industry, with the cross-industry differentials likely to be large: energy-intensive industries will generally incur the largest increase in cost. Even within the same industry, firms will have different increases in their costs, possibly due to differential energy intensities.

I assume that the increase in marginal cost for any one firm is uniform across all volumes of output and denote that increase as ΔMC_i, with the subscript allowing for firm heterogeneity. This increase in marginal cost depends upon the carbon emissions per unit of energy input, (tCO_2e/unit of energy), as well as the energy intensity per unit of output. In electricity production, for example, the change in marginal cost per megawatt hour of electricity produced, for a firm using coal as the energy input, would be based on the tCO_2e in a ton of coal as well as the amount of coal needed to produce a megawatt hour of electricity.

With marginal costs increasing, firms' average (total) costs also increase; a uniform increase in marginal cost causes average cost to increase by that same amount, ΔMC_i. Firm heterogeneity implies we will have a continuum of firms, with some experiencing a relatively low increase in marginal and average costs and some experiencing a relatively high increase.

These effects are depicted in Figure 17.4, Panel A. The increase in marginal cost ΔMC_i must be understood as representing conditions *after* firms' optimal adjustments to minimize costs due to the tax. This is an important consideration; without considering firms' adjustments the price increases implied by the carbon tax would be over-estimated (see below for more discussion, including a comparison to the sulfur-dioxide cap-and-trade program in the US in the 1990s).

Increased marginal costs of production in turn cause the short run industry supply curve, S_{SR}, to shift up, as depicted in Figure 17.4, Panel B. If all firms were identical, S_{SR} would shift up by the same amount as every firm's marginal cost; if firms are heterogeneous then the shift in S_{SR} can be more complex. If, for example, the carbon price affected higher-marginal cost firms more than lower-marginal cost firms, then S_{SR} will shift up by more at high industry output levels, for that is when the higher marginal cost firms would be operating.

With S_{SR} shifting upward, the short-run equilibrium price increases and the quantity produced by the industry decreases, with the quantitative changes being dependent upon both the supply and demand elasticities. For small changes, and assuming for simplicity that all firms are identical, the change in the equilibrium price for a small change in marginal cost can be written as $\left\{ \dfrac{dp^*}{dt} = \dfrac{\varepsilon^S}{(\varepsilon^S - \varepsilon^D)} \times dt \right\}$, where t is the marginal cost and ε^D and ε^S are elasticities of demand and supply. As is evident, the extent of short-run pass-through of the carbon tax in this downstream industry is complete $\left(\dfrac{dp^*}{dt} = 1 \right)$ if $\varepsilon^D = 0$, that is, a vertical or perfectly inelastic demand curve. If all firms are the same, and ε^D is negative, then short-run pass-through will always be incomplete in the sense that the increase in price will be less than the increase in marginal cost ($\dfrac{\varepsilon^S}{\left(\varepsilon^S - \varepsilon^D\right)}$ will be less than one).

Technological heterogeneity and innovation across firms, and energy input switching, makes for interesting and important differences in even the short-run implications. These considerations suggest that it is easy to over-estimate the increase in prices caused by a carbon tax. The experience of using a cap-and-trade policy to reduce sulfur dioxide emissions in the United States is illustrative. That program resulted in emissions reductions at costs lower than were anticipated; one reason cited is that electric utilities (the main entities being regulated under the policy) found it was relatively cheap to reduce emissions by switching to low-sulfur coal.[18] Rather than have prices for electricity increase by the implied increase in cost for utilities that burned high-sulfur coal, the price only had to increase by the implied emissions cost for burning low-sulfur coal, along with any higher market price for the low-sulfur input. With the current situation of the electric industry, suppose that a carbon price is imposed on fossil fuel inputs (focusing on coal and natural gas) for the electric power industry. Suppose that this price is sufficient to switch the relative market position of coal-based versus natural gas-based power producers, so that natural gas producers now become the overall low-cost firm and knock coal producers out of the market.[19] The increase in the price of electricity then consists of two effects, the first being the increase in marginal cost due to the carbon price for only the natural gas producers, which depends upon the carbon emissions per megawatt of natural gas producers, not of coal producers. The second reason for an increase in the price of electricity will be any higher marginal cost of producing electricity via natural gas rather than coal, not considering the carbon price. Given that coal was a dominant fuel before carbon pricing, this differential must exist but it need not be large.

The point here is that the cost curves of Figure 17.4 must reflect optimal adjustments to firms' operations, including energy switching. If there are easy technological/operational ways to reduce emissions in an industry, then the cost curves of Figure 17.4 would not increase by much and neither would the equilibrium price. Of course, as the sulfur dioxide experience shows, knowledge of such switching is imperative in forecasting the increase in price due to carbon pricing.[20]

Long-run analysis considers the impact of possible changes in profitability on entry and exit of fixed assets, and firms, from the industry. Without any firm-level heterogeneity, all firms will incur losses, with price now less than average total cost. This situation – which can be considered to be the base case scenario, with identical firms – is depicted in Figure 17.2, Panel B. Losses cause the exit of fixed assets and firms; this is shown with the shifting back of the short-run supply curve to S'_{SR}. In the base case with identical firms, there must be sufficient exit to cause the price to increase so as to bring the firms back to a zero economic-profit state. With identical firms and a uniform increase in marginal cost across volume, price must increase by the increase in marginal cost – which also equals the increase in average cost. In this sense, in the long run, consumers bear all of the cost of the carbon price: in the long run there has to be complete pass-through. The industry quantity demanded and therefore total output of this industry is lower at the higher long-run price: consumers have been induced to shift their purchases to sectors that are less affected by the carbon price.

With heterogeneity among existing firms, the firms and assets to exit will be the ones whose post-carbon pricing costs are the highest. Exit will continue until the resulting market price covers average total post-carbon pricing cost for the marginal firm. The new long run price will be determined by the carbon emissions per unit of output, $C*(tCO_2e$/unit of energy) and energy intensity, *for the firm that is now the marginal firm*. The new price must also be higher to the extent that the newly marginal firm has non-emissions costs that are higher than the firms that were displaced. Taking an extreme example, suppose the pre-carbon pricing output price in the industry was $5 and that a new technology is available to produce at $6 but is carbon-free. As long as the impact of the carbon price on the existing firms is greater than $1, then in the long run the carbon-free technology will come to dominate and the long run price will increase by only $1, the relative cost disadvantage of the clean technology.

It is easy to talk about exit of resources and firms in a clinical way, removed from the difficulties of the real world. A review of what is happening in the US electric industry illustrates the practical issues that arise when an industry is induced to shrink and change its production technology (see Fisher et al., 2023), and the disruptions caused by the closing of both coal and nuclear plants. Assets that have long physical lives remaining are now economically reduced in value – stranded assets. Investors lose as do employees. Consumers pay higher prices. This is exactly what the policy is meant to achieve, but for many the increases are particularly painful. Of course, government receipts from the carbon price may be large, and could be used to compensate those that are hurt by the policy.

Efficiency and Equalization of Marginal Costs

One last issue that is not transparent in the above analysis involves inter-firm efficiency of both energy usage and carbon abatement. This issue becomes especially salient in the case of internal carbon pricing so it is useful to review it now. Generally, firms affected by a carbon price are incentivized to reduce their emissions not only by reducing their output, but also directly by using different production or prevention technologies. While I focus here on emissions from energy sources, there are many production processes that create carbon emissions even without the burning of fossil fuels (e.g., from chemical reactions). In those cases, carbon emissions would need to be measured and a carbon price would need to be assessed directly on the firms using such processes. Assuming this is done, these firms would then have incentives to mitigate emissions and will do so until the marginal cost of reducing emissions equals the carbon price. Since all firms face the same carbon price, the marginal cost of reducing emissions would be equalized; this is a necessary condition for minimizing the total cost of any industry-wide mitigation objective and is generally held as one of the key attributes of a region-wide, uniform carbon pricing policy. It is an important principle: carbon pricing incentivizes those firms and sectors that can easily abate emissions to do so, while allowing hard-to-abate firms and sectors the flexibility to continue emitting, albeit paying a cost to do so.

In the case of an industry that only emits carbon from use of fossil fuels, this result on the equalization of marginal costs of mitigation is less obvious, but the result still holds. In the example of a downstream industry consuming a fossil fuel on which a carbon price has been assessed, the downstream firms will reduce emissions by reducing their energy usage to the point where the marginal cost of reducing energy usage equals the post-tax energy price. Since the post-tax price of energy is the same across all firms, the marginal cost of reducing energy usage is equalized. This is the necessary condition for productive efficiency; if the equality did not hold it would be cost-saving at the economy level to have firms with the lower marginal cost of reducing energy usage cut back on their energy consumption. Emphasizing the principle at work, firms that don't have low-cost ways to reduce energy consumption (airlines, perhaps) will not cut their consumption by very much, while those that have low-cost ways to reduce will do so.

We can re-cast this result in terms of reducing carbon emissions rather than in terms of just reducing energy usage. Let P_{NOCARB} be the before-carbon pricing energy price and $P_{AFTERCARB}$ be the price after. Since we are assuming full pass-through of the carbon price at the upstream level we will have

$$P_{AFTERCARB} = P_{NOCARB} + C$$

where C is the per-unit of energy carbon price (e.g., dollars per MCF if dealing with natural gas).

Let MC_{REDUCE} be the marginal cost of reducing energy usage at a given firm; this could reflect the cost of low-carbon energy or of reducing production levels. From the above discussion, we will have, at profit-maximizing firms,

$$MC_{REDUCE} = P_{AFTERCARB}$$

But since $P_{AFTERCARB} = P_{NOCARB} + C$, we can write

$$MC_{REDUCE} = P_{NOCARB} + C$$

or, re-arranging,

$$(MC_{REDUCE} - P_{NOCARB}) = C$$

The left-hand side term, $MC_{REDUCE} - P_{NOCARB}$, can be thought of as the net marginal cost of reducing carbon emissions: the MC of reducing energy consumption less the savings in pre-carbon pricing energy expenses. Firms will have this marginal cost of reducing carbon emissions equal to the carbon price; since all firms face the same carbon price, we will have the marginal cost of reducing carbon emissions equal at all firms – but some firms will reach that equality after significant reductions in emissions, while others will reach it after very little. This equality of marginal costs of emissions reduction is again a necessary condition for efficiency as to the economy-wide reduction of carbon.[21]

INTERNAL CARBON PRICING

What is Internal Carbon Pricing?

The logic of a region-wide carbon price is compelling, yet as documented above, there are still wide swaths of the global economy that either lack a region-wide carbon price entirely or have a price that is by many accounts lower than the efficient level. Perhaps encouraged by this void, corporations have begun deploying a policy of "internal carbon pricing."[22] This section covers what internal carbon pricing is, and it presents some of the rationale for why corporations might implement it. The section also assesses the likely economic efficiency of internal carbon pricing in a world with an existing, although possibly incomplete, regulatory regime for carbon emissions – an issue that has gotten little if any attention in the literature yet would seem to be of some importance. To jump to an extreme case, if there were an efficient region-wide carbon pricing policy in place, then there would be no economic justification to firms adopting internal carbon prices – the full cost of carbon emissions would already be represented in energy prices and firms would have efficient incentives for emissions reductions through those prices.[23] Any additional internal carbon price would be overkill.

Internal carbon pricing is simply the use by one company of a carbon price to guide some of the company's strategy and decisions. As discussed below, most often such internal prices are just shadow prices, used to indicate the cost of carbon emissions but not resulting in any actual financial charges to a business unit. According to data from the Carbon Disclosure Project almost half of the world's 500 largest companies use some kind of internal carbon price, with the overall median price in 2020 being $25t$CO_2$e.[24] Below, I take a deeper dive into the empirical evidence on internal carbon prices; in this section, I focus on the principles guiding the use of such prices and what the practice might accomplish. Detailed case studies of companies and other organizations adopting internal carbon pricing are scarce. In the boxes below, I discuss three organizations adopting carbon pricing that have publicly documented their experience: Microsoft, British Petroleum and Yale University.

BOX 17.1 MICROSOFT'S INTERNAL CARBON FEES

As part of its pledge to be carbon neutral, Microsoft Corporation implemented an internal carbon fee in July 2012. According to company documents, they didn't use estimates of the social marginal cost of carbon to establish the internal price but instead used the estimated cost of achieving their carbon emissions goal, divided by the current level of emissions. In 2019, the company reported that they were "nearly doubling" the internal carbon fee to $15 per metric ton, implying that the old fee was in the vicinity of $7–$8.[25] The company attributes carbon emissions to its various divisions and requires the divisions to pay into a central fund the implied cost of carbon emissions. The collected funds are then used for projects to further help reduce energy use and carbon emissions; in principle, given the way the internal fee is calculated, the collected funds should be enough to fund all the projects required to meet the emissions goal. Microsoft states that the internal carbon fee policy helps promote efficiency; helps guide responsible investments and creates consistency with its code of ethics; helps meet the demands of various environmentally-active stakeholders; and helps establish leadership in the corporate environmental sphere.[26]

In terms of what internal carbon pricing actually encompasses in terms of corporate behavior, a taxonomy is useful.[27] Four dimensions can be used to characterize a company's internal carbon pricing practice:

- *Width*: What carbon emissions in the company's value chains are covered.
- *Height*: The actual level of the internal carbon price.
- *Depth*: The extent of influence over business decisions and operations.
- *Time*: The historical and projected path for the first three dimensions.

In regard to the "width" dimension, most internal carbon pricing will apply only to the firm's direct emissions from within that company's facilities, or more properly, Scope 1 emissions. Firms might also use an internal carbon price to evaluate the full cost of energy purchased; in this case, Scope 2 emissions are being covered. Scope 3 emissions – including purchased goods and services and employee travel – might be the hardest to include under an internal carbon pricing policy. For purchased goods and services, a company would need to know the carbon content of all its purchases in order to assess a carbon charge to them.[28]

In regard to height, or the actual level of the internal carbon price, as the papers discussed in the fifth section document, there is great variation across countries and industries, as well as across companies in a given country and industry.

In regard to depth, there is also great variation in how internal carbon prices are used. In some cases there is an actual internal tax that divisions or units of the company are charged, based on their emissions (as with government sponsored carbon pricing regimes, there is then the question of what happens to the "tax" revenues). Alternatively, there might be only a shadow price that is used in certain business decisions and planning. Internal carbon prices might be used to evaluate energy supplies, with an internal carbon price tilting against carbon-intensive sources. Such internal prices might be used for new building construction, creating additional incentives for energy-efficient construction methods. They could be used for evaluating the profitability of new projects, such as new products or new market entry, with the profitability of such projects being reduced to the extent of their leading to greater carbon emissions. Taking the logic of internal carbon pricing to perhaps its deepest use, all capital budgeting analyses would include the negative cash flow implications from the project's carbon emissions multiplied by the appropriate internal carbon price for that year and region.[29]

Why do Firms use Internal Carbon Pricing and what Principles Support its Use?

Many reasons have been given for companies to implement internal carbon pricing; a relatively complete (and clearly overlapping) list would be:

1. As an efficient, decentralized means of achieving a corporate-wide reduction in carbon emissions.
2. As a tool for assessing the risk from possible future carbon taxes and regulations.
3. To generally prepare the organization for stronger governmental/external regulation of carbon emissions.
4. As a way to generally reduce risk from future carbon regulations; specifically as a way to reduce the risk from investing in assets that would lose value from stronger governmental regulation of carbon emissions.
5. To tilt the company's investments toward low-carbon activities and assets.
6. To achieve a competitive advantage.
7. To signal commitment to reducing carbon emissions and help meet the demands of employees, customers, suppliers, investors, politicians and activists.
8. To establish and showcase a leadership position in the arena of climate change.

BOX 17.2 BRITISH PETROLEUM'S EARLY USE OF INTERNAL CARBON PRICING

One of the earliest corporate uses of an internal carbon pricing regime is that of British Petroleum (BP), which initiated an internal carbon emissions trading regime in 1998. Unlike a direct internal carbon price, a trading regime requires allocation of emissions permits among business units. BP chose to do this using a grandfathering approach, with units receiving permits based on historical emissions. BP's initial goals for emissions were relatively moderate, just 1 percent off a business-as-usual forecast, yet internal permit prices were relatively high. This anomaly might have been caused by business units banking permits for the future, or simply because the internal market wasn't efficient. When prices got too high, the company tolerated non-compliance as a kind of "safety valve." The regime was cancelled in 2002, at least partly because of what were seen as inaccurate price signals and because the European Union had started a region-wide trading system, which BP would be a part of (see discussion below on possible redundancy of internal and region-wide trading systems). Victor and House (2006) conclude that the system was not a "textbook trading system," found "little evidence to support the hypothesis that pecuniary rewards – rooted in performance contracts – had much effect on (business unit) leaders", and that the biggest factors in emissions reductions over the period were not the trading system but instead corporate leadership that created awareness and measurement of emissions. They also discuss the substantial opportunity costs imposed by the system in the form of re-directed managerial time.

The World Bank's (2021a) *State and Trends of Carbon Pricing 2021* gives the main drivers of internal carbon pricing. The top three reasons for adopting internal carbon pricing most cited by firms are: (1) to drive low carbon investment; (2) to drive energy efficiency, and (3) to change internal behavior.[30]

Most of the published rationale for companies to implement internal carbon pricing appeals to the company and directly to its shareholders' interests: carbon pricing is a means to reduce risk and increase value. However, there seems little doubt that many advocates also believe that internal carbon pricing achieves societal goals as well. For instance, the "Business Leadership Criteria" – put out by Caring for Climate, and supported by the United Nations Environment Program, among several other NGOs – have the setting of an internal carbon price "high enough to materially affect investment decisions" as the first of three criteria to assess if companies are committed to the 2°C global target.[31] There appears to be an equating of internal carbon pricing to external, governmental carbon pricing, and if the latter represents an efficient means to reduce emissions so must the former – they are essentially the same tool. We turn now to address in more critical detail both the private and socially-oriented rationale for internal carbon pricing.

Starting with the first reason given above for companies, using internal carbon prices to efficiently reduce company-wide emissions is related to the so-called "bubble concept" developed for the control of sulfur dioxide emissions.[32] If a company wants to reduce its carbon emissions – taking that goal as a given – and if the company has multiple emissions sources, then an internal carbon price (or even an internal carbon emissions trading regime) might be an efficient way to achieve that goal. The key to efficiency in emissions reductions for a multi-source company is the same as for a region or country: make sure that the marginal cost of reducing emissions is the same across all sources.[33] If a company installs a company-wide carbon price, such that different units and divisions bear the actual cost of their emissions, then those units have incentives to reduce emissions up to where their marginal cost of cutting emissions equals the internal price. There are at least two critical conditions to make this scheme work. First, it must be that information on marginal cost of emissions reductions is held at the level of the unit and not centrally – otherwise, the command and control structure of a corporation would allow for efficient assigning of reductions across the units. Second, it must be that decision-makers at the unit level truly bear the cost of the internal carbon price. If a unit's profitability reflected carbon emissions at the internal carbon price, and if unit managers' pay schemes depended upon unit profitability, then decision-makers bear the cost of emissions. But for efficiency, marginal costs of emissions must be equated across all sources; this technically requires all managers' pay schemes to depend on the internal carbon price in the same way. In practice, it is not clear that very many companies' internal carbon pricing schemes meet these requirements.

Rationales 2–5 above are similar and can be categorized under the general theme of "using internal carbon pricing to help the company anticipate future carbon taxes or other regulations." Measuring and managing carbon emissions today can be an effective way to prepare for a future where a company's own carbon emissions are taxed, or where energy prices are significantly higher because of upstream carbon taxes. Measuring emissions – at whatever "scope" level seems most appropriate from a future regulatory view – is a good start, but to assess the financial liability a company faces from future taxes or regulations, it would be even more insightful to put a dollar value on those emissions. An internal carbon price satisfies that requirement. Knowledge of future financial risk can help the company prepare, by beginning to focus management time and effort on reducing emissions. Even without a binding internal carbon price, as for the first rationale, management can use the level of risk to begin moving the organization towards a lower-carbon future. This might mean including the cost of emissions in evaluation of new investments so as to reduce the likelihood of investing in assets that would suffer loss in value if carbon regulations were to significantly tighten. The European Bank for Reconstruction and Development (EBRD, 2019) specifically recommends a sensitivity analysis on possible carbon prices for all EBRD investment projects, including a breakeven carbon price that would change the investment decision.

Achieving and communicating (signaling) competitive advantage – rationales 6 and 7 above – are often cited as reasons for reducing carbon emissions generally and internal carbon pricing in particular.[34] The general idea is that if either future regulations tighten, or if important constituencies such as consumers, employees or investors increasingly care about a company's carbon emissions, then reducing emissions can convey a competitive advantage. Heterogeneity of companies makes this argument more compelling: those firms that have the greatest capability to reduce emissions at low cost can gain the most from reducing their emissions relative to "dirtier" competitors. This behavior fits the "raising rivals' costs" model of Salop and Scheffman (1983). Signaling models have also been developed to show how companies can gain from credibly conveying their emissions reduction capabilities; the key idea is that investors or other outside agents can infer a company's carbon risks from its emissions reductions behavior so long as the cost of such behavior is higher for some firms than others. Internal carbon prices can also be a means of credibly conveying to other constituencies, such as employees, that the company is committed to reducing carbon emissions. Credibility here is enhanced by actions that have real impact; measuring emissions is one thing but having those emissions priced internally and having decisions affected is another.[35]

The final rationale, establishing a leadership position on climate change, embodies all of the other rationales. The leadership rationale is cited by many as a desirable benefit of a strong corporate climate change strategy, including use of internal carbon prices.[36] The benefits of such leadership might accrue to the company and its shareholders, if such a leadership position signals a credible commitment to low-cost reductions of emissions, per the discussion above. Board members and other constituencies might bask in the publicity that such a leadership position conveys. It might also be that value from a leadership position accrues mostly to the company's executives, who might find enhanced job prospects at other companies whose boards want their own company to develop such a reputation for climate change leadership.

BOX 17.3 YALE UNIVERSITY'S CARBON PRICING[37]

Yale University in the United States began levying an internal carbon charge in 2015. The charge was initially set at $40/tonCO$_2$e, with 20 campus buildings in the initial pilot program. Each of the 20 buildings was assigned to one of four "treatments:" no carbon pricing but the building did receive a report on energy use and what their charge would have been if actually levied; carbon pricing with 20 percent of any revenue from the charge redistributed to energy efficiency programs; carbon pricing with revenue redistributed to buildings that were more efficient than in the past; and carbon pricing with revenue redistributed to buildings that achieved higher than average reductions in emissions. It is not clear how any revenues returned to buildings affected the building managers, who were responsible for any changes due to the carbon pricing. The buildings in all of the treatment groups all reduced emissions by more than the control group (all other campus buildings), with the greatest reductions in the last two treatment categories. One finding was that all the treatment building units reported higher levels of understanding and motivation for reducing energy use; this is similar to the finding in the BP case study above that cited the creation of awareness as one reason for successful emissions reductions. Yale has since moved to charging all of its administrative units an internal carbon price based on the carbon emissions of the unit's buildings.

The Social Value of Internal Carbon Pricing

The private benefits of internal carbon pricing – whether to the company or to individual executives – are difficult to argue. Companies should be forward-looking to anticipate future regulations, and they should make long-term investment decisions with future regulations in mind. Internal carbon prices can help companies efficiently meet internal emissions targets and can credibly convey a company's relative cost position in regard to emissions reductions. Leadership on climate issues certainly might be valuable to the company and to its executives.

But is a world where many if not most companies have a deep and broad policy of internal carbon pricing necessarily one of improved economic efficiency? Internal carbon pricing, following as it does from the well-established efficiency of region-wide carbon pricing, would seem to offer clear-cut social benefits. But could there be unintended consequences of internal carbon prices?

I argue that there are indeed several reasons to be concerned about the social value of internal carbon pricing, and I further argue that this issue has been neglected in most discussions of internal carbon pricing. I will focus on four possible sources of inefficiency.

1. Unless all companies use the same internal carbon price, there will be differences in the marginal cost of reducing emissions across companies, meaning that the societal-wide reduction is not being done at minimum cost.
2. It is possible that "cleaner" companies will be the ones to use internal carbon prices, and if the internal prices are binding on decisions, then the market-wide effect will cause "dirtier" companies to gain market share and emit more, possibly increasing total emissions.
3. There is a risk of redundancy of internal carbon prices with region-wide carbon regulations such as renewable portfolio standards for electric utilities or in regions that already have a carbon pricing regime, even an incomplete one.[38]
4. It will be more difficult to estimate downstream price increases associated with internal carbon pricing, possibly resulting in inefficient output and investment decisions.

Starting with the first point, one large benefit of region-wide carbon pricing is that, since all firms face the same price of carbon, all firms end up with the same marginal cost of reducing emissions. This is unlikely to be the case with internal carbon pricing, where the variation across companies in internal carbon prices is significant. Bento and Gianfrate (2020) report internal carbon prices ranging from $0.01 to almost $1000/t$CO_2$e. Having one firm incur $1000/t$CO_2$e to reduce emissions and another incur only $10/t$CO_2$e is a waste of resources; for efficiency, we want reductions from the low-cost sources. While this situation may not cause a net loss to society – so long as the marginal cost of emissions exceeds $1000/t$CO_2$e, reductions at even the high-cost firm are valuable – there is definitely an opportunity for further gains that is not being exercised. Indeed, internal carbon pricing really runs afoul of what might be considered as the core condition for economic efficiency in competitive markets: that all firms face the same prices for inputs and outputs.

That some firms will have higher or lower costs of reducing carbon emissions also plays into the second possible source of inefficiency and, in this case, possible net losses. It is possible, if not likely, that internal carbon pricing will be most profitable for "clean" firms that believe they can reduce emissions relatively easily. Firms that don't believe they have good options for reducing emissions have little reason to put in place a policy that attempts to do that. Another reason to think that firms with good mitigation options will be more likely to adopt internal carbon pricing is the "raising rivals' cost argument", where companies pursue regulations that harm their competitors more than themselves; adoption of internal carbon prices could serve to support a company's lobbying efforts for a region-wide carbon tax.

Suppose that firms that adopt internal carbon pricing do have greater potential for reducing emissions than those that don't. The firms adopting internal carbon pricing will be voluntarily increasing their costs. Following the logic of the previous discussion around Figure 17.4, this will shift the industry supply curve up and cause the market price of the industry output to increase. This higher price in turn gives firms that did not adopt internal carbon pricing incentive to increase their output, with the somewhat paradoxical result that a greater share of industry output is now accounted for by the firms not adopting carbon pricing. Depending upon elasticities of firms' outputs with respect to market prices and the output/emissions rates of different types of firms, it is possible that total emissions will increase. The carbon-price adopting firms will be, in effect, subsidizing the other firms; costs will be incurred but, with the net effect on emissions being unclear, the overall outcome could be worse than without any internal carbon pricing at any firms. This is identical to the problem of "leakage" in regional carbon pricing when pricing carbon in one region just causes emissions activity to move to a region where carbon is not priced.[39]

Redundancy is also a potential problem with internal carbon pricing, as my third point notes.[40] If an efficient region-wide carbon pricing policy is in place, and if economic efficiency is the goal, then there is no room for additional carbon pricing by other entities. An efficient region-wide carbon price will be set according to the principles of Figure 17.1, fully internalizing the carbon externality. Any enhancements to that price will over-penalize carbon emissions, with the result being too-high prices and too little production for carbon-intensive products. If a region-wide efficient carbon tax is placed upstream on energy, say electricity generation, such a tax would likely be invisible downstream; energy prices would simply be higher (the tax could be made apparent by putting it on invoices but that shouldn't have any additional effect). With other kinds of regulations upstream, such as portfolio standards for electricity generation, the lack of transparency will be greater. This makes it more possible that firms will overcompensate with internal carbon pricing.

One should consider the possibility that even moderate internal carbon prices could today be redundant, in some regions. Take California, for instance, which had a renewable portfolio standard for retail electricity sales of 20 percent in 2017 and which currently stands at 60 percent by 2030 and 100 percent carbon-free sources by 2045.[41] What is the implied carbon price that would induce electric utilities to have 60 percent of their energy be from renewable sources? One study from 2016 estimated that a 50 percent renewable standard would increase electricity prices by about 3 cents per kWh in California.[42] That is a large increase as compared with the 0.7 cents per kWh estimated as the increase in average US electric prices resulting from a $28/tCO$_2$e tax.[43] A very rough estimate from this, assuming linearity, is that a $120/tCO$_2$e tax would cause electricity prices to increase by 3 cents/kWh and therefore that the 50 percent standard is equivalent to a $120/tCO$_2$e tax.

Another study pegs the implicit cost per tCO$_2$e for California's renewable portfolio standards at over $100.[44]

On top of these regulations, California runs its own cap-and-trade system that also applies to electric utilities; the average auction settlement price for 2020 was in the vicinity of $17 per tCO$_2$e.[45]

While not necessarily the efficient tax level, a carbon tax of $35 (2017 US$) per tCO$_2$e in 2030 has been estimated to be sufficient for large country emitters to meet Paris Agreement pledges.[46]

It is not only possible then, but perhaps likely, that at least in the electricity-consuming sector in CA, the combination of policies already in place implies a carbon tax already well in excess of the efficient tax. This means that any internal carbon pricing on electricity usage by companies is redundant and in fact will lead to inefficient investment and production decisions.

My last point on the possible economic inefficiency of internal carbon pricing policies has to do with the downstream price impacts from internal carbon pricing. With an explicit region-wide carbon price, and for the simplest case with homogeneous firms in the downstream industries, the long-run downstream impact is to increase downstream prices by the full increase in average cost for all products impacted by the carbon price. Downstream price adjustments should affect companies' production and investment decisions; without including price increases due to carbon prices in decisions, companies would produce and invest too little, as they would only be looking at the increase in cost. A carbon price does force carbon-intensive industries to shrink, but in the long run the new equilibrium is simply a smaller industry with higher cost and commensurately higher prices. The firms that remain should be producing at the same rate as before, and they should be reinvesting to maintain those production levels.

With internal carbon pricing, there will be greater uncertainty about the downstream price impacts of companies' actions. Not all companies will institute internal carbon pricing, but among those that do the depth of their policies will vary. Some companies' policies will actually affect pricing, output and investment decisions while others will use internal carbon pricing merely to assess their liability. Companies that do use internal carbon prices to guide business decisions should not take output prices as fixed; as more companies adopt deeper internal carbon pricing policies those prices will increase. But uncertainty over the extent of internal carbon pricing makes such forecasts difficult. Take a company that institutes a $35/tonne CO_2e internal carbon tax, with that tax binding as a real cost on all the firm's decisions. With uncertainty over which firms will choose voluntarily to self-impose such a tax, what price increase should the company assume? If no price increase is assumed, then the firm will end up producing less and investing less than in a world with a similar but binding government-imposed tax of $35/tonne.

Take as an explicit example the EBRD (2019) recommended methodology for calculating the breakeven carbon tax for investment projects: given a project's cash flows, carbon emissions (relative to some base case) should be estimated and an internal carbon price applied to that to create another stream of project costs. The carbon price that makes the project's net present value zero is the breakeven carbon price – and that should presumably be compared with a standard, such as the efficient carbon tax, with the project being turned down if the breakeven carbon price is less than the relevant standard. The problem here is that the project's revenues have been taken as given, while if the project were being undertaken in a world with a mandated carbon tax, the prices associated with the project's output would be higher, and the net present value would support higher carbon taxes. Without assuming higher prices, too many projects will be turned down.

Overall, a regime where companies voluntarily adopt internal carbon pricing policies, while sounding similar to a regime of a government-mandated, region-wide carbon tax, is in fact very different. While there are clearly some benefits, ranging from heightened corporate awareness of carbon emissions liability and some reductions in emissions, there is a real possibility of unintended and inefficient consequences. These consequences are not unlike those from the "patchwork quilt" of state-level electric market regulations documented in Fisher et al. (2023) in this *Handbook*.

EMPIRICAL EVIDENCE ON INTERNAL CARBON PRICING: WHO, WHY, AND TO WHAT EXTENT?

Organizations such as the Carbon Disclosure Project, World Bank, World Business Council for Sustainable Development, and Bank for European Construction and Development, among others, have advocated for internal carbon pricing and, in the case of the Carbon Disclosure Project, collected data on use of carbon pricing by companies. Case studies on companies and even universities have also been written to document such experiences. But scholarly research on internal carbon pricing, using rich data on many companies and subjected to statistical testing, is in its early stages. Bento and Gianfrate (2020) and Bento, Gianfrate and Aldy (2021) use the Carbon Disclosure Project data to study the determinants of ICP by companies, and in particular to examine the effect of national carbon pricing policies on decisions by firms to implement ICP. As of 2021, these papers represent the most up-to-date statistical analysis of these issues.

Bento and Gianfrate (2020) estimate the following regression model on panel data from the Carbon Disclosure Project during the period 2015–2017:

$$ICP_{it} = \alpha + \beta_1 SIZE_{it} + \beta_2 ROIC_{it} + \beta_3 BOARDQ_{it} + \beta_4 GDPP_{it} + \beta_5 NCP_{it} + \beta_6 ENE_{it} + \varepsilon_{it}$$

where ICP is the natural log of a company's reported internal carbon price; $SIZE$ is the natural log of company revenues; $ROIC$ is return on invested capital, $BOARDQ$ are measures of corporate governance quality, $GDPP$ is the log of GDP per capita in the company's home country, NCP is a dummy variable if the home country has a climate tax or equivalent, and ENE is a dummy variable if the company is in the energy sector.

The main results are as follows: on a national level, companies headquartered in countries with higher GDP per capita, and with a national climate tax, report significantly higher internal carbon prices. On a corporate level the results are mixed. Companies in the energy sector report significantly higher internal carbon prices, but coefficients for the other variables vary in size and significance depending upon the set of variables used in the regression. With the national and company-level variables in the regression, firm profitability is positively and significantly associated with the level of internal carbon prices but size as measured by revenues is not. The authors use two variables for measures of corporate governance, the percentage of female directors and the percentage of independent directors. Both of these are positively and significantly related to the level of internal carbon prices, when entered individually, not both in the same model.

Bento, Gianfrate and Aldy (2021) focus more on the issue of how a national carbon pricing regime affects company internal carbon prices. They estimate statistical matching models that estimate the "treatment effect" of a given country implementing a national carbon pricing regime. The matching technique compares internal carbon pricing in countries that are similar except in their national carbon pricing regimes; just putting the national carbon price in as an independent variable in a standard regression would likely yield biased estimates of the effect of the national price on internal carbon prices. The first finding is internal carbon prices are about $27 per metric ton higher in countries with a national carbon tax to those in countries without such a national tax. Using a subset of the sample on just European firms, the authors additionally find some statistical support for the hypothesis that the level of internal carbon prices is positively related to the level of national carbon taxes (and not just the existence of a national carbon tax).

The positive association between national carbon regulations and internal company carbon prices is probably the most interesting, and curious, aspect of this empirical work. What might drive the positive relationship? As noted above, a prime motivation for companies to initiate internal carbon pricing is to prepare for a world with region-wide carbon pricing; the finding of a positive relationship between national carbon pricing and internal carbon pricing could be consistent with this motivation, if firms believe that the existing national policy is just the beginning of more aggressive regulations around carbon emissions, including non-price regulations. Second, nations that have a national carbon tax, or a relatively high one, must have a political system and an electorate that supports climate regulation; corporations are likely to be subject to similar political considerations and therefore more likely to have internal carbon pricing to meet the demands of their stakeholders.

While one can hypothesize what might drive the positive association between internal and region-wide carbon pricing, the more important question might be whether the relationship is indicative of wasteful redundancy. Bento et al. (2021) report an average internal carbon price of $48/t$CO_2$e in countries with a national carbon pricing regime. This by itself is in the range of $40 to $80/t$CO_2$e set by the Paris Agreement to meet a 2°C temperature goal. In Europe, during 2021, carbon prices in the European cap and trade system rose above $70, implying a total average carbon tax – internal and national – for EU companies with internal carbon prices of well over $100/t$CO_2$e, exceeding the Paris Agreement range.

CONCLUSION

Companies are increasingly taking independent actions to reduce their climate impact, with a significant number of companies opting for internal carbon pricing as an important part of their climate toolkit. Internal carbon pricing moves companies beyond just the physical measurement of carbon emissions to putting a dollar value on those emissions and even to using the assumed cost of carbon emissions to affect business decisions. The practice can help companies assess their potential liabilities; help to efficiently reduce emissions in multi-unit enterprises; help to communicate and motivate the company's climate strategy to employees, customers, investors and others; and help to avoid investing in projects that may end up being non-economic due to future carbon prices. One should be careful, however, when comparing the economy-wide effects of internal carbon pricing with those of government-imposed, region-wide carbon pricing. While the analogy seems appropriate, there are some large differences. First, since companies use widely varying internal prices, there is little chance that the firms with the lowest mitigation costs will be the ones that carry the brunt of any emission reductions. Second, the possibility of redundancy is real, especially as public policy increasingly moves to higher explicit and implicit carbon prices. If we get to a point where public policy has arrived at efficient carbon policies, will internal carbon pricing disappear, or will companies still be under pressure to go beyond policy mandates? Third, the problem of leakage, while typically thought of as occurring between countries with different carbon policies, can also occur between companies, with the possibility that aggregate emissions will actually increase as high-emitting companies without internal carbon pricing gain market share. Lastly, one of the desirable effects of region-wide carbon pricing is that heavy-emitting firms and industries will have large carbon bills and will be induced to shrink. It seems unlikely that profit-oriented firms will impose this on themselves and their shareholders but that remains to be seen.

Internal carbon pricing is part of what Fisher et al. (2023) refer to as the "patchwork quilt" of carbon regulations. The heterogeneity in corporate carbon pricing policies and especially in regard to the variance of assessed carbon prices should be disturbing to anyone schooled in the efficiency of a decentralized market-based pricing system. From a policy standpoint, it would be good to look at the totality of this patchwork quilt and think seriously about its overall efficiency, but that train may already be too far out of the station.

ACKNOWLEDGEMENT

I want to thank Anant Sundaram for reading earlier drafts and giving me excellent feedback for improving both the content and exposition.

NOTES

1. Throughout this chapter, the term "carbon emissions" or simply "carbon" is used as shorthand for greenhouse gases and emissions, and the symbol tCO_2e is used for tonnes of carbon dioxide equivalent emissions.
2. Economic efficiency is, technically, a state of affairs wherein nobody can be made better off without hurting someone else (Pareto optimality). For the purpose of this chapter, economic efficiency can be thought of as maximizing the net benefits from some activity.
3. The chapter generally uses the term *carbon price* rather than *carbon tax*. Emissions trading/cap-and-trade policies are implicitly included, as they result in an explicit carbon price and are on first approximation equivalent to an explicit carbon price. *Implicit* carbon prices can be inferred from any regulation that limits carbon emissions, either by dividing the cost of the regulations by the emissions reduction, or through calculation of the carbon price that would be necessary to obtain the emissions reduction. As the latter fall under more general regulatory schemes, I do not generally consider these to be carbon pricing schemes, although they are clearly important.
4. Fisher et al. (2023).
5. World Bank (2021a, 2021b).
6. Data in this paragraph from World Bank (2021a).
7. Note that the carbon prices across countries are not directly comparable due to differences in industry coverage and exemptions.
8. Carbon Disclosure Project (2021).
9. The literature on carbon pricing is voluminous, with contributions from academics, government, non-governmental organizations, and corporations. Among many others, see Pigou (1920), Coase (1960), Baumol (1972), Carlton and Loury (1980), Bovenberg and Golder (2002), Victor and House (2006), Tietenberg (2006, 2013), Nordhaus (2007), Weisbach and Metcalf (2009), Aldy and Stavins (2012), High-Level Commission on Carbon Prices (2017), Ahluwalia (2017), Stavins and Schmalensee (2019), Hafstead (2019), Addicott et al. (2019), Bento and Gianfrate (2020), Bento, Gianfrate and Aldy (2021), Fan et al. (2021), and Sundaram (2023).
10. Since removing carbon from the atmosphere is equivalent to reducing emissions, carbon capture should be included as a possible way to lower the marginal cost of reducing emissions.
11. See Weisbach and Metcalf (2009).
12. I focus on a competitive equilibrium as a representative model for many downstream industries and to permit discussion of differential effects across firms in the same industry.
13. This analysis should be understood as an *expected* set of outcomes, in the probabilistic sense, with actual outcomes possibly being more or less favorable. For example, higher energy prices could incentivize the search for better production technologies; if successful, those technologies could end up making firms' costs lower than without carbon pricing. There are many anecdotal stories of firms discovering efficiencies when faced with emissions regulations. But if this is the expected outcome, then firms could just pretend they are being regulated and force themselves to discover efficiencies: an unlikely situation.
14. This question of firm heterogeneity is important for assessing the impact of a carbon price on profitability. I incorporate the effects of heterogeneity in the analysis where that can be done without undue complexity. Inframarginal firms are firms that have unique assets giving them the ability to produce at lower cost; this allows them to earn abnormally high profits even in equilibrium.
15. This is sometimes referred to as allocative efficiency, to distinguish it from other aspects of economic efficiency inherent in the competitive equilibrium.
16. For example, a carbon price of $100/$tCO_2e$ would, with full pass-through, cause natural gas prices on a thousand cubic foot basis (mcf) to increase by $100*(0.055) = $5.50/mcf since there are about

0.055 tCO$_2$e in one MCF of gas. The average industrial price of natural gas in the US in September 2021 was $5.57/mcf.
17. As will become clear from analysis of the downstream industry, the short-run increase in the price of energy will be less than this, but in the long run the price per unit of energy should indeed increase by the full amount of $C*(tCO$_2$e/unit of energy).
18. See Stavins and Schmalensee (2013, 2019). The switch to low-sulfur coal was made more economic by a serendipitous deregulation of the rail industry; this lowered freight rates for low-sulfur coal from the Western US.
19. I abstract here from the details of wholesale electric markets, especially in regard to the minute-by-minute determination of prices. There may be times of high demand when high-emissions, high-cost producers would still be called on to produce, and at those times price will increase by the carbon tax effect given those producers' high emissions rates.
20. If all the technological innovations and switching possibilities were known, a command and control policy regime would, in principle, be implementable. It is uncertainty and lack of knowledge of specific conditions that makes a decentralized system such as a tax or cap-and-trade the efficient policy.
21. Again, if the downstream industry emitted carbon directly, not just through energy consumption, and if such emissions were taxed, then each firm would equate the marginal cost of directly reducing carbon emissions to the tax. This would result in the marginal cost of reducing carbon to be equal across all firms in the economy and therefore efficiency as to carbon reduction activity.
22. Below, I address evidence showing that internal carbon pricing appears to be more prevalent in countries that have economy-wide carbon pricing regimes. Thus, rather than "filling a void", internal carbon pricing might be responding to other factors, which also help determine public policy (citizens' preferences for climate action for example).
23. To the extent that a firm had carbon emissions unrelated to energy consumption, those emissions would also have to be regulated under the efficient region-wide regulatory scheme.
24. Carbon Disclosure Project (2021).
25. Smith (2019).
26. See DiCaprio (2013) for more detail.
27. See Aldy and Gianfrate (2019) and Ecofys, The Generation Foundation and Carbon Disclosure Project (2017).
28. For both Scope 2 and 3 emissions, the risk of duplicative carbon pricing is obvious.
29. European Bank for Reconstruction and Development (2019).
30. World Bank (2021a)
31. Caring for Climate (2014).
32. Stavins and Schmalensee (2013, 2019).
33. See Victor and House (2006, p. 2101) in regard to British Petroleum's motivation for using an internal carbon pricing mechanism: "Third, the firm sought a decentralized mechanism that would encourage business units to find the most advantageous cuts in emissions. As an organization, BP is geographically and operationally diverse, and business units have varying marginal costs of emissions reductions—exactly the attributes that have drawn many policymakers to use market instruments for cutting emissions rather than central command."
34. See Clarkson et al. (2023); Broadstock et al. (2018) and Matsumura et al. (2014).
35. These arguments assume that carbon regulations will tighten in the future. If they don't, any corporate value enhancement should disappear and the investments in emissions reductions will have been for naught.
36. See Argenti, Holmes and Smittenaar (2023) and Lawton and Kock (2023).
37. Yale University (2016).
38. See Fisher, Phillips and Scovic (2023).
39. Fowlie and Reguant (forthcoming).
40. Gundlach, Minsk and Kaufman (2019) analyze interactions between a Federal carbon tax and other policies, including state-level carbon pricing policies. They set up a useful continuum between complementary versus redundant policies. While not explicitly considering company-level policies such as internal carbon pricing, they place state-level carbon pricing far on the redundant side of the spectrum. Corporate-level policies would fall even further on the redundant side.

41. See Gill, Gutierrez and Weeks (2021).
42. Rouhani et al. (2016).
43. Cleary and Palmer (2020).
44. Greenstone and Nath (2020)
45. International Carbon Action Partnership (2021).
46. International Monetary Fund (2019).

REFERENCES

Addicott, E., Badahdah, A., Elder, L. & Tan, W. (2019). *Internal Carbon Pricing: Policy Framework and Case Studies*. Yale School of Forestry and Environmental Studies. https://cbey.yale.edu/sites/default/files/2019-09/Internal%20Carbon%20Pricing%20Report%20Feb%202019.pdf

Ahluwalia, M.B. (2017). *The Business of Pricing Carbon: How Companies are Pricing Carbon to Mitigate Risks and Prepare for a Low-Carbon Future*. Center for Climate and Energy Solutions. https://www.c2es.org/wp-content/uploads/2017/09/business-pricing-carbon.pdf

Aldy, J.E. & Gianfrate, G. (2019). Future-proof your climate strategy. *Harvard Business Review* (May/June), 86–97.

Aldy, J.E. & Stavins, R.N. (2012). The problems and promise of pricing carbon: Theory and experience. *Journal of Environment and Development*, *21*(2), 152–180. https://doi.org/10.1177/1070496512442508

Argenti, P., Holmes, P., and Smittenaar, M. (2023). Climate change communication strategies. In A. Sundaram and R. Hansen (Eds.), *The Handbook of Business and Climate Change*. Edward Elgar.

Auffhammer, M. (2018). Quantifying economic damages from climate change. *Journal of Economic Perspectives*, *32*(4), 33–52. https://doi.org/10.1257/jep.32.4.33

Baumol, W. (1972). On taxation and the control of externalities. *American Economic Review*, *62*(3), 307–322.

Bento, N., & Gianfrate, G. (2021). Determinants of internal carbon pricing. *Energy Policy*, *143*(C), 1–8. https://doi.org/10.1016/j.enpol.2020.111499

Bento, N., Gianfrate, G., & Aldy, J.E. (2020). National climate policies and corporate internal carbon pricing. *The Energy Journal*, *42*(5), 87–98. https://doi.org/10.5547/01956574.42.5.nben

Bovenberg, A.L. & Goulder, L.H. (2002). Environmental taxation and regulation. In A.J. Auerbach & M. Feldstein (Eds), *The Handbook of Public Economics* (pp. 1471–1545). North-Holland.

Broadstock, D.C., Collins, A., Hunt, L. & Vergos, K. (2018). Voluntary disclosure, greenhouse gas emissions and business performance: Assessing the first decade of reporting. *The British Accounting Review*, *50*, 48–59. http://dx.doi.org/10.1016/j.bar.2017.02.002

Carbon Disclosure Project. (2021). *Putting a Price on Carbon: The State of Internal Carbon Pricing by Corporates Globally*.

Caring for Climate. (2014). Business leadership criteria: carbon pricing. https://www.unglobalcompact.org/library/1051

Carleton, T. & Greenstone, M. (2021). *Updating the United States Government's Social Cost of Carbon*. Energy Policy Institute at the University of Chicago. https://epic.uchicago.edu/wp-content/uploads/2021/01/BFI_WP_202104_Final.pdf

Carlton, D.W. & Loury, G.C. (1980). The limitations of Pigouvian taxes as a long-run remedy for externalities. *The Quarterly Journal of Economics*, *95*(3), 559–566.

Clarkson, P., Grewal, J. & Richardson, G. (2023). The equity value relevance of carbon emissions. In A. Sundaram and R. Hansen (Eds.), *The Handbook of Business and Climate Change*. Edward Elgar.

Cleary, K. and Palmer, K. (2020). *Carbon Pricing 201: Pricing Carbon in the Electricity Sector*. Resources for the Future. https://www.rff.org/publications/explainers/carbon-pricing-201-pricing-carbon-electricity-sector/

Coase, R. (1960). The problem of social cost. *Journal of Law and Economics*, *III*, 1–44.

DiCaprio, T. (2013). *The Microsoft Carbon Fee: Theory and Practice*. Microsoft Corporation. https://download.microsoft.com/documents/en-us/csr/environment/microsoft_carbon_fee_guide.pdf

Ecofys, The Generation Foundation and Carbon Disclosure Project. (2017). How-to guide to corporate internal carbon pricing – Four dimensions to best practice approaches. http://b8f65cb373b1b7b15feb-c70d8ead6ced550b4d987d7c03fcdd1d.r81.cf3.rackcdn.com/cms/reports/documents/000/002/740/original/cpu-2017-how-to-guide-to-internal-carbon-pricing.pdf?1521554897

Enkvist, P, Naucler, T. & Rosander, J. (2007). A cost curve of greenhouse gas reduction. *The McKinsey Quarterly*, No. 1, 35 45. https://www.mckinsey.com/business-functions/sustainability/our-insights/a-cost-curve-for-greenhouse-gas-reduction

European Bank for Reconstruction and Development. (2019). Methodology for the economic assessment of EBRD projects with high greenhouse gas emissions. Technical Note. https://www.ebrd.com/news/publications/institutional-documents/methodology-for-the-economic-assessment-of-ebrd-projects-with-high-greenhouse-gasemissions.html

Fan, J., Rehm, W. & Siccardo, G. (2021). *The State of Internal Carbon Pricing*. McKinsey & Co. https://www.mckinsey.com/business-functions/strategy-and-corporate-finance/our-insights/the-state-of-internal-carbon-pricing

Fisher, S., Phillips, B. and Scovic, M. (2023). The patchwork quilt: business complexities of decarbonizing the electric sector. In A. Sundaram and R. Hansen (Eds), *The Handbook of Business and Climate Change*. Edward Elgar.

Fowlie, M.L. and Reguant, M. (forthcoming). Mitigating emissions leakage in incomplete carbon markets. *Journal of the Association of Environmental and Resource Economists*. https://www.journals.uchicago.edu/toc/jaere/0/ja

Gill, L., Gutierrez, A. and Weeks, T. (2021) *SB 100 Joint Agency Report, Achieving 100 Percent Clean Electricity in California: An Initial Assessment* (Publication CEC-200-2021-001). California Energy Commission. https://www.energy.ca.gov/publications/2021/2021-sb-100-joint-agency-report-achieving-100-percent-clean-electricity

Gillingham, K. & Stock, J.H. (2018). The cost of reducing greenhouse gas emissions. *The Journal of Economic Perspectives*, 32(4), 53–72. https://doi.org/10.1257/jep.32.4.53

Greenstone, M. and Nath, I. (2020). *Do Renewable Portfolio Standards Deliver Cost-Effective Carbon Abatement*. Energy Policy Institute (Working Paper No. 2019-62). University of Chicago.

Gundlach, J., Minsk, R. and Kaufman, N. (2019). *Interactions between a Federal Carbon Tax and Other Climate Policies*. Center on Global Energy Policy, Columbia University. https://www.energypolicy.columbia.edu/research/report/interactions-between-federal-carbon-tax-and-other-climate-policies

Hafstead, M. (2019). *Carbon Pricing 101*. Resources for the Future. https://media.rff.org/documents/Carbon_Pricing_Explainer.pdf

High-Level Commission on Carbon Prices. (2017). *Report of the High-Level Commission on Carbon Prices*. Washington, DC: World Bank. License: Creative Commons Attribution CC BY 3.0 IGO.

International Carbon Action Partnership (2021, August 9). USA – California Cap-and-Trade Program. https://icapcarbonaction.com/en/?option=com_etsmap&task=export&format=pdf&layout=list&systems[]=45

International Monetary Fund. (2019). *Fiscal Policies for Paris Climate Strategies – From Principle to Practice* (Policy Paper No. 19/010). https://www.imf.org/en/Publications/Policy-Papers/Issues/2019/05/01/Fiscal-Policies-for-Paris-Climate-Strategies-from-Principle-to-Practice-46826

Lawton, T.C. and Kock, C.J. (2023) Corporate strategy and climate change: A nonmarket approach to environmental advantage. In A. Sundaram and R. Hansen (Eds), *The Handbook of Business and Climate Change*. Edward Elgar.

Matsumura, E.M., Prakash, R., & Vera-Munoz, S.C. (2014). Firm-value effects of carbon emissions and carbon disclosures. *The Accounting Review*, 89(2), 695–724. http://doi.10.2308/accr-50629.

Nordhaus, W.D. (2007). To tax or not to tax: Alternative approaches to slowing global warming. *Review of Environmental Economics and Policy*, 1(1), 26–44. https://doi.org/10.1093/reep/rem008

Pigou, A.C. (1920). *The Economics of Welfare*. Macmillan & Co.

Rouhani, O.M., Niemeier, D., Gao, H.O. & Bel, G. (2016). Cost-benefit analysis of various California renewable portfolio standard targets: Is a 33% RPS optimal? *Renewable and Sustainable Energy Reviews*, 62, 1122–1132. https://doi.org/10.1016/j.rser.2016.05.049

Salop, S.C. and Scheffman, D.T. (1983). Raising rivals' costs. *American Economic Review Papers and Proceedings*, 73(2), 267–271. https://www.jstor.org/stable/1816853

Smith, B. (2019, April 15). We're increasing our carbon fee as we double down on sustainability. *Microsoft on the Issues*. https://blogs.microsoft.com/on-the-issues/2019/04/15/were-increasing-our-carbon-fee-as-we-double-down-on-sustainability/

Stavins, R.N. & Schmalensee, R. (2013). The SO_2 allowance trading system: The ironic history of a grand policy experiment. *Journal of Economic Perspectives*, *27*(1), 103–122. http://dx.doi.org/10.1257/jep.27.1.103

Stavins, R.N. & Schmalensee, R. (2019). Policy evolution under the clean air act. *Journal of Economic Perspectives*, *33*(4), 27–50. doi: 10.1257/jep.33.4.27

Sundaram, A.K. (2023). Business and climate change. In A. Sundaram and R. Hansen (Eds.), *The Handbook of Business and Climate Change*. Edward Elgar.

Tietenberg, T.H. (2006). *Emissions Trading: Principles and Practice* (2nd ed.). RFF Press.

Tietenberg, T.H. (2013). Reflections: Carbon pricing in practice. *Review of Environmental Economics and Policy*, *7*(2), 313–329. https://doi.org.10.1093/reep/ret008

Victor, D.G. & House, J.C. (2006). BP's emissions trading system. *Energy Policy*, *34*, 2100–2112. https://doi:10.1016/j.enpol.2005.02.014

Volvo Cars. (2021, November 10). Volvo Cars signs zero emission road transport declaration at COP26, reveals groundbreaking internal carbon pricing mechanism [Press release]. https://www.media.volvocars.com/global/en-gb/media/pressreleases/290035/volvo-cars-signs-zero-emission-road-transport-declaration-at-cop26-reveals-groundbreaking-internal-c

Weisbach D.A. & Metcalf, G.E. (2009). The design of a carbon tax. *Harvard Environmental Law Review*, *33*, 499–554.

World Bank. (2021a). *State and Trends of Carbon Pricing 2021*. Washington, DC: World Bank. https://openknowledge.worldbank.org/handle/10986/35620 License: CC BY 3.0 IGO."

World Bank. (2021b). *Carbon Pricing Dashboard*. Washington, DC: World Bank. https://carbonpricingdashboard.worldbank.org

Yale University. (2016). *Yale University's Carbon Charge: Preliminary Results from Learning by Doing*. https://carbon.yale.edu/resources/white-papers.

18. Shifting consumers' decisions towards climate-friendly behavior

Rishad Habib and Katherine White

INTRODUCTION

> It is human action that will determine the climate of the future, not systems beyond our control. (David Wallace-Wells, *The Uninhabitable Earth: Life After Warming*,

Climate change is arguably the most pressing issue of our time, one that may culminate in devastating results for humankind (IPCC, 2014). We have already begun to see the effects these changes can bring with raging wildfires (Alam, 2021), hurricanes, and floods (Brunner et al., 2021) recently occurring at unprecedented levels (Pruitt-Young, 2021). The scale and power of forces behind this shift and its consequences are tremendous, yet how we respond and adapt to climate change remains entirely in our hands. Fortunately, businesses and governments are stepping forward and embracing climate-friendly initiatives (Manel, 2020). This chapter summarizes research that addresses the challenges businesses, governments and other organizations face to effectively promote climate-friendly initiatives to consumers (Trudel, 2019; van Valkengoed & Steg, 2019; White, Habib, et al., 2019; White, Hardisty, et al., 2019). We focus on interventions that encourage individuals to behave in a more climate-friendly manner, given that individual behavior change, along with efforts from the business sector, government policies and other tools, are important levers to combat the growing climate crisis. Individual consumers can help spur change by purchasing more sustainable products, using, and disposing of them in a more sustainable manner, and supporting positive change through voting, advocacy, and collective action. Marketers that successfully appeal to consumers who care about sustainable offerings will be rewarded with faster growth and sales of their climate-friendly products and services. Throughout this chapter we will outline key principles from behavioral science that marketers and policymakers can leverage to encourage climate-friendly consumer behavior more effectively.

Climate-friendly consumer behaviors are those that reduce the negative impacts of climate change through the reduction of greenhouse gas emissions. We've already seen how consumer actions have the power to make a large difference. When people collectively reduced air travel and driving at the start of the COVID-19 pandemic, daily emissions decreased by 17 percent compared with the previous year (Le Quéré et al., 2020). The reduction was as high as 26 percent in some countries and could have been even higher if people had also changed food habits, household energy use, and other actions along with travel behaviors.

This chapter focuses on human consumption behaviors that are most influential in reducing climate change and offers recommendations for how marketers can shift consumer behaviors towards more climate-friendly options. We use the 'SHIFT' framework to categorize behavior-change strategies according to five psychological factors: Social influence, Habit, Individual self, Feelings and cognition, and Tangibility (White, Habib, et al., 2019). The SHIFT framework was developed via a systematic review of behavioral science research in marketing, psychology, and economics to determine the most impactful drivers of sustainable consumer behaviors. We then highlight examples of companies that have used the SHIFT principles to successfully encourage climate-friendly behavior across a variety of domains.

What are the Most Climate-friendly Consumer Behaviors?

The average American has a carbon footprint of 16.2 tons of CO_2 and CO_2-equivalents (tCO_2e) a year, with Canadians following closely behind with 15.6 tCO_2e. In contrast many European countries have emissions much closer to the global average of 4.8 tCO_2e (Ritchie, 2019). These variations are largely due to differences in lifestyles as is evident when comparing countries with similar GDP per capita but widely dissimilar carbon footprints. For instance, in 2018, the United States (US$63,064; 16.2 tCO_2e) had a much higher carbon footprint per capita than Denmark (US$61,592; 6.3 tCO_2e), and Canada (US$46,455; 15.6 tCO_2e) had a much higher carbon footprint per capita than Germany (US$47,950; 9.1 tCO_2e) (World Bank, 2021). If we are to truly make a dent in our greenhouse gas (GHG) emissions, it would be wise to concentrate on the areas where our impact can be greatest. Focusing on emissions in the US, research identifies five broad areas in which behavioral shifts can have the greatest potential to impact greenhouse gas reductions: transportation choices, food choices, energy conservation, reducing waste, and material purchases (Gifford & Nilsson, 2014; Stern, 2000).

1. *Transportation choices*: The US Environmental Protection Agency (EPA) estimates that transportation made up 28 percent of CO_2 output in 2018. For transportation, greenhouse gas emissions come mainly from burning petroleum-based fossil fuels to power vehicles (EPA, 2020). Commuting without a car and forgoing air travel have been shown to save 2.4 tCO_2e per person per year (i.e., 15 percent of the average American's carbon footprint) and 1.6 tCO_2e per roundtrip transatlantic flight (i.e., 10 percent of the average American's carbon footprint), respectively, making them two of the top three ways to reduce CO_2 emissions (Wynes et al., 2018; Wynes & Nicholas, 2017).
2. *Food choices*: Ten percent of US greenhouse gas emissions in 2018 came from agriculture, largely livestock such as cows and other animals (EPA, 2020). Switching to a plant-based diet decreases an individual's carbon footprint by 0.8 tCO_2e per year or 5 percent on average (Wynes & Nicholas, 2017).
3. *Energy conservation*: Twenty-seven percent of US greenhouse gas emissions in 2018 came from electricity usage (EPA, 2020). This is because electricity production and residential-use GHG emissions in the USA come mainly from burning natural gas and coal. Switching to renewable or non-CO_2-emitting sources or reducing high-emission activities can have a substantial impact on CO_2 emissions. For instance, using green energy can reduce up to 1.6 tCO_2e per year or 10 percent of the average yearly footprint. Other actions such as washing clothes in cold water or hanging dry clothes can reduce individual carbon footprints by 0.25 tCO_2e (1.5 percent) and 0.21 tCO_2e (1.2 percent) (Wynes & Nicholas, 2017).
4. *Reducing waste*: The EPA estimates that a family of four could reduce its carbon footprint by 0.97 tCO_2e per year (i.e., 6 percent of the average American's carbon footprint) by refusing, reducing, reusing, repairing, and recycling plastic, glass, textiles, and other materials (EPA, 2021).

5. *Material purchases*: One study of US household emissions finds that clothing, furnishing and housing supplies, and electronics and machinery, account for 12.1 percent, 3.9 percent, and 2.2 percent of total overseas emissions respectively (Song et al., 2019). Carbon emissions from producing plastic are estimated to be 56 gigatons in the next 30 years (Joyce, 2019) and the fashion industry contributes 10 percent of global carbon emissions every year (The World Bank, 2019). Extending the life of clothes, purchasing fewer products, or favoring products made from recycled materials (such as recycled plastic) can reduce carbon emissions.

These five behaviors can substantially reduce our carbon footprint and impact on the climate. In addition, consumers and companies can directly purchase carbon offsets which invest in green energy and climate positive projects to balance emissions.

What Makes Climate Change a Difficult Problem for Marketers?

Climate change and greenhouse gas reduction present thorny challenges for individuals and society. While consumers state high support for climate-friendly companies and actions, they do not always follow through with behaviors, leading to an intention–action gap (Auger & Devinney, 2007). There are four major characteristics of the climate crisis that make climate-friendly behavior more challenging for people to engage in – and for marketers to shift – than other consumption behaviors

1. *Uncertainty.* While there is broad scientific agreement on the fundamental science of climate change and the impact of greenhouse gas levels on temperature rise (IPCC, 2014), the specific consequences of climate change feel uncertain to the average consumer (Habib et al., 2021). Weather is a chaotic system, and it is very difficult to predict the exact effects of a rise in greenhouse gases and temperature on a specific location at a specific point in time. Thus, consumers may be uncertain about the impacts of climate change and how their own individual actions may contribute to the issue.
2. *Psychological distance.* The effects of climate change can feel far away both temporally and geographically. This is because future generations and people in developing countries are likely to feel the brunt of the blow (Reczek et al., 2018). This psychological distance from the effects of climate change makes it harder for people to prioritize addressing climate change.
3. *Invisibility.* A viral video of a plastic straw stuck in a turtle's nose can have such a strong emotional impact that states and countries ban plastic straws. However, unlike sustainability issues such as plastic pollution, the cause of climate change – greenhouse gases – is largely colorless gases, invisible to the human eye. We drive cars, use electricity, and purchase products without noticing the amount of carbon dioxide produced or released. As they say, "out of sight, out of mind". This makes it especially challenging to generate an emotional connection and make the numerous consequences of a heated atmosphere feel tangible.
4. *Need for collective action.* Climate change is a collective problem and requires a collective response. As difficult as it is to change our own behaviors, it can be even more challenging to believe that others will change theirs, reducing individual motivation to change. People may also decide to act as "free-riders," benefitting from the changes that others make while continuing to behave in the way that is most convenient for them, resulting in a collective

action problem also known as the tragedy of the commons, where people do not want to make adjustments or sacrifices for the common good (Olson, 2009; Shultz & Holbrook, 1999).

Following from these challenges, certain factors are particularly relevant in encouraging climate-friendly consumer behaviors. Social influence, habits, individual self, and cognitive factors can be effectively harnessed to encourage climate-friendly behaviors (van Valkengoed & Steg, 2019). However, due to the invisibility and the distant and abstract nature of climate change, tangibility and emotional connection (feelings) can be harder to leverage.

THE SHIFT FRAMEWORK

The SHIFT framework is based on a systematic literature review of over 300 academic articles and consists of five key psychological routes to sustainable consumer behavior change: Social influence, Habit, Individual self, Feelings and cognition, and Tangibility (White, Habib, et al., 2019). Confirming the relevance of these psychological factors, a recent meta-analysis of more than a hundred studies related to climate-change adaptation revealed that norms (Social influence), negative affect (Feelings and cognition), and self-efficacy (Individual self) were most strongly associated with behavior change in this area (van Valkengoed & Steg, 2019). The following sections delve into each of these factors.

Social Influence

The first factor of the framework is social influence, referring to the fact that people's thoughts, beliefs, and actions regarding climate change and consumption are influenced by other people. Below, we outline the important roles of social norms, social group membership, and social desirability and their effects on people's behaviors. We illustrate the role of social influence in Tide's Coldwater campaign and in BC Hydro's Team Power Smart program in a later section on marketing strategies.

Social norms
Social norms are beliefs about what others are doing (descriptive norms) and what they approve of (injunctive norms) in a given situation (Cialdini, 2001; Cialdini & Goldstein, 2004). People are more likely to engage in climate-friendly behaviors such as installing solar panels (Bollinger & Gillingham, 2012), conserving energy (Goldstein et al., 2008; Schultz et al., 2007), recycling (White & Simpson, 2013), and making climate-friendly transportation choices (Harland et al., 1999) when others are doing the same behavior and when they think that others approve of that behavior (Cialdini, 2007; Cialdini et al., 1991). Using positive descriptive and injunctive norms can strongly encourage climate-friendly behavior. However, this approach can be challenging since many climate-friendly behaviors are not yet the norm in terms of what people are currently doing. In such cases, messages emphasizing dynamic norms, or how such behaviors have become more common can be very effective. For instance, when people were informed that eating less meat has become more common over time, i.e. "over the last 5 years, 30 percent of Americans have started to make an effort to limit their meat consumption", they were more likely to change their own behavior and order a meatless

lunch (Sparkman & Walton, 2017). In such situations, people are "conforming in advance," as they try to adopt a behavior that is becoming more prevalent over time. Tide has used the power of social norms to promote the use of their cold-water laundry detergent by asking people not only to sign-up for free samples, but also to share it with their friends (see marketing strategies section for more details).

Social group memberships
In general, people conform to the behaviors of others, particularly when those others are part of a group that one is a member of or strongly identifies with (Goldstein et al., 2008; Han & Stoel, 2017). However, when people find out that a rival group is performing better in a positive behavior, they may be more likely to engage in climate-friendly action in an effort to make their own group look better. For instance, business school students were more likely to recycle when they learned that computer science students were recycling the most (White et al., 2014). Later in this chapter we discuss how BC Hydro, the largest electricity provider in British Columbia, has successfully created a social group around the idea of energy conservation.

Social desirability
Consumers tend to engage in climate-friendly actions that make a positive impression on others. This can be through actions that communicate status, such as owning an electric or hybrid car (Griskevicius et al., 2010). These effects are strongest when the behavior is publicly observable (Green & Peloza, 2014; Peloza et al., 2013). Climate consciousness can also be seen as undesirable; some male consumers avoid climate-friendly behaviors because they associate them with femininity. However, a more masculine messaging can increase the appeal of climate-friendly initiatives to men (Brough et al., 2016).

Habit

Habits are defined as automatic, relatively uncontrolled behaviors that arise after regularly encountered contextual cues (Kurz et al., 2015). They require little conscious processing to carry out and therefore feel effortless and easy. We all have habits that we carry out every day, such as brushing our teeth or tying our shoelaces, that take little effort or thought to accomplish. By encouraging climate-friendly habits, we can make it easier for people to behave in a climate-friendly manner. This is especially important for actions that are enacted on a regular basis and matters less for one-time or relatively infrequent behaviors. For instance, reducing driving and flying both have climate-positive impacts, but while driving is a regular activity, flights are typically occasional. Thus, the strategies in this section will apply more to driving, showering, shopping, and other routine actions, than to infrequent behaviors such as flying.

Over the years, many of us have developed unsustainable habits such as driving everywhere; this is understandable – it is convenient, relatively inexpensive, and easy to do. The first and most difficult step can therefore be breaking existing climate-unfriendly habits.

Moments of change
One line of research indicates that timing is extremely important; habits are easiest to change in moments when many other things are changing (Verplanken, 2011). This relies on the fact that shifts in our environment and routines can help break established contextual cues, making it more effortful to carry out old habits and easier to adopt new ones. For instance, moving to

a new city means figuring out new driving routes and is an optimum time to explore public transportation options (Verplanken et al., 2008; Verplanken & Roy, 2016). Marketers can thus target people who are undergoing life changes such as moving to a new city, graduating college, or starting a new job to direct them on a path of climate-friendly habits. One potential moment of change is the ongoing COVID-19 pandemic, which has already brought about substantial changes in how we live our lives. Once countries, organizations and individuals emerge from this, we will be able to choose how we build our "new normal." The reduction in greenhouse gases we've seen over this period came from lifestyle changes such as working from home and reduced air travel. We can take advantage of this moment of change to envision new climate-friendly habits for the future.

Penalties
Marketers can bestow penalties for climate-unfriendly behavior. One example is an additional charge for disposable coffee cups, a tactic that has been shown to be more effective in encouraging sustainable behavior than rewarding those who bring their own with a discount (Poortinga & Whitaker, 2018). Such changes can be effectively implemented by marketers in stores or other situations where outright bans on disposable cups or bags do not exist or are not feasible. Similarly, taxes, fines, and tariffs are penalties that can reduce climate-unfriendly behaviors such as the usage of gasoline-powered vehicles (Krause, 2009). One study looked at the performance of carbon taxes and greenhouse gas emissions trading systems (ETSs). The authors found that carbon taxes led to up to 6.5 percent reduction and ETSs led to an average of 6 percent reduction between 2008 and 2012 compared with business-as-usual emissions (Haites, 2018). Two other papers investigate specific carbon pricing strategies; one showed that a consumption charge of €80/tCO_2 could lead to a 10 percent reduction in carbon emissions in the EU by 2050 (Pollitt et al., 2020). Another showed that an economy wide $136/tCO2 tax would reduce emissions in the US by 20 percent (Choi et al., 2010). Companies can also look into carbon trading pricing to encourage climate-friendly behavior. However, we need to be cautious when using penalties, as they can be difficult to monitor and lead to negative reactions if they seem unreasonable (Bolderdijk et al., 2012; Fullerton & Kinnaman, 1995; Steg & Vlek, 2009).

In addition to breaking bad habits, organizations can use several strategies to build more climate-friendly habits, including making the climate-friendly choice easy, giving incentives for positive actions, placing prompts at the moment they are most influential, and providing real-time feedback.

Making it easy
There are a wide range of nudges that can help make climate friendly habits easier and thus reduce any barriers individuals face when carrying out positive actions. One of the best ways to make climate-friendly behavior easy is to make it the default option. A paper looking at 40 peer reviewed studies found that setting lower defaults for meat consumption saved an average of 51 kgCO_2 per individual per year (Wynes et al., 2018). In another case, an energy supplier in Europe who used green energy as the default option when customers chose between different options led to approximately 94 percent of people sticking with green electricity (Pichert & Katsikopoulos, 2008). Defaults work so well because they make climate-friendly actions easier, but they are not the only way to do so. We can also make it easy by placing recycling bins close to people (Brothers et al., 1994; Ludwig et al., 1998), or installing low-flow show-

erheads (Wong et al., 2016). In the marketing strategies section, we will go over how Loop, a company that works to reduce waste by increasing packaging reuse, makes it easy for companies and consumers to use high-quality, refillable containers rather than disposable plastic ones, by designing their system around existing consumer habits.

Incentives

Incentives have the opposite effect of penalties, encouraging desired behaviors. In a paper examining 40 studies involving 886,576 participants, the authors found that an incentive for driving less (e.g., employees could cash out parking spots) led to an average of 571 $kgCO_2$ saved per driver per year (Wynes et al., 2018). Incentives must be used carefully; marketers need to make sure that monetary incentives are not too small, and they should consider using non-monetary incentives such as gifts, lottery entries, or public social rewards when possible (Handgraaf et al., 2013; Hutton & McNeill, 1981).

Prompts

Building on the idea that habits involve contextual cues, prompts can be extremely effective in guiding behavior change if they are in the right place at the right time. Prompts are most effective when they are given close to when and where a climate-friendly behavior needs to be performed (Austin et al., 1993; Lehman & Geller, 2004), when they are clear and easy to follow (Werner et al., 1998), and when they are combined with other effective strategies (Delmas et al., 2013). In the marketing strategies section, we discuss how Modo, a cooperative car sharing company in British Columbia, Canada, successfully used prompts to encourage car-sharing.

Feedback

Technology has made it easier to provide clear, real-time feedback on energy use, driving behavior, and other activities related to greenhouse gas emissions. Such feedback can improve behavior, but may also backfire if consumers get the impression that they are already performing better than others (Schultz, 1999; Schultz et al., 2007).

Individual Self

Individuals are driven by a desire to maintain a positive view of themselves, to maximize their self-interests, to be consistent with themselves, and to feel like they can make a difference. In addition, there are variations among individuals that can direct them towards climate-friendly behavior.

Positive self-concept

People want to see themselves as good, moral individuals, and they can do this by engaging in climate-friendly consumption such as buying repurposed or green products (Kamleitner et al., 2019; Paharia, 2020) or buying from an organization with ethical leaders (Van Quaquebeke et al., 2019). In general, those with a greater concern for the environment and belief in climate change are more likely to engage in climate-friendly behavior (Anderson & Cunningham, 1972; Haws et al., 2014).

Self-interest

Organizations can appeal to people's self-interests by highlighting the personal benefits from engaging in climate-friendly behavior, such as energy savings or higher quality. Such messages can also help overcome barriers such as price and performance concerns (Luchs et al., 2010; Luchs & Kumar, 2017). They are most effective when presented in private and when consumers focus on their independent self (Green & Peloza, 2014; White & Simpson, 2013). However, presenting both personal and other-focused benefits together, such as cost savings alongside environmental benefits, can sometimes lower preference for climate-friendly products, as prosocial motivations are crowded out (Edinger-Schons et al., 2018). Self-interest appeals, such as economic benefits to consumers, have been found to be most effective when consumers are thinking concretely, and least effective when consumers are thinking abstractly (Goldsmith et al., 2016).

Self-consistency

People want to act consistently, and one climate-friendly action can lead to another, often referred to as a positive spillover effect (Lanzini & Thøgersen, 2014; Thøgersen & Ölander, 2003). It can thus be beneficial to encourage individuals to engage in an initial action or commitment and encourage more time-consuming or difficult actions later (Bodur et al., 2015). One large field study found that symbolic commitments, such as a lapel pin for committing to reuse towels, increased the number of towels hung rather than washed in a hotel by 40 percent (Baca-Motes et al., 2013). Consumers also respond positively and engage in more climate-friendly actions when companies also consistently engage in climate-friendly actions (Wang et al., 2016).

Self-efficacy

People want to believe that their actions will make a difference in order to feel a sense of self-efficacy. Thus, they are unlikely to act if they think that climate change is an insurmountable challenge. Messages emphasizing the impact individuals can have as well as those calling for collective action are effective in changing behavior (Schutte & Bhullar, 2017; Sparkman et al., 2020). Ikea's "One Little Thing" campaign discussed in the marketing strategies section emphasizes the ability of small actions to make a large collective difference.

Individual differences

Additionally, differences between people can make them more or less interested in climate-friendly action. Marketing can target specific demographic groups that are more likely to act in a climate-friendly manner such as women, liberals, and higher-income and more educated individuals (Davidson & Freudenburg, 1996; Gilg et al., 2005). Messaging can also be designed to appeal to those generally less inclined to act in a climate-friendly manner; this includes appealing to men by making messages more masculine (Brough et al., 2016), and to conservatives by focusing on values they care about such as duty and being part of a group (Kidwell et al., 2013).

Feelings and Cognition

People are responsive to messages that influence them in two different but related ways. The first is an emotional response that occurs immediately and is relatively automatic (often referred to as system 1), and the second is a cognitive, rational response that requires deliberation and time to decide (often referred to as system 2; Shiv & Fedorikhin, 1999; Tversky & Kahneman, 1974, 1981).

Feelings (system 1)
Messages that lead to emotional responses can be highly effective in encouraging climate-friendly behaviors. This is true, for example, when negative emotions such as guilt and fear are activated, since people are motivated to alter negative emotional states (Antonetti & Maklan, 2014; Banerjee et al., 1995; Paharia, 2020). One study found that people chose sustainable granola bars more when they were in a group because they anticipated stronger feelings of guilt (Peloza et al., 2013). However, marketers need to be careful that these feelings are not too negative, as that can lead to pushback or a sense that we cannot make a difference (Peloza et al., 2013). Similarly, marketers should be aware that the effects of emotional states can be temporary and are less likely to affect behaviors once they cool off. One effective strategy is to ask consumers for a behavioral commitment right after seeing an emotional message (Schwartz & Loewenstein, 2017).

Positive emotions can be effective for encouraging climate-friendly action, particularly the emotions of hope (Feldman & Hart, 2018), warm glow (Giebelhausen et al., 2016; Tezer & Bodur, 2020), and pride (Antonetti & Maklan, 2014; Bissing-Olson et al., 2016; Peter & Honea, 2012). One study shows that feelings of inspiration from transformed products led to more recycling and fewer products going to the trash (Winterich et al., 2019). Beyond Meat, a plant-based meat company, has effectively used hope in their consumer campaigns to create a sense of an ideal world that we are all working towards. We discuss this in more detail in the marketing strategies section.

Cognition (system 2)
Consumers may take a more cognitive route to climate-friendly behavior change, focusing on information labelling and the way messages are framed. A potential obstacle to initiating climate-friendly behaviors is that people often underestimate the climate impact of their actions, particularly air travel and meat consumption (Wynes et al., 2020). Clear informative labels, especially those involving third-party validation, can help people make better decisions consistent with their goals (Borin et al., 2011; Thøgersen, 2000; Wynes et al., 2018). One study shows that labelling food with the number of lightbulb hours used to produce it led consumers to shift purchases towards low-emission foods (Camilleri et al., 2019).

Marketers can encourage climate-friendly action by framing information so that it focuses on potential losses, rather than gains, such as energy costs rather than savings (Hardisty & Weber, 2009; J. Min et al., 2014). They can maximize the message's impact by using larger values, such as lifetime costs or cost per 100,000 miles (Camilleri & Larrick, 2014; Kallbekken et al., 2013). These messages are most effective when coupled with concrete information about desirable and beneficial climate-friendly behaviors (White et al., 2011). Last but not least, companies need to ensure that the information they share is accurate and genuine, as consumers react negatively to suspicions of greenwashing (Orazi & Chan, 2020; Pizzetti et al., 2021; Szabo & Webster, 2021).

Tangibility

Climate change can feel like an abstract, uncertain, and distant topic, concerning events that will occur in indeterminate ways to people very far away (McDonald et al., 2015; Van der Linden et al., 2015). One way for marketers to address this barrier to change is to ask consumers to think more abstractly and focus on future benefits, future generations, or one's own legacy (Reczek et al., 2018; Zaval et al., 2015). Another way is to make the consequences of actions feel more concrete, such as by communicating local and/or immediate impacts or sharing personal experiences, and leveraging publicity and interest around naturally occurring climate events (Li et al., 2011; Marx et al., 2007; Scannell & Gifford, 2010; Weber, 2010). Marketers can also highlight specific steps consumers should take (White et al., 2011), or use extreme weather events that make climate change seem closer, in order to influence behavior (Spence et al., 2011). Marketers can also increase consumers' feelings of psychological ownership towards public goods by communicating how they belong to them, such as saying "your park" instead of "the park" (Peck et al., 2021). One exciting approach to tangibility was taken by Offsetters, a carbon-management solutions company that allows individuals and organizations to purchase carbon offsets. They used guerilla marketing strategies and set up installations throughout the city of Vancouver during the winter Olympic Games to make the consequences of climate change feel closer and tangible to passersby. We discuss their strategies in a later section.

In another pro-environmental approach, marketers can shift consumers towards intangibles such as digital products, experiences, or services, such as watching movies, eating out, or getting a massage (Atasoy & Morewedge, 2018; Van Boven & Gilovich, 2003). They can try to reduce people's desire to own objects through promoting voluntary simplicity or by boosting the sharing economy (Donnelly et al., 2016; Eckhardt et al., 2019). Companies such as Rent the Runway for borrowing clothes and Modo for car ownership (see next section) have already changed many people's tendency to purchase goods towards a preference for sharing. Despite the material challenges of the tangibility route in capturing consumer awareness of climate change, creativity in this area of the SHIFT framework (such as these two businesses are demonstrating) will help move the current consumption model in a positive direction.

MARKETING STRATEGIES FOR CLIMATE-FRIENDLY BEHAVIOR

Since human behavior can be difficult to change and people need to work together to make a collective impact on climate emissions, it makes sense to focus behavior change on those areas where we can make the biggest difference. As mentioned at the start of the chapter, the five areas in which consumers can have the largest impact on greenhouse gas emissions by modifying their behavior are transportation choices, food choices, energy conservation, waste disposal, and material purchases (Gifford & Nilsson, 2014; Stern, 2000). In this section we share examples of how organizations have successfully encouraged climate-friendly consumer action in these five important areas as well as in the area of purchasing carbon offsets.[1]

Transportation Choices – Modo

Modo, one of the earliest car-sharing companies, is a member-owned co-operative with a mission to create healthier, more livable communities by reducing car ownership and usage, and encouraging more active and sustainable forms of transportation when possible. To make this vision a reality, Modo harnesses the Individual self and Habit factors from the SHIFT framework.

We will first dive into how Modo uses the Individual self factor by appealing to consumers' self-interest. The company realized early on that their customers valued affordability and convenience, so messaging often highlights these benefits to the individual first. Owning and maintaining a car can be very expensive, and living car-free can free up around 600 Canadian Dollars (CAD) a month, which customers can then spend on other things including rent, education, travel, etc. Modo's messaging also highlights the convenience of car sharing, where there is no need to take the vehicle in for repairs or to manage insurance. By emphasizing the benefits to the self in terms of money saved and convenience, Modo is able to attract a wide range of users and maximize reduction in car ownership and greenhouse gases.

Modo also builds its strategies around consumers' existing habits to make things easy and effectively introduces prompts to ensure consumers receive the right message at the right time. Modo simplifies car-sharing by placing cars where consumers are most likely to need them. One such place is ferry terminals. Letting consumers know that there are cars at ferry terminals, particularly at busy times such as holidays, gives them the option to walk onto ferries and grab a car on the other side conveniently and cheaply. This has been very well-received, with over half of their customers using Modo cars in conjunction with ferries at least once a year.

Modo also effectively uses prompts through a partnership with Poparide, an inter-city ridesharing platform. When Modo customers book a car for 24 hours or more, it serves as an indication that they are going on a longer trip. The company then sends an automatic prompt to ask if they want to make their trip even more affordable by posting it on Poparide and taking passengers with them. The same thing happens when someone is looking for a ride on Poparide; the site prompts them, asking if they would instead like to book a car with Modo and sell their seats. Sharing seats means money saved for the consumer and fewer cars on the road, reducing climate impacts. In fact, for each Modo car, approximately 9–13 private cars no longer use the streets and this reduces greenhouse gas emissions by up to 50 percent. These prompts have been extremely successful; Poparide has referred more new member joins at Modo than any other single promotion or collaboration by providing messages at the point at which customers were most likely to use them.

Energy Conservation – Tide

Tide, Proctor & Gamble's flagship laundry brand, recently announced its ambitions for 2030 (Business Wire, 2021). This includes a series of goals to reduce its carbon footprint, such as encouraging washing in cold water. Tide states that if three out of four loads were washed in cold water, it would reduce GHG emissions by 4.25 million metric tons a year – equivalent to taking a million cars off the road (Tide, 2021). Tide initiated its Coldwater Challenge campaign in 2005 after realizing that over two-thirds of the energy consumption during laundry is based on how consumers use their products. How did the company go about trying to encourage people to use cold water? Tide used messaging around Social influence through online campaigns, by getting people to sign up and share with their friends.

Tide gave out free samples of its new cold-water detergent, and whenever someone signed up for a sample, an orange dot showed up on their location on a US map on the Coldwater Challenge website. The company also encouraged people to share the product and challenge their friends to get them to try it. Whenever a friend signed up, the orange dots became connected by a line. The US map quickly became populated with orange dots connected to each other across states. This communicated to others that their friends and people around the country were switching to cold water, indicating strong social norms around cold water usage. This worked so well that over a million consumers pledged to try it and tell their friends, leading to more than 343 million consumer impressions (Neff, 2005), and showcases the power of social influence in changing consumer behaviors.

Energy Conservation – BC Hydro

BC Hydro is the main electricity provider for residential households in British Columbia, Canada. Besides providing electricity, it has also designed, implemented, and managed energy conservation programs and services in the region for over 30 years. A variety of programs target residential, commercial, and industrial customer segments. All programs have measurable outcomes that are established through rigorous impact evaluations.

A notable, even stand-out feature of BC Hydro is its rigor in testing and persistent focus on energy conservation. During the research phase they brainstorm with leading marketplace organizations, apply behavioral economic and social science principles, map customer journeys, and carry out large-scale testing. Their activities range from providing monetary and social rewards to sharing feedback on energy usage. A detailed impact evaluation process ensures that every program has a positive effect on consumers and energy conservation. One of their most effective strategies involves appealing to the power of the Individual self and Social norms.

BC Hydro encourages people to join Team Power Smart, a growing group of more than 160,000 households in British Columbia (i.e., 8.5 percent of households in the province) who truly care about energy conservation. After joining, the Team Power Smart challenge gives participants $50 if they can reduce their energy usage by 10 percent over 12 months. Once someone joins this group dedicated to energy conservation, they are encouraged to follow group norms around smart energy use. As people are intrinsically motivated to act like a member of the Team, they are interested in taking up challenges, participating in research, and becoming advocates. With an average per capita consumption of 12.9 MWh, the members of Team Power Smart have reduced their consumption by over 200 kWh annually (Government of Canada, 2021). By combining a valued individual identity with a social group that has norms around energy conservation, BC Hydro has created an effective strategy to encourage climate-friendly behavior.

Food Choices – Beyond Meat

Beyond Meat makes plant-based products that taste like meat. The company is arguably most famous for its Beyond Burger, a patty that mimics the texture, taste, and nutritional profile of a ground-beef patty. Part of what makes the Beyond Burger more successful than other plant-based meat alternatives is its appeal to Feelings and to the Individual self.

Beyond Meat appeals to consumer Feelings, particularly hope, in its messaging. A television advertisement asks, "What if?" and proceeds to unfold scenarios where taking cows off the table would make "us and the planet healthier", tying its product to the vision of a healthier, more sustainable, and humane future. By instilling hope in consumers, the campaign brings forward the idea that change and a better future for everyone is possible. Believing in this change is essential if we are to act in a way that brings it about.

By challenging consumers to "go beyond" and imagine a world where most proteins come from plants, using less land and energy, and releasing less greenhouse gases, Beyond Meat appeals to consumers' positive self-concept. The company emphasizes how its products can make the world a better place, associating consumption with more positive views of the self. This, combined with hope of a better world, can be extremely powerful as people are able to feel good about creating a more ideal society. Simultaneously, the brand appeals to consumers' self-interest by emphasizing the taste in every bite. This approach has enabled Beyond Meat to reach more than 100,000 stores, restaurants, hotels, and other organizations and to have an extremely successful initial public offering in 2019 – its stock price rose from $25 to $65 (i.e., 163 percent) on the day of its IPO, followed by a rise of 250 percent in the next two weeks (S. Min, 2019).

Waste Disposal – Loop

Loop is part of TerraCycle, a social enterprise (i.e., a revenue-generating organization with an objective to have social impact) whose mission is to eliminate waste. While TerraCycle has been fighting waste for nearly 20 years, Loop is a relatively new idea that launched in May 2019 with its first two markets in the Northeastern US and Paris, France. They are building a global platform for reuse, working with their many partner brands to move them toward high quality, reusable packaging and close the end-of-cycle loop between consumers and companies through collection, cleaning, and distribution. The concept of Loop is simple, particularly on the consumer end, and has been designed to fit with consumers' existing behaviors, successfully utilizing the power of Habits.

Loop's direct-to-consumer online store allows customers to order brands they are most familiar with using, such as Pantene from P&G, in aesthetically pleasing, high-quality reusable containers delivered in a reusable tote. For instance, Pantene shampoo and conditioner bottles on Loop.com are made with durable aluminum designed to be cleaned, refilled, and reused as part of the Loop system. When customers are ready, they can order their next batch of products and have them delivered and have their existing product packaging, such as shampoo bottles, taken away for refilling and reuse. This is similar to how the milkman operated with deliveries of glass bottles of milk – Loop is effectively modernizing the milkman.

While Loop's first phase has been as an online retailer, its eventual aim is to be in stores around the world. The company has built partnerships with large grocery stores such as Carrefour in Europe, Loblaws in Canada, Tesco in the UK, and Kroger, Walgreens, and Ulta Beauty in the US as well as quick service restaurants such as Burger King and Tim Hortons. Loop launched its first tests in ten Carrefour stores in Paris, France, in November 2020 with remarkable results. Customers shop as they always have and can stick to their existing habitual behaviors. At the point of purchase, they can choose a beautiful, reusable bottle of their favorite products, often made of glass, aluminum, or stainless steel, paying a small deposit for the container. The next time the consumer is in the store, they can return the empty bottle and get back their deposit. Loop then collects the containers and cleans, sanitizes, and returns them to the manufacturer for refilling and reuse, thus closing the loop in the last stage of the consumption cycle. The aim is for packaging in the Loop system to be used hundreds of times, diverting materials from the landfill, and reducing the demand for flimsy, one-time-use plastic products. So far over 90 percent of people participating in the program have returned their containers.

By turning product packaging from a cost passed to the consumer to a company asset, Loop incentivizes higher-quality, longer-lasting packaging over cheap, disposable products. This, in turn, ensures that containers are diverted from the landfill and that more reusable materials are used to create packaging.

Source: Loop.

Figure 18.1 Loop corner in a Carrefour store (Paris). From left to right: Collection point, return bag, Loop product shelf and fresh section

Material Purchases – IKEA

IKEA, the much beloved build-it-yourself furniture company, has set an ambitious goal to "become circular and climate positive", using only renewable or recycled materials by 2030. Its new "One Little Thing" campaign and advertisement use Feelings and Individual self to showcase how small individual efforts to live more sustainably can add up to a large impact and greater collective good. This is complemented by a series of targeted advertisements referring to various items from IKEA that encourage a sustainable lifestyle; for instance, one ad shows a reusable food container labeled "one little container can change the world", echoing the emotion and self-efficacy of the campaign. The company realizes that the pandemic recovery is a turning point for many consumers to build back better, and it hopes to encourage more environmentally friendly habits. As such, IKEA is considering the long-term use of products with a view to making them circular, thereby building reuse, repair, refurbishment, remanufacturing, reselling, and recycling into the design process. The company is also encouraging customers to return used items and purchase pre-loved items from its circular economy hub. As consumers move towards more reusable materials, IKEA aims to provide the quality products needed to meet its customers' sustainability goals as well as the company's own.

Carbon Offsets – Offsetters

Reducing total greenhouse gas emissions is ideal, but it may not always be possible to do so. In situations where it is difficult, undesirable, or even impossible, people can choose to offset their carbon emissions themselves. Offsetters is the leading provider of carbon-management solutions in Canada and allows individuals and organizations to offset their carbon footprint. Offsetters was the official provider of carbon offsets to the Vancouver 2010 Olympic and Paralympic Winter Games, making it the first carbon-neutral Olympic Games in history. Not only did Offsetters create an extensive portfolio of clean energy projects to offset the over 250,000 metric tons of carbon emitted during the Games, but they also engaged consumers in the process using the principles of Tangibility.

During the Games, Offsetters created attention grabbing, thoughtful installations throughout the city to make people aware that "without winter, there could be no winter games." To drive this point home, they created a fictional sport that would become a reality in a climate future: bobwheeling. This is just like bobsledding but without ice, making wheels a necessity. Throughout the city, they set up installations that forebode a future with climate change, including lifejackets on park benches and bus seats, lifeboats that were inflated off the sides of giant buildings and a lifeguard standing on duty in the middle of the street. The message to "stop global warming" was loud and clear and the guerilla marketing brought the idea of climate change to the forefront. This tangible display made consumers realize that this might be the situation if climate change continues unabated. The media campaign generated around 30 million media impressions overall and was covered worldwide. In addition, over 40 sponsors of the Olympic Games including corporate sponsors, governments and broadcasters also chose to offset their carbon emissions.

420 *Handbook of business and climate change*

Source: Rethink Communications.

Figure 18.2   Rethink campaign for Offsetters during the Vancouver 2010 Olympic and Paralympic Winter Games

DISCUSSION

Throughout this chapter, we've seen how the five SHIFT factors can be used to encourage climate-friendly behavior, both in theory and in practice. Social influence, Habit, Individual self, Feelings and cognition, and Tangibility are all effective psychological routes to changing behavior. However, the exact tools that are most appropriate will vary depending on the context. It is important to carefully consider the details of a specific situation: the behavior you are trying to change, the target market, their motivations, values, personalities, etc., the barriers and benefits to behavior change, the medium of communication, and other factors. Only after such careful thought will decision makers be able to select the tool that is most appropriate. For instance, if the target audience consists of people who are very concerned about climate change, relying on the self-concept, and using environmental messaging can be very effective. However, if the target audience has low concerns about climate change, a better approach might be self-interest and tactics that make the climate-friendly action easier, such as defaults.

Marketers can consider combining different aspects of the SHIFT framework to maximize the impact. This could be two or more tools from a single factor or a variety of tools from different factors. As we have seen in the cases in the previous section, organizations have successfully combined different parts of the framework to encourage climate-friendly action. For example, Beyond Meat's advertisements to "go beyond" envision that eating plant-based burgers can help change the world, thus inspiring feelings of hope and contributing to a positive self-concept. Marketers should remain cautious when combining factors and ideally test their strategies with a sample of customers to ensure that such messaging actually resonates with their target audience. Once they test their strategy, they can then confidently scale it to a broader audience.

The research in this chapter is a starting point for encouraging climate-friendly action. There are many directions future research can take to fill existing gaps. First, there is a lack of studies that measure real consumer actions, in both lab and field contexts. Behavioral measures in the lab can include whether people turn off the light when they leave, whether they choose to throw items in the recycling or trash, or what products they choose to take home. Behavioral measures in the field are even more important, as consumers are usually unaware that their actions are being studied and so are unlikely to change their behavior to appeal to experimenters. This can include data on energy use from meters, driving behaviors from smart cars, volume of recycling, and data on product purchases.

Second, we have little idea about the impact of many of these interventions in the long-run. Will advertisements that make the impacts of climate change more tangible still influence behavior after days, weeks, or even months? Understanding the long-term consequences can help marketers design strategies and reinforcement schedules to maximize behavior change. We encourage future work to look at these effects in order to better understand the true impact of each tool and factor. In this respect it can be valuable to partner with companies that track such metrics and that are passionate about making a lasting change. For instance, energy companies can help us understand the timing and frequency of messaging that maximizes energy conservation in the long-run.

Third, while some work has explored combining factors and tools, there is a wide scope for future research to delve into this. Which factors and tools are most useful when combined in a certain context? How can tools within a specific factor be used most effectively? Future work can address such questions to form a better understanding of how to shift consumers towards climate-friendly behavior.

Fourth, we do not fully understand when each technique will be most effective. In general, there are situations in which consumers process information using system 2, which is slow and rational, and other situations in which they make decisions using system 1, which is fast and intuitive. There are, however, many nuances that could be investigated further, such as the consumption situations, types of consumers and product categories in which cognitive information is more persuasive than affective information. There may also be situations where affective and cognitive strategies are more effective together and future research can look into this.

Fifth, future work can also examine changes in the way markets operate and the constraints of market failures in order to understand how consumers will respond. For instance, market failures might occur due to information complexities in making climate-smart purchasing decisions ("what does that number on the yellow Energy Star sticker actually mean?"), financing constraints (e.g., banks may be less willing or knowledgeable to lend for climate-related investments), etc., and might impact the likelihood that consumers will shift their behaviors (Gillingham & Sweeney, 2010). We also need a deeper understanding of how consumers interpret information and whether they perceive and respond differently to climate specific positioning compared with general environmentally friendly or "green" positioning. There is also scope to consider how consumers respond to climate-change related positioning compared with health, wellness, or other positioning. Additionally, technological developments such as the rise of social media, artificial intelligence, automation, surveillance, etc., can influence the ways in which consumers interact with marketing communications and inventions, influencing how they behave. For instance, consumers may respond differently to social media influencers than to traditional celebrities when they promote climate-friendly lifestyles, and they may have adverse reactions to tracking and surveillance tools that hamper their sense of privacy.

The factors and tools we have outlined, while effective at changing consumer behavior, form only a partial solution to the ongoing climate crisis. There are systemic actions that can have a wider influence on behavior, including, but not limited to, actions from government and policymakers, initiatives from large and small organizations, and global alliances. For example, regulations such as minimum standards for energy efficiency, bans on certain products, and national goals to reduce greenhouse gases can all help tackle and ideally rein in the most destructive effects of climate change.

One question we sometimes get asked is whether it makes sense to focus on influencing the actions of individual consumers, and whether, instead, bigger pushes from government and business are needed to meaningfully move the needle on climate change. However, this is not an "either/or" type of situation. Because climate change is such a large and multifaceted issue, it is most certainly going to need to be tackled from multiple angles. We need the individuals within society to SHIFT their actions to support sustainable companies, with their wallets, consumption, use, and disposal of products and services in a sustainable manner, as well as to vote and advocate for change. Behavioral science principles – like those we highlight in the SHIFT Framework – should be used alongside business, regulatory, and other policy levers to tackle the complex challenges that lay ahead (Howlett & Rawat, 2019). These are complementary tools that can be used together in a synergistic fashion. Behavioral tools like the SHIFT factors are not replacements for other system-wide levers, but they will be most effective when they are supported and reinforced by broader governmental policies, business practices, and societal trends. Furthermore, broader policy interventions can be better optimized when combined with these types of behavioral principles (Benartzi et al., 2017). For instance, firms and marketers can make policy levers such as carbon pricing more effective by leveraging behavioral science. Our evidence-based analysis suggests that the factors within the SHIFT Framework – Social influence, Habit, Individual self, Feelings and cognition, and Tangibility – are the most promising tools from the marketing and behavioral science literature to move us forward in this regard.

NOTE

1. We would like to thank Selena McLachlan, Director, Marketing & Communications at Modo; Annika Greve, Business Development Head North America, Loop; and Arien Korteland, Program manager, Residential programs at BC Hydro, for providing insights into successful marketing strategies they have used in their organizations.

REFERENCES

Alam, H. (2021, July 25). Climate change expected to bring longer wildfire seasons and more area burned. *Global News*. https://globalnews.ca/news/8058810/bc-wildfires-climate-change/

Anderson, W. T., & Cunningham, W. H. (1972). The socially conscious consumer. *Journal of Marketing*, *36*(3), 23–31. https://doi.org/10.1177/002224297203600305

Antonetti, P., & Maklan, S. (2014). Feelings that make a difference: How guilt and pride convince consumers of the effectiveness of sustainable consumption choices. *Journal of Business Ethics*, *124*(1), 117–134. https://doi.org/10.1007/s10551-013-1841-9

Atasoy, O., & Morewedge, C. K. (2018). Digital goods are valued less than physical goods. *Journal of Consumer Research, 44*(6), 1343–1357. https://doi.org/10.1093/jcr/ucx102

Auger, P., & Devinney, T. M. (2007). Do what consumers say matter? The misalignment of preferences with unconstrained ethical intentions. *Journal of Business Ethics, 76*(4), 361–383.

Austin, J., Hatfield, D. B., Grindle, A. C., & Bailey, J. S. (1993). Increasing recycling in office environments: The effects of specific, informative cues. *Journal of Applied Behavior Analysis, 26*(2), 247–253.

Baca-Motes, K., Brown, A., Gneezy, A., Keenan, E. A., & Nelson, L. D. (2013). Commitment and behavior change: Evidence from the field. *Journal of Consumer Research, 39*(5), 1070–1084. https://doi.org/10.1086/667226

Banerjee, S., Gulas, C. S., & Iyer, E. (1995). Shades of green: A multidimensional analysis of environmental advertising. *Journal of Advertising, 24*(2), 21–31.

Benartzi, S., Beshears, J., Milkman, K. L., Sunstein, C. R., Thaler, R. H., Shankar, M., Tucker-Ray, W., Congdon, W. J., & Galing, S. (2017). Should governments invest more in nudging? *Psychological Science, 28*(8), 1041–1055. https://doi.org/10.1177/0956797617702501

Bissing-Olson, M. J., Fielding, K. S., & Iyer, A. (2016). Experiences of pride, not guilt, predict pro-environmental behavior when pro-environmental descriptive norms are more positive. *Journal of Environmental Psychology, 45*(Supplement C), 145–153. https://doi.org/10.1016/j.jenvp.2016.01.001

Bodur, H. O., Duval, K. M., & Grohmann, B. (2015). Will you purchase environmentally friendly products? Using prediction requests to increase choice of sustainable products. *Journal of Business Ethics, 129*(1), 59–75.

Bolderdijk, J. W., Lehman, P. K., & Geller, E. S. (2012). Encouraging pro-environmental behaviour with rewards and penalties. *Environmental Psychology: An Introduction,* 233–242.

Bollinger, B., & Gillingham, K. (2012). Peer effects in the diffusion of solar photovoltaic panels. *Marketing Science, 31,* 900–912.

Borin, N., Cerf, D. C., & Krishnan, R. (2011). Consumer effects of environmental impact in product labeling. *Journal of Consumer Marketing, 28*(1), 76–86.

Brothers, K. J., Krantz, P. J., & McClannahan, L. E. (1994). Office paper recycling: A function of container proximity. *Journal of Applied Behavior Analysis, 27*(1), 153–160.

Brough, A. R., Wilkie, J. E. B., Ma, J., Isaac, M. S., & Gal, D. (2016). Is eco-friendly unmanly? The green-feminine stereotype and its effect on sustainable consumption. *Journal of Consumer Research, 43*(4), 567–582. https://doi.org/10.1093/jcr/ucw044

Brunner, M. I., Swain, D. L., Wood, R. R., Willkofer, F., Done, J. M., Gilleland, E., & Ludwig, R. (2021). An extremeness threshold determines the regional response of floods to changes in rainfall extremes. *Communications Earth & Environment, 2*(1), 1–11. https://doi.org/10.1038/s43247-021-00248-x

Business Wire. (2021, March 18). Tide reinvents clean on journey to decarbonize laundry with efforts to turn consumers to cold, explore carbon capture and reduce virgin plastic. *Financial Post.* https://financialpost.com/pmn/press-releases-pmn/business-wire-news-releases-pmn/tide-reinvents-clean-on-journey-to-decarbonize-laundry-with-efforts-to-turn-consumers-to-cold-explore-carbon-capture-and-reduce-virgin-plastic

Camilleri, A. R., & Larrick, R. P. (2014). Metric and scale design as choice architecture tools. *Journal of Public Policy & Marketing, 33*(1), 108–125. https://doi.org/10.1509/jppm.12.151

Camilleri, A. R., Larrick, R. P., Hossain, S., & Patino-Echeverri, D. (2019). Consumers underestimate the emissions associated with food but are aided by labels. *Nature Climate Change, 9*(1), 53–58. https://doi.org/10.1038/s41558-018-0354-z

Choi, J.-K., Bakshi, B. R., & Haab, T. (2010). Effects of a carbon price in the U.S. on economic sectors, resource use, and emissions: An input–output approach. *Energy Policy, 38*(7), 3527–3536. https://doi.org/10.1016/j.enpol.2010.02.029

Cialdini, R. B. (2001). *Influence: Science and Practice* (4th ed.). Allyn & Bacon.

Cialdini, R. B. (2007). Descriptive social norms as underappreciated sources of social control. *Psychometrika, 72*(2), 263. https://doi.org/10.1007/s11336-006-1560-6

Cialdini, R. B., & Goldstein, N. J. (2004). Social influence: Compliance and conformity. *Annual Review of Psychology, 55*(1), 591–621. https://doi.org/10.1146/annurev.psych.55.090902.142015

Cialdini, R. B., Kallgren, C. A., & Reno, R. R. (1991). A focus theory of normative conduct: A theoretical refinement and reevaluation of the role of norms in human behavior. *Advances in Experimental Social Psychology, 24*, 201–234.

Davidson, D. J., & Freudenburg, W. R. (1996). Gender and environmental risk concerns: A review and analysis of available research. *Environment and Behavior, 28*(3), 302–339. https://doi.org/10.1177/0013916596283003

Delmas, M. A., Fischlein, M., & Asensio, O. I. (2013). Information strategies and energy conservation behavior: A meta-analysis of experimental studies from 1975 to 2012. *Energy Policy, 61*, 729–739.

Donnelly, G. E., Lamberton, C., Reczek, R. W., & Norton, M. I. (2016). Social recycling transforms unwanted goods into happiness. *Journal of the Association for Consumer Research, 2*(1), 48–63. https://doi.org/10.1086/689866

Eckhardt, G. M., Houston, M. B., Jiang, B., Lamberton, C., Rindfleisch, A., & Zervas, G. (2019). Marketing in the sharing economy. *Journal of Marketing, 83*(5), 5–27. https://doi.org/10.1177/0022242919861929

Edinger-Schons, L. M., Sipilä, J., Sen, S., Mende, G., & Wieseke, J. (2018). Are two reasons better than one? The role of appeal type in consumer responses to sustainable products. *Journal of Consumer Psychology, 28*(4), 644–664. https://doi.org/10.1002/jcpy.1032

EPA. (2020). *U.S. greenhouse gas emissions and sinks (inventory) 1990-2018* (No. 430-R-20–002). https://www.epa.gov/sites/production/files/2020-04/documents/us-ghg-inventory-2020-main-text.pdf

EPA. (2021). *Carbon Footprint Calculator Climate Change* [Data & Tools]. https://www3.epa.gov/carbon-footprint-calculator

Feldman, L., & Hart, P. S. (2018). Is there any hope? How climate change news imagery and text influence audience emotions and support for climate mitigation policies. *Risk Analysis, 38*(3), 585–602. https://doi.org/10.1111/risa.12868

Fullerton, D., & Kinnaman, T. C. (1995). Garbage, recycling, and illicit burning or dumping. *Journal of Environmental Economics and Management, 29*(1), 78–91.

Giebelhausen, M., Chun, H. H., Cronin, J. J., & Hult, G. T. M. (2016). Adjusting the warm-glow thermostat: How incentivizing participation in voluntary green programs moderates their impact on service satisfaction. *Journal of Marketing, 80*(4), 56–71. https://doi.org/10.1509/jm.14.0497

Gifford, R., & Nilsson, A. (2014). Personal and social factors that influence pro-environmental concern and behaviour: A review. *International Journal of Psychology, 49*(3), 141–157. https://doi.org/10.1002/ijop.12034

Gilg, A., Barr, S., & Ford, N. (2005). Green consumption or sustainable lifestyles? Identifying the sustainable consumer. *Futures, 37*(6), 481–504. https://doi.org/10.1016/j.futures.2004.10.016

Gillingham, K., & Sweeney, J. (2010). Market failure and the structure of externalities. In B. Moselle, J. Padilla & R. Schmalensee (eds), *Harnessing Renewable Energy in Electric Power Systems: Theory, Practice, Policy*. New York: Taylor & Francis.

Goldsmith, K., Newman, G. E., & Dhar, R. (2016). Mental representation changes the evaluation of green product benefits. *Nature Climate Change, 6*(9), 847–850. https://doi.org/10.1038/nclimate3019

Goldstein, N. J., Cialdini, R. B., & Griskevicius, V. (2008). A room with a viewpoint: Using social norms to motivate environmental conservation in hotels. *Journal of Consumer Research, 35*(3), 472–482.

Government of Canada, C. E. R. (2021, March 17). *NEB – Provincial and Territorial Energy Profiles – British Columbia*. Canada Energy Regulator. https://www.cer-rec.gc.ca/en/data-analysis/energy-markets/provincial-territorial-energy-profiles/provincial-territorial-energy-profiles-british-columbia.html

Green, T., & Peloza, J. (2014). Finding the right shade of green: The effect of advertising appeal type on environmentally friendly consumption. *Journal of Advertising, 43*(2), 128–141.

Griskevicius, V., Tybur, J. M., & Bergh, B. V. den. (2010). Going green to be seen: Status, reputation, and conspicuous conservation. *Journal of Personality and Social Psychology, 98*(3), 392–404.

Habib, R., White, K., Hardisty, D. J., & Zhao, J. (2021). Shifting consumer behavior to address climate change. *Current Opinion in Psychology, 42*, 108–113.

Haites, E. (2018). Carbon taxes and greenhouse gas emissions trading systems: What have we learned? *Climate Policy, 18*(8), 955–966. https://doi.org/10.1080/14693062.2018.1492897

Han, T.-I., & Stoel, L. (2017). Explaining socially responsible consumer behavior: A meta-analytic review of theory of planned behavior. *Journal of International Consumer Marketing, 29*(2), 91–103.

Handgraaf, M. J. J., Van Lidth de Jeude, M. A., & Appelt, K. C. (2013). Public praise vs. private pay: Effects of rewards on energy conservation in the workplace. *Ecological Economics, 86*, 86–92. https://doi.org/10.1016/j.ecolecon.2012.11.008

Hardisty, D. J., & Weber, E. U. (2009). Discounting future green: Money versus the environment. *Journal of Experimental Psychology: General, 138*(3), 329.

Harland, P., Staats, H., & Wilke, H. A. (1999). Explaining proenvironmental intention and behavior by personal norms and the theory of planned behavior. *Journal of Applied Social Psychology, 29*(12), 2505–2528.

Haws, K. L., Winterich, K. P., & Naylor, R. W. (2014). Seeing the world through GREEN-tinted glasses: Green consumption values and responses to environmentally friendly products. *Journal of Consumer Psychology, 24*(3), 336–354.

Howlett, M., & Rawat, S. (2019). Behavioral Science and Climate Policy. In *Oxford Research Encyclopedia of Climate Science*. Oxford University Press. https://doi.org/10.1093/acrefore/9780190228620.013.624

Hutton, R. B., & McNeill, D. L. (1981). The value of incentives in stimulating energy conservation. *Journal of Consumer Research, 8*(3), 291–298.

IPCC. (2014). *AR5 climate change 2014: Impacts, adaptation, and vulnerability*. Cambridge University Press, Cambridge. https://www.ipcc.ch/report/ar5/wg2/

Joyce, C. (2019, July 9). Plastic has a big carbon footprint—But that isn't the whole story. *NPR.Org*. https://www.npr.org/2019/07/09/735848489/plastic-has-a-big-carbon-footprint-but-that-isnt-the-whole-story

Kallbekken, S., Sælen, H., & Hermansen, E. A. (2013). Bridging the energy efficiency gap: A field experiment on lifetime energy costs and household appliances. *Journal of Consumer Policy, 36*(1), 1–16.

Kamleitner, B., Thürridl, C., & Martin, B. A. S. (2019). A Cinderella story: How past identity salience boosts demand for repurposed products. *Journal of Marketing, 83*(6), 76–92. https://doi.org/10.1177/0022242919872156

Kidwell, B., Farmer, A., & Hardesty, D. M. (2013). Getting liberals and conservatives to go green: Political ideology and congruent appeals. *Journal of Consumer Research, 40*(2), 350–367. https://doi.org/10.1086/670610

Krause, R. M. (2009). Developing conditions for environmentally sustainable consumption: Drawing insight from anti-smoking policy. *International Journal of Consumer Studies, 33*(3), 285–292. https://doi.org/10.1111/j.1470-6431.2009.00769.x

Kurz, T., Gardner, B., Verplanken, B., & Abraham, C. (2015). Habitual behaviors or patterns of practice? Explaining and changing repetitive climate-relevant actions. *Wiley Inter- Disciplinary Reviews: Climate Change, 6*(1), 113–128. https://doi.org/10.1002/wcc.327

Lanzini, P., & Thøgersen, J. (2014). Behavioural spillover in the environmental domain: An intervention study. *Journal of Environmental Psychology, 40*, 381–390.

Le Quéré, C., Jackson, R. B., Jones, M. W., Smith, A. J. P., Abernethy, S., Andrew, R. M., De-Gol, A. J., Willis, D. R., Shan, Y., Canadell, J. G., Friedlingstein, P., Creutzig, F., & Peters, G. P. (2020). Temporary reduction in daily global CO_2 emissions during the COVID-19 forced confinement. *Nature Climate Change, 10*(7), 647–653. https://doi.org/10.1038/s41558-020-0797-x

Lehman, P. K., & Geller, E. S. (2004). Behavior analysis and environmental protection: Accomplishments and potential for more. *Behavior and Social Issues, 13*(1), 13.

Li, Y., Johnson, E. J., & Zaval, L. (2011). Local warming: Daily temperature change influences belief in global warming. *Psychological Science, 22*(4), 454–459. https://doi.org/10.1177/0956797611400913

Luchs, M. G., & Kumar, M. (2017). "Yes, but this other one looks better/works better": How do consumers respond to trade-offs between sustainability and other valued attributes? *Journal of Business Ethics, 140*(3), 567–584.

Luchs, M. G., Naylor, R. W., Irwin, J. R., & Raghunathan, R. (2010). The sustainability liability: Potential negative effects of ethicality on product preference. *Journal of Marketing, 74*(5), 18–31. https://doi.org/10.1509/jmkg.74.5.18

Ludwig, T. D., Gray, T. W., & Rowell, A. (1998). Increasing recycling in academic buildings: A systematic replication. *Journal of Applied Behavior Analysis, 31*(4), 683–686.

Manel, K. (2020, October 7). *Companies are making major climate pledges. Here's what they really mean*. HuffPost Canada. https://www.huffpost.com/entry/companies-climate-pledges-what-they-mean_n_5f7b6237c5b66fab25dc3079

Marx, S. M., Weber, E. U., Orlove, B. S., Leiserowitz, A., Krantz, D. H., Roncoli, C., & Phillips, J. (2007). Communication and mental processes: Experiential and analytic processing of uncertain climate information. *Global Environmental Change*, *17*(1), 47–58. https://doi.org/10.1016/j.gloenvcha.2006.10.004

McDonald, R. I., Chai, H. Y., & Newell, B. R. (2015). Personal experience and the 'psychological distance' of climate change: An integrative review. *Journal of Environmental Psychology*, *44*(44), 109–118. https://doi.org/10.1016/j.jenvp.2015.10.003

Min, J., Azevedo, I. L., Michalek, J., & de Bruin, W. B. (2014). Labeling energy cost on light bulbs lowers implicit discount rates. *Ecological Economics*, *97*, 42–50.

Min, S. (2019, May 16). Eat dirt, Uber! Beyond Meat is most successful IPO of 2019 so far. *CBS News*. https://www.cbsnews.com/news/beyond-meat-ipo-most-successful-initial-public-offering-of-2019-so-far/

Neff, J. (2005, November 7). *Tide Coldwater*. AdAge. https://adage.com/article/special-report-marketing-50/tide-coldwater/105098

Olson, M. (2009). *The Logic of Collective Action: Public Goods and the Theory of Groups*. Harvard University Press.

Orazi, D. C., & Chan, E. Y. (2020). "They did not walk the green talk!:" How information specificity influences consumer evaluations of disconfirmed environmental claims. *Journal of Business Ethics*, *163*(1), 107–123. https://doi.org/10.1007/s10551-018-4028-6

Paharia, N. (2020). Who receives credit or blame? The effects of made-to-order production on responses to unethical and ethical company production practices. *Journal of Marketing*, *84*(1), 88–104. https://doi.org/10.1177/0022242919887161

Peck, J., Kirk, C. P., Luangrath, A. W., & Shu, S. B. (2021). Caring for the commons: Using psychological ownership to enhance stewardship behavior for public goods. *Journal of Marketing*, *85*(2), 33–49. https://doi.org/10.1177/0022242920952084

Peloza, J., White, K., & Shang, J. (2013). Good and guilt-free: The role of self-accountability in influencing preferences for products with ethical attributes. *Journal of Marketing*, *77*(1), 104–119.

Peter, P. C., & Honea, H. (2012). Targeting social messages with emotions of change: The call for optimism. *Journal of Public Policy & Marketing*, *31*(2), 269–283. https://doi.org/10.1509/jppm.11.098

Pichert, D., & Katsikopoulos, K. V. (2008). Green defaults: Information presentation and pro-environmental behaviour. *Journal of Environmental Psychology*, *28*(1), 63–73.

Pizzetti, M., Gatti, L., & Seele, P. (2021). Firms talk, suppliers walk: Analyzing the locus of greenwashing in the blame game and introducing 'vicarious greenwashing.' *Journal of Business Ethics*, *170*(1), 21–38. https://doi.org/10.1007/s10551-019-04406-2

Pollitt, H., Neuhoff, K., & Lin, X. (2020). The impact of implementing a consumption charge on carbon-intensive materials in Europe. *Climate Policy*, *20*(sup1), S74–S89. https://doi.org/10.1080/14693062.2019.1605969

Poortinga, W., & Whitaker, L. (2018). Promoting the use of reusable coffee cups through environmental messaging, the provision of alternatives and financial incentives. *Sustainability*, *10*(3), 873. https://doi.org/10.3390/su10030873

Pruitt-Young, S. (2021, September 11). Climate change is making natural disasters worse—along with our mental health. *NPR*. https://www.npr.org/2021/09/11/1035241392/climate-change-disasters-mental-health-anxiety-eco-grief

Reczek, R. W., Trudel, R., & White, K. (2018). Focusing on the forest or the trees: How abstract versus concrete construal level predicts responses to eco-friendly products. *Journal of Environmental Psychology*, *57*, 87–98. https://doi.org/10.1016/j.jenvp.2018.06.003

Ritchie, H. (2019, October 4). *Where in the world do people emit the most CO_2?* Our World in Data. https://ourworldindata.org/per-capita-co2

Scannell, L., & Gifford, R. (2010). The relations between natural and civic place attachment and pro-environmental behavior. *Journal of Environmental Psychology*, *30*(3), 289–297. https://doi.org/10.1016/j.jenvp.2010.01.010

Schultz, P. W. (1999). Changing behavior with normative feedback interventions: A field experiment on curbside recycling. *Basic and Applied Social Psychology*, *21*(1). http://www.leaonline.com/doi/abs/10.1207/s15324834basp2101_3

Schultz, P. W., Nolan, J. M., Cialdini, R. B., Goldstein, N. J., & Griskevicius, V. (2007). The constructive, destructive, and reconstructive power of social norms. *Psychological Science*, *18*(5), 429–434. https://doi.org/10.1111/j.1467-9280.2007.01917.x

Schutte, N. S., & Bhullar, N. (2017). Approaching environmental sustainability: Perceptions of self-efficacy and changeability. *The Journal of Psychology*, *151*(3), 321–333. https://doi.org/10.1080/00223980.2017.1289144

Schwartz, D., & Loewenstein, G. (2017). The chill of the moment: Emotions and proenvironmental behavior. *Journal of Public Policy & Marketing*, *36*(2), 255–268. https://doi.org/10.1509/jppm.16.132

Shiv, B., & Fedorikhin, A. (1999). Heart and mind in conflict: The interplay of affect and cognition in consumer decision making. *Journal of Consumer Research*, *26*(3), 278–292.

Shultz, C. J., & Holbrook, M. B. (1999). Marketing and the tragedy of the commons: A synthesis, commentary, and analysis for action. *Journal of Public Policy & Marketing*, *18*(2), 218–229. https://doi.org/10.1177/074391569901800208

Song, K., Qu, S., Taiebat, M., Liang, S., & Xu, M. (2019). Scale, distribution and variations of global greenhouse gas emissions driven by U.S. households. *Environment International*, *133*, 105137. https://doi.org/10.1016/j.envint.2019.105137

Sparkman, G., & Walton, G. M. (2017). Dynamic norms promote sustainable behavior, even if it is counternormative. *Psychological Science*, *28*(11), 1663–1674. https://doi.org/10.1177/0956797617719950

Sparkman, G., Howe, L., & Walton, G. (2020). How social norms are often a barrier to addressing climate change but can be part of the solution. *Behavioural Public Policy*, 1–28. https://doi.org/10.1017/bpp.2020.42

Spence, A., Poortinga, W., Butler, C., & Pidgeon, N. F. (2011). Perceptions of climate change and willingness to save energy related to flood experience. *Nature Climate Change*, *1*(1), 46–49. https://doi.org/10.1038/nclimate1059

Steg, L., & Vlek, C. (2009). Encouraging pro-environmental behaviour: An integrative review and research agenda. *Journal of Environmental Psychology*, *29*(3), 309–317.

Stern, P. C. (2000). New environmental theories: Toward a coherent theory of environmentally significant behavior. *Journal of Social Issues*, *56*(3), 407–424.

Szabo, S., & Webster, J. (2021). Perceived greenwashing: The effects of green marketing on environmental and product perceptions. *Journal of Business Ethics*, *171*(4), 719–739. https://doi.org/10.1007/s10551-020-04461-0

Tezer, A., & Bodur, H. O. (2020). The greenconsumption effect: How using green products improves consumption experience. *Journal of Consumer Research*, *47*(1), 25–39. https://doi.org/10.1093/jcr/ucz045

Thøgersen, J. (2000). Psychological determinants of paying attention to eco-labels in purchase decisions: Model development and multinational validation. *Journal of Consumer Policy*, *23*(3), 285–313.

Thøgersen, J., & Ölander, F. (2003). Spillover of environment-friendly consumer behaviour. *Journal of Environmental Psychology*, *23*(3), 225–236.

Tide. (2021, March). *Tide Ambition 2030*. Loads of Good. https://tide.com/en-us/our-commitment/a-load-of-good

Trudel, R. (2019). Sustainable consumer behavior. *Consumer Psychology Review*, *2*(1), 85–96. https://doi.org/10.1002/arcp.1045

Tversky, A., & Kahneman, D. (1974). Judgment under uncertainty: Heuristics and biases. *Science*, *185*(4157), 1124–1131. https://doi.org/10.1126/science.185.4157.1124

Tversky, A., & Kahneman, D. (1981). The framing of decisions and the psychology of choice. *Science*, *211*(4481), 453–458.

Van Boven, L., & Gilovich, T. (2003). To do or to have? That is the question. *Journal of Personality and Social Psychology*, *85*, 1193–1202.

Van der Linden, S., Maibach, E., & Leiserowitz, A. (2015). Improving public engagement with climate change: Five "best practice" insights from psychological science. *Perspectives on Psychological Science*, *10*(6), 758–763. https://doi.org/10.1177/1745691615598516

Van Quaquebeke, N., Becker, J. U., Goretzki, N., & Barrot, C. (2019). Perceived ethical leadership affects customer purchasing intentions beyond ethical marketing in advertising due to moral identity self-congruence concerns. *Journal of Business Ethics*, *156*(2), 357–376. https://doi.org/10.1007/s10551-017-3577-4

van Valkengoed, A. M., & Steg, L. (2019). Meta-analyses of factors motivating climate change adaptation behaviour. *Nature Climate Change*, *9*(2), 158–163. https://doi.org/10.1038/s41558-018-0371-y

Verplanken, B. (2011). Old habits and new routes to sustainable behaviour. In L. Whitmarsh, I. Lorenzoni, & S. O'Neill (eds), *Engaging the Public with Climate Change: Behaviour Change and Communication*. New York: Taylor & Francis, 17–30.

Verplanken, B., & Roy, D. (2016). Empowering interventions to promote sustainable lifestyles: Testing the habit discontinuity hypothesis in a field experiment. *Journal of Environmental Psychology*, *45*(Supplement C), 127–134. https://doi.org/10.1016/j.jenvp.2015.11.008

Verplanken, B., Walker, I., Davis, A., & Jurasek, M. (2008). Context change and travel mode choice: Combining the habit discontinuity and self-activation hypotheses. *Journal of Environmental Psychology*, *28*(2), 121–127.

Wallace-Wells, D. (2020). *The Uninhabitable Earth: Life After Warming*. Crown/Archetype.

Wang, W., Krishna, A., & McFerran, B. (2016). Turning off the lights: Consumers' environmental efforts depend on visible efforts of firms. *Journal of Marketing Research*, *54*(3), 478–494. https://doi.org/10.1509/jmr.14.0441

Weber, E. U. (2010). What shapes perceptions of climate change? *Wiley Interdisciplinary Reviews: Climate Change*, *1*(3), 332–342.

Werner, C. M., Rhodes, M. U., & Partain, K. K. (1998). Designing effective instructional signs with schema theory: Case studies of polystyrene recycling. *Environment and Behavior*, *30*(5), 709–735.

White, K., & Simpson, B. (2013). When do (and don't) normative appeals influence sustainable consumer behaviors? *Journal of Marketing*, *77*(2), 78–95.

White, K., Habib, R., & Hardisty, D. J. (2019). How to SHIFT consumer behaviors to be more sustainable: A literature review and guiding framework. *Journal of Marketing*, *83*(3), 22–49. https://doi.org/10.1177/0022242919825649

White, K., Hardisty, D. J., & Habib, R. (2019, July 1). The elusive green consumer. *Harvard Business Review, July–August 2019*. https://hbr.org/2019/07/the-elusive-green-consumer

White, K., MacDonnell, R., & Dahl, D. W. (2011). It's the mind-set that matters: The role of construal level and message framing in influencing consumer efficacy and conservation behaviors. *Journal of Marketing Research*, *48*(3), 472–485.

White, K., Simpson, B., & Argo, J. J. (2014). The motivating role of dissociative out-groups in encouraging positive consumer behaviors. *Journal of Marketing Research*, *51*(4), 433–447.

Winterich, K. P., Nenkov, G. Y., & Gonzales, G. E. (2019). Knowing what it makes: How product transformation salience increases recycling. *Journal of Marketing*, *83*(4), 21–37. https://doi.org/10.1177/0022242919842167

Wong, L., Mui, K., & Zhou, Y. (2016). Impact evaluation of low flow showerheads for Hong Kong residents. *Water*, *8*(7), 305. https://doi.org/10.3390/w8070305

World Bank. (2019, September 23). *How Much Do Our Wardrobes Cost to the Environment?* World Bank. https://www.worldbank.org/en/news/feature/2019/09/23/costo-moda-medio-ambiente

World Bank. (2021). *GDP per Capita (current US$)*. The World Bank. https://data.worldbank.org/indicator/NY.GDP.PCAP.CD?most_recent_value_desc=true

Wynes, S., & Nicholas, K. A. (2017). The climate mitigation gap: Education and government recommendations miss the most effective individual actions. *Environmental Research Letters*, *12*(7), 074024. https://doi.org/10.1088/1748-9326/aa7541

Wynes, S., Nicholas, K. A., Zhao, J., & Donner, S. D. (2018). Measuring what works: Quantifying greenhouse gas emission reductions of behavioural interventions to reduce driving, meat consumption, and household energy use. *Environmental Research Letters*, *13*(11), 113002. https://doi.org/10.1088/1748-9326/aae5d7

Wynes, S., Zhao, J., & Donner, S. D. (2020). How well do people understand the climate impact of individual actions? *Climatic Change*, *162*(3), 1521–1534. https://doi.org/10.1007/s10584-020-02811-5

Zaval, L., Markowitz, E. M., & Weber, E. U. (2015). How will I be remembered? Conserving the environment for the sake of one's legacy. *Psychological Science*, *26*(2), 231–236.

PART V

CLIMATE FINANCE

19. Mainstreaming climate action in public and private investments: mobilizing finance towards sustainable investments through the bond markets

Heike Reichelt, David P. Allen and Scott M. Cantor

GLOSSARY

AfDB	African Development Bank
CBDR	Common But Differentiated Responsibilities and Respective Capabilities
CDM	Clean Development Mechanism
CDP	Carbon Disclosure Project
CER	Certified Emission Reductions
COP	Conference of Parties
EIB	European Investment Bank
ESG	Environmental, Social and Governance
EU	European Union
EU ETS	Emissions Trading System
GCF	Green Climate Fund
GDP	Gross Domestic Product
GHG	Greenhouse Gases
GRI	Global Reporting Initiative
ICMA	International Capital Markets Association
IFRS	International Financial Reporting Standard
IMF	International Monetary Fund
IPCC	Intergovernmental Panel on Climate Change
ITS	International Emission Trading
JI	Joint Implementation
MDBs	Multilateral Development Banks
NDCs	Nationally Determined Contributions
NGFS	Network of Central Banks and Supervisors for the Greening of the Financial System
NGO	Nongovernmental Organization
ODA	Official Development Assistance
OECD	Organization for Economic Co-operation and Development
RGGI	Regional Greenhous Gas Initiative
SASB	Sustainability Accounting Standards Board
SDGs	Sustainable Development Goals

SEB	Skandinaviska Enskilda Banken
SFDR	Sustainability Related Financial Disclosure Regulation
TCFD	Task Force on Climate-Related Financial Disclosures
UNCTAD	UN Commission on Trade and Development
UNEP	UN Environment Programme
UNFCCC	United Nations Framework Convention on Climate Change
WCI	Western Climate Initiative

INTRODUCTION

The compounding impacts of climate change are leading to a cascade of global tipping points that would cause irreversible change to the Earth's ecosystems. These negative impacts are well documented: rising sea levels, melting ice sheets, frequent and exacerbated extreme weather events, heat waves, flooding, droughts and crop failure, increased vector borne diseases and pests, among others (IPCC, 2021). These climate-related shocks disproportionally impact the poor who do not have the asset accumulation or access to adequate social safety nets. Each year, extreme natural disasters force 26 million people into poverty and reduce global annual consumption by $520 billion (Hallegatte et al., 2017). Without intervention, climate-related shocks could thrust an additional 100 million people into poverty by 2030, and result in over 200 million climate migrants by mid-century, as families and whole communities are driven to find more hospitable and climate-resilient places to reside (Clement et al., 2021).

To achieve mid-century decarbonization and improve the resiliency of vulnerable populations to climate impacts, finance needs to be channeled into low-carbon, climate-smart solutions. Extensive and rapid structural changes are required to prevent and to lessen the impact climate change will have on poverty (Hallegatte et al., 2019). In low and middle-income countries, natural disasters can cause $18 billion in damage repairs to the transport and power generation sectors alone. Damage from these events can be further intensified as communities face disruption of their day-to-day activities, costing households and firms at least $390 billion in low and middle-income countries (Hallegatte et al., 2019). Yet, the payback to investments are real and measurable; for example, investing in resilient infrastructure provides an overall net benefit of $4 for every $1 invested in low- and middle-income countries, resulting in $4.2 trillion over the lifetime of new infrastructure (Hallegatte et al., 2019). Yet, despite the return on investment and long-term environmental benefits, especially when external costs are included in financial models, the upfront costs may be considered too great to bear.

Closing the financing gap for low-carbon, climate-resilient solutions to meet the goals of the Paris Agreement and the Sustainable Development Goals (SDGs) will require substantial investment from both public and private sources. The Intergovernmental Panel on Climate Change (IPCC) estimates that climate investment will need to be $1.6 to $3.8 trillion annually between 2016 and 2050 to achieve a low-carbon transition (IPCC, 2018). Simultaneously, the Global Commission on Adaptation and the UN Environment Programme (UNEP) estimate that $300 billion is needed annually for adaptation costs to enable communities to keep up with the rapidly changing environment (Global Center on Adaptation, 2020). In response, as the largest provider of international climate investments in developing countries, Multilateral Development Banks (MDBs) are helping to catalyze other sources of climate finance and assist countries in integrating climate action into their core development agendas. In 2020, MDBs committed $38 billion to climate finance in developing and emerging economies, and $28 billion for high-income countries (totaling $66 billion) (European Bank for Reconstruction and Development, 2020). Approximately 76 percent of that amount went to mitigation efforts, while the remaining amounts went to adaptation efforts (European Bank for Reconstruction and Development, 2020). This disparity and need for an increase in adaptation and resilience finance by developing has been highlighted by the international community, including at COP26 with the launch of *The Glasgow – Sharm el-Sheikh Work Programme on the Global Goal on Adaptation* (UNFCC, 2021). For its part, the World Bank, the largest source of MDB financing, committed to making adaptation and resilient financing 50 percent of its climate finance as part of its updated Climate Action Plan in 2021 (World Bank, 2021a).

Despite the billions of dollars in climate investments mobilized by MDBs, current climate finance flows fall short of the amount of funding necessary to limit global warming to well below 2°C. In 2020, total global climate finance flows reached a two-year 2019/2020 average of $632 billion, representing an annual growth rate between 2013/2014 and 2019/2020 of 24 percent each period, but still short of the amount recommended by the IPCC (Buchner, 2021). In this context, public sector resources must be used in a transformative manner to incentivize the rapid scale up of private finance to achieve enhanced national climate targets and prompt a low-carbon transition across economies.

The COVID-19 pandemic has brought a tremendous toll to human life as well as caused an unprecedented level of economic hardships. In response, governments have had to focus on their immediate efforts to reinforce health systems, maintain economic growth, and alleviate job loss. This shift in development priorities at first seemed at odds with setting long-term targets and financing goals to address climate change. In 2020, countries were expected to act to limit global rising temperatures by submitting enhanced national climate strategies, known as Nationally Determined Contributions (NDCs), in line with the Paris Agreement. Yet despite the postponement of high-level climate dialogues, the pandemic offered an opportunity to build back in a way that fosters a more inclusive, and also green and resilient recovery, as countries try to stimulate their economies and align stimulus packages with their NDCs. This alignment of recovery packages with the Paris Agreement and NDCs can serve as an impetus for rebuilding equitable economies that are resilient to future health shocks and limit climate impacts.

While traditional climate investment efforts initially focused on mobilizing public finance, capital markets and commercial bank lending are increasingly recognized as important components of the climate finance architecture. Private firms and financial institutions can greatly affect climate outcomes by integrating climate metrics into business models and strategies through measuring and disclosing climate risks. Through initiatives such as the Task Force on Climate-Related Financial Disclosures (TCFD), firms can adopt a voluntary set of climate-related financial risk disclosures that can inform investors and the public about the risks companies face related to climate change. These disclosures contribute to greater understanding of climate risks and facilitate more sustainable investment. Similarly, as argued in this chapter, financing raised through green bonds, and the impact it has on investor behavior, can help shift the direction of climate investment towards low-carbon solutions. The two-year average for issuance of green bonds in 2019/2020 reached $287 billion, compared with just $44 billion in 2014/2015 five years earlier (Buchner, 2021). While this amount captures only less than one percent of the overall total bond market, the rapid issuance of green bonds has had an outsized effect on the market. All issuers of green bonds measure, track, and report on the social and environmental impact of their investments, creating a fundamental shift in the bond market as investors respond to initiatives and projects that provide both financial and social returns for their stakeholders. Overall, there is recognition that climate risk is a threat to the stability of the financial system, including in the United States and Europe, driving momentum towards integrating costs through data analysis of financial decisions, in order to reduce risks and channel more funds towards sustainable activities (Behnam et al., 2020).

This chapter develops the thesis that the green bond market has catalyzed a change in investor strategies. As investors have become more aware of the risks to their investments and increasingly seek to emphasize the sustainable impacts of their investments to foster systemic changes which, in turn, will have long-term benefit on the value of their assets, they can redirect their capital. In the wake of the economic downturn caused by COVID-19, it has become even clearer for investors in capital markets to equitably address the interconnected risks between climate, health systems, and the economy, and take a more holistic approach. Green bonds remain a crucial starting point for promoting transparency in the market and integrating environmental, social, and corporate governance (ESG) considerations into investments. This chapter will explore the existing global framework of climate finance, the development of the capital markets into this architecture, and the future of socially responsible investment in achieving the goals of the Paris Agreement and the 2030 Sustainable Development Goals (SDGs).

HISTORY OF PUBLIC CLIMATE FINANCE

The current flows and mechanisms of climate finance[1] have been influenced and driven by three main phases, each building on the progress of the previous phase before. These three phases were initialized by: (i) the signing of the United Nations Framework Convention on Climate Change and enforcement of the Kyoto Protocol; (ii) the $100 billion finance commitment and adoption of the Paris Agreement; and (iii) the introduction of the "billions to trillions" strategy and the active cultivation of the private sector. Throughout each phase, global ambition has steadily increased for mobilizing financing and investments that contribute to climate mitigation and adaption efforts, while accelerating low-carbon, resilient growth.

First Phase

The current climate finance architecture was started with the signing of the United Nations Framework Convention on Climate Change (UNFCCC), in 1992 at the Earth Summit in Rio. Originating with 154 signatories, member nations of the UNFCCC committed to reducing concentrations of greenhouse gases (GHG) with the goal of "preventing dangerous anthropogenic interference with Earth's climate system." Article 3 of the Convention introduced the principle of "common, but differentiated responsibilities and respective capabilities," (CBDR) in which countries' level of development and historical GHG emissions would dictate their share of the global burden in reducing GHG concentrations (United Nations Framework Convention on Climate Change, 1992). The Convention further delineated Parties into two different categories: 'Annex 1' parties for developed counties, and 'non-Annex 1' for developing countries. Under the UNFCCC, Annex 1 Parties were to aim to stabilize their GHG emissions at 1990 levels by 2000.

Subsequent annual meetings of the UNFCCC signatories led to the Conference of Parties (COP) process in which members would discuss actions for accomplishing the objective of the treaty. As the COP meetings progressed, UNFCCC parties concluded that the aim of Annex 1 countries in stabilizing their GHG emissions to 1990 levels was not adequate. Further discussion led to the adoption of the Kyoto Protocol at COP3 in 1997, which codified the delineation of country annex classifications and cemented the principle of CBDR (Rajamani, 2000). This bifurcation led to Annex I countries' emission reduction targets to be legally bound.

In order for Annex 1 Parties to meet their GHG emission reduction targets, three market mechanisms were included in the Kyoto Protocol: (i) International Emission Trading (IET); (ii) Clean Development Mechanism (CDM); (iii) and Joint Implementation (JI) (United Nations Framework Convention on Climate Change Kyoto Protocol, 1997). Of the three mechanisms, the CDM grew faster than anticipated, as it allowed developed countries to invest in projects that reduced GHG emissions in developing countries and count them against their own reduction targets, otherwise known as carbon offsetting (Bernstein et al., 2010). Credits generated through the CDM were called Certified Emission Reductions (CER) and were produced when a country implemented a project that reduced GHG emissions against a hypothetical baseline. This baseline was derived for each country based on the GHG emissions that would have been predicted to occur in the absence of a planned CDM project. This essentially allowed developing countries to receive funding by selling CERs for cutting emissions to Annex 1 countries seeking to meet their Kyoto Protocol target.

To meet its Kyoto target, the European Union launched its Emissions Trading System (EU ETS) in 2005, becoming the largest greenhouse gas emissions trading scheme in the world. While the EU ETS was created for EU members to meet their GHG reduction commitments under the Kyoto Protocol, it was independently established and enacted before the Kyoto Protocol's first commitment period began in 2008 (Ellerman and Joskow, 2008). Under the EU ETS, member states agreed on national emissions caps, approved by the European Commission. Emission allowances were allocated to participating firms, which could either surrender the allowances for each extra ton of carbon emitted, or purchase permits or sell excess allowances through the ETS. Eventually, the EU agreed to incorporate the three flexible mechanisms of the Kyoto Protocol, allowing EU firms covered under the ETS to use the Kyoto mechanisms as additional options for covering their emissions.

After the implementation of the EU ETS, several other regional and national ETSs were developed. Notably, the Regional Greenhouse Gas Initiative (RGGI), covering nine US states in New England,[2] and the Western Climate Initiative (WCI), incorporating several Canadian provinces[3] and California, are two other large regional ETS that have been operating since 2009 and 2007, respectively. These market-based approaches to deliver carbon financing[4] proliferated and, as of June 2020, 31 regional, national and subnational ETS have been implemented globally, and 30 carbon tax policies applied. Cumulatively, these carbon pricing initiatives cover approximately 22 percent of global GHG emissions (World Bank, 2020a).

Second Phase

In 2009 at COP15 in Copenhagen, developed countries promised to provide and to mobilize $100 billion a year by 2020, from a wide variety of public, private, and multilateral sources, to address the climate finance needs of developing countries (The Copenhagen Accord, 2012). The announcement was significant as it elevated the calls for developed countries to provide greater financial commitment given their historical contribution to overall emissions. The $100 billion commitment was envisioned to be administered and disbursed through the Green Climate Fund (GCF), a new funding institution that emerged from the annual meeting of the UNFCCC Conference of the Parties (COP) and acts as part of the UNFCCC's financial mechanism. However, in practice, this can come from a wide variety of sources, including public, private, multilateral, and alternative sources, causing different views regarding what financial sources counted towards the commitment. In 2018, the Organization for Economic Co-operation and Development (OECD) estimated that climate finance mobilized for developing countries totaled $79 billion, an increase of 11 percent from the year prior, primarily driven by public resources rather than private investment (Organization for Economic Cooperation and Development, 2020).

Formally established in 2010 at the 16th Conference of Parties, which took place in Cancun, Mexico (COP16), the GCF seeks to support developing countries in their efforts to respond to climate change through the provision of various loans, grants, and concessional financing for mitigation and adaptation projects (Lattanzio et al., 2019). The GCF currently is governed by a Board of 24 members and partners with 84 development and commercial banks, state agencies, and civil society groups to develop and implement a variety of projects (Rowling, 2019). As of March 2021, GCF has committed approximately $8.4 billion to over 173 projects with over $5 billion already disbursed to developing countries (Green Climate Fund, 2021). However, many observers of the GCF are concerned about the funding shortfall and overall inability to secure $100 billion by 2020 (Ives, 2018; World Resources Institute, 2019).

The 195 parties to the UNFCCC gathered in Paris, France, for COP21 in December 2015 to discuss a new climate agreement framework that would set forth a global agenda on climate change through 2020 and beyond. In preparation for the Paris Summit, UNFCCC parties publicly released post-2020 national climate action plans known as Intended Nationally Determined Contributions (INDCs).[5] The Paris Agreement aims to hold any increase in the global average temperature to below 2°C above pre-industrial levels, and to "pursue best efforts" to limit temperature increase to 1.5°C (Leggett, 2015). Parties further pledged to reach zero net emissions of GHGs by 2050. The Paris Agreement also laid out a framework in which member nations communicate more ambitious NDCs every five years, participate in a single transparency framework for communicating their GHG inventories, and submit progress reports for international review. Finally, the Paris Agreement reaffirmed the commitment to mobilize $100 billion a year in climate finance and agreed to continue mobilizing the level of financing until 2025 (Thwaites et al., 2015).

In contrast to the Kyoto Protocol, the Paris Agreement deliberately abandoned previous top-down approaches that sought to fairly distribute emission reduction responsibilities by collectively determining each country's obligation. While a second commitment period of the Kyoto Protocol was agreed in 2012, emerging economies were classified as non-Annex 1 countries, including several of the largest emitters, which limited the overall effectiveness of emission reductions. This was further hindered by the US not ratifying the treaty and the withdrawal of Canada. Ultimately, this process illustrated that any future agreement needed to be simplified in its approach and include participation from developing countries. In contrast, the Paris Agreement invited countries to develop their own commitments, considering their own economic condition and technical capabilities. The flexibility of the Paris Agreement allowed for 189 countries to submit NDCs, causing the agreement to be hailed as a major turning point in climate negotiations for uniting developed and developing countries under a common framework (Davenport, 2015).

Third Phase

Shortly prior to COP21 that took place in 2015 in Paris, France, the MDBs and the International Monetary Fund (IMF) jointly announced a plan to extend more than $400 billion in financing over the next three years to achieve the Sustainable Development Goals (SDGs) (World Bank, 2015a). The SDGs were adopted by 193 countries as an urgent call to action as part of a global partnership between the public and private sector. They emphasize that ending poverty and other development challenges must be tackled together with strategies that improve health and education, reduce inequality, and spur economic growth – all while ensuring a clean environment (United Nations Department of Economic and Social Affairs, 2015). The MDBs and IMF anticipated that the announcement of additional financing, coupled with the existing official development assistance (ODA) flows of $135 billion a year, would leverage and mobilize "trillions" of private sector sustainable investments. This financing strategy became known as "billions to trillions," and became a call globally for increasing public and private investments, as well as a general appeal for a change of stakeholder approaches and mindsets, to achieve the transformative vision of the SDGs (Multilateral Development Banks Joint Report, 2015).

The "Billion to Trillions" strategy became the focal point for the Third International Conference on Financing for Development, held in Addis Ababa in July 2015 (United Nations General Assembly, 2015). Delegates from 175 UN member states (including 24 heads of government), as well as participants from UN Agencies, the IMF, MDBs, the World Trade Organization, civil society leaders, and other private stakeholders, met to create a global policy framework to align sustainable development financing with nations' economic, social, and environmental priorities. This framework was codified in the Addis Ababa Action Agenda, which later helped to underpin the adoption of the SDGs and 2030 Agenda at the UN Special Summit in New York in September 2015. Concretely, members agreed to 100 measures that draw upon all "sources of finance, technology, innovation, trade, debt and data to support achievement of the SDGs" (United Nations General Assembly, 2015).

Despite these approaches, there remains a sizeable financing gap in achieving the SDGs according to the UN Commission on Trade and Development (UNCTAD) which found a $2.5–3 trillion funding gap in a 2014 study; IMF found a gap of $2.6 trillion in a 2019 study (United Nations, 2018). A major source of the gap, according to the UNCTAD report pointed to investment incentives focusing mostly "on economic performance objectives, less on sustainable development". It recommends a change in incentives by governments and the spending of public funds to be more closely aligned with the SDGs. The report pointed to other challenges such as the failures in global capital markets to price externalities such as carbon emissions or water use and the lack of transparency in corporate reporting (World Investment Report, 2014). To fill this disparity, several international organizations have called for the private sector to play a bigger role by joining their investments together with public sources in so-called "blended finance" for climate investments. Blended finance is an approach to boost private finance by combining concessional, or below market rate, funding with market rate or commercial funding. Blended finance can be provided by grant-making organizations such as government-backed programs, foundations, or grant equivalent elements of development banks (such as MDBs) and partners (Development Finance Institutions, multilateral climate funds, etc.) to mobilize private capital flows to emerging markets. This "blending" intends to overcome commercial investors' hesitation of investing in developing countries due to high risk and uncertainty of returns by sharing risk with public investors that will often end up taking any "first-losses". While blending of public sources of climate finance and private capital has been successful, it has not reached the scale needed (Group of Development Finance Institutions, 2019).

Over the past few years, there have been several key developments to systematically support climate action by the private sector by engaging the financial market participants. One such development includes the Network of Central Banks and Supervisors for the Greening of the Financial System (NGFS), which was established at the Paris "One Planet Summit" in December 2017 to help strengthen the global response required to meet the goals of the Paris Agreement. NGFS aims primarily to contribute to the analysis and management of climate and environment-related risks in the financial sector, but also to provide a framework that can help mobilize mainstream finance to support the transition toward a sustainable economy. The NGFS playing a leadership role among the voluntary group of central banks that participate, including through their 2019 Call for Action report, deem climate change as one of many sources of structural change affecting the financial system saying there is a need for collective leadership and globally coordinated action (Networking for Greening the Financial System, 2019). The United States Federal Reserve in 2021 established the Financial Stability Climate Committee to identify, access and address climate-related risks to financial stability (Brainard, 2021). In addition, the TCFD, which released its first climate-related finance disclosure set of recommendations in 2017 was established by the Financial Stability Board. Through TCFD, firms can adopt a voluntary set of climate-related financial risk disclosures that can inform investors and the public about the risks companies face related to climate change (Task Force on Climate-Related Financial Disclosure, 2021).

Other key developments include efforts by the G20 and United Nations to create a global roadmap for sustainable finance, as well as to outline regional options through taskforces such as the China G20 Green Finance Study Group (G20 Sustainable Finance Study Group, 2018). In Europe, major steps have been taken to maximize the EU's efforts in examining how to integrate sustainability considerations into its financial policy framework for mobilizing finance for sustainable growth. In 2018, the Sustainable Finance Action Plan was launched to support the adoption of a taxonomy that defines which activities should be considered as 'green' or 'sustainable (The Commission to the European Parliament, the European Council, the Council, the European Central Bank, the European Economic and Social Committee and the Committee of the Regions, 2018). The Plan was also used to inform the European Green Deal, announced in 2019, which is a growth strategy aiming to make Europe the first climate neutral continent by 2050. In the United States, the Commodity Futures Trading Commission report published in 2020 was also a step forward in terms of helping to manage climate risks in the financial system (Behnam et al., 2020).

BOX 19.1 IMPACT OF COVID-19 ON CLIMATE FINANCE

Prior to the COVID-19 pandemic, expectations were high for increasing ambition on climate finance ahead of COP26, originally scheduled for November 2020. However, this momentum was disrupted as countries shifted development priorities, as the crisis exposed the vulnerabilities of countries' social security, health systems, and economies, for both developed and developing countries alike. The conference was itself postponed until November 2021. The year 2020 was also the beginning of the ten-year countdown, according to the IPCC, for the world to achieve low-carbon transitions and limit temperature rise to below 1.5°C (Volz, 2020).

However, in the midst of rebuilding, the pandemic provides an opportunity for countries to align economic stimulus and recovery packages in a manner that is sustainable, equitable and inclusive. Furthermore, global political will is aligned for accelerating climate finance, as the United States re-enters the Paris Agreement and boosts other multilateral climate efforts. Yet, despite this international congruence, it may be difficult to adequately finance a green recovery as COVID-19 has dramatically worsened public finances. The IMF predicts that global public debt is likely to increase 101 percent of global Gross Domestic Product (GDP) in 2021, an increase of 19 percent from a year ago (International Monetary Fund, 2020). In the aftermath of the pandemic, it will be critical for countries and development institutions to align financial flows with the goals of the Paris Agreement and spur low-emission, climate-resilient development.

Recent Developments – COP26

Despite the ongoing COVID-19 pandemic, a record number of delegates (approximately 40,000) gathered for COP26 in Glasgow in early November 2021 in the hope of maintaining momentum on the goals of the Paris Agreement. By the end of the summit, 141 countries submitted new or enhanced NDCs to cut their emissions by 2030 (Climate Watch, 2021). Additionally, the parties at COP26 adopted the Glasgow Climate Compact, which committed signatories to double adaptation finance and called for more ambitious climate pledges ahead of COP27 in Egypt in 2022. In comparison to previous COP agreements, the Compact stressed the importance of the IPCC's findings from its Sixth Assessment Report and recognized that in order to limit warming to 1.5°C, carbon emissions must fall to 45 percent below 2010 levels by 2030, and to net-zero by 2050 (Evens et al., 2021).

In addition to the Glasgow Climate Compact, the US and China made a surprise announcement to cooperate on limiting emissions in line with the Paris Agreement by sharing policy and technology development, announcing new climate targets for 2035 by 2025, and establish a working group on enhancing climate action in the 2020s (Shapiro et al., 2021; US Department of State, 2021). While the US–China agreement lacked firm deadlines or specific commitments, cooperation between the two countries would account for nearly 40 percent of global carbon emissions. Beyond these two agreements, several parties gathered for COP26 made several additional collective commitments, including to: (i) reduce methane emissions; (ii) halt and reverse deforestation; (iii) align the finance sector with net-zero goals by 2050; (iv) abandon the use of internal combustion engines; (v) accelerate the phase-out of coal; and (vi) end financing for fossil fuels, among other goals (Mountford et al., 2021).

The summit also saw a surge of net-zero targets from countries promising to balance their emissions by 2050, bringing the total of sovereign net-zero pledges to 74 by the end of COP26. However, countries' net-zero pledges and enhanced NDCs are still forecast by external analysts to collectively miss the 2°C target of the Paris Agreement, and provide less than a 20 percent chance for the planet to keep temperature rise to 1.5°C (Kaplan et al., 2021). In fact, according to a UN analysis, current projections put the Earth on track to warm to 2.5°C, representing a large gap between countries' NDCs and long-term promises to zero out emissions (United Nations Environment Programme, 2021). Many observers were disappointed that this COP did not increase financial flows to developing countries to significantly boost their resilience to the unavoidable impacts of climate change. The Compact itself noted with "deep regret" that the $100 billion target in climate finance had not been met, given the 2020 deadline, and instead extended the deadline to 2025. This brings back the nearly two decade-long issue of the CBDR principle in the Kyoto Protocol and the expectation to transfer climate finance between developed (Annex I) and developing countries (non-Annex 1).

ENGAGING CAPITAL MARKETS TO FINANCE CLIMATE ACTION

While public sources of climate finance, blended concessional climate finance, and systemic actions of the financial sector steered private investment, the capital markets have simultaneously been evolving to better steer the trillions necessary for the sustainable future.

Capital markets, where savings meet investment, include both those who are saving money and those who seek funding by issuing bonds. A range of actors can participate in capital markets including individuals, companies, and governments – some directly and others through institutional investors such as asset managers and pension funds pooling investments. There are two primary types of capital markets: (i) stock markets, or equity markets, and (ii) bond markets, also known as fixed income markets as investors typically receive a consistent, usually fixed, stream of interest payments. Global equity markets total $60 trillion, while bond markets total approximately $110 trillion (Dealogic, 2021; World Bank, 2019c).[6,7] The remainder of the chapter focuses on changes to the bond markets that have facilitated climate financing and sustainable choices. In equity markets, investors typically own shares of companies and therefore have different levers and levels of influence over a company's climate strategy. In comparison, bond investors are lending money, and therefore influence over a bond issuer's climate strategy involves seeking information and transparency about the climate risks and opportunities, when deciding to purchase. In practice this means that investors can engage directly with bond issuers on their environmental, social, and governance policies before choosing to invest in a company or a sovereign government bond (World Bank, 2020b).

BOX 19.2 BUILDING SUSTAINABLE DEBT CAPITAL MARKETS HELPS CHANNEL INVESTMENTS TOWARDS MORE SUSTAINABLE ACTIVITIES

Investor behavior is changing. Investors are increasingly aware of global environmental and social challenges that impact the value of their investments (and their life) and are looking for environmental, social, and governance data to integrate into their investment strategies to mitigate risk and/or find investments that help contribute to a better society. Investor priorities differ. Some are more focused on analyzing ESG to reduce risk while others are intentionally seeking to influence how they are contributing to positive impact, and many pursue strategies that combine those approaches. Meanwhile, as investors are seeking and utilizing ESG data they are confronted with myriad ESG data sources, and face comparability and reliability issues with the data.

Issuer behavior is changing. Issuers are realizing that investors are looking for more robust ESG data and information and are increasing transparency around how they use investor funds.

Financial models are evolving. The financial system uses models that underestimate financial risks (external costs and potential regulatory changes that will have financial implications are not sufficiently included in asset prices) but are evolving with more transparency and requirements for sustainability reporting, better data from issuer/company disclosures required by regulators – including on climate stress-testing, more expertise on asset pricing and new technology.

Regulatory environments are shifting. Regulation is moving from one that is focused exclusively on financial disclosure to include sustainability metrics that go beyond short-term financial impact on the reporting entity, catalyzing market changes. Examples include the European Union's Taxonomy and Action Plan. There is also a proliferation of different standards and measures for reporting that will ultimately converge and harmonize but will remain sector-specific and will diverge considering local market situations.

Bond markets are evolving. The growth of the labeled bond market is key to the transition towards a holistic approach to sustainable investing that focuses on transparency for the entire bond market and all assets under management, not separate issuance programs or separate portfolios. The labeled bond market introduces transparency and intentional purpose of sustainable financing, with labels such as "green", "social" and "sustainability", indicating how the issuer intends to allocate proceeds.

Across this entire fixed income market, investor and issuer behavior is changing to reflect increased demand for climate action. Investors are increasingly aware of global environmental and social challenges that impact the value of their investments and are looking for ESG data to integrate into their investment strategies to mitigate risk and help contribute to a better society. To match investors' interest, bond issuers are changing: issuers are realizing that investors are looking for more robust ESG data and information and are increasing transparency around how they use investors' money. The creation of the green bond, and other so-called "labeled bonds" that followed, including social and sustainable bonds, has helped catalyze this change in investor behavior (World Bank, 2021b).

As of 2021, the total value of bonds that are associated with a label is more than $2.5 trillion. Issuers use labels for bonds, such as "green", "social", or "sustainable" as a shorthand way to explain to investors what they intend to use an equivalent amount of the bond proceeds for. Issuers provide documents to explain the details, including frameworks and impact reports. In October 2021, Bloomberg introduced a tagging system with icons for issuers' bonds that have labels – "green", "social" or "sustainable" for issuers supporting green and social projects. Out of the total market volume for labeled bonds, the value of green bonds is approximately $1.8 trillion. This market has grown from the first green bond issued by the World Bank in 2008 and was recognized early on as a product model that could help to mobilize more funds towards climate activities (Reichelt, 2010). While the size of the green bond market comprises only a tiny fraction of the global bond market ($110 trillion), the green bond market has grown tremendously, reaching a record of over $300 billion issued in 2020, which is a sizeable increase from $100 billion issued in 2016 and $7 billion in 2010.[8] These annual volumes alone underrepresent the change that green bonds have sparked in the global bond market, capital markets, and sustainable investing in general.

BOX 19.3 DEFINING GREEN BONDS

Green bonds are debt instruments that help issuers communicate their strategy and activities supporting climate change mitigation and adaptation activities (World Bank, 2015d). An equivalent amount to the proceeds raised using the green bond label is allocated to projects designed to deliver environmental benefits, according to pre-defined criteria. Green bonds have the same financial terms and risks as non-labeled bonds, but by providing information about specific eligible "green" projects through impact reporting, the issuer communicates to the investor about the expected impact of the projects that funds are allocated to. Green bond proceed amounts are allocated to projects in several categories that address climate change, for example:

- Renewable Energy
- Energy Efficiency
- Pollution Prevention and Control
- Biodiversity Conservation
- Clean Transportation
- Climate Change Adaptation
- Wastewater Management
- Solid Waste Management
- Sustainable Land Use
- Resilient Infrastructure

THE GREEN BOND: FROM EVOLUTION TO REVOLUTION

Origins

The World Bank pioneered the first labeled green bond when Scandinavian investors approached it in 2007 (Reichelt, 2010). Concerned that climate change posed a significant risk to their portfolios, a group of Swedish pension funds, through the bank Skandinaviska Enskilda Banken (SEB), was assessing its portfolio for climate risks and looking for opportunities that would help mitigate these risks and support climate-friendly investment. The investors wanted high quality, liquid bonds that would not carry project risk. They also wanted assurance that the selected projects would be identified and managed through a due diligence process that ensured good governance, adhered to environmental and social standards, and effectively helped countries mitigate and adapt to climate impacts. In addition, the Swedish investors were interested in issuer reporting to understand how their investments were helping to achieve positive impact. SEB worked with the World Bank and collaborated on the design of a new financial instrument for Swedish pension funds in Swedish Krona, which became the first labeled green bond (World Bank, 2008a).[9] This green bond issued by the World

442 *Handbook of business and climate change*

Bank created the blueprint for today's green bond market (World Bank, 2019a). It defined the criteria for projects eligible for green bond support, included a recognized climate research institution, CICERO, as a second opinion provider, and added impact reporting as an integral part of the process, highlighting the importance of transparency.

Figure 19.1   Tombstone for first green bond, issued in 2008

Green bonds have the same terms, credit and financial risks as other bonds. What separates a green bond from a conventional bond is the "use of proceeds" and an impact reporting pledge, showing that the issuer intends to use an equivalent amount of the funds raised for a specified purpose and commits to reporting back to investors. A green bond allocates an equivalent amount of the proceeds for certain types of projects with climate mitigation or adaptation benefits, as well as projects that have other environmental benefits. In most cases where issuers fund multiple activities and projects with their bond proceeds, the bond issuer does not transparently discuss what assets they fund using the proceeds from the bond. It was not until the issuance of the first green bond that issuers engaged with investors around the purpose of their investment and the expected social and environmental impacts.

To achieve this level of transparency, green bonds follow a standardized process that is encapsulated in the "Green Bond Principles", a voluntary set of guidelines created by market participants who collaborated to bring clarity to the market (International Capital Markets Association, 2021). The Green Bond Principles were originally drafted by a group of bankers, tasked with underwriting bonds with the label, keen to make sure there was a general understanding in the market among other underwriters, issuers and investors (Bank of America Corporation et al., 2014). The Green Bond Principles used the model set up by the World Bank with its first green bond and made it applicable to a broader range of issuers. They gave clear guidance on how the process could be adopted and adapted by different types of issuers such as corporates and sovereigns, based on their specific situation. When considering labeling a bond as green, an issuer describes the process it uses in a document called the green bond framework. This framework defines criteria the issuer intends to use to select the eligible projects, referred to as the green bond definition, and steps for monitoring and reporting back on them to investors. In the case of the World Bank, both climate mitigation projects, which reduce greenhouse gas emissions, and adaptation projects, which help people respond and build resiliency to the effects of climate change, are included in its green bond definition (World Bank, 2014). More broadly, any green bond issuer must have an established project selection process to match these criteria. Issuers also describe how they allocate the proceeds, like through a separate account to help track amounts of the proceeds raised and keep track as equivalent amounts are allocated to eligible projects. When allocating proceeds, it has become common practice for an issuer to obtain an independent third-party assessment of their framework to aid in providing investors additional assurance (International Capital Markets Association, 2021).

Transparency and Impact Reporting

The most important step of a green bond comes next, the reporting, which has had the most influence on the broader fixed income market as a whole. The World Bank collaborated with other major green bond issuers in the early years of the green bond market, including the African Development Bank (AfDB), the European Investment Bank (EIB), and the International Finance Corporation (IFC), to develop a harmonized reporting framework for green bonds (World Bank, 2015b). The World Bank's first Green Bond Impact Report has been widely recognized by the market as a standard and model for impact reporting (World Bank, 2015c). When the World Bank issued its first Green Bond Impact Report in 2015, the initial reporting template and set of indicators that were presented became a model for issuer reporting to investors on green bonds. This standardization benefitted from the efforts of multilateral development banks to harmonize metrics for GHG accounting and reporting on climate finance activities at the time, which has since continued to evolve (African Development Bank et al., 2015). As more diverse issuers, including by type, sector, and geography where they operate, have entered the market, the harmonization and comparability of impact reporting has become an additional challenge for investors.

Although according to the published guidelines it is voluntary, investors generally expect green bond issuers to measure, track, and report on the social and environmental impact of their investments. This level of transparency has also created accountability and lock-in within issuers' own organizations as the issuance of green bonds requires sustained and a long-term commitment by the organization, including to report on the positive impacts of the investments and the issuers' overall climate policies and targets. An impact working group that provides guidance on best practice includes a variety of green bond issuers, including those early pioneers, and is one of the working groups facilitated by the Green Bond Principles coordinated by the International Capital Markets Association (ICMA) (International Capital Markets Association, 2014b).

> ### BOX 19.4 IMPACT REPORTING (PROJECT EXAMPLES)[10]
>
> The World Bank applies a *country-driven* approach for directing program and project financing in the developing countries it serves. It supports countries with sustainable development financing for projects based on their own priorities in the areas where the World Bank can help the most. The institution works across the world in every sector related to economic and human development from health and education, to agriculture and critical infrastructure. And within these sectors, the World Bank has projects that are specifically focused on helping countries mitigate or adapt to climate change. Some examples that are included in the World Bank green bond portfolio include:
>
> - *China*. A project to promote renewable energy in educational institutions through rooftop solar panels installed in schools serving 650,000 children, which saves about 90,000 tons of emissions each year – equivalent to taking nearly 20,000 cars off the road each year.[11]
> - *Brazil*. A project that is closing open waste dumps and financing the construction and operation of environmentally safe landfills to reduce emissions which reduced nearly 800,000 tons of CO_2 emissions – equivalent to over 160,000 passenger vehicles being taken off the road.
> - *Indonesia*. Protecting and managing unique coral ecosystems covering 1.4 million hectares of marine areas and benefiting over 1000 people in local fishing communities.

International Community Takes Note – Green Bond Market Consolidation

With the sale of the World Bank's first green bond in 2008, several investors and banks took notice. A new development in the market was that investors were disclosing their purchase of the securities – this was not the case before the World Bank's green bond was issued in 2008 (World Bank, 2008b). In 2009, the first green bond denominated in US dollars was purchased by the California State Treasurer's Office (World Bank, 2009). In 2010, the World Bank collaborated with Nikko Asset Management to develop and launch a green bond fund that offered investors an opportunity to invest in green bonds issued by the World Bank in a variety of currencies (World Bank and Public-Private Infrastructure Advisory Facility, 2015 and World Bank, 2010). In 2011, the G20, IMF, and OECD formally recognized the potential of the market, and by 2012 the green bond was being suggested by the OECD to governments to finance their climate change solutions (G20 Finance Ministers, 2011). In this period of consolidation, several market stakeholders, such as the Climate Bonds Initiative (Climate Bonds Initiative, 2021), a non-governmental organization and early supporter of the growth of the green bond market, ESG research firms and others, including underwriters seeking to harmonize the understanding of what the green bond label described, called for rules or principles to be followed for what could be labeled *green* in the market. The banks most engaged at that point built from the World Bank's original approach, as well as the outcome summary of the Global Symposium on Green Bonds hosted by the World Bank in 2012, to create a draft version of the Green Bond Principles, later published by ICMA (World Bank, 2013; International Capital Markets Association, 2014b).

First published by ICMA in 2014, the Green Bond Principles outline criteria for labeling green bonds, selecting eligible projects, allocating proceeds, and reporting on project impacts in a transparent manner (International Capital Markets Association, 2014a). While the Green Bond Principles have been constructive in defining the green bond process in a standardized way, they do not attempt to define the activities that are considered *green*. Instead, the Green Bond Principles encourage issuers to transparently disclose their eligibility criteria and provide broad illustrative categories. The Green Bond Principles are a voluntary set of standards that encourage issuers to offer investors a second opinion to provide an independent assessment of the issuer's framework and activities, as a resource for investors. Others provided guidance in the early days of the green bond market, to clarify what they considered to be *green*. For example, in February 2015, an investor group convened by Ceres' Investor Network on Climate Risk (INCR), a network of major institutional investors, published a Statement of Investor Expectations to support the development of a consistent, durable framework for the green bond market to encourage its further growth and increase clean energy financing and other solutions to climate change (Reichelt and Keenan, 2017).

The Green Bond Market Matures

Until 2012, the green bond market was dominated by issuers such as the World Bank and EIB, who already had programs in place to finance climate action. Since 2013, green bond issuers include a broader variety of issuers including companies, such as Apple and Toyota as early green bond issuers, banks such as Bank of America and Deutsche Bank, US states such as Massachusetts, and several sovereign issuers. A growing number of corporations and national and sub-national governments also issue green bonds. In 2016, Poland became the first country to issue a green bond (Organization for Economic Cooperation and Development, 2017). A few months later, France issued the largest ever green bond, a €7 billion benchmark bond with a 22-year maturity (Environment Finance, 2017). In October 2017, Fiji became the first emerging market issuer of a green bond (Government of Fiji, 2021). Germany issued its first green bond in September 2020, introducing a new concept aimed at bringing transparency to the pricing of green bonds in the market (Government of Germany, Ministry of Finance, 2021). Since then, other issuers such as Belgium, Colombia, Egypt, Indonesia, Kenya, Jordan, Morocco, the Netherlands, Spain, Sweden, and the United Kingdom have also issued green bonds. Sovereign issuers decide to label their bonds for a variety of reasons. Some for signaling or political reasons to set an example and help move the market towards sustainable investing, or to show leadership in the debate around green finance, or a combination of other factors. In 2020, less than 5 percent of green bonds were issued by multilateral development banks and government agencies (e.g., World Bank, EIB and KfW), evidence that the market had expanded to other types of issuers.[12]

Beyond Green Labels

Since 2015, organizations such as the World Bank have started introducing bonds labeled as sustainable development bonds with issuer-level sustainability focus for all activities. This has influenced the growth of the labeled market and will continue to increase the volume of bonds supporting green and social activities. The United Kingdom issued its first green bond in September 2021, including for the first time for a sovereign green bond, a commitment to reporting on the social co-benefits such as job creation, from the expenditures that proceeds would be allocated to (United Kingdom Debt Management Office, 2021). These green gilts were referred to as "green+" in the market. Ahead of the UNFCCC Conference of Parties in Glasgow, Scotland, in November 2021, the World Bank launched an initiative highlighting to investors how climate is woven into all of its activities, including in sectors such as healthcare and education, which have not historically been associated with green financing (World Bank, 2021c). Such holistic approaches, including a mix of green and social in sustainable financing through the bond markets, have been accelerating.

There are more and more types of issuers using labeled green, social or sustainable bonds to articulate their intentions to allocate equivalent amounts of proceeds raised to specified types of projects that are intentionally designed to achieve positive impact. Across a range of investor types from commercial banks to asset managers, investors are increasingly adopting bond labels as a simplified approach to align their investment strategies with sustainability and climate action. Other investors are choosing to go further in their analysis and reviewing corporate data and creating their own ESG scoring and metrics to determine if an asset qualifies for the sustainable strategy (Strategy & Part of PwC Network, 2021).

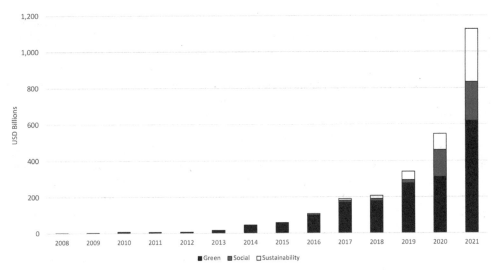

Source: Bloomberg and World Bank data.

Figure 19.2 Issuance by bond label 2008–2021

BOX 19.5 SUSTAINABILITY REPORTING AND REGULATION

Since the early days in the green bond market, there have been several milestones to add consistency in the green bond and sustainable bond market, including the launch by the European Commission of the Sustainable Finance Action Plan in 2018. The Action Plan included the proposal of an EU Taxonomy as well as other measures to channel more capital flows towards sustainable investment, to achieve sustainable and inclusive growth. Recognizing that "one size does not fit all" when defining if an industrial activity, product, process, etc. is sustainable, green, or social, and, in addition, attempting to assess the extent, there have been several other countries setting up their own taxonomies to address the need among financial market participants for clarity and transparency in what is understood and what qualifies as green (Hussain et al., 2020). There have been other significant developments in the area of disclosure and reporting, including the Sustainability Related Financial Disclosure Regulation (SFDR) that is being introduced by the EU, requiring fund managers and other market participants to evaluate and disclose sustainability-related data and policies. This comes in addition to other disclosure requirements at a company-level discussed earlier, such as TCFD, and more established sustainability reporting, such as the CDP (Carbon Disclosure Project), the GRI (Global Reporting Initiative), and SASB (Sustainability Accounting Standards Board). More recently, the IFRS (International Financial Reporting Standard) Foundation and new Sustainability Standards Board has started an initiative to consolidate and harmonize sustainability reporting.

As the remainder of this chapter will argue, the importance of green bonds is bigger than the size of the small yet growing market. The development of green bonds has created momentum for sustainable fixed income markets and promotes greater transparency from issuers with a focus on positive impact. Green bonds are 'connectors', linking (i) investors with the social and environmental purpose of their investment; (ii) climate policymakers and nongovernmental organizations (NGOs) with financial markets; (iii) bankers, issuers, and investors around a single topic and social challenge; and (iv) sustainable and responsible investors with mainstream investors. These interlinkages have encouraged mainstream investors to pursue ESG investing more systematically, and, in the case of sovereign green bonds, promoted intragovernmental coordination between different issuer agencies in separate ministries. Over a decade from their invention, green bonds are still a key entry point to setting up an investment process that integrates transparency around ESG aspects into the investment process and, in addition, supports a socially minded approach to fixed income.

BOX 19.6 THE WORLD BANK'S SUSTAINABLE DEVELOPMENT BOND APPROACH

The World Bank issues bonds to finance its sustainable development activities. All World Bank development operations are designed to achieve positive environmental and social impacts and outcomes consistent with the World Bank Group's twin goals of ending extreme poverty and advancing shared prosperity. The World Bank takes a portfolio approach to issuing sustainable bonds, which means it supports a variety of sustainable development sectors and the range of SDGs and does not allocate bond proceeds for individual sectors or projects.

The World Bank uses its issuance program to engage investors on the World Bank's development mandate, explain how the use of bond proceeds supports the financing of sustainable development activities, and describe how these activities contribute to the SDGs. While green bonds draw attention to the World Bank's climate activities, the World Bank uses its Sustainable Development bonds to raise awareness for certain SDGs and development challenges, using World Bank projects as examples. Dedicated themes have allowed for bonds to raise awareness around topics such as food loss and waste; gender equality/ health and nutrition of women and children; water and oceans; and sustainable cities.

MOVING TOWARDS A HOLISTIC 'SUSTAINABILITY' APPROACH

While this chapter has previously underscored the need to scale up climate investment to meet the objectives of the Paris Agreement's goal to limit global warming to well below 2°C, there is value to work on a broader and more holistic basis to achieve the Sustainable Development Goals (SDGs). It is estimated that $2.5 trillion needs to be invested annually by 2030 to achieve the SDGs, but this estimate was developed prior to the setbacks of the coronavirus pandemic (The United Nations Conference on Trade and Development, 2014). Capital markets, and specifically fixed income markets, play a crucial role in connecting global savings with global investment in this regard and will need to rely on both labeled bonds, such as green bonds, and non-labeled types of fixed income instruments to achieve this level of investment. As detailed earlier, given the relative size and growth of the labeled bond market to the overall bond market, the entire volume of financing from the capital markets will need to substantially shift and incorporate green priorities to achieve decarbonization and sustainability goals.

Integrating ESG and Impact in the Entire Portfolio

Bond investors are increasingly looking to understand the environmental, social and governance impacts of their investments and seeking greater opportunities for positive impact beyond labeled bonds. Their motivations vary from a narrow approach to understand ESG risks to protect the financial value of their immediate portfolio to meet the financial expectations of their shareholders and other stakeholders, to investors focused on values and interested in supporting society – in most cases also based on pressures from stakeholders. Most investors, however, are primarily focused on financial returns and integrate ESG for financial risk reasons. In the case of central bank reserve management, for example, an essential consideration is whether ESG is consistent with investment objectives (Bouyé et al., 2021). To meet the expectations for ESG disclosures, issuers are committing more and more resources and to analyze the information, investors are doing the same. The costs on both sides are increasing and becoming part of the investment process that is evolving to a more holistic approach. If done well, sustainable capital markets can be an important pathway for achieving the SDGs. For investors of sovereign bonds, there is also a need to assess ESG risks and opportunities by integrating ESG data, such as data provided on the World Bank's ESG Sovereign data portal.[13]

BOX 19.7 ABCS OF SUSTAINABLE INVESTMENT[14]

It is helpful to introduce a simple construct to understand how investors are considering broader sustainability in their decision making:

- **A**voiding certain securities, issuers or sectors, sometimes called "Socially Responsible Investing" (SRI); can be 'negative screening', 'best-in-class' etc.
- **B**enefiting stakeholders by focusing on minimizing ESG risks, often referred to as "ESG integration."
- **C**ontributing to society is a goal that impact investors pursue.

Investors typically use a combination of strategies that cover **A**, **B** and **C** and can be referred to as "Sustainable Investing" (Stewart and Reichelt, 2020).

Avoiding Certain Securities – Socially Responsible Investment: Socially responsible investment has its origins decades before the first green bond. Investors, typically in stocks, looked to screen out companies that were contrary to their moral and ethical understandings, such as weapons, alcohol, tobacco and nuclear energy (Stewart and Reichelt, 2020).

Benefitting Stakeholders – ESG Integration: ESG factors are becoming increasingly important in the investment decision processes for mainstream fixed income. This is driven by growing evidence that these factors influence credit risk and must be part of risk mitigation. Importantly, reporting on ESG and/or investment impact is becoming more common. Investors are increasingly managing ESG risks by explicitly and consistently considering ESG in the investment process (Bertolotti et al., 2019). An approach to integrate ESG into investment decision making is being confirmed by investors in government, or sovereign bonds. A World Bank survey of investors found that 65 percent of these investors are integrating ESG into their security selection. In response, the issuers of bonds must become increasingly adept at engaging with these investors and communicating the ESG risks and opportunities in their activities (Hussain et al., 2020). For government bonds specifically, the World Bank has stepped in to aggregate publicly available data sources. This has been compiled into a one-stop data portal, launched in 2019, to support investors in their decision-making for choosing which projects to finance (World Bank, 2021d).

Contributing to Society – Investing for Impact: Increasingly, investors are looking to the benefits to society of their investments, the intentional positive impact. Often using the Sustainable Development Goals as a framework, such strategies focus on certain themes (Operating Principles for Impact Management, 2019). Impacts are being measured and accounted.

Pandemic Puts Focus on "S" in ESG

The health and economic crisis brought by the coronavirus pandemic in 2020 has put an emphasis on the "Social" aspect of ESG investments. The attention to "S" in this period mirrors the origins of the creation of the green bond, mentioned earlier in this chapter, where investors were attracted to an instrument to minimize risks, yet quickly saw a larger opportunity for positive impact. Likewise, as the pandemic shuttered economies and increased market volatility in early 2020, investors quickly confronted the importance of minimizing risk in their portfolios from weakened healthcare systems and knock-on effects of the pandemic on food, education, and safety net systems. The "S" in ESG was blinking bright red as the interconnections between these systems and their sudden fragility became material for asset valuation. While some investors had already been saying "don't forget the S in ESG", with the pandemic it became clear to investors that global systems are only as strong as their weakest links and therefore addressing vulnerabilities has risen as a key part of the response by ESG focused investors.

And yet these social risks, interconnections, inequalities have always been present, just perhaps largely overlooked by all but a few socially responsible investors (Stewart and Reichelt, 2020). An ESG integration approach that appropriately captures and values these social risks is a good first step and is indeed accelerating. Yet the momentum among investors at the time of writing, one year from the start of the global pandemic, is on putting their funding towards addressing challenges and achieving impact more broadly. Therefore, the increased focus on social aspects, as a result of the pandemic, will likely lead to increased focus on environmental and climate impacts as well.

An area where investors are particularly focused to achieve positive impact with their funds is the reduction of inequality, including on certain groups of the population such as women and girls. For many investors, the pandemic has peeled away the layers of inequality that have always been present, presenting risks that have not been adequately incorporated into their financial models and they are keen to highlight this through their investments, such as in bonds issued by the World Bank (Nordea, 2021). The unequal impact of the pandemic on those most vulnerable in society has challenged investors to confront racial, gender and other biases, if not only just on their own merit, but because leaving inequalities unaddressed creates unsustainable investments.

And so, today the connections between E, S and G have never been clearer for capital market investors. For example, climate action ("E") and building resilience to the next health emergency and addressing persistent inequalities ("S") are seen more clearly as two sides of the same coin. And without strong governance ("G"), communities are left impotent to respond to challenges they face that cross city, jurisdictional or national boundaries. This leads investors back to a holistic view of sustainability, and the importance investors are placing on reviewing all the activities that they are funding through their investments. And by extension the importance to look at each individual investment, such as a company, or public entity, as a whole, rather than a sum of its parts.

CONCLUSION – WORKING TOWARDS A NEW FINANCIAL ARCHITECTURE: PUBLIC FUNDING AND SUSTAINABLE CAPITAL MARKETS

Systemic and lasting change in the allocation of capital is required to meet the financing needed to achieve the goals in the Paris Agreement and the 2030 Sustainable Development Goals. Public funds, as delivered through traditional climate finance described in this chapter, are playing a leading role, but will never be sufficient in volume to meet the investment needs. Beyond financing, the public sector's role is best designed to catalyze public policies, regulations, frameworks that lead to transparency and predictability for private sector investors.

The capital markets, fueled by investors' increasing demands to integrate ESG as part of their risk management as well as their desire to show their stakeholders how they are contributing to positive impact with their savings, have likewise catalyzed a shift towards sustainability.

The World Bank as a sustainable bond issuer in the markets has been playing a catalytic role in encouraging transparency in the bond markets, focusing on ESG criteria and social and environmental purpose. But even at over $2.5 trillion, the labeled bond market is a tiny fraction of the overall bond market. Through the inclusion of all climate-related costs and risks, integrating additional information provided through more reporting and transparency, more investments can be channeled towards sustainable activities – not just the labeled bond market. At the end of the day, investors are still measured by their financial returns. Integrating ESG criteria is critical from a risk consideration, including reputational risks based on pressures from stakeholders, but a new financial architecture is needed.

Many investors are seeking external guidance on what they can consider as 'green' in their investment process. Efforts to find what activities should be considered 'green' must incorporate a country's national context and capabilities. There is a need to be ambitious and push for better, cleaner technologies to prevent locking-in infrastructure assets that have a high carbon intensity over a long period of time. On the other hand, there are constraints that must be considered as with any financial decision, including social impacts. Activities that help move countries towards lower carbon activities should be supported considering each country context. There is a risk that focusing too much on making the labeled bond market perfect, or only considering the 'darkest green' activities, would result in losing sight of the goal to 'green' the entire market, resulting in channeling more funds towards sustainable activities. Labels are helpful as a starting point – once all climate risks and opportunities are integrated in financial decision-making and social risks and benefits are also more transparent and priced in, 'sustainable' investing will simply become 'investing.'

As the UNFCCC climate finance system was evolving, so too were early adopters of ESG approaches and the creation and demand for green, social and sustainable bonds in the capital markets. As the Paris Agreement has passed its five-year anniversary, both public and private financers are recognizing that climate action is critical for a sustainable future. A new financial architecture with increased transparency and a holistic mindset from policymakers, regulators, investors, and issuers will be necessary to achieve this sustainable future.

ACKNOWLEDGEMENTS

The authors would like to acknowledge valuable input to this chapter from many colleagues, including from the World Bank's Climate Change Group, Jonathan Coony, Senior Carbon Finance Specialist and Sandhya Srinivasan, Senior Climate Change Specialist, from the World Bank Treasury, Farah Hussain, Colleen Keenan, Concepcion Aisa Otin, and James Seward, Senior Financial Officers in the Capital Markets & Investments Department. The authors would also like to thank Anant K. Sundaram and Robert Hansen from the Tuck School of Business at Dartmouth for their support and thoughtful comments, and Christa Clapp, Co-Founder and Managing Partner of CICERO Shades of Green Ltd., for her feedback and review of the chapter.

The opinions expressed here are the authors' and do not necessarily reflect those of the World Bank or its stakeholders.

NOTES

1. As defined by the UNFCCC, "climate finance" refers to the transfer of public resources from developed to developing countries that aims to reduce emissions, enhance GHG sinks and limits vulnerability of human ecological systems to the negative impacts of climate change. In comparison, "climate investment" includes not only financial flows, but the allocation of public or private resources or assets that accelerate or scale up the adoption of technological solutions that addresses climate change in a manner that also provides a return. This chapter intends to provide the connection between climate finance and climate investment.
2. US states participating in RGGI include Connecticut, Delaware, Maine, Maryland, Massachusetts, New Hampshire, New Jersey, New York, Rhode Island, and Vermont. The Regional Greenhouse Gas Initiative (2021). www.rggi.org
3. Canadian provinces participating in WCI are British Columbia, Nova Scotia, and Quebec. The Western Climate Initiative. www.wci-inc.org
4. Carbon finance refers to the financial tools, such as an emissions trading system, that price carbon emissions in an effort to reduce GHG emissions and their impact on the environment.
5. Now more commonly known as Nationally Determined Contributions (NDC). Intended Nationally Determined Contributions (INDC) were converted into an NDC once a member nation formally submitted the INDC to the UNFCCC and ratified the Paris Agreement.
6. Data source: Dealogic. *Outstanding Bonds*, accessed October 2021.
7. Data source: World Bank. *Stocks Traded, Total Value (US $ 2019)*, accessed March 2021.
8. Data Source: Bloomberg and World Bank accessed October 2021.
9. The European Investment Bank (EIB) issued its first "Climate Awareness Bond" a year earlier. The contribution of this bond to the green bond market has also been recognized by the market, since it included the feature of allocating proceeds that has become standard for green bonds. However, that initial bond was a structured bond (linked to an equity index), didn't have the 'green bond' label as well as other key features such as the liquidity, second opinion, and originating institutional investor demand that the World Bank's first green bond had, which were all key to the scalability of the product.
10. For more project examples see: World Bank (2019b), *World Bank Green Bond Impact Report*. https://thedocs.worldbank.org/en/doc/790081576615720375-0340022019/original/IBRDGreenBondImpactReportFY2019.pdf
11. Carbon Dioxide equivalency results based on Greenhouse Gas Equivalences Calculator; United States Environmental Protection Agency accessed April 2021 at https://www.epa.gov/energy/greenhouse-gas-equivalencies-calculator
12. Data source: Bloomberg and World Bank accessed March 2021.
13. For more information see The World Bank's *Sovereign Environmental, Social and Governance Data Portal* https://datatopics.worldbank.org/esg/
14. The ABCs of Sustainable Investing are adapted from the Impact Management Project (2021), https://impactmanagementproject.com/impact-management/how-investors-manage-impact/

REFERENCES

African Development Bank, Asian Development Bank, The European Bank for Reconstruction and Development, The European Investment Bank, the Inter-American Development Bank and International Finance Corporation and the World Bank. (2015). *Joint Report on Multilateral Development Banks' Climate Finance*. https://www.worldbank.org/content/dam/Worldbank/document/Climate/mdb-climate-finance-2014-joint-report-061615.pdf

Bank of America Corporation, Citigroup Inc. Credit Agricole CIB, JP Morgan Chase, BNP Paribas, Daiwa, Deutsche Bank, Goldman Sachs, HSBC, Mizuho, Morgan Stanley, Rabobank and SEB. (2014, January 13). *Green Bond Principles Created to Help Issuers and Investors Deploy Capital for Green Projects* [Press release]. https://www.ca-cib.com/sites/default/files/2017-03/2014-01-13-cp-cacib-green-bond-principles-en-final.pdf

Behnam, R., Gillers, D., Litterman, B., Martinez-Diaz, L., Keenan, J., and Moch, S. (2020). *Managing Climate Risk in the U.S. Financial System: Report of the Climate-Related Market Risk Subcommittee, Market Risk Advisory Committee of the U.S. Commodity Futures Trading Commission.* Washington, DC, USA.

Bernstein, S., Betsill, M., Hoffmann, M., and Paterson, M. (2010). A tale of two Copenhagens: Carbon markets and climate governance. *Millennium: Journal of International Studies*, 39(1), 161–173.

Bertolotti, A., Fancy, T., Mateos y Lago, I., Pellegrini, G., Schulten, A., and Smith, J. (2019, February). Sustainability: The future of investing. *Blackrock Global Insights*.

Bouyé, E., Klingebiel, D., and Ruiz, M. (2021). *Environmental, Social, And Governance Investing. A Primer for Central Banks' Reserve Managers.* World Bank. https://openknowledge.worldbank.org/bitstream/handle/10986/36285/163525.pdf?sequence=4&isAllowed=y

Brainard, L. (2021). *Speech at CERES conference March 2021, Financial Stability Implications of Climate Change.* The Federal Reserve. https://www.federalreserve.gov/newsevents/speech/brainard20210323a.htm

Buchner, B. (2021). *Global Landscape of Climate Finance 2021*, Climate Policy Initiative. https://www.climatepolicyinitiative.org/publication/global-landscape-of-climate-finance-2021/

Clement, V., Rigaud, K., Sherbinin, A., Jones, A., Adamo, S., Schewe, J., Sadiq, N., and Shabahat, E. (2021). *Groundswell Part 2: Acting on Internal Climate Migration.* World Bank. https://openknowledge.worldbank.org/handle/10986/36248

Climate Bonds Initiative. (2021). *History.* https://www.climatebonds.net/standard/about/history

Climate Watch. (2021). *Explore Nationally Determined Contributions (NDCs).* https://www.climatewatchdata.org/2020-ndc-tracker

Davenport, C. (2015, 12 December). *Nations Approve Landmark Climate Accord in Paris.* The New York Times. https://www.nytimes.com/2015/12/13/world/europe/climate-change-accord-paris.html

Dealogic. (2021). *Fixed Income Database.* Available from httyp//www.dealogic.com

Ellerman, D., and Joskow, P. (2008). *The European Union's Emissions Trading System in Perspective.* Pew Center on Climate Change. https://www.c2es.org/site/assets/uploads/2008/05/european-union-emissions-trading-system.pdf

Environmental Finance. (2017). *Green Bonds: Review of 2017.* London, UK.

European Bank for Reconstruction and Development. (2020). *2020 Joint Report on Multilateral Development Banks' Climate Finance.* https://www.ebrd.com/news/2021/mdbs-climate-finance-rose-to-us-66-billion-in-2020-joint-report-shows.html

Evans, S., Gabbatiss, J., McSweeney, R., Chandrasekhar, A, Tandon, A., Viglione, G., Hausfather, Z., You, X., Goodman, J., and Hayes, S. (2021). *COP26: Key Outcomes Agreed at the UN Climate Talks in Glasgow.* The Carbon Brief. https://www.carbonbrief.org/cop26-key-outcomes-agreed-at-the-un-climate-talks-in-glasgow

G20 Finance Ministers. (2011). *Mobilizing Climate Finance.* https://www.imf.org/external/np/g20/pdf/110411c.pdf

G20 Sustainable Finance Study Group. (2018). *Sustainable Finance Synthesis Report.* https://unepinquiry.org/g20greenfinancerepositoryeng/

Global Center on Adaptation. (2020). *State and Trends in Adaptation Report 2020.* https://gca.org/report-category/flagship-reports/

Government of Fiji. (2021). *Climate Change Division.* http://fijiclimatechangeportal.gov.fj/

Government of Germany, Ministry of Finance. (2021). *The Bund's Green Twins: Green Federal Securities.* https://www.deutsche-finanzagentur.de/en/institutional-investors/federal-securities/green-federal-securities/

Green Climate Fund. (2021). *Green Climate Fund Portfolio Dashboard.* https://www.greenclimate.fund/projects/dashboard

Group of Development Finance Institutions. (2019). *DFI Working Group on Blended Concessional Finance for Private Sector Projects.* https://www.ifc.org/wps/wcm/connect/topics_ext_content/ifc_external_corporate_site/bf/bf-details/bf-dfi

Hallegatte, S., et al. (2017). *Unbreakable: Building the Resilience of the Poor in the Face of Natural Disasters.* The World Bank. https://openknowledge.worldbank.org/bitstream/handle/10986/25335/9781464810039.pdf

Hallegatte, S., Rentschler, J. and Rozenburg, J. (2019). *Lifelines: The Resilient Infrastructure Opportunity.* The World Bank. https://openknowledge.worldbank.org/handle/10986/31805

Hussain, F., Tlaiye, L., and Jordan, M. (2020). *Developing a National Green Taxonomy: A World Bank Guide.* The World Bank. http://documents1.worldbank.org/curated/en/953011593410423487/pdf/Developing-a-National-Green-Taxonomy-A-World-Bank-Guide.pdf

International Monetary Fund. (2020). *World Economic Outlook.* https://www.imf.org/-/media/Files/Publications/WEO/2020/Update/June/English/WEOENG202006.ashx

Impact Management Project. (2021). *How Investors Manage Impact.* https://impactmanagementproject.com/impact-management/how-investors-manage-impact/

International Capital Markets Association. (2014a). *Terms of Reference, Impact Working Group.* https://www.icmagroup.org/assets/documents/Regulatory/Green-Bonds/WG-Impact-Reporting-ToR-20202021-050121.pdf

International Capital Markets Association. (2014b). *Green Bond Principles.* https://www.icmagroup.org/assets/documents/Regulatory/Green-Bonds/Green-Bonds-Principles-2014.pdf

International Capital Markets Association. (2021). *Green Bond Principles (GBP 2021).* https://www.icmagroup.org/sustainable-finance/the-principles-guidelines-and-handbooks/green-bond-principles-gbp/

IPCC. (2021). Summary for policymakers. In V. Masson-Delmotte, P. Zhai, A. Pirani, S.L. Connors, C. Péan, S. Berger, N. Caud, Y. Chen, L. Goldfarb, M.I. Gomis, M. Huang, K. Leitzell, E. Lonnoy, J.B.R. Matthews, T.K. Maycock, T. Waterfield, O. Yelekçi, R. Yu and B. Zhou (eds.), *Climate Change 2021: The Physical Science Basis. Contribution of Working Group I to the Sixth Assessment Report of the Intergovernmental Panel on Climate Change.* Cambridge University Press.

IPCC. (2018). *Special Report: Global Warming of 1.5°C.* International Governmental Panel on Climate Change. https://www.ipcc.ch/sr15/

Ives, M. (2018, 9 September). Rich nations vowed billions for climate change. Poor countries are waiting. *The New York Times.* https://www.nytimes.com/2018/09/09/world/asia/green-climate-fund-global-warming.html?auth=login-email&login=email

Kaplan, S., and Birnbaum, M. (2021, 9 November). Despite COP26 pledges world still on track for dire warning. *Washington Post.* https://www.washingtonpost.com/climate-environment/2021/11/09/cop26-un-emissions-gap/

Lattanzio, R. (2019). *Paris Agreement: U.S. Climate Finance Commitments.* Congressional Research Service. https://fas.org/sgp/crs/misc/IF10763.pdf.

Leggett, J.A. (2015, 15 December). *Climate Change Pact Agreed in Paris.* The Congressional Research Service: CRS Insight. https://fas.org/sgp/crs/misc/IN10413.pdf

Mountford, H., et al. (2021) *COP26: Key Outcomes from the UN Climate Talks in Glasgow,* World Resources Institute, Washington, DC. https://www.wri.org/insights/cop26-key-outcomes-un-climate-talks-glasgow

Multilateral Development Bank Joint Report. (2015). *From Billions to Trillions: Transforming Development Finance.* The World Bank. http://pubdocs.worldbank.org/en/622841485963735448/DC2015-0002-E-FinancingforDevelopment.pdf

Networking for Greening the Financial System. (2019). *A Call to Action Climate Change as a Source of Financial Risk.* https://www.ngfs.net/sites/default/files/medias/documents/ngfs_first_comprehensive_report_-_17042019_0.pdf

Nordea. (2021). *Open Insights by Nordea: Shining a Light on Covid-19's Impact on Women.* https://insights.nordea.com/en/sustainability/world-bank-covid-19/

Operating Principles for Impact Management. (2019). *Impact Principles.* https://www.impactprinciples.org/

Organization for Economic Cooperation and Development. (2017). *Mobilising Bond Markets for a Low-Carbon Transition, Green Finance and Investment.* OECD Publishing, Paris.

Organization for Economic Cooperation and Development. (2020). *Climate Finance for Developing Countries Rose to USE 78.9 Billion in 2018.* https://www.oecd.org/newsroom/climate-finance-for-developing-countries-rose-to-usd-78-9-billion-in-2018oecd.htm

Rajamani, L. (2000). The principle of common but differentiated responsibility and the balance of commitments under the climate regime. *Review of European, Comparative & International Environmental Law*, 9(2), 120–131.

Regional Greenhouse Gas Initiative. (2021). www.rggi.org

Reichelt, H. (2010). Green bonds: a model to mobilise private capital to fund climate change mitigation and adaptation projects. *Euromoney Environmental Finance Handbook*. London, UK.

Reichelt, H. and Keenan, C. (2017). *The Green Bond Market: 10 Years Later and Looking Ahead*. Pension Fund Service, Knutsford, UK. https://pubdocs.worldbank.org/en/554231525378003380/publicationpensionfundservicegreenbonds201712-rev.pdf

Rowling, M. (2019, 3 May). *Green Climate Fund Must Take Risks in Warming Fight, Says New Head*. Reuters. https://www.reuters.com/article/us-global-climatechange-finance-intervie-idUSKCN1S91LL

Shapiro, A., et al. (2021). *U.S. and China Announce Surprise Climate Agreement at COP26 Summit*. National Public Radio. https://www.npr.org/2021/11/11/1054648598/u-s-and-china-announce-surprise-climate-agreement-at-cop26-summit

Stewart, F. and Reichelt, H. (2020). *Starting on a Sustainable Investing Journey*. Pension Fund Service. Knutsford, UK.

Strategy & Part of PwC Network. (2021). *Scaling the Sustainable Finance Market*. https://www.pwc.com/gx/en/insights/scaling-sustainable-finance-market.pdf

Task Force on Climate-Related Financial Disclosure. (2021). *About*. https://www.fsb-tcfd.org/

The Commission to the European Parliament, the European Council, the Council, the European Central Bank, the European Economic and Social Committee and the Committee of the Regions. (2018). *Action Plan: Financing Sustainable Growth*. https://eur-lex.europa.eu/legal-content/EN/TXT/?uri=CELEX:52018DC0097

The Copenhagen Accord. United Nations Framework Convention on Climate Change. (2012). https://unfccc.int/process-and-meetings/conferences/past-conferences/copenhagen-climate-change-conference-december-2009/copenhagen-climate-change-conference-december-2009

Thwaites, J., et al. (2015). *What Does the Paris Agreement Do for Finance?* The World Resources Institute. https://www.wri.org/blog/2015/12/what-does-paris-agreement-do-finance.

United Kingdom Debt Management Office. (2021). *Green Gilt Issuance*. https://www.dmo.gov.uk/responsibilities/green-gilts

United Nations. (2018, 1 January). *UN Secretary-General's Strategy for Financing the 2030 Agenda*. https://www.un.org/sustainabledevelopment/sg-finance-strategy/

United Nations Conference on Trade and Development. (2014). *World Investment Report 2014 Investing in the SDGs: An Action Plan*. Geneva, Switzerland.

United Nations Department of Economic and Social Affairs. (2015). *Sustainable Development Goals*. https://sdgs.un.org/goals

United Nations Environment Programme. (2021). *Addendum to the Emissions Gap Report 2021*. https://wedocs.unep.org/bitstream/handle/20.500.11822/37350/AddEGR21.pdf

United Nations Framework Convention on Climate Change. (1992). The United Nations General Assembly. https://unfccc.int/resource/docs/convkp/conveng.pdf

United Nations Framework Convention on Climate Change, Kyoto Protocol. (1997). https://unfccc.int/resource/docs/convkp/kpeng.pdf

United Nations Framework Convention on Climate Change. (2021). *COP26 Outcomes: Finance for Climate Adaptation*. https://unfccc.int/process-and-meetings/the-paris-agreement/the-glasgow-climate-pact/cop26-outcomes-finance-for-climate-adaptation#eq-1

United Nations General Assembly. (2015). *Outcome of the Third International Conference on Financing for Development*. UN Report of the Secretary General, New York, USA. https://undocs.org/A/70/320

US Department of State. (2021). *U.S.–China Joint Glasgow Declaration on Enhancing Climate Action in the 2020s*. https://www.state.gov/u-s-china-joint-glasgow-declaration-on-enhancing-climate-action-in-the-2020s/

US Environmental Protection Agency. (2021). *Carbon Dioxide Equivalency Results Based on Greenhouse Gas Equivalences Calculator*. https://www.epa.gov/energy/greenhouse-gas-equivalencies-calculator

Volz, U. (2020). *Investing in a Green Recovery*. International Monetary Fund. https://www.imf.org/external/pubs/ft/fandd/2020/09/investing-in-a-green-recovery-volz.htm

World Bank. (2008a, November 6). *World Bank and SEB partner with Scandinavian Institutional Investors to Finance "Green" Projects* [Press release]. https://www.worldbank.org/en/news/press

-release/2008/11/06/world-bank-and-seb-partner-with-scandinavian-institutional-investors-to-finance-green-projects.
World Bank. (2008b, November 14). *World Bank "Green Bonds" Increased to SEK 2.7 Billion* [Press Release]. https://www.worldbank.org/en/news/press-release/2008/11/14/world-bank-green-bonds-increased-to-sek-2-7-billion
World Bank. (2009, April 24). *State of California Buys USD 300 Million World Bank Green Bonds to Boost Global Solutions to Climate Change* [Press release]. https://www.worldbank.org/en/news/press-release/2009/04/24/state-of-california-buys-usd-300-million-world-bank-green-bonds-to-boost-global-solutions-to-climate-change
World Bank. (2010, February 22). *Nikko Asset Management Set to Launch Green Fund with World Bank Bonds* [Press release]. https://www.worldbank.org/en/news/press-release/2010/02/22/nikko-asset-management-set-to-launch-green-fund-with-world-bank-bonds
World Bank. (2013), *Green Bond Symposium.* http://pubdocs.worldbank.org/en/980521525116735167/Green-Bond-Symposium-Summary.pdf
World Bank. (2014). *Green Bond Process Implementation Guidelines.* https://pubdocs.worldbank.org/en/217301525116707964/Green-Bond-Implementation-Guidelines.pdf
World Bank. (2015a, July 10). *International Financial Institutions Announce $400 Billion to Achieve Sustainable Development Goals* [Press release]. https://www.worldbank.org/en/news/press-release/2015/07/10/international-financial-institutions-400-billion-sustainable-development-goals
World Bank. (2015b). *Green Bonds Working Towards a Harmonized Framework for Impact Reporting.* https://pubdocs.worldbank.org/en/760481552506496044/report-green-bonds-working-towards-harmonized-framework-2015-03.pdf.
World Bank. (2015c). *World Bank Green Bond Impact Report.* https://thedocs.worldbank.org/en/doc/275171507751972339-0340022017/original/reportimpactgreenbond2015.pdf
World Bank. (2015d). *What are Green Bonds?* https://openknowledge.worldbank.org/handle/10986/22791
World Bank. (2019a). *10 Years of Green Bonds: Creating the Blueprint for Sustainability across Capital Markets.* https://www.worldbank.org/en/news/immersive-story/2019/03/18/10-years-of-green-bonds-creating-the-blueprint-for-sustainability-across-capital-markets.
World Bank. (2019b). *World Bank Green Bond Impact Report.* https://thedocs.worldbank.org/en/doc/790081576615720375-0340022019/original/IBRDGreenBondImpactReportFY2019.pdf
World Bank. (2019c). *Stocks Traded, Total Value (US $)* [data set]. World Bank. https://data.worldbank.org/indicator/CM.MKT.TRAD.CD.
World Bank. (2020a). *State and Trends of Carbon Pricing 2020.* World Bank, Washington, DC.
World Bank. (2020b). *Engaging with Investors on Environmental, Social and Governance (ESG) Issues – A World Bank Guide for Sovereign Debt Managers.* https://pubdocs.worldbank.org/en/375981604591250621/World-Bank-ESG-Guide-2020-FINAL-11-5-2020.pdf
World Bank. (2021a). *World Bank Group Climate Change Action Plan 2021–2025: Supporting Green, Resilient, and Inclusive Development.* World Bank, Washington, DC.
World Bank. (2021b). *What You Need to Know About Sustainable Development Bonds.* https://www.worldbank.org/en/news/feature/2021/09/28/what-you-need-to-know-about-sustainable-development-bonds
World Bank. (2021c, September 23). *The World Bank Launches Initiative to Issue USD 10 Billion in Sustainable Development Bonds while Highlighting the Urgency of Mainstreaming Climate Action* [Press release]. https://www.worldbank.org/en/news/press-release/2021/09/23/world-bank-launches-initiative-to-issue-usd-10-billion-in-sustainable-development-bonds-while-highlighting-the-urgency-o
World Bank. (2021d). *Sovereign Environmental, Social and Governance.* [data set]. World Bank. https://datatopics.worldbank.org/esg/
World Bank and Public-Private Infrastructure Advisory Facility. (2015). *What Are Green Bonds.* https://documents1.worldbank.org/curated/en/400251468187810398/pdf/99662-REVISED-WB-Green-Bond-Box393208B-PUBLIC.pdf
World Investment Report. (2014). *Investing in the SDG: An Action Plan.* United Nations Conference on Trade and Development, New York, USA.

World Resources Institute. (2019, October 25). *Green Climate Fund Replenishment Pledges a Positive Step, But Far More Finance Needed* [Statement]. https://www.wri.org/news/2019/10/statement-green-climate-fund-replenishment-pledges-positive-step-far-more-finance

20. Green bonds: investor, issuer and climate perspectives

Christa Clapp, Keith Lee and Anouk Brisebois

INTRODUCTION

Striving to meet the appetite of responsible investors, the green bond market has been on the rise for over a decade. Demand for green-labeled financial products from investors spurred the start of the green bond market, with the Swedish pension fund pushing for the first World Bank issued green bond in 2008. Since then, the green bond market has grown to over one trillion USD in cumulative issuances, from multilateral development banks, corporations, municipalities, and sovereign governments.

Green bonds are debt instruments that specifically target climate or environmental projects (see Chapter 19 by Reichelt, Allen, and Cantor in this volume). In most cases, the use of proceeds of the bond are ring-fenced for projects with positive climate or environment benefits. Green bond proceeds primarily target climate mitigation projects that reduce net levels of greenhouse gas emissions. Owing to this predominant use, they have sometimes been referred to as 'climate' bonds, but the most commonly used term is 'green bond'.

The 'green' bond label is conferred by an independent reviewer. Such labels can help investors identify bonds and other financial products that channel funds either directly or indirectly to projects with positive climate and environmental impact. By producing a framework document that outlines selection criteria for eligible green projects, assets or activities, and how the environmental impact will be reported back to investors, a green bond issuer can establish a closer link between issuers and investors.

From an investor's perspective, green bonds can help reduce climate risk exposure in their portfolio and can signal a company's intent on environmental targets (Flammer, 2020). From an issuer's perspective, green bonds can result in a lower cost of financing than conventional bonds because the reduced interest paid to investors exceeds the costs of getting third-party certification or verification (Schumacher, 2020). While it is difficult to trace specific climate impact from green bonds, they can support a positive climate shift via transparent and robust labelling to help investors identify opportunities.

While green bonds have a role to play in the financial sector's contribution to a low-carbon shift, they are not a substitute for climate policy. Carbon taxes or emission trading schemes provide a more direct route to climate action via the 'polluter pays' principle, and typically provide the underlying economic conditions for a project or investment to occur in the first place. But transparency requirements related to green bonds can support improved identification of issuers involved in green sectors – as well as greater disclosure on climate risk in the financial sector, which can help shifting financial flows in a supporting direction for the climate, particularly when coupled with strong climate policy signals.

Reflecting on over a decade of green bond market growth, this chapter provides an overview of the market characteristics today, including an examination of the 'green' label, explores the changing regulatory landscape, reflects on both investors and issuers perspectives, and discusses the current lack of evidence for the direct impact of green bonds on the climate. The chapter ultimately proposes some trends to watch going forward.

MARKET CHARACTERISTICS

Driven by increasing investor demand, green bond issuances have grown significantly over the past five years. Green bond issuance in 2020 was valued at 290 billion USD, which is nearly seven times the value of issuance in 2015 (Climate Bonds Initiative, 2015, 2021b). Cumulative total green bond issuance from the beginning of the market in 2008 to the end of 2020 was 1.1 trillion USD (Climate Bonds Initiative, 2021b). Even during the Covid-19 crisis, green bonds issuance remained steady, with 2020 issuance up 9 percent from the previous year. The label 'green' is the most widely used label in the bond market, and thus the focus of this chapter, with notable issuers including Apple,[1] Fannie Mae,[2] and the Industrial and Commercial Bank of China[3] (see also Box 20.1 for a discussion on other labels).

Looking across the universe of green and sustainability labeled bonds, and including socially targeted bonds[4], total issuances in 2020 were over 600 billion USD, indicating growing investor interest in addressing environmental and social issues (Environmental Finance, 2021). Beyond the labeled bond universe, there can also be bonds that result in positive environmental impact but are not represented in these figures owing to their lack of a label.

Despite strong growth trends, green-labeled bonds remain a niche portion of the global bond market, comprising approximately 1% of outstanding global bonds issued by both public and private sector issuers (Climate Bonds Initiative, 2021b; International Capital Market Association, 2020). Given the portion of unlabeled bonds in the global market, there could be significant opportunity to scale up green bonds. However, it remains to be seen how much of the market could ultimately be eligible for a green label.

BOX 20.1 PROLIFERATION OF LABELS

There has been a proliferation of multiple other labels in the market, including but not limited to climate bonds, blue bonds, climate resilience bonds, sustainable development bonds, transition, and environmental, social and governance (ESG) bonds. Many can be considered green bonds in that they aim to generate positive environmental impact. Blue bonds can be used to describe bonds that encompass ocean-related activities, although some green bonds also include blue elements. These include Grieg Seafood's green bond, which finances sustainable fish farming, and Cadeler's green bond, which finances shipping vessels that service offshore wind turbines. Some labels, such as ESG and sustainable development, are used for bonds that encompass projects with both environmental and social impact, such as sustainable low-income housing, and can be generally labeled as 'sustainability' bonds. Notable sustainability bond issuers include the Agricultural Development Bank of China and the Government of the Grand-Duchy of Luxembourg. Many of these labels have also been applied in Islamic finance to the issuance of sukuk, beginning with the world's first green sukuk in 2017 by Tadau Energy. Major issuances have since included the Republic of Indonesia's green sukuk in 2018 and the Islamic Development Bank's sustainability sukuk in 2021. A more recent development has been sustainability-linked bonds, such as issued by

H&M Group, which tie bond interest rates, or other financial characteristics, to an issuer's sustainability performance. Transition bonds refer to bonds that support emission reductions in typically heavy-emitting sectors as a first step on a low-carbon pathway.

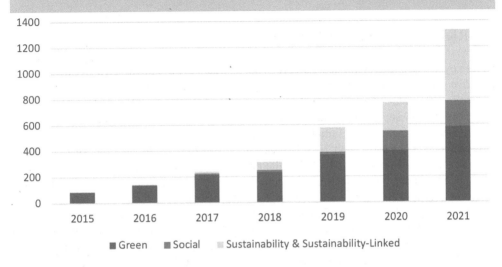

Figure 20.1   Green and sustainable bond value in 2015–2020 (million USD)

Issuer Types

From the initial creation of the green bond market by the World Bank (for more details, see Chapter 19 by Reichelt, Allen, and Cantor in this volume), followed by green bond issuances from other multilateral development banks, the market today is represented by a diverse group of issuers. Issuers include a wide range of public and private sector actors, such as sovereigns (e.g., Kenya[5], the United Kingdom[6]), municipalities (e.g. the City of Reykjavik[7], the Province of Quebec[8]), corporations (e.g. Daimler[9], Electrolux[10]), financial institutions (e.g. HSBC[11], the Ontario Teachers' Pension Plan Fund[12]), as well as development banks (e.g. the Asian Development Bank[13]). Corporations are the fastest growing issuer type. Issuers in 2020 were primarily corporations with 42 percent, followed by financial institutions with 17 percent (government agencies also with 17 percent), sovereign governments with 12 percent, and a smaller value of issuances from municipalities and supranational organizations, including multilateral development banks (Environmental Finance, 2021).

Issuers based in North America and Europe have been the most active in the green bond market. The largest green bond deals in 2020 originated in the US, the Netherlands and France (Environmental Finance, 2021). Asia is witnessing a rapidly-growing market, with issuers in many emerging market countries being active in the green bond market, including public and private sectors issuers across Brazil, China, India, Korea, Mexico and South Africa (Markandya et al., 2017). Green bond issuances in emerging markets grew to 168 billion USD in 2019, a 21 percent increase from the previous year (IFC, 2020).

Project Categories

The most popular project categories for green bonds issued in 2020 were renewable energy (102 billion USD), green buildings (76 billion USD), and sustainable transportation (58 billion USD) (Climate Bonds Initiative, 2020a, 2021b) . For example, Ørsted,[14] a Danish energy company, uses its green bond to finance offshore wind power generation. Volvo Cars'[15] green bond finances the development of electric vehicles. And Hungary's[16] sovereign green bond finances electric railways.

Climate adaptation projects, such as measures in buildings to accommodate heat stress or early warning and drainage systems for flooding, are funded to a far lesser extent. This could be in part due to the smaller and more spread out nature of some types of adaptation projects, which could be eligible for green bond funding if the projects are bankable and can be structured in a fixed-income instrument (Kato, Ellis, & Clapp, 2015). Examples of issuers that have included climate-adaptation related projects into their framework are, among others, the Republic of Kenya,[17] and the Province of Ontario.[18]

ESTABLISHING A GREEN LABEL

Providing transparency to investors on how the use of proceeds will be applied to green projects is the aim of a robust labelling system. This requires both a high level of transparency, but also some level of common understanding on green definitions that helps build trust in the market and support comparability across bonds. Recent voluntary and regulatory trends sustain the green bond market, but some gaps remain for future market development.

Best Practice

For most green bonds, the label 'green' is conferred by an independent external reviewer. This model of the external review was established in 2008 in the first green-labeled bond deal, whereby the investors requested a so-called 'second opinion', with the 'first opinion' provided by the issuer. The earliest such opinion was provided by CICERO (Center for International Climate Research), based in Norway, to give investors assurance that the World Bank's intended use of proceeds was indeed 'green'.[19] This established a model for transparency to investors that has guided the market since, with external reviewers providing a check against what green activities the issuer intends to finance with the bond. The Green Bond Principles (GBP), voluntary guidance that has established best practice that is widely followed in the market, recommends the use of an external reviewer to check the alignment of green bonds at the stage of issuance with the pillars of the transparency outlined by the GBP (International Capital Market Association, 2021) (Box 20.2). In 2020, 89 percent of green labeled bonds globally used an external reviewer to confirm their bonds' green credentials (Climate Bonds Initiative, 2021b).

462 *Handbook of business and climate change*

BOX 20.2 GREEN BOND PRINCIPLES: BEST PRACTICE FOR ISSUERS

The Green Bond Principles (GBP) outline voluntary guidelines for issuing green bonds that are considered best practice. The GBP were established in 2014 by a group of underwriting banks to set out guidelines to encourage transparency and protect the integrity of the green bond market. The GBP are independently hosted by the International Capital Market Association.

The GBP outline four pillars for a green label that encourage transparency on (1) the use of proceeds; (2) the issuer's process for project evaluation and selection; (3) the management of proceeds from an accounting perspective; and (4) reporting back to investors. Issuers provide this transparency by making a green bond framework document publicly available that describes their approach to these four pillars.

While the GBP do not provide an exhaustive or detailed list of eligible green project types, they do provide examples focused on climate mitigation and adaptation projects, in sectors such as energy efficiency, land use, and circular economy, and also include environmental pollution prevention and control. The GBP encourage the use of an external reviewer to provide an independent view on the pre-issuance alignment with the four core components of the GBP.

External reviews come in several different forms and at different points during the bond issuance process. Second opinions are the most common type of review, which are short reports provided by independent reviewers that check the green bond framework against the Green Bond Principles (see Box 20.2) and offer an opinion on whether the bond is green. External reviewers currently active in the market are a diverse set of actors, comprised of sustainability rating agencies such as Sustainalytics, credit rating agencies such as Moody's, accounting firms such as KPMG, or derived from research organizations such as CICERO Shades of Green. External reviewers can voluntarily commit to following the Guidelines for External Reviewers established by the GBP, but are currently not subject to any regulations for qualifications (ICMA, 2021). In addition to the GBP, a voluntary standard for the types of projects that can be financed by a green bond has been established by the Climate Bonds Initiative (CBI), a non-profit organization. Approximately 15 percent of the green bond market has been certified according to the CBI standard at the time of issuance by CBI-approved certifiers (Climate Bonds Initiative, 2021b).

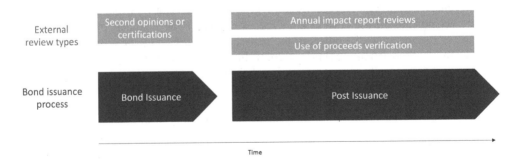

Figure 20.2   *Types of green bond external reviews*

To maintain credibility in the green bond market, the labels need to be conferred by credible actors based on clear and ambitious definitions of green. While concerns about 'greenwashing' have persisted in the green bond market for several years (Shislov et al., 2018), the fear of reputational risk to both the issuer and investors has held this in check to some degree (Hoepner et al., 2017). A Climate Bonds Initiative report (2021a) suggests that greenwashing is rare based on an observation that the majority of bond issuances analyzed in the report was accompanied by use of proceeds reporting, i.e., there was assurance that issuers were actually financing green projects and assets.

After a green bond is issued, the follow-up by the issuer and external reviewers is less uniform. The issuer is encouraged under the GBP to provide annual reports to investors on the environmental impact of the investments, but not all issuers do so, perhaps in part depending on how engaged the investors are in asking for this information. While 77 percent of green bond issuers report on allocation of use-of-proceeds, fewer (59 percent) report on environmental impact (Climate Bonds Initiative, 2021a). Some issuers choose to have the impact reports reviewed or verified by external reviewers. While this is not common, it is a growing practice and an area where investors are paying more attention. It is also possible to have an accounting auditor verify that the investment was used for its intended purpose. While there are no specific penalties that can be imposed based on the current regulatory system, there are legal (Doran & Tanner, 2019) and reputational risks at stake. For sustainability-linked bonds (SLBs), the financial terms are tied to the sustainability performance criteria of the issuer's design, but impact is highly dependent on the ambition of the performance targets. The World Bank[20] was one of the early movers in developing its green bond impact report, which includes project examples and metrics such as megawatts of renewable energy generated, the number of people with improved access to electricity, and the number of hectares of improved irrigation or drainage.

One of the challenges in post-issuance reporting is the lack of consistent data, providing a barrier to aggregating impact across a green bond portfolio for investors. Post-issuance impact reporting for green bonds is patchy and inconsistent, with variations in metrics and methodologies employed (Environmental Finance, 2020). However, several voluntary coordination efforts are pushing for greater consistency in reporting (ICMA, 2020; Nordic Public Sector Issuers, 2019).

Defining 'Green'

The GBPs' ascendancy to market best practice has improved the credibility of the green bond market, particularly in terms of process and transparency (see Box 20.2). However, the GBP's guidance on activities that are eligible for financing with green bond proceeds is indicative and leaves much room for interpretation. Consequently, one of the most important checks conducted by external reviewers is their evaluation of whether an issuer's stated use of proceeds is 'green' or not, and external reviewers have developed varying methodologies for doing so. This often means making decisions about what types of environmental projects are acceptable or layering in additional sustainability criteria for certain project types that may have more questionable environmental impact. Variation in these methodologies, alongside bespoke frameworks from issuers, has meant a degree of heterogeneity in green bonds to date, and the green credentials of transactions are occasionally questioned in the media and public discourse.

Financial regulators have recognized the importance of creating a more common language, defining 'green' to support the further growth of the market while protecting against greenwashing and improving comparability. The EU Commission, among other regulators, is developing a classification system for sustainable economic activities (see below), which shifts some of the burden of defining green to the government authorities for green bonds that adopt this approach, while still leaving room for interpretation by external reviewers.

Once the initial external review is provided at the time of issuance, other market actors that consolidate or list green bonds can influence what is considered green by how the bonds are presented to investors. In response to increasing investor demand, a plethora of indices have cropped up in recent years, focusing on low-carbon securities such as green bonds (UN PRI, 2018). Indices such as S&P's Green Bond Index, and exchange traded funds such as BlackRock Green Bond ETF, each have their own selection criteria for what types of green bonds can be included. For example, the Bloomberg Barclays MSCI Green Bond Indices require additional standards to be met for large hydropower projects to be included (Bloomberg Barclays MSCI, 2019). Exchanges are also entering the labeled security arena by creating separate listings for green labeled bonds and equities help investors identify credible green financial products. From the first green bond listing on the Oslo Stock Exchange in 2015, separate green or sustainability bond listing segments grew to 15 in 2018 (SSE, 2018).

Beyond the market, international organizations, NGOs and think tanks (such as the Anthropocene Fixed Income Institute, Climate-KIC, UNEP-FI and WWF) also play a role, whether in shaping definitions of green, or acting as watchdogs against greenwashing.

The discourse around definitions of green has attracted increasing attention for a good reason – there is no simple approach to defining an environmentally sustainable activity. What constitutes a positive impact, or robust risk management, depends on varying factors, chief among which are the timeframe, context, and assessment scope – and the picture becomes more complex when trying to factor in and balance primary objectives, e.g., climate mitigation, against other environmental concerns, let alone interlinked social ones. In response to this challenge, the need for criteria to be 'science-based' has gained increasing traction in approaches to assessing greenness. In the context of climate change, this most commonly refers to the concept of 'Paris-alignment', i.e., whether an economic activity or actor is compatible with achieving the global warming target of the Paris Agreement or not. Specifying Paris-alignment helps define the timeframe for assessing the use of proceeds (does it help limit climate warming to well-below 2°C by 2100 from pre-industrial levels and is it resilient to climate impact at this level of warming?) and the context for their assessment (the level of climate mitigation and adaptation entailed by the Paris Agreement varies across different sectors and countries).

Given the challenges in timeframe and context, definitions of green do not need to be binary. Investments leading to short-term emission reductions or other environmental benefits are often necessary in the transition to a low-carbon and climate-resilient future, and there is a case for calling them green as they do not lock in emissions by extending the lifetime of emissions-intensive assets, e.g., by improving their economic viability. This necessitates definitions of green that allow for these nuances to be captured in traditionally more heavy-emitting sectors, while also rewarding those that are already shifting away from fossil fuels in line with the Paris Agreement.[21]

Shifting Regulatory Landscape

The norms and regulations affecting the green bond market, and more broadly the field of sustainable finance, have shifted dramatically in recent years. Generally, regulatory initiatives have attempted to address two main areas: (1) developing the market for green finance, including by tackling definitional issues, and (2) the threats from climate risk to financial stability.

Developing the market for green finance

The norms of the green bond market largely established via the GBP are, in part, beginning to evolve to reflect new regulatory developments. Regulatory initiatives aimed at supporting continued green bond growth are emerging in several different regions (see Box 20.3). Some aim to provide clarity to market actors on eligible green activities and improve transparency, while others aim at supporting some of the transaction costs of issuing a green bond.

BOX 20.3 GLOBAL GREEN BOND REGULATORY HIGHLIGHTS

Several initiatives in different regions of the world support increased green bond issuance by helping to define green activities and providing incentives or subsidies to green bond issuers. The two most comprehensive green bond regulatory initiatives and central bank support mechanisms have developed in China and the EU. China and the EU recently participated in an exercise convened by the International Platform on Sustainable Finance to create a common ground taxonomy that compares activities at the sector level.

The Chinese Green Industry Guidance Catalogue was the first of such initiatives to define green activities for green bonds. The latest iteration in 2020 defines eligibility criteria across several environmental sectors including transportation, energy, and adaptation. Green bonds issued in the domestic Chinese market must align with the catalogue. To further incentivize green bond issuance, the People's Bank of China (PBoC) offers macro prudential assessment points for banks issuing green bonds to improve financial terms on bank deposits and includes green bonds as eligible collateral for re-lending at below-market rates. Local governments also offer subsidies for green bond issuers and underwriters in China.

The EU Taxonomy Regulation, currently in draft form but expected to be finalized in 2021, defines activities by industrial code for climate mitigation, adaptation, and other sustainability objectives. The accompanying draft legislation for an EU Green Bond Standard would establish a voluntary EU green bond label where issuers must disclose alignment with the EU Taxonomy. The taxonomy is also accompanied by the Corporate Sustainability Reporting Directive, which requires companies to disclose climate risk and report alignment with the taxonomy beginning in 2021. The European Central Bank has also announced its intent to invest in a green bond investment fund and accept sustainability-linked loans as collateral.

There are many other initiatives that seek to establish a supporting environment for sustainable finance and green bonds. Reaching across ten national securities regulators in southeast Asia, the ASEAN Green Bond Standards build on ICMA's Green Bond Principles. Both Hong Kong and Singapore provide incentives to offset the cost of external reviews.

The European Union has introduced one of the most comprehensive sustainable finance initiatives, raising the bar for precision and transparency for green bonds – as well as overall corporate disclosure of sustainable taxonomy-aligned activities. The EU legislative package includes a taxonomy of green and sustainable activities and an accompanying voluntary EU Green Bond Standard (European Commission, 2020). Although the draft Green Bond Standard will be voluntary, it includes requirements to meet the standard, and some investors may be looking for verification against this standard as a sign of credibility. This will inevitably raise the burden of proof for new issuers. Some market actors informally lamented that many existing green bond issuers would not qualify, while others see it as a necessary step to guard the integrity of the green bond market (Cooper, 2020; Hurley, 2019).

The EU Taxonomy intends to create a common language on what 'green' means, by laying out detailed environmental requirements by industrial activity code, focused on several different environmental objectives including climate mitigation and adaptation. For an activity to be aligned with the Taxonomy, it must pass through three checkpoints (European Commission, 2021a):

1. Substantially contribute to at least one of the six environmental objectives (of which climate mitigation and climate adaptation have been detailed to date), as measured by specified environmental thresholds;
2. 'Do no significant harm' to any of the other five environmental objectives, as guided by the considerations specified in the regulatory annex; and
3. Comply with minimum social safeguards, as measured by alignment with the OECD Guidelines for Multinational Enterprises, the UN Guiding Principles on Business and Human Rights, the ILO core conventions, and the International Bill of Human Rights.

For green bonds, the draft EU Green Bond Standard would require transparency showing 100 percent alignment with the EU Taxonomy to qualify for the label (European Commission, 2021b).

Adding more administrative and reporting requirements can help protect the integrity of the market and provide improved transparency on climate risk but can also come with a transaction cost that could be a deterrent for some green bond issuers (see the discussion in the Issuer Perspective section later in this chapter). Further, a taxonomy does not completely eliminate the need for interpretation and assessment by external reviewers (Clapp, 2020). Although the Taxonomy specifies green activities, it does not provide transparency on the direction the company or organization is headed in, on sustainability. Transparency on corporate level environmental targets and strategies can be important to investors, particularly in the case of a company transitioning from greenhouse gas (GHG)-intensive activities to greener activities.

Further, a green bond that aligns with such a taxonomy (e.g., Île-de-France Mobilités,[22] SpareBank 1 Østlandet[23]) can indicate green activity level but does not provide a complete measurement of climate risk (Cai et al., 2021). In the EU Taxonomy, there are separate lists of climate mitigation and adaptation with the 'do no significant harm' principle applying across these objectives. However, this is not the same as a holistic view on climate risk across both transition and physical risk. For example, a newly constructed building that meets the energy efficiency requirements under the mitigation criteria will also have to avoid significant harm on adaptation, i.e., not aggravate adaptation by causing further soil erosion that contributes to flooding. But this is a one-lens approach versus a holistic view that considers both the building's energy efficiency and its use of building materials that are adapted for managing physical impact from climate change, e.g., heat stress.

Regulatory initiatives will also impact post-issuance reporting on green bonds. Annual impact reports are encouraged under the Green Bond Principles, and are seen as best practice, but the methodology, level of detail, and validation ranges to a large degree (Environmental Finance, 2020). Although green bonds have been around for over a decade, this is a relatively new area of focus as investors start to ask more questions about the environmental benefits resulting from their investments. As one example, the draft EU Green Bond Standard legislation would require annual allocation and impact reporting but would only require verification on the allocation reports (European Commission, 2020).

Threats from climate risk to financial stability
There has been a shift in financial discourse on climate change from an ethical or CSR framing to a risk framing (Cai et al., 2021), driven by a recognition that the financial sector needs to actively manage risk in the face of a changing climate and policy landscape to safeguard their financial returns (Clapp & Sillmann, 2019). Investment, underwriting, and lending portfolios are exposed via sectors that are vulnerable to changing carbon prices and via regions that are vulnerable to more frequent and severe physical climate impact (Battiston et al., 2017). The Financial Stability Board (FSB), an international body that monitors the global financial system, has been influential in placing climate risk on the agenda of central bankers, financial institutions, and companies. A useful framing of climate risk into three main categories faced by financial actors is provided by the FSB's Task Force on Climate-Related Financial Disclosures (TCFD) (TCFD, 2017). Physical risk results from the physical impact of the changing climate, such as extreme weather patterns and flooding, which are already resulting in significant damage costs. Transition risk results from transition to a low-carbon economy via changing policy, societal preferences, and technologies. Liability risk results from the legal responsibility for the impact of climate change.

To support both the active redirection of capital to achieve climate goals and the management of climate risk in portfolios, one of the current focuses of the financial sector is on increasing transparency and disclosure on climate risk. France was the first country to mandate climate risk disclosure from financial institutions (under Article 173 of the Energy and Climate Law), followed by an EU-wide push for climate risk disclosure for companies and financial institutions under the Corporate Sustainability Reporting Directive (CSRD).[24] In Europe, such financial sector directives provide additional disclosure coverage on top of the EU Emissions Trading Scheme, which governs the trading of emissions permits by companies. Elsewhere, authorities in Brazil, Hong Kong, New Zealand, Switzerland, and the UK, have introduced mandatory requirements specifically for financial institutions to disclose in accordance with the TCFD recommendations, while others, e.g. Australia, Japan and Malaysia, reference TCFD as best practice in voluntary guidelines and other communications around expectations to the financial sector.

The balancing act between climate risk disclosure regulation and avoiding undue burden to green issuers will likely continue for some time. However, in at least one instance, climate risk disclosure regulation has been linked with evidence of increasing investor appetite for green investments, and a decline in fossil fuel investments as evidenced by French institutional investors following the implementation of the Energy Transition and Green Growth law implemented in France in 2016 (Mésonnier & Nguyen, 2021). Although climate risk disclosure regulation in the financial sector is not a full substitute for environmental regulation, it could provide a signaling effect via transparency on climate risk that investors incorporate into their portfolio management.

INVESTOR PERSPECTIVE

A robust green label can facilitate investors to identify low-carbon investment opportunities. The number of investors with mandates that include a focus on climate change is growing. While some investors are more focused on the environmental impact of their portfolios, others are focused on reducing climate risk. These investors' strategies result in rising demand for green-labeled financial products.

Benefits for Investors

Investors, many of which self-identify as responsible or sustainable investors, are increasingly seeking investment opportunities that can help reduce their climate risk exposure and/ or support their contribution to the achievement of national and international net zero carbon goals, whether as part of their risk/return considerations, or to meet beneficiary/client expectations, voluntary commitments, and regulatory obligations. Responsible and/or sustainable investors are also increasingly engaging with issuers to understand and influence the impact of their investments.

The plethora of investor initiatives targeting low- and zero-carbon portfolios reflects the growing interest from investors in actively contributing to climate action. One example is the UN-convened Net Zero Asset Owners Alliance with 42 members, representing over 6 trillion USD in assets under management, setting and reporting on emissions targets for their portfolios. Another example is the European-based Institutional Investors Group on Climate Change (IIGCC) with over 200 financial institution members including pension funds and asset managers, which has developed the Net Zero Investment Framework, where signatories work together on low carbon investment frameworks and collaboratively shape demand for low-carbon financial products.

As more and more investors identify with low-carbon initiatives, portfolio managers can incorporate green labeled financial products into their existing investment strategy to mitigate climate risk or achieve impact. Green bonds with robust labels can signal investors towards lower climate risk issuers, assuming they are taking steps to mitigate loss of value from climate risk (Flammer, 2020), e.g., from higher prices for commodities and fossil fuel-based energy, or increasing climate-related extreme weather events such as heat waves and flooding. Further, the additional information green bond issuers provide around environmental strategies, use of proceeds, and anticipated impact is valuable for investors, who can use it to improve their investment strategies and facilitate assessment and reporting on their portfolios' impact (Shishlov et al., 2016).

An estimated 50 percent of green bond investors are either responsible or impact investors (CBI, 2020a), and a growing number of dedicated green bond funds and portfolios that incorporate green bonds illustrate the demand for financial products that address climate change. A total of 49 funds hold more than 50 percent of their portfolio assets in green bonds, with fund managers such as Amundi and Mirova in Europe, and Calvert and TIAA-CREF in the US (Environmental Finance, 2020).

As another indicator of investor appetite, green bond deals have been frequently over-subscribed, where the investor demand is greater than the amount of the bond issued. For example, Nauman (2020) notes that green bond offerings have been oversubscribed by three times on average, while another source noted that green bonds are five times oversubscribed compared with three times for non-labeled bonds (Gore & Berrospi, 2019), although specific statistics on timing and scope of analysis were not available. A survey of global treasurers revealed that demand for green bonds was higher than non-labeled bonds for 70 percent of the respondents (Climate Bonds Initiative, 2020b).

Challenges for Investors

Green bond investors focused on environmental impact are interested in claiming the impact attributed to their investments, but this impact is not only difficult to trace (see the discussion in the Climate Impact section) but can be subject to double counting. Investors holding green bonds purchased in the secondary market may also be making claims to impact that overlaps the attribution in the primary market. Further, impact reporting at the bond level is patchy and may have different methodologies for attributing impact to the funding, including a potential lack of transparency in the case of multiple funding sources for a specific climate-oriented project.

The structure of the bond also determines, to some degree, the ability to trace the funding to specific project-level impact. Green bonds structured as asset-backed securities (ABS) or project bonds channel funding directly to the project level, whereas most green bonds follow the use-of-proceeds model. Use-of-proceeds bonds can be issued by project developers, owners, or financiers, that essentially ring-fence funding for green projects. This can free up financing for other assets that may not be green, depending on the overall issuer context (Nicol et al., 2018).

Another complication in tracing environmental impact to financing is that green bonds are frequently used to re-finance existing projects. This means that for some projects the green labelling is new, but the project to be financed may not be (CICERO & CPI, 2015). In essence, a portion of green bonds may be replacing existing debt financing with green labels, rather than representing new or additional green project financing. Projects financed by green bonds may therefore only partially represent 'new' green financed projects. Indeed, surveys and interviews with issuers indicate that projects financed by green bonds would have gone ahead regardless (Gyura, 2020; Maltais & Nykvist, 2020). Assessing additionality is difficult due to a lack of transparency from issuers on the portion of proceeds going to refinancing (Tuhkanen & Vulturius, 2020) and the generated impact that is additional (Tolliver et al., 2019).

ISSUER PERSPECTIVE

There are both tangible and intangible benefits for an issuer of green bonds in addition to several challenges.

470 *Handbook of business and climate change*

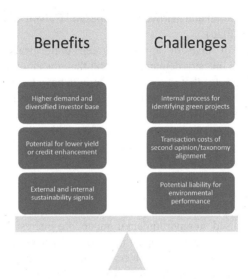

Figure 20.3   Benefits and challenges for green bond issuers

Benefits for Issuers

On the benefit side of the equation, a boost in demand with a diversified investor base, that in many cases is willing to accept lower yields in exchange for the green label, can be a big driver for issuance. In addition, the more intangible benefit of integrating sustainability into the internal decision-making structure of a company or organization can have an important external signaling effect.

In addition to attracting more investor demand (see above), one of the most cited benefits to issuing a green bond is increased access to diversified investors (Climate Bonds Initiative, 2020b). While this evidence is difficult to measure uniformly given the confidential investor data on many deals, this aspect has been often cited anecdotally. Attracting more investors that identify as 'responsible' investors can help orient the profiling of the issuer and potentially provide incentives for an issuer to grow their green-labeled debt offerings in the future, which also aligns with the signaling argument of environmental commitment from a company issuing a green bond (Flammer, 2020).

The process of issuing a green bond can also help an organization align internally across financial and environmental departments, which in some cases can be seen as a competitive advantage (see Box 20.4 for example).[25] While this can be time consuming and potentially involve organizational changes, sustainability alignment internally can prepare an organization to address future environmental targets in addition to providing a foundation for a stronger sustainability profile externally. Stronger corporate governance has been associated with green bond issuance in at least one study (Schneeweiss, 2019), however green bond issuance is not always tied with stronger climate targets at a corporate level (Tuhkanen & Vulturius, 2020), reflecting the variation in internal strategies and alignment.

Following from this, green bonds may serve as a credible signal of the company's commitment toward the environment, as well as the ability to undertake investments in green projects and to improve the company's environmental footprint (Flammer, 2020). Such a signal can be valuable, as investors may lack sufficient information about the company's environmental commitment and the level of ambition. Therefore, companies issuing green bonds may be better able to attract long term investors sensitive to climate risk. If corporate green bonds provide a credible signal of companies' commitment to the environment, then the announcement of a green bond issuance can result in a positive market reaction, with stronger responses for independently reviewed green bonds and first-time issuers.

> ### BOX 20.4 CASE STUDY: VASAKRONAN'S INTERNAL ALIGNMENT FOR GREEN FINANCING
>
> One of the immeasurable benefits of issuing a green bond is the opportunity to align internally on sustainability. Treasuries and sustainability departments need to work together to identify and finance green projects. The case of Vasakronan, a Swedish real estate company, highlights the internal shifts that can accompany green financing strategies.
>
> As the first corporate green bond issuer in the world, Vasakronan recognized the benefit of accessing responsible investors, and used the opportunity to organize internally to link environmental performance with financial decisions. Since its first issuance in 2013, it has become an experienced issuer of green bonds. Projects financed include new construction and major renovation projects with excellent building certificate classifications in addition to energy performance requirements of at least 25 percent below Swedish building regulations. In addition to the ambition of the energy performance of their buildings, Vasakronan applies a life cycle approach to building design, underscoring a holistic approach to sustainability. Environmental impacts are considered during material use, construction, property management, and demolition. Transport infrastructure, waste, and water management are also taken into consideration. Vasakronan has established targets to shift to 100 percent renewable energy to power its portfolio of properties, achieve climate neutrality throughout its value chain by 2030, and energy self-sufficiency of all its properties.
>
> Vasakronan's green bond framework was independently evaluated by CICERO Shades of Green and received their highest rating of Dark Green. The full second opinion is available at https://pub.cicero.oslo.no/cicero-xmlui/bitstream/handle/11250/2720386/vasakronan_2018.pdf.
>
> In parallel to, and in support of, the green financing developments at Vasakronan, sustainability and finance have become more closely aligned. At the time it issued its first green bond, it had a separate sustainability department, like most companies. Now, it incorporates sustainability across all departments and no longer has a carved-out sustainability department. The integration of sustainability across its internal decision-making processes reflects a significant internal shift that began with the issuance of its first green bond.

Lower cost of capital

Recently there has been a focus on the potential to achieve a premium, with investors accepting a lower yield, for a green bond issuance over a non-labeled bond issuance. There is some consensus across a survey of the most recent literature, with 56 percent of studies finding the existence of a premium for green bonds in the primary market, and 70 percent of the studies finding the existence of a premium in the secondary market (MacAskill et al., 2021). Premiums have been observed for green bonds issued by government entities, for those that are investment grade, and for those that followed defined governance and reporting practices, e.g., the Green Bond Principles.

The existence of a premium varies across regions, with mixed evidence in the USD-denominated market, and stronger evidence in the EUR-denominated market. However, there is a wide variance of the amount of premium calculated, with an average premium of –1 to –9 basis points in the secondary market (MacAskill et al., 2021), which can be attributed in part to how the pairings between green and non-green bonds are made, including differences in tenor or the sample size and characteristics. Importantly, investors have been shown to reward credible green labelling (Dorfleitner et al., 2021), with premiums for green bonds with external reviews, which are recommended by the Green Bond Principles.

In the US municipal bond market, some studies find evidence of a premium for the issuer, or a lower yield, for green bonds in the primary or secondary market (Baker et al., 2018; Karpf & Mandel, 2018; Partridge & Medda, 2020), while others find no premium (Hyun et al., 2019; Larcker & Watts, 2020).

In the European market, more evidence has been found of a premium for green bonds, perhaps reflecting the preferences of European 'responsible' investors. One of the most recent studies shows a premium in the primary market for 58 percent of green bonds denominated in EUR, and 46 percent for green bonds in USD (Climate Bonds Initiative, 2020a). The premium carries forward into the secondary EUR market, even if the amount of the premium is lower (Gianfrate & Peri, 2019). On the other hand, a study by Hachenberg & Schiereck (2018) did not find a significant influence of issuance size, maturity or currency on pricing for green bonds, although they did find that government issued green bonds showed a marginal premium.

Credit ratings can show an improvement following issuance, leading to a lower cost of capital for green bond issuers. Equity market spillover effects of increased stock prices may also be, in part, tied to green bond issuance (Tang & Zhang, 2020). Sovereign green bond issuance may be associated with positive spillover effects on credit market performance, as modelled by Monasterolo & Raberto (2018).

If there is a link between green labeled bonds and reduced climate risk, then over time the willingness of investors to accept a lower yield may increase. This in part depends also on the regulatory landscape for climate policy and climate risk disclosure in the finance sector (see the section on the regulatory landscape). Acceptance of a lower yield for green bonds also implies a robust green label (see section on the subjectiveness of green).

Challenges for Issuers

Yet there are also challenges that issuers face, as the burden of proof lies with issuers wishing to label their debt green. This means taking the time to align internally across financial and sustainability departments, as well as paying the transaction cost to prove green alignment and accept potential liability.

Green bond issuers face a transaction cost when they issue. There are internal time and capacity considerations for alignment across financial and environmental departments to identify and follow up funded green activities. In addition, there is a fee for an external review, which is generally less than the cost of, for instance, an engineering study, and thus is insignificant compared with the green bond deal size, especially with consideration of potentially achieving a lower cost of financing (Schumacher, 2020). However, transaction costs could rise to some extent with the additional detailed assessment necessary to align taxonomies.

Some potential issuers have also expressed concerns about their liability related to environmental performance, for example if they are not achieving the expected impact from a green bond, or if they are misrepresenting the use of proceeds (Shislov et al., 2018). While there have not to date been any lawsuits related to green bonds and misrepresenting the use of proceeds, as market regulation develops this could be an area for increasing scrutiny as taxonomies provide more detail of what constitutes a green project. One case of accused fraud has been filed with the US Securities and Exchange Commission for misrepresentation of fossil fuel involvement by an issuer of a sustainability-linked bond.[26] It may also be a consideration for green bond issuance in the United States, which has different norms than in Europe, regarding what is included in the legal prospectus for a green bond issuance.

CLIMATE IMPACT

While the causal chain leading to environmental impact is difficult to prove, it is premature to dismiss green bonds as not making a positive shift for the environment, considering green bonds' broad signaling effect on climate finance. Some benefits of green bonds have been touched upon in this chapter already, such as greater internal alignment on sustainability within issuing corporations and other indirect impacts such as improved internal profiles for sustainability teams and higher sustainability ambitions within issuers and investors alike, greater public awareness around climate change and sustainability, advancing policy dialogue on green finance, and contributing generally to the greater 'investability' of climate-focused and green activities (Maltais & Nykvist, 2020; TEG 2019). All of these indirect impacts can help to establish a foundation for identifying and shifting to smarter green investments over time. In the long term, more consistent reporting and methodologies could help establish stronger impact links.

Patchy Climate Impact Chain

Studies exploring emission reductions associated with green bonds have been focused on the corporate emission reporting level. This may be due, in part, to current inconsistent impact reporting from green bond issuers, but also in part to the challenge of linking green bond financing with project funding (see also the discussion in the previous section).

It remains challenging to trace evidence for impact, whether due in part to lack of data, or lack of direct impact tied to a green bond's use of proceeds. Impact reporting remains patchy and difficult to compare across bonds (Environmental Finance, 2020). However, market initiatives such as ICMA have taken some steps towards harmonizing impact reporting metrics that could be strengthened in the future to support impact measurement and further support green bond market credibility.

The few studies available show mixed evidence of environmental impact linked with green bonds to date. For instance, one study did not find any strong evidence that green bond issuance was followed by decreases in issuers' carbon intensities in the three years post-issuance (Ehlers et al., 2020), although there could be cases of companies issuing green bonds after the transition to green has begun. Another study covering 20 European-based green bond issuers found only 14 had identified climate emission reduction targets, with varying approaches and levels of ambition (Tuhkanen & Vulturius, 2020). Further, Flammer (2020) found that corporate emissions decreased after a green bond issuance, although the author notes that green bonds may not be directly responsible for significant improvements at the firm level due to the small size the average green bond comprises in relation to an average issuer's total assets. Rather, the author suggests the emissions reductions reflect the issuer's commitment to improving its environmental performance, with green bond issuances serving as a signal of that commitment (the previously discussed 'signaling' argument).

Fungibility of Financing

The positive impacts from a green bond may be undermined if an issuer uses proceeds to invest in green projects while investing in environmentally counterproductive projects via other parts of its organization (Ehlers & Packer, 2017). A recent example is the case of the State Bank of India, a green bond issuer that became subject to public criticism, NGO campaigns, and the divestment of its green bond by several asset managers after its plans to finance the controversial Carmichael coal mine in Australia surfaced (Kirakosian, 2020). The case highlights the risk posed by fungibility of capital, namely that issuers can use green bond proceeds to free up financing for other, potentially environmentally damaging purposes, and at a lower cost of capital should they enjoy any 'greeniums'. This underscores the importance of transparency as well as the need for investors to assess a firm's broader sustainability governance, policies and targets alongside the considerations that are specific to the green bond.

THE GREEN BOND MARKET LOOKING FORWARD

The increasing demand for green bonds illustrates rising investor interest in sustainability issues, environmental impact, and climate risk. While green bonds are not a silver bullet to solving the climate crisis, they can offer an opportunity for issuers to build internal capacity and link financial decisions with environmental considerations. Further work is needed to consider how the green bond issuance process, including internal alignment of sustainability and financial decisions, impacts issuers and the market in the long-term. The findings could have impact that supports more climate-smart decisions for decades. Given the size of the global bond market and increasing investor focus on sustainability, we can expect the portion of green labeled bonds to continue to increase. This may particularly be the case for corporate issuers as they also face regulatory pressure on climate risk disclosure. We can also expect more issuances of social and sustainability focused bonds in response to the Covid-19 pandemic.

Given investors' focus on environmental impact, we can also expect more focus on impact reporting from green bonds. Over time, with industry guidance, impact reporting could move towards more consistent reporting to aid portfolio managers in considering aggregate impacts and provide more credibility to the green bond market. Further, investors currently lack any legal recourse should issuers fail to comply with their stated use of proceeds, or a 'green default,' which could be a growing source of risk for investors, as the integration of green criteria into investment mandates becomes more widespread. Legal experts indicate this could lead to the introduction of contractual protections for investors against green defaults, with one expected mechanism being the linkage of green defaults to margin ratchets (Corke et al.,

2019). Other possibilities include covenants requiring as-yet voluntary best practices, e.g., regular third-party assured reporting and ring-fencing of green bond proceeds, or put options allowing bondholders to accelerate or redeem their bonds in the event of a green default (Czerniecki & Saunders, 2016).

As the regulatory landscape with respect to climate risk transparency evolves, a more common language on green will help to align investors and issuers in the green bond market but may in fact increase the level of transaction and compliance cost as issuers face more requirements to issue a green bond. The forthcoming regulatory pressure in the EU is on issuers and companies that pursue green activities. Exemplifying a top-down approach, the EU's Corporate Sustainability Reporting Directive (CSRD) takes a step towards requiring disclosure for all large companies but is still based on the EU Taxonomy of green and transition activities. This again places the burden on those proving they are doing green activities. It may have been more politically feasible to begin with a green taxonomy than one for activities with high climate risk, but this still leaves an uneven playing field until a taxonomy of GHG emitting activities is included. In contrast, the existing norms based on the Green Bond Principles have been more of bottom-up norms, with some regional differences, with reputational risk for both issuers and investors operating as a check against misuse of the green label.

The increase in regulatory requirements translates into higher transaction costs for bond issuers involved in green activities than for those involved in GHG-emitting activities. These transaction costs will need to be weighed against the benefits of internal alignment for continuing pressure on sustainable corporate management and potential premiums, or lower yields, for green labels. Some regional initiatives that provide incentives for external reviews by covering the cost of pre-issuance external review, such as those implemented in Hong Kong and Singapore, can also help bridge this gap before a level playing field is reached. Although these transaction costs can be balanced out by the benefits of issuing a green bond as outlined in this chapter, it makes it more difficult to imagine a global bond market that is mostly green.

In the long run, as climate and environmental policy tightens, as is the expectation under the Paris Agreement, and as climate risk disclosure policy spreads globally, we might expect to reach a tipping point where most of the bonds issued globally help mitigate climate risk. Responsible investors may be increasingly willing to accept a lower yield for financial products with lower climate risk. As central banks and financial supervisors increasingly push for climate risk transparency from financial institutions (NGFS, 2019), who in turn raise their expectations from companies, we may see the burden of proof shifting over time not just from companies proving they are green, but to all companies and issuers. The EU Corporate Sustainability Reporting Directive that requires large companies to disclose their EU Taxonomy alignment is a step in this direction. This may become easier for issuers if governments tighten environmental and climate policy at the same time. Over time, if companies that are not incorporating climate mitigation and resiliency planning into their financial decisions face a higher burden of proof, this would help reset this balance. This could also incite a broad range of investors, not just the so-called responsible or sustainable investors, to put pressure on bond as well as equity issuers to disclose climate risk.

At the same time, stronger climate policy can reinforce investors' preferences to reduce climate risk. Tightening the price or implied price on carbon, while raising pressure to disclose climate risk, would place the financial sector in a strong position to support a shift to a low carbon and climate resilient economy.

NOTES

1. https://www.apple.com/newsroom/2021/03/apples-four-point-seven-billion-green-bond-spend-is-helping-to-create-one-point-two-gigawatts-of-clean-power/
2. https://multifamily.fanniemae.com/media/document/pdf/multifamily-green-bond-framework
3. https://www.government.se/49bcc9/contentassets/ed959d7b700e429a98cc85bdb64ef1af/swedens-sovereign-green-bond-framework.pdf
4. Some of which may have been issued in response to the Covid-19 crisis, which has resulted in massive growth in socially targeted bond issuances.
5. https://www.greenbondskenya.co.ke/
6. https://assets.publishing.service.gov.uk/government/uploads/system/uploads/attachment_data/file/1002578/20210630_UK_Government_Green_Financing_Framework.pdf
7. https://reykjavik.is/sites/default/files/reykjavik_green_bond_framework_2019_-_baeklingur.pdf
8. http://www.finances.gouv.qc.ca/documents/Autres/en/AUTEN_Green_Bond_Framework.pdf
9. https://www.daimler.com/investors/reports-news/financial-news/20200903-green-bond.html
10. https://www.electroluxgroup.com/en/green-bond-framework-29317/
11. https://www.hsbc.com/investors/fixed-income-investors/green-and-sustainability-bonds
12. https://www.otpp.com/documents/10179/792185/OTPP+Green+Bond+Framework.pdf/a14845e6-634d-4b53-ae4b-6e0b58cbc94d
13. https://www.adb.org/sites/default/files/adb-green-bonds-framework.pdf
14. https://orsted.com/en/investors/debt/green-financing
15. https://www.volvogroup.com/en/investors/debt-information/green-financing.html
16. https://www.akk.hu/download?path=64709b3f-e69d-4969-b271-9d1db8f469bd.pdf
17. https://www.fsdafrica.org/wp-content/uploads/2021/06/Kenya-Sovereign-Green-Bond-Framework.pdf
18. https://www.ofina.on.ca/pdf/green_bond_framework.pdf
19. Second opinions are now provided through CICERO's subsidiary company CICERO Shades of Green.
20. https://treasury.worldbank.org/en/about/unit/treasury/ibrd/ibrd-green-bonds
21. The Shades of Green methodology developed by CICERO Shades of Green allows for three shades of green to indicate the level of climate ambition of the green bond. See https://cicero.green/green-bonds
22. https://www.iledefrance-mobilites.fr/medias/portail-idfm/403255b5-fdcc-4744-a42c-1e28a8276546_FrameworkGreenbund_EN-110521.pdf
23. https://www.sparebank1.no/content/dam/SB1/bank/ostlandet/omoss/investor/Rapporter2021/SpareBank1-Ostlandet-Green-Bond-Framework_vF.pdf
24. The European Commission website explains the EU Taxonomy, EU Green Bond Standard, and corporate disclosure of climate-related information: https://ec.europa.eu/info/business-economy-euro/banking-and-finance/sustainable-finance_en.
25. As a green bond external reviewer, CICERO Shades of Green has observed many conversations during the issuance process where an issuer's CFOs and sustainability officers need to cooperate, perhaps for the first time, on identifying and following up green projects eligible for funding.
26. https://www.globalcapital.com/article/297sitz2boxhpl0ffm29s/sri/banks-hit-by-fraud-complaint-to-sec-over-adani-slb-coal-links

REFERENCES

Baker, M. P., Bergstresser, D. B., Serafeim, G., & Wurgler, J. A. (2018). Financing the response to climate change: The pricing and ownership of U.S. green bonds. *SSRN Electronic Journal*. https://doi.org/10.2139/ssrn.3275327

Battiston, S., Mandel, A., Monasterolo, I., Schütze, F., & Visentin, G. (2017). A climate stress-test of the financial system. *Nature Climate Change*, 7(4), 283–288. https://doi.org/10.1038/nclimate3255

Bloomberg Barclays MSCI. (2019). *Bloomberg Barclays MSCI Green Bond Indices*.

Cai, W., Clapp, C., Das, I., Perkins-Kirkpatrick, S., Thomas, A., & Tierney, J. E. (2021). Reflections on weather and climate research. *Nature Reviews Earth & Environment*, 2(1), 9–14. https://doi.org/10.1038/s43017-020-00123-x

CICERO & CPI. (2015). *Background Report on Long-term Climate Finance prepared for the German G7 Presidency 2015 by CICERO and Climate Policy Initiative.*

Clapp, C. (2020). Green bonds: Thinking beyond standardization. *InvestESG*.

Clapp, C., & Sillmann, J. (2019). Facilitating climate-smart investments. *One Earth*, 1(1), 57–61. https://doi.org/10.1016/j.oneear.2019.08.009

Climate Bonds Initiative. (2015). *2015 Green Bond Market Roundup*.

Climate Bonds Initiative. (2020a). *Green Bond Pricing in the Primary Market: Report Highlights*.

Climate Bonds Initiative. (2020b). Green bond treasurer survey. In *Green Bond Treasurer Survey 2020*.

Climate Bonds Initiative. (2021a). *Post-issuance Reporting in the Green Bond Market*.

Climate Bonds Initiative. (2021b). *Sustainable Debt: Global State of the Market 2020*.

Cooper, G. (2020). Investors and verifiers welcome EU Green Bond Standard. *Environmental Finance*, March.

Corke, C., Myers, J., & Busch, C. (2019, October 17). *Green Bonds Series: Part 4 – When 'green' bonds go brown*. Lexology. https://www.lexology.com/library/detail.aspx?g=0a6503d3-d4ff-44fc-ab2b-5166c157f630

Czerniecki, K., & Saunders, S. (2016, February 11). *Green Bonds: An Introduction and Legal Considerations*. Bloomberg Law. https://news.bloomberglaw.com/environment-and-energy/green-bonds-an-introduction-and-legal-considerations

Doran, M., & Tanner, J. (2019). *Green Bonds – An Overview*. https://www.bakermckenzie.com/en/-/media/files/insight/publications/2019/05/green-bonds--an-overview--may-2019.pdf

Dorfleitner, G., Utz, S., & Zhang, R. (2021). The pricing of green bonds: External reviews and the shades of green. *SSRN Electronic Journal*. https://doi.org/10.2139/SSRN.3594114

Ehlers, T., & Packer, F. (2017). Green bond finance and certification. In *BIS Quarterly Review*, September, 89–104. https://www.bis.org/publ/qtrpdf/r_qt1709h.htm

Ehlers, T., Mojon, B., & Packer, F. (2020). Green bonds and carbon emissions: Exploring the case for a rating system at the firm level. *BIS Quarterly Review*, September, https://ssrn.com/abstract=3748440

Environmental Finance. (2020). *Green Bond Funds – Impact Reporting Practices 2020*.

Environmental Finance. (2021). *Sustainable Bonds Insight*.

European Commission. (2020). *Consultation document – Targeted consultation on the establishment on an EU Green Bond Standard*.

European Commission. (2021a). *Annex to the Commission Delegated Regulation Establishing Technical Criteria for Climate Change Mitigation*. EU.

European Commission. (2021b). *Proposal for a Regulation on European Green Bonds*. EU.

Flammer, C. (2020). Corporate green bonds. *SSRN Electronic Journal*. https://doi.org/10.2139/ssrn.3125518

Gianfrate, G., & Peri, M. (2019). The green advantage: Exploring the convenience of issuing green bonds. *Journal of Cleaner Production*, 219. https://doi.org/10.1016/j.jclepro.2019.02.022

Gore, G., & Berrospi, M. (2019). Rise of controversial transition bonds leads to call for industry standards. *Reuters*, September.

Gyura, G. (2020). Green bonds and green bond funds: The quest for the real impact. *The Journal of Alternative Investments*, 23(1), 71–79. https://doi.org/10.3905/JAI.2020.1.098

Hachenberg, B., & Schiereck, D. (2018). Are green bonds priced differently from conventional bonds? *Journal of Asset Management*, 19(6). https://doi.org/10.1057/s41260-018-0088-5

Hoepner, A.G.F. Dimatteo, S., Schaul, J., Yu, P.-S. & Musolesi, M. (2017). Tweeting about sustainability: Can emotional nowcasting discourage greenwashing? *Corporate Finance*, 8(3&4), https://ssrn.com/abstract=2924088 or http://dx.doi.org/10.2139/ssrn.2924088

Hurley, M. (2019, August). Just 17% of bonds in MSCI green bond index would satisfy EU GBS criteria. *Environmental Finance*. https://www.environmental-finance.com/content/news/just-17-of-bonds-in-msci-green-bond-index-would-satisfy-eu-gbs-criteria.html

Hyun, S., Park, D., & Tian, S. (2019). The price of going green: the role of greenness in green bond markets. *Accounting and Finance*. https://doi.org/10.1111/acfi.12515

ICMA. (2020). *Harmonized Framework for Impact Reporting*.

ICMA. (2021). *Guidelines for Green, Social* February).
IFC. (2020). *Emerging Market Green Bonds Report 2019.*
International Capital Market Association. (2020, August). *Bond Market Size.*
International Capital Market Association. (2021). *Green Bond Principles.*
Karpf, A., & Mandel, A. (2018). The changing value of the 'green' label on the US municipal bond market, *Nature Climate Change, 8*(2), 161–165, doi:10.1038/s41558-017-0062-0.
Kato, T., Ellis, J., Clapp, C. (2015). The role of the 2015 agreement in enhancing adaptation to climate change. In *Climate Change Expert Group Paper OECD* (Vol. 2015, Issue 2014).
Kirakosian, M. (2020, December 18). *Amundi Axes State Bank of India Green Bonds over Coal Mine Financing - Citywire.* Citywire Selector. https://citywireselector.com/news/amundi-axes-state-bank-of-india-green-bonds-over-coal-mine-financing/a1441102
Larcker, D. F., & Watts, E. M. (2020). Where's the greenium? *Journal of Accounting and Economics, 69*(2–3). https://doi.org/10.1016/j.jacceco.2020.101312
MacAskill, S., Roca, E., Liu, B., Stewart, R. A., & Sahin, O. (2021). Is there a green premium in the green bond market? Systematic literature review revealing premium determinants. *Journal of Cleaner Production, 280*, 124491. https://doi.org/10.1016/j.jclepro.2020.124491
Maltais, A., & Nykvist, B. (2020). Understanding the role of green bonds in advancing sustainability. *Journal of Sustainable Financial Investments, 11*(3), 233–252. https://doi.org/10.1080/20430795.2020.1724864
Markandya, A., Galarraga, I., & Rübbelke, D. (2017). *Climate Finance* (Vol. 2). World Scientific. https://doi.org/10.1142/9433
Mésonnier, J.-S., & Nguyen, B. (2021). *Showing Off Cleaner Hands: Mandatory Climate-related Disclosure by Financial Institutions and the Financing of Fossil Energy.* (November 18, 2020). SSRN: https://ssrn.com/abstract=3733781 or http://dx.doi.org/10.2139/ssrn.3733781
Monasterolo, I., & Raberto, M. (2018). The EIRIN flow-of-funds behavioural model of green fiscal policies and green sovereign bonds. *Ecological Economics, 144*, 228–243. https://doi.org/10.1016/j.ecolecon.2017.07.029
Nauman, B. (2020). Green bonds set to keep flying off shelves in 2020. *Financial Times.* https://www.ft.com/content/61631c2c-1a65-11ea-9186-7348c2f183af
NGFS. (2019). *Network for Greening the Financial System First Comprehensive Report A Call for Action. Climate Change as a Source of Financial Risk.* NGFS.
Nicol, M., Shishlov, I., & Cochran, I. (2018). *Green Bonds: Improving their Contribution to the Low-carbon and Climate Resilient Transition.* Institute for Climate Economics. https://www.i4ce.org/wp-content/uploads/2022/07/I4CE-GreenBondsProgram-Contribution-Energy-Transition-web-5-1.pdf
Nordic Public Sector Issuers. (2019). *Position Paper on Green Bonds Impact Reporting.* Nordic Public Sector Issuers
Partridge, C., & Medda, F. R. (2020). The evolution of pricing performance of green municipal bonds. *Journal of Sustainable Finance & Investment, 10*(1), 44–64. https://doi.org/10.1080/20430795.2019.1661187
Schneeweiss, A. (2019). *Great Expectations Credibility and Additionality of Green Bonds.* https://doi.org/10.13140/RG.2.2.15563.64808
Schumacher, K. (2020). The shape of green fixed income investing to come. *The Journal of Environmental Investing, 10*(1). https://doi.org/10.2139/ssrn.3663308
Shishlov, I., Morel, R., & Cochran, I. (2016). Beyond transparency: unlocking the full potential of green bonds. In *I4CE – Institute for Climate Economics* (Issue June). https://www.cbd.int/financial/greenbonds/i4ce-greenbond2016.pdf
Shislov, I., Nicol, M., & Cochran, I. (2018). *Environmental Integrity of Green Bonds: Stakes, Status and Next Steps.* Paris: Institute for Climate Economics. https://www.i4ce.org/wp-content/uploads/I4CE-GreenBondsProgram-Environmental-Integrity-web.pdf
SSE. (2018). *2018 Report on Progress.* SSE.
Tang, D. Y., & Zhang, Y. (2020). Do shareholders benefit from green bonds? *Journal of Corporate Finance, 61.* https://doi.org/10.1016/j.jcorpfin.2018.12.001
TCFD. (2017). *Recommendations of the Task Force on Climate-related Financial Disclosures i Letter from Michael R. Bloomberg.*

Technical Expert Group on Sustainable Finance. (2019). *Report on EU Green Bond Standard.*
Tolliver, C., Keeley, A. R., & Managi, S. (2019). Green bonds for the Paris agreement and sustainable development goals. *Environmental Research Letters, 14*(6), 064009. https://doi.org/10.1088/1748-9326/AB1118
Tuhkanen, H., & Vulturius, G. (2020). Are green bonds funding the transition? Investigating the link between companies' climate targets and green debt financing. *Journal of Sustainable Finance and Investment.* https://doi.org/10.1080/20430795.2020.1857634
UN PRI. (2018). *How to Invest in the Low-carbon Economy: An Institutional Investors' Guide.* UN.

21. Cost of capital and climate risks
Gianfranco Gianfrate, Dirk Schoenmaker and Saara Wasama

INTRODUCTION

As climate change and societal issues are addressed by tougher regulation, emerging technologies, and shifts in consumer behaviors, global companies and investors are increasingly treating environmental and social risks as a key aspect when making investment and financing decisions, pricing financial assets, and deciding the allocation of their investment portfolios. As a consequence, the quantification of the impact of social and environmental performance on firms' cost of capital has become a popular and policy-relevant research topic.

At the same time, amid rising concerns over the environmental impact of businesses and the responsibility that businesses have in addressing environmental and social issues caused by – or relevant to – their operations, companies and practitioners have started to explore ways to incorporate sustainability considerations in the estimation of cost of capital (Bianchini and Gianfrate, 2018).

Since cost of capital is a measure of investors' perceptions on what is a relevant determinant of firms' market value, the growing global interest in sustainability suggests that ESG and corporate social responsibility (CSR) practices might be priced by investors. In this chapter, we present a literature review of studies that examine the relationship between environmental performance (EP) and firms' cost of capital (COC). To our knowledge, no comparable systematic analysis of the literature on this branch of research exists. Previous works by Benlemlih (2017), Izzo (2014) and Ambec and Lanoie (2008) review existing literature on the general relationship between corporate social performance (CSP) and corporate financial performance (CFP). However, none of these reviews focuses solely on cost of capital. Artiach and Clarkson (2011) review existing literature on the relationship between cost of equity, corporate disclosure and choice of accounting policy. The disclosure aspect of this review focuses on a broad-based disclosure measure, not environmental or social performance only. Huang and Watson (2015) review existing literature on CSR research in accounting, discussing the relation between CSR and financial performance among other topics, but not focusing solely on it, nor cost of capital.

Although empirical results on the relationship between sustainability and cost of capital exhibit variance, the majority of the research finds a statistically significant negative relationship between the two variables, suggesting that sustainability performance is value-relevant and that the valuation effect is – at least partially – realized through lower cost of capital. Because firms with a better sustainability footprint are expected to be less vulnerable to climate-related risks, they would be less risky than otherwise similar companies. Lower risk translates into lower return to be expected from investors, implying a lower cost of capital for firms. Testing such a relationship empirically has delivered mixed evidence. The findings differ across geographies, time periods, choice of method, and choice of sustainability measure. Additionally, a number of country-, firm- and, for cost of debt, loan-specific external variables play a role in explaining the strength of the relationship between environmental performance and cost of capital. The heterogeneity in the measures of sustainability used, apparent flaws in some of the most commonly used sustainability metrics, and the reliability issues inherent in sustainability reporting by companies in the first place limit the generalizability of these results.

Our focus is on climate risks but many of the contributions surveyed do not deal with firms' exposure to climatic risks in isolation but they have investigated holistically the spectrum of risks associated with firm sustainability. In particular, much of the literature in the field has privileged a comprehensive environmental, social and governance (ESG) perspective of corporate sustainability and only a relatively limited number of studies is specifically exploring the relation between climatic risks and cost of capital. This chapter will therefore also survey papers that empirically have investigated sustainability more broadly.

The remainder of the paper is structured as follows. The next section introduces the theoretical insights and describes the methodology and the sustainability measures used in existing research. The third section discusses existing research on the relationship between exposure to climate risks and cost of equity. The fourth section introduces empirical research on cost of debt, and the fifth section concludes, presenting our main findings and avenues for future research.

THEORETICAL, METHODOLOGICAL AND MEASUREMENT ISSUES

Theoretical Perspective

Firms finance their operations using capital they obtain from investors. The capital received is in the form of equity or debt, and each source of capital has its own "cost"; hence, firms usually bear a "cost of equity" and a "cost of debt". For many financial management applications, firms and investors consider a comprehensive cost of capital that is a (weighted) average of the cost of debt and equity. The estimation of cost of capital is essential for many corporate decisions, including what operating investments to make. It is also essential for managers and investors to measure the value of firms. In such contexts, cost of capital is used as a discount rate applied to future expected cash flows. The climate footprint is likely to materialize in firms' cash-flow in the form, for example, of additional costs due to carbon emission taxes: as a consequence, the future cash-flow of projects and/or companies will diminish. They are also likely to affect the cost of capital, as the additional riskiness brought by climatic exposures is a source of risks that investors would expect to be compensated for.

The cost of capital reflects the capital markets' perception of what is value-relevant for businesses, including the corporate sustainability footprint. Because markets are segmented to a certain degree (meaning that local investors mostly hold local securities), the relationship between exposure to climate risks and cost of capital is likely to be geography-dependent and to be affected by mandatory requirements for corporate disclosure as well as by cultural and institutional elements. However, regardless of the geographical context, several scholars have hypothesized that environmental risk management reduces a firm's riskiness which, in turn, should reduce expected cost of capital and *lower* returns going forward. Indeed, Pastor et al. (2021) show that sustainable firms eventually deliver lower stock returns while, from the point of view of corporate debt, firms that are more exposed to climate risks do appear to be perceived as more likely to default (Capasso et al., 2020).

To frame the discussion, it is of paramount importance to introduce a distinction between climate-related transition and physical risks. Climate Transition Risks (CTR) arise from changes in policy (i.e., carbon taxes, carbon allowances markets) and new technologies, such as the growth of renewable energy. Climate Physical Risks (CPR) are environmental events such as floods or storms that can damage and disrupt productive assets and processes (Gianfrate, 2018). Figure 21.1 shows that there are three possible channels through which exposure to both CTR and CPR can theoretically affect a firm's cost of capital.

First, climate risks could contribute to defining the cost of equity. Using a multi-factor model, the cost of equity for firm i can be estimated by considering all the sources of systemic risk the company is exposed to, besides the risk-free return rate (r_f). For each risk factor one should consider the Risk Premium (RP) and the associated β which measures how sensitive the firm is to the risk factor. In theory, CTR should generate a CTR Premium (or extra-return) and a β^{CTR}. Analogously, CPR is associated with a CPR Premium and to a β^{CPR} which expresses how sensitive the firm is to the extra-returns related to climate physical risks. In the seminal literature on the topic β^{CTR} is often referred to as "carbon beta" (e.g., Huij et al., 2021). The intuition is that investors should be compensated for the risk they bear because they hold stocks that are exposed to CTR and β^{CTR} measures how much a certain security's return is exposed to such risk.

Second, both CTR and CPR could affect the creditworthiness of individual companies, as reflected in credit default spreads added to the risk-free rate. For example, Capasso et al. (2020) show that firms most exposed to CTR are perceived as more likely to default.

Finally, climate risks could introduce some limitation about how much debt capital companies can raise to finance their assets. Exposure to climate risks therefore can impact the degree of leverage firms implement. Empirical support to this view is provided for example by Nguyen and Phan (2020) but we will not cover this channel here.

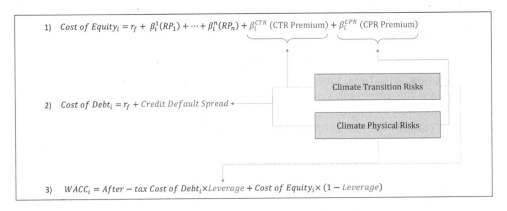

Figure 21.1 Climate risks and cost of capital

Methodology

For the purpose of this review, studies that focus on a measure of sustainability performance and/or a measure of sustainability disclosure and its impact on cost of capital have been included. The choice to include measures of sustainability disclosure as well as performance is based on the notion that for sustainability disclosure to be value-relevant, investors must, by default, consider sustainability performance value-relevant as well. Therefore, any study that identifies a link between environmental disclosure and cost of capital inherently supports the notion that sustainability performance is linked to cost of capital as well. Although some of the articles focus explicitly on cost of (debt or equity) capital, many also take a broader approach, examining the role of sustainability in explaining firm value, expected cash flows, or the systematic risk of a company. These studies have also been included, with emphasis placed on their findings regarding the link between environmental performance and cost of capital.

Existing empirical literature on the link between environmental performance and cost of capital exhibits considerable heterogeneity in the choice of measure for sustainability and cost of capital. The majority of existing research focuses on developed markets, particularly the United States. More recently, acknowledging the importance of the institutional and cultural context in which companies and investors operate for the cost of capital–sustainability relationship, more research using global samples has emerged. Even so, emerging markets are underrepresented in the existing empirical literature, which, given their growing economic and environmental importance globally, is an important limitation in current research.

Measuring the Environmental Performance

Measuring corporate environmental performance (EP) varies considerably across studies and over time. Du et al. (2017) recognize four major approaches for EP measurement in existing literature: (i) analysis of publicly available environmental disclosures; (ii) using the toxic release inventory (TRI) database maintained by the US Environmental Performance Agency (EPA); (iii) using data from social performance databases, such as the KLD (also known as MSCI ESG STATS); and (iv) using a content analysis framework created by Clarkson et al. (2008) and based on the Global Reporting Initiative (GRI)[1] reporting guidelines. In the research reviewed for this chapter, two additional commonly used measures of environmental performance were firms' carbon intensity and different researcher-constructed or publicly administered CSR performance surveys or analysis frameworks.

The KLD database uses a binary scoring system across six categories: environment, community, product, diversity, employee relations, human rights and corporate governance. Firms are given strength and concern scores in these categories. Additionally, the database has exclusionary screens, which measure firm participation in controversial businesses: alcohol, gambling, firearms, military, nuclear power, and tobacco (El Ghoul et al., 2011). The KLD database coverage begins in 1991 with approximately 600 firms, growing up to circa 2,400 US firms by 2003. Since 2013, the database has also rated the 2,600 largest non-US firms (Huang and Watson, 2015). The KLD database provides independent and consistent social ratings of a large sample of firms (Graves and Waddock, 1994). Plumlee et al. (2015) credit the database for measuring both positive and negative environmental performance, as opposed to the TRI, which only documents negative actions. Additionally, the researchers note that the KLD database covers a broad range of environmental issues beyond pollution, thus being a more comprehensive measure of environmental performance than simple measures of emissions intensity.

The KLD database is broadly used in research on firms' environmental performance, but it has also come under extensive criticism since its inception. Sharfman (1996) criticizes the measure for the lack of an underlying theory about the choice of variables, while Graves and Waddock (1994) note that the equal weighting of different CSR dimensions is an arbitrary choice not supported by theory. Mattingly and Berman (2006) posit that the data are better suited to measure corporate social action rather than corporate social performance (i.e. the presumed outcome of action). Semenova and Hassel (2015) argue that KLD concern scores are good measures of the environmental impact of industrial activities, but the strength scores are not and the commonly used method of netting KLD scores is, therefore, not a valid measure of a firm's overall environmental performance. Goss and Roberts (2011) echo Semenova and Hassel's conclusions in stating that net KLD scores are not representative of firms' CSR performance. Furthermore, the assumption of cardinality for ordinal CSR measures such as KLD is problematic: assigning a score of +2 to one company and +1 to another implies the former is better, but this does not mean that it performs twice as well as the other one (Surroca et al., 2010; Goss and Roberts, 2011).

The TRI database is maintained by the US EPA and it includes annual data on the toxic chemicals released by US companies in selected industries, mainly mining, utilities, manufacturing and hazardous waste processing. The program covers over 675 toxic chemicals and 22,000 facilities. Although a widely used measure of environmental performance, this database has come under criticism for (i) data accuracy due to lack of routine validation checks of company-reported data; (ii) changes in US EPA instructions in 1994, which caused a large reduction in reported emissions; and (iii) data reliability, as some of the data in TRI are estimated rather than measured (Toffel and Marshall, 2004). Furthermore, the TRI measures chemical emissions and as such cannot be considered a comprehensive indicator of a firm's total environmental performance (Ambec and Lanoie, 2008). In particular, the latter argument can easily be extended to the use of carbon emissions as a measure of environmental performance, which, although not yet common, has grown in popularity with the introduction of carbon pricing in different countries.

Another measure of environmental performance commonly used is the Sustainability Asset Management Group GmbH (SAM) ESG ratings database. SAM provides ESG research for the Dow Jones Sustainability Indices and has received numerous industry awards for the quality of its ESG research. It assesses firms' opportunities and risks across a number of general and industry-specific ESG criteria. The data are derived from an annual ESG company questionnaire, complemented with additional information from company-published information. SAM calculates a total CSP score from 0 to 100 for each firm from the ESG dimensions for both general and industry-specific criteria (RobecoSAM, 2016). Humphrey, Lee and Shen (2012) credit SAM's ratings methodology for (i) differentiating between firms engaged in the same activities; (ii) considering firms' industry membership and hence not downgrading a well-performing firm in an environmentally loaded industry such as mining, while upgrading a poor performer in a less environmentally loaded industry such as software development; and (iii) having a balanced scoring method across categories as opposed to the unequal number of strengths and concerns in each criterion of the KLD database.

Outside the use of external CSR ratings and emissions databases, a number of studies reviewed in this paper use a self-constructed measure of environmental performance, typically based on a CSR checklist or content analysis framework that relies on the Global Reporting Initiative (GRI) guidelines (see, for example, Plumlee et al., 2015). Additionally, a number of researchers rely on publicly administered CSR surveys or databases provided by local authorities in the markets studied (see, for example, Pae and Choi, 2011). These measures are typical especially in research on emerging markets, such as China, as most of the existing databases for CSR or ESG ratings do not cover these markets (Du et al., 2017). The issue of comparability is, naturally, even higher with these studies, as the measure of environmental performance varies study per study.

The difficulty in generalizing results of studies that use different measures of environmental performance has also been studied empirically. Research on environmental performance measures has revealed a high degree of heterogeneity between different ratings providers, particularly between ratings from different countries. Sharfman (1996) compares the KLD database CSR scores with the *Fortune* magazine corporate reputation survey, finding that the correlations between rating scores explain less than 50 percent of the variance in the variables compared. On the other hand, Semenova and Hassel (2015) identify correlations above 50 percent for a comparison of the KLD, Asset4 and GES databases. Chatterji et al. (2016) study the validity of six major social rating databases: KLD, Asset4, Calvert, FTSE4Good, DJSI and Innovest and find, first, that the theoretical definition underlying the CSR scores differs across databases, and, second, that even when the CSR construct is defined in the same way, different rating agencies measure it differently. The mean pairwise correlations between indices vary from 0.13 to 0.52 across all indices studied, indicating low validity of the CSR measures used. When taking the rating agency's country into consideration, the researchers find a much higher correlation between ratings, suggesting that location plays an important role in determining the definition of CSR. Taking an analogy from the home bias in the investment literature (Chan, Covrig and Ng, 2005), the higher correlation also suggests that rating agencies have more information on CSR practices close to home.

Rowley and Berman (2000) argue that corporate social performance (CSP) as a whole is not a viable concept theoretically or operationally and should, therefore, be considered in specific operational settings only. The authors note that the proliferation of ways in which CSR performance is measured renders it impossible to make valid comparisons between the results of different studies. Furthermore, most studies on the relationship between CSP and financial performance (CFP) involve multiple industries and we cannot assume that a given dimension of CSR performance in one industry is relevant for another. Most notably, the authors argue strongly in favor of building a theoretical framework to explain under which conditions a relationship between corporate social and financial performance should be expected. This criticism is still valid today, as, apart from one notable exception (Heinkel, Kraus and Zechner, 2001, discussed in the following section), the vast majority of research on this topic engages in "scanning data for statistically significant results linking CSP and FP", instead of considering the theoretical aspects of this relationship.

Given the range of issues with existing measurements of environmental performance, one must approach research on the relationship between corporate environmental and financial performance with caution. In particular, it is important not to generalize the results of one study too easily.

CORPORATE SOCIAL PERFORMANCE AND COST OF EQUITY

The majority of the research on the relationship between corporate environmental/social performance (CSP) and cost of capital (COC) is empirical, trying to document a statistically significant relationship between the two variables of interest. However, an early research paper by Heinkel et al. (2001) introduces an equilibrium model to provide a theoretical explanation for how ethical investing can influence a firm's cost of capital. The framework includes two types of risk-averse investors: neutral and green, and three types of firms: polluting, green and reformed. Green investors will only invest in green and reformed firms, while neutral investors have no preference. All firms act to maximize their share price. Two variables are key to determining whether exclusionary investing affects corporate behavior: the proportion of total investors that are green and the cost of reform. Using empirically reasonable variables, Heinkel et al. (2001) show that about 25 percent of investors need to be green to induce polluting firms to reform. At the time of this research, approximately 10 percent of investable funds were reportedly invested in a socially responsible manner. This proportion is not enough to induce reform in polluting firms, but it does increase the cost of capital for all firms. The increase in cost of capital results from a reduced demand for polluting firms' stocks, which decreases diversification possibilities between technologies and thus raises all firms' cost of capital. The pioneering work by Heinkel et al. (2001) can be seen as the theoretical backdrop for later empirical research on the relationship between CSP and COC.

The first empirical study on the relationship between a firm's environmental performance and cost of capital was conducted by Sharfman and Fernando (2008), who examine the impact of environmental risk management on the cost of debt and equity capital in a sample of publicly listed US firms. The researchers use the Capital Asset Pricing Model (CAPM)[2] to estimate each firm's cost of equity and the Bloomberg database to estimate each firm's marginal cost of borrowing. The results point to a statistically significant negative relationship between environmental risk management and both cost of equity and weighted average cost of capital.

In the research that follows Sharfman and Fernando (2008), two methods of studying the relationship between sustainability and cost of equity have emerged: (i) using, like Sharfman and Fernando, an asset pricing model to estimate firms' cost of equity in portfolios/individual firms sorted on a measure of environmental performance; and (ii) regressing cost of equity on a measure of environmental performance and control variables. In the latter method, an implied cost of equity is typically used.

Three studies represent the first strand of research. Koch and Bassen (2013) rely on the CAPM to estimate cost of equity capital but augment it with a carbon risk factor. The researchers find that carbon price movements are a statistically significant risk factor for utility companies with extremely high-emitting fuel mix, leading to a higher cost of equity. Gregory, Tharyan and Whittaker (2014) show that markets value several dimensions of CSR and that the majority of this effect is attributable to higher expected cash flows, while the overall implied cost of capital differences between green and non-green portfolios are small. A later study by Gregory, Whittaker and Yan (2016) finds that the lower cost of capital of high CSR firms is attributable to industry membership and that the valuation effect of CSR performance is driven by earnings persistence. This result is restricted to the short term, as the researchers only examine the impact of CSR on returns one period ahead.

The second strand of research, which regresses a measure of cost of equity on sustainability and control variables is considerably more popular among researchers. The first study of such kind is attributable to El Ghoul et al. (2011), who study the relationship between corporate social responsibility (CSR) and cost of equity capital across a broad sample of US firms before the financial crisis. The researchers examine CSR performance in more detail and show that out of the six qualitative issues measured in the KLD database, employee relations, environmental policies and product strategies have a statistically significant negative relationship with cost of equity capital. The researchers also document a positive relationship between cost of equity and firm participation in two controversial business areas: nuclear power and tobacco. Findings from further research are inconclusive: Salama, Anderson and Toms (2011) document a negative but economically meaningless relationship between a company's environmental performance and its systematic financial risk in the UK. Humphrey, Lee and Shen (2012), also in the UK, consider both general and industry-specific ESG criteria to measure sustainability performance. The researchers find no evidence of a difference in the risk-adjusted performance of high- and low-ESG firms, regardless of the measure of ESG performance used. Li, Eddie and Liu (2014) find a statistically significant positive relationship between emissions intensity and cost of debt in Australia, but not between emissions intensity and cost of equity.

On the other hand, Girerd-Potin, Jimenez-Garces and Louvet (2014) show that French investors require a risk premium for holding non-socially responsible stocks (namely stocks of firms operating in sectors such as tobacco, gambling, weapons). Later, Li, Liu, Tang and Xiong (2017) confirm this result in the Chinese market: the researchers find a negative and statistically significant relationship between carbon information disclosure (CID) and cost of equity. The researchers note a negative and statistically significant (10 percent) relationship between non-financial CID and cost of equity, as well as a negative and statistically significant (5 percent) relationship between financial CID and cost of equity. Suto and Takehara (2017) also find a negative relationship between corporate social performance and cost of equity in Japan and note that institutional ownership of a firm impacts this relationship and reduces cost of equity.

The empirical literature has also focused on two specific sources of CSP that weighted on the magnitude of the cost of equity, namely, the "sin premium" (Hong and Kacperczyk, 2009) and the "carbon premium" (Bolton and Kacperczyk, 2022). The former derives from investors increasingly choosing to exclude stocks from their portfolios because of nonfinancial considerations, traditional targets for exclusion being firms with morally questionable business models, such as the tobacco, weapons, alcohol, gambling, and adult entertainment industries. The latter refers to the evidence that carbon emissions do affect stock returns in most geographic areas of the world (Bolton and Kacperczyk, 2022). Such premium appears to be economically sizable with respect to both direct emissions (scope 1), indirect emissions from consumption of purchased electricity, heat, or steam (scope 2), and other indirect emissions from the production of purchased materials, product use, waste disposal, and other supply chain activities (scope 3). Interestingly, what matters is both the total level of CO_2 emissions produced by companies as well as their annual changes, with the emissions volume emerging as a long-term risk projection, given that emissions are highly persistent, while the year-by-year change is a short-term projection of the risk. No premium appears to be associated with emission intensity (the ratio of emission volume to revenues), which is a common measure used by policymakers and climate scientists, thus suggesting that absolute levels of carbon emissions are more important metrics of carbon risk.

Empirical results on the COE–CSP relationship are therefore inconclusive, though mostly in favor of a negative relationship between the two variables. Some studies indicate that a statistically significant negative relationship exists between cost of equity and environmental performance, while others find no, or only marginal, statistical significance between the variables of interest or find a statistically significant but economically negligible relationship. As discussed before, the choice of environmental performance metric is likely to play a large role in the heterogeneity of existing empirical results. In this regard, recent work by Pastor, Stambaugh, and Taylor (2021) shows that "green" assets have lower expected returns than more polluting ones: this is because environmentally-concerned investors derive utility from holding greener assets and because such assets allow investors to hedge against climate risk. The next section discusses findings from studies where environmental disclosure has been chosen as the explanatory variable instead of environmental performance.

Disclosure and Cost of Equity

An early contribution on the link between non-financial disclosure and cost of equity was conducted by Orens, Aerts and Cormier (2010), who study the link between web-based non-financial disclosure and a firm's cost of debt and implied cost of equity capital in a sample of European and North American companies. Using a broad disclosure measure, the researchers show that cross-sectional levels of non-financial web-based disclosure are negatively associated with cross-sectional differences in the implied cost of equity capital. These findings are supported by later research from Gruning (2011), who studies the interaction between disclosure, market liquidity and cost of equity in listed German firms and finds a reduction in investors' return expectations associated with higher disclosure levels. Embong, Mohd-Saleh and Hassan (2012) also find a statistically significant negative relationship between disclosure and cost of equity capital for large firms in Malaysia. On the other hand, Qiu, Shaukat and Tharyan (2016) document a positive and statistically significant relationship between a firm's overall environmental and social disclosure score and its stock price in the UK, but find no significant relationship between firms' cost of equity and disclosure scores. These results are not conclusive since all of the above-mentioned studies use a broad disclosure measure, rendering it difficult to draw conclusions regarding the impact of environmental disclosure alone on the cost of equity capital.

Later research with a stronger focus on sustainability performance is mostly supportive of a negative relationship between environmental disclosure and cost of capital. Reverte (2012) posits that higher CSR disclosure quality is associated with lower cost of equity capital in Spanish firms, and that this effect is more pronounced for firms in environmentally sensitive industries. Ng and Rezaee (2015) find a statistically significant negative relationship between economic sustainability disclosure and cost of equity capital and note that the relationship is more pronounced when ESG performance is strong. Chauhan and Kumar (forthcoming) study the impact of nonfinancial ESG disclosure on firm value in Indian firms, finding a statistically significant negative relationship between ESG disclosure and cost of equity, but not cost of debt. However, in a recent study of the Malaysian market, Atan et al. (2018) find a statistically significant positive relationship between a firm's total ESG score and its weighted average cost of capital (WACC). It should be noted, however, that the sample in the study is limited to only 54 firms, which makes the results more susceptible to bias.

As before, the variance in empirical results can partially be explained with differences in the measure of disclosure used: many of the studies cited here rely on specifically prepared disclosure surveys or checklists, hand-collected disclosure data, or disclosure data published by public authorities in the local market. Although the measures themselves may be sound, their heterogeneity makes it difficult to generalize any one set of results (Rowley and Berman, 2000). There are also external factors such as the social and cultural context a firm operates in, the legal requirements with respect to corporate disclosure, and the role of third-party assurance on corporate disclosure, which play a role in explaining the strength and direction of the COE-CSP relationship. The following three subsections discuss empirical findings from each of these factors in detail.

Culture, Institutional Context and the COE–CSP Relationship

Dhaliwal et al. (2014) note that the importance of CSR disclosure depends on the extent to which the culture and laws of a country support non-shareholder stakeholders' interest in a firm's operations and disclosure. Taking note of these factors, much of the more recent research on the relationship between sustainability and cost of capital has emphasized the role of cultural, economic, geographical and institutional factors in supporting or weakening this relationship. Dhaliwal et al. (2014) specifically examine the impact of country-level stakeholder orientation and financial transparency on the relationship between corporate social responsibility disclosure and cost of equity capital. Studying a global sample prior to the financial crisis, the researchers conclude that the reduction in cost of capital as a result of higher CSR disclosure is stronger in more stakeholder-oriented countries and in less financially transparent countries.

Later research supports Dhaliwal et al.'s findings: Feng, Wang and Huang (2015) identify a statistically significant negative relationship between CSR performance and cost of equity in North America and Europe, but not in Asia. Martínez-Ferrero, Banerjee and García-Sánchez (2016) find an economically and statistically significant negative relationship between CSR practices and cost of capital, noting also that CSR performance shields cost of capital from the punitive impact of earnings management, an effect which is stronger in countries with strong investor protection and weaker in countries with strong commitment to CSR. Matthiesen and Salzmann (2017) find a negative relationship between CSR performance and cost of equity and show that this relationship is stronger in countries with lower levels of assertiveness, higher levels of humane orientation, and higher institutional collectivism. Gupta (2018) finds a negative, statistically significant relationship between environmental performance and cost of equity in a broad international sample spanning 43 countries. The relationship is stronger in countries with weaker country-level governance.

On the other hand, El Ghoul et al. (2018) show that the negative relationship between corporate environmental responsibility (CER) and cost of equity holds across different legal, economic and geographic settings and is robust to alternative statistical techniques, and to accounting for noise in analyst forecasts. The strength of the results may be driven by the choice of sample, as El Ghoul et al. limit their study to a global sample of only manufacturing firms. Industry membership has been shown to have a strong effect on the relationship between sustainability disclosure and cost of capital (see, for example, Reverte, 2012, or Gregory et al., 2016) and it is conceivable that investors across geographical and economic settings would be more attuned to environmental risks in a notably emissions-intensive industry. These results find support from an earlier study by Cajias, Fuerst and Bienert (2014), who record significant differences in the strength and significance of the negative relationship between CSR performance and cost of equity capital across industries in the US market, particularly between consumer- and asset-oriented industries.

Some research also supports the notion that company culture can impact the strength of the relationship between sustainable performance and cost of capital. Pae and Choi (2011) examine the relationship between corporate governance, ethical commitment and cost of equity capital in the Korean market, finding that a statistically significant negative relationship exists between corporate governance and cost of equity capital, and that this relationship is more pronounced for companies with weak corporate ethical commitment.

These findings show the importance of economic, institutional, industrial, cultural and geographic context in determining whether higher sustainability performance leads to lower cost of (equity) capital. Furthermore, the findings are strongly consistent in supporting the negative relationship between environmental performance and cost of equity, suggesting that the omissions of country-level and cultural factors may have driven the variance in results from earlier studies.

Voluntary versus Mandatory Disclosure and the COE–CSP Relationship

Within the literature on the relationship between sustainability disclosure and cost of equity, a specific strand of research exists, which focuses on the differences between voluntary and mandatory disclosure (see Clarkson et al., this volume, Chapter 14). Voluntary environmental disclosure presumably has a signaling role in the economy, with good CSR performers producing direct voluntary disclosures that cannot be easily replicated by poor CSR performers. This should theoretically increase firm value and potentially lower a firm's cost of capital due to reduced information asymmetry.

The first study to specifically focus on voluntary disclosure was presented by Clarkson, Fang and Richardson (2013), who study the role of voluntary environmental disclosure in explaining firm valuation and, in particular, the channel through which this valuation effect is realized (cost of capital or future profitability). Focusing on the US market and five highly-polluting industries in 2003 and 2006, the researchers document a positive and statistically significant relationship between environmental performance and cost of capital, but no such relationship between voluntary environmental disclosure and cost of capital. Voluntary environmental disclosure does have an incremental valuation effect, but this effect is realized through higher future profitability, not cost of capital.

Li and Foo (2015) find similar results in the Chinese market, studying a broad sample of listed Chinese firms in 2008–2012. The researchers identify a statistically significant negative relationship between CSR report quality and ex-ante implied cost of equity, and a statistically significant difference in cost of equity for firms that disclose CSR information (lower cost of capital) and firms that do not (higher cost of capital). However, there seems to be no difference between the impact of mandatory and voluntary disclosure. Kim, An and Kim (2015) find that carbon intensity is positively related to the cost of equity capital in Korea, but whether firms disclose sustainability reports voluntarily or by law does not impact the relationship. It should be noted here that third-party verification of carbon emissions is compulsory in the Korean market. Research on the role of assurance in sustainability disclosure suggests that third-party verification strengthens the connection between sustainability disclosure and cost of capital (see Martínez-Ferrero and García-Sánchez, 2017, discussed below). In this instance, it may have contributed to reducing the difference between mandatory and voluntary disclosure.

Other studies have identified a stronger impact on cost of capital in one type of disclosure over the other. Plumlee et al. (2015) show that firms with better relative environmental performance have lower cost of equity capital than their less prudent peers, and that higher quality positive voluntary disclosure is negatively associated with cost of equity and vice versa. Harjoto and Jo (2015) specifically examine the differential impact of mandatory versus voluntary CSR expenditure on the implied cost of equity capital in publicly listed US firms between 1993 and 2009, showing that the negative effect of CSR on cost of capital is driven by mandatory CSR spending. Voluntary CSR spending has the opposite effect initially, but results in lower COC in the longer term.

External Assurance and the COE–CSP Relationship

Another relevant strand of research considers the role of third-party assurance in the relationship between sustainability disclosure and cost of capital. Martínez-Ferrero and García-Sánchez (2017) use a global sample of companies from 16 different countries and nine activity sectors in 2007–2014 to study the interaction between sustainability assurance, type of assurance provider and cost of capital. The researchers find, first, that the issuance of a sustainability report has a statistically significant negative relationship with a firm's cost of capital. Second, that securing external assurance for a sustainability report has the same effect, and third, that the decrease in cost of capital is larger when assurance is provided by one of the Big Four accounting firms.

This strand of research is promising as it confirms the validity of the criticism (Toffel and Marshall, 2004) expressed against the TRI as a measure of environmental performance due to the lack of validity checks on company-reported data by the US EPA. It also raises concern over the measures of environmental performance that rely entirely on company-published, unverified environmental disclosures, which are common, especially in research covering emerging markets, as explained in the second section. Intuitively, one would expect the role of external assurance to be more important in markets where the institutional setting and CSR requirements are not yet fully developed, which provides an additional incentive to examine this phenomenon in the context of emerging markets.

This section provided an overview of current research into the relationship between corporate social and/or environmental performance and cost of equity. The majority of the studies surveyed support the notion that better environmental performance is associated with lower cost of equity. However, the heterogeneity of the samples and measures of environmental performance, as well as the issues inherent in existing measures of environmental performance make it difficult to generalize any specific set of results. These issues also partially explain the variance in results over time. Interesting recent contributions to research in this field come from studies on the role of country-level factors, voluntary and mandatory disclosure, and external assurance on the COE–CSP relationship. Future research should also focus more on emerging markets, given their growing importance in economic and environmental terms globally. The following section gives an overview of current research into the relationship between CSP and cost of debt.

CORPORATE SOCIAL PERFORMANCE AND COST OF DEBT

Sharfman and Fernando's (2008) paper on the relationship between environmental performance and cost of debt and equity capital in a sample of US firms, is one of the early studies in this field. The researchers hypothesize that environmental risk management reduces a firm's expected cost of financial distress and the probability of value-relevant exposure to extreme environmental events, therefore leading to lower cost of debt. Although the relationship between environmental risk management, as measured by KLD and TRI scores, and cost of equity and weighted average cost of capital is found to be negative, the researchers find no support for their hypothesis with respect to the cost of debt.

After Sharfman and Fernando's (2008) seminal research article, numerous papers studying the relationship between different measures of social and/or environmental performance and cost of capital have emerged, the majority of them focusing on cost of equity capital. The first study on the CSP–COC relationship that focused entirely on cost of debt was conducted by Menz (2010). Menz examines the standards of CSR and the valuation of European corporate bonds in a sample of 498 bonds observed over 38 months between 2007 and 2010. Based on the thesis that socially responsible firms are regarded as more commercially successful and less risky, Menz hypothesizes that firms with higher CSR scores, as measured by SAM Research, should have lower cost of debt. The results, however, do not support a statistically significant relationship between cost of debt, as measured by credit spreads, and the social performance of firms. Menz concludes that credit ratings are more important for bond investors than CSR ratings. This is an interesting finding given that more recent research (see, for example, Ge and Liu, 2015, and La Rosa et al., 2018) suggests that credit ratings are one of the channels through which corporate social performance (CSP) lowers a firm's cost of debt (COD). These findings are discussed in more detail later.

The findings from later studies largely contradict those of Sharfman and Fernando (2008) and Menz (2010) with respect to cost of debt. Schneider (2011) studies 48 US companies active in the pulp and paper and chemical industries in 1994–2004 and identifies a highly statistically significant negative relationship between environmental performance (EP), as measured by TRI, and yield spreads. Again, the importance of credit ratings is highlighted, as the EP-COD relationship is strongest for low-rated bonds and not significant for bonds rated A- or higher. Goss and Roberts (2011) study a much broader sample covering 1,265 US companies between 1991 and 2006, and examine the KLD strength and concern scores separately, as recommended by Semonova (2010). They find that poor CSR performers pay a higher interest on bank loans than their more responsible peers, but higher CSR performance is not rewarded. The effect is stronger for loans issued without security, but insignificant for loans with it. Oikonomou, Brooks and Pavelin (2014) build on Menz (2010) and Goss and Roberts (2011), examining the impact of corporate social performance (CSP) on the pricing of corporate debt and the credit quality of bond issues. The findings show that support for local communities, higher levels of product safety and quality, and good employee relations all materially decrease cost of debt. The aggregate measures of CSP show that good CSR performance is rewarded with a lower cost of debt, while transgressions in this respect are penalized. The relationship is stronger for longer-maturity bonds. In one of the rare contributions from emerging markets, Du et al. (2017) examine the role of environmental performance, as measured by company-published environmental information analyzed according to the GRI guidelines, on cost of debt in privately-owned Chinese firms. The researchers find an economically and statistically significant negative relationship between corporate environmental performance and interest rates on debt.

In addition to studies using the KLD and TRI databases or company-published information, a few studies exist that use historical carbon emissions as a measure of environmental performance. Li, Eddie and Liu (2014) study the potential effect of carbon emissions intensity on companies' cost of debt and equity capital in Australia. Using a sample of 210 listed Australian firms in 2006–2010, the researchers find a statistically significant positive relationship between emissions intensity and cost of debt, but not between emissions intensity and cost of equity. Jung, Herbohn and Clarkson (2018) focus more closely on risk awareness by studying the interaction between a firm's historical carbon emissions, awareness of carbon risk exposure, and cost of public debt in a sample of 255 firm-year observations of Australian firms from eight different industries in 2009–2013. The results show that firms with higher carbon risk and lower risk awareness pay between 38 and 62 basis points more for their loans than companies with demonstrated carbon awareness. The result is robust across different measures of carbon awareness and across industries, though more important for companies in high emitting industries.

Even though the majority of research on the CSP–COD relationship after Sharfman and Fernando's (2008) and Menz's (2010) initial findings supports the notion that better environmental performance leads to lower cost of debt, opposite results have also been found. Suto and Takehara (2017) find a positive relationship between cost of debt and CSP in a sample of Japanese firms in 2008–2013. The relationship is statistically significant only in 2008–2010, suggesting that debtors became more aware of CSP when estimating firm risk after the financial crisis. Magnanelli and Izzo (2017) study the CSP–cost of bank debt relationship in an international sample of 332 firms across different industries from 2005 to 2009. Measuring CSP with the CSR ratings provided by the Dow Jones Sustainability Index and RobecoSAM, the researchers find a statistically and economically positive relationship between CSP and cost of bank debt, suggesting that higher CSR performance is not interpreted by banks as reducing risk but as a costly diversion of firm resources. Magnanelli and Izzo's result may be driven by sample selection, as the majority of the companies selected operate in the US (132 out of 332 firms), where prior research has not identified lender support for CSR performance, presumably due to lower stakeholder-centricity of the economy (see the third section). Furthermore, it should be noted that Magnanelli and Izzo, like Menz (2010), include financial companies in their sample – something most researchers exclude as their inclusion would reduce cross-industrial variance due to the amount of bonds they issue. Finally, the measure of environmental performance used, the RobecoSAM, is the same employed by Menz (2010).

Disclosure and Cost of Debt

As with studies on the CSP–COE relationship, some of the studies on CSP–COD focus on a measure of environmental disclosure instead of environmental performance. The research on CSR disclosure and cost of debt is largely focused on emerging markets, especially China. This is due to data availability: as mentioned in the second section, the most commonly used sustainability measures in existing research are the CSR scores provided by the KLD or a comparable social performance database, and the TRI database maintained by the US EPA. Most of these measures are not available for Chinese companies (Du et al., 2017), so research must focus on company-published information instead.

Using a sample of 668 European and North American companies, Orens et al. (2010) find that web-based non-financial disclosure is negatively related to the rate of interest paid on bonds, but this relationship differs across geographies: it is statistically significant for Continental European firms, but not for North American firms. In general, the relationship is stronger in countries where the mandatory requirements for corporate disclosure are lower, which supports the notion that cultural and institutional context in which companies and investors operate impacts the relationship between environmental performance (or disclosure) and cost of capital. The findings of this study may not be generalizable to the broader CSP–COD debate since the disclosure measure used is very broad. Furthermore, the sample is limited to large listed companies, which may benefit more from CSP activities than their smaller counterparts (see, for example, Embong Mohd-Saleh and Hassan 2012). Finally, Orens et al. do not distinguish between negative and positive disclosure – an approach which has been criticized, for example, by Semenova and Hassel (2015) due to the difference in the ability of negative and positive disclosure to capture environmental performance.

Subsequent evidence is mixed: Gong, Xu and Gong (2018) study Chinese industrial companies in 2010–2013 and find that high CSR disclosure quality is associated with lower cost of corporate bonds. This relationship is weaker in firms located in areas with weak institutional environment, providing further support for the notion that cultural factors may impact the CSR–cost of capital relationship (see above). The researchers also contribute to research on the difference in investors' reaction to mandatory and voluntary CSR disclosure – a phenomenon studied by, for example, Plumlee et al. (2015) regarding cost of equity. In another study of the Chinese market, Fonseka, Rajapakse and Richardson (2019) examine the impact of environmental information disclosure (EID) and energy product type on the cost of debt in a sample of Chinese energy firms in 2008–2014. The researchers document a statistically significant negative relationship between environmental disclosure and cost of debt. Including the type of energy product in the analysis shows that the EID–COD relationship is only negative for solar and wind power firms, but positive for gas, thermal power and hydro firms. This shows that energy firms that produce less polluting energy products benefit from more favorable lending terms. On the other hand, Chauhan and Kumar (forthcoming) find no statistically significant relationship between ESG disclosure and cost of debt in the Indian market.

Taken together, the results suggest that environmental disclosure is associated with a lower cost of debt, but the strength and direction of this relationship varies across geographies. These results are in accordance with previous studies conducted on the impact of the cultural context on the CSP–COE relationship (see the third section). As shown by the limited number of articles reviewed in this section, this phenomenon ought to be more extensively studied for the relationship between CSP and COD in the future.

In recent years, more focused research has emerged on the COD–CSP relationship. Namely, researchers focus on (i) the relevance of country-level factors for the COD–CSP relationship, (ii) the role of credit risk and credit ratings as a channel through which the impact of CSP on COD is realized, and (iii) the potential nonlinearities in the COD–CSP relationship. The following three subsections discuss these recent findings in more detail.

Culture, Institutional Context and the COD–CSP Relationship

The role of country-level factors in moderating or strengthening the relationship between cost of capital and sustainability has not yet received much attention in the literature on cost of debt capital. Two interesting contributions to the literature come from Cheung, Tan and Wang (2015) and Hoepner et al. (2016).

Cheung, Tan and Wang (2015) study an international sample of 1,462 loan facilities issued by 622 firms in 20 countries and consider the impact of national stakeholder orientation on the relationship between CSR performance and cost of debt. Arguing that stakeholder groups have higher expectations with respect to a firm's CSR performance and are more likely to publicly scrutinize a firm's CSR performance in more stakeholder-oriented countries, the researchers hypothesize that the relationship between CSR performance, as measured by FTSE4Good ESG ratings, and cost of debt should be stronger in these countries. The researchers find no support for a relationship between CSR performance and cost of debt in the overall sample. However, the relationship is negative and highly statistically significant when stakeholder orientation is accounted for, suggesting that the failure of Goss and Roberts (2011) to document a lower cost of debt associated with higher CSR performance may have been driven by the sample being restricted to the US market.

Hoepner et al. (2016) find similar results. Studying a sample of 470 loan agreements from 28 countries between 2005 and 2012, the researchers document an economically and statistically significant relationship between country-level sustainability and cost of bank debt: a one-unit increase in a country's sustainability score decreases cost of debt by 64 basis points on average. The effect is driven by environmental performance, being twice as impactful as the social dimension. The researchers fail to find support for the impact of firm-level sustainability on the cost of bank loans. The researchers suggest that firm-level sustainability may only be priced through country-level sustainability. It should be noted, however, that the sustainability measure used in this study, from Oekom research, provides only a single rating and does lend itself to separating positive and negative performance. Furthermore, the sample consists mostly of financially robust European and North American firms. Since previous research has shown that the relationship between non-financial disclosure and cost of capital is less important for investment grade companies (see, for example, Schneider, 2011), it is possible that the lack of company-level sustainability effect on cost of debt is driven by the financial robustness of the sample companies.

These results suggest that the COD–CSP relationship is strongly influenced by country-level factors. Given the preponderance of US samples in existing literature, future research should focus on comprehensive international samples, which render comparisons between countries possible. Furthermore, special emphasis should be placed on emerging markets, as they are currently severely underrepresented in existing research and the difference in their cultural, social, economic and institutional context is likely to significantly influence the relationship between environmental performance and cost of capital.

Credit Ratings, Credit Risk and the COD–CSP Relationship

Although a growing body of literature addresses the connection between corporate social performance (CSP) and cost of capital, the link between CSP and credit risk has received considerably less attention. As discussed in the fourth section, many existing studies have documented the importance of credit ratings for the CSP–COD relationship (see, for example, Menz, 2010; Schneider, 2011; Goss and Roberts, 2011). However, most of these studies use credit ratings only as a control variable or at most an interaction variable. More recent research on the CSP–COD relationship has begun to consider the role of credit risk and credit ratings as a channel through which the impact of CSP on cost of debt is realized. This section gives an overview of the existing research on this topic.

The first paper in which the relationship between CSP and credit ratings is considered as part of the CSP–COD relationship is from Ge and Liu (2015). Building on the work by Menz (2010) and Oikoonomou et al. (2014), Ge and Liu contribute to existing literature by using multiple dimensions of CSR performance and by considering the explanatory power of CSP over credit ratings, a known determinant of cost of debt. Using a large sample of US firms and the KLD strength and concern scores for environmental performance, the researchers find that higher overall CSR performance is associated with better credit ratings and, after controlling for credit ratings, with lower bond yield spreads. CSR strengths are negatively, and CSR concerns positively, associated with bond yield spreads. A closer examination of the components of CSR performance shows that overall scores in community, product, employee and governance dimensions are associated with lower cost of debt.

As stated before, research on this front is still limited, but Ge and Liu's (2015) findings are validated by an earlier study from Jiraporn et al. (2014), as well as by later research from Erragragui (2018) and La Rosa et al. (2018). Jiraporn et al. (2014) find that a firm's geographical location is a significant determinant of CSR performance in the US, even more so than the widely studied industry effect (see, for example, Reverte, 2012, or Gregory et al., 2016). Using this relationship, Jiraporn et al. show that CSR performance has a highly significant positive relationship with credit ratings. Although the researchers do not study the relationship between CSR performance and COD directly, they conclude that better CSR performers enjoy lower COD due to the negative relationship between higher credit ratings and cost of debt. Erragragui (2018) finds a statistically significant positive relationship between environmental concerns and perceived default risk for firms in the US. He also finds support for the negative relationship between governance and environmental strengths, and cost of debt. La Rosa et al. (2018) use a sample of European listed companies included in the S&P Europe 350 index in 2005–2012 and find a statistically significant negative relationship between a firm's social performance and cost of debt. The researchers also document a positive relationship between corporate social performance and credit ratings.

The relationship between CSP and credit risk is an interesting new avenue of research moving on from considering credit ratings as an external determinant of cost of debt to studying their role as a channel through which part of the influence of CSP on cost of debt materializes. Indicative findings suggest that the relationship holds in the US and in Europe, though more research is needed to confirm the initial results. Furthermore, given the importance of country-level factors on the CSP–cost of capital relationship (see above), future research should focus on international samples that allow comparison across geographical areas, as well as on emerging markets, where the institutional context can differ greatly from developed markets.

Nonlinearity in the COD–CSP Relationship

Prior research has focused solely on the relationship between environmental performance and cost of capital, not considering the evolution of this relationship as environmental performance or CSR spending increases or decreases. Three papers have, so far, identified a U-shaped relationship between CSR performance and cost of debt, two of them studying the Chinese market and one the US market. The proposition that the CSP–COD relationship is U-shaped implies that an optimal level of CSR spending exists: low levels of CSR are ineffective but at the same time over-spending in CSR activities is detrimental.

Ye and Zhang (2011) and Zhou et al. (2018) both find a U-shaped relationship between CSR performance and cost of debt in China. Ye and Zhang (2011) study publicly listed Chinese companies in 2007 and 2008. Measuring CSR performance as the ratio of charitable donations to annual sales, Ye and Zhang document a highly statistically significant negative relationship between CSR and cost of debt, which only holds when the CSR ratio is below 0.357. For higher CSR spending, the relationship is reversed. Thus, firms with extremely low or extremely high CSR spending are punished with higher cost of debt. Zhou et al. (2018) study the relationship between carbon risk, as measured by a scale of 0–3 based on whether a company has been punished for carbon emissions, and cost of debt financing. The sample covers 191 publicly listed Chinese companies from high-carbon industries between 2011 and 2015. Like Ye and Zhang (2011), Zhou et al. also find a U-shaped relationship between carbon risk and COD, noting that this relationship is mainly driven by private (as opposed to state-owned) firms.

These results are supported by a recent paper from Bae, Chang and Yi (forthcoming), who document the same non-linear relationship between corporate social responsibility and the cost of bank debt in an extensive sample of 5,810 private syndicated bank loans issued by US companies between 1991 and 2008. Bae et al. use the strength and concern scores from the KLD database to rank the firms in their sample according to CSR performance. The authors show that as a borrower marginally increases CSR engagement, the cost of debt is reduced at a decreasing rate. This relationship is strongest in firms with moderate levels of CSR concerns.

These preliminary results suggest that the relationship exists at least in China and the US. Given the difference of these two markets in terms of their level of development, economic and social context, as well as culture, one might assume that the U-shape of the COD–CSP relationship may be universal. Future research should focus on verifying the conclusions of these initial findings, but also to extend this study to other markets and other measures of cost of capital.

This section provided an overview of current research into the relationship between corporate social and/or environmental performance and cost of debt. Most studies surveyed support a negative relationship between better environmental performance and cost of debt. However, the results are plagued by the same issues as research on CSP and cost of equity, namely the heterogeneity of the samples and measures of environmental performance, as well as the issues inherent in existing measures of environmental performance. Furthermore, the strength and direction of the relationship between CSP and COD is influenced by maturity of loans (Oikonomou et al., 2014), issuance of security for a loan (Goss and Roberts, 2011), industry membership (Jung, Herbohn and Clarkson, 2018) and large external events, such as the financial crisis (Suto and Takehara, 2017). Interesting recent contributions and potential avenues of future research in this field come from studies on (i) the role of country-level factors in explaining the COD–CSP relationship; (i) credit ratings and credit risk as transitory mechanisms through which CSP impacts COD; and (iii) the potential non-linearity of the COE–CSP relationship.

CONCLUSION AND AVENUES FOR FUTURE RESEARCH

Research on corporate social/environmental performance and cost of capital is, in general, supportive of a negative relationship between these two variables, as theory on investor perceptions of risk that matter would broadly – and intuitively – predict. However, a number of external factors have a considerable effect on this relationship.

Country-level factors, such as the cultural, social, economic and institutional context in which investors and businesses operate, play a significant role in determining the strength, direction and sometimes even the existence of the CSP–COC relationship for both cost of equity and cost of capital. Country-level factors that have been shown to impact the CSP–COC relationship are stakeholder orientation, financial transparency, the degree of investor protection and commitment to CSR, the level of assertiveness, humane orientation and institutional collectivism, country-level governance, institutional environment and country-level sustainability. Additionally, some studies support the notion that culture on company-level may impact the COC–CSP relationship.

In addition to country-level factors, industry membership is a significant determinant of the CSP–COC relationship, with research typically identifying a more prominent reduction in cost of capital in industries with high emissions intensity. Some scholars note a particularly strong difference between consumer and asset-oriented industries. Some research, particularly on the Chinese market has documented a difference in the way cost of capital reacts to environmental performance in private and state-owned firms, private ownership being typically associated with a larger reduction in cost of capital.

In terms of research on the cost of debt capital, a number of debt characteristics have been linked to the COD–CSP relationship. Credit ratings, loan maturity, issuance of security and significant external events have all been linked to the relationship between COD and CSP. The relationship is more pronounced in lower rated firms, for long-term maturities, for loans issued without security, and following catastrophic events. In general, the conclusion could be made that factors which increase loan- or firm-specific risk also heighten the importance of CSR performance for debtors.

Future research on the CSP–COC relationship should focus more on emerging markets, given (i) their growing importance in environmental and economic terms globally; (ii) the role that country-level factors play in the CSP–COC relationship; and (iii) the clear underrepresentation of these markets in current research. Another interesting avenue of future research is the theory behind the relationship between CSP and COC. To date, only one paper has attempted to create a comprehensive theoretical framework to explain the impact of socially responsible investing on cost of equity capital. The apparent lack of a theoretical framework is a serious flaw which renders existing empirical research a simple data analysis exercise without contributing to a better understanding of the relationship between CSP and COC. Moreover, research on the cost of equity, the role of external assurance of environmental disclosure, and performance measures is a recent addition to existing research and ought to be explored further. In research that focuses on cost of debt, the impact of environmental performance on credit risk merits further attention both in terms of theoretical and empirical contributions. To date, most researchers refer to the role of environmental performance in reducing information asymmetry and risk of financial distress when explaining the theory underpinning their hypothesized relationship between the two variables. Empirical evidence to support these assumptions, as well as a more robust theoretical framework would both be interesting additions to current research. Additionally, the U-shaped relationship documented between CSR performance and cost of debt in the Chinese and US markets is an interesting finding, which ought to be explored further in different geographical contexts as well as with respect to the cost of equity.

NOTES

1. GRI is an independent standards organization that supports businesses, governments and other organizations to understand and communicate their impacts to various stakeholders on issues such as climate change, human rights and corruption. Because of their modularity, GRI Standards have been adopted globally by numerous firms and other organizations.
2. CAPM is a model that describes the relationship between market risk and expected return for assets, particularly stocks. CAPM is widely used in finance for pricing risky securities and estimating firm cost of capital.

REFERENCES

Ambec, S. and Lanoie, P. (2008). Does it pay to be green? A systematic overview. *Academy of Management Perspectives*, 22(4), 45–62.

Atan, R., Alam, M., Said, J. and Zamri, M. (2018). The impacts of environmental, social, and governance factors on firm performance: Panel study of Malaysian companies. *Management of Environmental Quality: An International Journal*, 29(2), 182–194.

Artiach, T.C. & Clarkson, P.M. (2011). Disclosure, conservatism and the cost of equity capital: A review of the foundation literature. *Accounting and Finance*, 51(1), 2–49.

Bae, S.C., Chang, K. and Yi, H-C. (forthcoming). Are more corporate social investments better? Evidence of non-linearity effect on costs of U.S. bank loans. *Global Finance Journal*, in press.

Benlemlih, M. (2017). Corporate social responsibility and firm financing decisions: A literature review. *Journal of Multinational Financial Management*, 42–43, 1–10.

Bianchini, R. and Gianfrate, G. (2018). Climate risk and the practice of corporate valuation. In D.J. Cummings et al. (eds), *Research Handbook of Finance and Sustainability*. London: Routledge.

Bolton, P. and M. T. Kacperczyk (2022). Global pricing of carbon-transition risk. Journal of Finance, Available at SSRN: https://ssrn.com/abstract=3550233 or http://dx.doi.org/10.2139/ssrn.3550233

Cajias, M., Fuerst, F. and Bienert, S. (2014). Can investing in corporate social responsibility lower a company's cost of capital? *Studies in Economics and Finance*, 31(2), 202–222.

Capasso, G., Gianfrate, G. and Spinelli, M. (2020). Climate change and credit risk. *Journal of Cleaner Production*, 266, September.

Chan, K., Covrig, V. and Ng, L. (2005). What determines the domestic bias and foreign bias? Evidence from mutual fund equity allocations worldwide. *The Journal of Finance*, 60(3), 1495–1534.

Chatterji, A.K., Durand, R., Levine, D.I. and Touboul, S. (2016). Do ratings of firms converge? Implications for managers, investors and strategy researchers. *Strategic Management Journal*, 37, 1597–1614.

Chauhan, Y. and Kumar, S.B. (forthcoming). Do investors value the nonfinancial disclosure in emerging markets? *Emerging Markets Review*, in press.

Cheung, Y-L., Tan, W. and Wang, W. (2015). National stakeholder orientation, corporate social responsibility, and bank loan cost. *Journal of Business Ethics*, 1–20.

Clarkson, P.M., Li, Y., Richardson, G.D. and Vasvari, F.P. (2008). Revisiting the relation between environmental performance and environmental disclosure: An empirical analysis. *Accounting, Organizations and Society*, 33(4-5), 303–327. doi: 10.1016/j.aos.2007.05.003

Clarkson, P.M., Fang, X., Li, Y. and Richardson, G. (2013). The relevance of environmental disclosures: Are such disclosures incrementally informative? *Journal of Accounting and Public Policy*, 32, 410–431.

Dhaliwal, D., Li, O.Z., Tsang, A. and Yang, Y.G. (2014). Corporate social responsibility disclosure and the cost of equity capital: The roles of stakeholder orientation and financial transparency. *Journal of Accounting Public Policy*, 33, 328–355.

Du, X., Weng, J., Zeng, Q., Chang, Y. and Pei, H. (2017). Do lenders applaud corporate environmental performance? Evidence from Chinese private-owned firms. *Journal of Business Ethics*, 143, 179–207.

El Ghoul, S., Guedhami, O., Kwok, C.C.Y. and Mishra, D.R. (2011). Does corporate social responsibility affect the cost of capital? *Journal of Banking & Finance*, 35, 2388–2406.

El Ghoul, S., Guedhami, O., Kim, H. and Park, K. (2018). Corporate environmental responsibility and the cost of capital: International evidence. *Journal of Business Ethics*, 149, 335–361.

Embong, Z., Mohd-Saleh, N. and Hassan, M.S. (2012). Firm size, disclosure and cost of equity capital. *Asian Review of Accounting*, 20(2), 119–139.

Erragragui, E. (2018). Do creditors price firms' environmental, social and governance risks? *Research in International Business and Finance*, 45, 197–207.

Feng, Z-Y., Wang, M-L. and Huang, H-W. (2015). Equity financing and social responsibility: Further international evidence. *The International Journal of Accounting*, 50, 247–280.

Fonseka, M., Rajapakse, T. and Richardson, G. (2019). The effect of environmental information disclosure and energy product type on the cost of debt: Evidence from energy firms in China. *Pacific-Basin Finance Journal*, 54, 159–182.

Ge, W. and Liu, M. (2015). Corporate social responsibility and the cost of corporate bonds. *Journal of Accounting Public Policy*, 34, 597–624.

Gianfrate, G. (2018). "Designing carbon-neutral investment portfolios." In *Designing a Sustainable Financial System*. Cham: Palgrave Macmillan.

Girerd-Potin, I., Jimenez-Garces, S. and Louvet, P. (2014). Which dimensions of social responsibility concern financial investors? *Journal of Business Ethics*, 121, 559–576.

Gong, G., Xu, S. and Gong, X. (2018). On the value of corporate social responsibility disclosure: An empirical investigation of corporate bond issues in China. *Journal of Business Ethics*, 150, 227–258.

Goss, A. and Roberts, G.S. (2011). The impact of corporate social responsibility on the cost of bank loans. *Journal of Banking & Finance*, 35, 1794–1810.

Graves, S.B. and Waddock, S.A. (1994). Institutional owners and corporate social performance. *The Academy of Management Journal*, 37(4), 1034–1046.

Gregory, A., Tharyan, R. and Whittaker, J. (2014). Corporate social responsibility and firm value: Disaggregating the effects on cash flow, risk and growth. *Journal of Business Ethics*, 124, 633–657.

Gregory, A., Whittaker, J. and Yan, X. (2016). Corporate social performance, competitive advantage, earnings persistence and firm value. *Journal of Business Finance & Accounting*, 43 (1), 3–30.

Gruning, M. (2011). Capital market implications of corporate disclosure: German evidence. *BuR – Business Research*, 4(1), 48–72.

Gupta, K. (2018). Environmental sustainability and implied cost of equity: International evidence. *Journal of Business Ethics*, 147, 343–365.

Harjoto, M.A. and Jo, H. (2015). Legal vs. normative CSR: Differential impact on analyst dispersion stock return volatility, cost of capital, and firm value. *Journal of Business Ethics*, 128, 1–20.

Heinkel, R., Kraus, A. and Zechner, J. (2001). The effect of green investment on corporate behavior. *Journal of Financial and Quantitative Analysis*, 36(4), 431–450.

Hoepner, A., Oikonomou, I., Scholtens, B. and Schroder, M. (2016). The effects of corporate and country sustainability characteristics on the cost of debt: An international investigation. *Journal of Business Finance & Accounting*, 43(1 and 2), 158–190.

Hong, H. and Kacperczyk, M, (2009). The price of sin: The effects of social norms on markets. *Journal of Financial Economics*, 93(1), 15–36.

Huang, X.B. and Watson, L. (2015). Corporate social responsibility research in accounting. *Journal of Accounting Literature*, 34, 1–16.

Huij, J., Laurs, D., Stork, P.A. and Zwinkels, R.C.J. (2021). Carbon Beta: A market-based measure of climate risk. Available at SSRN: https://ssrn.com/abstract=3957900

Humphrey, J.E., Lee, D.D. and Shen, Y. (2012). Does it cost to be sustainable? *Journal of Corporate Finance*, 18, 626–639.

Izzo, M.F. (2014). Bringing theory to practice: How to extract value from corporate social responsibility. *Journal of Global Responsibility*, 5(1), 22–44.

Jiraporn, P., Jiraporn, N., Boeprasert, A. and Chang, K. (2014). Does Corporate Social Responsibility (CSR) improve credit ratings? Evidence from geographic identification. *Financial Management*, Fall, 505–531.

Jung, J., Herbohn, K. and Clarkson, P. (2018). Carbon risk, carbon risk awareness and the cost of debt financing. *Journal of Business Ethics*, 150, 1151–1171.

Kim, Y-B., An, H.T. and Kim, J.D. (2015). The effect of carbon risk on the cost of equity capital. *Journal of Cleaner Production*, 93, 279–287.

Koch, N. and Bassen, A. (2013). Valuing the carbon exposure of European utilities. The role of fuel mix, permit allocation and replacement investments. *Energy Economics*, 36, 431–441.

La Rosa, F., Liberatore, G., Mazzi, F. and Terzani, S. (2018). The impact of corporate social performance on the cost of debt and access to debt financing for listed European non-financial firms. *European Management Journal*, 36(4), 1–11.

Li, L., Liu, Q., Tang, D. and Xiong, J. (2017). Media reporting, carbon information disclosure, and the cost of equity financing: evidence from China. *Environmental Science and Pollution Research*, 24, 9447–9459.

Li, Y. and Foo, C.T. (2015). A sociological theory of corporate finance: Societal responsibility and cost of equity in China. *Chinese Management Studies*, 9(3), 269–294.

Li, Y., Eddie, I. and Liu, J. (2014). Carbon emissions and the cost of capital: Australian evidence. *Review of Accounting and Finance*, 13 (4), 400–420.

Magnanelli, B.S. and Izzo, M.F. (2017). Corporate social performance and cost of debt: The relationship. *Social Responsibility Journal*, 13(2), 250–265.

Martínez-Ferrero, J. and García-Sánchez, I-M. (2017). Sustainability assurance and cost of capital: Does assurance impact on credibility of corporate social responsibility information? *Business Ethics: A European Review*, 26, 223–239.

Martinez-Ferrero, J., Banerjee, S. and García-Sánchez, I.M. (2016). Corporate social responsibility as a strategic shield against costs of earnings management practices. *Journal of Business Ethics*, 133, 305–324.

Matthiesen, M-L. and Salzmann, A.J. (2017). Corporate social responsibility and firms' cost of equity: How does culture matter? *Cross Cultural & Strategic Management*, 24(1), 105–124.

Mattingly, J.E. and Berman, S.L. (2006). Measurement of corporate social action: Discovering taxonomy in the Kinder Lydenberg Domini ratings data. *Business & Society*, 45(1), 20-46.

Menz, K-M. (2010). Corporate social responsibility: Is it rewarded by the corporate bond market? A critical note. *Journal of Business Ethics*, 96, 117–134.

Ng, A.C. and Rezaee, Z. (2015). Business sustainability performance and cost of equity capital. *Journal of Corporate Finance*, 34, 128–149.

Nguyen, J.H., and Phan, H.V. (2020). Carbon risk and corporate capital structure. *Journal of Corporate Finance*, 64, 101713. https://doi.org/10.1016/j.jcorpfin.2020.101713

Oikonomou, I., Brooks, C. and Pavelin, S. (2014). The effects of corporate social performance on the cost of corporate debt and credit ratings. *The Financial Review*, 49, 49–75.

Orens, R., Aerts, W. and Cormier, D. (2010). Web-based non-financial disclosure and cost of finance. *Journal of Business Finance & Accounting*, 37(9 and 10), 1057–1093.

Pae, J. and Choi, T.H. (2011). Corporate governance, commitment to business ethics, and firm valuation: Evidence from the Korean stock market. *Journal of Business Ethics*, 100, 323–348.

Pastor, L., Stambaugh, R.F. and Taylor, L.A. (2021). Sustainable investing in equilibrium. *Journal of Financial Economics*, 142(2), 550–571.

Plumlee, M., Brown, D., Hayes, R.M. and Marshall, R.S. (2015). Voluntary environmental disclosure quality and firm value: Further evidence. *Journal of Accounting and Public Policy*, 34, 336–361.

Qiu, Y., Shaukat, A. and Tharyan, R. (2016). Environmental and social disclosures: Link with corporate financial performance. *The British Accounting Review*, 48, 102–116.

RobecoSAM (2016). RobecoSAM's Corporate sustainability assessment methodology, available from: http://www.sustainability-indices.com/images/corporate-sustainability-assessment-methodology-guidebook.pdf (accessed June 14, 2018).

Reverte, C. (2012). The impact of better corporate social responsibility disclosure on the cost of equity capital. *Corporate Social Responsibility and Environmental Management*, 29, 253–272.

Rowley, T. and Berman, S. (2000). A brand new brand of corporate social performance. *Business & Society*, 39(4), 397–418. https://doi.org/10.1177/000765030003900404

Salama, A., Anderson, K. and Toms, J.S. (2011). Does community and environmental responsibility affect firm risk? Evidence from UK panel data 1994-2006. *Business Ethics: A European Review*, 20(2), 192–204.

Schneider, T.E. (2011). Is environmental performance a determinant of bond pricing? Evidence from the U.S. pulp and paper and chemical industries. *Contemporary Accounting Research*, 28(5), 1537–1561.

Semenova, N. (2010). Corporate environmental performance: Consistency of metrics and identification of drivers. Proceedings of the PRI Academic Conference.

Semenova, N. and Hassel, L.G. (2015). On the validity of environmental performance metrics. *Journal of Business Ethics*, 135(2), 249–258.

Sharfman, M. (1996). The construct validity of the Kinder, Lydenberg & Domini social performance ratings data. *Journal of Business Ethics*, 15, 287–296.

Sharfman, M.P. and Fernando, C.S. (2008). Environmental risk management and the cost of capital. *Strategic Management Journal*, 29, 569–592.

Surroca, J., Tribo, J.A. and Waddock, S. (2010). Corporate responsibility and financial performance: The role of intangible resources. *Strategic Management Journal*, 31, 463–490.

Suto, M. and Takehara, H. (2017). CSR and cost of capital: evidence from Japan. *Social Responsibility Journal*, 13(4), 798–816.

Toffel, M.W. and Marshall, J.D. (2004). Improving environmental performance assessment. A comparative analysis of weighting methods used to evaluate chemical release inventories. *Journal of Industrial Ecology*, 8(1-2), 143–172.

Ye, K. and Zhang, R. (2011). Do lenders value corporate social responsibility? Evidence from China. *Journal of Business Ethics*, 104, 197–206.

Zhou, Z., Zhang, T., Wen, K., Zeng, H. and Chen, X. (2018). Carbon risk, cost of debt financing and the moderation effect of media attention: Evidence from Chinese companies operating in high-carbon industries. *Business Strategy and the Environment*, 1–14.

22. ESG investing

Anant K. Sundaram

INTRODUCTION

Equity investments in assets with attractive environmental (including climate), social, and governance – ESG – characteristics are now ubiquitous in global capital markets.[1] During the past decade-and-a-half, ESG investing, also commonly referred to as 'responsible investing' or 'sustainable investing', has become a central topic of conversation amongst asset managers, pension & mutual funds, boardrooms, stock exchanges, financial media, academic researchers and even the public-at-large. A variety of investing styles, driven by non-pecuniary goals such as ethical and faith-based motives (for which investors may be willing to forego returns to invest virtuously) on the one end, all the way to pecuniary goals that screen for attractive environmental, social and governance attributes of firms and portfolios (as a means to earn excess market returns) at the other end, characterize the arena of ESG investing.

For example, "interest over time" in the acronym "ESG" revealed by Google web searches in the US shows a modest popularity score of 9 percent (relative to peak popularity)[2] in January 2010. It then takes almost nine years for that score to exceed 25 percent. However, within less than three years afterwards, US public interest in ESG *quadruples*, with peak Google web-search popularity attaining a score of 100 percent in September 2021 (see Figure 22.1).

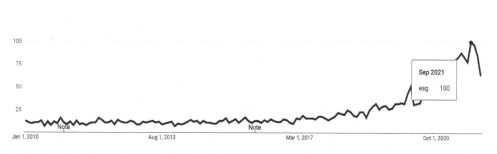

Source: Google Trends. See Note 2 for a definition of "interest over time."

Figure 22.1 "Interest over time" in "ESG" from Google web searches

Similarly, an analysis of approximately 10,000 earnings call transcripts of global companies between May 2005 to May 2021 done by PIMCO (Brown & Sundstrom 2021), an investment management firm with more than $2 trillion in assets, reveals impressive increases in ESG mentions in earnings calls in 2021 compared with even just a couple of years prior: from less than 1 percent of earnings calls mentioning ESG in 2005 (with that proportion staying at or near that level for well over a decade), it was not until mid-2019 that the proportion of mentions exceeded 5 percent. However, just two years later, by mid-2021, the proportion of earning calls mentioning ESG had increased nearly four-fold, to almost 20 percent (Carlson 2021; see Figure 22.2).

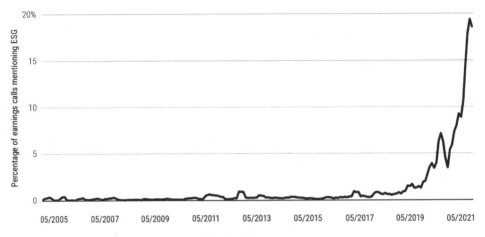

Source: Brown and Sundstrom (2021); used with permission.

Figure 22.2   *Percentage of earnings calls mentioning ESG*

ESG investing has equally rapidly acquired its share of ardent evangelists and detractors, as well as those with all shades of views in-between. For example, at the evangelist end of the spectrum, the CEO of the largest asset management firm in the world, BlackRock's Larry Fink, calls it a "tectonic shift of capital" (Fink 2022), noting that access to capital is not a right for companies, and that failing key ESG and carbon mitigation tests could lead BlackRock to sell its holdings in a company (Sorkin et al. 2022). At the other end, in response to a question from the *Wall Street Journal* on "Where will ESG investing be five years from now?" a scholar on retirement research, Alicia Munnell of Boston College, responded, "Dead, I hope," adding, "…ESG funds are a marketing ploy by financial services firms to repackage actively-managed investments … in a trendy wrapper" (*Wall Street Journal* 2021).

The literature on this topic – with many thousands of papers written in just the past few decades – can be daunting and sometimes confusing, as it often tends to conflate different aspects of ESG into one amorphous, co-mingled whole. Right at the outset, it is important to clarify that there are two sides to the ESG coin. One side is 'corporate' ESG investments whereby firms invest in building (or gaining) attractive ESG characteristics with a view to improving their stock market appeal (e.g., by being seen as less risky and thus lowering their cost of capital) or to do better financially (e.g., on metrics such as return on assets, return on equity, profit margins, and cash flows). The other side of the ESG coin consists of investors seeking out investment portfolios that comprise assets with attractive ESG characteristics, or, equally, avoiding portfolios with unattractive ESG characteristics. Essentially, we can think of the former – which I will call the 'corporate' side of ESG investing – as the assets that are then bundled into portfolios into which investors put their money (which I will call the 'portfolio' side of ESG investing). Much of this chapter will focus on the latter, although it will also discuss how corporate ESG investments are defined, characterized, and so forth (see Chapter 2 in this volume for a more detailed discussion of the corporate side of things, especially on corporate climate-related and low-carbon investments, and whether/how those can impact financial or stock market performance).

The goal of this chapter is to take a closer look at the ESG investing phenomenon, with a view to synthesizing and unpacking the considerable evidence and controversies that have emerged. It will first look at the origins, definitions, key characteristics, and components of how firms – the entities that comprise ESG assets in an investment portfolio – frame ESG. It then looks at the growth in, and current size of the market for ESG investing in equities globally and in the US in particular, and lists the investing strategies that are typically said to constitute this form of investing. Following that, it will summarize the main findings of a vast academic literature that now exists on ESG scores and ratings, as well as on the link between corporate ESG investments and financial or stock price performance, and the links between ESG portfolios and their return/risk characteristics.

The chapter will close with some thoughts for companies seeking to enhance their ESG rankings and scores, for portfolio investors who seek out and invest in ESG assets and portfolios, and will reflect on the opportunities (and potential perils) for sellers of this form of portfolio investing.

ESG: ORIGINS, DEFINITIONS, CHARACTERISTICS

ESG Origins

While the use of the phrase "environmental, social and governance" and its acronym 'ESG' are only a decade-and-a-half old (see the next paragraph for the origins of the phrase), such investing approaches have existed in various forms for at least a half-century, under different nomenclatures (Starks 2021). The phrase "socially responsible investing" or SRI[3] was often used, with references to "ethics in investing" or "ethical investing" being prevalent (see, e.g., Hall 1986; Domini & Kinder 1986). The focus was typically on divestment from, for example, exclusion of stocks of companies whose practices and policies were seen as inconsistent with an investor's ethical preferences, or with their definition of what constitutes socially responsible business behavior. The first ratings agency to start scoring companies, called "Ethical Investment Research and Information Services" (EIRIS) was created in the UK in 1983 (Dimson et al. 2020a), and a similar company "KLD Research & Analytics" was founded in in the US in 1989.

The term ESG is the result of the efforts of the UN Global Compact in 2004 which persuaded CEOs of many major financial institutions to consider how to integrate environmental, social and governance issues into capital market decisions (Kell 2018). These efforts produced a report titled *Who Cares Wins* (The Global Compact 2004), which asserted:[4]

> Companies that perform better with regard to these issues can increase shareholder value by, for example, properly managing risks, anticipating regulatory action or accessing new markets, while at the same time contributing to the sustainable development of the societies in which they operate.

Working with the United Nations Environment Program, these efforts resulted in the creation of "Principles for Responsible Investment" (or PRI) in 2006, an entity founded with a goal to advance the integration of ESG into investment analysis and decision-making. Today, PRI claims to represent over 600 asset managers globally with $120 trillion in assets under management (AUM),[5] and its 'PRI principles' are widely-accepted by financial market participants worldwide. The 'Principles' cover: incorporation of ESG issues into investment analysis and decision-making, making efforts towards being active owners, ESG disclosure/reporting, and promoting acceptance of ESG and its implementation.

Alongside these developments, the creation of the Global Reporting Initiative (GRI) and the Carbon Disclosure Project (CDP) in 2000, as well as the International Integrated Reporting Council (IIRC) in 2011 and the Sustainability Accounting Standards Board (SASB) in 2013, combined with rapid growth of data aggregators and providers comprising an ESG ratings industry, provided a further fillip to the incorporation of a broad range of ESG metrics, scores, rankings and indexes in asset managers' investment evaluation decisions. Corporate reporting decisions are also currently in a great deal of ferment over ESG reporting (see Sundaram 2023, Chapter 2 in this volume).

Definitions and Characteristics of ESG

The *Who Cares Wins* report initially identified 15 broad sets of indicators as examples of ESG attributes for investors to consider in analysis of assets for inclusion in portfolios (The Global Compact 2004, p. 6). *Environmental* issues included climate, toxic releases, waste, and environmental liability; *social* issues included workplace health and safety, community relations, human rights, and government/community relations; *governance* issues addressed board structure and accountability, audit committee structures and independence, accounting and disclosure practices, executive compensation, and corruption/bribery.

As time went by, with numerous additional constituencies and participants joining in and the conversation growing, the list of ESG concerns steadily expanded: by 2013, SASB listed 26 different areas of ESG concern (Boffo & Patalano 2020). By 2022, MSCI, a leading provider of ESG scores and rankings, listed 37 different ESG metrics as meriting concern (MSCI 2022). Key additions in the last decade-and-a-half since the *Who Cares Wins* report have included management of energy, water, wastewater, materials, and product end-of-life under 'E'; diversity, equity and inclusion, employee engagement, anticompetitive practices, board diversity, customer privacy, data security, and access/affordability of products and services under 'S' and 'G.' A summary of the main issues that now commonly fall under the rubric of ESG concerns is shown in Table 22.1.

As is evident from the list, there is little universal agreement among the major ESG service providers and consultants on what these key components of ESG should be, and whether and how they are material from one industry to another or one geographic location to another. For example, toxic emissions are important to a firm in the mining industry to address, but immaterial for the operations of a bank or a consulting firm; DEI issues around ethnicity may be more salient in ethnically diverse countries such as the US, while they may be less so in ethnically homogeneous countries such as Japan (or DEI concerns may be around issues other than that of ethnicity).

While the variables in the 'E' category are easier to define and measure (e.g., direct (Scope 1) and indirect purchased (Scope 2) emissions, energy and water use, toxic releases, waste generated) many, if not most, of the variables that populate the 'S' and the 'G' categories are difficult to define, let alone reliably measure. As discussed further in the fourth section, there

Table 22.1   Key issues that constitute ESG concerns

Environmental (E)	• Addressing externalities from extraction/exploitation of natural resources
	• Corporate responsibility to address:
	  • Energy
	  • Emissions
	  • Water
	• Material use
	• Biodiversity impacts
	• Toxic releases
	• Waste
	• Product end-of-life
	• Circular economy

Social (S)	• Employee rights; employee safety • Employee engagement • Diversity, equity, inclusion • Government & community relations • Consumer health and privacy • Access to and affordability of products/services • Human rights • Corporate philanthropy • Anticompetitive practices

Governance (G)	• Board independence
	• Board accountability (shareholders v. stakeholders)
	• Board diversity
	• Executive compensation and pay-for-performance
	• Board committee structures
	• Corporate codes of conduct:
	  • Ethics
	  • Disclosure/transparency
	  • Risk management
	  • Compliance

is an eye-of-the-beholder quality to the latter type of metrics. Thus, there is concern about how, say, 15 or 26 or 37 different metrics are aggregated into a combined asset-level cardinal ESG score, not to mention the fact that the scores are then synthesized into a ranking with the system of weighting the different variables often being opaque. Then there is the question of the appropriate benchmark to use in assessing ESG performance of funds and individual companies (e.g., using an index, or relative to industry peers). The plethora of hard-to-define and hard-to-measure characteristics in the 'S' and 'G' categories of ESG result in "…dissimilarity in ratings reflecting firm-specific attributes, differing terminologies, metrics and units of measurement" (Dimson et al. 2020a, p. 78). Further, the differences in benchmarks used to compare firms' ESG performance, weightings used to aggregate different metrics, and perhaps most problematically, how missing data are either not filled-in (hence not scored) or filled-in using unclear methodologies (hence unreliably scored) raise numerous additional concerns (see the next section for more on this issue).

Concomitantly, there has been dramatic growth in the number and specificity of ESG indexes to use as benchmarks or for index investing. According to an analysis by BlackRock, the first ESG index was the "MSCI KLD 400 Social Index" that was created in 1990 – but less than three decades later, by 2019, there were over 1,000 ESG indexes (Kjellberg et al. 2020). Figure 22.3, adapted and updated from that report, summarizes the creation timeline of the more prominent ones from 1990–2021.

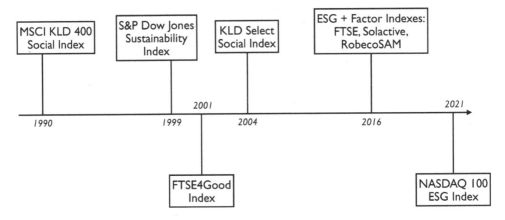

Source: Adapted from Kjellberg et al. (2020) and updated by author.

Figure 22.3   *Evolution of key ESG indexes 1990–2021*

As noted elsewhere in that report, availability of data, metrics, and benchmarks on the ESG front went from "…scarcity to superabundance" (Kjellberg et al. 2020, p. 7).

ESG INVESTING: SIZE, SCOPE AND STRATEGIES

Size of Market for ESG Investing

What is the size of the market for ESG investing, or more precisely, the value of assets under management (AUM) that would qualify for the 'ESG' label? The answer is not so straightforward. Mirroring the incertitude of definitions, categories, and components that comprise ESG, it appears that the assessment of AUM which constitutes ESG investing in the US and around the world is somewhat all-over-the-map. Consider just four examples.

A February 2020 report from BlackRock Investment Institute pegged the AUM of ESG investments in equity, fixed income and mixed allocation securities put together as being collectively less than $1 trillion in 2019 in the US (Hildebrand et al. 2020, p. 4). While the BlackRock report does not display the actual data behind the bar charts, eyeballing the chart suggests that the size of ESG AUM nearly tripled during 2011–2019.[6] Then, in his annual "Letter to CEOs" in January 2022, the CEO of BlackRock notes that "sustainable investments"[7] had reached $4 trillion by mid-to-end of 2021 globally (Fink 2022).[8] If we assume that the two terms are interchangeable *and* that "investments" mean the same as the assets defined in Hildebrand et al. (2020), this would be a quadrupling of market size in two years, which seems unlikely.

The Forum for Sustainable and Responsible Investing (also called US-SIF), a membership-based non-profit organization that advocates for sustainable investing worldwide (and claims to represent $5 trillion in assets), estimates that sustainable investing in the US in 2020 – defined as 'ESG incorporation, shareholder advocacy, and overlapping strategies' – as having assets worth between $16 trillion and $18 trillion, with just the 'ESG incorporation' portion of it accounting for between $14 and $16 trillion.[9] Similarly, the Global Sustainable Investing Alliance or GSIA – a membership-based international collaboration of sustainable investment firms – reports that sustainable investing assets totaled $35.3 trillion in 2020 in the US/Canada, the EU, Australasia, and Japan, growing at 11.5 percent per year starting from a base of $22.8 trillion in 2016 (GSIA 2021).[10]

The bottom line is, based in data from widely used industry sources, there is not a great deal of precision that can be attached to estimates of the size of the ESG market. Terms and definitions used to characterize the market are vague, and the composition of assets that constitute the data for AUM is unclear. My own view, is that, an AUM number in the vicinity of $4 trillion globally, across all asset classes, may be closest to the actual number.

In summary, ESG clearly appears to be an asset class in the investment management industry that is growing at double-digit percentages. While the rapid growth in asset values across-the-board during the past decade is certainly a factor driving that increase in value of ESG assets as well, there appears to be little room for doubt that net fund inflows into ESG investing have grown faster than into conventional funds.

ESG Investing Strategies

What are the investing strategies or styles that characterize ESG investing? It turns out, while there isn't a precise, commonly defined list – such as, say, active, passive, value or growth investing – the profession is gravitating towards one. Depending on the protagonist, the list of ESG investing strategies ranges from just three, to many multiples of that. Many strategies have a tendency to overlap, and hence, the boundaries of where one ends and the other begins are not always clear. As a result, assets may fit into multiple style categories.

Early manifestations of ESG, which took the form of SRI investing or ethical investing, were typically focused on divesting or avoiding assets in sectors or industries such as tobacco, alcohol, adult businesses and defense, i.e., those that did not comport with particular investor preferences. Such strategies are now referred to as "exclusionary" ESG strategies or "negative screening".[11] It was expected (and understood) that an investor adopting such strategies had non-pecuniary goals driving investment decisions, and would settle for lower-than-optimal returns, since, after all, they were consciously limiting their investment opportunity set. However, this was seen as the necessary price to pay for investing virtuously, and consistent with the standard folk wisdom in finance that "there is no free lunch".[12] In other words, by owning just a narrower set of assets that consists of "ethical" companies, an investor was willing to give up returns, i.e., incur a cost to stick to ethical principles. As Asness (2017) notes, "…accepting a lower return is not just an unfortunate ancillary consequence…, it's precisely the point."

On the flip side, what about the returns to those investors who seek out these 'fallen' assets? The logic goes thus: when virtuous investors divest such assets, their stock prices would be expected to fall, and hence expected future returns would be *higher* for those investors buying the assets at the lower prices. Thus, not entirely paradoxically (but perhaps ironically), virtue is penalized while vice is rewarded. Academic research does, indeed, confirm that investments in such fallen sectors have tended to produce higher-than expected – or 'excess' – returns in the long run (e.g., Hong and Kacperczyk 2009; Dimson et al. 2020b).

In the past decade-and-a-half, as ESG investing gained more prominence, the focus shifted to "inclusionary" strategies whereby assets are actively chosen to be *included* in portfolios for their positive ESG attributes, hence also called "positive screening". Similarly, "ESG integration" has become a buzzword in the profession: according to BlackRock, this is a means to "arming portfolio managers with tools and opportunities..., enhancing the investment process..., and making investment decisions that take financially material ESG information into account." In other words, it is *not* seen as a values-based exercise or negative screening, but rather a conscious strategy to seek higher returns by investing in high-ESG-quality firms.[13] More recently, ESG styles have embraced "impact" investing strategies, which is the process of approaching investing with a goal to create measurable and verifiable positive environmental, social and governance outcomes,[14] and "thematic" investing, which approaches investing from the standpoint of structural social, environmental or demographic shifts rather than in a particular sector or industry (examples of such "themes" would include climate change, aging populations, increasing resource scarcity, access to safe drinking water, and so forth).

Table 22.2 provides examples of ESG investing style categorizations used by four well-known firms in the ESG arena, The World Bank, BlackRock, the membership-based Global Sustainable Investment Alliance (GSIA), and the data/ratings provider MSCI.

Table 22.2   Examples of ESG investment style classifications

The Word Bank	• Avoid (Negative screening) • Benefit (ESG integration) • Contribute (Impact investing)
BlackRock	• Avoid (Negative screening) • Advance: ESG integration • Advance: Thematic • Advance: Impact
GSIA	• Negative screening/exclusionary • ESG integration • Corporate engagement • Norms-based screening • Positive screening • Sustainability investing • Impact investing

MSCI	• Integration
	  • Bottom-up
	  • Top-down
	  • Best-in-class
	  • Active ownership
	  • Thematic
	• Values
	  • Best-in-class
	  • Exclusionary/negative
	  • Active ownership
	  • Socially responsible
	  • Faith-based
	• Impact
	  • Impact
	  • Thematic
	  • Active ownership
	  • Mission-related

Characteristics of ESG Portfolios and ESG Funds

What do typical ESG equity portfolios look like? Do ESG attributes really provide the market with *additional* information about expected returns and risk in portfolio management, thereby leading to better portfolio performance, or is it more likely that these portfolios reflect a systematic tilt towards – or away from – some well-known and commonly-accepted return-generating factors, i.e., do they proxy for well-known risk factors such as size, value, profitability, volatility, etc.?[15] While the answer is not definitive, evidence increasingly points to the latter.

First, researchers have observed that higher ESG scores are titled towards larger market capitalization-firms. Drempetic et al. (2020) hypothesize that there should be a positive association between firm size and data availability as well as firm size and ESG scores since, after all, larger firms have more resources. Their analysis does, indeed, find a "highly significant" influence of size on ESG scores. They suggest that this finding implies that larger companies more often have the resources for ESG data disclosure, and that "…more resources lead to more available data in the ESG database, and more available data, regardless of whether it is directly positive or negative, raises the overall sustainability assessment of the company" (Drempetic et al. 2020, p. 335). LeBella et al. (2019) also find similarly, noting that a portfolio built around companies with the highest ESG scores typically contains a higher proportion of larger market capitalization firms than the benchmark.[16]

Why does the size tilt matter? Recalling that much of the interest in – and funds flows into – ESG investing is a phenomenon of the last decade-and-a-half, any portfolio that is skewed towards holding larger firms would naturally have performed better, since large market capitalization firms have handily outperformed small market capitalization firms during this period: During the period 2007–2021, stocks of firms in the top quintile of market value outperformed the bottom quintile by 1.73 percent per year on a value-weighted basis.[17] Put differently, $1.00 invested at the beginning of 2007 in a portfolio of the top 20 percent of firms by size (and re-formed annually) would have been worth $5.90 at the end of 2021, while doing so with the bottom 20 percent of firms by size would have resulted in only $4.68. In other words, returns attributed to ESG characteristics may perhaps be explained, at least in part, by not much more than the tilt towards larger firms during a particularly fortuitous period of returns for this category of stocks.

Second, research looking in more detail into whether there is a "growth" tilt seems to be indicating that might, indeed, be the case. Schnellner (2021) notes, for example that, compared with the average market-to-book ratio of 4.17 for firms in the Russell 3000 Index (a stock market index of the 3,009 largest firms in the US by market capitalization), an aggregate of sustainability funds in the US in their dataset have a market-to-book ratio of 5.33, i.e., 28 percent higher. Since "growth" and "value" are most commonly screened on the basis of firms' market-to-book ratios, this clearly implies a tilt towards growth assets in sustainability funds. As with the tilt towards larger firms, why might this matter for returns?[18] During the period 2007–2021, stocks of firms in the top quintile of market-to-book ratios outperformed the bottom quintile by 2.99 percent per year on a value-weighted basis. Thus, $1.00 invested at the beginning of 2007 in a portfolio of the top 20 percent of firms by market-to-book ratios representing "growth" firms (and re-formed annually) would have been worth $9.16 at the end of 2021, while doing so with the bottom 20 percent of firms by market-to-book ratios (representing "value" firms) would have resulted in only $6.19.[19] Returns attributed to ESG characteristics therefore appear to be explained by their bias towards "growth" assets even more so than with size, again during a fortuitous period of returns for this category of stocks.

Third, research has similarly noted that there is a bias in ESG holdings towards stocks in the more advanced economies (see, e.g., LaBella et al. 2019), and here again, given a great deal of evidence which shows that emerging market stocks have substantially underperformed those in developed markets during the past decade-and-a-half, excess return attributions made to ESG stocks may be merely the result of luck rather than skill, arising from non-ESG asset tilts.

The importance of not conflating propitious sector tilts with excess returns to ESG characteristics is lent further credence by the observation that equity returns have a documented tendency to mean-revert over longer horizons (Poterba & Summers 1988; Fama & French 1988). Colloquially, in equity markets, it implies that "what goes up tends to come down." If history is any guide, the gains in returns to larger, growth firms in advanced economies observed in the past 15 years being misidentified with ESG excess returns may merit caution when looking ahead: mean-reversion in these asset categories can easily return, wiping away any gains (mis) attributed to ESG, perhaps even turning them into losses as time goes on.

Detailed evidence on the collective of ESG funds and their fund characteristics does not yet seem to be available, but some tentative contours are emerging. Murugaboopathy & Mann (2021) report, based on Morningstar data, that there are nearly 7,500 sustainable funds worldwide, with total AUM of $3.9 trillion, with nearly 88 percent of sustainable AUM located in Europe, followed by 8.5 percent in the US, and 3.5 percent in the rest of the world combined. Data from the Investment Company Institute (ICI 2021), the leading industry association representing regulated funds globally, claims that as of 2020 there were 592 ESG mutual funds and exchange-traded funds (ETFs) in the US, with $465 billion in net AUM.

Stankiewicz (2022), based on Morningstar data, estimates that 60 percent of US sustainable AUM are in active funds (as opposed to 'passive' which are funds that mimic their asset holdings to a chosen market index benchmark). This is in line with the proportions of active versus passive for *all* AUM in US funds, and hence not unusual. But average expense ratios, i.e., fund fees paid by investors, are about 50 percent higher with sustainable funds: Johnson & DiBenedetto (2021) estimate that sustainable funds in the US have 0.61 percent in asset-weighted expense ratios compared with 0.41 percent for conventional peers. In an analysis of 196 ESG funds spread across 17 different ESG investing strategies and $214 billion in AUM, Schneller (2021) finds that ESG funds in their data include a total of 2,870 firms of the 3,009 firms in the Russell 3000 Index. In other words, there is a 95 percent chance that an investment in a randomly selected stock of the highest-valued 3,009 firms in the US (which, in turn, account for over 95 percent of US stock market capitalization) would qualify as an ESG investment!

ARE ESG RATINGS CREDIBLE? DOES ESG INVESTING RESULT IN BETTER PERFORMANCE?

Credibility of ESG Scores and Rankings

During the recent past, we also have witnessed impressive rates of growth in the number of data-, scoring-, and rankings-providers in the ESG arena. Concomitantly, academic research assessing the quality of their outputs has rapidly grown. The literature raises troubling concerns about the quality and consistency of ESG data, scores and ratings. Concerns are also beginning to be simultaneously raised about potential conflicts of interest in the data provider industry.

As of the time of writing, there are at least eight major providers of such services (see also the fifth section of Chapter 2 in this volume, Sundaram 2023). There are two main sources of *actual* reported data, namely Carbon Disclosure Project (CDP) and the Global Reporting Initiative (GRI), to which companies voluntarily disclose. In addition to CDP and GRI, there are now at least seven or eight large data aggregators and scoring/rankings providers, who *estimate* ESG data points for thousands of non-reporting companies by extrapolating values – typically based on metrics such as revenue- or employee-based multiples from those firms that report – to companies in the entire industry/sector: FTSE Russell, ISS, MSCI, Morningstar, Refinitiv, RobecoSAM, Sustainalytics and Trucost.[20] The high ratios of estimated-to-reported data are concerning. For example, ISS has coverage of *reported* data on GHG emissions from about 4,000 companies, but their *total* (i.e., estimated plus reported) coverage is approximately 25,000 companies: in other words, nearly 85 percent of their data is based on estimated values. Similarly, numbers for Trucost are 1,800 and 14,000 respectively, i.e., over 85 percent based on estimated values (Busch et al. 2018, Table 1).

Estimation methods that providers use to impute values for non-reporting companies are opaque, with few actual examples or explanations or calculations for a typical company or industry shown. Moreover, it is unclear how they address change over time in the number of reporting firms: for example, in 2009, CDP had about 2,600 companies reporting data (CDP 2009), a number that grew to over 5,600 in 2015 (estimated from BusinessGreen Staff 2020),[21] but by 2021, over 13,000, or 5× the number relative to 2009, were reporting (CDP 2021).[22] How exactly is this rapid change over time in the ratio of firms that report to those that don't, incorporated into the estimation methods over time in provider databases? The answer is not clear.

As discussed above, such values, often comprising dozens of hard-to-measure E, S, and G indicators are aggregated into a cardinal ESG score, which then becomes the basis for ranking companies according to their ESG 'performance.' As a result, we should not be surprised if there are inconsistencies and disagreements in reported ESG scores across different providers – often for the same company – arising from what Kosantonis and Serafeim (2019, p. 51) and Dimson et al. (2020a, p. 78) list as data discrepancies, benchmarks used (e.g., is the benchmark for comparisons with other firms in the industry or the entire index?), data imputation and estimation methods, weighting schemes for aggregation to one 'ESG score', and information overload. Indeed, a considerable amount of research now shows precisely such an outcome.

Dimson et al. (2020a) compare ratings across three ESG rankings providers, FTSE Russell, MSCI, and Sustainalytics, for some large, well-known companies with mega-market capitalizations. They note, for example, that Tesla gets a high 'AA' rating from MSCI, whereas FTSE ranks it low, and Sustainalytics puts the company in the middle. In the case of Facebook's 'E' attribute – the easier, more precise attributes to measure, relative to S and G – Sustainalytics places the company in the first percentile, while MSCI ranks it in the 96th percentile. Many other such examples are provided. Berg et al. (2022) show, for example, that there are substantial rater disagreements on a simple (and, again, seemingly easy-to-measure) binary variable such as whether the jobs of Chairman and CEO of a company are combined or separate.

Evidence is now fairly substantial on unacceptably low correlations between ratings firms on each three of the E, S, and G attributes. Dimson et al. (2020b) show a correlation coefficient of 0.30 to 0.59 on the aggregate ESG measure across the three ratings firms they analyze.[23] The correlation coefficients for the G attribute are so low as to practically render it meaningless for an investor: the correlation coefficient between FTSE and Sustainalytics for G is 0.07; between MSCI and FTSE, 0.00; and between MSCI and Sustainalytics, –0.02, i.e., a negative number (even if likely not significantly different from zero). Berg et al. (2022), Boffo & Patalano (2020),[24] Bilio et al. (2020) and Brandon et al. (2021) document similar results across pretty much all of the eight ratings providers listed prior, between them.

Such disagreements can have negative consequences for not just investors looking at ESG assets ("Which firms should I invest in for their attractive ESG attributes?"), but also firms that are looking to assess and communicate their ESG performance ("Am I doing well or poorly relative to my peers?"). Additionally, academic research will find it difficult, if not impossible to come to convincing conclusions regarding the impact of ESG performance on corporate financial or stock market performance, since the variable on the right-hand side that purports to explain profitability or returns – the ESG score for the company – is, simply put, not credible. At best, the evidence is provider-dependent and unless everyone can agree on provider quality, the measurement impact of the ESG on performance becomes suspect, or at the very least, in the eye of the beholder.

Even worse, there could be provider-induced negative market consequences in the form of a cost of equity capital-penalty for some firms due to no fault of their own: in a study based on ESG scores from seven ratings providers for S&P 500 firms, Brandon et al. (2021) find that there is a positive association between stock returns and ratings disagreements. In other words, their analysis implies that the market expects a higher risk premium for stocks with greater ratings disagreement. Complicating matters, Christensen et al. (2022) find that, the greater the amount of disclosure from a company, the greater the ratings disagreement. Further, that study finds that greater ratings disagreement is associated with higher return volatility, larger absolute price movements, and lower likelihood of the firm raising external financing. If such evidence holds up under further scrutiny, it would not be out-of-line to say that the current state of the ESG scoring/ranking industry runs the risk of imposing a negative externality on firms' financial decisions.

Finally, there is emerging concern from multiple fronts regarding potential conflicts of interest inherent to the business model of the ESG ratings provider industry, not unlike what was observed with the credit ratings industry (implicated as a contributing factor to the global financial crisis of 2008–2009) and the governance/proxy-voting advisory industry (regarding which the US Securities and Exchange Commission, or SEC, is currently in the process of drafting rules). A recent report from the staff of the SEC (Securities and Exchange Commission 2022, p. 8) says:

> NRSROs[25] and their affiliates have developed and are offering an increasing number of ESG-related products and services. Development in this area has grown rapidly, and competition has increased among NRSRO providers and non-NRSRO providers, leading the Staff to identify several areas of potential risk to NRSROs. These include the risks that, in incorporating ESG factors into ratings determinations, NRSROs may not adhere to their methodologies or policies and procedures, consistently apply ESG factors, make adequate disclosure regarding the use of ESG factors applied in ratings actions, or maintain effective internal controls involving the use of ratings of ESG-related data from affiliates or unaffiliated third parties. The Staff also identifies the potential risk for conflicts of interest if an NRSRO offers ratings and non-ratings products and services.

In the popular press, for example, Eaglesham (2022) notes in *The Wall Street Journal*:

> The market for helping companies with corporate ESG reporting alone is worth an estimated $1.6 billion globally, and forecast to increase by 21% a year over the next six years... In many cases, firms that rate or evaluate companies on things like climate risk also sell services to help companies address these issues. Many of the firms providing these ratings, such as credit raters and auditors, are already managing deep conflicts of interest because they are paid by the companies they judge.... One new set of potential conflicts springs from the widespread practice of selling ESG ratings alongside consulting and other services.

On the academic research front, focus on this issue is early and tentative, but emerging work points to concerns: for example, Tang et al. (2021) find that firms held by the same owners as the ratings firms (what they call "sister firms") receive higher ESG ratings, and that the larger the stakes common owners have in the sister firms, the higher the firm's ESG ratings.

Such across-the-board concerns imply that we should not be surprised to see significant legislative and regulatory efforts emerging to forestall potential ESG ratings-provider conflicts of interest, in the next few years.

Evidence on ESG and Corporate Financial/Stock Market Performance

As of this point, *thousands* of papers – both published and unpublished – have been written on the links between ESG (or their variants) scores or ratings and financial/stock market performance of companies, and similarly for ESG funds and their performance against benchmarks or against conventional funds. Fortunately, however, there have not only been many meta-studies of these papers, but there are now *meta*-meta-studies that allow us to summarize the findings of this vast literature. In a meta-meta-study done by Whelan et al. (2021) of the ESG-performance literature during just the period 2015–2020, the authors document 13 meta-studies covering 1,272 unique studies "with a quantitative approach" on firm-level links between ESG and performance, and two meta-studies covering 107 unique studies. Two other meta-studies often mentioned in academic reviews of the literature are Margolis et al. (2009) which covers 35 years of research and covers 214 unique papers, and Friede et al. (2015), which covers more than 2,200 unique studies.

What do these meta- and meta-meta-studies find? Margolis et al. (2009), who look only at the link between corporate social and financial performance, do not find meaningful increases in performance from companies investing in ESG attributes. While they do find an overall non-negative effect, it is small in the aggregate. Friede et al. (2015) split their assessment into two, namely corporate versus portfolio investments in ESG. On corporate ESG and financial performance, with 90 percent of the studies finding a positive association, they conclude that there is "...clearly a business case for ESG investing" (p. 226), but on ESG attributes of portfolios and portfolio performance, results are mixed or neutral. Whelan et al. (2021, pp. 2, 4) find that: (i) in "corporate" studies, 58 percent show a positive impact, 13 percent neutral impact, and 21 percent mixed impact, and 8 percent negative impact; (ii) in "portfolio" studies, when compared with conventional investment approaches, these proportions were 33 percent, 26 percent, 28 percent, and 14 percent, respectively; i.e., positive effects were less strong with the corporate evidence; and (iii) from their meta-meta-analysis, they find "... consistent positive correlations between ESG and corporate financial performance" but that "...ESG investing returns [are] generally indistinguishable from conventional investing."

Thus, taken together in total, assessments of the thousands of papers on the link between ESG and performance collectively find that the association is non-negative at worst, and positive at best, with stronger results on the corporate side, and weakly-positive to no results on the portfolio side.

A well-known report from BlackRock (Deese et al. 2019, p. 6) summarizes evidence on indistinguishability of returns from ESG versus conventional ("traditional") investing during the period 2012–2018: Table 22.3 provides an (adapted) summary. As we can observe, ESG stocks have risk-return characteristics that are little different from those of conventional stocks both in the US and in the rest of the world. Essentially, it appears that such investments do no worse than conventional investing.[26]

Table 22.3   Conventional versus ESG Stock performance 2012–2018

	US Conventional	US ESG	World (excluding US) Conventional	World (excluding US) ESG
Annual return	14.4%	14.5%	7.7%	8.1%
Volatility	9.7%	9.8%	11.5%	14.4%
Sharpe Ratio*	1.42	1.42	0.62	0.64
Maximum monthly drawdown**	–13.9%	–13.9%	–23.3%	–22.7%
ESG score	5.4	6.5	6.6	7.8
Number of stocks	621	313	1,012	453

Source: Adapted by the author from Deese et al. (2019, p. 6).
*The Sharpe Ratio is a measure of risk-adjusted return, defined as the ratio of mean excess returns of an asset during a particular time period (relative to the risk-free rate of return) divided by the standard deviation of the excess returns.
** 'Maximum drawdown' is the peak-to-trough value decline of an investment during a specific period.

Simply put, the aggregate evidence on ESG and performance appears to imply that we can have it all: we can invest virtuously and do well. With a finding like that, an obvious question to ask would be, *what's not to like*? There are at least three worrisome issues that have to be considered.

The first worrisome aspect is regarding how "doing well" and "doing good" are related for companies. As with a great deal of evidence in the social sciences, we must address issues of causality before we can make statements such as "X results in Y". In other words, a central question is, does desirable ESG performance result in better corporate performance (i.e., "does doing good lead to doing well?")? Or is the causality the other way around: is it good corporate performance that leads to better ESG performance (i.e., "is it doing well that leads to doing good?")? The answer matters since, if society's goal is to encourage firms to become more ESG-friendly, public policies and market signals that reward managerial actions that lead them to being successful and profitable first should be unambiguously preferred, since such actions will then naturally be expected to lead companies to make more ESG investments. Equally – and perhaps more crucially – such a finding would also imply that encouraging or requiring companies to simply behave better on the ESG front may have little to do with enabling them to become successful or profitable. In other words, the approach of "doing good leads to doing well" may not lead investors to the rewards that the literature says ESG brings, but may just end up putting the cart before the horse. Moreover, for the numerous reasons and the considerable evidence on imperfectness and noisiness in company ESG scores and ratings discussed above (see the fourth section which describes the current state of affairs on that front), how can we be certain that conclusions we come to are not spurious?

Similarly, on the portfolio front, does the analysis control for all the asset pricing factors that matter *and* are the econometrics designed to tease out causality (i.e., it's not just a matter of running regressions, even with some controls, as is commonly done)? Otherwise, just as with corporate actions – and even leaving aside the problems with scores and ratings that lead to noisy measurements of the key variable that purports to explain a "good" company or a "good" portfolio – we cannot conclude much, if anything, from merely observing associations between ESG ratings and stock (or portfolio) returns.

The second worrisome aspect arises from how to interpret the portfolio return findings. Suppose high-ESG-rated companies' stocks or high-ESG-rated portfolios do, indeed, perform better. Is that a good thing or a bad thing for an investor? Well, it depends. If we assume that markets are efficient and they reward companies that perform well on ESG characteristics with higher stocks prices, then the positive effects of ESG performance are, by definition, already embedded in the price of the asset. If investors bid up the price of a company's stock on the basis of expectations of better performance on an attribute – any attribute – that they think drives returns, then the expected return to an investor buying the asset at the higher price is *lower*, going forward. Similarly, if investors exit a company's stock on the basis of expectations of poorer performance on an attribute they think drives returns, thus driving the stock price down, then the expected return for the investor buying the asset at the lower price is higher, going forward. Thus, in an efficient market, there are no *excess* risk-adjusted returns to be had for an investor regardless of whether they are invested in a good ESG or a bad ESG asset. Investors simply get what they pay for given the risks associated with the cash flows of that asset, no more, no less. Thus, the premise that investing in portfolios of high-ESG-rated firms is the path to greater portfolio riches does not stand up to logical scrutiny in a well-functioning market.

The only way to achieve such excess returns is for an investor to be presciently skilled in identifying good ESG firms or portfolios *before* the rest of the market does so, then hope that others in the market see, in sufficiently large numbers, what the investor foresaw, thus raising its market value to produce excess returns for the early investor. Moreover, unless they were merely lucky in identifying such an asset, sustaining such excess returns requires the investor to be able to find such assets (and asset classes) systematically and repeatedly.

A simple, yet formidable framework to think about this comes from Damodaran (2020); see Table 22.4. He argues, "Whatever your beliefs may be on whether ESG increases or decreases value, you have to start with a fresh slate, and incorporate market behavior, to make judgments on whether investors will benefit from ESG investing." Focusing on the first row of Table 22.4, suppose ESG *does* increase value. What would happen if markets overreacted to the company's goodness, as can sometimes happen? The company's stock price increases too much, yielding *negative* excess returns to the investor. Even more interestingly, what if the impact of ESG investing on value is negative, and markets overreact? In that situation, investing in good ESG firms would yield *positive* excess returns. In other words, as Damodaran notes, "… studies that conclude that ESG investing earns positive (or negative) returns tell us nothing or very little about underlying benefits of ESG, since the market acts as an intervening variable."

Table 22.4 Impact of ESG-performance association and market pricing on investor returns

Value effect	Market pricing	Investor returns to ESG
ESG increases value	Markets overreact, pushing *up* prices too much	*Negative excess returns* for investors in good ESG firms
ESG decreases value	Markets overreact, pushing *down* prices too much	*Positive excess returns* for investors in good ESG firms
ESG increases value	Markets underreact, with prices going *up* too little	*Positive excess returns* for investors in good ESG firms
ESG decreases value	Markets underreact, with prices going *down* too little	*Negative excess returns* for investors in good ESG firms
ESG increases value	Markets react correctly, with prices *increasing* to reflect value	*Zero excess returns* for investors in good ESG firms
ESG decreases value	Markets react correctly, with prices *decreasing* to reflect value	*Zero excess returns* for investors in good ESG firms

Source: Damodaran (2020); used with permission.

The third, and perhaps most vexing question is, if evidence on the corporate front is so unassailable that ESG investing entails no performance sacrifices but rather, provides opportunities for performance gains, why doesn't every company invest in ESG at all times since doing so will apparently improve its financial and stock market performance? Similarly on the portfolio front, we might ask, why does any rational investor put their money into a conventional portfolio at all? The simple fact that we observe neither to be the case – i.e., we don't observe every company going overboard on such investments, and we do not see investors abandoning conventional funds in droves in favor of ESG funds – suggests that there surely are costs as well as benefits to ESG investing, and that in the aggregate, the costs perhaps outweigh the benefits.

Consider the case of companies and the types of ESG investments they make. Investing in ESG and maintaining that investment over time is not costless. ESG-related data must be gathered, disclosures must be verified and made, internal systems – e.g., capital budgeting systems, employee hiring/training/retention/promotion processes, performance-incentive systems, supply chains, product and process R&D, even governing boards – must be retooled and redesigned, all with the hope that the benefits from doing so outweigh the costs. And people must be employed to enable all this to happen. In most countries of the world, these investments are, certainly at the moment, being made voluntarily by companies in the absence of any substantial policies – whether such policies are designed to reward good behavior (subsidies) or penalties for bad behavior (taxes) – which require them to do so. There are a number of sensible reasons why this could happen, including brand- and reputation-building, raising rivals' costs, sending a credible (even if costly) signal of quality, enhancing the firm's ability to attract and retain better talent, forestalling regulatory risk, and so forth. In other words, becoming an ESG-friendly company is far from costless.

What about the benefits from such investing? In my estimation it is more likely than not that the benefits are perhaps overstated, since many – certainly not all – ESG-related outcomes are hard to measure, or they are typically measured on inputs and processes associated with ESG rather than on impacts achieved. At a minimum, unless we can agree on the key attributes to consider, can reliably define, measure and verify those key attributes, then achieve consensus on aggregation rules to convert them into ESG scores and ratings, and finally, identify the paths through which ESG affect cash flows and risks of the company *as well as* the impact achieved, it is difficult to make any claim about benefits realized.

CONCLUSION

A closer look at this rapidly emerging landscape of ESG investing reveals that it is, at best, a work-in-progress and, at worst, an investing arena in a state of turmoil from perhaps being still in early stages of evolution. Confusing definitions (to the point where participants define the size of the market to be anywhere from $1 trillion to $37 trillion globally), imprecise and overlapping investing strategies, fortuitous asset tilts, data- and ratings-related obfuscations and inconsistencies, and the possibility that a lot of it may be a sales pitch by interested parties (even if well-intentioned), abound. Until the dust settles, we cannot rule out the possibility that a state of noise overwhelms this space. Additionally, the absence of clear causality attributions, the lack of consideration of financial markets as intervening variables in setting asset prices (see Table 22.4), and the "we-can-have-it-all" aspect of how investment opportunities are characterized and sold, raise concerns. We might wonder whether the quest to find an 'ESG investing factor' that explains the purported higher (or the same) returns for the same (or lower) risk profile might be difficult, if not impossible.

For companies, there are two implications. One, there is little benefit in trying to manage ESG ratings. It may be a futile exercise to try to work with every major data provider to get them to agree on every major ESG attribute, its definitions and its measurements, in a manner that reflects what is material to the firm. A better strategy for firms would be to continue to focus on doing well what they already do, and to voluntarily disclose as much sustainability-related information as best (and at as reasonable a cost) as they can, letting the rest take care of itself. Two, companies might consider focusing more attention on measuring the true costs of ESG investments and, equally, measuring their impact. While this may be easier said than done, it would be useful to be armed with as much relevant data as possible on costs versus benefits of investments in ESG, to be able to make the case that what the company does on sustainability is neither free nor inexpensive, at the same time being able to make the case for how it all matters in terms of impact.

For sellers of these investment opportunities, i.e., the industry players and their advisors, it is in their long-term interest to make a deeper case for how the direction of causality works between 'doing well' and 'doing good'. Perhaps, breaking with past practice, the best opportunities here may be to draw investors' attention to traditional, old-line sectors such as energy, materials, heavy industry and transportation, which tend to be the ones that are most often shunned by purveyors of ESG portfolios. If we are to get in front of difficult-to-solve issues such as climate change, these are precisely the sectors in which change must happen, particularly in relation to the 'E' variable. These may also be the very sectors where investors still have opportunities to identify assets that are in the early stages of making potentially value-creating change happen. As a result, opportunities for screening to generate excess returns for investors may exist with greater likelihood in such sectors than in the larger, US-focused, growth firms that seem to populate many ESG portfolios. Given their skills and access to data and models, asset managers have a unique opportunity in this regard. Further, ESG investing has reached a stage whereby it is necessary to see major players in the space quickly coalesce around a clear, commonly-accepted, concise set of attributes, definitions and rules that characterize investing in this arena, so as to minimize confusion. Equally importantly, problems arising from lack of reliable, carefully measured, externally-verified data based on what is *reported* by companies (rather than estimates based on multiples and such) cannot be wished away, and need to be urgently addressed.

For investors, at least for the moment, much of the evidence marshalled here reinforces the need to remind themselves of the age-old, common-sense advice of *caveat emptor*, especially when it comes to what seems like, and is sold as a 'free lunch.'

Without such developments being visibly – and authentically – led by the ESG investing and advising community, and unless it all starts to happen sooner rather than later, we should not be surprised to see regulators and rule-makers stepping into the void. This will likely happen in a substantive manner over the next few years; indeed, it is already starting to happen. For example, major disclosure-related regulation is in the works in the EU, under the "Corporate Sustainability Reporting Directive" (CSRD) and its emphasis on the so-called "EU Taxonomy" for green investments and the unusual notion of "double-materiality", which is the idea that publicly-traded firms must start to disclose information not only on how non-financial attributes such as ESG materially affect their business model, but also the reverse, i.e., how the firm's operations may materially affect ESG attributes and society-at-large. In the US, the SEC is very much in the process of examining similar disclosure-related rulemaking, although it is not nearly as far ahead as regulators in the EU are (see the fifth section of Chapter 2 in this volume, Sundaram 2023, for more on this issue).

In closing, it is abundantly obvious that ESG investing is here to stay and will likely continue to grow over time. The first decade-and-a-half has been something of a free-for-all, with relatively lax oversight. One can quite reasonably speculate that the next decade-and-a-half will not offer the industry that luxury.

ACKNOWLEDGEMENT

The author is deeply grateful to Bob Hansen for his incisive comments.

NOTES

1. For ease of exposition and to keep things manageable, this chapter will not specifically or separately discuss climate-related and low-carbon investments, noting that the ESG space includes a plethora of those. The main insights and conclusions we draw for the broader ESG space will directly apply to the climate-related and low-carbon investment subset as well.
2. https://trends.google.com/trends/explore?date=2010-01-01%202022-01-01&geo=US&q=%22ESG%22, where "interest over time" is defined as the percentage of searches relative to the highest point on the chart for the given region and time. A value of '100' is the peak popularity for the term. A value of 50 means that the term is half as popular, and so forth.
3. Today, the acronym SRI typically refers to "sustainable and responsible investing."
4. Other than for a survey of investors and the reports of a consulting firm (see The Global Compact 2004, pp. 11–12), the *Who Cares Wins* report did not independently assess evidence – academic or otherwise – on the claimed association between ESG characteristics and shareholder value-creation.
5. See https://www.unpri.org/about-us/about-the-pri
6. This would imply a 14–15 percent compounded annual growth rate (CAGR) in AUM during this period.

ESG investing 521

7. There may be differences between the definition of what constitutes "sustainable" versus how "ESG" is defined, but increasingly, the two terms are used interchangeably in both the practitioner and academic literatures. In any event, for the rest of this chapter, I will use them in an interchangeable fashion, but attempt to stick to the term used in the actual report or data being cited.
8. For example, Morningstar, one of the sources he cites in the *Letter*, estimates the size of sustainable equity investments *globally* at $3.9 trillion in October 2021 (Murugaboopathy and Mann 2021), a number close to the CEO's estimate of $4 trillion.
9. https://www.ussif.org/sribasics
10. If we assume that US AUM are 55 percent of global AUM (Dimson et al. 2021, p. 8, estimates it to be 55.9 percent), GSIA's estimate would imply that US sustainable investing assets under management equal 0.55×35.3 = $19.4 trillion, which would put it close to the $16–$18 trillion estimated by US-SIF.
11. It should be noted that this investing style has made a strong comeback recently, in the form of a steadily-expanding 'divest fossil fuels' movement in many parts of the Western world, especially amongst universities, foundation-investors, and public pension funds.
12. For a brief (and amusing) history of the origins of this phrase – one that people often associate with economist Milton Friedman – see https://quoteinvestigator.com/2016/08/27/free-lunch/
13. https://www.blackrock.com/institutions/en-us/solutions/sustainable-investing/esg-integration
14. Arguably, the need for a separate "impact" category is somewhat superfluous, since we should expect that the goal of *any* type of ESG investing must be to create measurable and verifiable impact on characteristics of environmental, social or governance investor interest. This applies even to firms we wish to avoid/exclude, since after all, the idea of exiting from their stocks would be to raise their cost of capital for 'undesirable' ESG activities in which they engage.
15. The fourth section discusses the link between ESG assets and return performance in more detail.
16. Damodaran (2021) drily notes, "It is entirely possible that big companies are better corporate citizens than smaller ones, but it is also just as plausible that big companies have the resources to play the ESG scoring game."
17. Calculated using data from Ken French's website, https://mba.tuck.dartmouth.edu/pages/faculty/ken.french/data_library.html).
18. 'Value' firms dominate sectors such as energy, utilities, materials, and heavy industry, which tend to be the most *directly* polluting and emissions-intensive sectors in the economy. 'Growth' firms, which dominate sectors such as technology, communication services, health care and biotechnology, tend to have much of their pollution and emissions occur *indirectly*, i.e., in the upstream (e.g., supply chain) and downstream (e.g., consumer use and end-of life treatment) portions of their value chains. Damodaran (2020) also points to ESG funds having a tech tilt.
19. Calculated using data from Ken French's website, https://mba.tuck.dartmouth.edu/pages/faculty/ken.french/data_library.html).
20. See Busch et al. (2018, Table 1) for the coverage of number of companies with actual and estimated GHG emissions data in the ISS, MSCI, Refinitiv, Sustainalytics and Trucost databases. See also Kosantonis and Serafeim (2019), who pick just one 'S' metric, "Employee Health and Safety," and based on hand-collected data from a few dozen companies, work through the complications that arise when trying to compare across companies. A comparison similar to that by Busch et al. (2018) for other major E, S, and G indicators across the major providers appears to be lacking, as of now.
21. https://www.greenbiz.com/article/cdp-reporting-record-almost-10000-companies-disclose-environmental-data-2020
22. https://www.cdp.net/en/articles/media/cdp-reports-record-number-of-disclosures-and-unveils-new-strategy-to-help-further-tackle-climate-and-ecological-emergency
23. By way of comparison, the correlation coefficient in bond credit ratings between Moody's and S&P is 99 percent (Collins-Dean & Geffroy 2021).
24. In the Boffo & Patalano (2020) analysis, not only did the authors find low ESG correlation coefficients, but they found, quite unusually, a *positive* correlation between 'E' scores and carbon emissions in the case of two out of the three providers they looked at.
25. NRSROs refer to "Nationally Recognized Statistical Ratings Organizations" which are credit rating organizations approved by the SEC. The two largest credit raters are now owners of ESG service providers as well – S&P through its acquisition of Trucost in 2015, and Moody's which acquired

RMS, a climate and natural disaster analytics firm, in 2021, which likely explains the SEC Staff's concerns.
26. It is interesting to note that, despite their generally higher ESG scores, the annualized dollar returns of stocks in the rest-of-the-world were substantially lower than that of US stocks during the period 2012–2018.

REFERENCES

Asness, C. (2017, May 18). Virtue is its own reward: Or, one man's ceiling is another man's floor. *Cliff's Perspective.* https://www.aqr.com/Insights/Perspectives/Virtue-is-its-Own-Reward-Or-One-Mans-Ceiling-is-Another-Mans-Floor

Berg, F., Koebel, J. F., & Rigobon, R. (2022). Aggregate confusion: The divergence of ESG ratings. SSRN: https://ssrn.com/abstract=3438533. http://dx.doi.org/10.2139/ssrn.3438533

Bilio, M., Costola, M., Hristova, I., Latino, C., & Pelizzon L. (2020). Inside the ESG ratings: (Dis)agreement and performance. *Department of Economics Working Paper.* Ca'Foscari University of Venice. https://papers.ssrn.com/sol3/papers.cfm?abstract_id=3659271

Boffo, R., & Patalano, R. (2020). *ESG investing: Practices, progress and challenges.* OECD Paris. https://www.oecd.org/finance/ESG-Investing-Practices-Progress-Challenges.pdf

Brandon, R. G., Krueger P., & Schmidt, P. S. (2021). ESG ratings disagreements and stock returns. *Finance Working Paper No. 651/2020.* European Corporate Governance Institute, August. https://ecgi.global/sites/default/files/working_papers/documents/gibsonkruegerschmidtfinal_1.pdf

Brown, E., & and Sundstrom, G. (2021). Mid-cycle investing: Time to get selective. *Asset Allocation Outlook.* PIMCO, July. https://global.pimco.com/en-gbl/insights/economic-and-market-commentary/global-markets/asset-allocation-outlook/mid-cycle-investing-time-to-get-selective

Busch, T., Johnson, M., Pioch, T., and Kopp, M. (2018). *Consistency of Corporate Carbon Emissions Data.* University of Hamburg/WWF. https://ec.europa.eu/jrc/sites/default/files/paper_timo_busch.pdf

BusinessGreen Staff. (2020, November 10). CDP: Almost 10,000 companies worldwide disclosed environmental data in 2020. https://www.businessgreen.com/news/4023045/cdp-companies-worldwide-disclosed-environmental-2020

Carlson, D. (2021, July 19). Mentions of 'ESG' and sustainability are being made on thousands of corporate earnings calls. *Marketwatch.* https://www.marketwatch.com/story/mentions-of-esg-and-sustainability-are-being-made-on-thousands-of-corporate-earnings-calls-11626712848

CDP. (2009). OECD presentation 1 July 2010 by Paul Simpson, Chief Operating Officer. Carbon Disclosure Project. https://www.oecd.org/daf/inv/mne/45634581.pdf

CDP. (2021, October 14). CDP reports record number of disclosures and unveils new strategy to help further tackle climate and ecological emergency. Carbon Disclosure Project. https://www.cdp.net/en/articles/media/cdp-reports-record-number-of-disclosures-and-unveils-new-strategy-to-help-further-tackle-climate-and-ecological-emergency

Christensen, M. D., Serafeim, G., & Sikochi, A. (2022). Why is corporate virtue in the eye of the beholder? The case of ESG ratings. *The Accounting Review*, 97(1), 147–175. DOI: 10.2308/TAR-2019-0506

Collins-Dean, W., & Geffroy, E. (2021). Do ESG ratings get high marks? Dimensional Research. https://www.dimensional.com/se-en/insights/do-esg-ratings-get-high-marks

Damodaran, A. (2020, September 3). Sounding good or doing good? A skeptical look at ESG. *Musings on Markets.* https://aswathdamodaran.blogspot.com/2020/09/sounding-good-or-doing-good-skeptical.html

Damodaran, A. (2021, September 14). The ESG movement: The "goodness" gravy train rolls on! *Musing on Markets.* https://aswathdamodaran.blogspot.com/2021/09/the-esg-movement-goodness-gravy-train.html

Deese, B., Hildebrand, P., Kushel, R., Novick, B., & Wiseman, M. (2019). Sustainability: The future of investing. Global Insights. *BlackRock Investment Institute.* https://www.blackrock.com/us/individual/literature/whitepaper/bii-sustainability-future-investing-jan-2019.pdf

Dimson, E., Marsh, P., & Staunton, M. (2020a). Divergent ESG ratings. *Journal of Portfolio Management*, November, 75-86.

Dimson, E., Marsh, P., & Staunton, M. (2020b). *Summary Edition: Credit Suisse Global Investment Yearbook 2020*. Credit Suisse, February.

Dimson, E., Marsh, P., & Staunton, M. (2021). *Summary Edition: Credit Suisse Global Investment Yearbook 2021*. Credit Suisse, March

Domini, A., & Kinder, P. D. (1986). *Ethical Investing*. 1st edn. Addison Wesley Publishing.

Drempetic, S., Klein, C., & Zwergel, B. (2020). The influence of firm size on ESG score: Corporate sustainability ratings under review. *Journal of Business Ethics*, 167, 333–360. https://doi.org/10.1007/s10551-019-04164-1

Eaglesham, J. (2022, January 29). Wall Street's green push exposes new conflicts of interest. *The Wall Street Journal*. https://www.wsj.com/articles/wall-streets-green-push-exposes-new-conflicts-of-interest-11643452202

Fama, E., & French, K. (1988). Permanent and temporary components of stock prices. *Journal of Political Economy*, 96(1988), 246–273.

Fink, L. (2022). The power of capitalism. *2022 Letter to CEOs*. BlackRock. https://www.blackrock.com/corporate/investor-relations/larry-fink-ceo-letter

Friede, G., Busch, T., & Bassen, A. (2015). ESG and financial performance: Aggregated evidence from more than 2,000 empirical studies. *Journal of Sustainable Finance and Investment*, 5(4), 210–233. https://doi.org/10.1080/20430795.2015.1118917

GSIA. (2021). *Global Sustainable Investment Review 2020*. Global Sustainable Investment Alliance. http://www.gsi-alliance.org/wp-content/uploads/2021/08/GSIR-20201.pdf

Hall, J. P. (1986). Ethics in investment: Divestment. *Financial Analysts Journal*, 42(4), 7–30.

Hildebrand, P., Polk, C., Deese, B., & Bolvin, J. (2020). Sustainability: The tectonic shift transforming investing. *BlackRock Investment Institute*. https://www.blackrock.com/us/individual/insights/blackrock-investment-institute/sustainability-in-portfolio-construction

Hong, H., & Kacperczyk, M. (2009). The price of sin: The effects of social norms on markets. *Journal of Financial Economics*, 93(1), 15–36. https://doi.org/10.1016/j.jfineco.2008.09.001

ICI. (2021). 2021 investment company factbook: A review of trends and activities in the investment company industry. Investment Company Institute. https://www.ici.org/system/files/2021-05/2021_factbook.pdf

Johnson, B., & DiBenedetto, G. (2021). *2020 US Fund Fee Study: Fees Keep Falling*. Morningstar Manager Research, August. https://www.morningstar.com/lp/annual-us-fund-fee-study

Kell, G. (2018, July 11). The remarkable rise of ESG. *Forbes*. https://www.forbes.com/sites/georgkell/2018/07/11/the-remarkable-rise-of-esg/?sh=228ab1091695

Kjellberg, S., Pradhan, T., & Kuh, T. (2020). *An Evolution in ESG Indexing*. iShares Blackrock. https://www.chinaesg-pa2f.com/upload/file/20210313/20210313211157575757.pdf

Kosantonis, S., & Serafeim, G. (2019). Four things no one will tell you about ESG data. *Journal of Applied Corporate Finance*, 31(2), 50–58.

LaBella, M. J., Sullivan, L., Russell, J., & Novikov, D. (2019). *The Devil is in the Details: The Divergence in ESG Data and Implications for Sustainable Investing*. QS Investors, September. https://qsinvestorsproduction.blob.core.windows.net/media/Default/PDF/The%20Devil%20is%20in%20the%20Details_Divergence%20in%20ESG%20Data.pdf

Margolis, J. D., Elfenbein, H. A., & Walsh, J. P. (2009). Does it pay to be good... and does it matter? A meta-analysis of the relationship between corporate social and financial performance. https://papers.ssrn.com/sol3/papers.cfm?abstract_id=1866371

MSCI. (2022). ESG 101: What is environmental, social, and governance? Explore the basic fundamentals of ESG investing. https://www.msci.com/esg-101-what-is-esg

Murugaboopathy, P., & Mann, A. (2021, October 29). Global sustainable fund assets hit record $3.9 trillion in Q3, says Morningstar. Reuters. https://www.reuters.com/business/sustainable-business/global-sustainable-fund-assets-hit-record-39-trillion-q3-says-morningstar-2021-10-29/

Poterba, J. M., & Summers, L. H. (1988). Mean reversion in stock prices: Evidence and implications. *Journal of Financial Economics*, 22(1), 27–59. https://doi.org/10.1016/0304-405X(88)90021-9

Schneller, W. (2021, August 12). *Beyond the Label, ESG Funds May Miss Their Mark*. Dimensional Research. https://www.dimensional.com/us-en/insights/beyond-the-label-esg-funds-may-miss-their-mark

Securities and Exchange Commission. (2022). *Staff Report on Nationally Recognized Statistical Rating Organizations*. Office of Credit ratings, January.

Sorkin, A. R., Karaian, J., Kessler, S., Gandel, S., de la Merced, M. J., Hirsch, L., & Livni, E. (2022, January 18). Larry Fink defends shareholder capitalism. *The New York Times*. https://www.nytimes.com/2022/01/18/business/dealbook/fink-blackrock-woke.html

Stankiewicz, A. (2022). Sustainable fund flows dip for the fourth quarter but peak for the year. *Sustainability Matters*. Morningstar. https://www.morningstar.com/articles/1076648/sustainable-fund-flows-dip-for-the-quarter-but-peak-for-the-year

Starks, L. T. (2021). Environmental, social, and governance issues and the Financial Analysts Journal. *Financial Analysts Journal*, 77(4), 5–21. https://doi.org/10.1080/0015198X.2021.1947024

Sundaram, A. K. (2023). Business and climate change. In A. K. Sundaram & R. G. Hansen (eds.), *Handbook of Business and Climate Change*. Edward Elgar Publishing.

Tang, D. Y., Yan, J., & Yao, Y. (2021). The determinants of ESG ratings: rater ownership matters. https://papers.ssrn.com/sol3/papers.cfm?abstract_id=3889395

The Global Compact. (2004). *Who Cares Wins: Connecting Financial Markets to a Changing World*. https://www.unepfi.org/fileadmin/events/2004/stocks/who_cares_wins_global_compact_2004.pdf

Wall Street Journal. (2021, November 18). Where will ESG investing be in five years? *Journal Reports: Wealth Management*. https://www.wsj.com/articles/esg-investing-in-five-years-11637161312

Whelan, T., Atz, U., Van Holt, T., & Clark, C. (2021). *ESG and Financial Performance: Uncovering the Relationship by Aggregating Evidence from 1,000 Plus Studies Published Between 2015 and 2020*. NYU Stern Center for Sustainable Business. https://www.stern.nyu.edu/sites/default/files/assets/documents/NYU-RAM_ESG-Paper_2021%20Rev_0.pdf

PART VI

THE FUTURE

23. Reflections on the future
Arranged and edited by Anant K. Sundaram and Robert G. Hansen

NOTE FROM THE EDITORS

For this final chapter, we asked the *Handbook* contributors to reflect on what the future might bring for business and climate change. Our request to them was the following.

> We would like to pull together a final chapter that would include each of our thoughts on the future of the "climate economy." There is no doubt that you have thought a lot about what the changing climate means for the world of business – contributing to this chapter will be a chance to reflect more on that, and to put your ideas to paper.
>
> Here are some suggestions to get you thinking: What might be one or two developments – these could be technological, market-based, policy-based, something else altogether – that will transform our ability to deal with climate change in the next decade or two? Is there something specific, around the corner, that makes you feel optimistic, or pessimistic? How do you think the world of regulation and corporate practice will play out: Will we reach a global, coordinated policy solution or will we end up with a hodgepodge of individual corporate action, taxes, trading schemes, subsidies, etc.?

What follows are the reflections that we received, edited only very lightly, and in alphabetical order.

EDWIN ANDERSON, ILYA KHAYKIN, ALBAN PYANET AND TIL SCHUERMANN

We've seen the world swing from doing almost nothing on climate to a vanguard of mobilization around governments and leading public companies pledging to drive improvements in risk transparency and global emissions. There is a distinct risk that momentum will swing back the other way as dirty assets head to private hands and many governments fail to make the tough decisions to lower emissions. Yet the increase in awareness of the rising risks from climate change and actions by enough governments together provides pressure on the global economy to address these issues. This pressure motivates investment into new technologies and changes in ways of doing business but is still well short of what is needed to stop a two-degree increase.

As climate change causes more and more disruptions in businesses and economies, banks and insurers – pressured by regulators – weather the storm better than most. Other parts of the financial landscape, using alternative investment mechanisms to gain promised higher returns in the private capital world, run into trouble and create some problems for the financial institutions that lend to them. But the financial fallout is limited, and the premium on projects with lower climate risks and lower emissions rises in a series of abrupt steps as the equity markets, which are often "stupid about risk until they are not," create unpleasant market swings for certain market participants.

The economics of tackling climate change become stronger and stronger, but too late to avoid damage that emotionally resonates with voters in democracies around the world and creates a chaotic outcry of voices demanding action. Non-economic insurance systems, set up by state and municipalities to protect their citizens from paying the higher fair insurance rates that have arisen, begin to crack and fail, demanding bailouts with results varying by the political calculus of local areas. Although an impressive amount of green power is put online, the power transmission infrastructure in most countries evolves far too slowly, forcing the continued use of more carbon-heavy power than might be expected.

Carbon border tariffs eventually start to drag the lagging governments and companies forward, forcing more and more change. Generational change further fuels the move to sustainability in a broad way, and things move towards a greener and better world -- too late to avoid economic damage, loss, and population displacement, but enough to avoid the worst of it. There are clear losers in the disruptions ahead, but there are also winners in companies that moved definitively and for governments that address the risks in their financial systems and infrastructure early.

PAUL ARGENTI, POSIE HOLMES AND MARLOES SMITTENAAR

Based on the observations we gleaned conducting research for our chapter focused on communication, we see three developments that could transform our ability to deal with climate change in the next 15–20 years.

First, organizations will need to position their brands strategically to align their reputations with how people are thinking about climate change. Companies such as Shell and Chevron in the energy space and Unilever and General Mills in consumer package goods will need to rebrand and reposition themselves as interest in climate change continues to grow, particularly among the younger generations. Both Shell and Unilever have tried this already, but they still have one foot in the past and another in the future.

Second, as energy companies try to offload fossil fuels and move to renewables, will the reputational benefit offset the potential financial loss? These companies would like incremental change to continue reaping profits while creating a sustainable future state and that is not something that works with the timeline we need. If you already know you need to do this, why not rip the Band-Aid off now and capture the first mover goodwill? This reputational leap will create a huge advantage for those that are bold enough to try.

And, finally, how can companies deal with the more basic needs of developing nations with huge demands for traditional food and energy while also satisfying the demand of more progressive thinking from highly developed countries? This will present huge communication, recruiting, and brand challenges in the years ahead.

CHRISTA CLAPP

In an optimistic vision of future capital markets, corporations with operations linked to fossil fuel will need to showcase how they are actively managing climate risk to achieve lower costs of financing. The current burden of proof for lower financing costs is on corporations with green activities. Under existing regulatory regimes, such as in the EU, organizations wishing to label green activities have the highest transaction costs. It is more work for a corporation that needs to prove what they are doing is green to raise debt capital than for heavy-emitting industries. While those transaction costs may be balanced out by the benefits of the signaling effects of corporate management alignment on sustainability, it leaves organizations that do not seek a green label with an easy way out.

To create a more even playing field, central banks and national governments could continue to put more pressure on corporate climate risk disclosure, for all economic activities and not just the green ones. Transaction costs would then shift more towards companies with heavy-emitting activities since they face higher climate risk. Companies that are not incorporating climate mitigation and resiliency planning into their financial decisions would face a higher burden of proof. The EU Corporate Sustainability Reporting Directive requires large companies to report against the taxonomy of green and transition activities. The next step could be to also require reporting on activities with high climate risk such as those dependent on fossil fuels. This, coupled with stronger carbon pricing signals via taxes or emission trading schemes, would ultimately shift capital markets in a greener direction.

VINCENT ETCHEBEHERE

Decarbonization is probably the greatest challenge aviation has yet to face in its relatively short history. The necessary ecological transition will alter consumers' attitudes towards flying and transform the aviation supply chain in the short term with new energy sources, and in the longer-term with new aircraft propulsion technology. The aviation decarbonization challenge perfectly illustrates the two driving factors of the climate economy: *energy efficiency* and *sobriété*, i.e., moderation. These two dimensions can be perceived as opposed by techno-optimists on the one hand, and by advocates for extreme 'degrowth' on the other. They are, however, interlinked: as long as efforts for energy efficiency face physical limits, moderation in our production and consumption habits will be required. Conversely, the more *sobriété* we collectively achieve, the greater the access we will have to resources that drive energy efficiency. In aviation, for example, greater feedstock availability could be allocated to sustainable aviation fuels (SAFs). The big question is how should moderation emerge? Can it solely rely on consumers' environmental consciousness, despite wide variations across regions? Can it be framed through global public policies that avoid competitive distortions or carbon leakages?

A climate economy requires a socio-economic transformation based on shared values that respect the environment's physical limits. Paving the way for such an economy requires us to address both social and environmental expectations from society, to identify and deal with physical constraints, and to build a vision of a desirable future upon which we can collectively embark as consumers, as employees, employers, or shareholders, but above all as citizens of our planet.

Looking back at where we were – collectively – 10 years ago, I believe we have come a long way towards becoming aware of the ecological urgency and starting to act upon it. Leading up to my 2006 business school graduation, I had not once heard the words *climate change* or *ecological transition* during my studies at a top Paris school. Now, students in most French engineering or business schools dedicate their first day to climate. Ecological transition has become a compulsory topic of study. Within companies, sustainability has moved from being a reporting-based, standalone, activity to one that is an integral part of corporate strategy. Those developments make me quite optimistic about our collective ability to address the ecological crisis in a meaningful manner. I am distraught nevertheless every time a new IPCC report reminds us that the climate crisis is intensifying, and that we are not acting fast enough. I personally can't be swayed by waves of optimism or pessimism, and instead feel we all need to *act* with determination in our respective fields of work, and in our personal lives.

GIANFRANCO GIANFRATE

Recent evidence sheds light on the magnitude and pervasiveness of climate risks exposure for global banks and financial institutions. The magnitude of climate-change-related risks is substantial and similar to (and likely bigger than) the ones that started the financial crisis of 2008. Besides measuring climate risks and acting on them (adaptation), the global financial system will be crucial in channeling capitals towards *new* green assets (mitigation) especially for clean power generation. By greening their investments portfolios, investors could, therefore, jointly reduce vulnerability to the consequences of climate change and contribute to the reduction of pollutant emissions.

Sustainable finance can mitigate climate change and help adaptation strategies in several ways. In terms of mitigation, finance can contribute to the reduction of GHG emissions by pricing efficiently the social cost of carbon, by reflecting the transition risks in the valuation of financial assets, and by channeling investments in low-carbon technologies. At the same time, there is a growing need to spur greater public and private capital into climate adaptation and resilience. Typically, adaptation activities are location and project-specific. Addressing specific climate vulnerabilities requires pricing and insuring effectively catastrophic risks, supporting local utilities to realize resilient infrastructure, and financing the development of adaptation technologies and climate-resilient agriculture.

On the other hand, the green mandates across the investment chain and investment definitions of some of the asset classes are still too ambiguous and poorly defined. The "greenness" label for bonds is all too often stretched to encompass financing facilities of issuers that are opportunistically misrepresenting the actual environmental footprint of their operations (so-called risk of "greenwashing"). Even in cases where the bonds' proceeds are actually used to finance green projects, investors remain nevertheless exposed to both the green and "brown" assets of the issuers. In general, measurement and reporting of an environmental footprint is, in most cases, partial, unaudited, and mostly left to the discretion of asset managers and issuers. The heterogeneity of metrics and rating methodologies – along inherent conflict of interests between issuers, investors and score/rating providers – results in inconsistent and unreliable quantification of the actual environmental footprint of corporate and sovereign issuers. In order to steer private capital towards issuers and projects that can deliver on the transition towards a low-carbon economy, the actual accountability of issuers, along with transparency and comparability of metrics for investors and rating providers is necessary.

In order to promote financial climate-related disclosures for companies and financial intermediaries, the financial system could play a key role in pricing carbon and in allocating capital toward slower emitting companies. Stable and predictable carbon-pricing regimes would significantly contribute to fostering financial innovation that can help further accelerate the de-carbonization of the global economy even in jurisdictions that are more lenient in implementing climate mitigation actions. While pension funds, insurance companies and other long-term oriented institutional investors are actively pursuing a greater standardization of green mandates and seeking reliable climate reporting metrics, there is still a clear need to unleash financial engineering in order to move from mere disclosure to adequate management of climate risks.

A growing number of financial regulators are intensifying the efforts to enhance the climate-related disclosure of financial actors. Importantly, central banks are considering the possibility of steering or tilting the allocation of their assets to favor the less polluting issuers. This, in turn, would translate into lower cost of capital for cleaner sectors, significantly accelerating the greening of the real economy.

MARYAM GOLNARAGHI

As we look ahead, a number of factors will drive our future socio-economic transformation; for example, the risks and opportunities associated with addressing climate change and large-scale nature-based issues linked to new technologies needed for transitioning;[1] the next wave of information and communication revolution; socio-economic reconstruction post COVID; and rethinking globalization and supply-chain related issues. No one person, no government, no company, no sector can do this alone. Enabling and incentivizing such transformation would need collective action at an unprecedented level and new innovative thinking. For example:

1. Breaking down the traditional disciplinary and institutional silos to enable more system-based thinking and integrated business models by governments and companies in various sectors.
2. Deeper cross-organization and cross-sectoral collaboration to drive out-of-the-box-thinking, for example, involving carbon-intensive industries, technology and engineering companies, the investor community, insurance industry and governments.
3. Scaling up technological development and innovation for decarbonizing key sectors, while considering environmental and nature-related sustainability.
4. Managing new risks associated with the wide range of untested new technologies, new processes, new industries and new infrastructure systems needed for decarbonization. For example, new technologies come with myriad untested risks, particularly when deployed at scale, spanning operational and safety risks, construction and maintenance risks, environmental and disposal risks, litigation and liability risks and weather-related risks, which need to be managed with consideration for the entire life cycle.
5. Leveraging public and private investments, aligning risk/return profiles, to enable more sustained financing of the socio-economic transition.
6. Rethinking of public policies and regulatory frameworks to enable incentives for new market development and to boost the supply of research and innovation for transitioning – as well as greening – of public infrastructure systems.
7. Investing in, building and repurposing human resources.
8. Influencing public behavior and consumption choices.

SUZANNE GREENE

As global corporations contend with the need to decarbonize their operations and supply chains, the reality of who pays for these upgrades and who can register the credit on their "GHG bottom line" becomes a critical element of the discussion. While some decarbonization activities, such as fuel efficiency, may save a company's money, other gains will come from costly upgrades, such as new equipment, energy sources, or carbon capture technologies. Both corporate procurement and sales teams, as well as non-governmental organizations and industry groups, are developing innovative strategies on how to calculate, validate and integrate new carbon reduction financing systems. For example, companies may be keen to pass some decarbonization costs along to their customers in the form of "green premiums" charged for lower carbon products. How to allocate sustainability-related upgrade costs to, for many companies, a complex portfolio of products or services, remains a new frontier that will push sales, marketing, and procurement teams to think creatively and dynamically. In another example, novel schemes such as carbon insets, book and claim systems, and other certificate-based programs are emerging as funding mechanisms outside the sales and procurement process, allowing operators to pass some decarbonization costs along to the customer without raising product prices – instead, placing costs into carbon offset or sustainability budgets.

In order for these systems to become truly effective within a corporate climate strategy, companies need to be able to integrate emissions reductions from low carbon products or certificates into the GHG bottom line, effectively tracking progress towards climate goals while avoiding any double counting. Work is needed to expand existing systems to make innovative supply chain collaboration strategies work within carbon accounting principles and climate targets programs such as the Science-Based Targets initiative. This likely requires new tools or methods for emissions calculations, allocation and sharing, allowing another opportunity to think innovatively and proactively about solutions that can accelerate climate progress. New and creative ideas should be carefully considered and advanced today to make sure decarbonization actions can be taken swiftly and in an economically feasible way.

RISHAD HABIB AND KATHERINE WHITE

As consumer awareness of the impacts of climate change increases and demand for climate-positive products and services grows, marketers will find more ways to incorporate carbon reductions into their offerings. Marketers will likely offer more ways to help consumers identify and calculate the carbon footprints of goods and services (such as offering apps and additional forms of certification), produce more carbon-neutral options, or offer opportunities to offset carbon more efficiently (such as including it in pricing or during checkout). One interesting possibility is that consumers might be offered product options that can absorb or have captured carbon themselves – think jewelry or clothing created through a carbon capture process! In addition, consumers will gravitate towards more climate-friendly ways of consuming such as buying less or second-hand, sharing with others, repairing what they own and reselling, as interest in the circular economy grows. More and more companies will grow in this space and existing organizations will provide options that help make their products more circular, such as a resale market, lifetime repair, or transforming old products into new ones.

Efforts by marketers will likely be complemented by policy changes that encourage climate-friendly consumption, incentivize making homes and transportation climate-friendly, and discourage damaging behaviors. As more people move to dense cities, there will be greater investments in public transport, bike lanes, and walkable urban cores. As the consequences of climate change become more concrete and affect countries around the world, support for such policies will increase and climate change will be at the forefront of electoral agendas. In addition, technological innovations will drive the move towards greener solutions including carbon capture technologies, solar home systems, electric vehicles, plant-based meats, home composting, etc. The success of early-mover companies in these spaces will encourage others to join in.

ROBERT G. HANSEN

With regard to the future of business and climate change, I admit to being of two minds, one pessimistic and one optimistic.

Starting with the pessimistic side, I lament the abandonment of core environmental economics principles in the carbon abatement policy arena. In the US, municipalities and states commonly mandate standards for the share of electricity from renewable sources, with little cost/benefit analysis or consideration of alternative policies. Flexibility in letting emitters achieve emissions reductions – a lesson learned in earlier environmental episodes – has been replaced by inflexible mandates for renewable energy no matter the cost. The world's companies all have their own carbon mitigation programs, with little if any consideration for whether they are low-cost mitigators. Even companies and industries with a marginal cost of reducing carbon far exceeding reasonable estimates of the damages are pushed to reduce, nonetheless. Relatively clean and safe natural gas is labeled as unsustainable and subjected to myriad regulatory burdens. Nuclear is often off the table. Cap-and-trade systems, or equivalent tax/pricing regimes, are nascent at best, despite having been effective in reducing sulfur dioxide from coal plants, phasing out lead in gasoline and in reducing CFCs.

One finds little consideration for the impact of higher energy prices on heating, cooling, transportation, communication, cooking and indeed all the essential activities of daily life, especially for disadvantaged populations and developing economies.

Yet there is room for optimism. Market economies are incredibly adaptable and can and will find ways to deal with regulatory inefficiencies in, yes, an efficient fashion. There will be costs that could be avoided with wiser regulations but those costs will be minimized. Wind and solar technologies have advanced rapidly, coming way down in cost, and deeper penetration into the electrical grid will happen with developments in storage. Nuclear is being reconsidered in some parts of the world, and use of relatively-clean and cheap natural gas continues to grow. Hydrogen and fusion are promising. All these developments are being supported through investments from governments and even the most profit-oriented private equity firms, while graduates of top MBA programs seek careers in businesses that are core to the energy transition. Capital, both human and financial, is flowing into the energy transition. The potential for much of our energy being produced at close to zero marginal cost and therefore to be sold at a negligible marginal price, while posing a dilemma for our current pricing systems, brings the potential for tremendous expansion of energy consumption and related economic growth.

While the next 20 years or so is likely to be a difficult and costly transition, our overall future remains bright.

DAVID HONE

A recent paper[2] I was fortunate to co-author with a Shell colleague and two researchers from the MIT Joint Program, analyzed the notion that the world is now heading towards a (net) zero emissions outcome irrespective of the pathway the energy system takes; so the real question had become when, not if.

We concluded that the climate issue is now a bounded problem rather than the wicked unbounded problem that existed a decade ago. The analysis points to an outcome somewhere between 1.5°C and 2.8°C depending on the pace of change. This represented a substantial change in the framing of the climate issue, even compared with the IPCC Fifth Assessment Report where the authors contrasted impacts in a 2° and a 4°C world over dozens of pages, charts and tables. We would argue that the 4°C world isn't even a possibility now. While 2.8°C of warming isn't a cheerful story from an adaptation perspective, the reframing that has become apparent in just the last few years is a reason to be optimistic. As change accelerates I am confident that these boundaries will continue to narrow.

ANDREW INKPEN

As the saying goes 'prediction is very difficult, especially about the future', so take this with the usual caveats. While there are many skeptics who believe mankind in a warming world is doomed, I take a different view. I am optimistic about the future and mankind's ability to innovate and adapt in order to manage climate change. Climate change will bring unanticipated consequences and these will not all be negative (unfortunately, some of the anticipated negative consequences will happen and hopefully they will jolt the deniers out of their complacency). Transformational technologies such as e-vehicles, batteries for homes and businesses, modular nuclear reactors, and scalable bio-fuels will change the way energy is produced, stored, and used. These technologies will complement the remarkable progress that is already happening with wind and solar energy.

New technologies will be driven by entrepreneurs and market forces, just as the innovative technologies of the past have been created. While our research into the strategies of oil and gas firms raises questions about energy firms and their commitment to energy transition, I am confident that industry leaders will emerge in various sectors to play a key role in the transition. Some oil and gas firms will play an important role in the transition and others will fall behind. Over more than a century the oil and gas majors have developed world-leading skills in project management, construction, engineering, strategic alliance management, and capital discipline. These skills will be central to developing scalable business models that support the energy transition.

As the energy transition evolves, markets and investors will support the leaders in environmental innovation – look at the valuations of e-vehicle companies versus those of the legacy internal combustion engine auto companies. Investors and consumers want the new technologies and governments can play a role in nudging and motivating change. However, waiting for governments to make commitments that will change the course of climate change will be a long and ultimately fruitless wait.

THOMAS C. LAWTON

Building on Lawton and Kock's chapter on corporate strategy and climate change, we advance some final thoughts on how business strategy around climate change, and environmental issues more generally, might look in the future. In doing so, the intent is to contribute to strategic foresight, and not to attempt strategic forecasting.

In reflecting on the future of the climate economy, Lawton and Kock emphasize the preponderance of heterogeneous corporate responses, which is also evident in other chapters. While homogeneity is unlikely, and perhaps even undesirable, we ponder if there is at least more convergence likely on the horizon. The most plausible answer is, "it depends". It depends, first, on size: smaller companies will need more time and support to adapt their strategies and structures to the requirements, and opportunities, of the climate economy. And, second, on location, as we see a potentially even greater divide emerge between firms based in advanced economies – where climate action oriented public policies and associated regulations are accelerating – and those coming from emerging economies – where state and societal demands may prioritize economic growth and prosperity over climate action. This divide is further aggravated by the economic legacy of COVID-19, which early indicators suggest has had a much greater cumulative negative impact on emerging economies. Of course, when we speak of emerging economies, it is important to delineate between at least three categories: first, those countries that constitute the most advanced developing economies, such as Mexico, Malaysia, and Thailand; second, those that are less advanced but display strong upward trajectories on most growth indices, such as Colombia, Morocco, and the Philippines; and third, lesser developed and higher risk, but often fast growing, new frontier economies such as Nigeria, Kenya, and Vietnam. Preliminary data and insights suggest that we are likely to also see a tiered or staggered approach to climate action from emerging economy-based firms, particularly those headquartered in lesser developed economies with nascent public discourse and policy on climate action, and arbitrary or lax enforcement of extant environmental regulation. This is without mentioning those, often populous, countries that remain far off any classification of developing, and are closer to being defined as failed or failing states, such as the Democratic Republic of Congo, Libya, Iraq, or Sudan.

In highlighting the fragmented nature of the global economy, and a future likely to be dominated by the so-called Global South, we emphasize the difficulty, if not impossibility, of achieving any form of corporate convergence on climate action and sustainable strategies. Nevertheless, glimmers of hope are evident, as China begins to address air pollution and invest in renewable energy, and as emerging market champions such as CEMEX and Gruma in Mexico, Natura in Brazil, Sintesa Group in Indonesia, and Godrej Group and Infosys from India, outpace governments on sustainability initiatives and lead the way in setting and delivering on social and environmental goals.

FRANK O'BRIEN-BERNINI

Whenever envisioning the future, and certainly when imagining a carbon-free economy, the one thing that we know for sure is that we will be wrong. However, that fact cannot stop us from developing and testing scenarios, and then doing our very best to advocate and drive action toward the scenarios we want (or need) to bet on.

We know that what we bet on needs to be big... and it needs to be fast. That is, it needs to be rapidly scalable, commensurate with both the magnitude and the urgency of the challenge.

So, here are the three bets I'm placing with my thoughts and with my actions:

1. Mechanisms of corporate standardization and accountability, such as SBTi, and commitments to decarbonize, or achieve net-zero, will drive material progress enabling massive portions of our economy to operate at sustainable carbon emissions levels – at least 90 percent lower than today.
2. The buzz around the concept of the Circular Economy will translate into real action as more companies realize that it will be impossible to reach their decarbonization goals without seriously decarbonizing their raw materials (i.e., their Scope 3 emissions), and one of the most powerful tools available to them is (carbon-smart) circularity.
3. Getting the right data into the right people's hands, at the right time, to enable the right decisions, is an evergreen backbone of business. With that in mind, my final bet is on actionable transparency of embodied carbon, of everything: raw materials, products, applications, entire assembled goods (e.g., buildings, cars, consumer goods). I have seen evidence, and I have great faith, that if people are sufficiently informed they will make planet-sustaining choices. This demand–creation feedback loop will drive further and faster decarbonization progress throughout our global supply chains.

BRUCE A. PHILLIPS AND SCOTT G. FISHER

As of early 2022, there are several reasons not to be optimistic about the prospects for achieving today's ambitious net-zero carbon emissions goals. The 26th United Nations Climate Change Conference held in November 2021 did not meet all the expectations of many climate advocates. Federal policy in the US is advancing more slowly than hoped with most conservatives and a few Democrats opposing the most progressive initiatives. And stepping back from recent events, the 50 to 100 years or more required to complete the historic energy transitions from wood to coal and then from coal to oil and gas mean the goal of net-zero global emissions within 30 years was always a daunting challenge, even under the best of circumstances. Together, these suggest the US and the rest of the world will, in the coming years, lag behind today's ambitious decarbonization targets and miss the mid-century decarbonization goal.

Yet, in other respects, the world has never been in such a promising position to eventually decarbonize the global energy system. Corporate support for ambitious climate goals and investment funding is growing rapidly. Solar and wind technologies continue to achieve important technical and cost milestones. Meaningful advancements are also being made with other technology families, including zero-carbon natural gas fired generation, zero-carbon liquid fuels, advanced nuclear and carbon capture. While not all these will be successful on a widespread basis, breakthroughs with one or more of them in the next 5 to 10 years could spur more rapid deployment even with inconsistent government policy and continued public reluctance to pay a large premium for clean electricity. As the competitiveness of clean energy technologies continues to improve, the constraints to deployment will shift to social considerations and a key question will be whether local communities hosting new clean energy infrastructure will see that as a threat to their sense of place and community or as an opportunity for economic development.

HEIKE REICHELT AND SCOTT M. CANTOR

The findings of the Sixth Assessment Report of the Intergovernmental panel on Climate Change in August 2021 served to drive home the urgency to meaningful progress on climate action – once again. Human influence has warmed the climate at a rate that is unprecedented in at least the last 200 years. The result of the political process at the UNFCCC COP26 in Glasgow, Scotland, at the end of this same year drew global attention to the opportunities and gaps to achieve our collective goals towards a world that aims to limit global warming to 1.5°C. It also highlighted the recognition that even a 1.5° warmer world is one that is dramatically altered, in which pressures will fall on the most vulnerable in society, requiring renewed focus on adaptation investments as part of climate action. While motivating individuals and the systems they lead to transform to fully incorporate climate action remains the key challenge, a necessary condition will be better data on how to assess climate risks and opportunities – as well as the experience and expertise to interpret these data properly. Only with better, harmonized data and the corresponding investments in analysis and interpretation will the financial system continue to remain at the forefront of the transition.

Armed with better data and analysis, financial decision makers will naturally incorporate long-term sustainability into decision making. For example, how well-suited is a company, or an investment, to thrive in a world where climate change mitigation and adaptation are integrated into all aspects of its business? These questions of sustainability will not just be asked of companies in low-carbon or climate resilient sectors such as renewable energy, battery storage or sustainable transport, but across all sectors and all investments. The financial system should approach the question of sustainability holistically – what are the positive and negative impacts of an investment, not only in terms of climate, but impacts to ecosystems, biodiversity, water, and other key systems? We are encouraged by the emergence of this holistic mindset – as encapsulated by the 'green taxonomies' with 'Do No Significant Harm' regulations – which has roots in the safeguarding policies of Multilateral Development Banks, which have, for decades, incorporated environmental and social issues into the design and implementation of all of their investments in developing countries.

This integration of sustainability and climate into financial markets takes two forms, both of which are underpinned by data and transparency. First, a focus on recognizing, disclosing, and working towards mitigating climate risks helps steer private capital to necessary adaptation and resilience investments. A strategy that identifies and avoids risks helps capital shift away from carbon-intensive industries and processes as well as reward resiliency. Second, investing in opportunities for positive climate impact makes the low-carbon transition possible because financial markets, again through increased transparency and the availability of more data, will allow for the application of a holistic approach to sustainable investing to see the opportunities that the new climate economy will generate, and integrate ESG costs in financial models.

In fixed income markets we have seen these trends of data and transparency accelerate at a rapid pace. While there are still tremendous gaps to overcome before the bond markets can transition to channel all finance for sustainable activities, they have demonstrated what is achievable in financial markets if trends continue. The first labeled green bond issued by the World Bank in 2008 catalyzed this transparency revolution – giving investors for the first time a window into the use of bond proceeds and expected impacts for equivalent amounts raised. As a result, the expectations of bond investors have been forever changed, unlocking the mindset that they can receive information on the positive and negative impacts of activities financed by issuers they support with their investments. This transformation is seen with the expansion of the green, social, and sustainable bonds, now topping $2.5 trillion in issuance, demonstrating that investors are demanding, ultimately, more information about how their money is supporting a sustainable future. But it goes further than labeled bonds. To achieve sustainable financial markets, specifically in the fixed income space, the end goal is not that all bonds are labeled but that all bonds share the same transparency and reporting features that are the truly innovative pieces.

As we look forward, the next step towards increased data and transparency around climate risks and opportunities is their standardization across financial markets and including ways to consider external costs in financial models. This is happening in real time across a variety of platforms, including through national regulation and via international disclosure and standard setting. It is the authors' view that the sheer demand from investors for transparency, data and better analysis will combine with these policy and harmonization tools to give them the tools they need to seek out the long-term value and steer away from risks. The future of sustainable finance should be that it becomes just 'finance'.

ANANT K. SUNDARAM

An economically- and socially-profound climate and energy transition is underway. Although a few loud – and exaggerated – pockets of negativism will persist, they will not matter a great deal in the final analysis, as this juggernaut will not be easily stopped. Yet, vexing questions about how fast, how expensively, and with what distributional impacts to constituencies in disparate socio-economic groups and diverse levels of economic development will arise soon enough. Addressing these questions may be neither easy nor uneventful.

In the process, the transition will not happen in a straight line, but rather, in a saw-toothed fashion (albeit with an upward slope). While I firmly believe that the goal of net-zero will be achieved, it will be achieved with delay, perhaps around three-quarters of the way through this century rather than by 2050. We will thus have a planet that is warmer by something closer to 2°C than 1.5°C. However, I have little doubt that as-yet unseen, unthought-of, unforeseeable technological innovations will surely arise in the next half-century to help cope with the consequences of the incremental 0.5°C.[3]

Most of the current policy discussions focus on what needs to be done now to avoid future damages from climate change, surely an issue of massive importance: this is, of course, the key social *benefit* of addressing climate change. But allow me to push a bit on the flip side of the coin, the social *costs*. Unduly speeding up the transition may be the approach we take in the next decade. Whatever form that approach takes, I worry that it will require policy interventions which ultimately raise the price of incumbent energy in some form or other, energy that is currently both affordable and reliable (if neither clean nor safe). Three issues will arise: (1) there will be an inevitable increase in energy poverty and energy insecurity that, in turn, invites social dislocations and civil strife; (2) hostilities between poorer countries at the bottom of the energy ladder and high-income countries will increase, arising from the fact that the latter achieved their economic success in significant part because of the low-cost 'carbon party' they were able to throw for a couple of centuries, thereby constraining the energy-choices that poorer countries now have to make; and (3) the costs, associated with spending too much too soon on solutions that subsequent innovations render suboptimal, as is inevitable in rapidly-changing technology settings, could be non-trivial. These social costs, their impacts, and how to forestall them – as well as how to trade-off the costs against benefits – need to be dealt with more urgently and more universally than seems to be happening currently.

I see a transition in which far-reaching nuclear technologies (e.g., fourth generation and beyond; small modular reactors; thorium-based; and who knows, even fusion, someday), storage technologies, carbon capture & reuse technologies (especially nature-based solutions and air capture) will play a massive role. Far more will happen in the realm of energy- and carbon-efficiency innovations than we can dream of today. This transition will be led by businesses, the constituency that will play a *central* role, since they will have to be the ones performing the R&D, deploying the talent and the technologies, and financing its development and rollout. But if we don't balance the expected benefits from solving the problem of climate change with the costs mentioned above, it will be a bumpy road ahead, indeed.

Perhaps therein lies the mega-opportunity for businesses to fuse profit with purpose. And for policymakers (and climate elites) to balance the need for speed and spending on climate solutions with what that entails for intra- and inter-country climate justice.

NOTES

1. Such as biodiversity loss, extensive degradation of natural ecosystems and large-scale air, water and soil pollution.
2. Morris, Hone, Haigh, Sokolov and Paltsev, "Future energy: in search of a scenario reflecting current and future pressures and trends", *Environmental Economics and Policy Studies*, February 2022.
3. To put things into perspective, think about all that is normal and widespread as innovations that exist today, which enhance (and inform) how we live, work and play, compared with, say, what we dreamed possible in 1972.

Index

Abernathy, W. J. 262
absolute carbon emissions 328
Acharya, V. V. 81
adaptation 304, 305, 312, 314
Advanced Research Projects Agency – Energy (ARPA-E) program 124
Aerts, W. 488
aggregate corporate emissions 22
agricultural sector 181
agriculture 360
Airbus A220 199
aircraft lift 192
Air France 3, 15, 187, 189, 191, 195, 197, 200, 204,
Air France-KLM Group 196
Aldy, J. E. 398
Allen, David B. 5, 251
Alphabet 21
Amazon 21
Ambec, S. 480
ambition 350
ambitious policies 127
American Clean Energy and Security Act 97, 128
American Economic Association 82
American Society of Testing and Materials (ASTM) 198
American Wind Energy Association 103
Amoco 209
An, H. T. 490
Andersen, T. T. 217
Anderson, Edwin 526
Anderson, K. 487
anthropogenic activities 180
Apple 16, 17, 18, 22, 232
    Apple's carbon strategy 16
Aragón-Correa, J. A. 258
Aravind, D. 260
ARCO 209
Arctic National Refuge 234
Arctic sea ice area 11
Argenti, Paul 3, 527
ARPA-E program 127, 137
Arrhenius, S. 8, 38
Artiach, T. C. 480
Asian Development Bank (ADB) 314
Asness, C. 509
Assessment Report Working Group 42
asset-backed securities (ABS) 469
asset-liability management (ALM) 150

assets 150
assets under management (AUM) 508
asset value 75
Atan, R. 488
atmospheric warming 31
A Tree for You (ATFY) 200
Audit Committee Charter 275
Auffhammer 68, 383
Australia 38
Australian Securities Exchange (ASX) 328
aviation 3
    employees 195
    sector
        actors' transformation 204
        emissions evolution 189

Bach, D. 251
Bae, S. C. 497
Banerjee, S. 489
Bank of America 234
Bank of England 79, 81
bank risk management 61
banks 58, 193
    bank assets 66
Barnett, M. 70
Barney, J. 252
Baron, D. 251
Basel Committee for Banking Supervision (BCBS) 59
Bassen, A. 486
Bateman, Alexis 4
batteries 135
    commercialization 95
Bay Area Council 312
BC Hydro 416
beach protection 308
BECCS 51, 52
benchmarking 326, 327, 333, 334, 335, 336
beneficial electrification 125
Benlemlih, M. 480
Bento, N. 398, 399
Berg, F. 514
Berman, S. L. 484, 485
Berner, R. 81
Bernstein, A. 66
Biden Administration 232
Bienert, S. 490
Bilio, M. 514
biofuels 198, 359

539

BlackRock 239, 316, 508, 510, 516
Boffo, R. 514, 522
Bolton, P. 28, 66, 79, 80
bond investors 448
bond markets 430, 439, 440, 446, 450
Bradt, J. T. 66
Braidwood Dispatch and Mining Journal 38
Brandon, R. G. 514
Brisebois, Anouk 5
Brock, W. 70
Brooks, C. 492
"brown" technologies 64
Brundtland Commission Report 273
Build Back Better Act 128
build barriers 261
building buy-in 289
business 30
    and climate change 1, 4, 6, 532
    -as-usual (BAU) emissions 18
    challenges 115, 121, 136
    complexities 89, 403
    continuity 314
    continuity plans 301, 302
    implications 133
    opportunities 115
    performance 74
    resilience 294
    risk 296
businesses 114, 300
    challenges for 115
    commercialization challenge for 136
    thematic challenges for 110
Buysse, K. 259, 260

Cai, Y. 70
Cajias, M. 490
Caldecott, B. 67
California Low Carbon Fuel Standard (CA-LCFS) 202
California Public Utilities Commission 101
Cantor, Scott M. 5, 536
capacity markets 91
'cap-and-trade' system 11, 105, 532
Capasso, G. 482
CapEx 75
Capital Asset Pricing Model (CAPM) 486, 499
capital
    costs 115, 116
    expenditure 76, 274
    -intensive technologies
        investment risks for 113
    markets 450, 451
Capron, L. 263
carbon
    allowances 337

beta 482
border 527
budget 38, 41, 42, 46, 52, 54, 55
capture and sequestration (CCS) 110, 131
capture and storage (CCS) 54, 73, 132
    and alternative use technologies (CCSU) 13
capture utilization and storage (CCUS) 200
economy 58
efficiency technologies 15
emissions 4, 123, 138, 287, 326, 351
    data, methodologies 345
    reductions 334, 335
footprint 58, 238, 344
-free generation 140
-free technology 388
information disclosure (CID) 487
insets 352
-intensive businesses 162
-intensive sectors 178
leakage 111
offsets 199, 352, 419
    Carbon Offsetting and Reduction Scheme for International Aviation (CORSIA) 192
party 538
premium 487
pricing 2, 5, 104, 396, 528
    downstream industries 384
    economic theory 381
    energy expenses 389
reduction
    commitments 342
    strategies 27
related damages 2
sinks 139, 199, 225
tax 104, 337, 380, 386, 387, 395, 396, 397, 398, 399, 400, 401, 402, 458
carbon dioxide (CO_2) 6, 8, 31, 38, 41, 43, 47, 51, 52, 136
carbon disclosure 334, 338
Carbon Disclosure Project (CDP) 14, 23, 33, 284, 305, 329, 330, 349, 350, 380, 390, 398, 513
Carroll, A. B. 256
cash flow 25, 26
Catastrophic (Cat) bonds 170
cause-and-effect association 8
Center for Climate and Energy Solutions (C2ES) 295, 310
Cevik, S. 67
Chambers, B. 289
Chang, K. 497
Chapple, L. 327, 328, 329
Chatain, O. 263

Chatterji, A. K. 485
Chauhan, Y. 488, 494
Chen, X. 497
Cheung, Y.-L. 495
Chevron 218, 221, 261, 527
Chiaro, P. 314
China 11, 12
Cho, C. 260
Choi, T. H. 490
Christa, C. 527
Christensen, C. 219, 223
Christensen, M. D. 514
Christmann, P. 258, 260
CICERO (Center for International Climate Research) 442
circular economy 290, 535
Citigroup 328
Clapp, Christa 5
Clarkson, Peter M. 4, 326, 327, 331, 332, 333, 333, 334, 336, 480, 490, 493
Clean Cargo 348
Clean Development Mechanism (CDM) 434
clean electricity 51, 54
Clean Energy Standard (CES) 102, 123
Clean Power Plan 128
'cliff effect' 162
climate
    communication 247
    economy 8, 13, 30, 526, 528, 534, 536
    finance 5, 433, 529, 537
        capital markets 439
        first phase 433
        second Phase 434
    footprint 481
    -friendly behavior 405
        climate-friendlier companies 27
    goals 350
        achievement of 110
    impact 295, 314, 473
        reduction strategies 350
    Minsky moment 79, 80
    Physical Risks (CPR) 482
    preparedness 294
    regulation 24
    resilience 3, 311
    risk assessment 152, 153, 154, 166, 174, 176, 185
    scenario analysis 62, 71, 72, 84
    science 8, 154
    stress testing 78, 80
    targets 195
    technologies 154, 173, 182
    transition 192
        Climate Transition Risks (CTR) 482
climate adaptation 461

Climate Bonds Initiative (CBI) 462
    Climate Bonds Initiative report 463
climate challenge 204
climate change 8, 25, 30, 58, 59, 60, 61, 63, 64, 65, 67, 68, 69, 70, 71, 72, 73, 74, 77, 79, 80, 81, 82, 83, 121, 122, 128, 141
    aviation sector's impact on 187
    communication strategies 231
    impact of 73
        on aviation 192
    communications 236
        objective of 233
        strategy 237
    dynamics 60
    litigation 155
    risks 152, 153, 155, 157, 158, 162, 177
coal 76
    consumption 38
    -fired
        electric capacity 138
        generation 142
        power 94, 224
Coalition for Climate Resilient Investment (CCRI) 312
coal mining companies 77
cobalt 46
Cochrane, P. L. 256
COD-CSP relationship 498
COD-DSP relationship 494
co-equity investments 16
cognition 405, 408, 412, 413, 419, 422
cognitive factors 408
Cohen, J. 302
cold extremes 11
Coleman Jr, H. J. 254
collaboration 352, 367
collateral impacts 10
collective action 265, 407
Colombia 18
combined cycle technology 95
combustion 38
commercial debt 66
commercial insurance 148
commercial technologies 133
    deploying 120
commodity prices 90
Common But Differentiated Responsibilities and Respective Capabilities (CBDR) 433
communication strategies 247
Community Independent Transaction Log (CITL) 331
company management profiles 368
compounded annual growth rate (CAGR) 521
"Conference of Parties" (COP) 31
constituencies 242

employees and 242
constituents' messages 243
consumer acceptance 129
consumer behavior 405, 408, 421
consumers 194
Consumer Technology Association 314
continuity planning 301, 305, 309
COP16 435
COP21 192, 436
COP26 145, 154, 438
core competencies 141
Cormier, D. 488, 494
CorpComm 231, 233, 236, 237
corporate
    carbon emissions 344
    climate strategy 3
    Communication (CorpComm) 231
    communications 232
    competitiveness 251
    customers 194
    disclosure 25
    emissions 18, 19, 20
        management 8
        measurement and reporting 8
    environmental responsibility (CER) 489
    environmental strategy 253
    financial 515
        performance (CFP) 480
    handprint 271, 278, 282
    resilience 312
    social performance (CSP) 480, 485
    strategy 251
corporate social responsibility 167, 231, 252, 480, 483, 484, 485, 486, 487, 488, 489, 490, 491, 492, 493, 494, 495, 496, 497, 498, 499
Corporate Sustainability Reporting Directive (CSRD) 475, 520
corporate value 25
corporations 237, 238, 240, 241, 243, 247
cost-benefit analyses 316
cost declines 95
cost of capital (COC) 5, 6, 25, 26, 28, 33, 134, 471, 472, 474, 480–502, 504, 530
    estimation of 481
    multi-factor model 482
cost of debt 481
cost of equity 481
COVID-19 43, 47, 82, 190, 197, 432, 437
Crane, C. 233
credibility 394
credit risk 64, 73, 74
CRO Forum 157
crop insurance 170
cross-border mechanisms 112

CSR *see* corporate social responsibility

Damodaran, A. 33, 517, 518, 521
data
    accuracy 484
    challenges 78
    coding 212
    extraction 212
    quality issues 20
    reliability 484
decarbonization 3, 133, 136, 141, 172, 187, 189, 190, 196, 197, 198, 199, 201, 202, 203, 204, 205, 282, 285, 290, 340, 354, 431, 528, 530, 531, 535
    pathway 132
    transition 145
Dell 21
Delmas, M. A. 257
deployment
    scaling 129
Deployment
    policies 110
    rates 54
Dhaliwal, D. 489
DiBenedetto, G. 512
DICE modeling framework 70
Dimson, E. 513, 514
direct air capture (DAC) 200
direct air capture of carbon dioxide (DACCS) 51
direct emissions 19
direct investments 16
direct subsidies 107
disaster insurance 172
disclosure landscape 23, 24
disclosure programs 349
disorderly transition scenarios 76
distributed energy resources (DERs) 96, 135, 136, 141
diversified electric companies 140
Dodds, R. 253, 261
domestic insurance market 172
DONG (Denmark) 217
Doss-Gollin, J. 302
double-materiality 520
Dow Jones Sustainability World Index 285
Downar, B. 335
downstream price 397
"dragon eggs" 317
Drempetic, S. 510
Drucker, P. 32
Du, X. 483, 492

Eaglesham, J. 515
EBITA (earnings before interest, taxes, depreciation, and amortization) 217

ecological droughts 11
economic development 140
economic inefficiency 397
economic variables 70
economy-wide decarbonization 130
Eddie, I. 487, 493
Edison Electric Institute (EEI) 111
efficiency 379, 382, 385, 388, 389, 390, 392, 393, 394, 395, 396, 400, 401
efficient carbon price 383
e-fuels 198, 202
"eGRID" initiative 21
electric
    businesses 133
    energy 120
    generation 125, 132
    grid 91
    industry 2, 120, 121, 122, 123, 141
    generation 42, 93
    loads 130
    reliability 141
    sector v, 89, 115, 121, 141
        carbon emissions 93, 123
    system reliability 121
    utilities 142
    vehicles (EV) 59
Electric Reliability Council of Texas (ERCOT) 92
electricity 49, 89, 91, 97, 101, 117, 118, 119, 217
electronic waste (e-waste) recycling initiative 263
El Ghoul, S. 487, 489
El Niño phenomena 39
Elving, W. J. 237
Embong, Z. 488
emission reductions 123
emissions 10
    corporate reporting of 11, 12
    Emissions Gap Report 31
    and firm value 26, 28, 29, 326, 329, 330, 331, 332,
    management 19, 261
    measurement 18, 19
    reductions 14
    reporting 19, 21, 22, 343, 350
emissions trading systems (ETS) 22, 104, 328, 410, 434
Employee Health and Safety 521
Enbridge 215, 216
end-of-life solutions 290
energy 297
    non-carbon sources of 17
energy
    conservation 406, 415, 416
    consumption 32
    demand 52, 53, 54
    efficiency 15, 17, 528
    intensities 385
    policy 96, 98, 112
    sector 224
    services 53
    system 50, 54
        system technologies 46
    transition 38, 41, 46, 52, 90, 121, 127, 208, 210, 225
        transition metals 76
"energy market" auctions 91
Energy Policy Act (EPACT) 92, 105
Engle, R. 67, 81
Enkvist, P. 384
enterprise commitment 288
environmental
    capabilities 251, 259, 260, 261, 262, 263, 266
    competitive advantage 251
    footprint reduction 271
        footprint reduction goals 279
    information disclosure (EID) 494
    issues 506
    laggard 264
    management systems (EMS) 257, 258, 259
    proactive firms 262
    performance (ER) 6, 480, 481, 483, 486, 487, 488, 489, 490, 491, 492, 493, 494, 495, 496, 497, 498
    proactivity 258, 260
    product declaration (EPD) 286
    strategy archetypes 254
Environmental Protection Agency (EPA) 21, 23
environmental, social and governance (ESG) 6, 25, 341, 448, 450, 481, 521
    characteristics 506, 512
    coin 504
    corporate financial 515
    and corporate performance 516
    credibility of 513
    definitions 506
    funds 510
    incorporation 509
    index 507
    integration 449, 450, 510
    investing 503
        benefits 518
        corporate 504, 505
        size of market for 508
        strategies 509
    investing styles and strategies 510
    origins 505
    performance 513
    portfolios 510, 520

performance 510, 516
　　rankings 6, 505, 513
　　scores and ratings 505, 517, 519
　　space 520
　　stock market performance 515
EOG Resources Inc. 21
Equinor of Norway 213
equity investments 503
equity market spill over effects 472
equity value relevance 326
Erragragui, E. 496
ESG *see* environmental, social and governance
estimation methods 23
Etchebehere, Vincent 15, 528
ethical investing 505
Ethical Investment Research and Information Services (EIRIS) 505
EU 12, 24
EU Corporate Sustainability Reporting Directive 528
EU-ETS 24, 33, 333
EU Green Bond Standard 466
European Bank for Reconstruction and Development (EBRD) 393, 397
European Central Bank (ECB) 81
European Climate Law 232
European Investment Bank (EIB) 452
European Union (EU) 11, 326
EU Taxonomy 466, 520
exchange-traded funds (ETFs) 512
exponential deployment 43
exposure at default (EAD) 74
external carbon pricing 380
Exxon 209, 227, 228, 261
ExxonMobil 214, 218

Facebook 21
Fang, X. 490
fast-growing commitments 195
federal and state policies 112
federal ecosystem 137
Federal Energy Regulatory Commission (FERC) 92
feedback 411
feed-in tariffs 102, 103
feelings 405, 408, 413, 414, 420, 422
FEMA (Federal Emergency Management Agency) 305
Feng, Z-Y. 489
FERC (Federal Energy Regulatory Commission) 92
　　FERC Order 2222 108, 142
Fernando, C.S. 486, 492, 493
F-gases 31
fiberglass insulation 4

fiduciary duties 150
financial
　　intermediaries 63
　　markets 233
　　models 440
　　risks 90
　　services' regulators 152
　　stability 467
　　sustainability 172
Financial Stability Board (FSB) 59
financial stability board's (FSB) task force on climate-related financial disclosure (TCFD) 152
Fink, L. 240, 316
firm value 26, 28, 29
Fisher, Scott G. 400, 535
fishing sector 181
Flammer, C. 474
fleet renewal 197
flight operations optimization 197
flooding 192
flygskam movement 194
FOAMULAR® NGX™ insulation 282
Fonseka, M. 494
Foo, C. T. 490
food choices 406, 417
Foote, E. 8
footprint reduction goals 275
Fortune magazine 485
Forum for Sustainable and Responsible Investing (US-SIF) 509
fossil-fired generation 114
fossil-fuel based primary energy 283
fossil fuels 2, 98, 208
Fourier, Joseph 8
Fourth National Climate Assessment (NCA4) 302
Fouts, P. A. 256, 257
fracking technology 95
fragmented policy 304
free cash flows 26
Friede, G. 515
fuel combustion 3
fuel efficiency levers 203
Fuerst, F. 490
Fungibility of financing 474

Garcia-Sanchez, I. M. 489, 491
gas-fired electric generation 139
General Mills 527
general risk management 150
generation fleets 121, 137
generation mix, changing 93
Geneva Association 146, 152, 153, 154, 156, 158, 175, 176, 178, 179, 180

Geneva Association Task Force 176
'geo-engineering' solutions 32
geothermal power generation 223
Germany 38
Ge, W. 496
GHGProtocol 19, 23
GHG reporting program ('GHGRP') 23
Gianfrate, Gianfranco 398, 529
Giglio, H. 68
Gillingham, K. 69, 70, 84
Girerd-Potin, I. 487
glacier melting 83
Glasgow Climate Pact 42
Glasgow Financial Alliance for Net Zero (GFANZ) 177
Global Association of Risk Professionals (GARP) 300
Global Change Analysis Model (GCAM) 66
Global Climate Strike 235
Global Commission on Adaptation (GCA) 313
global
  deforestation 360
  deployment 43
  economy 30
  energy 208
    demand 54
Global Logistics Emissions Council (GLEC) 346
global palm oil trade 362
Global Reporting Initiative (GRI) 23, 349, 370, 483, 485, 499, 506, 513
Global South 534
global supply chains 303, 340, 342
Global Sustainable Investing Alliance (GSIA) 509, 521
global warming 9, 10, 90
"global warming potential" (GWP) 6, 31
Glozer, S. 233
Goldsmith-Pinkham, P. S. 67
Golnaraghi, Maryam 3, 530
good bank / bad bank scenario 64
Goss, A. 484, 492
government investments 316
government-mandated portfolio standards 101
Graves, S.B. 484
"Green Beauty Contest" (GBC) 286, 287
green bonds 5, 180, 181, 432, 433, 440, 441, 442, 443, 444, 445, 447, 448, 452, 458, 461, 470
  Index 464
  investors 469
  market 445, 447, 462, 474
  Principles (GBP) 461, 463, 467
Green Climate Fund (GCF) 435
Greene, Suzanne 4, 531
green finance 465, 473

green gilts 446
greenhouse gas (GHG) 31, 60
  accounting 359
  bottom line 531
  emissions 8, 10, 12, 13, 15, 17, 19, 21, 24, 28, 33, 90, 116, 121, 146, 281, 355, 359
    reductions of 341
  footprint 354
Greenhouse Gas Protocol 344, 346
Greenhouse Gas Protocol Initiative 18
green label 461
greenness 529
Greenpeace 244
green premiums 531
green taxonomies 536
greenwashing 200, 237, 244, 245, 260, 529
Gregory, A. 486
Grewal, Jody 4, 333, 335, 336
grid-connected electricity 98
Griffin, P. 327, 328, 329, 332
growth 511, 512, 521
Grubb, M. 337
Gruning, M. 488
Gulf Coast Carbon Collaborative 311
Gundlach, J. 402
Gupta, K. 234, 489
Gustafson, M. 66

Habib, Rishad 531
habit 405, 408, 409, 415, 419, 422
Hachenberg, B. 472
Hansen, Robert G. 5, 14, 70, 83, 379, 451, 520, 532
Harjoto, M. A. 491
Hart, S. L. 256, 258, 259
Hartmann, Julia 3, 208
Hassan, M. S. 488
Hassel, L. G. 484, 485, 494
heat extremes 11
heavy precipitation events 11
Heinkel, R. 486
Herbohn, K. 493
highly active companies 368
Hildebrand. P. 508
Hill, C. W. 258
Hoepner, A. 495
Holmes, Posie 3, 527
Hone, David 31, 533
Hopper, Abigail Ross 114
House, J. C. 392
Huang, H-W. 489
Huang, X. B. 480
Huber, Kristiane 4, 27, 294
Hugonnet, R. 83

human-induced climate change 11
Humphrey, J. E. 484, 487
Huynh, T. D. 67
hydraulic fracturing 95
hydrogen 136, 142
Hydrogen Energy of California project 111

Icarus 2
Idimetsu Kosan 213
Iizzo 480
IKEA 418
Ilhan, E. 66, 309
incentives 3, 411
incremental fuel efficiency improvements 203
indirect impacts 192
individual differences 412
individual self 405, 408, 411, 422
Industrial Revolution 38
industry alliances 176
industry-led initiatives 176
industry restructuring 91
in-flight fuel consumption optimization 197
infrastructure deployment 120
Infrastructure Investment and Jobs Act (IIJA)
    legislation 124
infrastructure systems 173
Inkpen, Andrew 3, 533
innovation process
    integrating sustainability 287
Institutional Shareholder Services 284
insurance 3, 77, 170, 298, 308
    companies 150
    industry 173, 175, 179
        value-chain 148
    policy 149
    prices 180
integrated assessment models (IAMs) 64, 65, 69,
    70, 73, 74, 75
Interagency Working Group (IWG) 65
Intercontinental Exchange Inc. (ICE) 20
Intergovernmental Panel on Climate Change
    (IPCC) 9, 39, 121, 241, 295, 536
internal carbon pricing 379, 380, 381, 382, 388,
    390, 391, 392, 393, 395, 396, 397, 398,
    399, 401, 402
    social value of 394, 395
internal combustion engines (ICE) 59
International Air Transport Association
    (IATA) 187
International Association of Credit Portfolio
    Managers (IACPM) 59
International Capital Markets Association
    (ICMA) 444
International Integrated Reporting Council
    (IIRC) 506

international rating agencies 153
International Standards Organization 351
interstate transmission 135
Institutional Investors Group on Climate Change
    (IIGCC) 468
investing style 521
Investment Company Institute (ICI) 512
investment opportunities 120, 126, 127, 135, 141
investment strategies 64, 174, 179
investments 209, 508
Investment Tax Credits (ITCs) 105
investors 150, 193, 458, 459, 461, 463, 464, 466,
    467, 468, 469, 470, 471, 472, 473, 474,
    475, 476
IPCC 10, 11, 55, 65, 71, 82, 151, 187, 241, 242,
    297, 431, 432
IRRC Institute and Trucost 332
issuer behavior 440
issuer perspective 469
issuers
    challenges for 472
IWB 263
Izzo, M. F. 493

Jalles, J. 67
Jiang, F. 67
Jimenez-Garces, S. 487
Jiraporn, P. 496
Jo, H. 491
Johnson, B. 512
joint ventures 16
Jones, T. M. 258
Jung, H. 81
Jung, J. 493

Kacperczyk, M. T. 28, 66
Kahn, M. E. 66
Kaufman, N. 402
Keenan, J. M. 66
key performance indicators (KPIs) 243
Keys, B. J. 66
Khaykin, Ilya 2, 58, 526
Kim, J. D. 490
Kim, Y-B. 490
KLD database 483, 484
Kleist, Dale 272
Koch, N. 486
Kock, Carl J. 251, 253, 254, 258, 534,
KPMG 238
Kumar, S. B. 488, 494
Kyoto Protocol 24, 434, 435

labeled market 446
LafargeHolcim 314
Lanoie, P. 480

large-scale battery storage systems 116
large-scale deployment 172
La Rosa, F. 496
latent carbon liability 326, 327, 331, 332, 333, 334, 336
Lavallee, Anthea 30
Lawrence Berkeley National Laboratory report 102
Lawton, Thomas C. 534
leadership 3
Lee, D. D. 484, 487
Lee, Keith 5, 24, 458
Lee, Yinjin 4, 5, 345, 359, 361, 364, 368, 370, 373
legacy infrastructure 2
levelized cost 116
Levis, William 272
Lewis, R. 66
liability 169
liability-driven investment strategy 150
Li, C. W. 67
life cycle assessments (LCAs) 285, 286, 344
life insurers 150
life re/insurers 162
life versus non-life (or P&C) insurance 148
Li, L. 487
lithium 46
Little, Arthur D. 199
Liu, J. 487, 493
Liu, M. 496
Liu, Q. 487
Li, Y. 487, 490, 493
long-term contracts 103
    competitive solicitations for 103, 104
Lontzek, T. S. 70
loss given default (LGD) 74
Louvet, P. 487
low-carbon design 17
low-carbon economy 152
low carbon transitioning 174
low-cost financing arrangements 107
Lukoil 215
Lund, Susan 301
Lyon 266

Magnanelli, B. S. 493
Magnan, M. 260
management profiles
    factors influencing 370
Mancur Olson 299
mandated carbon emissions disclosure 327, 334
mandatory reporting 327
Mann, A. 512
marginal cost 383, 384, 385, 386, 387, 388, 389, 390, 393, 395, 401

Margolis, J. D. 515
market analysis 233
market-based insurance 151
market-based strategies 4
market characteristics 459
marketers 407, 410, 413, 414, 420
marketing 406, 408, 409, 411, 412, 413, 414, 419, 421, 422
market revenues 114
market risk premium 33
market uncertainty 114
market value penalty 30
Marshall, R. S. 491, 494
Martínez-Ferrero, J. 489, 491
materiality 292
Matías-Reche, F. 258
Matsumura, E. 327, 328, 329, 330, 332, 337
Matthiesen, M-L. 489
Mattingly, J. E. 484
McCollum, D. 64
McKinsey Global Institute 317
McLennan, Marsh 305
Meadows, D. H. 205
Meadows report 205
Meehan, Amanda 4, 15, 16, 271
Menz, K.-M. 492, 493, 496
Methane (CH4) 6, 21, 31, 43, 61,
Meyer, A. D. 254
micro-insurance products 170
microprudential banking regulation 79
Microsoft 22
Minimum Offer Price Rule (MOPR) 112
Minsk, R. 402
Mishra, D. R. 487
MIT 'Growing Pressures' scenario 49
mitigation 383, 529
MIT Joint Program on the Science and Policy of Global Change (MITJP) 48
Mobil 209, 227, 228, 229
Modified Accelerated Cost Recovery System (MACRS) treatment 106
Mohd-Saleh, N. 488
monopsony 112
mortgages 66
MSCI KLD 400 Social Index 507
Mulder, P. 66
Multilateral Development Banks (MDBs) 431
multi-lateral initiatives 176
multinomial logistics regression 374
multi-stakeholder standards 364
municipal debt 66
Murugaboopathy, P. 512

NatCat 146
    -centric research 158

National Greenhouse and Energy Reporting Act 2007 (NGER Act) 330
Nationally Determined Contribution (NDC) 192, 432, 452
Nationally Recognized Statistical Ratings Organizations (NRSROs) 515, 522
National Oceanic and Atmospheric Administration (NOAA) 294
natural capital 181
natural catastrophe 146
natural gas 94, 139
natural gas-fired electric generation 94
Naucler, T. 384
negative market clearing 106
Neste 223
net carbon footprint 217
net emissions 9
net metering 108
Network for Greening the Financial System (NGFS) 59, 63, 66, 72, 74, 80
Net-Zero Banking Alliance (NZBA) 59
net-zero emissions 38, 41, 46, 47, 48, 49, 50, 51, 52, 55, 59, 195, 201
New Deal legislation 91
New York Stock Exchange 272
Ng, A.C. 488
Nguyen, J. H. 482
nickel 46
Nike Supply Chains 308
Nisbet, Matthew C. 241
nitrous oxide (N2O) 6, 21, 31
non-CO2 impacts impacts 190
non-economic insurance systems 527
non-financial disclosure 488, 494, 495
Non-Financial Reporting Directive (NFRD) 25
non-governmental organizations (NGOs) 1, 340, 372, 373
non-hydropower renewables 93
non-life (P&C) insurers 158
nonmarket approach 251
nonmarket strategy
    environmental activism in 253
non-utility developers 134
Nordhaus, William 82
nuclear reactors 138

O'Brien-Bernini, Frank 15, 271, 272, 273, 277, 285, 289, 291, 535
Ocasio-Cortez, Alexandria 245
ocean acidification 11
offsetters 419
Oikonomou, I. 492
oil and gas 208
oil and gas industry
    oil and gas companies

    and renewable energy mini-cases 214
    and the energy transition landscape 213
oil and gas firms
    renewables' commitments 219
    and GHG emissions 210
    renewable energy strategy in 211
Oliver Wyman 59
Olsen, Mancur 265
"operational" emissions 32
optimization process 91
Orens, R. 488, 494
Ørsted, Hans Christian 227
Ouasad, A. 66
oversight 275
Owens Corning 271
    alignment 273
    brief history of 272
    environmental footprint and sustainable products 272
    GHG emissions & energy 280
    managerial performance evaluation 274
    materiality assessments 277
    stakeholders 277
oxy-combustion technologies 139

Pae, J. 490
palm oil 5
    industry 359
    supply chain 360
    sustainability 362
parametric insurance 3
Paris Agreement 2, 39, 40, 41, 42, 48, 54, 55, 62, 72, 97, 124, 151, 155, 156, 192, 210, 233, 351, 432, 435, 437, 451
Paris conference (COP21) 340
Park, K. 489
PAROC® Natura™ 282
passive management profile 368
Pastor, L. 481, 488
Patagonia 238, 241, 255
Patalano, R. 514, 522
"patchwork quilt" 90, 97, 115, 380, 400
    policies 101
Pavelin, S. 492
Peace, Janet 4, 27, 294, 295
Pei, H. 483, 492
penalties 410
personal insurance 148
pharmaceutical sector 181
Phillips, Bruce A. 2, 16, 89, 120, 127, 131, 535
physical climate risks 297
physical risk 27, 61, 58, 67, 73, 77, 79, 177, 296
Pigou, A. C. 381
Pigouvian tax 14, 32
PJM 107, 111, 112

plastic and packaging sector 181
Plumlee, M. 491, 494
policy developments 79
policy goals 10
policy uncertainty 114
population densities 98
positive emotions 413
positive screening 510
positive self-concept 411
Poulsen, Henrik 217
power capacity 116
power generation 89
power purchase agreements (PPAs) 103, 104, 134, 282
pre-carbon pricing 385
primary insurers 171
principle component analysis 374
Principles for Responsible Investment (PRI) 505
proactive engagement 256
probability of default (PD) 74
product innovation 255
Production Tax Credits (PTCs) 105, 106
project categories 461
Project Gigaton 266
Prudential Regulation Authority (PRA) 62
psychological distance 407
public funding 450
publicly-traded corporations 23
public opinion 194
Public-Private Partnerships (PPPs) 172, 177
pure public goods 267
PURPA 102
Pyanet, Alban 2, 58, 526

Qian, Y. 67
Qiu, Y. 488
qualifying facilities (QFs) 92
quality management approach 258
quilt 90, 97

Raberto 472
radiative forcing (RF) 191, 192
Rajapakse, T. 494
Ramaswamy, K. 3, 208,
reactive compliance 260
reactive-reactive pairing 264
reactors 255
real estate credit 66
real estate lending 58
reduction governance
   GHG emissions 192
redundancy 384, 392, 395, 396, 399
Regional Greenhouse Gas Initiative (RGGI) 122
regulatory developments 79, 80
regulatory environments 440

regulatory landscape 465
regulatory pressure 227
Reichelt, Heike 5, 430, 440, 441, 445, 449, 450, 458, 460, 536
re/insurers 166, 174
renewable energy 3, 5, 208, 209, 210, 221, 532, 534, 536
   developers 134
   firm-level commitment to 223
Renewable Energy Act 103
renewable generation projects 103
Renewable Portfolio Standard (RPS) policies 122
representative concentration pathways (RCPs) 71, 72, 73
reputation 233, 234, 237, 244, 245, 247
residential mortgages 63
residential solar generation 109
residential space heating 125
residual annual fluctuations 149
resilience 151, 294
resilience bonds 317
resilience partnerships 311
resilient low carbon economy 145, 146
resilient sustainable infrastructure 173
retail ratemaking 108
retrocession 149
Reverte, C. 488
Rezaee, Z. 488
Riahi, K. 72
Richardson, Gordon D. 4, 22, 326, 490, 494
Richardson, M. 81
risk
   -based capital charges 175
   disclosure 309, 310
   framework 235
   management 114
   transferring and carrying 149
Risk Premium (RP) 482
Roberts, G. S. 484, 492
Rockström, Johan 63
Rodrigue, M. 260
Rosander, J. 384
Roundtable of Sustainable Palm Oil (RSPO) 362, 364, 372
Rowley 485
Royal Dutch Shell 216
Russo 256, 257

Sabatier 265
Sachs, Goldman 314
SAF 198, 201, 202, 203
Salama, A. 487
Salzmann, A. J. 489
Sautner, Z. 66, 232, 233
SBTi Net Zero referential 196

scenario analysis 153
scenarios 47, 50, 51, 52, 54, 55
Schneider Electric 254, 261
Schneider, T. E. 492
Schneller, W. 512
Schoenmaker, Dirk 5, 28, 480
Schroder, M. 495
Schuermann, Til 2, 58, 526
Science Based Targets Initiative (SBTi) 195, 292
Scovic, Mark W. 2, 16, 89, 120,
secondary market 471
sectoral decarbonization 154, 172
sectoral transitioning 172
Securities and Exchange Commission (SEC) 514
securitization 150
selection bias 337
Selective Opportunities 114
self-consistency 412
self-efficacy 412
self-interests 412
Semenova, N. 484, 485, 494
Senise-Barrio, M. E. 258
shale rock formations 95
shared socio-economic pathways (SSPs) 10, 71
Sharfman, M. P. 484, 485, 486, 492, 493
Shaukat, A. 488
Shell 224, 225, 244, 245, 527
Shen, Y. 484, 487
SHIFT framework 408, 422
short-term supply deficits 96
signaling models 394
sinks 41, 51, 52, 55
Sky 1.5 54, 55
Smale, R. 337
small- and medium-sized businesses 303
Smittenaar, Marloes 3, 231, 527
Sneader, Kevin 301
Snow, C. C. 254
sobriété 528
social benefit 537
social costs 537
social desirability 409
social group memberships 409
social influence 405, 408, 422
social insurance 148
socially responsible investing (SRI) 449, 505, 520
social norms 408
social pressures 225
societal behaviors 178
socio-economic resilience 146, 151, 169, 177
socioeconomic transformation 530
solar
    commercialization 95
solar capacity 133

solar generation technologies 95
Solar REC (SREC) 107
sovereign debt 66
S&P 500 32
specialized energy business units 170
S&P Trucost 23
SRISK methodology 81
Stambaugh, R. F. 488
standard risk management activities 307
Stankiewicz, A. 512
state of emissions 2
state regulations 92
Steffen, W. 60
Stock, Corey 30
Stock, J. H. 69, 384
stock market performance 6, 515
STOXX 600 Europe Index 20
stranded assets 75
strategic management 252, 254, 265, 266
strategic stance 254
stress testing 80
structure decision 240
substantial investment opportunities 90
Sundaram, Anant K. 1, 2, 6, 8, 21, 28, 83, 400, 451, 503, 506, 513, 520, 526, 537
supervisory bodies 152
Supplier Clean Energy Program 18
supplier engagement 352
supply chain
    climate accounting 347
    climate goals 340
    decarbonization 353
    emissions 343
    GHG emissions 343
    management 6, 359, 373
        commitment 363
        compliance instruments 365
        components 362
        external standards 363
    risks 309
    sustainability 4, 352, 355, 372
    Sustainability Report 359
    traceability 366
    transparency 347, 354
sustainability 213, 244, 275, 291, 480, 481, 483, 486, 487, 488, 489, 490, 491, 493, 494, 495, 498, 536
    goals 271, 274, 275, 276, 285, 287, 288, 289, 290, 291, 292
    leadership 292
    -linked bonds (SLBs) 463
    momentum 286
    rankings 285
    reporting 447
    Summit 289

thematics 251
trajectory 196
Sustainability Accounting Standards Board (SASB) 506
Sustainability Asset Management Group GmbH (SAM) 484
sustainable 521
   aviation fuels 3
   bonds 440, 446, 447
   business models 177
   debt capital markets 439
   development bond approach 447
   finance 437, 465, 466
      frameworks 152
      initiatives 152
   infrastructure 173,
   investing 509
   policies 127
   sourcing 351, 362, 373
   sourcing
      and procurement 351
Sustainable Aviation Fuels (SAF) 194, 196, 198, 528
Sustainable Development Goals (SDGs) 431, 448, 449
Suto, M. 487, 493
Swiss Re's research institute 294
synthetic 'drop-in' hydrocarbon fuels 51
Synthetic Genomics 214
"systematic" risk 33

Takehara, H. 487, 493
Tan, C. K. 242
Tan, W. 495
Tang, D. Y. 515
tangibility 405, 408, 414, 419, 422
Taskforce on Climate-related Financial Disclosure (TCFD) 2, 27, 59, 80, 82, 152, 176, 310, 350, 432, 437, 467
tax credit 105, 116
tax depreciation 106
tax incentive 137, 142
Taylor, L. A. 488
technological
   developments 155
   disruption 199, 203
   heterogeneity 387
   innovations 69
   pathways 130
technology companies 21
technology pathways 130
technology-specific capacity factor 116
temperature scenarios 10
TerraCycle 417
Terzani, S. 496

Tesla 263
Texas Clean Energy Project (TCEP) 137
Thaker, Jagadish 233
Thaman, Mike 273, 274
Tharyan, R. 486, 488
thematic investing 510
The Nature Conservancy (TNC) 308
third-party verifications 242
Thunberg, Greta 245
Tide 415
Toms, J. S. 487
Total Productive Maintenance (TPM) 274
Touboul, S. 485
tourism sector 182
toxic air emissions 278
toxic release inventory (TRI) 483
Toyota 264
traceability 366
traditional portfolio data 78
transaction costs 528
transformational technologies 533
transition 41, 46
   bonds 181
   risk 27, 58, 60, 61, 62, 63, 65, 66, 68, 71, 73, 74, 76, 78, 81
transmission system 135
transparency 443
transportation 15
Transportation and Climate Initiative (TCI) 122
tropical cyclone activity 11
Trucost 337
trust 245
Tushman 255
Tyndall, John 8

uncertainties 70, 120, 126, 127, 133
underwriting 173
uneven climate policy 111
UN Global Compact 505
Unilever 527
United National Environment Program - Financial Initiative's Principles for Sustainable Insurance (UNEP-FI PSI) 177
United Nations Environment Programme (UNEP) 392, 505
   United Nations Environment Programme – Finance Initiative (UNEP-FI) 59
United Nations Framework Convention on Climate Change (UNFCCC) 11, 12, 50, 145, 155, 433, 434, 435, 451
U.S. Energy Information Administration project 96
U.S. Equity Market Liability 28
U.S. municipal bond market 472

U.S. Securities and Exchange Commission (SEC) 23
U.S. Sustainable Skies Act 202

valuation 480, 486, 490, 492
valuation penalty 326, 336
value 511, 521
value-chain emissions 17
value-creation process 8
value relevance 326
variable renewable-dominant systems 131
Verified Emissions Reductions (VERs) 283
VicSuper 328
Victor, D. G. 392
Vilkov, G. 66
"virtual net metering" 109
virtual power purchase agreement (VPPA) 282
vulnerability assessments 306, 315
vulnerability screening tools 307

Waddock, S.A. 484
Wall Street Journal 504
Walmart 243
Wang, M-L. 489
Wang, W. 495
Wartick, S. L. 256
Wasama, Saara 5, 28, 480
waste disposal 417
Watson, L. 480
weighted average cost of capital (WACC) 488
weight reductions 197
Whelan, T. 515, 516
White, Katherine 5, 405, 408, 409, 412, 413, 414, 531
Whittaker, J. 486
Who Cares Wins 505, 506, 521
wholesale market design 107
Wilson, C. 67

wind
    commercialization 95
    capacity 133
    changes 192
    energy production 282
    generation 124
    power development 98
Woodlawn Associates 106
World Bank 5, 314, 392, 432, 434, 436, 439, 440, 441, 443, 444, 445, 446, 447, 448, 449, 450, 452
World Business Council for Sustainable Development (WBCSD) 19, 307
World Economic Forum (WEF) 155, 297
World Wildlife Fund (WWF) 367
Wyoming's tax 107

Xia, Y. 67
Xi Jinping 232
Xiong, J. 487
Xu, S. 494

Yale University 394
Yan, X. 486
Ye, K. 497
Yi, H-C. 497

Zamri, M. 488
Zechner, J. 486
zero-carbon
    energy 131
    fuels 130
    generating technologies 122
    investments 134
zero emission electricity 51
Zhang, R. 497
Zhou, X. 67
Zhou, Z. 497